Laplace Transform Table

Laplace Transform $F(s)$	Time Function $f(t)$
$\dfrac{\omega_n^2}{s(s^2 + \omega_n^2)}$	$1 - \cos \omega_n t$
$\dfrac{\omega_n^2(s + \alpha)}{s^2 + \omega_n^2}$	$\omega_n \sqrt{\alpha^2 + \omega_n^2} \, \sin(\omega_n t + \theta)$ where $\theta = \tan^{-1}(\omega_n/\alpha)$
$\dfrac{\omega_n}{(s + \alpha)(s^2 + \omega_n^2)}$	$\dfrac{\omega_n}{\alpha^2 + \omega_n^2} e^{-\alpha t} + \dfrac{1}{\sqrt{\alpha^2 + \omega_n^2}} \sin(\omega_n t - \theta)$ where $\theta = \tan^{-1}(\omega_n/\alpha)$
$\dfrac{\omega_n^2}{s^2 + 2\zeta\omega_n s + \omega_n^2}$	$\dfrac{\omega_n}{\sqrt{1-\zeta^2}} e^{-\zeta\omega_n t} \sin \omega_n \sqrt{1-\zeta^2}\, t \quad (\zeta < 1)$
$\dfrac{\omega_n^2}{s(s^2 + 2\zeta\omega_n s + \omega_n^2)}$	$1 - \dfrac{1}{\sqrt{1-\zeta^2}} e^{-\zeta\omega_n t} \sin\left(\omega_n \sqrt{1-\zeta^2}\, t + \theta\right)$ where $\theta = \cos^{-1}\zeta \quad (\zeta < 1)$
$\dfrac{s\omega_n^2}{s^2 + 2\zeta\omega_n s + \omega_n^2}$	$\dfrac{-\omega_n^2}{\sqrt{1-\zeta^2}} e^{-\zeta\omega_n t} \sin\left(\omega_n \sqrt{1-\zeta^2}\, t - \theta\right)$ where $\theta = \cos^{-1}\zeta \quad (\zeta < 1)$
$\dfrac{\omega_n^2(s + \alpha)}{s^2 + 2\zeta\omega_n s + \omega_n^2}$	$\omega_n \sqrt{\dfrac{\alpha^2 - 2\alpha\zeta\omega_n + \omega_n^2}{1 - \zeta^2}} e^{-\zeta\omega_n t} \sin\left(\omega_n \sqrt{1-\zeta^2}\, t + \theta\right)$ where $\theta = \tan^{-1}\dfrac{\omega_n \sqrt{1-\zeta^2}}{\alpha - \zeta\omega_n} \quad (\zeta < 1)$
$\dfrac{\omega_n^2}{s^2(s^2 + 2\zeta\omega_n s + \omega_n^2)}$	$t - \dfrac{2\zeta}{\omega_n} + \dfrac{1}{\omega_n \sqrt{1-\zeta^2}} e^{-\zeta\omega_n t} \sin\left(\omega_n \sqrt{1-\zeta^2}\, t + \theta\right)$ where $\theta = \cos^{-1}(2\zeta^2 - 1) \quad (\zeta < 1)$

9TH EDITION

Automatic Control Systems

FARID GOLNARAGHI
Simon Fraser University

BENJAMIN C. KUO
University of Illinois at Urbana-Champaign

JOHN WILEY & SONS, INC.

VP & Executive Publisher   *Don Fowley*
Associate Publisher   *Daniel Sayre*
Senior Production Editor   *Nicole Repasky*
Marketing Manager   *Christopher Ruel*
Senior Designer   *Kevin Murphy*
Production Management Services   *Elm Street Publishing Services*
Editorial Assistant   *Carolyn Weisman*
Media Editor   *Lauren Sapira*
Cover Photo   *Science Source/Photo Researchers*

This book was set in Times Roman by Thomson Digital and printed and bound by Quebecor/Versailles. The cover was printed by Quebecor/Versailles.

This book is printed on acid free paper. ∞

Copyright © 2010, 2003, 2000, 1991, 1987, 1982, 1975, 1967, 1962 John Wiley & Sons, Inc. All rights reserved. No part of this publication may be reproduced, stored in a retrieval system or transmitted in any form or by any means, electronic, mechanical, photocopying, recording, scanning or otherwise, except as permitted under Sections 107 or 108 of the 1976 United States Copyright Act, without either the prior written permission of the Publisher, or authorization through payment of the appropriate per-copy fee to the Copyright Clearance Center, Inc., 222 Rosewood Drive, Danvers, MA 01923, website www.copyright.com. Requests to the Publisher for permission should be addressed to the Permissions Department, John Wiley & Sons, Inc., 111 River Street, Hoboken, NJ 07030-5774, (201)748-6011, fax (201)748-6008, website www.wiley.com/go/permissions.

To order books or for customer service, please call 1-800-CALL WILEY (225-5945).

MATLAB® and Simulink® are trademarks of The MathWorks, Inc. and are used with permission. The MathWorks does not warrant the accuracy of the text or exercises in this book. This book's use or discussion of MATLAB® software or related products does not constitute endorsement or sponsorship by The MathWorks of a particular pedagogical approach or particular use of the MATLAB® software.

ISBN-13   978-0470-04896-2

Printed in the United States of America

10 9 8 7 6 5 4 3 2

To my wife, Mitra, and to Sophia and Carmen, the joys of my life.
—M. Farid Golnaraghi

Preface (Readme)

This is the ninth edition of the text but the first with Farid Golnaraghi as the lead author. For this edition, we increased the number of examples, added MATLAB®[1] toolboxes, and enhanced the MATLAB GUI software, **ACSYS**. We added more computer-aided tools for students and teachers. The prepublication manuscript was reviewed by many professors, and most of the relevant suggestions have been adopted. In this edition, Chapters 1 through 4 are organized to contain all background material, while Chapters 5 through 10 contain material directly related to the subject of control.

In this edition, the following materials have been moved into appendices on this book's Web site at www.wiley.com/college/golnaraghi.

    Appendix A: Elementary Matrix Theory and Algebra

    Appendix B: Difference Equations

    Appendix C: Laplace Transform Table

    Appendix D: z-Transform Table

    Appendix E: Properties and Construction of the Root Loci

    Appendix F: General Nyquist Criterion

    Appendix G: **ACSYS** 2008: Description of the Software

    Appendix H: Discrete-Data Control Systems

In addition, the Web site contains the MATLAB files for **ACSYS**, which are software tools for solving control-system problems, and PowerPoint files for the illustrations in the text.

The following paragraphs are aimed at three groups: professors who have adopted the book or who we hope will select it as their text; practicing engineers looking for answers to solve their day-to-day design problems; and, finally, students who are going to live with the book because it has been assigned for the control-systems course they are taking.

To the Professor: The material assembled in this book is an outgrowth of senior-level control-system courses taught by the authors at their universities throughout their teaching careers. The first eight editions have been adopted by hundreds of universities in the United States and around the world and have been translated into at least six languages. Practically all the design topics presented in the eighth edition have been retained.

This text contains not only conventional MATLAB toolboxes, where students can learn MATLAB and utilize their programming skills, but also a graphical MATLAB-based software, **ACSYS**. The **ACSYS** software added to this edition is very different from the software accompanying any other control book. Here, through extensive use of MATLAB GUI programming, we have created software that is easy to use. As a result, students will need to focus only on learning control problems, not programming! We also have added two new applications, SIMLab and Virtual Lab, through which students work on realistic problems and conduct speed and position control labs in a software environment. In SIMLab, students have access to the system parameters and can alter them (as in any simulation). In Virtual Lab, we have introduced a black-box approach in which the students

---

[1] MATLAB® is a registered trademark of The MathWorks, Inc.

have no access to the plant parameters and have to use some sort of system identification technique to find them. Through Virtual Lab we have essentially provided students with a realistic online lab with all the problems they would encounter in a real speed- or position-control lab—for example, amplifier saturation, noise, and nonlinearity. We welcome your ideas for the future editions of this book.

Finally, a sample section-by-section for a one-semester course is given in the *Instructor's Manual*, which is available from the publisher to qualified instructors. The *Manual* also contains detailed solutions to all the problems in the book.

To Practicing Engineers: This book was written with the readers in mind and is very suitable for self-study. Our objective was to treat subjects clearly and thoroughly. The book does not use the theorem–proof–Q.E.D. style and is without heavy mathematics. The authors have consulted extensively for wide sectors of the industry for many years and have participated in solving numerous control-systems problems, from aerospace systems to industrial controls, automotive controls, and control of computer peripherals. Although it is difficult to adopt all the details and realism of practical problems in a textbook at this level, some examples and problems reflect simplified versions of real-life systems.

To Students: You have had it now that you have signed up for this course and your professor has assigned this book! You had no say about the choice, though you can form and express your opinion on the book after reading it. Worse yet, one of the reasons that your professor made the selection is because he or she intends to make you work hard. But please don't misunderstand us: what we really mean is that, though this is an easy book to study (in our opinion), it is a no-nonsense book. It doesn't have cartoons or nice-looking photographs to amuse you. From here on, it is all business and hard work. You should have had the prerequisites on subjects found in a typical linear-systems course, such as how to solve linear ordinary differential equations, Laplace transform and applications, and time-response and frequency-domain analysis of linear systems. In this book you will not find too much new mathematics to which you have not been exposed before. What is interesting and challenging is that you are going to learn how to apply some of the mathematics that you have acquired during the past two or three years of study in college. In case you need to review some of the mathematical foundations, you can find them in the appendices on this book's Web site. The Web site also contains lots of other goodies, including the **ACSYS** software, which is GUI software that uses MATLAB-based programs for solving linear control systems problems. You will also find the Simulink®[2]-based SIMLab and Virtual Lab, which will help you to gain understanding of real-world control systems.

This book has numerous illustrative examples. Some of these are deliberately simple for the purpose of illustrating new ideas and subject matter. Some examples are more elaborate, in order to bring the practical world closer to you. Furthermore, the objective of this book is to present a complex subject in a clear and thorough way. One of the important learning strategies for you as a student is not to rely strictly on the textbook assigned. When studying a certain subject, go to the library and check out a few similar texts to see how other authors treat the same subject. You may gain new perspectives on the subject and discover that one author may treat the material with more care and thoroughness than the others. Do not be distracted by written-down coverage with oversimplified examples. The minute you step into the real world, you will face the design of control systems with nonlinearities and/or time-varying elements as well as orders that can boggle your mind. It

---

[2] Simulink® is a registered trademark of The MathWorks, Inc.

may be discouraging to tell you now that strictly linear and first-order systems do not exist in the real world.

Some advanced engineering students in college do not believe that the material they learn in the classroom is ever going to be applied directly in industry. Some of our students come back from field and interview trips totally surprised to find that the material they learned in courses on control systems is actually being used in industry today. They are surprised to find that this book is also a popular reference for practicing engineers. Unfortunately, these fact-finding, eye-opening, and self-motivating trips usually occur near the end of their college days, which is often too late for students to get motivated.

There are many learning aids available to you: the MATLAB-based **ACSYS** software will assist you in solving all kinds of control-systems problems. The SIMLab and Virtual Lab software can be used for simulation of virtual experimental systems. These are all found on the Web site. In addition, the Review Questions and Summaries at the end of each chapter should be useful to you. Also on the Web site, you will find the errata and other supplemental material.

We hope that you will enjoy this book. It will represent another major textbook acquisition (investment) in your college career. Our advice to you is not to sell it back to the bookstore at the end of the semester. If you do so but find out later in your professional career that you need to refer to a control systems book, you will have to buy it again at a higher price.

Special Acknowledgments: The authors wish to thank the reviewers for their invaluable comments and suggestions. The prepublication reviews have had a great impact on the revision project.

The authors thank Simon Fraser students and research associates Michael Ages, Johannes Minor, Linda Franak, Arash Jamalian, Jennifer Leone, Neda Parnian, Sean MacPherson, Amin Kamalzadeh, and Nathan (Wuyang) Zheng for their help. Farid Golnaraghi also wishes to thank Professor Benjamin Kuo for sharing the pleasure of writing this wonderful book, and for his teachings, patience, and support throughout this experience.

M. F. Golnaraghi,
Vancouver, British Columbia,
Canada

B. C. Kuo,
Champaign, Illinois, U.S.A.

2009

Contents

Preface iv

▶ **CHAPTER 1**
Introduction 1

1-1 Introduction 1
    1-1-1 Basic Components of a Control System 2
    1-1-2 Examples of Control-System Applications 2
    1-1-3 Open-Loop Control Systems (Nonfeedback Systems) 5
    1-1-4 Closed-Loop Control Systems (Feedback Control Systems) 7
1-2 What Is Feedback, and What Are Its Effects? 8
    1-2-1 Effect of Feedback on Overall Gain 8
    1-2-2 Effect of Feedback on Stability 9
    1-2-3 Effect of Feedback on External Disturbance or Noise 10
1-3 Types of Feedback Control Systems 11
    1-3-1 Linear versus Nonlinear Control Systems 11
    1-3-2 Time-Invariant versus Time-Varying Systems 12
1-4 Summary 14

▶ **CHAPTER 2**
Mathematical Foundation 16

2-1 Complex-Variable Concept 16
    2-1-1 Complex Numbers 16
    2-1-2 Complex Variables 18
    2-1-3 Functions of a Complex Variable 19
    2-1-4 Analytic Function 20
    2-1-5 Singularities and Poles of a Function 20
    2-1-6 Zeros of a Function 20
    2-1-7 Polar Representation 22
2-2 Frequency-Domain Plots 26
    2-2-1 Computer-Aided Construction of the Frequency-Domain Plots 26
    2-2-2 Polar Plots 27
    2-2-3 Bode Plot (Corner Plot or Asymptotic Plot) 32
    2-2-4 Real Constant K 34
    2-2-5 Poles and Zeros at the Origin, $(j\omega)^{\pm p}$ 34
    2-2-6 Simple Zero, $1 + j\omega T$ 37
    2-2-7 Simple Pole, $1/(1 + j\omega T)$ 39
    2-2-8 Quadratic Poles and Zeros 39
    2-2-9 Pure Time Delay, $e^{-j\omega T_d}$ 42
    2-2-10 Magnitude-Phase Plot 44
    2-2-11 Gain- and Phase-Crossover Points 46
    2-2-12 Minimum-Phase and Nonminimum-Phase Functions 47
2-3 Introduction to Differential Equations 49
    2-3-1 Linear Ordinary Differential Equations 49
    2-3-2 Nonlinear Differential Equations 49
    2-3-3 First-Order Differential Equations: State Equations 50
    2-3-4 Definition of State Variables 50
    2-3-5 The Output Equation 51
2-4 Laplace Transform 52
    2-4-1 Definition of the Laplace Transform 52
    2-4-2 Inverse Laplace Transformation 54
    2-4-3 Important Theorems of the Laplace Transform 54
2-5 Inverse Laplace Transform by Partial-Fraction Expansion 57
    2-5-1 Partial-Fraction Expansion 57
2-6 Application of the Laplace Transform to the Solution of Linear Ordinary Differential Equations 62
    2-6-1 First-Order Prototype System 63
    2-6-2 Second-Order Prototype System 64
2-7 Impulse Response and Transfer Functions of Linear Systems 67
    2-7-1 Impulse Response 67
    2-7-2 Transfer Function (Single-Input, Single-Output Systems) 70
    2-7-3 Proper Transfer Functions 71
    2-7-4 Characteristic Equation 71
    2-7-5 Transfer Function (Multivariable Systems) 71
2-8 Stability of Linear Control Systems 72
2-9 Bounded-Input, Bounded-Output (BIBO) Stability—Continuous-Data Systems 73
2-10 Relationship between Characteristic Equation Roots and Stability 74
2-11 Zero-Input and Asymptotic Stability of Continuous-Data Systems 74
2-12 Methods of Determining Stability 77
2-13 Routh-Hurwitz Criterion 78

		2-13-1	Routh's Tabulation 79
		2-13-2	Special Cases when Routh's Tabulation Terminates Prematurely 80
2-14	MATLAB Tools and Case Studies 84
	2-14-1	Description and Use of Transfer Function Tool 84
	2-14-2	MATLAB Tools for Stability 85
2-15	Summary 90

► CHAPTER 3
Block Diagrams and Signal-Flow Graphs   104

3-1 Block Diagrams   104
    3-1-1 Typical Elements of Block Diagrams in Control Systems   106
    3-1-2 Relation between Mathematical Equations and Block Diagrams   109
    3-1-3 Block Diagram Reduction   113
    3-1-4 Block Diagram of Multi-Input Systems—Special Case: Systems with a Disturbance   115
    3-1-5 Block Diagrams and Transfer Functions of Multivariable Systems   117

3-2 Signal-Flow Graphs (SFGs)   119
    3-2-1 Basic Elements of an SFG   119
    3-2-2 Summary of the Basic Properties of SFG   120
    3-2-3 Definitions of SFG Terms   120
    3-2-4 SFG Algebra   123
    3-2-5 SFG of a Feedback Control System   124
    3-2-6 Relation between Block Diagrams and SFGs   124
    3-2-7 Gain Formula for SFG   124
    3-2-8 Application of the Gain Formula between Output Nodes and Noninput Nodes   127
    3-2-9 Application of the Gain Formula to Block Diagrams   128
    3-2-10 Simplified Gain Formula   129

3-3 MATLAB Tools and Case Studies   129
3-4 Summary   133

► CHAPTER 4
Theoretical Foundation and Background Material: Modeling of Dynamic Systems   147

4-1 Introduction to Modeling of Mechanical Systems   148
    4-1-1 Translational Motion   148
    4-1-2 Rotational Motion   157
    4-1-3 Conversion between Translational and Rotational Motions   161
    4-1-4 Gear Trains   162
    4-1-5 Backlash and Dead Zone (Nonlinear Characteristics)   164

4-2 Introduction to Modeling of Simple Electrical Systems   165
    4-2-1 Modeling of Passive Electrical Elements   165
    4-2-2 Modeling of Electrical Networks   165

4-3 Modeling of Active Electrical Elements: Operational Amplifiers   172
    4-3-1 The Ideal Op-Amp   173
    4-3-2 Sums and Differences   173
    4-3-3 First-Order Op-Amp Configurations   174

4-4 Introduction to Modeling of Thermal Systems   177
    4-4-1 Elementary Heat Transfer Properties   177

4-5 Introduction to Modeling of Fluid Systems   180
    4-5-1 Elementary Fluid and Gas System Properties   180

4-6 Sensors and Encoders in Control Systems   189
    4-6-1 Potentiometer   189
    4-6-2 Tachometers   194
    4-6-3 Incremental Encoder   195

4-7 DC Motors in Control Systems   198
    4-7-1 Basic Operational Principles of DC Motors   199
    4-7-2 Basic Classifications of PM DC Motors   199
    4-7-3 Mathematical Modeling of PM DC Motors   201

4-8 Systems with Transportation Lags (Time Delays)   205
    4-8-1 Approximation of the Time-Delay Function by Rational Functions   206

4-9 Linearization of Nonlinear Systems   206
    4-9-1 Linearization Using Taylor Series: Classical Representation   207
    4-9-2 Linearization Using the State Space Approach   207

4-10 Analogies   213
4-11 Case Studies   216
4-12 MATLAB Tools   222
4-13 Summary   223

► CHAPTER 5
Time-Domain Analysis of Control Systems   253

5-1 Time Response of Continuous-Data Systems: Introduction   253
5-2 Typical Test Signals for the Time Response of Control Systems   254
5-3 The Unit-Step Response and Time-Domain Specifications   256
5-4 Steady-State Error   258

Contents ◄ ix

	5-4-1	Steady-State Error of Linear Continuous-Data Control Systems 258
	5-4-2	Steady-State Error Caused by Nonlinear System Elements 272
5-5	Time Response of a Prototype First-Order System 274	
5-6	Transient Response of a Prototype Second-Order System 275	
	5-6-1	Damping Ratio and Damping Factor 277
	5-6-2	Natural Undamped Frequency 278
	5-6-3	Maximum Overshoot 280
	5-6-4	Delay Time and Rise Time 283
	5-6-5	Settling Time 285
5-7	Speed and Position Control of a DC Motor 289	
	5-7-1	Speed Response and the Effects of Inductance and Disturbance-Open Loop Response 289
	5-7-2	Speed Control of DC Motors: Closed-Loop Response 291
	5-7-3	Position Control 292
5-8	Time-Domain Analysis of a Position-Control System 293	
	5-8-1	Unit-Step Transient Response 294
	5-8-2	The Steady-State Response 298
	5-8-3	Time Response to a Unit-Ramp Input 298
	5-8-4	Time Response of a Third-Order System 300
5-9	Basic Control Systems and Effects of Adding Poles and Zeros to Transfer Functions 304	
	5-9-1	Addition of a Pole to the Forward-Path Transfer Function: Unity-Feedback Systems 305
	5-9-2	Addition of a Pole to the Closed-Loop Transfer Function 307
	5-9-3	Addition of a Zero to the Closed-Loop Transfer Function 308
	5-9-4	Addition of a Zero to the Forward-Path Transfer Function: Unity-Feedback Systems 309
5-10	Dominant Poles and Zeros of Transfer Functions 311	
	5-10-1	Summary of Effects of Poles and Zeros 313
	5-10-2	The Relative Damping Ratio 313
	5-10-3	The Proper Way of Neglecting the Insignificant Poles with Consideration of the Steady-State Response 313
5-11	Basic Control Systems Utilizing Addition of Poles and Zeros 314	
5-12	MATLAB Tools 319	
5-13	Summary 320	

► CHAPTER 6

The Control Lab 337

6-1	Introduction 337	
6-2	Description of the Virtual Experimental System 338	
	6-2-1	Motor 339
	6-2-2	Position Sensor or Speed Sensor 339
	6-2-3	Power Amplifier 340
	6-2-4	Interface 340
6-3	Description of SIMLab and Virtual Lab Software 340	
6-4	Simulation and Virtual Experiments 345	
	6-4-1	Open-Loop Speed 345
	6-4-2	Open-Loop Sine Input 347
	6-4-3	Speed Control 350
	6-4-4	Position Control 352
6-5	Design Project 1—Robotic Arm 354	
6-6	Design Project 2—Quarter-Car Model 357	
	6-6-1	Introduction to the Quarter-Car Model 357
	6-6-2	Closed-Loop Acceleration Control 359
	6-6-3	Description of Quarter Car Modeling Tool 360
	6-6-4	Passive Suspension 364
	6-6-5	Closed-Loop Relative Position Control 365
	6-6-6	Closed-Loop Acceleration Control 366
6-7	Summary 367	

► CHAPTER 7

Root Locus Analysis 372

7-1	Introduction 372	
7-2	Basic Properties of the Root Loci (RL) 373	
7-3	Properties of the Root Loci 377	
	7-3-1	$K = 0$ and $K = \pm\infty$ Points 377
	7-3-2	Number of Branches on the Root Loci 378
	7-3-3	Symmetry of the RL 378
	7-3-4	Angles of Asymptotes of the RL: Behavior of the RL at $|s| = \infty$ 378
	7-3-5	Intersect of the Asymptotes (Centroid) 379
	7-3-6	Root Loci on the Real Axis 380
	7-3-7	Angles of Departure and Angles of Arrival of the RL 380
	7-3-8	Intersection of the RL with the Imaginary Axis 380
	7-3-9	Breakaway Points (Saddle Points) on the RL 380
	7-3-10	The Root Sensitivity 382

7-4 Design Aspects of the Root Loci 385
  7-4-1 Effects of Adding Poles and Zeros to $G(s)H(s)$ 385
7-5 Root Contours (RC): Multiple-Parameter Variation 393
7-6 MATLAB Tools and Case Studies 400
7-7 Summary 400

▶ **CHAPTER 8**
Frequency-Domain Analysis 409

8-1 Introduction 409
  8-1-1 Frequency Response of Closed-Loop Systems 410
  8-1-2 Frequency-Domain Specifications 412
8-2 M_r, ω_r, and Bandwidth of the Prototype Second-Order System 413
  8-2-1 Resonant Peak and Resonant Frequency 413
  8-2-2 Bandwidth 416
8-3 Effects of Adding a Zero to the Forward-Path Transfer Function 418
8-4 Effects of Adding a Pole to the Forward-Path Transfer Function 424
8-5 Nyquist Stability Criterion: Fundamentals 426
  8-5-1 Stability Problem 427
  8-5-2 Definition of Encircled and Enclosed 428
  8-5-3 Number of Encirclements and Enclosures 429
  8-5-4 Principles of the Argument 429
  8-5-5 Nyquist Path 433
  8-5-6 Nyquist Criterion and the $L(s)$ or the $G(s)H(s)$ Plot 434
8-6 Nyquist Criterion for Systems with Minimum-Phase Transfer Functions 435
  8-6-1 Application of the Nyquist Criterion to Minimum-Phase Tranfer Functions That Are Not Strictly Proper 436
8-7 Relation between the Root Loci and the Nyquist Plot 437
8-8 Illustrative Examples: Nyquist Criterion for Minimum-Phase Transfer Functions 440
8-9 Effects of Adding Poles and Zeros to $L(s)$ on the Shape of the Nyquist Plot 444
8-10 Relative Stability: Gain Margin and Phase Margin 449
  8-10-1 Gain Margin (GM) 451
  8-10-2 Phase Margin (PM) 453
8-11 Stability Analysis with the Bode Plot 455
  8-11-1 Bode Plots of Systems with Pure Time Delays 458
8-12 Relative Stability Related to the Slope of the Magnitude Curve of the Bode Plot 459
  8-12-1 Conditionally Stable System 459
8-13 Stability Analysis with the Magnitude-Phase Plot 462
8-14 Constant-M Loci in the Magnitude-Phase Plane: The Nichols Chart 463
8-15 Nichols Chart Applied to Nonunity-Feedback Systems 469
8-16 Sensitivity Studies in the Frequency Domain 470
8-17 MATLAB Tools and Case Studies 472
8-18 Summary 472

▶ **CHAPTER 9**
Design of Control Systems 487

9-1 Introduction 487
  9-1-1 Design Specifications 487
  9-1-2 Controller Configurations 489
  9-1-3 Fundamental Principles of Design 491
9-2 Design with the PD Controller 492
  9-2-1 Time-Domain Interpretation of PD Control 494
  9-2-2 Frequency-Domain Interpretation of PD Control 496
  9-2-3 Summary of Effects of PD Control 497
9-3 Design with the PI Controller 511
  9-3-1 Time-Domain Interpretation and Design of PI Control 513
  9-3-2 Frequency-Domain Interpretation and Design of PI Control 514
9-4 Design with the PID Controller 528
9-5 Design with Phase-Lead Controller 532
  9-5-1 Time-Domain Interpretation and Design of Phase-Lead Control 534
  9-5-2 Frequency-Domain Interpretation and Design of Phase-Lead Control 535
  9-5-3 Effects of Phase-Lead Compensation 554
  9-5-4 Limitations of Single-Stage Phase-Lead Control 555
  9-5-5 Multistage Phase-Lead Controller 555
  9-5-6 Sensitivity Considerations 559
9-6 Design with Phase-Lag Controller 561
  9-6-1 Time-Domain Interpretation and Design of Phase-Lag Control 561
  9-6-2 Frequency-Domain Interpretation and Design of Phase-Lag Control 563
  9-6-3 Effects and Limitations of Phase-Lag Control 574
9-7 Design with Lead–Lag Controller 574
9-8 Pole-Zero-Cancellation Design: Notch Filter 576
  9-8-1 Second-Order Active Filter 579
  9-8-2 Frequency-Domain Interpretation and Design 580

9-9 Forward and Feedforward Controllers 588
9-10 Design of Robust Control Systems 590
9-11 Minor-Loop Feedback Control 601
    9-11-1 Rate-Feedback or Tachometer-Feedback Control 601
    9-11-2 Minor-Loop Feedback Control with Active Filter 603
9-12 A Hydraulic Control System 605
    9-12-1 Modeling Linear Actuator 605
    9-12-2 Four-Way Electro-Hydraulic Valve 606
    9-12-3 Modeling the Hydraulic System 612
    9-12-4 Applications 613
9-13 Controller Design 617
    9-13-1 P Control 617
    9-13-2 PD Control 621
    9-13-3 PI Control 626
    9-13-4 PID Control 628
9-14 MATLAB Tools and Case Studies 631
9-15 Plotting Tutorial 647
9-16 Summary 649

CHAPTER 10

State Variable Analysis 673

10-1 Introduction 673
10-2 Block Diagrams, Transfer Functions, and State Diagrams 673
    10-2-1 Transfer Functions (Multivariable Systems) 673
    10-2-2 Block Diagrams and Transfer Functions of Multivariable Systems 674
    10-2-3 State Diagram 676
    10-2-4 From Differential Equations to State Diagrams 678
    10-2-5 From State Diagrams to Transfer Function 679
    10-2-6 From State Diagrams to State and Output Equations 680
10-3 Vector-Matrix Representation of State Equations 682
10-4 State-Transition Matrix 684
    10-4-1 Significance of the State-Transition Matrix 685
    10-4-2 Properties of the State-Transition Matrix 685
10-5 State-Transition Equation 687
    10-5-1 State-Transition Equation Determined from the State Diagram 689
10-6 Relationship between State Equations and High-Order Differential Equations 691
10-7 Relationship between State Equations and Transfer Functions 693
10-8 Characteristic Equations, Eigenvalues, and Eigenvectors 695
    10-8-1 Characteristic Equation from a Differential Equation 695
    10-8-2 Characteristic Equation from a Transfer Function 696
    10-8-3 Characteristic Equation from State Equations 696
    10-8-4 Eigenvalues 697
    10-8-5 Eigenvectors 697
    10-8-6 Generalized Eigenvectors 698
10-9 Similarity Transformation 699
    10-9-1 Invariance Properties of the Similarity Transformations 700
    10-9-2 Controllability Canonical Form (CCF) 701
    10-9-3 Observability Canonical Form (OCF) 703
    10-9-4 Diagonal Canonical Form (DCF) 704
    10-9-5 Jordan Canonical Form (JCF) 706
10-10 Decompositions of Transfer Functions 707
    10-10-1 Direct Decomposition 707
    10-10-2 Cascade Decomposition 712
    10-10-3 Parallel Decomposition 713
10-11 Controllability of Control Systems 714
    10-11-1 General Concept of Controllability 716
    10-11-2 Definition of State Controllability 716
    10-11-3 Alternate Tests on Controllability 717
10-12 Observability of Linear Systems 719
    10-12-1 Definition of Observability 719
    10-12-2 Alternate Tests on Observability 720
10-13 Relationship among Controllability, Observability, and Transfer Functions 721
10-14 Invariant Theorems on Controllability and Observability 723
10-15 Case Study: Magnetic-Ball Suspension System 725
10-16 State-Feedback Control 728
10-17 Pole-Placement Design Through State Feedback 730
10-18 State Feedback with Integral Control 735
10-19 MATLAB Tools and Case Studies 741
    10-19-1 Description and Use of the State-Space Analysis Tool 741
    10-19-2 Description and Use of tfsym for State-Space Applications 748
10-20 Summary 751

INDEX 773

Appendices can be found on this book's companion Web site: www.wiley.com/college/golnaraghi.

APPENDIX A

Elementary Matrix Theory and Algebra A-1

A-1 Elementary Matrix Theory A-1
    A-1-1 Definition of a Matrix A-2

A-2 Matrix Algebra A-5
    A-2-1 Equality of Matrices A-5
    A-2-2 Addition and Subtraction of Matrices A-6
    A-2-3 Associative Law of Matrix (Addition and Subtraction) A-6
    A-2-4 Commutative Law of Matrix (Addition and Subtraction) A-6
    A-2-5 Matrix Multiplication A-6
    A-2-6 Rules of Matrix Multiplication A-7
    A-2-7 Multiplication by a Scalar k A-8
    A-2-8 Inverse of a Matrix (Matrix Division) A-8
    A-2-9 Rank of a Matrix A-9
A-3 Computer-Aided Solutions of Matrices A-9

▶ APPENDIX B
Difference Equations B-1

B-1 Difference Equations B-1

▶ APPENDIX C
Laplace Transform Table C-1

▶ APPENDIX D
z-Transform Table D-1

▶ APPENDIX E
Properties and Construction of the Root Loci E-1

E-1 $K = 0$ and $K = \pm\infty$ Points E-1
E-2 Number of Branches on the Root Loci E-2
E-3 Symmetry of the Root Loci E-2
E-4 Angles of Asymptotes of the Root Loci and Behavior of the Root Loci at $|s| = \infty$ E-4
E-5 Intersect of the Asymptotes (Centroid) E-5
E-6 Root Loci on the Real Axis E-8
E-7 Angles of Departure and Angles of Arrival of the Root Loci E-9
E-8 Intersection of the Root Loci with the Imaginary Axis E-11
E-9 Breakaway Points E-11
    E-9-1 (Saddle Points) on the Root Loci E-11
    E-9-2 The Angle of Arrival and Departure of Root Loci at the Breakaway Point E-12
E-10 Calculation of K on the Root Loci E-16

▶ APPENDIX F
General Nyquist Criterion F-1

F-1 Formulation of Nyquist Criterion F-1
    F-1-1 System with Minimum-Phase Loop Transfer Functions F-4
    F-1-2 Systems with Improper Loop Transfer Functions F-4
F-2 Illustrative Examples—General Nyquist Criterion Minimum and Nonminimum Transfer Functions F-4
F-3 Stability Analysis of Multiloop Systems F-13

▶ APPENDIX G
ACSYS 2008: Description of the Software G-1

G-1 Installation of ACSYS G-1
G-2 Description of the Software G-1
    G-2-1 tfsym G-2
    G-2-2 Statetool G-3
    G-2-3 Controls G-3
    G-2-4 SIMLab and Virtual Lab G-4
G-3 Final Comments G-4

▶ APPENDIX H
Discrete-Data Control Systems H-1

H-1 Introduction H-1
H-2 The z-Transform H-1
    H-2-1 Definition of the z-Transform H-1
    H-2-2 Relationship between the Laplace Transform and the z-Transform H-2
    H-2-3 Some Important Theorems of the z-Transform H-3
    H-2-4 Inverse z-Transform H-5
    H-2-5 Computer Solution of the Partial-Fraction Expansion of $Y(z)/z$ H-7
    H-2-6 Application of the z-Transform to the Solution of Linear Difference Equations H-7
H-3 Transfer Functions of Discrete-Data Systems H-8
    H-3-1 Transfer Functions of Discrete-Data Systems with Cascade Elements H-12
    H-3-2 Transfer Function of the Zero-Order-Hold H-13
    H-3-3 Transfer Functions of Closed-Loop Discrete-Data Systems H-14
H-4 State Equations of Linear Discrete-Data Systems H-16
    H-4-1 Discrete State Equations H-16
    H-4-2 Solutions of the Discrete State Equations: Discrete State-Transition Equations H-18
    H-4-3 z-Transform Solution of Discrete State Equations H-19
    H-4-4 Transfer-Function Matrix and the Characteristic Equation H-20
    H-4-5 State Diagrams of Discrete-Data Systems H-22
    H-4-6 State Diagrams for Sampled-Data Systems H-23
H-5 Stability of Discrete-Data Systems H-26
    H-5-1 BIBO Stability H-26
    H-5-2 Zero-Input Stability H-26
    H-5-3 Stability Tests of Discrete-Data Systems H-27
H-6 Time-Domain Properties of Discrete-Data

Systems H-31
- H-6-1 Time Response of Discrete-Data Control Systems H-31
- H-6-2 Mapping between s-Plane and z-Plane Trajectories H-34
- H-6-3 Relation between Characteristic-Equation Roots and Transient Response H-38

H-7 Steady-State Error Analysis of Discrete-Data Control Systems H-41

H-8 Root Loci of Discrete-Data Systems H-45

H-9 Frequency-Domain Analysis of Discrete-Data Control Systems H-49
- H-9-1 Bode Plot with the w-Transformation H-50

H-10 Design of Discrete-Data Control Systems H-51
- H-10-1 Introduction H-51
- H-10-2 Digital Implementation of Analog Controllers H-52
- H-10-3 Digital Implementation of the PID Controller H-54
- H-10-4 Digital Implementation of Lead and Lag Controllers H-57

H-11 Digital Controllers H-58
- H-11-1 Physical Realizability of Digital Controllers H-58

H-12 Design of Discrete-Data Control Systems in the Frequency Domain and the z-Plane H-61
- H-12-1 Phase-Lead and Phase-Lag Controllers in the w-Domain H-61

H-13 Design of Discrete-Data Control Systems with Deadbeat Response H-68

H-14 Pole-Placement Design with State Feedback H-70

CHAPTER 1

Introduction

▶ 1-1 INTRODUCTION

The main objectives of this chapter are:

1. To define a control system.
2. To explain why control systems are important.
3. To introduce the basic components of a control system.
4. To give some examples of control-system applications.
5. To explain why feedback is incorporated into most control systems.
6. To introduce types of control systems.

One of the most commonly asked questions by a novice on a control system is: What is a control system? To answer the question, we can say that in our daily lives there are numerous "objectives" that need to be accomplished. For instance, in the domestic domain, we need to regulate the temperature and humidity of homes and buildings for comfortable living. For transportation, we need to control the automobile and airplane to go from one point to another accurately and safely. Industrially, manufacturing processes contain numerous objectives for products that will satisfy the precision and cost-effectiveness requirements. A human being is capable of performing a wide range of tasks, including decision making. Some of these tasks, such as picking up objects and walking from one point to another, are commonly carried out in a routine fashion. Under certain conditions, some of these tasks are to be performed in the best possible way. For instance, an athlete running a 100-yard dash has the objective of running that distance in the shortest possible time. A marathon runner, on the other hand, not only must run the distance as quickly as possible, but, in doing so, he or she must control the consumption of energy and devise the best strategy for the race. The means of achieving these "objectives" usually involve the use of control systems that implement certain control strategies.

• Control systems are in abundance in modern civilization.

In recent years, control systems have assumed an increasingly important role in the development and advancement of modern civilization and technology. Practically every aspect of our day-to-day activities is affected by some type of control system. Control systems are found in abundance in all sectors of industry, such as quality control of manufactured products, automatic assembly lines, machine-tool control, space technology and weapon systems, computer control, transportation systems, power systems, robotics, Micro-Electro-Mechanical Systems (MEMS), nanotechnology, and many others. Even the control of inventory and social and economic systems may be approached from the theory of automatic control.

Figure 1-1 Basic components of a control system.

1-1-1 Basic Components of a Control System

The basic ingredients of a control system can be described by:

1. Objectives of control.
2. Control-system components.
3. Results or outputs.

The basic relationship among these three components is illustrated in Fig. 1-1. In more technical terms, the **objectives** can be identified with **inputs**, or **actuating signals**, u, and the results are also called **outputs**, or **controlled variables**, y. In general, the objective of the control system is to control the outputs in some prescribed manner by the inputs through the elements of the control system.

1-1-2 Examples of Control-System Applications

Intelligent Systems

Applications of control systems have significantly increased through the development of new materials, which provide unique opportunities for highly efficient actuation and sensing, thereby reducing energy losses and environmental impacts. State-of-the-art actuators and sensors may be implemented in virtually any system, including biological propulsion; locomotion; robotics; material handling; biomedical, surgical, and endoscopic; aeronautics; marine; and the defense and space industries. Potential applications of control of these systems may benefit the following areas:

- **Machine tools.** Improve precision and increase productivity by controlling chatter.
- **Flexible robotics.** Enable faster motion with greater accuracy.
- **Photolithography.** Enable the manufacture of smaller microelectronic circuits by controlling vibration in the photolithography circuit-printing process.
- **Biomechanical and biomedical.** Artificial muscles, drug delivery systems, and other assistive technologies.
- **Process control.** For example, on/off shape control of solar reflectors or aerodynamic surfaces.

Control in Virtual Prototyping and Hardware in the Loop

The concept of virtual prototyping has become a widely used phenomenon in the automotive, aerospace, defense, and space industries. In all these areas, pressure to cut costs has forced manufacturers to design and test an entire system in a computer environment before a physical prototype is made. Design tools such as **MATLAB** and **Simulink** enable companies to design and test controllers for different components (e.g., suspension, ABS, steering, engines, flight control mechanisms, landing gear, and specialized devices) within the system and examine the behavior of the control system on the virtual prototype in real time. This allows the designers to change or adjust controller parameters online before the actual hardware is developed. Hardware in the loop terminology is a new approach of testing individual components by attaching them to the virtual and controller prototypes. Here the physical controller hardware is interfaced with the computer and replaces its mathematical model within the computer!

Smart Transportation Systems
The automobile and its evolution in the last two centuries is arguably the most transformative invention of man. Over years innovations have made cars faster, stronger, and aesthetically appealing. We have grown to desire cars that are "intelligent" and provide maximum levels of comfort, safety, and fuel efficiency. Examples of intelligent systems in cars include climate control, cruise control, anti-lock brake systems (ABSs), active suspensions that reduce vehicle vibration over rough terrain or improve stability, air springs that self-level the vehicle in high-G turns (in addition to providing a better ride), integrated vehicle dynamics that provide yaw control when the vehicle is either over- or understeering (by selectively activating the brakes to regain vehicle control), traction control systems to prevent spinning of wheels during acceleration, and active sway bars to provide "controlled" rolling of the vehicle. The following are a few examples.

Drive-by-wire and Driver Assist Systems The new generations of intelligent vehicles will be able to understand the driving environment, know their whereabouts, monitor their health, understand the road signs, and monitor driver performance, even overriding drivers to avoid catastrophic accidents. These tasks require significant overhaul of current designs. Drive-by-wire technology replaces the traditional mechanical and hydraulic systems with electronics and control systems, using electromechanical actuators and human–machine interfaces such as pedal and steering feel emulators—otherwise known as haptic systems. Hence, the traditional components—such as the steering column, intermediate shafts, pumps, hoses, fluids, belts, coolers, brake boosters, and master cylinders—are eliminated from the vehicle. Haptic interfaces that can offer adequate transparency to the driver while maintaining safety and stability of the system. Removing the bulky mechanical steering wheel column and the rest of the steering system has clear advantages in terms of mass reduction and safety in modern vehicles, along with improved ergonomics as a result of creating more driver space. Replacing the steering wheel with a haptic device that the driver controls through the sense of touch would be useful in this regard. The haptic device would produce the same sense to the driver as the mechanical steering wheel but with improvements in cost, safety, and fuel consumption as a result of eliminating the bulky mechanical system.

Driver assist systems help drivers to avoid or mitigate an accident by sensing the nature and significance of the danger. Depending on the significance and timing of the threat, these on-board safety systems will initially alert the driver as early as possible to an impending danger. Then, they will actively assist or, ultimately, intervene in order to avert the accident or mitigate its consequences. Provisions for automatic over-ride features when the driver loses control due to fatigue or lack of attention will be an important part of the system. In such systems, the so-called advanced vehicle control system monitors the longitudinal and lateral control, and by interacting with a central management unit, it will be ready to take control of the vehicle whenever the need arises. The system can be readily integrated with sensor networks that monitor every aspect of the conditions on the road and are prepared to take appropriate action in a safe manner.

Integration and Utilization of Advanced Hybrid Powertrains Hybrid technologies offer improved fuel consumption while enhancing driving experience. Utilizing new energy storage and conversion technologies and integrating them with powertrains would be prime objectives of this research activity. Such technologies must be compatible with current platforms and must enhance, rather than compromise, vehicle function. Sample applications would include developing plug-in hybrid technology, which would enhance the vehicle cruising distance based on using battery power alone, and utilizing sustainable energy resources, such as solar and wind power, to charge the batteries. The smart plug-in vehicle can be a part of an integrated smart home and grid energy system of the future, which would utilize smart energy metering devices for optimal use of grid energy by avoiding peak energy consumption hours.

High Performance Real-time Control, Health Monitoring, and Diagnosis Modern vehicles utilize an increasing number of sensors, actuators, and networked embedded computers. The need for high performance computing would increase with the introduction of such revolutionary features as drive-by-wire systems into modern vehicles. The tremendous computational burden of processing sensory data into appropriate control and monitoring signals and diagnostic information creates challenges in the design of embedded computing technology. Towards this end, a related challenge is to incorporate sophisticated computational techniques that control, monitor, and diagnose complex automotive systems while meeting requirements such as low power consumption and cost effectiveness.

The following represent more traditional applications of control that have become part of our daily lives.

Steering Control of an Automobile
As a simple example of the control system, as shown in Fig. 1-1, consider the steering control of an automobile. The direction of the two front wheels can be regarded as the controlled variable, or the output, y; the direction of the steering wheel is the actuating signal, or the input, u. The control system, or process in this case, is composed of the steering mechanism and the dynamics of the entire automobile. However, if the objective is to control the speed of the automobile, then the amount of pressure exerted on the accelerator is the actuating signal, and the vehicle speed is the controlled variable. As a whole, we can regard the simplified automobile control system as one with two inputs (steering and accelerator) and two outputs (heading and speed). In this case, the two controls and two outputs are independent of each other, but there are systems for which the controls are coupled. Systems with more than one input and one output are called **multivariable systems**.

Idle-Speed Control of an Automobile
As another example of a control system, we consider the idle-speed control of an automobile engine. The objective of such a control system is to maintain the engine idle speed at a relatively low value (for fuel economy) regardless of the applied engine loads (e.g., transmission, power steering, air conditioning). Without the idle-speed control, any sudden engine-load application would cause a drop in engine speed that might cause the engine to stall. Thus the main objectives of the idle-speed control system are (1) to eliminate or minimize the speed droop when engine loading is applied and (2) to maintain the engine idle speed at a desired value. Fig. 1-2 shows the block diagram of the idle-speed control system from the standpoint of inputs–system–outputs. In this case, the throttle angle α and the load torque T_L (due to the application of air conditioning, power steering, transmission, or power brakes, etc.) are the inputs, and the engine speed ω is the output. The engine is the controlled process of the system.

Sun-Tracking Control of Solar Collectors
To achieve the goal of developing economically feasible non-fossil-fuel electrical power, the U.S. government has sponsored many organizations in research and development of solar power conversion methods, including the solar-cell conversion techniques. In most of

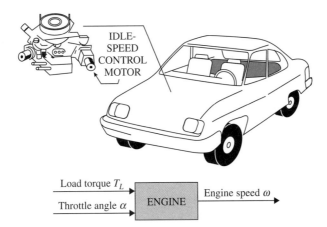

Figure 1-2 Idle-speed control system.

Figure 1-3 Solar collector field.

these systems, the need for high efficiencies dictates the use of devices for sun tracking. Fig. 1-3 shows a solar collector field. Fig. 1-4 shows a conceptual method of efficient water extraction using solar power. During the hours of daylight, the solar collector would produce electricity to pump water from the underground water table to a reservoir (perhaps on a nearby mountain or hill), and in the early morning hours, the water would be released into the irrigation system.

One of the most important features of the solar collector is that the collector dish must track the sun accurately. Therefore, the movement of the collector dish must be controlled by sophisticated control systems. The block diagram of Fig. 1-5 describes the general philosophy of the sun-tracking system together with some of the most important components. The controller ensures that the tracking collector is pointed toward the sun in the morning and sends a "start track" command. The controller constantly calculates the sun's rate for the two axes (azimuth and elevation) of control during the day. The controller uses the sun rate and sun sensor information as inputs to generate proper motor commands to slew the collector.

1-1-3 Open-Loop Control Systems (Nonfeedback Systems)

- Open-loop systems are economical but usually inaccurate.

The idle-speed control system illustrated in Fig. 1-2, shown previously, is rather unsophisticated and is called an **open-loop control system**. It is not difficult to see that the system as shown would not satisfactorily fulfill critical performance requirements. For instance, if the throttle angle α is set at a certain initial value that corresponds to a certain

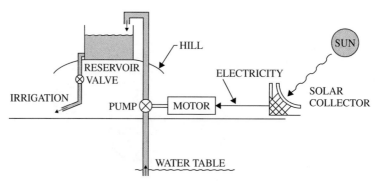

Figure 1-4 Conceptual method of efficient water extraction using solar power.

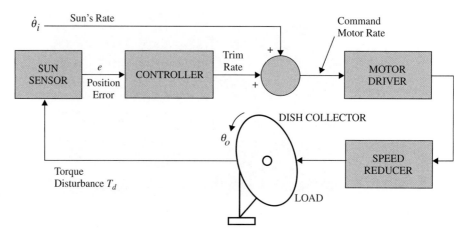

Figure 1-5 Important components of the sun-tracking control system.

engine speed, then when a load torque T_L is applied, there is no way to prevent a drop in the engine speed. The only way to make the system work is to have a means of adjusting α in response to a change in the load torque in order to maintain ω at the desired level. The conventional electric washing machine is another example of an open-loop control system because, typically, the amount of machine wash time is entirely determined by the judgment and estimation of the human operator.

The elements of an open-loop control system can usually be divided into two parts: the **controller** and the **controlled process**, as shown by the block diagram of Fig. 1-6. An input signal, or command, r, is applied to the controller, whose output acts as the actuating signal u; the actuating signal then controls the controlled process so that the controlled variable y will perform according to some prescribed standards. In simple cases, the controller can be

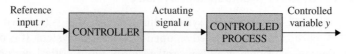

Figure 1-6 Elements of an open-loop control system.

an amplifier, a mechanical linkage, a filter, or other control elements, depending on the nature of the system. In more sophisticated cases, the controller can be a computer such as a microprocessor. Because of the simplicity and economy of open-loop control systems, we find this type of system in many noncritical applications.

1-1-4 Closed-Loop Control Systems (Feedback Control Systems)

What is missing in the open-loop control system for more accurate and more adaptive control is a link or feedback from the output to the input of the system. To obtain more accurate control, the controlled signal y should be fed back and compared with the reference input, and an actuating signal proportional to the difference of the input and the output must be sent through the system to correct the error. A system with one or more feedback paths such as that just described is called a **closed-loop system**.

- Closed-loop systems have many advantages over open-loop systems.

A closed-loop idle-speed control system is shown in Fig. 1-7. The reference input ω_r sets the desired idling speed. The engine speed at idle should agree with the reference value ω_r, and any difference such as the load torque T_L is sensed by the speed transducer and the error detector. The controller will operate on the difference and provide a signal to adjust the throttle angle α to correct the error. Fig. 1-8 compares the typical performances of open-loop and closed-loop idle-speed control systems. In Fig. 1-8(a), the idle speed of the open-loop system will drop and settle at a lower value after a load torque is applied. In Fig. 1-8 (b), the idle speed of the closed-loop system is shown to recover quickly to the preset value after the application of T_L.

The objective of the idle-speed control system illustrated, also known as a **regulator system**, is to maintain the system output at a prescribed level.

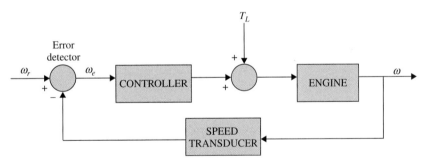

Figure 1-7 Block diagram of a closed-loop idle-speed control system.

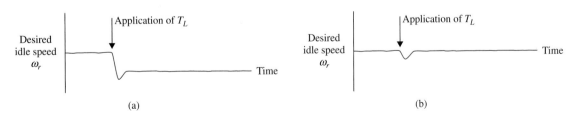

Figure 1-8 (a) Typical response of the open-loop idle-speed control system. (b) Typical response of the closed-loop idle-speed control system.

1-2 WHAT IS FEEDBACK, AND WHAT ARE ITS EFFECTS?

The motivation for using feedback, as illustrated by the examples in Section 1-1, is somewhat oversimplified. In these examples, feedback is used to reduce the error between the reference input and the system output. However, the significance of the effects of feedback in control systems is more complex than is demonstrated by these simple examples. The reduction of system error is merely one of the many important effects that feedback may have upon a system. We show in the following sections that feedback also has effects on such system performance characteristics as **stability**, **bandwidth**, **overall gain**, **impedance**, and **sensitivity**.

- Feedback exists whenever there is a closed sequence of cause-and-effect relationships.

To understand the effects of feedback on a control system, it is essential to examine this phenomenon in a broad sense. When feedback is deliberately introduced for the purpose of control, its existence is easily identified. However, there are numerous situations where a physical system that we recognize as an inherently nonfeedback system turns out to have feedback when it is observed in a certain manner. In general, we can state that whenever a closed sequence of **cause-and-effect relationships** exists among the variables of a system, feedback is said to exist. This viewpoint will inevitably admit feedback in a large number of systems that ordinarily would be identified as nonfeedback systems. However, control-system theory allows numerous systems, with or without physical feedback, to be studied in a systematic way once the existence of feedback in the sense mentioned previously is established.

We shall now investigate the effects of feedback on the various aspects of system performance. Without the necessary mathematical foundation of linear-system theory, at this point we can rely only on simple static-system notation for our discussion. Let us consider the simple feedback system configuration shown in Fig. 1-9, where r is the input signal; y, the output signal; e, the error; and b, the feedback signal. The parameters G and H may be considered as constant gains. By simple algebraic manipulations, it is simple to show that the input–output relation of the system is

$$M = \frac{y}{r} = \frac{G}{1 + GH} \qquad (1\text{-}1)$$

Using this basic relationship of the feedback system structure, we can uncover some of the significant effects of feedback.

1-2-1 Effect of Feedback on Overall Gain

- Feedback may increase the gain of a system in one frequency range but decrease it in another.

As seen from Eq. (1-1), feedback affects the gain G of a nonfeedback system by a factor of $1 + GH$. The system of Fig. 1-9 is said to have **negative feedback**, because a minus sign is assigned to the feedback signal. The quantity GH may itself include a minus sign, so the *general effect of feedback is that it may increase or decrease the gain* G. In a practical control system, G and H are functions of frequency, so the magnitude of $1 + GH$ may be

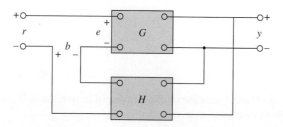

Figure 1-9 Feedback system.

greater than 1 in one frequency range but less than 1 in another. Therefore, *feedback could increase the gain of system in one frequency range but decrease it in another.*

1-2-2 Effect of Feedback on Stability

• A system is unstable if its output is out of control.

Stability is a notion that describes whether the system will be able to follow the input command, that is, be useful in general. In a nonrigorous manner, *a system is said to be unstable if its output is out of control.* To investigate the effect of feedback on stability, we can again refer to the expression in Eq. (1-1). If $GH = -1$, the output of the system is infinite for any finite input, and the system is said to be unstable. Therefore, we may state that *feedback can cause a system that is originally stable to become unstable.* Certainly, feedback is a double-edged sword; when it is improperly used, it can be harmful. It should be pointed out, however, that we are only dealing with the static case here, and, in general, $GH = -1$ is not the only condition for instability. The subject of system stability will be treated formally in Chapters 2, 5, 7, and 8.

It can be demonstrated that one of the advantages of incorporating feedback is that it can stabilize an unstable system. Let us assume that the feedback system in Fig. 1-9 is unstable because $GH = -1$. If we introduce another feedback loop through a negative feedback gain of F, as shown in Fig. 1-10, the input–output relation of the overall system is

$$\frac{y}{r} = \frac{G}{1 + GH + GF} \tag{1-2}$$

• Feedback can improve stability or be harmful to stability.

It is apparent that although the properties of G and H are such that the inner-loop feedback system is unstable, because $GH = -1$, the overall system can be stable by properly selecting the outer-loop feedback gain F. In practice, GH is a function of frequency, and the stability condition of the closed-loop system depends on the **magnitude** and **phase** of GH. The bottom line is that *feedback can improve stability or be harmful to stability if it is not properly applied.*

Sensitivity considerations often are important in the design of control systems. Because all physical elements have properties that change with environment and age, we cannot always consider the parameters of a control system to be completely stationary over the entire operating life of the system. For instance, the winding resistance of an electric motor changes as the temperature of the motor rises during operation. Control systems with electric components may not operate normally when first turned on because

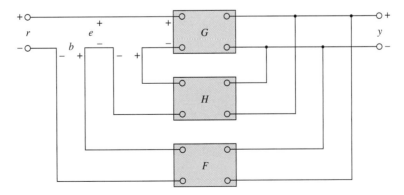

Figure 1-10 Feedback system with two feedback loops.

of the still-changing system parameters during warmup. This phenomenon is sometimes called "morning sickness." Most duplicating machines have a warmup period during which time operation is blocked out when first turned on.

In general, a good control system should be very insensitive to parameter variations but sensitive to the input commands. We shall investigate what effect feedback has on sensitivity to parameter variations. Referring to the system in Fig. 1-9, we consider G to be a gain parameter that may vary. The sensitivity of the gain of the overall system M to the variation in G is defined as

$$S_G^M = \frac{\partial M/M}{\partial G/G} = \frac{\text{percentage change in } M}{\text{percentage change in } G} \quad (1\text{-}3)$$

- Note: Feedback can increase or decrease the sensitivity of a system.

where ∂M denotes the incremental change in M due to the incremental change in G, or ∂G. By using Eq. (1-1), the sensitivity function is written

$$S_G^M = \frac{\partial M}{\partial G} \frac{G}{M} = \frac{1}{1 + GH} \quad (1\text{-}4)$$

This relation shows that if GH is a positive constant, the magnitude of the sensitivity function can be made arbitrarily small by increasing GH, provided that the system remains stable. It is apparent that, in an open-loop system, the gain of the system will respond in a one-to-one fashion to the variation in G (i.e., $S_G^M = 1$). Again, in practice, GH is a function of frequency; the magnitude of $1 + GH$ may be less than unity over some frequency ranges, so feedback could be harmful to the sensitivity to parameter variations in certain cases. In general, the sensitivity of the system gain of a feedback system to parameter variations depends on where the parameter is located. The reader can derive the sensitivity of the system in Fig. 1-9 due to the variation of H.

1-2-3 Effect of Feedback on External Disturbance or Noise

All physical systems are subject to some types of extraneous signals or noise during operation. Examples of these signals are thermal-noise voltage in electronic circuits and brush or commutator noise in electric motors. External disturbances, such as wind gusts acting on an antenna, are also quite common in control systems. Therefore, control systems should be designed so that they are insensitive to noise and disturbances and sensitive to input commands.

- Feedback can reduce the effect of noise.

The effect of feedback on noise and disturbance depends greatly on where these extraneous signals occur in the system. No general conclusions can be reached, but in many situations, *feedback can reduce the effect of noise and disturbance on system performance*. Let us refer to the system shown in Fig. 1-11, in which r denotes the command

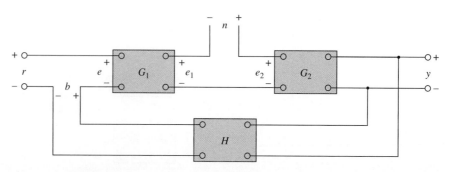

Figure 1-11 Feedback system with a noise signal.

signal and n is the noise signal. In the absence of feedback, that is, $H = 0$, the output y due to n acting alone is

$$y = G_2 n \qquad (1\text{-}5)$$

With the presence of feedback, the system output due to n acting alone is

$$y = \frac{G_2}{1 + G_1 G_2 H}\, n \qquad (1\text{-}6)$$

- Feedback also can affect bandwidth, impedance, transient responses, and frequency responses.

Comparing Eq. (1-6) with Eq. (1-5) shows that the noise component in the output of Eq. (1-6) is reduced by the factor $1 + G_1 G_2 H$ if the latter is greater than unity and the system is kept stable.

In Chapter 9, the feedforward and forward controller configurations are used along with feedback to reduce the effects of disturbance and noise inputs. In general, feedback also has effects on such performance characteristics as bandwidth, impedance, transient response, and frequency response. These effects will be explained as we continue.

▶ 1-3 TYPES OF FEEDBACK CONTROL SYSTEMS

Feedback control systems may be classified in a number of ways, depending upon the purpose of the classification. For instance, according to the method of analysis and design, control systems are classified as **linear** or **nonlinear**, and **time-varying** or **time-invariant**. According to the types of signal found in the system, reference is often made to **continuous-data** or **discrete-data** systems, and **modulated** or **unmodulated** systems. Control systems are often classified according to the main purpose of the system. For instance, a **position-control system** and a **velocity-control system** control the output variables just as the names imply. In Chapter 9, the **type** of control system is defined according to the form of the open-loop transfer function. In general, there are many other ways of identifying control systems according to some special features of the system. It is important to know some of the more common ways of classifying control systems before embarking on the analysis and design of these systems.

1-3-1 Linear versus Nonlinear Control Systems

- Most real-life control systems have nonlinear characteristics to some extent.

This classification is made according to the methods of analysis and design. Strictly speaking, linear systems do not exist in practice, because all physical systems are nonlinear to some extent. Linear feedback control systems are idealized models fabricated by the analyst purely for the simplicity of analysis and design. When the magnitudes of signals in a control system are limited to ranges in which system components exhibit linear characteristics (i.e., the principle of superposition applies), the system is essentially linear. But when the magnitudes of signals are extended beyond the range of the linear operation, depending on the severity of the nonlinearity, the system should no longer be considered linear. For instance, amplifiers used in control systems often exhibit a saturation effect when their input signals become large; the magnetic field of a motor usually has saturation properties. Other common nonlinear effects found in control systems are the backlash or dead play between coupled gear members, nonlinear spring characteristics, nonlinear friction force or torque between moving members, and so on. Quite often, nonlinear characteristics are intentionally introduced in a control system to improve its performance or provide more effective control. For instance, to achieve minimum-time control, an on-off (bang-bang or relay) type controller is used in many missile or spacecraft control systems. Typically in these systems, jets are mounted on the sides of the vehicle to provide reaction torque for attitude control. These jets are often controlled in a full-on or full-off fashion, so a fixed amount of air is applied from a given jet for a certain time period to control the attitude of the space vehicle.

For linear systems, a wealth of analytical and graphical techniques is available for design and analysis purposes. A majority of the material in this text is devoted to the analysis and design of linear systems. Nonlinear systems, on the other hand, are usually difficult to treat mathematically, and there are no general methods available for solving a wide class of nonlinear systems. It is practical to first design the controller based on the linear-system model by neglecting the nonlinearities of the system. The designed controller is then applied to the nonlinear system model for evaluation or redesign by computer simulation. The Virtual Lab introduced in Chapter 6 is mainly used to model the characteristics of practical systems with realistic physical components.

- There are no general methods for solving a wide class of nonlinear systems.

1-3-2 Time-Invariant versus Time-Varying Systems

When the parameters of a control system are stationary with respect to time during the operation of the system, the system is called a time-invariant system. In practice, most physical systems contain elements that drift or vary with time. For example, the winding resistance of an electric motor will vary when the motor is first being excited and its temperature is rising. Another example of a time-varying system is a guided-missile control system in which the mass of the missile decreases as the fuel on board is being consumed during flight. Although a time-varying system without nonlinearity is still a linear system, the analysis and design of this class of systems are usually much more complex than that of the linear time-invariant systems.

Continuous-Data Control Systems

A continuous-data system is one in which the signals at various parts of the system are all functions of the continuous time variable t. The signals in continuous-data systems may be further classified as ac or dc. Unlike the general definitions of ac and dc signals used in electrical engineering, ac and dc control systems carry special significance in control systems terminology. When one refers to an **ac control system**, it usually means that the signals in the system are *modulated* by some form of modulation scheme. A **dc control system**, on the other hand, simply implies that the signals are *unmodulated*, but they are still ac signals according to the conventional definition. The schematic diagram of a closed-loop dc control system is shown in Fig. 1-12. Typical waveforms of the signals in response to a step-function input are shown in the figure. Typical components of a dc control system are potentiometers, dc amplifiers, dc motors, dc tachometers, and so on.

Figure 1-13 shows the schematic diagram of a typical ac control system that performs essentially the same task as the dc system in Fig. 1-12. In this case, the signals in the system are modulated; that is, the information is transmitted by an ac carrier signal. Notice that the output controlled variable still behaves similarly to that of the dc system. In this case, the modulated signals are demodulated by the low-pass characteristics of the ac motor. Ac control systems are used extensively in aircraft and missile control systems in which noise and disturbance often create problems. By using modulated ac control systems with carrier frequencies of 400 Hz or higher, the system will be less susceptible to low-frequency noise. Typical components of an ac control system are synchros, ac amplifiers, ac motors, gyroscopes, accelerometers, and so on.

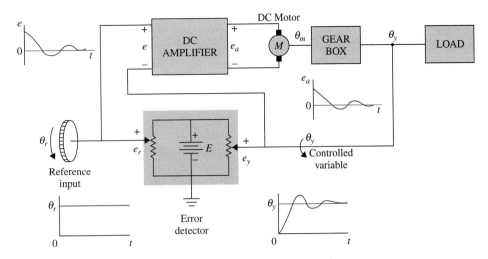

Figure 1-12 Schematic diagram of a typical dc closed-loop system.

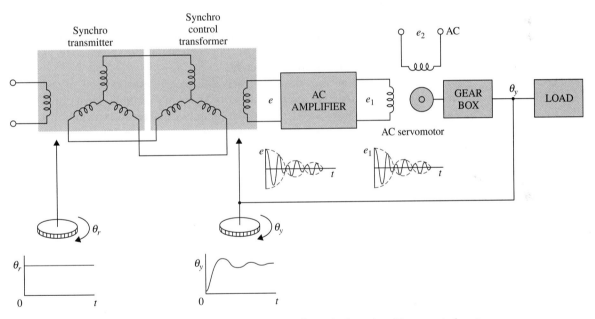

Figure 1-13 Schematic diagram of a typical ac closed-loop control system.

In practice, not all control systems are strictly of the ac or dc type. A system may incorporate a mixture of ac and dc components, using modulators and demodulators to match the signals at various points in the system.

Discrete-Data Control Systems

Discrete-data control systems differ from the continuous-data systems in that the signals at one or more points of the system are in the form of either a pulse train or a digital code. Usually, discrete-data control systems are subdivided into **sampled-data** and **digital control systems**. Sampled-data control systems refer to a more general class of

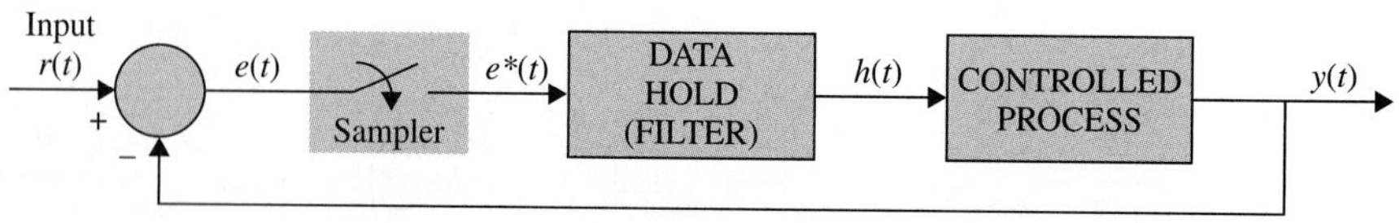

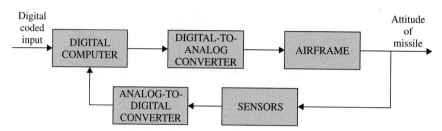

Figure 1-14 Block diagram of a sampled-data control system.

Figure 1-15 Digital autopilot system for a guided missile.

discrete-data systems in which the signals are in the form of pulse data. A digital control system refers to the use of a digital computer or controller in the system so that the signals are digitally coded, such as in binary code.

In general, a sampled-data system receives data or information only intermittently at specific instants of time. For example, the error signal in a control system can be supplied only in the form of pulses, in which case the control system receives no information about the error signal during the periods between two consecutive pulses. Strictly, a sampled-data system can also be classified as an ac system, because the signal of the system is pulse modulated.

Figure 1-14 illustrates how a typical sampled-data system operates. A continuous-data input signal $r(t)$ is applied to the system. The error signal $e(t)$ is sampled by a sampling device, the **sampler**, and the output of the sampler is a sequence of pulses. The sampling rate of the sampler may or may not be uniform. There are many advantages to incorporating sampling into a control system. One important advantage is that expensive equipment used in the system may be time-shared among several control channels. Another advantage is that pulse data are usually less susceptible to noise.

- Digital control systems are usually less susceptible to noise.

Because digital computers provide many advantages in size and flexibility, computer control has become increasingly popular in recent years. Many airborne systems contain digital controllers that can pack thousands of discrete elements into a space no larger than the size of this book. Figure 1-15 shows the basic elements of a digital autopilot for guided-missile control.

▶ 1-4 SUMMARY

In this chapter, we introduced some of the basic concepts of what a control system is and what it is supposed to accomplish. The basic components of a control system were described. By demonstrating the effects of feedback in a rudimentary way, the question of why most control systems are closed-loop systems was also clarified. Most important, it was pointed out that feedback is a double-edged sword—it can benefit as well as harm the system to be controlled. This is part of the challenging task of designing a control system, which involves consideration of such performance criteria as stability, sensitivity, bandwidth, and accuracy. Finally, various types of control systems were categorized according to the system signals, linearity, and control objectives. Several typical control-system examples were given to illustrate the analysis and design of control systems. Most systems encountered in real life are nonlinear and time-varying to some extent. The concentration on the studies of linear systems is due primarily to the availability of unified and simple-to-understand analytical methods in the analysis and design of linear systems.

▶ REVIEW QUESTIONS

1. List the advantages and disadvantages of an open-loop system.
2. List the advantages and disadvantages of a closed-loop system.
3. Give the definitions of ac and dc control systems.
4. Give the advantages of a digital control system over a continuous-data control system.
5. A closed-loop control system is usually more accurate than an open-loop system. (T) (F)
6. Feedback is sometimes used to improve the sensitivity of a control system. (T) (F)
7. If an open-loop system is unstable, then applying feedback will always improve its stability. (T) (F)
8. Feedback can increase the gain of a system in one frequency range but decrease it in another. (T) (F)
9. Nonlinear elements are sometimes intentionally introduced to a control system to improve its performance. (T) (F)
10. Discrete-data control systems are more susceptible to noise due to the nature of their signals. (T) (F)

Answers to these review questions can be found on this book's companion Web site: www.wiley.com/college/golnaraghi.

CHAPTER 2

Mathematical Foundation

The studies of control systems rely to a great extent on applied mathematics. One of the major purposes of control-system studies is to develop a set of analytical tools so that the designer can arrive with reasonably predictable and reliable designs without depending solely on the drudgery of experimentation or extensive computer simulation.

In this chapter, it is assumed that the reader has some level of familiarity with these concepts through earlier courses. Elementary matrix algebra is covered in Appendix A. Because of space limitations, as well as the fact that most subjects are considered as review material for the reader, the treatment of these mathematical subjects is not exhaustive. The reader who wishes to conduct an in-depth study of any of these subjects should refer to books that are devoted to them.

The main objectives of this chapter are:

1. To introduce the fundamentals of complex variables.
2. To introduce frequency domain analysis and frequency plots.
3. To introduce differential equations and state space systems.
4. To introduce the fundamentals of Laplace transforms.
5. To demonstrate the applications of Laplace transforms to solve linear ordinary differential equations.
6. To introduce the concept of transfer functions and how to apply them to the modeling of linear time-invariant systems.
7. To discuss stability of linear time-invariant systems and the Routh-Hurwitz criterion.
8. To demonstrate the MATLAB tools using case studies.

▶ 2-1 COMPLEX-VARIABLE CONCEPT

To understand complex variables, it is wise to start with the concept of complex numbers and their mathematical properties.

2-1-1 Complex Numbers

A complex number is represented in **rectangular form** as

$$z = x + jy \qquad (2\text{-}1)$$

where, $j = \sqrt{-1}$ and (x, y) are real and imaginary coefficients of z respectively. We can treat (x, y) as a point in the **Cartesian** coordinate frame shown in Fig. 2-1. A point in a

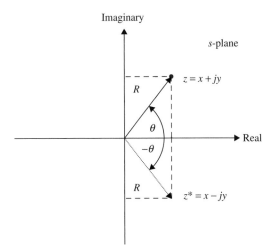

Figure 2-1 Complex number z representation in rectangular and polar forms.

rectangular coordinate frame may also be defined by a vector R and an angle θ. It is then easy to see that

$$x = R\cos\theta \\ y = R\sin\theta \tag{2-2}$$

where,

R = magnitude of z

θ = phase of z and is measured from the x axis. Right-hand rule convention: positive phase is in counter clockwise direction.

Hence,

$$R = \sqrt{x^2 + y^2} \\ \theta = \tan^{-1}\frac{x}{y} \tag{2-3}$$

Introducing Eq. (2-2) into Eq. (2-1), we get

$$z = R\cos\theta + jR\sin\theta \tag{2-4}$$

Upon comparison of Taylor series of the terms involved, it is easy to confirm

$$e^{j\theta} = R\cos\theta + jR\sin\theta \tag{2-5}$$

Eq. (2-5) is also known as the **Euler formula**. As a result, Eq. (2-1) may also be represented in **polar form** as

$$z = Re^{j\theta} = R\angle\theta \tag{2-6}$$

We define the **conjugate** of the complex number z in Eq. (2-1) as

$$z^* = x - jy \tag{2-7}$$

Or, alternatively,

$$z^* = R\cos\theta - jR\sin\theta = Re^{-j\theta} \tag{2-8}$$

Note:

$$zz^* = R^2 = x^2 + y^2 \tag{2-9}$$

Table 2-1 shows basic mathematical properties of complex numbers.

TABLE 2-1 Basic Properties of Complex Numbers

Addition	$\begin{cases} z_1 = x_1 + jy_1 \\ z_2 = x_2 + jy_2 \end{cases}$ $\rightarrow z = (x_1 + x_2) + j(y_1 + y_2)$	$\begin{cases} z_1 = R_1 e^{j\theta_1} \\ z_2 = R_2 e^{j\theta_2} \end{cases}$ $\rightarrow z = (R_1 + R_2) e^{j(\theta_1 + \theta_2)}$ $\rightarrow z = (R_1 + R_2) \angle (\theta_1 + \theta_2)$
Subtraction	$\begin{cases} z_1 = x_1 + jy_1 \\ z_2 = x_2 + jy_2 \end{cases}$ $\rightarrow z = (x_1 + x_2) - j(y_1 + y_2)$	$\begin{cases} z_1 = R_1 e^{j\theta_1} \\ z_2 = R_2 e^{j\theta_2} \end{cases}$ $\rightarrow z = (R_1 - R_2) e^{j(\theta_1 - \theta_2)}$ $\rightarrow z = (R_1 - R_2) \angle (\theta_1 - \theta_2)$
Multiplication	$\begin{cases} z_1 = x_1 + jy_1 \\ z_2 = x_2 + jy_2 \end{cases}$ $\rightarrow z = (x_1 x_2 - y_1 y_2) - j(x_1 y_2 + x_2 y_1)$ $j^2 = -1$	$\begin{cases} z_1 = R_1 e^{j\theta_1} \\ z_2 = R_2 e^{j\theta_2} \end{cases}$ $\rightarrow z = (R_1 R_2) e^{j(\theta_1 + \theta_2)}$ $\rightarrow z = (R_1 R_2) \angle (\theta_1 + \theta_2)$
Division	$\begin{cases} z_1 = x_1 + jy_1 \\ z_2 = x_2 + jy_2 \end{cases}$ $\begin{cases} z_1^* = x_1 - jy_1 \\ z_2^* = x_2 - jy_2 \end{cases}$ Complex Conjugate $\rightarrow z = \dfrac{z_1}{z_2}$ $\rightarrow z = \dfrac{z_1 z_2^*}{z_2 z_2^*} = \dfrac{(x_1 x_2 + y_1 y_2) + j(x_1 y_2 + x_2 y_1)}{x_2^2 + y_2^2}$	$\begin{cases} z_1 = R_1 e^{j\theta_1} \\ z_2 = R_2 e^{j\theta_2} \end{cases}$ $\begin{cases} z_1^* = R_1 e^{-j\theta_1} \\ z_2^* = R_2 e^{-j\theta_2} \end{cases}$ $\rightarrow z = \left(\dfrac{R_1}{R_2}\right) e^{j(\theta_1 - \theta_2)}$ $\rightarrow z = \left(\dfrac{R_1}{R_2}\right) \angle (\theta_1 - \theta_2)$

▶ **EXAMPLE 2-1-1** Find j^3 and j^4.

$$j = \sqrt{-1} = \cos\frac{\pi}{2} + j\sin\frac{\pi}{2} = e^{j\frac{\pi}{2}}$$

$$j^3 = \sqrt{-1}\sqrt{-1}\sqrt{-1} = -\sqrt{-1} = -j$$

$$j^3 = e^{j\frac{3\pi}{2}} = e^{-j\frac{\pi}{2}}$$

$$j^4 = j^3 j = -j^2 = 1$$

◀

▶ **EXAMPLE 2-1-2** Find z^n using Eq. (2-6).

$$z^n = \left(R e^{j\theta}\right)^n = R^n e^{jn\theta} = R^n \angle n\theta \tag{2-10}$$

◀

2-1-2 Complex Variables

A complex variable s has two components: a real component σ and an imaginary component ω. Graphically, *the real component of* s *is represented by a* σ *axis* in the horizontal direction, and *the imaginary component is measured along the vertical* $j\omega$ *axis*, in the complex s-plane. Fig. 2-2 illustrates the complex s-plane, in which any arbitrary point $s = s_1$ is defined by the coordinates $\sigma = \sigma_1$, and $\omega = \omega_1$, or simply $s_1 = \sigma_1 + j\omega_1$.

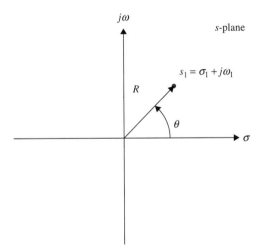

Figure 2-2 Complex s-plane.

2-1-3 Functions of a Complex Variable

The function $G(s)$ is said to be a function of the complex variable s if, for every value of s, there is one or more corresponding values of $G(s)$. Because s is defined to have real and imaginary parts, the function $G(s)$ is also represented by its real and imaginary parts; that is,

$$G(s) = \text{Re}[G(s)] + j\,\text{Im}[G(s)] \tag{2-11}$$

where $\text{Re}[G(s)]$ denotes the real part of $G(s)$, and $\text{Im}[G(s)]$ represents the imaginary part of $G(s)$. The function $G(s)$ is also represented by the complex $G(s)$-plane, with $\text{Re}[G(s)]$ as the real axis and $\text{Im}[G(s)]$ as the imaginary axis. If for every value of s there is only one corresponding value of $G(s)$ in the $G(s)$-plane, $G(s)$ is said to be a **single-valued function**, and the mapping from points in the s-plane onto points in the $G(s)$-plane is described as **single-valued** (Fig. 2-3). If the mapping from the $G(s)$-plane to the s-plane is also single-valued, the mapping is called **one-to-one**. However, there are many functions for which the mapping from the function plane to the complex-variable plane is not single-valued. For instance, given the function

$$G(s) = \frac{1}{s(s+1)} \tag{2-12}$$

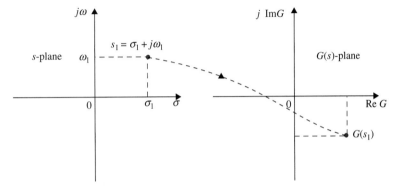

Figure 2-3 Single-valued mapping from the s-plane to the $G(s)$-plane.

it is apparent that, for each value of s, there is only one unique corresponding value for $G(s)$. However, the inverse mapping is not true; for instance, the point $G(s) = \infty$ is mapped onto two points, $s = 0$ and $s = -1$, in the s-plane.

2-1-4 Analytic Function

A function $G(s)$ of the complex variable s is called an **analytic function** in a region of the s-plane if the function and all its derivatives exist in the region. For instance, the function given in Eq. (2-12) is analytic at every point in the s-plane except at the points $s = 0$ and $s = -1$. At these two points, the value of the function is infinite. As another example, the function $G(s) = s + 2$ is analytic at every point in the finite s-plane.

2-1-5 Singularities and Poles of a Function

The **singularities** of a function are the points in the s-plane at which the function or its derivatives do not exist. A pole is the most common type of singularity and plays a very important role in the studies of classical control theory.

The definition of a **pole** can be stated as: If a function $G(s)$ is analytic and single-valued in the neighborhood of point p_i, it is said to have a pole of order r at $s = p_i$ if the limit $\lim_{s \to p_i} [(s - p_i)^r G(s)]$ has a finite, nonzero value. In other words, the denominator of $G(s)$ must include the factor $(s - p_i)^r$, so when $s = p_i$, the function becomes infinite. If $r = 1$, the pole at $s = p_i$ is called a **simple pole**. As an example, the function

$$G(s) = \frac{10(s+2)}{s(s+1)(s+3)^2} \tag{2-13}$$

has a pole of order 2 at $s = -3$ and simple poles at $s = 0$ and $s = -1$. It can also be said that the function $G(s)$ is analytic in the s-plane except at these poles. See Fig. 2-4 for the graphical representation of the finite poles of the system.

2-1-6 Zeros of a Function

The definition of a **zero** of a function can be stated as: If the function $G(s)$ is analytic at $s = z_i$, it is said to have a zero of order r at $s = z_i$ if the limit $\lim_{s \to z_i} [(s - z_i)^{-r} G(s)]$ has a finite, nonzero value. Or, simply, $G(s)$ has a zero of order r at $s = z_i$ if $1/G(s)$ has an rth-order pole at $s = z_i$. For example, the function in Eq. (2-13) has a simple zero at $s = -2$.

If the function under consideration is a rational function of s, that is, a quotient of two polynomials of s, the total number of poles equals the total number of zeros, counting the multiple-order poles and zeros and taking into account the poles and zeros at infinity. The function in Eq. (2-13) has four finite poles at $s = 0, -1, -3$, and -3; there is one finite zero at $s = -2$, but there are three zeros at infinity, because

$$\lim_{s \to \infty} G(s) = \lim_{s \to \infty} \frac{10}{s^3} = 0 \tag{2-14}$$

Therefore, the function has a total of four poles and four zeros in the entire s-plane, including infinity. See Fig. 2-4 for the graphical representation of the finite zeros of the system.

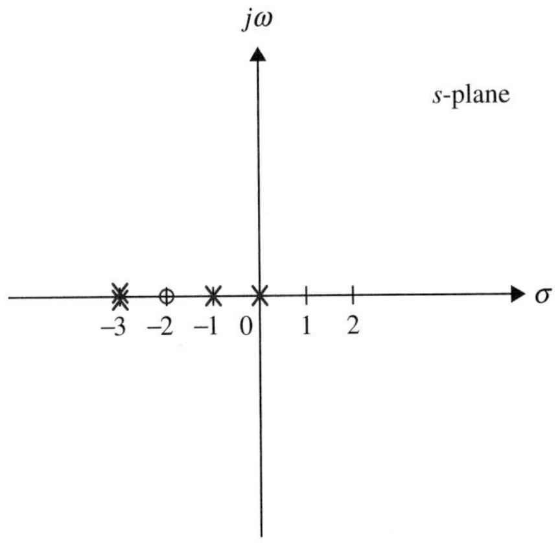

Figure 2-4 Graphical representation of $G(s) = \frac{10(s+2)}{s(s+1)(s+3)^2}$ in the s-plane: × poles and O zeros.

Toolbox 2-1-1

For Eq. (2-13), use "zpk" to create zero-pole-gain models by the following sequence of MATLAB functions

```
>> G = zpk([-2],[0 -1 -3 -3],10)

Zero/pole/gain:
     10 (s + 2)
---------------
s (s + 1) (s + 3)^2
```

Convert the transfer function to polynomial form

```
>> Gp = tf(G)

Transfer function:
         10 s + 20
--------------------------
s^4 + 7 s^3 + 15 s^2 + 9 s
```

Alternatively use:

```
>> clear all
>> s = tf('s');
>> Gp = 10*(s + 2)/(s*(s + 1)*(s + 3)^2)

Transfer function:
         10 s + 20
--------------------------
s^4 + 7 s^3 + 15 s^2 + 9 s
```

Use "pole" and "zero" to obtain the poles and zeros of the transfer function

```
>> pole(Gp)

ans =
     0
    -1
    -3
    -3

>> zero(Gp)

ans =
    -2
```

Convert the transfer function Gp to zero-pole-gain form

```
>> Gzpk = zpk(Gp)

Zero/pole/gain:
     10 (s + 2)
---------------
s (s + 3)^2 (s + 1)
```

2-1-7 Polar Representation

To find the **polar representation** of $G(s)$ in Eq. (2-12) at $s = 2j$, we look at individual components. That is

$$G(s) = \frac{1}{s(s+1)} \tag{2-15}$$

$$s = 2j = Re^{j\theta} = 2e^{j\frac{\pi}{2}}$$

$$\begin{cases} s+1 = 2j+1 = Re^{j\theta} \\ R = \sqrt{2^2+1} = \sqrt{5} \\ \theta = \tan^{-1}\frac{1}{2} = 0.46\,rad\,(= 26.57°) \end{cases} \tag{2-16}$$

$$G(2j) = \frac{1}{2j(2j+1)} = \frac{1}{2}e^{-j\frac{\pi}{2}}\frac{1}{\sqrt{5}}e^{-j\tan^{-1}\frac{1}{2}} = \frac{1}{2\sqrt{5}}e^{-j\left(\frac{\pi}{2}+\tan^{-1}\frac{1}{2}\right)} \tag{2-17}$$

See Fig. 2-5 for a graphical representation of $s_1 = 2j+1$ in the s-plane.

▶ **EXAMPLE 2-1-3** Find the polar representation of $G(s)$ given below for $s = j\omega$, where ω is a constant varying from zero to infinity.

$$G(s) = \frac{16}{s^2+10s+16} = \frac{16}{(s+2)(s+8)} \tag{2-18}$$

To evaluate Eq. (2-18) at $s = j\omega$, we look at individual components. Thus,

$$j\omega + 2 = \sqrt{2^2+\omega^2}\,e^{j\phi_1} \tag{2-19}$$

$$\omega = R_1 \sin\phi_1 \tag{2-20}$$

$$2 = R_1 \cos\phi_1 \tag{2-21}$$

$$R_1 = \sqrt{2^2+\omega^2} \tag{2-22}$$

$$\phi_1 = \tan^{-1}\frac{\omega/R_1}{2/R_1} \tag{2-23}$$

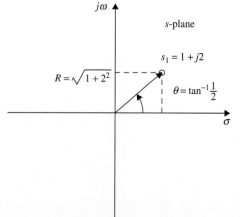

Figure 2-5 Graphical representation of $s_1 = 2j+1$ in the s-plane.

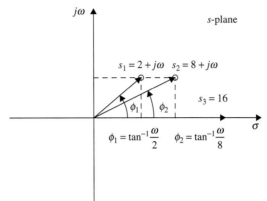

Figure 2-6 Graphical representation of components of $\frac{16}{(\omega j + 2)(\omega j + 8)}$.

$$j\omega + 2 = R_1(j\sin\phi_1 + \cos\phi_1) \tag{2-24}$$

$$j\omega + 2 = R_1 e^{j\phi_1} \tag{2-25}$$

$$j\omega + 8 = \sqrt{8^2 + \omega^2}\, e^{j\phi_2} \tag{2-26}$$

$$\phi_2 = \tan^{-1}\frac{\omega/R_2}{8/R_2} \tag{2-27}$$

$$16 = 16\, e^0 \tag{2-28}$$

See Fig. 2-6 for a graphical representation of components of $\frac{16}{(\omega j + 2)(\omega j + 8)}$.

Hence,

$$\begin{aligned}\frac{1}{j\omega + 2} &= \frac{1}{\sqrt{2^2 + \omega^2}\, e^{j\phi_1}} \\ \frac{1}{j\omega + 8} &= \frac{1}{\sqrt{8^2 + \omega^2}\, e^{j\phi_2}}\end{aligned} \tag{2-29}$$

As a result, $G(s = j\omega)$ becomes:

$$G(j\omega) = \frac{16}{\sqrt{2^2 + \omega^2}\sqrt{8^2 + \omega^2}}\, e^{-j(\phi_1 + \phi_2)} = |G(j\omega)| e^{j\phi} \tag{2-30}$$

where

$$R = G(\omega) = |G(j\omega)| = \frac{16}{\sqrt{(\omega^2 + 4)(\omega^2 + 64)}} \tag{2-31}$$

Similarly, we can define

$$\phi = \tan^{-1}\frac{\operatorname{Im} G(j\omega)}{\operatorname{Re} G(j\omega)} = \angle G(s = j\omega) = -\phi_1 - \phi_2 \tag{2-32}$$

Table 2-2 describes different R and ϕ values as ω changes. As shown, the magnitude decreases as the frequency increases. The phase goes from 0° to −180°.

TABLE 2-2 Numerical Values of Sample Magnitude and Phase of the System in Example 2-1-3

ω rad/s	R	ϕ
0.1	0.999	−3.58
1	0.888	−33.69
10	0.123	−130.03
100	0.0016	−174.28

Alternative Approach: If we multiply both numerator and denominator of Eq. (2-18) by the complex conjugate of the denominator, i.e. $\dfrac{(-j\omega + 2)(-j\omega + 8)}{(-j\omega + 2)(-j\omega + 8)} = 1$, we get

$$\begin{aligned}
G(j\omega) &= \frac{16(-j\omega + 2)(-j\omega + 8)}{(\omega^2 + 2^2)(\omega^2 + 8^2)} \\
&= \frac{16}{(\omega^2 + 4)(\omega^2 + 64)}\left[(16 - \omega^2) - j10\omega\right] \\
&= \text{Real} + \text{Imaginary} \\
&= \frac{16\sqrt{(16 - \omega^2)^2 + (10\omega)^2}}{(\omega^2 + 4)(\omega^2 + 64)}e^{j\phi} \\
&= \frac{16}{\sqrt{(\omega^2 + 4)(\omega^2 + 64)}}e^{j\phi} \\
&= R e^{j\phi}
\end{aligned} \qquad (2\text{-}33)$$

where $\phi = \tan^{-1}\dfrac{-10\omega/R}{(16 - \omega^2)/R} = \dfrac{\text{Im}(G(j\omega))}{\text{Re}(G(j\omega))}$

See Fig. 2-7 for a graphical representation of $\dfrac{16}{(\omega j + 2)(\omega j + 8)}$ for a fixed value of ω. So as you have noticed, the frequency response can be determined graphically. Consider the following second order system:

$$G(s) = \frac{K}{(s + p_1)(s + p_2)} \qquad (2\text{-}34)$$

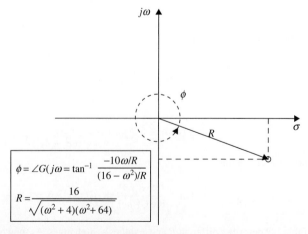

Figure 2-7 Graphical representation of $\dfrac{16}{(\omega j + 2)(\omega j + 8)}$ for a fixed value of ω.

Toolbox 2-1-2

Here are MATLAB commands to treat complex variables:

Z = complex (a,b)

creates a complex output, Z, from the two real inputs $Z = a + bi$

ZC = conj (Z)

returns the complex conjugate of the elements of Z

X = real (Z)

returns the real part of the elements of the complex array Z

Y = imag (Z)

returns the imaginary part of the elements of array Z

R = abs (Z)

returns the complex modulus (magnitude), which is the same as

R = sqrt(real(Z).^2 + imag(Z).^2)

theta = angle(Z)

returns the phase angles, in radians, for each element of complex array Z

The angles lie between the "real axis" in the *s*-plane and the magnitude *R*

Z = R.*exp(i*theta)

converts back to the original complex Z

```
>> Z = complex(3,2)
Z =
   3.0000 + 2.0000i
>> ZC = conj (Z)
ZC =
   3.0000 - 2.0000i
>> R = abs(Z)
R =
     3.6056
>> theta = angle(Z)
theta =
   0.5880
>> ZRT = R.*exp(i*theta)
ZRT =
   3.0000 + 2.0000i
```

26 ▶ Chapter 2. Mathematical Foundation

where $(-p_1)$ and $(-p_2)$ are poles of the function $G(s)$. By definition, if $s = j\omega$, $G(j\omega)$ is the **frequency response function** of $G(s)$, because ω has a unit of frequency (rad/s):

$$G(s) = \frac{K}{(j\omega + p_1)(j\omega + p_2)} \tag{2-35}$$

The magnitude of $G(j\omega)$ is

$$R = |G(j\omega)| = \frac{K}{|j\omega + p_1||j\omega + p_2|} \tag{2-36}$$

and the phase angle of $G(j\omega)$ is

$$\phi = \angle G(j\omega) = \angle K - \angle j\omega + p_1 - \angle j\omega + p_2$$
$$= -\phi_1 - \phi_2 \tag{2-37}$$

For the general case, where

$$G(s) = K \frac{\sum_{k=1}^{m}(s + z_k)}{\sum_{i=1}^{n}(s + p_i)} \tag{2-38}$$

The magnitude and phase of $G(s)$ are as follows

$$R = |G(j\omega)| = K \frac{|j\omega + z_1|\cdots|j\omega + z_m|}{|j\omega + p_1|\cdots|j\omega + p_n|} \tag{2-39}$$
$$\phi = \angle G(j\omega) = (\psi_1 + \cdots + \psi_m) - (\phi_1 + \cdots + \phi_n)$$

▶ 2-2 FREQUENCY-DOMAIN PLOTS

Let $G(s)$ be the forward-path **transfer function**[1] of a linear control system with unity feedback. The frequency-domain analysis of the closed-loop system can be conducted from the frequency-domain plots of $G(s)$ with s replaced by $j\omega$.

The function $G(j\omega)$ is generally a complex function of the frequency ω and can be written as

$$G(j\omega) = |G(j\omega)|\angle G(j\omega) \tag{2-40}$$

where $|G(j\omega)|$ denotes the magnitude of $G(j\omega)$, and $\angle G(j\omega)$ is the phase of $G(j\omega)$.

The following frequency-domain plots of $G(j\omega)$ versus ω are often used in the analysis and design of linear control systems in the frequency domain.

1. **Polar plot.** A plot of the magnitude versus phase in the polar coordinates as ω is varied from zero to infinity
2. **Bode plot.** A plot of the magnitude in decibels versus ω (or $\log_{10}\omega$) in semilog (or rectangular) coordinates
3. **Magnitude-phase plot.** A plot of the magnitude (in decibels) versus the phase on rectangular coordinates, with ω as a variable parameter on the curve

2-2-1 Computer-Aided Construction of the Frequency-Domain Plots

The data for the plotting of the frequency-domain plots are usually quite time consuming to generate if the computation is carried out manually, especially if the function is of high order. In this textbook, we use MATLAB and the **ACSYS** software for this purpose.

---
[1] For the formal definition of a "transfer function," refer to Section 2-7-2.

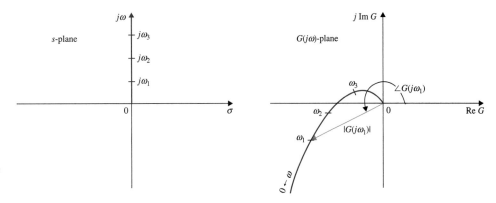

Figure 2-8 Polar plot shown as a mapping of the positive half of the $j\omega$-axis in the s-plane onto the $G(j\omega)$-plane.

From an analytical standpoint, the analyst and designer should be familiar with the properties of the frequency-domain plots so that proper interpretations can be made on these computer-generated plots.

2-2-2 Polar Plots

The polar plot of a function of the complex variable s, $G(s)$, is a plot of the magnitude of $G(j\omega)$ versus the phase of $G(j\omega)$ on polar coordinates as ω is varied from zero to infinity. From a mathematical viewpoint, the process can be regarded as the mapping of the positive half of the imaginary axis of the s-plane onto the $G(j\omega)$-plane. A simple example of this mapping is shown in Fig. 2-8. For any frequency $\omega = \omega_1$, the magnitude and phase of $G(j\omega_1)$ are represented by a vector in the $G(j\omega)$-plane. In measuring the phase, counterclockwise is referred to as positive, and clockwise is negative.

▶ **EXAMPLE 2-2-1** To illustrate the construction of the polar plot of a function $G(s)$, consider the function

$$G(s) = \frac{1}{1 + Ts} \tag{2-41}$$

where T is a positive constant. Setting $s = j\omega$, we have

$$G(j\omega) = \frac{1}{1 + j\omega T} \tag{2-42}$$

In terms of magnitude and phase, Eq. (2-42) is rewritten as

$$G(j\omega) = \frac{1}{\sqrt{1 + \omega^2 T^2}} \angle -\tan^{-1} \omega T \tag{2-43}$$

When ω is zero, the magnitude of $G(j\omega)$ is unity, and the phase of $G(j\omega)$ is at $0°$. Thus, at $\omega = 0$, $G(j\omega)$ is represented by a vector of unit length directed in the $0°$ direction. As ω increases, the magnitude of $G(j\omega)$ decreases, and the phase becomes more negative. As ω increases, the length of the vector in the polar coordinates decreases and the vector rotates in the clockwise (negative) direction. When ω approaches infinity, the magnitude of $G(j\omega)$ becomes zero, and the phase reaches $-90°$. This is presented by a vector with an infinitesimally small length directed along the $-90°$-axis in the $G(j\omega)$-plane. By substituting other finite values of ω into Eq. (2-43), the exact plot of $G(j\omega)$ turns out to be a semicircle, as shown in Fig. 2-9.

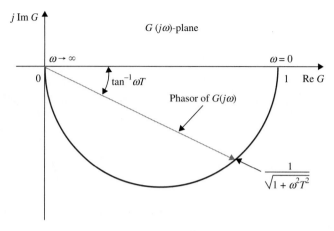

Figure 2-9 Polar plot of $G(j\omega) = \frac{1}{(1+j\omega T)}$.

▶ **EXAMPLE 2-2-2** As a second illustrative example, consider the function

$$G(j\omega) = \frac{1 + j\omega T_2}{1 + j\omega T_1} \tag{2-44}$$

where T_1 and T_2 are positive real constants. Eq. (2-44) is re-written as

$$G(j\omega) = \sqrt{\frac{1 + \omega^2 T_2^2}{1 + \omega^2 T_1^2}} \angle (\tan^{-1} \omega T_2 - \tan^{-1} \omega T_1) \tag{2-45}$$

The polar plot of $G(j\omega)$, in this case, depends on the relative magnitudes of T_1 and T_2. If T_2 is greater than T_1, the magnitude of $G(j\omega)$ is always greater than unity as ω is varied from zero to infinity, and the phase of $G(j\omega)$ is always positive. If T_2 is less than T_1, the magnitude of $G(j\omega)$ is always less than unity, and the phase is always negative. The polar plots of $G(j\omega)$ of Eq. (2-45) that correspond to these two conditions are shown in Fig. 2-10.

The general shape of the polar plot of a function $G(j\omega)$ can be determined from the following information.

1. The behavior of the magnitude and phase of $G(j\omega)$ at $\omega = 0$ and $\omega = \infty$.
2. The intersections of the polar plot with the real and imaginary axes, and the values of ω at these intersections

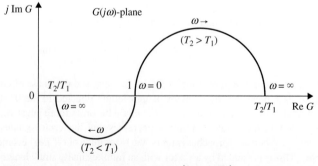

Figure 2-10 Polar plots of $G(j\omega) = \frac{(1 + j\omega T_2)}{(1 + j\omega T_1)}$.

Toolbox 2-2-1

The Nyquist diagram for Eq. (2-44) for two cases is obtained by the following sequence of MATLAB functions:

```
T1 = 10;
T2 = 5;
num1 = [T2 1];
den1 = [T1 1];
G1 = tf(num1,den1);
nyquist(G1);
hold on;
num2 = [T1 1];
den2 = [T2 1];
G2 = tf (num2,den2);
nyquist (G2);
title ('Nyquist diagram of G1 and G2')
```

Note: The "nyquist" function provides a complete polar diagram, where ω is varying from $-\infty$ to $+\infty$.

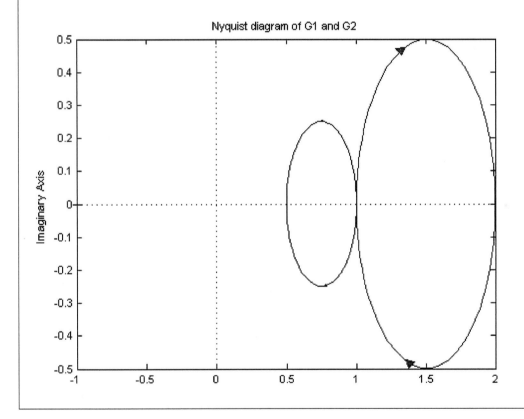

Comparing the results in Toolbox 2-2-1 and Fig. 2-10, it is clear that the polar plot reflects only a portion of the Nyquist diagram. In many control-system applications, such as the Nyquist stability criterion (see Chapter 8), an exact plot of the frequency response is not essential. Often, a rough sketch of the polar plot of the transfer function is adequate for stability analysis in the frequency domain.

> **EXAMPLE 2-2-3** In frequency-domain analyses of control systems, often we have to determine the basic properties of a polar plot. Consider the following transfer function:

$$G(s) = \frac{10}{s(s+1)} \tag{2-46}$$

By substituting $s = j\omega$ in Eq. (2-46), the magnitude and phase of $G(j\omega)$ at $\omega = 0$ and $\omega = \infty$ are computed as follows:

$$\lim_{\omega \to 0} |G(j\omega)| = \lim_{\omega \to 0} \frac{10}{\omega} = \infty \tag{2-47}$$

$$\lim_{\omega \to 0} \angle G(j\omega) = \lim_{\omega \to 0} \angle 10/j\omega = -90° \tag{2-48}$$

$$\lim_{\omega \to \infty} |G(j\omega)| = \lim_{\omega \to \infty} \frac{10}{\omega^2} = 0 \tag{2-49}$$

$$\lim_{\omega \to \infty} \angle G(j\omega) = \lim_{\omega \to \infty} \angle 10/j\omega^2 = -180° \tag{2-50}$$

Thus, the properties of the polar plot of $G(j\omega)$ at $\omega = 0$ and $\omega = \infty$ are ascertained. Next, we determine the intersections, if any, of the polar plot with the two axes of the $G(j\omega)$-plane. If the polar plot of $G(j\omega)$ intersects the real axis, at the point of intersection, the imaginary part of $G(j\omega)$ is zero; that is,

$$\text{Im}[G(j\omega)] = 0 \tag{2-51}$$

To express $G(j\omega)$ as the sum of its real and imaginary parts, we must rationalize $G(j\omega)$ by multiplying its numerator and denominator by the complex conjugate of its denominator. Therefore, $G(j\omega)$ is written

$$\begin{aligned} G(j\omega) &= \frac{10(-j\omega)(-j\omega+1)}{j\omega(j\omega+1)(-j\omega)(-j\omega+1)} = \frac{-10\omega^2}{\omega^4+\omega^2} - j\frac{10\omega}{\omega^4+\omega^2} \\ &= \text{Re}[G(j\omega)] + j\,\text{Im}[G(j\omega)] \end{aligned} \tag{2-52}$$

When we set $\text{Im}[G(j\omega)]$ to zero, we get $\omega = \infty$, meaning that the $G(j\omega)$ plot intersects only with the real axis of the $G(j\omega)$-plane at the origin.

Similarly, the intersection of $G(j\omega)$ with the imaginary axis is found by setting $\text{Re}[G(j\omega)]$ of Eq. (2-52) to zero. The only real solution for ω is also $\omega = \infty$, which corresponds to the origin of the $G(j\omega)$-plane. The conclusion is that the polar plot of $G(j\omega)$ *does not* intersect any one of the axes at any finite nonzero frequency. Under certain conditions, we are interested in the properties of the $G(j\omega)$ at infinity, which corresponds to $\omega = 0$ in this case. From Eq. (2-52), we see that $\text{Im}[G(j\omega)] = \infty$ and $\text{Re}[G(j\omega)] = -10$ at $\omega = 0$. Based on this information as well as knowledge of the angles of $G(j\omega)$ at $\omega = 0$ and $\omega = \infty$, the polar plot of $G(j\omega)$ is easily sketched without actual plotting, as shown in Fig. 2-11.

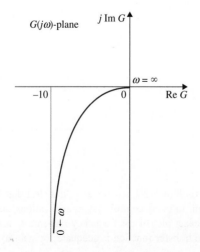

Figure 2-11 Polar plot of $G(s) = \frac{10}{s(s+1)}$.

▶ **EXAMPLE 2-2-4** Given the transfer function

$$G(s) = \frac{10}{s(s+1)(s+2)} \quad (2\text{-}53)$$

we want to make a rough sketch of the polar plot of $G(j\omega)$. The following calculations are made for the properties of the magnitude and phase of $G(j\omega)$ at $\omega = 0$ and $\omega = \infty$:

$$\lim_{\omega \to 0} |G(j\omega)| = \lim_{\omega \to 0} \frac{5}{\omega} = \infty \quad (2\text{-}54)$$

$$\lim_{\omega \to 0} \angle G(j\omega) = \lim_{\omega \to 0} \angle 5/j\omega = -90° \quad (2\text{-}55)$$

$$\lim_{\omega \to \infty} |G(j\omega)| = \lim_{\omega \to \infty} \frac{10}{\omega^3} = 0 \quad (2\text{-}56)$$

To find the intersections of the $G(j\omega)$ plot on the real and imaginary axes of the $G(j\omega)$-plane, we rationalize $G(j\omega)$ to give

$$G(j\omega) = \frac{10(-j\omega)(-j\omega+1)(-j\omega+2)}{j\omega(j\omega+1)(j\omega+2)(-j\omega)(-j\omega+1)(-j\omega+2)} \quad (2\text{-}57)$$

After simplification, the last equation is written

$$G(j\omega) = \text{Re}[G(j\omega)] + j\text{Im}[G(j\omega)] = \frac{-30}{9\omega^2 + (2-\omega^2)^2} - \frac{j10(2-\omega^2)}{9\omega^3 + \omega(2-\omega^2)^2} \quad (2\text{-}58)$$

Setting $\text{Re}[G(j\omega)]$ to zero, we have $\omega = \infty$, and $G(j\infty) = 0$, which means that the $G(j\omega)$ plot intersects the imaginary axis only at the origin. Setting $\text{Im}[G(j\omega)]$ to zero, we have $\omega = \pm\sqrt{2}$ rad/sec. This gives the point of intersection on the real axis at

$$G(\pm j\sqrt{2}) = -5/3 \quad (2\text{-}59)$$

The result, $\omega = -\sqrt{2}$ rad/sec, has no physical meaning, because the frequency is negative; it simply represents a mapping point on the negative $j\omega$-axis of the s-plane. In general, if $G(s)$ is a rational function of s (a quotient of two polynomials of s), the polar plot of $G(j\omega)$ for negative values of ω is the mirror image of that for positive ω, with the mirror placed on the real axis of the $G(j\omega)$-plane. From Eq. (2-58), we also see that $\text{Re}[G(j0)] = \infty$ and $\text{Im}[G(j0)] = \infty$. With this information, it is now possible to make a sketch of the polar plot for the transfer function in Eq. (2-53), as shown in Fig. 2-12.

Although the method of obtaining the rough sketch of the polar plot of a transfer function as described is quite straightforward, in general, for complicated transfer functions that may have multiple crossings on the real and imaginary axes of the transfer-function plane, the algebraic manipulation may again be quite involved. Furthermore, the polar plot is basically a tool for analysis; it is somewhat awkward for design purposes. We shall show in the next section that approximate information on the polar plot can always be obtained from the Bode plot, which can be sketched

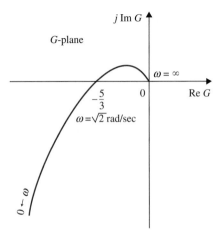

Figure 2-12 Polar plot of $G(s) = \frac{10}{s(s+1)(s+2)}$.

without any calculations. Thus, for more complicated transfer functions, sketches of the polar plots can be obtained with the help of the Bode plots, unless MATLAB is used.

2-2-3 Bode Plot (Corner Plot or Asymptotic Plot)

The Bode plot of the function $G(j\omega)$ is composed of two plots, one with the amplitude of $G(j\omega)$ in decibels (dB) versus $\log_{10}\omega$ or ω and the other with the phase of $G(j\omega)$ in degrees as a function of $\log_{10}\omega$ or ω. A Bode plot is also known as a **corner plot** or an **asymptotic plot** of $G(j\omega)$. These names stem from the fact that the Bode plot can be constructed by using straight-line approximations that are asymptotic to the actual plot.

In simple terms, the Bode plot has the following features:

1. Because the magnitude of $G(j\omega)$ in the Bode plot is expressed in dB, product and division factors in $G(j\omega)$ became additions and subtractions, respectively. The phase relations are also added and subtracted from each other algebraically.

2. The magnitude plot of the Bode plot of $G(j\omega)$ can be approximated by straight-line segments, which allow the simple sketching of the plot without detailed computation.

Because the straight-line approximation of the Bode plot is relatively easy to construct, the data necessary for the other frequency-domain plots, such as the polar plot and the magnitude-versus-phase plot, can be easily generated from the Bode plot.

Consider the function

$$G(s) = \frac{K(s+z_1)(s+z_2)\cdots(s+z_m)}{s^j(s+p_1)(s+p_2)\cdots(s+p_n)} e^{-T_d s} \qquad (2\text{-}60)$$

where K and T_d are real constants, and the z's and the p's may be real or complex (in conjugate pairs) numbers. In Chapter 7, Eq. (2-60) is the preferred form for root-locus construction, because the poles and zeros of $G(s)$ are easily identified. For constructing the Bode plot manually, $G(s)$ is preferably written in the following form:

$$G(s) = \frac{K_1(1+T_1 s)(1+T_2 s)\cdots(1+T_m s)}{s^j(1+T_a s)(1+T_b s)\cdots(1+T_n s)} e^{-T_d s} \qquad (2\text{-}61)$$

where K_1 is a real constant, the T's may be real or complex (in conjugate pairs) numbers, and T_d is the real time delay. If the Bode plot is to be constructed with a computer program, then either form of Eq. (2-60) or Eq. (2-61) can be used.

Because practically all the terms in Eq. (2-61) are of the same form, then without loss of generality, we can use the following transfer function to illustrate the construction of the Bode diagram.

$$G(s) = \frac{K(1+T_1 s)(1+T_2 s)}{s(1+T_a s)\left(1+2\zeta s/\omega_n + s^2/\omega_n^2\right)} e^{-T_d s} \qquad (2\text{-}62)$$

where K, T_d, T_1, T_2, T_a, ζ, and ω_n are real constants. It is assumed that the second-order polynomial in the denominator has complex-conjugate zeros.

The magnitude of $G(j\omega)$ in dB is obtained by multiplying the logarithm (base 10) of $|G(j\omega)|$ by 20; we have

$$|G(j\omega)|_{\text{dB}} = 20\log_{10}|G(j\omega)|$$

$$= 20\log_{10}|K| + 20\log_{10}|1+j\omega T_1| + 20\log_{10}|1+j\omega T_2|$$

$$- 20\log_{10}|j\omega| - 20\log_{10}|1+j\omega T_a| - 20\log_{10}\left|1+j2\zeta\omega - \omega^2/\omega_n^2\right| \quad (2\text{-}63)$$

The phase of $G(j\omega)$ is

$$\angle G(j\omega) = \angle K + \angle(1 + j\omega T_1) + \angle(1 + j\omega T_2) - \angle j\omega - \angle(1 + j\omega T_a)$$
$$- \angle(1 + 2\zeta\omega/\omega_n - \omega^2/\omega_n^2) - \omega T_d \quad \text{rad} \qquad (2\text{-}64)$$

In general, the function $G(j\omega)$ may be of higher order than that of Eq. (2-62) and have many more factored terms. However, Eqs. (2-63) and (2-64) indicate that additional terms in $G(j\omega)$ would simply produce more similar terms in the magnitude and phase expressions, so the basic method of construction of the Bode plot would be

---

Toolbox 2-2-2

The Bode plot for Example 2-1-3, using the MATLAB "bode" function, is obtained by the following sequence of MATLAB functions.

Approach 1

```
num = [16];
den = [1 10 16];
G = tf(num,den);
bode(G);
```

Approach 2

```
s = tf('s');
G=16/(s^2 + 10*s + 16);
bode(G);
```

The "bode" function computes the magnitude and phase of the frequency response of linear time invariant models. The magnitude is plotted in decibels (dB) and the phase in degrees. Compare the results to the values in Table 2-2.

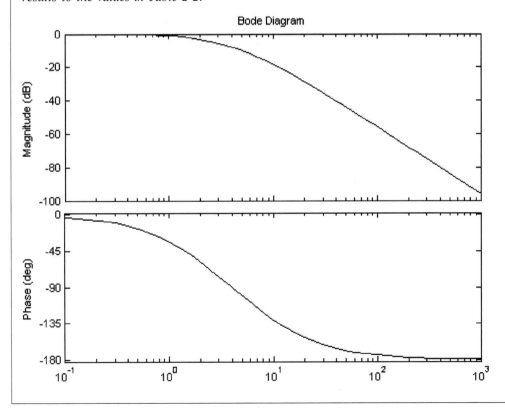

the same. We have also indicated that, in general, $G(j\omega)$ can contain just five simple types of factors:

1. Constant factor: K
2. Poles or zeros at the origin of order p: $(j\omega)^{\pm p}$
3. Poles or zeros at $s = -1/T$ of order q: $(1 + j\omega T)^{\pm q}$
4. Complex poles and zeros of order r: $\left(1 + j2\zeta\omega/\omega_n - \omega^2/\omega_n^2\right)^{\pm r}$
5. Pure time delay $e^{-j\omega T_d}$, where T_d, p, q, and r are positive integers

Eqs. (2-63) and (2-64) verify one of the unique characteristics of the Bode plot in that each of the five types of factors listed can be considered as a separate plot; the individual plots are then added or subtracted accordingly to yield the total magnitude in dB and the phase plot of $G(j\omega)$. The curves can be plotted on semilog graph paper or linear rectangular-coordinate graph paper, depending on whether ω or $\log_{10}\omega$ is used as the abscissa.

We shall now investigate sketching the Bode plot of different types of factors.

2-2-4 Real Constant K

Because

$$K_{\text{dB}} = 20\log_{10} K = \text{constant} \tag{2-65}$$

and

$$\angle K = \begin{cases} 0° & K > 0 \\ 180° & K < 0 \end{cases} \tag{2-66}$$

the Bode plot of the real constant K is shown in Fig. 2-13 in semilog coordinates.

2-2-5 Poles and Zeros at the Origin, $(j\omega)^{\pm p}$

The magnitude of $(j\omega)^{\pm p}$ in dB is given by

$$20\log_{10}\left|(j\omega)^{\pm p}\right| = \pm 20 p \log_{10}\omega \quad \text{dB} \tag{2-67}$$

for $\omega \geq 0$. The last expression for a given p represents a straight line in either semilog or rectangular coordinates. The slopes of these lines are determined by taking the derivative of Eq. (2-67) with respect to $\log_{10}\omega$; that is,

$$\frac{d}{d\log_{10}\omega}(\pm 20 p \log_{10}\omega) = \pm 20 p \quad \text{dB/decade} \tag{2-68}$$

These lines pass through the 0-dB axis at $\omega = 1$. Thus, a unit change in $\log_{10}\omega$ corresponds to a change of $\pm 20 p$ dB in magnitude. Furthermore, a unit change in $\log_{10}\omega$ in the rectangular coordinates is equivalent to one **decade** of variation in ω, that is, from 1 to 10, 10 to 100, and so on, in the semilog coordinates. Thus, the slopes of the straight lines described by Eq. (2-68) are said to be $\pm 20 p$ dB/decade of frequency.

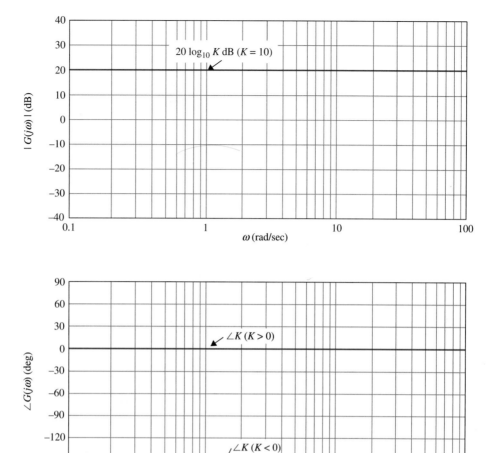

Figure 2-13 Bode plot of constant K.

Instead of decades, sometimes **octaves** are used to represent the separation of two frequencies. The frequencies ω_1 and ω_2 are separated by one octave if $\omega_2/\omega_1 = 2$. The number of decades between any two frequencies ω_1 and ω_2 is given by

$$\text{number of decades} = \frac{\log_{10}(\omega_2/\omega_1)}{\log_{10} 10} = \log_{10}\left(\frac{\omega_2}{\omega_1}\right) \qquad (2\text{-}69)$$

Similarly, the number of octaves between ω_2 and ω_1 is

$$\text{number of octaves} = \frac{\log_{10}(\omega_2/\omega_1)}{\log_{10} 2} = \frac{1}{0.301} \log_{10}\left(\frac{\omega_2}{\omega_1}\right) \qquad (2\text{-}70)$$

Thus, the relation between octaves and decades is

$$\text{number of octaves} = 1/0.301 \text{ decades} = 3.32 \text{ decades} \qquad (2\text{-}71)$$

Substituting Eq. (2-71) into Eq. (2-67), we have

$$\pm 20 p \text{ dB/decade} = \pm 20 p \times 0.301 \cong 6 p \quad \text{dB/octave} \tag{2-72}$$

For the function $G(s) = 1/s$, which has a simple pole at $s = 0$, the magnitude of $G(j\omega)$ is a straight line with a slope of -20 dB/decade, and it passes through the 0-dB axis at $\omega = 1$ rad/sec.

The phase of $(j\omega)^{\pm p}$ is written

$$\angle (j\omega)^{\pm p} = \pm p \times 90° \tag{2-73}$$

The magnitude and phase curves of the function $(j\omega)^{\pm p}$ are shown in Fig. 2-14 for several values of P.

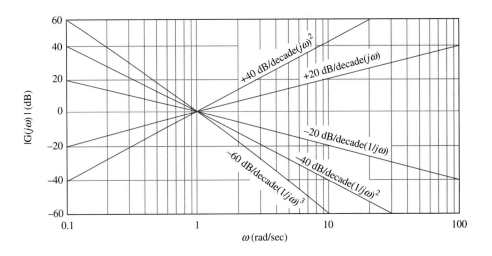

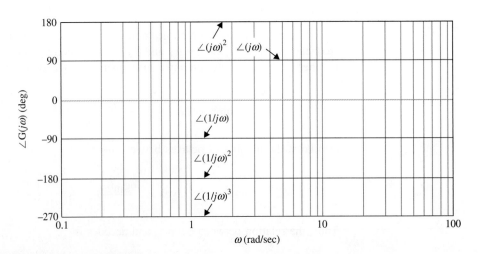

Figure 2-14 Bode plots of $(j\omega)^p$.

2-2-6 Simple Zero, $1 + j\omega T$

Consider the function

$$G(j\omega) = 1 + j\omega T \tag{2-74}$$

where T is a positive real constant. The magnitude of $G(j\omega)$ in dB is

$$|G(j\omega)|_{\text{dB}} = 20\log_{10}|G(j\omega)| = 20\log_{10}\sqrt{1 + \omega^2 T^2} \tag{2-75}$$

To obtain asymptotic approximations of $|G(j\omega)|_{\text{dB}}$, we consider both very large and very small values of ω. At very low frequencies, $\omega T \ll 1$, Eq. (2-75) is approximated by

$$|G(j\omega)|_{\text{dB}} \cong 20\log_{10} 1 = 0 \quad \text{dB} \tag{2-76}$$

because $\omega^2 T^2$ is neglected when compared with 1.

At very high frequencies, $\omega T \gg 1$, we can approximate $1 + \omega^2 T^2$ by $\omega^2 T^2$; then Eq. (2-75) becomes

$$|G(j\omega)|_{\text{dB}} \cong 20\log_{10}\sqrt{\omega^2 T^2} = 20\log_{10}\omega T \tag{2-77}$$

Eq. (2-76) represents a straight line with a slope of 20 dB/decade of frequency. The intersect of these two lines is found by equating Eq. (2-76) to Eq. (2-77), which gives

$$\omega = 1/T \tag{2-78}$$

This frequency is also the intersect of the high-frequency approximate plot and the low-frequency approximate plot, which is the 0-dB axis. The frequency given in Eq. (2-78) is also known as the **corner frequency** of the Bode plot of Eq. (2-74), because the asymptotic plot forms the shape of a corner at this frequency, as shown in Fig. 2-15. The actual $|G(j\omega)|_{\text{dB}}$ plot of Eq. (2-74) is a smooth curve and deviates only slightly from the straight-line approximation. The actual values and the straight-line approximation of $|1 + j\omega T|_{\text{dB}}$ as functions of ωT are tabulated in Table 2-3. The error between the actual magnitude curve and the straight-line asymptotes is symmetrical with respect to the corner frequency $\omega = 1/T$. It is useful to remember that the error is 3 dB at the corner frequency, and it is 1 dB at 1 octave above ($\omega = 2/T$) and 1 octave below ($\omega = 1/2T$) the corner frequency. At 1 decade above and below the corner frequency, the error is dropped to approximately 0.3 dB. Based on these facts, the procedure of drawing $|1 + j\omega T|_{\text{dB}}$ is as follows:

1. Locate the corner frequency $\omega = 1/T$ on the frequency axis.
2. Draw the 20-dB/decade (or 6-dB/octave) line and the horizontal line at 0 dB, with the two lines intersecting at $\omega = 1/T$.
3. If necessary, the actual magnitude curve is obtained by adding the errors to the asymptotic plot at the strategic frequencies. Usually, a smooth curve can be sketched simply by locating the 3-dB point at the corner frequency and the 1-dB points at 1 octave above and below the corner frequency.

The phase of $G(j\omega) = 1 + j\omega T$ is

$$\angle G(j\omega) = \tan^{-1}\omega T \tag{2-79}$$

Similar to the magnitude curve, a straight-line approximation can be made for the phase curve. Because the phase of $G(j\omega)$ varies from 0° to 90°, we can draw a line from 0° at 1 decade below the corner frequency to 90° at 1 decade above the corner frequency. As shown in Fig. 2-15, the maximum deviation between the straight-line approximation and the actual curve is less than 6°. Table 2-3 gives the values of $\angle(1 + j\omega T)$ versus ωT.

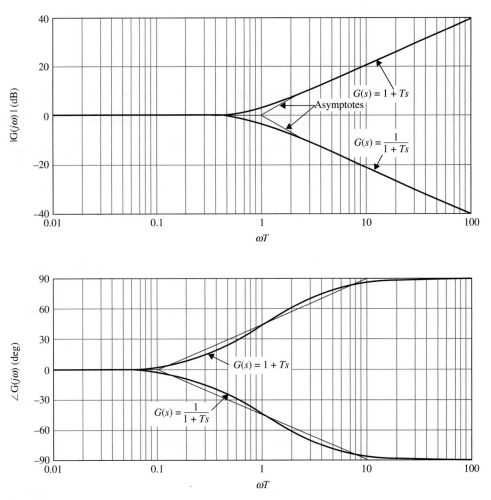

Figure 2-15 Bode plots of $G(s) = 1 + Ts$ and $G(s) = \frac{1}{(1+Ts)}$.

TABLE 2-3 Values of $\angle(1 + j\omega t)$ versus ωT

Straight-Line		Approximation			Error $\angle(1 + j\omega T)$	
ωT	$\log_{10}\omega T$	$\lvert 1 + j\omega T \rvert$	$\lvert 1 + j\omega T \rvert_{dB}$	$\lvert 1 + j\omega T \rvert_{dB}$	(dB)	(deg)
0.01	-2	1.0	0.000043	0	0.00043	0.5
0.10	-1	1.04	0.043	0	0.043	5.7
0.50	-0.3	1.12	1	0	1	26.6
0.76	-0.12	1.26	2	0	2	37.4
1.00	0	1.41	3	0	3	45.0
1.31	0.117	1.65	4.3	2.3	2	52.7
2.00	0.3	2.23	7.0	6.0	1	63.4
10.00	1.0	10.4	20.043	20.0	0.043	84.3
100.00	2.0	100.005	40.00043	40.0	0.00043	89.4

2-2-7 Simple Pole, $1/(1+j\omega T)$

For the function

$$G(j\omega) = \frac{1}{1+j\omega T} \qquad (2\text{-}80)$$

the magnitude, $|G(j\omega)|$ in dB, is given by the negative of the right side of Eq. (2-75), and the phase $\angle G(j\omega)$ is the negative of the angle in Eq. (2-79). Therefore, it is simple to extend all the analysis for the case of the simple zero to the Bode plot of Eq. (2-80). The asymptotic approximations of $|G(j\omega)|_{\text{dB}}$ at low and high frequencies are

$$\omega T \ll 1 \quad |G(j\omega)|_{\text{dB}} \cong 0\,\text{dB} \qquad (2\text{-}81)$$

$$\omega T \gg 1 \quad |G(j\omega)|_{\text{dB}} \cong -20\log_{10}\omega T \qquad (2\text{-}82)$$

Thus, the corner frequency of the Bode plot of Eq. (2-80) is still at $\omega = 1/T$, except that at high frequencies the slope of the straight-line approximation is $-20\,\text{dB/decade}$. The phase of $G(j\omega)$ is 0 degrees at $\omega = 0$, and $-90°$ when $\omega = \infty$. The magnitude in dB and phase of the Bode plot of Eq. (2-80) are shown in Fig. 2-15. The data in Table 2-3 are still useful for the simple-pole case if appropriate sign changes are made to the numbers. For instance, the numbers in $|1+j\omega T|_{\text{dB}}$, the straight-line approximation of $|1+j\omega T|_{\text{dB}}$, the error (dB), and the $\angle(1+j\omega T)$ columns should all be negative. At the corner frequency, the error between the straight-line approximation and the actual magnitude curve is $-3\,\text{dB}$.

2-2-8 Quadratic Poles and Zeros

Now consider the second-order transfer function

$$G(s) = \frac{\omega_n^2}{s^2 + 2\zeta\omega_n s + \omega_n^2} = \frac{1}{1 + (2\zeta/\omega_n)s + (1/\omega_n^2)s^2} \qquad (2\text{-}83)$$

We are interested only in the case when $\zeta \leq 1$, because otherwise $G(s)$ would have two unequal real poles, and the Bode plot can be obtained by considering $G(s)$ as the product of two transfer functions with simple poles.

By letting $s = j\omega$, Eq. (2-83) becomes

$$G(j\omega) = \frac{1}{\left[1 - (\omega/\omega_n)^2\right] + j2\zeta(\omega/\omega_n)} \qquad (2\text{-}84)$$

The magnitude of $G(j\omega)$ in dB is

$$|G(j\omega)| = 20\log_{10}|G(j\omega)| = -20\log_{10}\sqrt{\left[1-(\omega/\omega_n)^2\right]^2 + 4\zeta^2(\omega/\omega_n)^2} \qquad (2\text{-}85)$$

At very low frequencies, $\omega/\omega_n \ll 1$, Eq. (2-85) can be approximated as

$$|G(j\omega)|_{\text{dB}} = 20\log_{10}|G(j\omega)| \cong -20\log_{10}1 = 0 \quad \text{dB} \qquad (2\text{-}86)$$

Thus, the low-frequency asymptote of the magnitude plot of Eq. (2-83) is a straight line that lies on the 0-dB axis. At very high frequencies, $\omega/\omega_n \gg 1$, the magnitude in dB of $G(j\omega)$ in Eq. (2-83) becomes

$$|G(j\omega)|_{dB} \cong -20\log_{10}\sqrt{(\omega/\omega_n)^4} = -40\log_{10}(\omega/\omega_n) \quad dB \quad (2\text{-}87)$$

This equation represents a straight line with a slope of -40 dB/decade in the Bode-plot coordinates. The intersection of the two asymptotes is found by equating Eq. (2-86) to Eq. (2-87), yielding the corner frequency at $\omega = \omega_n$. The actual magnitude curve of $G(j\omega)$ in this case may differ strikingly from the asymptotic curve. The reason for this is that the amplitude and phase curves of the second-order $G(j\omega)$ depend not only on the corner frequency ω_n but also on the damping ratio ζ, which does not enter the asymptotic curve. The actual and the asymptotic curves of $|G(j\omega)|_{dB}$ are shown in Fig. 2-16 for several values of ζ. The errors between the two sets of curves are shown in Fig. 2-17 for the same set of values of ζ. The standard procedure of constructing the second-order $|G(j\omega)|_{dB}$ is to first locate the corner frequency ω_n and -40-dB/decade line to the right of ω_n. The actual curve is obtained by making corrections to the asymptotes by using either the data from the error curves of Fig. 2-17 or the curves in Fig. 2-16 for the corresponding ζ.

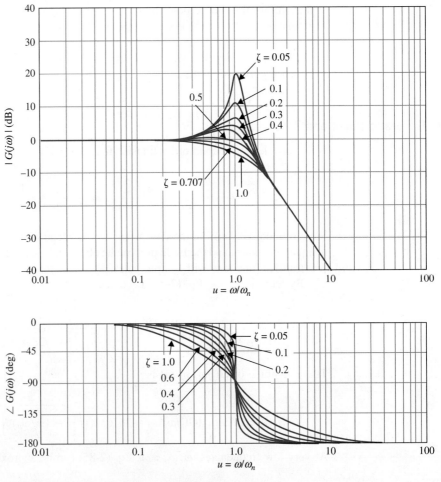

Figure 2-16 Bode plot of $G(s) = \dfrac{1}{1+2\zeta(s/\omega_n)+(s/\omega_n)^2}$.

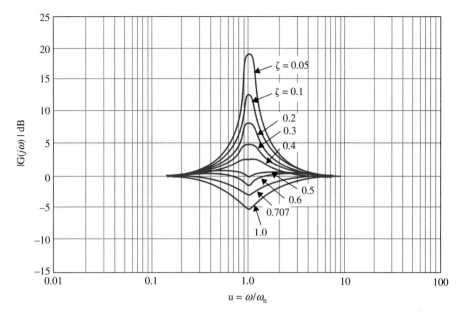

Figure 2-17 Errors in magnitude curves of Bode plots of $G(s) = \frac{1}{1+2\zeta(s/\omega_n)+(s/\omega_n)^2}$.

The phase of $G(j\omega)$ is given by

$$\angle G(j\omega) = -\tan^{-1}\left\{\frac{2\zeta\omega}{\omega_n}\left[1 - \left(\frac{\omega}{\omega_n}\right)^2\right]\right\} \qquad (2\text{-}88)$$

and is plotted as shown in Fig. 2-16 for various values of ζ.

The analysis of the Bode plot of the second-order transfer function of Eq. (2-83) can be applied to the second-order transfer function with two complex zeros. For

$$G(s) = 1 + \frac{2\zeta}{\omega_n}s + \frac{1}{\omega_n^2}s^2 \qquad (2\text{-}89)$$

the magnitude and phase curves are obtained by inverting those in Fig. 2-16. The errors between the actual and the asymptotic curves in Fig. 2-17 are also inverted.

Toolbox 2-2-3

The Bode plot for Fig. 2-17 when $\zeta = 0.05$ and $\omega = 1$, using the MATLAB "bode" function, is obtained by the following sequence of MATLAB functions.

Approach 1

```
num = [1];
den = [1 .1 1];
G = tf(num,den);
bode(G);
```

Approach 2

```
s = tf('s');
G = 1/(s^2 + .1*s + 1);
bode(G);
```

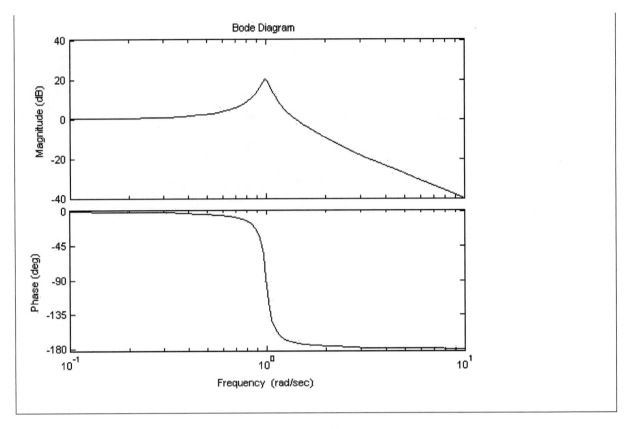

2-2-9 Pure Time Delay, $e^{-j\omega T_d}$

The magnitude of the pure time delay term is equal to unity for all values of ω. The phase of the pure time delay term is

$$\angle e^{-j\omega T_d} = -\omega T_d \tag{2-90}$$

which decreases linearly as a function of ω. Thus, for the transfer function

$$G(j\omega) = G_1(j\omega)e^{-j\omega T_d} \tag{2-91}$$

the magnitude plot $|G(j\omega)|_{dB}$ is identical to that of $|G_1(j\omega)|_{dB}$. The phase plot $\angle G(j\omega)$ is obtained by subtracting ωT_d radians from the phase curve of $G_1(j\omega)$ at various ω.

▶ **EXAMPLE 2-2-5** As an illustrative example on the manual construction of the Bode plot, consider the function

$$G(s) = \frac{10(s+10)}{s(s+2)(s+5)} \tag{2-92}$$

The first step is to express $G(s)$ in the form of Eq. (2-61) and set $s = j\omega$ (keeping in mind that, for computer plotting, this step is unnecessary); we have

$$G(j\omega) = \frac{10(1+j0.1\omega)}{j\omega(1+j0.5\omega)(1+j0.2\omega)} \tag{2-93}$$

Eq. (2-92) shows that $G(j\omega)$ has corner frequencies at $\omega = 2$, 5, and 10 rad/sec. The pole at $s = 0$ gives a magnitude curve that is a straight line with a slope of -20 dB/decade, passing through the $\omega = 1$ rad/sec point on the 0-dB axis. The complete Bode plot of the magnitude and phase of $G(j\omega)$ is obtained by adding the component curves together, point by point, as shown in Fig. 2-18. The actual curves can be obtained by a computer program and are shown in Fig. 2-18.

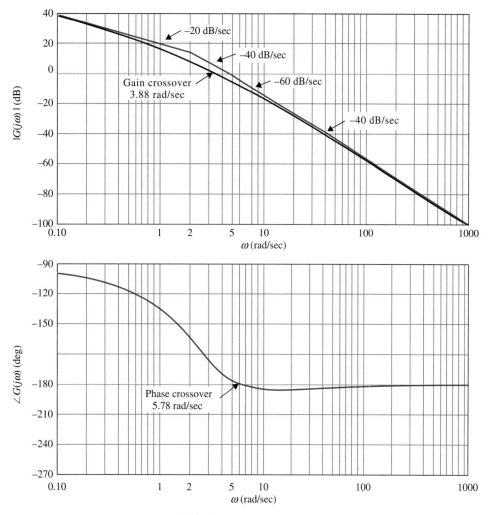

Figure 2-18 Bode plot of $G(s) = \frac{10(s+10)}{s(s+2)(s+5)}$.

Toolbox 2-2-4

The Bode plot for Eq. (2-93), using the MATLAB "bode" function, is obtained by the following sequence of MATLAB functions.

```
num = [1 10];
den = [.1 .7 1 0];
G = tf(num,den);
bode(G);
```

The result is a graph similar to Fig. 2-18.

2-2-10 Magnitude-Phase Plot

The magnitude-phase plot of $G(j\omega)$ is a plot of the magnitude of $G(j\omega)$ in dB versus its phase in degrees, with ω as a parameter on the curve. One of the most important applications of this type of plot is that, when $G(j\omega)$ is the forward-path transfer function of a unity-feedback control system, the plot can be superposed on the Nichols chart (see Chapter 8) to give information on the relative stability and frequency response of the system. When constant coefficient K of the transfer function varies, the plot is simply raised or lowered vertically according to the value of K in dB. However, in the construction of the plot, the property of adding the curves of the individual components of the transfer function in the Bode plot does not carry over to this case. Thus, it is best to make the magnitude-phase plot by computer or transfer the data from the Bode plot.

▶ **EXAMPLE 2-2-6** As an illustrative example, the polar plot and the magnitude-phase plot of Eq. (2-92) are shown in Fig. 2-19 and Fig. 2-20, respectively. The Bode plot of the function is already shown in Fig. 2-18. The relationships among these three plots are easily identified by comparing the curves in Figs. 2-18, 2-19, and 2-20.

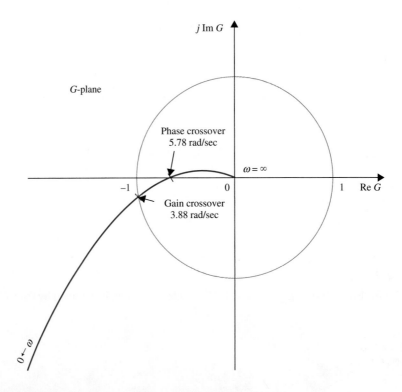

Figure 2-19 Polar plot of $G(s) = \frac{10(s+10)}{s(s+2)(s+5)}$.

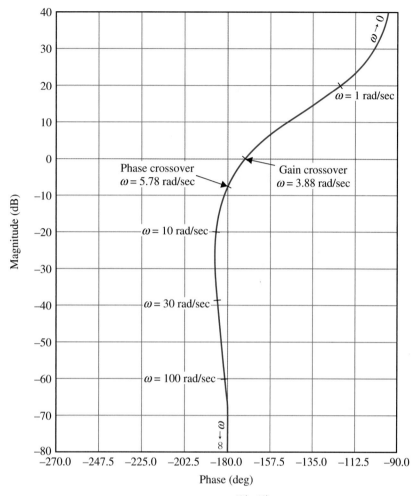

Figure 2-20 Magnitude-phase plot of $G(s) = \frac{10(s+10)}{s(s+2)(s+5)}$.

Toolbox 2-2-5

The magnitude and phase plot for Example 2-2-6 may be obtained using the MATLAB "nichols" function, by the following sequence of MATLAB functions.

```
>> G = zpk([-10],[0 -2 -5],10)

Zero/pole/gain:

  10 (s + 10
  -----------
  s (s + 2) (s + 5)

>> nichols(G)
```

See Fig. 2-20.

Toolbox 2-2-6

The phase and gain margins for Eq. (2-92) are obtained by the following sequence of MATLAB functions.

Approach 1
```
num = [10 100];
den = [1 7 10 0];
G1 = tf(num,den);
margin(G1);
```

Approach 2
```
s = tf('s');
G1=(10*s + 100)/(s^3 + 7*s^2 + 10*s);
margin(G1);
```

"Margin" produces a Bode plot and displays the margins on this plot.

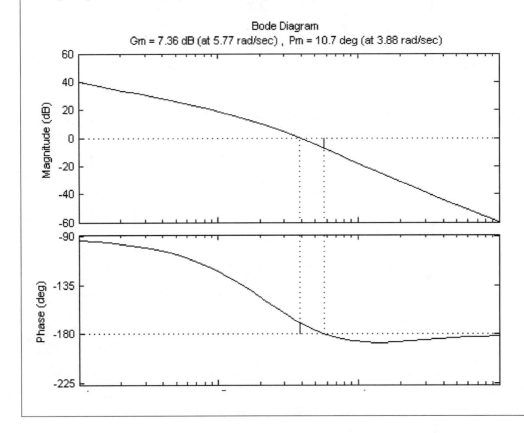

2-2-11 Gain- and Phase-Crossover Points

Gain- and phase-crossover points on the frequency-domain plots are important for analysis and design of control systems. These are defined as follows.

- **Gain-crossover point.** The gain-crossover point on the frequency-domain plot of $G(j\omega)$ is the point at which $|G(j\omega)| = 1$ or $|G(j\omega)|_{dB} = 0$ dB. The frequency at the gain-crossover point is called the **gain-crossover frequency** ω_g.
- **Phase-crossover point.** The phase-crossover point on the frequency-domain plot of $G(j\omega)$ is the point at which $\angle G(j\omega) = 180°$. The frequency at the phase-crossover point is called the **phase-crossover frequency** ω_p.

The gain and phase crossovers are interpreted with respect to three types of plots:

- **Polar plot.** The gain-crossover point (or points) is where $|G(j\omega)| = 1$. The phase-crossover point (or points) is where $\angle G(j\omega) = 180°$ (see Fig. 2-19).
- **Bode plot.** The gain-crossover point (or points) is where the magnitude curve $|G(j\omega)|_{dB}$ crosses the 0-dB axis. The phase-crossover point (or points) is where the phase curve crosses the 180° axis (see Fig. 2-18).
- **Magnitude-phase plot.** The gain-crossover point (or points) is where the $G(j\omega)$ curve crosses the 0-dB axis. The phase-crossover point (or points) is where the $G(j\omega)$ curve crosses the 180° axis (see Fig. 2-20).

2-2-12 Minimum-Phase and Nonminimum-Phase Functions

A majority of the process transfer functions encountered in linear control systems do not have poles or zeros in the right-half s-plane. This class of transfer functions is called the **minimum-phase transfer function**. When a transfer function has either a pole or a zero in the right-half s-plane, it is called a **nonminimum-phase transfer function**.

Minimum-phase transfer functions have an important property in that their magnitude and phase characteristics are uniquely related. In other words, given a minimum-phase function $G(s)$, knowing its magnitude characteristics $|G(j\omega)|$ completely defines the phase characteristics, $\angle G(j\omega)$. Conversely, given $\angle G(j\omega)$, $|G(j\omega)|$ is completely defined.

Nonminimum-phase transfer functions do not have the unique magnitude-phase relationships. For instance, given the function

$$G(j\omega) = \frac{1}{1 - j\omega T} \qquad (2\text{-}94)$$

the magnitude of $G(j\omega)$ is the same whether T is positive (nonminimum phase) or negative (minimum phase). However, the phase of $G(j\omega)$ is different for positive and negative T.

Additional properties of the minimum-phase transfer functions are as follows:

- For a minimum-phase transfer function $G(s)$ with m zeros and n poles, excluding the poles at $s = 0$, if any, when $s = j\omega$ and as ω varies from ∞ to 0, the total phase variation of $G(j\omega)$ is $(n - m)\pi/2$.
- The value of a minimum-phase transfer function cannot become zero or infinity at any finite nonzero frequency.
- A nonminimum-phase transfer function will always have a more positive phase shift as ω is varied from ∞ to 0.

▶ **EXAMPLE 2-2-7** As an illustrative example of the properties of the nonminimum-phase transfer function, consider that the zero of the transfer function of Eq. (2-92) is in the right-half s-plane; that is,

$$G(s) = \frac{10(s - 10)}{s(s + 2)(s + 5)} \qquad (2\text{-}95)$$

The magnitude plot of the Bode diagram of $G(j\omega)$ is identical to that of the minimum-phase transfer function in Eq. (2-92), as shown in Fig. 2-18. The phase curve of the Bode plot of $G(j\omega)$ of Eq. (2-95) is shown in Fig. 2-21(a), and the polar plot is shown in Fig. 2-21(b). Notice that the nonminimum-phase function has a net phase shift of 270° (from $-180°$ to $+90°$) as ω varies from ∞ to 0, whereas the minimum-phase transfer function of Eq. (2-92) has a net phase change of only 90° (from $-180°$ to $-90°$) over the same frequency range.

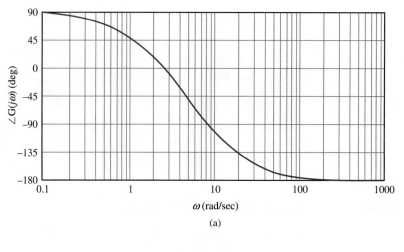

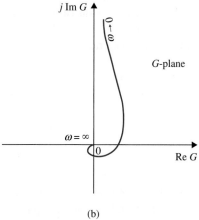

Figure 2-21 (a) Phase curve of the Bode plot. (b) Polar plot. $G(s) = \frac{10(s-10)}{s(s+2)(s+5)}$.

Care should be taken when using the Bode diagram for the analysis and design of systems with nonminimum-phase transfer functions. For stability studies, the polar plot, when used along with the Nyquist criterion discussed in Chapter 8, is more convenient for nonminimum-phase systems. Bode diagrams of nonminimum-phase forward-path transfer functions *should not* be used for stability analysis of closed-loop control systems. The same is true for the magnitude-phase plot.

Here are some important notes:

- A Bode plot is also known as a corner plot or an asymptotic plot.
- The magnitude of the pure time delay term is unity for all ω.
- The magnitude and phase characteristics of a minimum-phase function are uniquely related.
- Do not use the Bode plot and the gain-phase plot of a nonminimum-phase transfer function for stability studies.

The topic of frequency response has a special importance in the study of control systems and is revisited later in Chapter 8.

2-3 INTRODUCTION TO DIFFERENTIAL EQUATIONS

A wide range of systems in engineering are modeled mathematically by differential equations. These equations generally involve derivatives and integrals of the dependent variables with respect to the independent variable—usually time. For instance, a series electric *RLC* (resistance-inductance-capacitance) network can be represented by the differential equation:

$$Ri(t) + L\frac{di(t)}{dt} + \frac{1}{C}\int i(t)dt = e(t) \tag{2-96}$$

where R is the resistance; L, the inductance; C, the capacitance; $i(t)$, the current in the network; and $e(t)$, the applied voltage. In this case, $e(t)$ is the forcing function; t, the independent variable; and $i(t)$, the dependent variable or unknown that is to be determined by solving the differential equation.

Eq. (2-96) is referred to as a second-order differential equation, and we refer to the system as a **second-order system**. Strictly speaking, Eq. (2-96) should be referred to as an integrodifferential equation, because an integral is involved.

2-3-1 Linear Ordinary Differential Equations

In general, the differential equation of an *n*th-order system is written

$$\frac{d^n y(t)}{dt^n} + a_{n-1}\frac{d^{n-1}y(t)}{dt^{n-1}} + \cdots + a_1\frac{dy(t)}{dt} + a_0 y(t) = f(t) \tag{2-97}$$

which is also known as a **linear ordinary differential equation** if the coefficients $a_0, a_1, \ldots, a_{n-1}$ are not functions of $y(t)$.

A first-order linear ordinary differential equation is therefore in the general form:

$$\frac{dy(t)}{dt} + a_0 y(t) = f(t) \tag{2-98}$$

and the second-order general form of a linear ordinary differential equation is

$$\frac{d^2 y(t)}{dt^2} + a_1 \frac{dy(t)}{dt} + a_0 y(t) = f(t) \tag{2-99}$$

In this text, because we treat only systems that contain lumped parameters, the differential equations encountered are all of the ordinary type. For systems with distributed parameters, such as in heat-transfer systems, partial differential equations are used.

2-3-2 Nonlinear Differential Equations

Many physical systems are nonlinear and must be described by nonlinear differential equations. For instance, the following differential equation that describes the motion of a pendulum of mass *m* and length *l*, later discussed in this chapter, is

$$m\ell\frac{d^2\theta(t)}{dt^2} + mg\sin\theta(t) = 0 \tag{2-100}$$

Because $\theta(t)$ appears as a sine function, Eq. (2-100) is nonlinear, and the system is called a **nonlinear system**.

2-3-3 First-Order Differential Equations: State Equations[2]

In general, an nth-order differential equation can be decomposed into n first-order differential equations. Because, in principle, first-order differential equations are simpler to solve than higher-order ones, first-order differential equations are used in the analytical studies of control systems. For the differential equation in Eq. (2-96), if we let

$$x_1(t) = \int i(t) dt \tag{2-101}$$

and

$$x_2(t) = \frac{dx_1(t)}{dt} = i(t) \tag{2-102}$$

then Eq. (2-96) is decomposed into the following two first-order differential equations:

$$\frac{dx_1(t)}{dt} = x_2(t) \tag{2-103}$$

$$\frac{dx_2(t)}{dt} = -\frac{1}{LC}x_1(t) - \frac{R}{L}x_2(t) + \frac{1}{L}e(t) \tag{2-104}$$

In a similar manner, for Eq. (2-97), let us define

$$\begin{aligned} x_1(t) &= y(t) \\ x_2(t) &= \frac{dy(t)}{dt} \\ &\vdots \\ x_n(t) &= \frac{d^{n-1}y(t)}{dt^{n-1}} \end{aligned} \tag{2-105}$$

then the nth-order differential equation is decomposed into n first-order differential equations:

$$\begin{aligned} \frac{dx_1(t)}{dt} &= x_2(t) \\ \frac{dx_2(t)}{dt} &= x_3(t) \\ &\vdots \\ \frac{dx_n(t)}{dt} &= -a_0 x_1(t) - a_1 x_2(t) - \cdots - a_{n-2} x_{n-1}(t) - a_{n-1} x_n(t) + f(t) \end{aligned} \tag{2-106}$$

Notice that the last equation is obtained by equating the highest-ordered derivative term in Eq. (2-97) to the rest of the terms. In control systems theory, the set of first-order differential equations in Eq. (2-106) is called the **state equations**, and $x_1, x_2, \ldots, x_n$ are called the **state variables**.

2-3-4 Definition of State Variables

The state of a system refers to the past, present, and future conditions of the system. From a mathematical perspective, it is convenient to define a set of state variables and state equations to model dynamic systems. As it turns out, the variables $x_1(t)$, $x_2(t), \ldots, x_n(t)$ defined in Eq. (2-105) are the state variables of the nth-order system

---

[2] Please refer to Chapter 10 for more in-depth study of State Space Systems.

described by Eq. (2-97), and the n first-order differential equations are the state equations. In general, there are some basic rules regarding the definition of a state variable and what constitutes a state equation. The state variables must satisfy the following conditions:

- At any initial time $t = t_0$, the state variables $x_1(t_0), x_2(t_0), \ldots, x_n(t_0)$ define the **initial states** of the system.
- Once the inputs of the system for $t \geq t_0$ and the initial states just defined are specified, the state variables should completely define the future behavior of the system.

The state variables of a system are defined as a **minimal set** of variables, $x_1(t), x_2(t), \ldots, x_n(t)$, such that knowledge of these variables at any time t_0 and information on the applied input at time t_0 are sufficient to determine the state of the system at any time $t > t_0$. Hence, the **space state form** for n state variables is

$$\dot{\mathbf{x}}(\mathbf{t}) = \mathbf{A}\mathbf{x}(\mathbf{t}) + \mathbf{B}\mathbf{u} \qquad (2\text{-}107)$$

where $\mathbf{x(t)}$ is the state vector having n rows,

$$\mathbf{x(t)} = \begin{bmatrix} x_1(t) \\ x_2(t) \\ \vdots \\ x_n(t) \end{bmatrix} \qquad (2\text{-}108)$$

and $\mathbf{u(t)}$ is the input vector with p rows,

$$\mathbf{u(t)} = \begin{bmatrix} u_1(t) \\ u_2(t) \\ \vdots \\ u_p(t) \end{bmatrix} \qquad (2\text{-}109)$$

The coefficient matrices $\mathbf{A}$ and $\mathbf{B}$ are defined as:

$$\mathbf{A} = \begin{bmatrix} a_{11} & a_{12} & \cdots & a_{1n} \\ a_{21} & a_{22} & \cdots & a_{2n} \\ \vdots & \vdots & \ddots & \vdots \\ a_{n1} & a_{n2} & \cdots & a_{nn} \end{bmatrix} \quad (n \times n) \qquad (2\text{-}110)$$

$$\mathbf{B} = \begin{bmatrix} b_{11} & b_{12} & \cdots & b_{1p} \\ b_{21} & b_{22} & \cdots & b_{2p} \\ \vdots & \vdots & \ddots & \vdots \\ b_{n1} & b_{n2} & \cdots & b_{np} \end{bmatrix} \quad (n \times p) \qquad (2\text{-}111)$$

2-3-5 The Output Equation

One should not confuse the state variables with the outputs of a system. An **output** of a system is a variable that *can be measured*, but a state variable does not always satisfy this requirement. For instance, in an electric motor, such state variables as the winding current, rotor velocity, and displacement can be measured physically, and these variables all qualify as output variables. On the other hand, magnetic flux can also be regarded as a state variable in an electric motor, because it represents the past, present, and future states of the motor, but it cannot be measured directly during operation and therefore does not ordinarily qualify as an output variable. In general, an output variable can be expressed as an algebraic combination of the state variables. For the system described by Eq. (2-97), if $y(t)$ is designated as the output, then the output equation is simply $y(t) = x_1(t)$. In general,

$$\mathbf{y}(t) = \begin{bmatrix} y_1(t) \\ y_2(t) \\ \vdots \\ y_q(t) \end{bmatrix} = \mathbf{Cx(t) + Du} \qquad (2\text{-}112)$$

$$\mathbf{C} = \begin{bmatrix} c_{11} & c_{12} & \cdots & c_{1n} \\ c_{21} & c_{22} & \cdots & c_{2n} \\ \vdots & \vdots & \ddots & \vdots \\ c_{q1} & c_{q2} & \cdots & c_{qn} \end{bmatrix} \qquad (2\text{-}113)$$

$$\mathbf{D} = \begin{bmatrix} d_{11} & d_{12} & \cdots & d_{1p} \\ d_{21} & d_{22} & \cdots & d_{2p} \\ \vdots & \vdots & \ddots & \vdots \\ d_{q1} & d_{q2} & \cdots & d_{qp} \end{bmatrix} \qquad (2\text{-}114)$$

We will utilize these concepts in the modeling of various dynamical systems.

▶ 2-4 LAPLACE TRANSFORM

The Laplace transform is one of the mathematical tools used to solve linear ordinary differential equations. In contrast with the classical method of solving linear differential equations, the Laplace transform method has the following two features:

1. The homogeneous equation and the particular integral of the solution of the differential equation are obtained in one operation.
2. The Laplace transform converts the differential equation into an algebraic equation in s-domain. It is then possible to manipulate the algebraic equation by simple algebraic rules to obtain the solution in the s-domain. The final solution is obtained by taking the inverse Laplace transform.

2-4-1 Definition of the Laplace Transform

Given the real function $f(t)$ that satisfies the condition

$$\int_0^\infty \left| f(t)e^{-\sigma t} \right| dt < \infty \qquad (2\text{-}115)$$

for some finite, real σ, the Laplace transform of $f(t)$ is defined as

$$F(s) = \int_{0^-}^\infty f(t)e^{-st} dt \qquad (2\text{-}116)$$

or

$$F(s) = \text{Laplace transform of } f(t) = \mathcal{L}[f(t)] \qquad (2\text{-}117)$$

The variable s is referred to as the **Laplace operator**, which is a complex variable; that is, $s = \sigma + j\omega$, where σ is the real component and ω is the imaginary component. The defining equation in Eq. (2-117) is also known as the **one-sided Laplace transform**, as the integration is evaluated from $t = 0$ to ∞. This simply means that all information contained in $f(t)$ prior to $t = 0$ is ignored or considered to be zero. This assumption does not impose any limitation on the applications of the Laplace transform to linear systems, since in the usual time-domain studies, time reference is often chosen at $t = 0$. Furthermore, for a physical system when an input is applied at $t = 0$, the response of the system does not start sooner than $t = 0$; that is, response does not precede excitation. Such a system is also known as being **causal** or simply **physically realizable**.

Strictly, the one-sided Laplace transform should be defined from $t = 0^-$ to $t = \infty$. The symbol $t = 0^-$ implies the limit of $t \to 0$ is taken from the left side of $t = 0$. This limiting process will take care of situations under which the function $f(t)$ has a jump discontinuity or an impulse at $t = 0$. For the subjects treated in this text, the defining equation of the Laplace transform in Eq. (2-117) is almost never used in problem solving, since the transform expressions encountered are either given or can be found from the Laplace transform table, such as the one given in Appendix C. Thus, the fine point of using 0^- or 0^+ never needs to be addressed. For simplicity, we shall simply use $t = 0$ or $t = t_0 (\geq 0)$ as the initial time in all subsequent discussions.

The following examples illustrate how Eq. (2-117) is used for the evaluation of the Laplace transform of $f(t)$.

▶ **EXAMPLE 2-4-1** Let $f(t)$ be a unit-step function that is defined as

$$f(t) = u_s(t) = 1 \quad t \geq 0$$
$$= 0 \quad t < 0 \quad (2\text{-}118)$$

The Laplace transform of $f(t)$ is obtained as

$$F(s) = \mathcal{L}[u_s(t)] = \int_0^\infty u_s(t) e^{-st} dt = -\frac{1}{s} e^{-st} \Big|_0^\infty = \frac{1}{s} \quad (2\text{-}119)$$

Eq. (2-119) is valid if

$$\int_0^\infty |u_s(t) e^{-\sigma t}| dt = \int_0^\infty |e^{-\sigma t}| dt < \infty \quad (2\text{-}120)$$

which means that the real part of s, σ, must be greater than zero. In practice, we simply refer to the Laplace transform of the unit-step function as $1/s$, and rarely do we have to be concerned with the region in the s-plane in which the transform integral converges absolutely. ◀

▶ **EXAMPLE 2-4-2** Consider the exponential function

$$f(t) = e^{-\alpha t} \quad t \geq 0 \quad (2\text{-}122)$$

where α is a real constant. The Laplace transform of $f(t)$ is written

$$F(s) = \int_0^\infty e^{-\alpha t} e^{-st} dt = \frac{e^{-(s+\alpha)t}}{s+\alpha} \Big|_0^\infty = \frac{1}{s+\alpha} \quad (2\text{-}122)$$
◀

Toolbox 2-4-1

Use the MATLAB symbolic toolbox to find the Laplace transforms.

```
>> syms t
>> f = t^4

f =

t^4

>> laplace(f)

ans =

24/s^5
```

2-4-2 Inverse Laplace Transformation

Given the Laplace transform $F(s)$, the operation of obtaining $f(t)$ is termed the inverse Laplace transformation and is denoted by

$$f(t) = \text{Inverse Laplace transform of } F(s) = \mathcal{L}^{-1}[F(s)] \qquad (2\text{-}123)$$

The inverse Laplace transform integral is given as

$$f(t) = \frac{1}{2\pi j} \int_{c-j\infty}^{c+j\infty} F(s) e^{st} ds \qquad (2\text{-}124)$$

where c is a real constant that is greater than the real parts of all the singularities of $F(s)$. Eq. (2-124) represents a line integral that is to be evaluated in the s-plane. For simple functions, the inverse Laplace transform operation can be carried out simply by referring to the Laplace transform table, such as the one given in Appendix C and on the inside back cover. For complex functions, the inverse Laplace transform can be carried out by first performing a partial-fraction expansion (Section 2-5) on $F(s)$ and then using the Transform Table from Appendix D. You may also use the **ACSYS** "Transfer Function Symbolic" Tool, **Tfsym**, for partial-fraction expansion and inverse Laplace transformation.

2-4-3 Important Theorems of the Laplace Transform

The applications of the Laplace transform in many instances are simplified by utilization of the properties of the transform. These properties are presented by the following theorems, for which no proofs are given here.

■ **Theorem 1.** *Multiplication by a Constant*

Let k be a constant and $F(s)$ be the Laplace transform of $f(t)$. Then

$$\mathcal{L}[kf(t)] = kF(s) \qquad (2\text{-}125)$$

■ **Theorem 2.** *Sum and Difference*

Let $F_1(s)$ and $F_2(s)$ be the Laplace transform of $f_1(t)$ and $f_2(t)$, respectively. Then

$$\mathcal{L}[f_1(t) \pm f_2(t)] = F_1(s) \pm F_2(s) \qquad (2\text{-}126)$$

■ **Theorem 3.** *Differentiation*

Let $F(s)$ be the Laplace transform of $f(t)$, and $f(0)$ is the limit of $f(t)$ as t approaches 0. The Laplace transform of the time derivative of $f(t)$ is

$$\mathcal{L}\left[\frac{df(t)}{dt}\right] = sF(s) - \lim_{t \to 0} f(t) = sF(s) - f(0) \qquad (2\text{-}127)$$

In general, for higher-order derivatives of $f(t)$,

$$\begin{aligned}\mathcal{L}\left[\frac{d^n f(t)}{dt^n}\right] &= s^n F(s) - \lim_{t \to 0}\left[s^{n-1} f(t) + s^{n-2}\frac{df(t)}{dt} + \cdots + \frac{d^{n-1} f(t)}{dt^{n-1}}\right] \\ &= s^n F(s) - s^{n-1} f(0) - s^{n-2} f^{(1)}(0) - \cdots - f^{(n-1)}(0)\end{aligned} \qquad (2\text{-}128)$$

where $f^{(i)}(0)$ denotes the ith-order derivative of $f(t)$ with respect to t, evaluated at $t = 0$.

■ Theorem 4. *Integration*

The Laplace transform of the first integral of $f(t)$ with respect to t is the Laplace transform of $f(t)$ divided by s; that is,

$$\mathcal{L}\left[\int_0^t f(\tau)d\tau\right] = \frac{F(s)}{s} \tag{2-129}$$

For nth-order integration,

$$\mathcal{L}\left[\int_0^{t_n}\int_0^{t_{n-1}}\cdots\int_0^{t_1} f(t)d\tau dt_1 dt_2 \cdots dt_{n-1}\right] = \frac{F(s)}{s^n} \tag{2-130}$$

■ Theorem 5. *Shift in Time*

The Laplace transform of $f(t)$ delayed by time T is equal to the Laplace transform $f(t)$ multiplied by e^{-Ts}; that is,

$$\mathcal{L}[f(t-T)u_s(t-T)] = e^{-Ts}F(s) \tag{2-131}$$

where $u_s(t-T)$ denotes the unit-step function that is shifted in time to the right by T.

■ Theorem 6. *Initial-Value Theorem*

If the Laplace transform of $f(t)$ is $F(s)$, then

$$\lim_{t \to 0} f(t) = \lim_{s \to \infty} sF(s) \tag{2-132}$$

if the limit exists.

■ Theorem 7. *Final-Value Theorem*

If the Laplace transform of $f(t)$ is $F(s)$, and if $sF(s)$ is analytic (see Section 2-1-4 on the definition of an analytic function) on the imaginary axis and in the right half of the s-plane, then

$$\lim_{t \to \infty} f(t) = \lim_{s \to 0} sF(s) \tag{2-133}$$

The final-value theorem is very useful for the analysis and design of control systems, because it gives the final value of a time function by knowing the behavior of its Laplace transform at $s = 0$. The final-value theorem is *not* valid if $sF(s)$ contains any pole whose real part is zero or positive, which is equivalent to the analytic requirement of $sF(s)$ in the right-half s-plane, as stated in the theorem. The following examples illustrate the care that must be taken in applying the theorem.

▶ **EXAMPLE 2-4-3** Consider the function

$$F(s) = \frac{5}{s(s^2 + s + 2)} \tag{2-134}$$

Because $sF(s)$ is analytic on the imaginary axis and in the right-half s-plane, the final-value theorem may be applied. Using Eq. (2-133), we have

$$\lim_{t \to \infty} f(t) = \lim_{s \to 0} sF(s) = \lim_{s \to 0} \frac{5}{s^2 + s + 2} = \frac{5}{2} \tag{2-135}$$

◀

▶ **EXAMPLE 2-4-4** Consider the function

$$F(s) = \frac{\omega}{s^2 + \omega^2} \qquad (2\text{-}136)$$

which is the Laplace transform of $f(t) = \sin \omega t$. Because the function $sF(s)$ has two poles on the imaginary axis of the s-plane, the final-value theorem *cannot* be applied in this case. In other words, although the final-value theorem would yield a value of zero as the final value of $f(t)$, the result is erroneous. ◀

Theorem 8. *Complex Shifting*

The Laplace transform of $f(t)$ multiplied by $e^{\mp \alpha t}$, where α is a constant, is equal to the Laplace transform $F(s)$, with s replaced by $s \pm \alpha$; *that is,*

$$\mathcal{L}[e^{\mp \alpha t} f(t)] = F(s \pm \alpha) \qquad (2\text{-}137)$$

TABLE 2-4 Theorems of Laplace Transforms

Multiplication by a constant	$\mathcal{L}[kf(t)] = kF(s)$
Sum and difference	$\mathcal{L}[f_1(t) \pm f_2(t)] = F_1(s) \pm F_2(s)$
Differentiation	$\mathcal{L}\left[\dfrac{df(t)}{dt}\right] = sF(s) - f(0)$
	$\mathcal{L}\left[\dfrac{d^n f(t)}{dt^n}\right] = s^n F(s) - s^{n-1} f(0) - s^{n-2} f(0)$
	where $f^{(k)}(0) = \left.\dfrac{d^k f(t)}{dt^k}\right|_{t=0}$
Integration	$\mathcal{L}\left[\displaystyle\int_0^t f(\tau)d\tau\right] = \dfrac{F(s)}{s}$
	$\mathcal{L}\left[\displaystyle\int_0^{t_n}\int_0^{t_{n-1}}\cdots\int_0^{t_1} f(t)d\tau dt_1 dt_2 \cdots dt_{n-1}\right] = \dfrac{F(s)}{s^n}$
Shift in time	$\mathcal{L}[f(t-T)u_s(t-T)] = e^{-Ts} F(s)$
Initial-value theorem	$\displaystyle\lim_{t \to 0} f(t) = \lim_{s \to \infty} sF(s)$
Final-value theorem	$\displaystyle\lim_{t \to \infty} f(t) = \lim_{s \to 0} sF(s)$ if $sF(s)$ does not have poles on or to the right of the imaginary axis in the s-plane.
Complex shifting	$\mathcal{L}[e^{\mp \alpha t} f(t)] = F(s \pm \alpha)$
Real convolution	$F_1(s)F_2(s) = \mathcal{L}\left[\displaystyle\int_0^t f_1(\tau) f_2(t-\tau) d\tau\right]$
	$= \mathcal{L}\left[\displaystyle\int_0^t f_2(\tau) f_1(t-\tau) d\tau\right] = \mathcal{L}[f_1(t) * f_2(t)]$
Complex convolution	$\mathcal{L}[f_1(t) f_2(t)] = F_1(s) * F_2(s)$

Theorem 9. *Real Convolution (Complex Multiplication)*

Let $F_1(s)$ and $F_2(s)$ be the Laplace transforms of $f_1(t)$ and $f_2(t)$, respectively, and $f_1(t) = 0$, $f_2(t) = 0$, for $t < 0$, then

$$F_1(s)F_2(s) = \mathcal{L}[f_1(t) * f_2(t)]$$
$$= \mathcal{L}\left[\int_0^t f_1(\tau)f_2(t-\tau)d\tau\right] \quad (2\text{-}138)$$
$$= \mathcal{L}\left[\int_0^t f_2(\tau)f_1(t-\tau)d\tau\right]$$

where the symbol $*$ denotes **convolution** in the time domain.

Eq. (2-138) shows that multiplication of two transformed functions in the complex s-domain is equivalent to the convolution of two corresponding real functions of t in the t-domain. An important fact to remember is that *the inverse Laplace transform of the product of two functions in the s-domain is **not** equal to the product of the two corresponding real functions in the t-domain*; that is, in general,

$$\mathcal{L}^{-1}[F_1(s)F_2(s)] \neq f_1(t)f_2(t) \quad (2\text{-}139)$$

There is also a dual relation to the real convolution theorem, called the **complex convolution**, or **real multiplication**. Essentially, the theorem states that multiplication in the real t-domain is equivalent to convolution in the complex s-domain; that is,

$$\mathcal{L}[f_1(t)f_2(t)] = F_1(s) * F_2(s) \quad (2\text{-}140)$$

where $*$ denotes complex convolution in this case. Details of the complex convolution formula are not given here. Table 2-4 summarizes the theorems of the Laplace transforms represented.

▶ 2-5 INVERSE LAPLACE TRANSFORM BY PARTIAL-FRACTION EXPANSION

In a majority of the problems in control systems, the evaluation of the inverse Laplace transform does not rely on the use of the inversion integral of Eq. (2-124). Rather, the inverse Laplace transform operation involving rational functions can be carried out using a Laplace transform table and partial-fraction expansion, both of which can also be done by computer programs.

2-5-1 Partial-Fraction Expansion

When the Laplace transform solution of a differential equation is a rational function in s, it can be written as

$$G(s) = \frac{Q(s)}{P(s)} \quad (2\text{-}141)$$

where $P(s)$ and $Q(s)$ are polynomials of s. It is assumed that *the order of $P(s)$ in s is greater than that of $Q(s)$*. The polynomial $P(s)$ may be written

$$P(s) = s^n + a_{n-1}s^{n-1} + \cdots + a_1 s + a_0 \quad (2\text{-}142)$$

where $a_0, a_1, \ldots, a_{n-1}$ are real coefficients. The methods of partial-fraction expansion will now be given for the cases of simple poles, multiple-order poles, and complex-conjugate poles of $G(s)$.

G(s) Has Simple Poles If all the poles of $G(s)$ are simple and real, Eq. (2-117) can be written as

$$G(s) = \frac{Q(s)}{P(s)} = \frac{Q(s)}{(s+s_1)(s+s_2)\cdots(s+s_n)} \qquad (2\text{-}143)$$

where $s_1 \neq s_2 \neq \cdots \neq s_n$. Applying the partial-fraction expansion, Eq. (2-143) is written

$$G(s) = \frac{K_{s1}}{s+s_1} + \frac{K_{s2}}{s+s_2} + \cdots + \frac{K_{sn}}{s+s_n} \qquad (2\text{-}144)$$

The coefficient $K_{si}(i=1,2,\ldots,n)$ is determined by multiplying both sides of Eq. (2-143) by the factor $(s+s_i)$ and then setting s equal to $-s_i$. To find the coefficient K_{s1}, for instance, we multiply both sides of Eq. (2-143) by $(s+s_1)$ and let $s=-s_1$. Thus,

$$K_{s1} = \left[(s+s_1)\frac{Q(s)}{P(s)}\right]_{s=-s_1} = \frac{Q(-s_1)}{(s_2-s_1)(s_3-s_1)\cdots(s_n-s_1)} \qquad (2\text{-}145)$$

▶ **EXAMPLE 2-5-1** Consider the function

$$G(s) = \frac{5s+3}{(s+1)(s+2)(s+3)} = \frac{5s+3}{s^3+6s^2+11s+6} \qquad (2\text{-}146)$$

which is written in the partial-fraction expanded form:

$$G(s) = \frac{K_{-1}}{s+1} + \frac{K_{-2}}{s+2} + \frac{K_{-3}}{s+3} \qquad (2\text{-}147)$$

The coefficients K_{-1}, K_{-2}, and K_{-3} are determined as follows:

$$K_{-1} = [(s+1)G(s)]\big|_{s=-1} = \frac{5(-1)+3}{(2-1)(3-1)} = -1 \qquad (2\text{-}148)$$

$$K_{-2} = [(s+2)G(s)]\big|_{s=-2} = \frac{5(-2)+3}{(1-2)(3-2)} = 7 \qquad (2\text{-}149)$$

$$K_{-3} = [(s+3)G(s)]\big|_{s=-3} = \frac{5(-3)+3}{(1-3)(2-3)} = -6 \qquad (2\text{-}150)$$

Thus, Eq. (2-146) becomes

$$G(s) = \frac{-1}{s+1} + \frac{7}{s+2} - \frac{6}{s+3} \qquad (2\text{-}151)$$

Toolbox 2-5-1

For Example 2-5-1, Eq. (2-146) is a ratio of two polynomials.

```
>> b = [5 3] % numerator polynomial coefficients
>> a = [1 6 11 6] % denominator polynomial coefficients
```

You can calculate the partial fraction expansion as

```
>> [r, p, k] = residue(b,a)
r =
    -6.0000
     7.0000
    -1.0000
```

```
p =
    -3.0000
    -2.0000
    -1.0000

k =
    []
```

Now, convert the partial fraction expansion back to polynomial coefficients.

```
>> [b,a] = residue(r,p,k)

b =
    0.0000    5.0000    3.0000

a =
    1.0000    6.0000   11.0000    6.0000
```

Note that the result is normalized for the leading coefficient in the denominator.

G(s) Has Multiple-Order Poles If r of the n poles of $G(s)$ are identical, or we say that the pole at $s = -s_i$ is of multiplicity r, $G(s)$ is written

$$G(s) = \frac{Q(s)}{P(s)} = \frac{Q(s)}{(s+s_1)(s+s_2)\cdots(s+s_{n-r})(s+s_i)^r} \qquad (2\text{-}152)$$

$(i \neq 1, 2, \ldots, n-r)$, then $G(s)$ can be expanded as

$$G(s) = \frac{K_{s1}}{s+s_1} + \frac{K_{s2}}{s+s_2} + \cdots + \frac{K_{s(n-r)}}{s+s_{n-r}}$$
$$|\leftarrow n-r \text{ terms of simple poles} \rightarrow| \qquad (2\text{-}153)$$
$$+ \frac{A_1}{s+s_i} + \frac{A_2}{(s+s_i)^2} + \cdots + \frac{A_r}{(s+s_i)^r}$$
$$|\leftarrow r \text{ terms of repeated poles} \rightarrow|$$

Then $(n-r)$ coefficients, $K_{s1}, K_{s2}, \ldots, K_{s(n-r)}$, which correspond to simple poles, may be evaluated by the method described by Eq. (2-145). The determination of the coefficients that correspond to the multiple-order poles is described as follows.

$$A_r = \left[(s+s_i)^r G(s)\right]\Big|_{s=-s_i} \qquad (2\text{-}154)$$

$$A_{r-1} = \frac{d}{ds}\left[(s+s_i)^r G(s)\right]\Big|_{s=-s_i} \qquad (2\text{-}155)$$

$$A_{r-2} = \frac{1}{2!}\frac{d^2}{ds^2}\left[(s+s_i)^r G(s)\right]\Big|_{s=-s_i} \qquad (2\text{-}156)$$

$$\vdots$$

$$A_1 = \frac{1}{(r-1)!}\frac{d^{r-1}}{ds^{r-1}}\left[(s+s_i)^r G(s)\right]\Big|_{s=-s_i} \qquad (2\text{-}157)$$

EXAMPLE 2-5-2 Consider the function

$$G(s) = \frac{1}{s(s+1)^3(s+2)} = \frac{1}{s^5 + 5s^4 + 9s^3 + 7s^2 + 2s} \qquad (2\text{-}158)$$

By using the format of Eq. (2-153), $G(s)$ is written

$$G(s) = \frac{K_0}{s} + \frac{K_{-2}}{s+2} + \frac{A_1}{s+1} + \frac{A_2}{(s+1)^2} + \frac{A_3}{(s+1)^3} \qquad (2\text{-}159)$$

The coefficients corresponding to the simple poles are

$$K_0 = \left[sG(s)\right]\Big|_{s=0} = \frac{1}{2} \qquad (2\text{-}160)$$

$$K_{-2} = \left[(s+2)G(s)\right]\Big|_{s=-2} = \frac{1}{2} \qquad (2\text{-}161)$$

and those of the third-order pole are

$$A_3 = \left[(s+1)^3 G(s)\right]\Big|_{s=-1} = -1 \qquad (2\text{-}162)$$

$$A_2 = \frac{d}{ds}\left[(s+1)^3 G(s)\right]\Big|_{s=-1} = \frac{d}{ds}\left[\frac{1}{s(s+2)}\right]\Big|_{s=-1} = 0 \qquad (2\text{-}163)$$

$$A_1 = \frac{1}{2!}\frac{d^2}{ds^2}\left[(s+1)^3 G(s)\right]\Big|_{s=-1} = \frac{1}{2}\frac{d^2}{ds^2}\left[\frac{1}{s(s+2)}\right]\Big|_{s=-1} = -1 \qquad (2\text{-}164)$$

The completed partial-fraction expansion is

$$G(s) = \frac{1}{2s} + \frac{1}{2(s+2)} - \frac{1}{s+1} - \frac{1}{(s+1)^3} \qquad (2\text{-}165)$$

Toolbox 2-5-2

For Example 2-5-2, Eq. (2-158) is a ratio of two polynomials.

```
>> clear all
>> a = [1 5 9 7 2] % coefficients of polynomial s^4 + 5*s^3 + 9*s^2 + 7*s + 2
a =
     1     5     9     7     2
>> b = [1] % polynomial coefficients
b =
     1
>> [r, p, k] = residue(b,a) % b is the numerator and a is the denominator
r =
   -1.0000
    1.0000
   -1.0000
    1.0000
```

```
p =
    -2.0000
    -1.0000
    -1.0000
    -1.0000
k =
    []
>> [b,a] = residue(r,p,k)% Obtain the polynomial form
b =
    -0.0000    -0.0000    -0.0000    1.0000

a =
    1.0000    5.0000    9.0000    7.0000    2.0000
```

G(s) Has Simple Complex-Conjugate Poles The partial-fraction expansion of Eq. (2-144) is valid also for simple complex-conjugate poles. Because complex-conjugate poles are more difficult to handle and are of special interest in control system studies, they deserve special treatment here.

Suppose that $G(s)$ of Eq. (2-117) contains a pair of complex poles:

$$s = -\sigma + j\omega \quad \text{and} \quad s = -\sigma - j\omega$$

The corresponding coefficients of these poles are found by using Eq. (2-145),

$$K_{-\sigma+j\omega} = (s + \sigma - j\omega)G(s)\big|_{s=-\sigma+j\omega} \tag{2-166}$$

$$K_{-\sigma-j\omega} = (s + \sigma + j\omega)G(s)\big|_{s=-\sigma-j\omega} \tag{2-167}$$

▶ **EXAMPLE 2-5-3** Consider the **second-order prototype** function

$$G(s) = \frac{\omega_n^2}{s^2 + 2\zeta\omega_n s + \omega_n^2} \tag{2-168}$$

Let us assume that the value of ζ is less than one, so that the poles of $G(s)$ are complex. Then, $G(s)$ is expanded as follows:

$$G(s) = \frac{K_{-\sigma+j\omega}}{s + \sigma - j\omega} + \frac{K_{-\sigma-j\omega}}{s + \sigma + j\omega} \tag{2-169}$$

where

$$\sigma = \zeta\omega_n \tag{2-170}$$

and

$$\omega = \omega_n\sqrt{1 - \zeta^2} \tag{2-171}$$

The coefficients in Eq. (2-169) are determined as

$$K_{-\sigma+j\omega} = (s+\sigma-j\omega)G(s)\Big|_{s=-\sigma+j\omega} = \frac{\omega_n^2}{2j\omega} \quad (2\text{-}172)$$

$$K_{-\sigma-j\omega} = (s+\sigma+j\omega)G(s)\Big|_{s=-\sigma-j\omega} = -\frac{\omega_n^2}{2j\omega} \quad (2\text{-}173)$$

The complete partial-fraction expansion of Eq. (2-168) is

$$G(s) = \frac{\omega_n^2}{2j\omega}\left[\frac{1}{s+\sigma-j\omega} - \frac{1}{s+\sigma+j\omega}\right] \quad (2\text{-}174)$$

Taking the inverse Laplace transform on both sides of the last equation gives

$$g(t) = \frac{\omega_n^2}{2j\omega}e^{-\sigma t}\left(e^{j\omega t} - e^{-j\omega t}\right) \quad t \geq 0 \quad (2\text{-}175)$$

Or,

$$g(t) = \frac{\omega_n}{\sqrt{1-\zeta^2}}e^{-\zeta\omega_n t}\sin\omega_n\sqrt{1-\zeta^2}\,t \quad t \geq 0 \quad (2\text{-}176)$$

◀

▶ 2-6 APPLICATION OF THE LAPLACE TRANSFORM TO THE SOLUTION OF LINEAR ORDINARY DIFFERENTIAL EQUATIONS

As we see later, the mathematical models of most components of control systems are represented by first- or second-order differential equations. In this textbook, we primarily study **linear ordinary differential equations** with constant coefficients such as the first-order linear system:

$$\frac{dy(t)}{dt} + a_0 y(t) = f(t) \quad (2\text{-}177)$$

or the second-order linear system:

$$\frac{d^2y(t)}{dt^2} + a_1\frac{dy(t)}{dt} + a_0 y(t) = f(t) \quad (2\text{-}178)$$

Linear ordinary differential equations can be solved by the Laplace transform method with the aid of the theorems on Laplace transform given in Section 2-4, the partial-fraction expansion, and the table of Laplace transforms. The procedure is outlined as follows:

1. Transform the differential equation to the s-domain by Laplace transform using the Laplace transform table.
2. Manipulate the transformed algebraic equation and solve for the output variable.
3. Perform partial-fraction expansion to the transformed algebraic equation.
4. Obtain the inverse Laplace transform from the Laplace transform table.

Let us examine two specific cases, first- and second-order prototype systems.

2-6-1 First-Order Prototype System

Consider Eq. (2-177), which may also be represented by the **first-order prototype** form:

$$\frac{dy(t)}{dt} + \frac{1}{\tau}y(t) = f(t) \tag{2-179}$$

where, t is known as the **time constant** of the system, which is a measure of how fast the system responds to initial conditions of external excitations.

▶ **EXAMPLE 2-6-1** Find the solution of the first-order differential Eq. (2-179).

SOLUTION For a unit step input

$$f(t) = u_s(t) = \begin{cases} 0, & t < 0, \\ 1, & t \geq 0 \end{cases} \tag{2-180}$$

Eq. (2-179) is written as

$$u_s(t) = \tau \dot{y}(t) + y(t) \tag{2-181}$$

If $y(0) = \dot{y}(0) = 0$, $\mathcal{L}(u_s(t)) = \frac{1}{s}$ and $L(y(t)) = Y(s)$, we have

$$\frac{1}{s} = s\tau Y(s) + Y(s) \tag{2-182}$$

or

$$Y(s) = \frac{1}{s} \frac{1}{\tau s + 1} \tag{2-183}$$

Notice that the system has a pole at $s = -1/\tau$.

Using partial fractions, Eq. (2-183) becomes

$$Y(s) = \frac{K_0}{s} + \frac{K_{-1/\tau}}{\tau s + 1} \tag{2-184}$$

where, $K_0 = 1$ and $K_{-1/\tau} = -1$. Applying the inverse Laplace transform to Eq. (2-184), we get the time response of Eq. (2-179).

$$v_o(t) = 1 - e^{-t/\tau} \tag{2-185}$$

where t is the time for $y(t)$ to reach 63% of its final value of $\lim_{t \to \infty} y(t) = 1$.

Typical unit-step responses of $y(t)$ are shown in Fig. 2-22 for a general value of t. As the value of time constant τ decreases, the system response approaches faster to the final value.

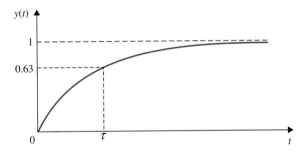

Figure 2-22 Unit-step response of a first-order *RC* circuit system.

Toolbox 2-6-1

The inverse Laplace transform for Eq. (2-183) is obtained using the MATLAB Symbolic Toolbox by the following sequence of MATLAB functions.

```
>> syms s tau;
>> ilaplace(1/(tau*s^2 + s));
```

The result is Eq. (2-185).

Note, the sym command lets you construct symbolic variables and expressions, and the command:

```
>> syms s tau;
```

is equivalent to:

```
>> s = sym('s');
>> tau = sym('tau');
```

Time response of Eq. (2-183), shown in Fig. 2-22, for a given value r = 0.1 is obtained using

```
>> clear all;
>> t = 0:0.01:1;
>> tau = 0.1;
>> plot(1-exp(-t/tau));
```

You may wish to confirm that at t = 0.1, y(t) = 0.63.

2-6-2 Second-Order Prototype System

Similarly, for the **second-order prototype** of the form:

$$\frac{d^2y(t)}{dt^2} + 2\zeta\omega_n\frac{dy(t)}{dt} + \omega_n^2 y(t) = f(t) \tag{2-186}$$

where ζ is known as the damping ratio, and ω_n is the natural frequency of the system. The prototype forms of differential equations provide a common format of representing various components of a control system. The significance of this representation will become more evident when we study the **time response** of control systems in Chapter 5.

▶ **EXAMPLE 2-6-2** Consider the differential equation

$$\frac{d^2y(t)}{dt^2} + 3\frac{dy(t)}{dt} + 2y(t) = 5u_s(t) \tag{2-187}$$

where $u_s(t)$ is the unit-step function. The initial conditions are $y(0) = -1$ and $y^{(1)}(0) = dy(t)/dt|_{t=0} = 2$. To solve the differential equation, we first take the Laplace transform on both sides of Eq. (2-153):

$$s^2 Y(s) - sy(0) - y^{(1)}(0) + 3sY(s) - 3y(0) + 2Y(s) = 5/s \tag{2-188}$$

Substituting the values of the initial conditions into the last equation and solving for $Y(s)$, we get

$$Y(s) = \frac{-s^2 - s + 5}{s(s^2 + 3s + 2)} = \frac{-s^2 - s + 5}{s(s+1)(s+2)} \tag{2-189}$$

Eq. (2-189) is expanded by partial-fraction expansion to give

$$Y(s) = \frac{5}{2s} - \frac{5}{s+1} + \frac{3}{2(s+2)} \quad (2\text{-}190)$$

Taking the inverse Laplace transform of Eq. (2-190), we get the complete solution as

$$y(t) = \frac{5}{2} - 5e^{-t} + \frac{3}{2}e^{-2t} \quad t \geq 0 \quad (2\text{-}191)$$

The first term in Eq. (2-191) is the steady-state solution or the particular integral; the last two terms represent the transient or homogeneous solution. Unlike the classical method, which requires separate steps to give the transient and the steady-state responses or solutions, the Laplace transform method gives the entire solution in one operation.

If only the magnitude of the steady-state solution of $y(t)$ is of interest, the final-value theorem of Eq. (2-133) may be applied. Thus,

$$\lim_{t \to \infty} y(t) = \lim_{s \to 0} sY(s) = \lim_{s \to 0} \frac{-s^2 - s + 5}{s^2 + 3s + 2} = \frac{5}{2} \quad (2\text{-}192)$$

where, in order to ensure the validity of the final-value theorem, we have first checked and found that the poles of function $sY(s)$ are all in the left-half s-plane.

▶ **EXAMPLE 2-6-3** Consider the linear differential equation

$$\frac{d^2y(t)}{dt^2} + 34.5\frac{dy(t)}{dt} + 1000y(t) = 1000u_s(t) \quad (2\text{-}193)$$

The initial values of $y(t)$ and $dy(t)/dt$ are zero. Taking the Laplace transform on both sides of Eq. (2-193), and solving for $Y(s)$, we have

$$Y(s) = \frac{1000}{s(s^2 + 34.5s + 1000)} = \frac{\omega_n^2}{s(s^2 + 2\zeta\omega_n s + \omega_n^2)} \quad (2\text{-}194)$$

where, using the second-order prototype representation, $\zeta = 0.5455$ and $\omega_n = 31.62$. The inverse Laplace transform of Eq. (2-194) can be executed in a number of ways. The Laplace transform table in Appendix C provides the time-response expression in Eq. (2-194) directly. The result is

$$y(t) = 1 - \frac{e^{-\zeta\omega_n t}}{\sqrt{1-\zeta^2}}\sin\left(\omega_n\sqrt{1-\zeta^2}t + \theta\right) \quad t \geq 0 \quad (2\text{-}195)$$

where

$$\theta = \cos^{-1}\zeta = 0.9938 \text{ rad}\left(= 56.94° \frac{\pi \text{ rad}}{180°}\right) \quad (2\text{-}196)$$

Thus,

$$y(t) = 1 - 1.193e^{-17.25t}\sin(26.5t + 0.9938) \quad t \geq 0 \quad (2\text{-}197)$$

Eq. (2-197) can be derived by performing the partial-fraction expansion of Eq. (2-194), knowing that the poles are at $s = 0$, $-\sigma + j\omega$, and $-\sigma - j\omega$, where

$$\sigma = \zeta\omega_n = 17.25 \quad (2\text{-}198)$$

$$\omega = \omega_n\sqrt{1-\zeta^2} = 26.5 \quad (2\text{-}199)$$

The partial-fraction expansion of Eq. (2-194) is written

$$Y(s) = \frac{K_0}{s} + \frac{K_{-\sigma+j\omega}}{s+\sigma-j\omega} + \frac{K_{-\sigma-j\omega}}{s+\sigma+j\omega} \quad (2\text{-}200)$$

where

$$K_0 = sY(s)\Big|_{s=0} = 1 \quad (2\text{-}201)$$

$$K_{-\sigma+j\omega} = (s+\sigma-j\omega)Y(s)\Big|_{s=-\sigma+j\omega} = \frac{e^{-j\phi}}{2j\sqrt{1-\zeta^2}} \quad (2\text{-}202)$$

$$K_{-\sigma-j\omega} = (s+\sigma+j\omega)Y(s)\Big|_{s=-\sigma-j\omega} = \frac{-e^{-j\phi}}{2j\sqrt{1-\zeta^2}} \quad (2\text{-}203)$$

The angle ϕ is given by

$$\phi = 180° - \cos^{-1}\zeta \quad (2\text{-}204)$$

and is illustrated in Fig. 2-23.

The inverse Laplace transform of Eq. (2-200) is now written

$$\begin{aligned} y(t) &= 1 + \frac{1}{2j\sqrt{1-\zeta^2}} e^{-\zeta\omega_n t}\left[e^{j(\omega t - \phi)} - e^{-j(\omega t - \phi)}\right] \\ &= 1 + \frac{1}{\sqrt{1-\zeta^2}} e^{-\zeta\omega_n t} \sin\left[\omega_n\sqrt{1-\zeta^2}\,t - \phi\right] \end{aligned} \quad (2\text{-}205)$$

Substituting Eq. (2-204) into Eq. (2-205) for ϕ, we have

$$y(t) = 1 - \frac{1}{\sqrt{1-\zeta^2}} e^{-\zeta\omega_n t} \sin\left[\omega_n\sqrt{1-\zeta^2}\,t + \cos^{-1}\zeta\right] \quad t \geq 0 \quad (2\text{-}206)$$

or

$$y(t) = 1 - 1.193 e^{-17.25t} \sin(26.5t + 0.9938) \quad t \geq 0 \quad (2\text{-}207)$$

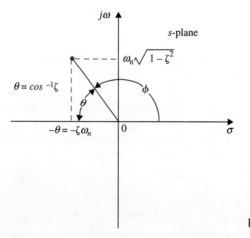

Figure 2-23 Root location in the *s*-plane.

Toolbox 2-6-2

Time response of Eq. (2-194) for a unit-step input may also be obtained using

```
                          Alternatively:
num = [1000];
den = [1 34.5 1000];      s = tf ('s');
G = tf (num,den);         G=1000/(s^2+34.5*s+1000);
step(G);                  step (G);
title ('Step Response')   title ('Step Response')
xlabel ('Time (sec')      xlabel ('Time(sec')
ylabel ('Amplitude')      ylabel ('Amplitude')
```

"step" produces the time response of a function for a unit-step input.

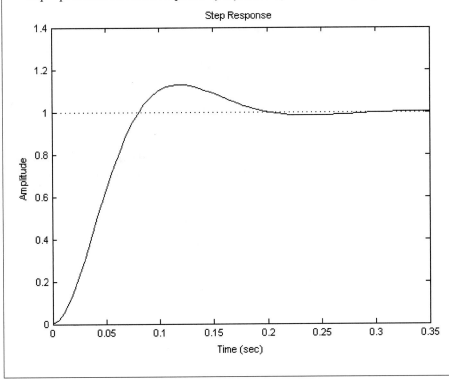

2-7 IMPULSE RESPONSE AND TRANSFER FUNCTIONS OF LINEAR SYSTEMS

The classical way of modeling linear time-invariant systems is to use **transfer functions** to represent input–output relations between variables. One way to define the transfer function is to use the impulse response, which is defined as follows.

2-7-1 Impulse Response

Consider that a linear time-invariant system has the input $u(t)$ and output $y(t)$. As shown in Fig. 2-24, a rectangular pulse function $u(t)$ of a very large magnitude $\hat{u}/2\varepsilon$ becomes

68 ▶ Chapter 2. Mathematical Foundation

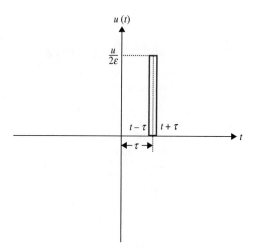

Figure 2-24 Graphical representation an impulse function.

an impulse function for very small durations as $\varepsilon \to 0$. The equation representing Fig. 2-24 is

$$u(t) = \begin{cases} 0 & t \leq \tau - \varepsilon \\ \dfrac{\hat{u}}{2\varepsilon} & \tau - \varepsilon < t < \tau + \varepsilon \\ 0 & t \geq \tau + \varepsilon \end{cases} \qquad (2\text{-}208)$$

For $\hat{u} = 1$, $u(t) = \delta(t)$ is also known as unit impulse or **Dirac delta** function. For $t = 0$ in Eq. (2-208), using Eq. (2-116) and noting the actual limits of the integral are defined from $t = 0^-$ to $t = \infty$, it is easy to verify that the **Laplace transform of $\delta(t)$ is unity**, i.e. $\mathcal{L}[\delta(t)] = 1$ as $\varepsilon \to 0$.

The important point here is that *the response of any system can be characterized by its impulse response* g(t), *which is defined as the output when the input is a unit-impulse function* $\delta(t)$. Once the impulse response of a linear system is known, *the output of the system* y(t), *with any input,* u(t), *can be found by using the transfer function*. We define

$$G(s) = \frac{\mathcal{L}(y(t))}{\mathcal{L}(u(t))} = \frac{Y(s)}{F(s)} \qquad (2\text{-}209)$$

as the **transfer function** of the system.

▶ **EXAMPLE 2-7-1** For the second-order prototype system Eq. (2-186), shown in Example 2-5-3 as:

$$\frac{d^2y(t)}{dt^2} + 2\zeta\omega_n \frac{dy(t)}{dt} + \omega_n^2 y(t) = \omega_n^2 u(t) \qquad (2\text{-}210)$$

Hence,

$$G(s) = \frac{\mathcal{L}(y(t))}{\mathcal{L}(u(t))} = \frac{\omega_n^2}{s^2 + 2\zeta\omega_n s + \omega_n^2} \qquad (2\text{-}211)$$

is the transfer function of the system in Eq. (2-210). Similar to Example 2-5-3, given zero initial conditions, the impulse response $g(t)$ is

$$g(t) = \frac{\omega_n}{\sqrt{1-\zeta^2}} e^{-\zeta\omega_n t} \sin \omega_n \sqrt{1-\zeta^2}\, t \quad t \geq 0 \qquad (2\text{-}212)$$

For a unit-step input $u(t) = u_s(t)$, using the convolution properties of Laplace transforms,

$$\mathcal{L}[y(t)] = \mathcal{L}[u_s * g(t)]$$
$$= \mathcal{L}\left[\int_0^t u_s g(t-\tau)d\tau\right] = \frac{G(s)}{s} \qquad (2\text{-}213)$$

From the inverse Laplace transform of Eq. (2-213), the output $y(t)$ is therefore

$$\int_0^t u_s g(t-\tau)d\tau$$

or

$$y(t) = 1 - \frac{e^{-\zeta\omega_n t}}{\sqrt{1-\zeta^2}} \sin\left(\omega_n\sqrt{1-\zeta^2}\,t + \theta\right) \quad t \geq 0 \qquad (2\text{-}214)$$

where, $\theta = \cos^{-1}\zeta$.

◂

Toolbox 2-7-1

The unit impulse response of Eq. (2-194) may be obtained using

```
num = [1000];
den = [1 34.5 1000];
G = tf(num,den);
impulse(G);
```

Alternatively:
```
s = tf('s');
G = 1000/(s^2+34.5*s+1000);
impulse(G);
```

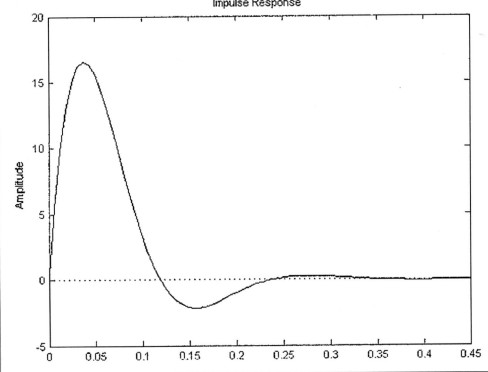

2-7-2 Transfer Function (Single-Input, Single-Output Systems)

The transfer function of a linear time-invariant system is defined as the Laplace transform of the impulse response, with all the initial conditions set to zero.

Let $G(s)$ denote the transfer function of a single-input, single-output (SISO) system, with input $u(t)$, output $y(t)$, and impulse response $g(t)$. The transfer function $G(s)$ is defined as

$$G(s) = \mathcal{L}[g(t)] \tag{2-215}$$

The transfer function $G(s)$ is related to the Laplace transform of the input and the output through the following relation:

$$G(s) = \frac{Y(s)}{U(s)} \tag{2-216}$$

with all the initial conditions set to zero, and $Y(s)$ and $U(s)$ are the Laplace transforms of $y(t)$ and $u(t)$, respectively.

Although the transfer function of a linear system is defined in terms of the impulse response, in practice, the input–output relation of a linear time-invariant system with continuous-data input is often described by a differential equation, so it is more convenient to derive the transfer function directly from the differential equation. Let us consider that the input–output relation of a linear time-invariant system is described by the following nth-order differential equation with constant real coefficients:

$$\begin{aligned} \frac{d^n y(t)}{dt^n} + a_{n-1}\frac{d^{n-1} y(t)}{dt^{n-1}} + \cdots + a_1 \frac{dy(t)}{dt} + a_0 y(t) \\ = b_m \frac{d^m u(t)}{dt^m} + b_{m-1}\frac{d^{m-1} u(t)}{dt^{m-1}} + \cdots + b_1 \frac{du(t)}{dt} + b_0 u(t) \end{aligned} \tag{2-217}$$

The coefficients $a_0, a_1, \ldots, a_{n-1}$ and $b_0, b_1, \ldots, b_m$ are real constants. Once the input $u(t)$ for $t \geq t_0$ and the initial conditions of $y(t)$ and the derivatives of $y(t)$ are specified at the initial time $t = t_0$, the output response $y(t)$ for $t \geq t_0$ is determined by solving Eq. (2-217). However, from the standpoint of linear-system analysis and design, the method of using differential equations exclusively is quite cumbersome. Thus, differential equations of the form of Eq. (2-217) are seldom used in their original form for the analysis and design of control systems. It should be pointed out that, although efficient subroutines are available on digital computers for the solution of high-order differential equations, *the basic philosophy of linear control theory is that of developing analysis and design tools that will avoid the exact solution of the system differential equations*, except when computer-simulation solutions are desired for final presentation or verification. In classical control theory, even computer simulations often start with transfer functions, rather than with differential equations.

To obtain the transfer function of the linear system that is represented by Eq. (2-145), we simply take the Laplace transform on both sides of the equation and assume **zero initial conditions**. The result is

$$(s^n + a_{n-1}s^{n-1} + \cdots + a_1 s + a_0)Y(s) = (b_m s^m + b_{m-1}s^{m-1} + \cdots + b_1 s + b_0)U(s) \tag{2-218}$$

The transfer function between $u(t)$ and $y(t)$ is given by

$$G(s) = \frac{Y(s)}{U(s)} = \frac{b_m s^m + b_{m-1}s^{m-1} + \cdots + b_1 s + b_0}{s^n + a_{n-1}s^{n-1} + \cdots + a_1 s + a_0} \tag{2-219}$$

The properties of the transfer function are summarized as follows:

- The transfer function is defined only for a linear time-invariant system. It is not defined for nonlinear systems.
- The transfer function between an input variable and an output variable of a system is defined as the Laplace transform of the impulse response. Alternately, the transfer function between a pair of input and output variables is the ratio of the Laplace transform of the output to the Laplace transform of the input.
- All initial conditions of the system are set to zero.
- The transfer function is independent of the input of the system.
- The transfer function of a continuous-data system is expressed only as a function of the complex variable s. It is not a function of the real variable, time, or any other variable that is used as the independent variable. For discrete-data systems modeled by difference equations, the transfer function is a function of z when the z-transform is used (refer to Appendix D).

2-7-3 Proper Transfer Functions

The transfer function in Eq. (2-147) is said to be strictly proper if the order of the denominator polynomial is greater than that of the numerator polynomial (i.e., $n > m$). If $n = m$, the transfer function is called proper. The transfer function is improper if $m > n$.

2-7-4 Characteristic Equation

The characteristic equation of a linear system is defined as the equation obtained by setting the denominator polynomial of the transfer function to zero. Thus, from Eq. (2-147), the characteristic equation of the system described by Eq. (2-145) is

$$s^n + a_{n-1}s^{n-1} + \cdots + a_1 s + a_0 = 0 \qquad (2\text{-}220)$$

Later we shall show that the stability of linear, single-input, single-output systems is completely governed by the roots of the characteristic equation.

2-7-5 Transfer Function (Multivariable Systems)

The definition of a transfer function is easily extended to a system with multiple inputs and outputs. A system of this type is often referred to as a multivariable system. In a multivariable system, a differential equation of the form of Eq. (2-145) may be used to describe the relationship between a pair of input and output variables, when all other inputs are set to zero. Because the principle of superposition is valid for linear systems, the total effect on any output due to all the inputs acting simultaneously is obtained by adding up the outputs due to each input acting alone.

In general, if a linear system has p inputs and q outputs, the transfer function between the jth input and the ith output is defined as

$$G_{ij}(s) = \frac{Y_i(s)}{R_j(s)} \qquad (2\text{-}221)$$

with $R_k(s) = 0$, $k = 1, 2, \ldots, p$, $k \neq j$. Note that Eq. (2-151) is defined with only the jth input in effect, whereas the other inputs are set to zero. When all the p inputs are in action, the ith output transform is written

$$Y_i(s) = G_{i1}(s)R_1(s) + G_{i2}(s)R_2(s) + \cdots + G_{ip}(s)R_p(s) \qquad (2\text{-}222)$$

It is convenient to express Eq. (2-152) in matrix-vector form:

$$\mathbf{Y}(s) = \mathbf{G}(s)\mathbf{R}(s) \quad (2\text{-}223)$$

where

$$\mathbf{Y}(s) = \begin{bmatrix} Y_1(s) \\ Y_2(s) \\ \vdots \\ Y_q(s) \end{bmatrix} \quad (2\text{-}224)$$

is the $q \times 1$ transformed output vector;

$$\mathbf{R}(s) = \begin{bmatrix} R_1(s) \\ R_2(s) \\ \vdots \\ R_p(s) \end{bmatrix} \quad (2\text{-}225)$$

is the $p \times 1$ transformed input vector; and

$$\mathbf{G}(s) = \begin{bmatrix} G_{11}(s) & G_{12}(s) & \cdots & G_{1p}(s) \\ G_{21}(s) & G_{22}(s) & \cdots & G_{2p}(s) \\ \cdot & \cdot & \cdots & \cdot \\ G_{q1}(s) & G_{q2}(s) & \cdots & G_{qp}(s) \end{bmatrix} \quad (2\text{-}226)$$

is the $q \times p$ transfer-function matrix.

▶ 2-8 STABILITY OF LINEAR CONTROL SYSTEMS

From the studies of linear differential equations with constant coefficients of SISO systems, we learned that the homogeneous solution that corresponds to the transient response of the system is governed by the roots of the characteristic equation. Basically, the design of linear control systems may be regarded as a problem of arranging the location of the poles and zeros of the system transfer function such that the system will perform according to the prescribed specifications.

Among the many forms of performance specifications used in design, the most important requirement is that the system must be stable. An unstable system is generally considered to be useless.

When all types of systems are considered—linear, nonlinear, time-invariant, and time-varying—the definition of stability can be given in many different forms. We shall deal only with the stability of linear SISO time-invariant systems in the following discussions.

For analysis and design purposes, we can classify stability as **absolute stability** and **relative stability**. Absolute stability refers to whether the system is stable or unstable; it is a *yes* or *no* answer. Once the system is found to be stable, it is of interest to determine how stable it is, and this degree of stability is a measure of relative stability.

In preparation for the definition of stability, we define the two following types of responses for linear time-invariant systems:

- **Zero-state response.** The zero-state response is due to the input only; all the initial conditions of the system are zero.

- **Zero-input response.** The zero-input response is due to the initial conditions only; all the inputs are zero.

From the principle of superposition, when a system is subject to both inputs and initial conditions, the total response is written

Total response = zero-state response + zero-input response

The definitions just given apply to continuous-data as well as discrete-data systems.

▶ 2-9 BOUNDED-INPUT, BOUNDED-OUTPUT (BIBO) STABILITY—CONTINUOUS-DATA SYSTEMS

Let $u(t)$, $y(t)$, and $g(t)$ be the input, output, and the impulse response of a linear time-invariant system, respectively. *With zero initial conditions, the system is said to be BIBO* (bounded-input, bounded-output) *stable, or simply stable, if its output* y(t) *is bounded to a bounded input* u(t).

The convolution integral relating $u(t)$, $y(t)$, and $g(t)$ is

$$y(t) = \int_0^\infty u(t-\tau)g(\tau)d\tau \qquad (2\text{-}227)$$

Taking the absolute value of both sides of the equation, we get

$$|y(t)| = \left|\int_0^\infty u(t-\tau)g(\tau)d\tau\right| \qquad (2\text{-}228)$$

or

$$|y(t)| \leq \int_0^\infty |u(t-\tau)||g(\tau)|d\tau \qquad (2\text{-}229)$$

If $u(t)$ is bounded,

$$|u(t)| \leq M \qquad (2\text{-}230)$$

where M is a finite positive number. Then,

$$|y(t)| \leq M \int_0^\infty |g(\tau)|d\tau \qquad (2\text{-}231)$$

Thus, if $y(t)$ is to be bounded, or

$$|y(t)| \leq N < \infty \qquad (2\text{-}232)$$

where N is a finite positive number, the following condition must hold:

$$M \int_0^\infty |g(\tau)|d\tau \leq N < \infty \qquad (2\text{-}233)$$

Or, for any finite positive Q,

$$\int_0^\infty |g(\tau)|d\tau \leq Q < \infty \qquad (2\text{-}234)$$

The condition given in Eq. (2-234) implies that the area under the $|g(\tau)|$–versus–τ-curve must be finite.

▶ 2-10 RELATIONSHIP BETWEEN CHARACTERISTIC EQUATION ROOTS AND STABILITY

To show the relation between the roots of the characteristic equation and the condition in Eq. (2-234), we write the transfer function $G(s)$, according to the Laplace transform definition, as

$$G(s) = \mathcal{L}[g(\tau)] = \int_0^\infty g(t)e^{-st}dt \qquad (2\text{-}235)$$

Taking the absolute value on both sides of the last equation, we have

$$|G(s)| = \left|\int_0^\infty g(t)e^{-st}dt\right| \leq \int_0^\infty |g(t)||e^{-st}|dt \qquad (2\text{-}236)$$

Because $|e^{-st}| = |e^{-\sigma t}|$, where σ is the real part of s, when s assumes a value of a pole of $G(s)$, $G(s) = \infty$, Eq. (2-236) becomes

$$\infty \leq \int_0^\infty |g(t)||e^{-\sigma t}|dt \qquad (2\text{-}237)$$

If one or more roots of the characteristic equation are in the right-half s-plane or on the $j\omega$-axis, $\sigma \geq 0$, then

$$|e^{-\sigma t}| \leq M = 1 \qquad (2\text{-}238)$$

Eq. (2-237) becomes

$$\infty \leq \int_0^\infty M|g(t)|dt = \int_0^\infty |g(t)|dt \qquad (2\text{-}239)$$

which violates the BIBO stability requirement. Thus, *for BIBO stability, the roots of the characteristic equation, or the poles of* $G(s)$, *cannot be located in the right-half* s-*plane or on the* $j\omega$-*axis; in other words, they must all lie in the left-half* s-*plane. A system is said to be unstable if it is not BIBO stable.* When a system has roots on the $j\omega$-axis, say, at $s = j\omega_0$ and $s = -j\omega_0$, if the input is a sinusoid, $\sin\omega_0 t$, then the output will be of the form of $t\sin\omega_0 t$, which is unbounded, and the system is unstable.

▶ 2-11 ZERO-INPUT AND ASYMPTOTIC STABILITY OF CONTINUOUS-DATA SYSTEMS

In this section, we shall define zero-input stability and asymptotic stability and establish their relations with BIBO stability.

Zero-input stability refers to the stability condition when the input is zero, and the system is driven only by its initial conditions. We shall show that the zero-input stability also depends on the roots of the characteristic equation.

Let the input of an nth-order system be zero and the output due to the initial conditions be $y(t)$. Then, $y(t)$ can be expressed as

$$y(t) = \sum_{k=0}^{n-1} g_k(t)y^{(k)}(t_0) \qquad (2\text{-}240)$$

where

$$y^{(k)}(t_0) = \left.\frac{d^k y(t)}{dt^k}\right|_{t=t_0} \qquad (2\text{-}241)$$

and $g_k(t)$ denotes the zero-input response due to $y^{(k)}(t_0)$. The zero-input stability is defined as follows: *If the zero-input response y(t), subject to the finite initial conditions, $y^{(k)}(t_0)$, reaches zero as t approaches infinity, the system is said to be zero-input stable, or stable; otherwise, the system is unstable.*

Mathematically, the foregoing definition can be stated: *A linear time-invariant system is zero-input stable if, for any set of finite $y^{(k)}(t_0)$, there exists a positive number* M, *which depends on $y^{(k)}(t_0)$, such that*

1.

$$|y(t)| \leq M < \infty \quad \text{for all } t \geq t_0 \qquad (2\text{-}242)$$

and

2.

$$\lim_{t \to \infty} |y(t)| = 0 \qquad (2\text{-}243)$$

Because the condition in the last equation requires that the magnitude of $y(t)$ reaches zero as time approaches infinity, the zero-input stability is also known at the **asymptotic stability**.

Taking the absolute value on both sides of Eq. (2-240), we get

$$|y(t)| = \left|\sum_{k=0}^{n-1} g_k(t) y^{(k)}(t_0)\right| \leq \sum_{k=0}^{n-1} |g_k(t)| \left|y^{(k)}(t_0)\right| \qquad (2\text{-}244)$$

Because all the initial conditions are assumed to be finite, the condition in Eq. (2-242) requires that the following condition be true:

$$\sum_{k=0}^{n-1} |g_k(t)| < \infty \qquad \text{for all } t \geq 0 \qquad (2\text{-}245)$$

Let the n characteristic equation roots be expressed as $s_i = \sigma_i + j\omega_i$, $i = 1, 2, \ldots, n$. Then, if m of the n roots are simple, and the rest are of multiple order, $y(t)$ will be of the form:

$$y(t) = \sum_{i=1}^{m} K_i e^{s_i t} + \sum_{i=0}^{n-m-1} L_i t^i e^{s_i t} \qquad (2\text{-}246)$$

where K_i and L_i are constant coefficients. Because the exponential terms $e^{s_i t}$ in the last equation control the response $y(t)$ as $t \to \infty$, to satisfy the two conditions in Eqs. (2-242) and (2-243), the real parts of s_i must be negative. In other words, the roots of the characteristic equation must all be in the left-half s-plane.

From the preceding discussions, we see that, *for linear time-invariant systems, BIBO, zero-input, and asymptotic stability all have the same requirement that the roots of the characteristic equation must all be located in the left-half* s-*plane. Thus, if a system is BIBO stable, it must also be zero-input or asymptotically stable.* For this reason, we shall simply refer to the stability condition of a linear system as **stable** or **unstable**. The latter

TABLE 2-5 Stability Conditions of Linear Continuous-Data Time-Invariant SISO Systems

Stability Condition	Root Values
Asymptotically stable or simply stable	$\sigma_i < 0$ for all i, $i = 1, 2, \ldots, n$. (All the roots are in the left-half s-plane.)
Marginally stable or marginally unstable	$\sigma_i = 0$ for any i for simple roots, and no $\sigma_i > 0$ For $i = 1, 2, \ldots, n$ (at least one simple root, no multiple-order roots on the $j\omega$-axis, and n roots in the right-half s-plane; note exceptions)
Unstable	$\sigma_i > 0$ for any i, or $\sigma_i = 0$ for any multiple-order root; $i = 1, 2, \ldots, n$ (at least one simple root in the right-half s-plane or at least one multiple-order root on the $j\omega$-axis)

condition refers to the condition that at least one of the characteristic equation roots is not in the left-half s-plane. For practical reasons, we often refer to the situation in which the characteristic equation has simple roots on the $j\omega$-axis and none in the right-half plane as **marginally stable** or **marginally unstable**. *An exception to this is if the system were intended to be an integrator (or, in the case of control systems, a velocity control system); then the system would have root(s) at $s = 0$ and would be considered stable.* Similarly, if the system were designed to be an oscillator, the characteristic equation would have simple roots on the $j\omega$-axis, and the system would be regarded as stable.

Because the roots of the characteristic equation are the same as the eigenvalues of **A** of the state equations, the stability condition places the same restrictions on the eigenvalues.

Let the characteristic equation roots or eigenvalues of **A** of a linear continuous-data time-invariant SISO system be $s_i = \sigma_i + j\omega_i$, $i = 1, 2, \ldots, n$. If any of the roots is complex, it is in complex-conjugate pairs. The possible stability conditions of the system are summarized in Table 2-5 with respect to the roots of the characteristic equation.

The following example illustrates the stability conditions of systems with reference to the poles of the system transfer functions that are also the roots of the characteristic equation.

▶ **EXAMPLE 2-11-1** The following closed-loop transfer functions and their associated stability conditions are given.

$$M(s) = \frac{20}{(s+1)(s+2)(s+3)} \qquad \text{BIBO or asymptotically stable (or, simply, stable)}$$

$$M(s) = \frac{20(s+1)}{(s-1)(s^2+2s+2)} \qquad \text{Unstable due to the pole at } s = 1$$

$$M(s) = \frac{20(s-1)}{(s+2)(s^2+4)} \qquad \text{Marginally stable or marginally unstable due to } s = \pm j2$$

$$M(s) = \frac{10}{(s^2+4)^2(s+10)} \qquad \text{Unstable due to the multiple-order pole at } s = \pm j2$$

$$M(s) = \frac{10}{s^4+30s^3+s^2+10s} \qquad \text{Stable if the pole at } s = 0 \text{ is placed intentionally}$$

◀

2-12 METHODS OF DETERMINING STABILITY

The discussions in the preceding sections lead to the conclusion that the stability of linear time-invariant SISO systems can be determined by checking on the location of the roots of the characteristic equation of the system. For all practical purposes, there is no need to compute the complete system response to determine stability. The regions of stability and instability in the s-plane are illustrated in Fig. 2-25. When the system parameters are all known, the roots of the characteristic equation can be found using MATLAB as demonstrated in various MATLAB Toolbox windows discussed earlier in this chapter. The Transfer Function Symbolic Tool (**tfsym**) developed for this chapter may also be utilized to find the transfer function poles and zeros. See the end of this chapter for some examples. These programs are discussed in detail in Appendix G. For design purposes, there will be unknown or variable parameters imbedded in the characteristic equation, so a Routh-Hurwitz stability routine has also been developed for this textbook (**tfrouth**), which is discussed at the end of this chapter.

The methods outlined in the following list are well known for determining the stability of linear continuous-data systems without involving root solving.

1. **Routh-Hurwitz criterion.** This criterion is an algebraic method that provides information on the absolute stability of a linear time-invariant system that has a characteristic equation with constant coefficients. The criterion tests whether any of the roots of the characteristic equation lie in the right-half s-plane. The number of roots that lie on the $j\omega$-axis and in the right-half s-plane is also indicated.

2. **Nyquist criterion.** This criterion is a semi-graphical method that gives information on the difference between the number of poles and zeros of the closed-loop transfer function that are in the right-half s-plane by observing the behavior of the Nyquist plot of the loop transfer function. This topic is discussed in detail in Chapter 8, and the concepts of loop transfer function and close-loop systems are discussed in Chapter 3.

3. **Bode diagram.** This diagram is a plot of the magnitude of the loop transfer function $G(j\omega)H(j\omega)$ in dB and the phase of $G(j\omega)H(j\omega)$ in degrees, all versus frequency ω. The concepts of loop transfer function and closed-loop systems are

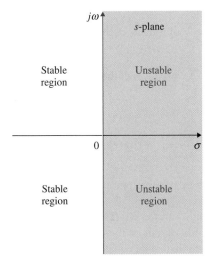

Figure 2-25 Stable and unstable regions in the s-plane.

discussed in Chapter 3. The stability of the closed-loop system can be determined by observing the behavior of these plots. This topic is discussed in detail in Chapter 8.

Thus, as will be evident throughout the text, most of the analysis and design techniques on control systems represent alternate methods of solving the same problem. The designer simply has to choose the best analytical tool, depending on the particular situation.

Details of the Routh-Hurwitz stability criterion are presented in the following section.

▶ 2-13 ROUTH-HURWITZ CRITERION

The Routh-Hurwitz criterion represents a method of determining the location of zeros of a polynomial with constant real coefficients with respect to the left half and right half of the s-plane, without actually solving for the zeros. *Because root-finding computer programs can solve for the zeros of a polynomial with ease, the value of the Routh-Hurwitz criterion is at best limited to equations with at least one unknown parameter.*

Consider that the characteristic equation of a linear time-variant SISO system is of the form

$$F(s) = a_n s^n + a_{n-1} s^{n-1} + \cdots + a_1 s + a_0 = 0 \qquad (2\text{-}247)$$

where all the coefficients are real. To ensure the last equation does not have roots with positive real parts, it is *necessary (but not sufficient)* that the following conditions hold:

1. All the coefficients of the equation have the same sign.
2. None of the coefficients vanish.

These conditions are based on the laws of algebra, which relate the coefficients of Eq. (2-247) as follows:

$$\frac{a_{n-1}}{a_n} = -\sum \text{all roots} \qquad (2\text{-}248)$$

$$\frac{a_{n-2}}{a_n} = \sum \text{products of the roots taken two at a time} \qquad (2\text{-}249)$$

$$\frac{a_{n-3}}{a_n} = -\sum \text{products of the roots taken three at a time} \qquad (2\text{-}250)$$

$$\vdots$$

$$\frac{a_0}{a_n} = (-1)^n \text{products of all the roots} \qquad (2\text{-}251)$$

Thus, all these ratios must be positive and nonzero unless at least one of the roots has a positive real part.

The two necessary conditions for Eq. (2-247) to have no roots in the right-half s-plane can easily be checked by inspection of the equation. However, these conditions are not sufficient, for it is quite possible that an equation with all its coefficients nonzero and of the same sign still may not have all the roots in the left half of the s-plane.

2-13-1 Routh's Tabulation

The Hurwitz criterion gives the necessary and sufficient condition for all roots of Eq. (2-247) to lie in the left half of the s-plane. The criterion requires that the equation's n Hurwitz determinants must all be positive.

However, the evaluation of the n Hurwitz determinants is tedious to carry out. But Routh simplified the process by introducing a tabulation method in place of the Hurwitz determinants.

The first step in the simplification of the Hurwitz criterion, now called the Routh-Hurwitz criterion, is to arrange the coefficients of the equation in Eq. (2-247) into two rows. The first row consists of the first, third, fifth, ..., coefficients, and the second row consists of the second, fourth, sixth, ..., coefficients, all counting from the highest-order term, as shown in the following tabulation:

$$\begin{array}{cccc} a_n & a_{n-2} & a_{n-4} & a_{n-6} \cdots \\ a_{n-1} & a_{n-3} & a_{n-5} & a_{n-7} \cdots \end{array}$$

The next step is to form the following array of numbers by the indicated operations, illustrated here for a sixth-order equation:

$$a_6 s^6 + a_5 s^5 + \cdots + a_1 s + a_0 = 0 \tag{2-252}$$

s^6	a_6	a_4	a_2	a_0
s^5	a_5	a_3	a_1	0
s^4	$\dfrac{a_5 a_4 - a_6 a_3}{a_5} = A$	$\dfrac{a_5 a_2 - a_6 a_1}{a_5} = B$	$\dfrac{a_5 a_0 - a_6 \times 0}{a_5} = a_0$	0
s^3	$\dfrac{A a_3 - a_5 B}{A} = C$	$\dfrac{A a_1 - a_5 a_0}{A} = D$	$\dfrac{A \times 0 - a_5 \times 0}{A} = 0$	0
s^2	$\dfrac{BC - AD}{C} = E$	$\dfrac{C a_0 - A \times 0}{C} = a_0$	$\dfrac{C \times 0 - A \times 0}{C} = 0$	0
s^1	$\dfrac{ED - C a_0}{E} = F$	0	0	0
s^0	$\dfrac{F a_0 - E \times 0}{F} = a_0$	0	0	0

This array is called the **Routh's tabulation** or **Routh's array**. The column of s's on the left side is used for identification purposes. The reference column keeps track of the calculations, and the last row of the Routh's tabulation should always be the s^0 row.

Once the Routh's tabulation has been completed, the last step in the application of the criterion is to investigate the *signs* of the coefficients in the *first column* of the tabulation, which contains information on the roots of the equation. The following conclusions are made:

> The roots of the equation are all in the left half of the s-plane if all the elements of the first column of the Routh's tabulation are of the same sign. The number of changes of signs in the elements of the first column equals the number of roots with positive real parts, or those in the right-half s-plane.

The following examples illustrate the applications of the Routh-Hurwitz criterion when the tabulation terminates without complications.

▶ **EXAMPLE 2.13.1** Consider the equation

$$2s^4 + s^3 + 3s^2 + 5s + 10 = 0 \tag{2-253}$$

Because the equation has no missing terms and the coefficients are all of the same sign, it satisfies the necessary condition for not having roots in the right-half or on the imaginary axis of the s-plane. However, the sufficient condition must still be checked. Routh's tabulation is made as follows:

$$
\begin{array}{c|ccc}
s^4 & 2 & 3 & 10 \\
s^3 & 1 & 5 & 0 \\
s^2 & \frac{(1)(3)-(2)(5)}{1}=-7 & 10 & 0 \\
s^1 & \frac{(-7)(5)-(1)(10)}{-7}=6.43 & 0 & 0 \\
s^0 & 10 & 0 & 0
\end{array}
$$

Sign change (at s^2 row)

Sign change (at s^1 row)

Because there are two sign changes in the first column of the tabulation, the equation has two roots in the right half of the s-plane. Solving for the roots of Eq. (2-253), we have the four roots at $s = -1.005 \pm j0.933$ and $s = 0.755 \pm j1.444$. Clearly, the last two roots are in the right-half s-plane, which cause the system to be unstable. ◀

Toolbox 2-13-1

The roots of the polynomial in Eq. (2-253) are obtained using the following sequence of MATLAB functions.

```
>> clear all
>> p = [2 1 3 5 10] % Define polynomial 2*s^4+s^3+3*s^2+5*s+10

p =
    2    1    3    5   10
>> roots(p)
ans =
   0.7555 + 1.4444i
   0.7555 - 1.4444i
  -1.0055 + 0.9331i
  -1.0055 - 0.9331i
```

2-13-2 Special Cases when Routh's Tabulation Terminates Prematurely

The equations considered in the two preceding examples are designed so that Routh's tabulation can be carried out without any complications. Depending on the coefficients of the equation, the following difficulties may occur, which prevent Routh's tabulation from completing properly:

1. The first element in any one row of Routh's tabulation is zero, but the others are not.
2. The elements in one row of Routh's tabulation are all zero.

In the first case, if a zero appears in the first element of a row, the elements in the next row will all become infinite, and Routh's tabulation cannot continue. To remedy the situation, we *replace the zero element in the first column by an arbitrary small positive number ε, and then proceed with Routh's tabulation*. This is illustrated by the following example.

▶ **EXAMPLE 2-13-2** Consider the characteristic equation of a linear system

$$s^4 + s^3 + 2s^2 + 2s + 3 = 0 \tag{2-254}$$

Because all the coefficients are nonzero and of the same sign, we need to apply the Routh-Hurwitz criterion. Routh's tabulation is carried out as follows:

$$\begin{array}{c|ccc} s^4 & 1 & 2 & 3 \\ s^3 & 1 & 2 & 0 \\ s^2 & 0 & 3 & \end{array}$$

Because the first element of the s^2 row is zero, the elements in the s^1 row would all be infinite. To overcome this difficulty, we replace the zero in the s^2 row with a small positive number ε, and then proceed with the tabulation. Starting with the s^2 row, the results are as follows:

$$\begin{array}{c|ccc} & s^2 & \varepsilon & 3 \\ \text{Sign change} & s^1 & \dfrac{2\varepsilon - 3}{\varepsilon} \cong -\dfrac{3}{\varepsilon} & 0 \\ \text{Sign change} & s^0 & 3 & 0 \end{array}$$

Because there are two sign changes in the first column of Routh's tabulation, the equation in Eq. (2-254) has two roots in the right-half s-plane. Solving for the roots of Eq. (2-254), we get $s = -0.091 \pm j0.902$ and $s = 0.406 \pm j1.293$; the last two roots are clearly in the right-half s-plane.

It should be noted that the ε-method described may not give correct results if the equation has pure imaginary roots.

In the second special case, when all the elements in one row of Routh's tabulation are zeros before the tabulation is properly terminated, it indicates that one or more of the following conditions may exist:

1. The equation has at least one pair of real roots with equal magnitude but opposite signs.
2. The equation has one or more pairs of imaginary roots.
3. The equation has pairs of complex-conjugate roots forming symmetry about the origin of the s-plane; for example, $s = -1 \pm j1$, $s = 1 \pm j1$.

The situation with the entire row of zeros can be remedied by using the **auxiliary equation** $A(s) = 0$, which is formed from the coefficients of the row just above the row of zeros in Routh's tabulation. The auxiliary equation is always an even polynomial; that is, only even powers of s appear. *The roots of the auxiliary equation also satisfy the original equation.* Thus, by solving the auxiliary equation, we also get some of the roots of the original equation. To continue with Routh's tabulation when a row of zero appears, we conduct the following steps:

1. Form the auxiliary equation $A(s) = 0$ by using the coefficients from the row just preceding the row of zeros.
2. Take the derivative of the auxiliary equation with respect to s; this gives $dA(s)/ds = 0$.

3. Replace the row of zeros with the coefficients of $dA(s)/ds = 0$.
4. Continue with Routh's tabulation in the usual manner with the newly formed row of coefficients replacing the row of zeros.
5. Interpret the change of signs, if any, of the coefficients in the first column of the Routh's tabulation in the usual manner.

▶ **EXAMPLE 2.13.3** Consider the following equation, which may be the characteristic equation of a linear control system:

$$s^5 + 4s^4 + 8s^3 + 8s^2 + 7s + 4 = 0 \quad (2\text{-}255)$$

Routh's tabulation is

$$\begin{array}{cccc} s^5 & 1 & 8 & 7 \\ s^4 & 4 & 8 & 4 \\ s^3 & 6 & 6 & 0 \\ s^2 & 4 & 4 & \\ s^1 & 0 & 0 & \end{array}$$

Because a row of zeros appears prematurely, we form the auxiliary equation using the coefficients of the s^2 row:

$$A(s) = 4s^2 + 4 = 0 \quad (2\text{-}256)$$

The derivative of $A(s)$ with respect to s is

$$\frac{dA(s)}{ds} = 8s = 0 \quad (2\text{-}257)$$

from which the coefficients 8 and 0 replace the zeros in the s^1 row of the original tabulation. The remaining portion of the Routh's tabulation is

$$\begin{array}{ccc} s^1 & 8 & 0 \quad \text{coefficients of } dA(s)/ds \\ s^0 & 4 & \end{array}$$

Because there are no sign changes in the first column of the entire Routh's tabulation, the equation in Eq. (2-257) does not have any root in the right-half s-plane. Solving the auxiliary equation in Eq. (2-256), we get the two roots at $s = j$ and $s = -j$, which are also two of the roots of Eq. (2-255). Thus, the equation has two roots on the $j\omega$-axis, and the system is marginally stable. These imaginary roots caused the initial Routh's tabulation to have the entire row of zeros in the s^1 row.

Because all zeros occurring in a row that corresponds to an odd power of s creates an auxiliary equation that has only even powers of s, the roots of the auxiliary equation may all lie on the $j\omega$-axis. For design purposes, we can use the all-zero-row condition to solve for the marginal value of a system parameter for system stability. The following example illustrates the realistic value of the Routh-Hurwitz criterion in a simple design problem. ◀

▶ **EXAMPLE 2-13-4** Consider that a third-order control system has the characteristic equation

$$s^3 + 3408.3s^2 + 1,204,000s + 1.5 \times 10^7 K = 0 \quad (2\text{-}258)$$

The Routh-Hurwitz criterion is best suited to determine the critical value of K for stability, that is, the value of K for which at least one root will lie on the $j\omega$-axis and none in the right-half s-plane. Routh's tabulation of Eq. (2-258) is made as follows:

$$\begin{array}{ccc} s^3 & 1 & 1,204,000 \\ s^2 & 3408.3 & 1.5 \times 10^7 K \\ s^1 & \dfrac{410.36 \times 10^7 - 1.5 \times 10^7 K}{3408.3} & 0 \\ s^0 & 1.5 \times 10^7 K & \end{array}$$

Toolbox 2-13-2

Refer to Section 2-14-2 for the MATLAB symbolic tool to solve this problem.

For the system to be stable, all the roots of Eq. (2-258) must be in the left-half s-plane, and, thus, all the coefficients in the first column of Routh's tabulation must have the same sign. This leads to the following conditions:

$$\frac{410.36 \times 10^7 - 1.5 \times 10^7 K}{3408.3} > 0 \qquad (2\text{-}259)$$

and

$$1.5 \times 10^7 K > 0 \qquad (2\text{-}260)$$

From the inequality of Eq. (2-259), we have $K < 273.57$, and the condition in Eq. (2-260) gives $K > 0$. Therefore, the condition of K for the system to be stable is

$$0 < K < 273.57 \qquad (2\text{-}261)$$

If we let $K = 273.57$, the characteristic equation in Eq. (2-258) will have two roots on the $j\omega$-axis. To find these roots, we substitute $K = 273.57$ in the auxiliary equation, which is obtained from Routh's tabulation by using the coefficients of the s^2 row. Thus,

$$A(s) = 3408.3 s^2 + 4.1036 \times 10^9 = 0 \qquad (2\text{-}262)$$

which has roots at $s = j1097$ and $s = -j1097$, and the corresponding value of K at these roots is 273.57. Also, if the system is operated with $K = 273.57$, the zero-input response of the system will be an undamped sinusoid with a frequency of 1097.27 rad/sec.

▶ **EXAMPLE 2.13.5** As another example of using the Routh-Hurwitz criterion for simple design problems, consider that the characteristic equation of a closed-loop control system is

$$s^3 + 3Ks^2 + (K+2)s + 4 = 0 \qquad (2\text{-}263)$$

It is desired to find the range of K so that the system is stable. Routh's tabulation of Eq. (2-263) is

$$\begin{array}{c|cc}
s^3 & 1 & K+2 \\
s^2 & 3K & 4 \\
s^1 & \dfrac{3K(K+2)-4}{3K} & 0 \\
s^0 & 4 &
\end{array}$$

From the s^2 row, the condition of stability is $K > 0$, and from the s^1 row, the condition of stability is

$$3K^2 + 6K - 4 > 0 \qquad (2\text{-}264)$$

or

$$K < -2.528 \quad \text{or} \quad K > 0.528 \qquad (2\text{-}265)$$

Toolbox 2-13-3

Refer to Section 2-14-2 for the MATLAB symbolic tool to solve this problem.

When the conditions of $K > 0$ and $K > 0.528$ are compared, it is apparent that the latter requirement is more stringent. Thus, for the closed-loop system to be stable, K must satisfy

$$K > 0.528 \tag{2-266}$$

The requirement of $K < -2.528$ is disregarded because K cannot be negative.

It should be reiterated that the Routh-Hurwitz criterion is valid only if the characteristic equation is algebraic with real coefficients. If any one of the coefficients is complex, or if the equation is not algebraic, for example, containing exponential functions or sinusoidal functions of s, the Routh-Hurwitz criterion simply cannot be applied.

Another limitation of the Routh-Hurwitz criterion is that it is valid only for the determination of roots of the characteristic equation with respect to the left half or the right half of the s-plane. The stability boundary is the $j\omega$-axis of the s-plane. The criterion *cannot* be applied to any other stability boundaries in a complex plane, such as the unit circle in the z-plane, which is the stability boundary of discrete-data systems (Appendix H).

▶ 2-14 MATLAB TOOLS AND CASE STUDIES

2-14-1 Description and Use of Transfer Function Tool

If you have access to the MATLAB Symbolic Toolbox, you may use the **ACSYS** Transfer Function Symbolic Tool by pressing the appropriate button in the **ACSYS** window or by typing in **tfsym** in the MATLAB command window. The Symbolic Tool window is shown in Fig. 2-26. Click the "Help for 1st Time User" button to see the instructions on how to use the toolbox. The instructions appear in a Help Dialog window, as shown in Fig. 2-27. As instructed, press the "Transfer Function and Inverse Laplace" button to run the program. You must run this program within the MATLAB command window. Enter the transfer function, as shown in Fig. 2-28, to get the time response.

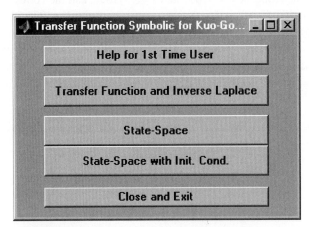

Figure 2-26 The Transfer Function Symbolic window.

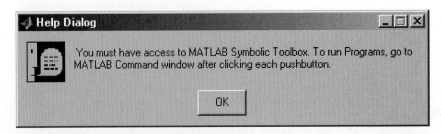

Figure 2-27 The Symbolic Help Dialog window.

> Transfer Function Symbolic. © Kuo & Golnaraghi 8th Edition, John Wiley & Sons. e.g., Use the following input format: (s+2)*(s^3+2*s+1)/(s*(s^2+2*s+1))
>
> Enter G=5*(s+0.6)/((s+1)*(s+2)*(s+3))
>
> $$\frac{5s+3}{(s+1)\ (s+2)\ (s+3)}$$
>
> G in polynomial form:
>
> Transfer function:
>
> $$\frac{5s+3}{s^3+6s^2+11s+6}$$
>
> G factored:
>
> Zero/pole/gain:
>
> $$\frac{5(s+0.6)}{(s+3)\ (s+2)\ (s+1)}$$
>
> Inverse Laplace transform:
>
> Gtime =
>
> −exp(−t)+7*exp(−2*t)−6*exp (−3*t)

Figure 2-28 The inverse Laplace transform of Eq. (2-267) for an impulse input, in the MATLAB command window.

▶ **EXAMPLE 2-14-1** Find the inverse Laplace transform of the transfer function

$$G(s) = \frac{5s+3}{(s+1)(s+2)(s+3)} = \frac{5s+3}{s^3+6s^2+11s+6} \quad (2\text{-}267)$$

You can do this either by using the *ilaplace* command in the MATLAB command window, as we demonstrated in Toolbox 2-5-1 for Example 2-5-1, or by utilizing the **tfsym** function, as shown in Fig. 2-28.

To find the time representation of Eq. (2-267) for a different input function such as a step or a sinusoid, the user may combine the input transfer function (e.g. 1/s for a unit-step input) with the transfer function in the TFtool input window. So to obtain Eq. (2-267) time representation for a unit-step input, use the following transfer function:

$$G(s) = \frac{5s+3}{s(s+1)(s+2)(s+3)} \quad (2\text{-}268)$$

and repeat the previous steps.

Similarly, for the transfer function

$$Y(s) = \frac{1000}{s(s^2+34.5s+1000)} = \frac{\omega_n^2}{(s^2+2\zeta\omega_n s+\omega_n^2)} \quad (2\text{-}269)$$

using the **tfsym** tool, the time representation of this system is obtained as

$$1+(-1/2+13/40^*i)^* \exp((-69/4-53/2^*i))^*t + (-1/2-13/40^*i)^* \exp((-69/4+53/2^*i)^*t)$$

◀

2-14-2 MATLAB Tools for Stability

The easiest way to assess stability of known transfer functions is to find the location of the poles. For that purpose, the MATLAB code that appears in Toolbox 2-13-1 is the easiest way for finding the roots of the characteristic equation polynomial—i.e., the poles of the system. However, many of the other tools within **ACSYS** software may also be used to find the poles of the closed-loop system transfer function, including the "Transfer Function Symbolic" (**tfsym**) and the "Transfer Function Calculator" (**tfcal**). You may also conduct a more thorough stability study of your system using the root locus and phase and gain margin concepts utilizing the "Controller Design Tool," respectively. These topics will be thoroughly discussed in Chapter 9.

In this section, we introduce the **tfrouth** tool, which may be used to find the Routh array, and more importantly it may be utilized for controller design applications where it is important to assess the stability of a system for a controller gain, say k.

The steps involved in setting up and then solving a given stability problem using tfrouth are as follows.

1. Type "tfrouth" in the MATLAB command module within the "tfsymbolic" directory.
2. Enter the characteristic polynomial in symbolic (e.g., s^3+s^2+s+1) or in vectorial (e.g. [1 1 1 1]) forms.
3. Press the "Routh-Hurwitz" button and check the results in the MATLAB command window.
4. In case you wish to assess the stability of the system for a design parameter, enter it in the box designated as "Enter Symbolic Parameters." For example, for s^3 + k1*s^2 + k2*s + 1, you need to enter "k1 k2" in the "Enter Symbolic Parameters" box, followed by entering the polynomial s^3 + k1∗s^2 + k2∗s + 1 in the "Characteristic Equation" box.
5. Press the "Routh-Hurwitz" button to form the Routh table and conduct the Routh-Hurwitz stability test.

To better illustrate how to use tfrouth, let us solve some of the earlier examples in this chapter.

▶ **EXAMPLE 2-14-2** Recall Example 2-13-1; let's use tfrouth for the following polynomial:

$$2s^4 + s^3 + 3s^2 + 5s + 10 = 0 \tag{2-270}$$

In the MATLAB command module, type in "tfrouth" and enter the characteristic Eq. (2-270) in polynomial form, followed by clicking the "Routh-Hurwitz" button to get the Routh-Hurwitz matrix, as shown in Fig. 2-29.

The results match Example 2-14-2. The system is therefore unstable because of two positive poles. The Routh's array first column also shows two sign changes to confirm this result. To see the complete Routh table, the user must refer to the MATLAB command window, as shown in Fig. 2-30. ◀

▶ **EXAMPLE 2-14-3** Consider Example 2-13-2 for characteristic equation of a linear system:

$$s^4 + s^3 + 2s^2 + 2s + 3 = 0 \tag{2-271}$$

After entering the transfer function characteristic equation using tfrouth and pressing the "Routh-Hurwitz" button, we get the results shown in Fig. 2-31.

As a result, because of the final two sign changes, we expect to see two unstable poles. ◀

▶ **EXAMPLE 2-14-4** Revisiting Example 2-13-3, use tfrouth to study the following characteristic equation:

$$s^5 + 4s^4 + 8s^3 + 8s^2 + 7s + 4 = 0 \tag{2-272}$$

to get the results shown in Fig. 2-32.

2-14 MATLAB Tools and Case Studies ◄ 87

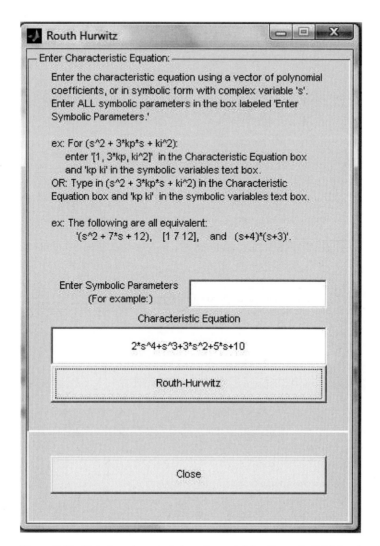

Figure 2-29 Entering characteristic polynomial for Example 2-14-2 using the tfrouth module.

Routh-Hurwitz Matrix:

$$\begin{matrix} [\ 2 & 3 & 10 &] \\ [\ 1 & 5 & 0 &] \\ [\ -7 & 10 & 0 &] \\ [45/7 & 0 & 0 &] \\ [\ 10 & 0 & 0 &] \end{matrix}$$

There are two sign changes in the first column.

Figure 2-30 Stability results for Example 2-14-2, after using the Routh-Hurwitz test.

88 ▶ Chapter 2. Mathematical Foundation

```
First element of row3 is zero. Epsilon is used.
Routh-Hurwitz Matrix:

              [  1        2       3    ]
              [                        ]
              [  1        2       0    ]
              [                        ]
              [ eps       3       0    ]
              [                        ]
              [-3 + 2 eps              ]
              [----------  0      0    ]
              [   eps                  ]
              [                        ]
              [  3        0       0    ]

There are two sign changes in the first column.
```

Figure 2-31 Stability results for Example 2-14-3, after using the Routh-Hurwitz test.

```
Row of zeros found at row5. Auxiliary polynomial is used.
Routh-Hurwitz Matrix:

              [1         8       7     ]
              [                        ]
              [4         8       4     ]
              [                        ]
              [6         6       0     ]
              [                        ]
              [4         4       0     ]
              [                        ]
              [8         0       0     ]
              [                        ]
              [4         0       0     ]

There are two sign changes in the first column.
```

Figure 2-32 Stability results for Example 2-14-4, after using the Routh-Hurwitz test.

In this case, the program has automatically replaced the whole row of zeros in the fifth row with the coefficients of the polynomial formed from the derivative of an auxiliary polynomial formed from the fourth row. As a result, the system is unstable. Further, because of the final zero sign changes, we expect to see no additional unstable poles. The unstable poles of the system may be obtained directly by obtaining the roots of the auxiliary polynomial:

$$A(s) = 4s^2 + 4 = 0 \tag{2-273}$$

◀

▶ **EXAMPLE 2-14-5** Considering the characteristic equation of a closed-loop control system

$$s^3 + 3Ks^2 + (K+2)s + 4 = 0 \tag{2-274}$$

It is desired to find the range of K so that the system is stable. See Figs. 2-33, 2-34, and 2-35 for more details.

In the end, the user is encouraged to make use of the software to solve examples and problems appearing in this chapter.

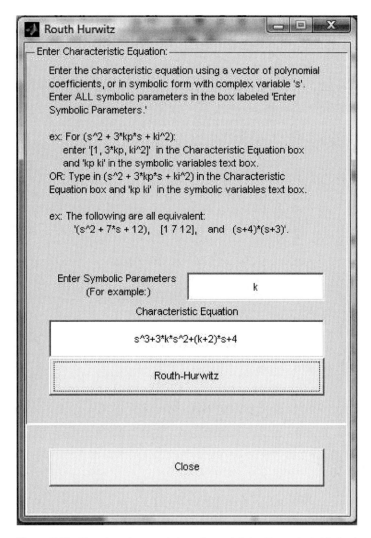

Figure 2-33 Entering characteristic polynomial for Example 2-14-5 using the tfrouth module.

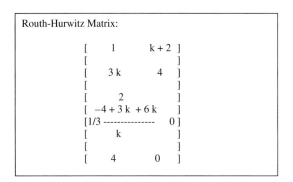

Figure 2-34 The Routh's array for Example 2-14-5.

```
>> k=.4;
>> RH

RH =

[          1,          k+2  ]
[         3*k,          4   ]
[ 1/3*(-4+3*k^2+6*k)/k,  0  ]
[          4,           0   ]

>> eval(RH)

ans =

  1.0000    2.4000
  1.2000    4.0000
 -0.9333         0
  4.0000         0
```

There are two sign changes in the first column.

```
>> k=1;

>> eval(RH)

ans =

  1.0000    3.0000
  3.0000    4.0000
  1.6667         0
  4.0000         0
```

There are no sign changes in the first column.

Figure 2-35 The Routh's array for Example 2-14-5.

2-15 SUMMARY

In this chapter, we presented some fundamental mathematics required for the study of linear control systems. Specifically, we started with complex numbers and their basic properties leading to frequency domain mathematics and plots. The Laplace transform is used for the solution of linear ordinary differential equations. This transform method is characterized by first transforming the real-domain equations into algebraic equations in the transform domain. The solutions are first obtained in the transform domain by using the familiar methods of solving algebraic equations. The final solution in the real domain is obtained by taking the inverse transform. For engineering problems, the transform tables and the partial-fraction expansion method are recommended for the inverse transformation.

In this chapter, the definitions of BIBO, zero-input, and asymptotic stability of linear time-invariant continuous-data and discrete-data systems are given. It is shown that the condition of these types of stability is related directly to the roots of the characteristic equation. For a continuous-data system to be stable, the roots of the characteristic equation must all be located in the left half of the s-plane.

The necessary condition for a polynomial $F(s)$ to have no zeros on the $j\omega$-axis and in the right half of the s-plane is that all its coefficients must be of the same sign and none can vanish. The necessary and sufficient conditions of $F(s)$ to have zeros only in the left half of the s-plane are checked with the Routh-Hurwitz criterion. The value of the Routh-Hurwitz criterion is diminished if the characteristic equation can be solved using MATLAB.

REVIEW QUESTIONS

1. Give the definitions of the poles and zeros of a function of the complex variable s.

2. What are the advantages of the Laplace transform method of solving linear ordinary differential equations over the classical method?

3. What are state equations?

4. What is a causal system?

5. Give the defining equation of the one-sided Laplace transform.

6. Give the defining equation of the inverse Laplace transform.

7. Give the expression of the final-value theorem of the Laplace transform. What is the condition under which the theorem is valid?

8. Give the Laplace transform of the unit-step function, $u_s(t)$.

9. What is the Laplace transform of the unit-ramp function, $tu_s(t)$?

10. Give the Laplace transform of $f(t)$ shifted to the right (delayed) by T_d in terms of the Laplace transform of $f(t)$, $F(s)$.

11. If $\mathcal{L}[f_1(t)] = F_1(s)$ and $\mathcal{L}[f_2(t)] = F_2(s)$, then find $\mathcal{L}[f_1(t)f_2(t)]$ in terms of $F_1(s)$ and $F_2(s)$.

12. Do you know how to handle the exponential term in performing the partial-fraction expansion of

$$F(s) = \frac{10}{(s+1)(s+2)} e^{-2s}$$

13. Do you know how to handle the partial-fraction expansion of a function whose denominator order is not greater than that of the numerator, for example,

$$F(s) = \frac{10(s^2 + 5s + 1)}{(s+1)(s+2)}$$

14. In trying to find the inverse Laplace transform of the following function, do you have to perform the partial-fraction expansion?

$$F(s) = \frac{1}{(s+5)^3}$$

15. Can the Routh-Hurwitz criterion be directly applied to the stability analysis of the following systems?

(a) Continuous-data system with the characteristic equation

$$s^4 + 5s^3 + 2s^2 + 3s + 2e^{-2s} = 0$$

(b) Continuous-data system with the characteristic equation

$$s^4 - 5s^3 + 3s^2 + Ks + K^2 = 0$$

16. The first two rows of Routh's tabulation of a third-order system are

$$\begin{array}{ccc} s^3 & 2 & 2 \\ s^2 & 4 & 4 \end{array}$$

Select the correct answer from the following choices:

(a) The equation has one root in the right-half s-plane.

(b) The equation has two roots on the $j\omega$-axis at $s = j$ and $-j$. The third root is in the left-half s-plane.

(c) The equation has two roots on the $j\omega$-axis at $s = 2j$ and $s = -2j$. The third root is in the left-half s-plane.

(d) The equation has two roots on the $j\omega$-axis at $s = 2j$ and $s = -2j$. The third root is in the right-half s-plane.

17. If the numbers in the first column of Routh's tabulation turn out to be all negative, then the equation for which the tabulation is made has at least one root not in the left half of the s-plane. **(T)** **(F)**

18. The roots of the auxiliary equation, $A(s) = 0$, of Routh's tabulation of a characteristic equation must also be the roots of the latter. **(T)** **(F)**

19. The following characteristic equation of a continuous-data system represents an unstable system because it contains a negative coefficient.

$$s^3 - s^2 + 5s + 10 = 0$$
(T) **(F)**

20. The following characteristic equation of a continuous-data system represents an unstable system because there is a zero coefficient.

$$s^3 + 5s^2 + 4 = 0$$
(T) **(F)**

21. When a row of Routh's tabulation contains all zeros before the tabulation ends, this means that the equation has roots on the imaginary axis of the s-plane. **(T)** **(F)**

Answers to these review questions can be found on this book's companion Web site: www.wiley.com/college/golnaraghi.

REFERENCES

Complex Variables, Laplace Transforms

1. F. B. Hildebrand, *Methods of Applied Mathematics*, 2nd Ed., Prentice Hall, Englewood Cliffs, NJ, 1965.
2. B. C. Kuo, *Linear Networks and Systems*, McGraw-Hill Book Company, New York, 1967.
3. C. R. Wylie, Jr., *Advanced Engineering Mathematics*, 2nd Ed., McGraw-Hill Book Company, New York, 1960.
4. K. Ogata, *Modern Control Engineering*, 4th ed., Prentice-Hall, NJ, 2002.
5. J. J. Distefano, A. R. Stubberud, and I. J. Williams, *Feedback and Control Systems*, 2nd ed., McGraw-Hill, 1990.
6. "ChE-400: Applied Chemical Engineering Calculations," http://www.ent.ohiou.edu/che/che400/Handouts%20and%20Class%20notes.htm.

Partial-Fraction Expansion

7. C. Pottle, "On the Partial Fraction Expansion of a Rational Function with Multiple Poles by Digital Computer," *IEEE Trans. Circuit Theory*, Vol. CT-11, pp. 161–162, Mar. 1964.
8. B. O. Watkins, "A Partial Fraction Algorithm," *IEEE Trans. Automatic Control*, Vol. AC-16, pp. 489–491, Oct. 1971.

Additional References

9. W. J. Palm, III, *Modeling, Analysis, and Control of Dynamic Systems*, 2nd Ed., John Wiley & Sons, New York, 1999.
10. K. Ogata, *Modern Control Engineering*, 4th Ed., Prentice Hall, NJ, 2002.
11. I. Cochin and W. Cadwallender, *Analysis and Design of Dynamic Systems*, 3rd Ed., Addison-Wesley, 1997.
12. A. Esposito, *Fluid Power with Applications*, 5th Ed. Prentice Hall, NJ, 2000.
13. H. V. Vu and R. S. Esfandiari, *Dynamic Systems*, Irwin/McGraw-Hill, 1997.
14. J. L. Shearer, B. T. Kulakowski, and J. F. Gardner, *Dynamic Modeling and Control of Engineering Systems*, 2nd Ed., Prentice Hall, NJ, 1997.
15. R. L. Woods and K. L. Lawrence, *Modeling and Simulation of Dynamic Systems*, Prentice Hall, NJ, 1997.

Stability

18. F. R. Gantmacher, *Matrix Theory*, Vol. II, Chelsea Publishing Company, New York, 1964.
19. K. J. Khatwani, "On Routh-Hurwitz Criterion," *IEEE Trans. Automatic Control*, Vol. AC-26, p. 583, April 1981.
20. S. K. Pillai, "The ε Method of the Routh-Hurwitz Criterion," *IEEE Trans. Automatic Control*, Vol. AC-26, 584, April1981.

▶ PROBLEMS

PROBLEMS FOR SECTION 2-1

2-1. Find the poles and zeros of the following functions (including the ones at infinity, if any). Mark the finite poles with × and the finite zeros with o in the *s*-plane.

(a) $G(s) = \dfrac{10(s+2)}{s^2(s+1)(s+10)}$

(b) $G(s) = \dfrac{10s(s+1)}{(s+2)(s^2+3s+2)}$

(c) $G(s) = \dfrac{10(s+2)}{s(s^2+2s+2)}$

(d) $G(s) = \dfrac{e^{-2s}}{10s(s+1)(s+2)}$

2-2. Poles and zeros of a function are given; find the function:

(a) Simple poles: 0, −2; poles of order 2: −3; zeros: −1, ∞
(b) Simple poles: −1, −4; zeros: 0
(c) Simple poles: −3, ∞; poles of order 2: 0,−1; zeros: ±j, ∞

2-3. Use MATLAB to find the poles and zeros of the functions in Problem 2-1.

PROBLEMS FOR SECTION 2-2

2-4. Find the polar representation of $G(s)$ given in Problem 2-1 for $s = j\omega$, where ω is a constant varying from zero to infinity.

2-5. Find the polar plot of the following functions:

(a) $G(j\omega) = \dfrac{10}{(j\omega - 2)}$

(b) $G(j\omega) = \dfrac{1}{1 + 2\zeta\left(j\dfrac{\omega}{\omega_n}\right) + \left(j\dfrac{\omega}{\omega_n}\right)^2} \quad 0 < \zeta < 1$

(c) $G(j\omega) = \dfrac{1}{1 + 2\zeta\left(j\dfrac{\omega}{\omega_n}\right) + \left(j\dfrac{\omega}{\omega_n}\right)^2} \quad \zeta > 1$

(d) $G(j\omega) = \dfrac{1}{j\omega(jT\omega + 1)}$

(e) $G(j\omega) = \dfrac{e^{-j\omega L}}{(jT\omega + 1)}$

2-6. Use MATLAB to find the polar plot of the functions in Problem 2-5.

2-7. Draw the Bode plot of the following functions:

(a) $G(j\omega) = \dfrac{2000(j\omega + 0.5)}{j\omega(j\omega + 10)(j\omega + 50)}$

(b) $G(j\omega) = \dfrac{25}{j\omega(j\omega - 2.5\omega^2 + 10)}$

(c) $G(j\omega) = \dfrac{(j\omega - 100\omega^2 + 100)}{-\omega^2(j\omega - 25\omega^2 + 100)}$

(d) $G(j\omega) = \dfrac{1}{1 + 2\zeta\left(j\dfrac{\omega}{\omega_n}\right) + \left(j\dfrac{\omega}{\omega_n}\right)^2} \quad 0 \leq \zeta \leq 1$

(e) $G(j\omega) = \dfrac{0.03(e^{j\omega t} + 1)^2}{(e^{j\omega t} - 1)(3e^{j\omega t} + 1)(e^{j\omega t} + 0.5)}$

2-8. Use MATLAB to draw the Bode plot of the functions in Problem 2-7.

PROBLEMS FOR SECTION 2-3

2-9. Express the following set of first-order differential equations in the vector-matrix form of $\dfrac{d\mathbf{x}(t)}{dt} = \mathbf{Ax}(t) + \mathbf{Bu}(t)$.

(a)
$\dfrac{dx_1(t)}{dt} = -x_1(t) + 2x_2(t)$
$\dfrac{dx_2(t)}{dt} = -2x_2(t) + 3x_3(t) + u_1(t)$
$\dfrac{dx_3(t)}{dt} = -x_1(t) - 3x_2(t) - x_3(t) + u_2(t)$

(b)
$\dfrac{dx_1(t)}{dt} = -x_1(t) + 2x_2(t) + 2u_1(t)$
$\dfrac{dx_2(t)}{dt} = 2x_1(t) - x_3(t) + u_2(t)$
$\dfrac{dx_3(t)}{dt} = 3x_1(t) - 4x_2(t) - x_3(t)$

PROBLEMS FOR SECTION 2-4

2-10. Prove theorem 3 in Section 2-4-3.

2-11. Prove the integration theorem 4 in Section 2-4-3.

2-12. Prove the shift-in-time theorem, which is

$$\mathcal{L}[g(t - T)u_s(t - T)] = e^{-Ts}G(s)$$

2-13. Prove the convolution theorem in both time and s domain, which is

$$\mathcal{L}[g_1(t) * g_2(t)] = G_1(s)G_2(s)$$
$$\mathcal{L}[g_1(t)g_2(t)] = G_1(s) * G_2(s)$$

2-14. Prove theorems 6 and 7.

2-15. Use MATLAB to obtain $\mathcal{L}\{\sin^2 2t\}$. Then, calculate $\mathcal{L}\{\cos^2 2t\}$ when you know $\mathcal{L}\{\sin^2 2t\}$. Verify your answer by calculating $\mathcal{L}\{\cos^2 2t\}$ in MATLAB.

2-16. Find the Laplace transforms of the following functions. Use the theorems on Laplace transforms, if applicable.

(a) $g(t) = 5te^{-5t}u_s(t)$
(b) $g(t) = (t\sin 2t + e^{-2t})u_s(t)$
(c) $g(t) = 2e^{-2t}\sin 2t\, u_s(t)$
(d) $g(t) = \sin 2t \cos 2t\, u_s(t)$
(e) $g(t) = \sum_{k=0}^{\infty} e^{-5kT}\delta(t-kT)$ where $\delta(t)$ = unit-impulse function

2-17. Use MATLAB to solve Problem 2-16.

2-18. Find the Laplace transforms of the functions shown in Fig. 2P-18. First, write a complete expression for $g(t)$, and then take the Laplace transform. Let $gT(t)$ be the description of the function over the basic period and then delay $gT(t)$ appropriately to get $g(t)$. Take the Laplace transform of $g(t)$ to get the following:

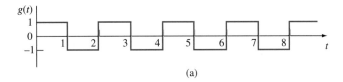

(a)

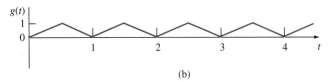

(b)

Figure 2P-18

2-19. Find the Laplace transform of the following function.

$$g(t) = \begin{cases} t+1 & 0 \le t < 1 \\ 0 & 1 \le t < 2 \\ 2-t & 2 \le t < 3 \\ 0 & t \ge 3 \end{cases}$$

2-20. Find the Laplace transform of the periodic function in Fig. 2P-20.

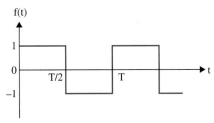

Figure 2P-20

2-21. Find the Laplace transform of the function in Fig. 2P-21.

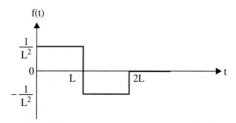

Figure 2P-21

2-22. Solve the following differential equations by means of the Laplace transform.

(a) $\dfrac{d^2 f(t)}{dt^2} + 5\dfrac{df(t)}{dt} + 4f(t) = e^{-2t} u_s(t)$ Assume zero initial conditions.

(b) $\begin{cases} \dfrac{dx_1(t)}{dt} = x_2(t) \\ \dfrac{dx_2(t)}{dt} = -2x_1(t) - 3x_2(t) + u_s(t) \end{cases}$ $x_1(0) = 1,\ x_2(0) = 0$

(c) $\dfrac{d^3 y(t)}{dt^3} + 2\dfrac{d^2 y(t)}{dt^2} + \dfrac{dy(t)}{dt} + 2y(t) = -e^{-t} u_s(t)$

$\dfrac{d^2 y}{dt^2}(0) = -1 \quad \dfrac{dy}{dt}(0) = 1 \quad y(0) = 0$

2-23. Use MATLAB to find the Laplace transform of the functions in Problem 2-22.

2-24. Use MATLAB to solve the following differential equation:

$$\dfrac{d^2 y}{dt^2} - y = e^t \quad \text{(Assuming zero initial conditions)}$$

2-25. A series of a three-reactor tank is arranged as shown in Fig. 2P-25 for chemical reaction.

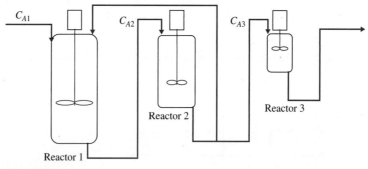

Figure 2P-25

The state equation for each reactor is defined as follows:

$$R1: \dfrac{dC_{A1}}{dt} = \dfrac{1}{V_1}[1000 + 100C_{A2} - 1100C_{A1} - k_1 V_1 C_{A1}]$$

$$R2: \frac{dC_{A2}}{dt} = \frac{1}{V_2}[1100C_{A1} - 1100C_{A2} - k_2V_2C_{A2}]$$

$$R3: \frac{dC_{A3}}{dt} = \frac{1}{V_3}[1000C_{A2} - 1000C_{A3} - k_3V_3C_{A3}]$$

when V_i and k_i represent the volume and the temperature constant of each tank as shown in the following table:

Reactor	V_i	k_i
1	1000	0.1
2	1500	0.2
3	100	0.4

Use MATLAB to solve the differential equations assuming $C_{A1} = C_{A2} = C_{A3} = 0$ at $t = 0$.

PROBLEMS FOR SECTION 2-5

2-26. Find the inverse Laplace transforms of the following functions. First, perform partial-fraction expansion on $G(s)$; then, use the Laplace transform table.

(a) $G(s) = \dfrac{1}{s(s+2)(s+3)}$

(b) $G(s) = \dfrac{10}{(s+1)^2(s+3)}$

(c) $G(s) = \dfrac{100(s+2)}{s(s^2+4)(s+1)} e^{-s}$

(d) $G(s) = \dfrac{2(s+1)}{s(s^2+s+2)}$

(e) $G(s) = \dfrac{1}{(s+1)^3}$

(f) $G(s) = \dfrac{2(s^2+s+1)}{s(s+1.5)(s^2+5s+5)}$

(g) $G(s) = \dfrac{2 + 2se^{-s} + 4e^{-2s}}{s^2+3s+2}$

(h) $G(s) = \dfrac{2s+1}{s^3+6s^2+11s+6}$

(i) $G(s) = \dfrac{3s^3+10s^2+8s+5}{s^4+5s^3+7s^2+5s+6}$

2-27. Use MATLAB to find the inverse Laplace transforms of the functions in Problem 2-26. First, perform partial-fraction expansion on $G(s)$; then, use the inverse Laplace transform.

2-28. Given the state equation of the system, convert it to the set of first-order differential equation.

(a) $A = \begin{bmatrix} 0 & -1 & 2 \\ 1 & 0 & 1 \\ -1 & -2 & 1 \end{bmatrix}$ $B = \begin{bmatrix} 0 & -1 \\ 1 & 0 \\ 0 & 0 \end{bmatrix}$

(b) $A = \begin{bmatrix} 3 & 1 & -2 \\ -1 & 2 & 2 \\ 0 & 0 & 1 \end{bmatrix}$ $B = \begin{bmatrix} -1 \\ 0 \\ 2 \end{bmatrix}$

2-29. The following differential equations represent linear time-invariant systems, where $r(t)$ denotes the input and $y(t)$ the output. Find the transfer function $Y(s)/R(s)$ for each of the systems. (Assume zero initial conditions.)

(a) $\dfrac{d^3y(t)}{dt^3} + 2\dfrac{d^2y(t)}{dt^2} + 5\dfrac{dy(t)}{dt} + 6y(t) = 3\dfrac{dr(t)}{dt} + r(t)$

(b) $\dfrac{d^4y(t)}{dt^4} + 10\dfrac{d^2y(t)}{dt^2} + \dfrac{dy(t)}{dt} + 5y(t) = 5r(t)$

(c) $\dfrac{d^3y(t)}{dt^3} + 10\dfrac{d^2y(t)}{dt^2} + 2\dfrac{dy(t)}{dt} + y(t) + 2\int_0^t y(\tau)d\tau = \dfrac{dr(t)}{dt} + 2r(t)$

(d) $2\dfrac{d^2y(t)}{dt^2} + \dfrac{dy(t)}{dt} + 5y(t) = r(t) + 2r(t-1)$

(e) $\dfrac{d^2y(t+1)}{dt^2} + 4\dfrac{dy(t+1)}{dt} + 5y(t+1) = \dfrac{dr(t)}{dt} + 2r(t) + 2\int_{-\infty}^t r(\tau)d\tau$

(f) $\dfrac{d^3y(t)}{dt^2} + 2\dfrac{d^2y(t)}{dt^2} + \dfrac{dy(t)}{dt} + 2y(t) + 2\int_{-\infty}^t y(\tau)d\tau = \dfrac{dr(t-2)}{dt} + 2r(t-2)$

2-30. Use MATLAB to find $Y(s)/R(s)$ for the differential equations in Problem 2-29.

2-31. Use MATLAB to find the partial-fraction expansion to the following functions.

(a) $G(s) = \dfrac{10(s+1)}{s^2(s+4)(s+6)}$

(b) $G(s) = \dfrac{(s+1)}{s(s+2)(s^2+2s+2)}$

(c) $G(s) = \dfrac{5(s+2)}{s^2(s+1)(s+5)}$

(d) $G(s) = \dfrac{5e^{-2s}}{(s+1)(s^2+s+1)}$

(e) $G(s) = \dfrac{100(s^2+s+3)}{s(s^2+5s+3)}$

(f) $G(s) = \dfrac{1}{s(s^2+1)(s+0.5)^2}$

(g) $G(s) = \dfrac{2s^3+s^2+8s+6}{(s^2+4)(s^2+2s+2)}$

(h) $G(s) = \dfrac{2s^4+9s^3+15s^2+s+2}{s^2(s+2)(s+1)^2}$

2-32. Use MATLAB to find the inverse Laplace transforms of the functions in Problem 2-31.

PROBLEMS FOR SECTIONS 2-7 THROUGH 2-13

2-33. Without using the Routh-Hurwitz criterion, determine if the following systems are asymptotically stable, marginally stable, or unstable. In each case, the closed-loop system transfer function is given.

(a) $M(s) = \dfrac{10(s+2)}{s^3 + 3s^2 + 5s}$

(b) $M(s) = \dfrac{s-1}{(s+5)(s^2+2)}$

(c) $M(s) = \dfrac{K}{s^3 + 5s + 5}$

(d) $M(s) = \dfrac{100(s-1)}{(s+5)(s^2+2s+2)}$

(e) $M(s) = \dfrac{100}{s^3 - 2s^2 + 3s + 10}$

(f) $M(s) = \dfrac{10(s+12.5)}{s^4 + 3s^3 + 50s^2 + s + 10^6}$

2-34. Use the ROOTS command in MATLAB to solve Problem 2-33.

2-35. Using the Routh-Hurwitz criterion, determine the stability of the closed-loop system that has the following characteristic equations. Determine the number of roots of each equation that are in the right-half s-plane and on the $j\omega$-axis.

(a) $s^3 + 25s^2 + 10s + 450 = 0$

(b) $s^3 + 25s^2 + 10s + 50 = 0$

(c) $s^3 + 25s^2 + 250s + 10 = 0$

(d) $2s^4 + 10s^3 + 5.5s^2 + 5.5s + 10 = 0$

(e) $s^6 + 2s^5 + 8s^4 + 15s^3 + 20s^2 + 16s + 16 = 0$

(f) $s^4 + 2s^3 + 10s^2 + 20s + 5 = 0$

(g) $s^8 + 2s^7 + 8s^6 + 12s^5 + 20s^4 + 16s^3 + 16s^2 = 0$

2-36. Use MATLAB to solve Problem 2-35.

2-37. Use MATLAB Toolbox 2-13-1 to find the roots of the following characteristic equations of linear continuous-data systems and determine the stability condition of the systems.

(a) $s^3 + 10s^2 + 10s + 130 = 0$

(b) $s^4 + 12s^3 + s^2 + 2s + 10 = 0$

(c) $s^4 + 12s^3 + 10s^2 + 10s + 10 = 0$

(d) $s^4 + 12s^3 + s^2 + 10s + 1 = 0$

(e) $s^6 + 6s^5 + 125s^4 + 100s^3 + 100s^2 + 20s + 10 = 0$

(f) $s^5 + 125s^4 + 100s^3 + 100s^2 + 20s + 10 = 0$

2-38. For each of the characteristic equations of feedback control systems given, use MATLAB to determine the range of K so that the system is asymptotically stable. Determine the value of K so that the system is marginally stable and determine the frequency of sustained oscillation, if applicable.

(a) $s^4 + 25s^3 + 15s^2 + 20s + K = 0$

(b) $s^4 + Ks^3 + 2s^2 + (K+1)s + 10 = 0$

(c) $s^3 + (K+2)s^2 + 2Ks + 10 = 0$

(d) $s^3 + 20s^2 + 5s + 10K = 0$

(e) $s^4 + Ks^3 + 5s^2 + 10s + 10K = 0$

(f) $s^4 + 12.5s^3 + s^2 + 5s + K = 0$

2-39. The loop transfer function of a single-loop feedback control system is given as

$$G(s)H(s) = \frac{K(s+5)}{s(s+2)(1+Ts)}$$

The parameters K and T may be represented in a plane with K as the horizontal axis and T as the vertical axis. Determine the regions in the T-versus-K parameter plane where the closed-loop system is asymptotically stable and where it is unstable. Indicate the boundary on which the system is marginally stable.

2-40. Given the forward-path transfer function of unity-feedback control systems, apply the Routh-Hurwitz criterion to determine the stability of the closed-loop system as a function of K. Determine the value of K that will cause sustained constant-amplitude oscillations in the system. Determine the frequency of oscillation.

(a) $G(s) = \dfrac{K(s+4)(s+20)}{s^3(s+100)(s+500)}$

(b) $G(s) = \dfrac{K(s+10)(s+20)}{s^2(s+2)}$

(c) $G(s) = \dfrac{K}{s(s+10)(s+20)}$

(d) $G(s) = \dfrac{K(s+1)}{s^3 + 2s^2 + 3s + 1}$

2-41. Use MATLAB to solve Problem 2-40.

2-42. A controlled process is modeled by the following state equations.

$$\frac{dx_1(t)}{dt} = x_1(t) - 2x_2(t) \quad \frac{dx_2(t)}{dt} = 10x_1(t) + u(t)$$

The control $u(t)$ is obtained from state feedback such that

$$u(t) = -k_1 x_1(t) - k_2 x_2(t)$$

where k_1 and k_2 are real constants. Determine the region in the k_1-versus-k_2 parameter plane in which the closed-loop system is asymptotically stable.

2-43. A linear time-invariant system is described by the following state equations.

$$\frac{d\mathbf{x}(t)}{dt} = \mathbf{A}\mathbf{x}(t) + \mathbf{B}u(t)$$

where

$$\mathbf{A} = \begin{bmatrix} 0 & 1 & 0 \\ 0 & 0 & 1 \\ 0 & -4 & -3 \end{bmatrix} \quad \mathbf{B} = \begin{bmatrix} 0 \\ 0 \\ 1 \end{bmatrix}$$

The closed-loop system is implemented by state feedback, so that $u(t) = -\mathbf{K}\mathbf{x}(t)$, where $\mathbf{K} = [k_1 \ k_2 \ k_3]$ and $k_1, k_2,$ and k_3 are real constants. Determine the constraints on the elements of $\mathbf{K}$ so that the closed-loop system is asymptotically stable.

2-44. Given the system in state equation form,

$$\frac{d\mathbf{x}(t)}{dt} = \mathbf{A}\mathbf{x}(t) + \mathbf{B}u(t)$$

(a) $\mathbf{A} = \begin{bmatrix} 1 & 0 & 0 \\ 0 & -3 & 0 \\ 0 & 0 & -2 \end{bmatrix} \quad \mathbf{B} = \begin{bmatrix} 1 \\ 0 \\ 1 \end{bmatrix}$

(b) $\mathbf{A} = \begin{bmatrix} 1 & 0 & 0 \\ 0 & -2 & 0 \\ 0 & 0 & 3 \end{bmatrix} \quad \mathbf{B} = \begin{bmatrix} 0 \\ 1 \\ 1 \end{bmatrix}$

Can the system be stabilized by state feedback $u(t) = -\mathbf{K}\mathbf{x}(t)$, where $\mathbf{K} = [k_1\ k_2\ k_3]$?

2-45. Consider the open-loop system in Fig. 2P-45(a).

$F(s) \longrightarrow \boxed{G(s)} \longrightarrow Y(s)$

Figure 2P-45a

where $\dfrac{d^2 y}{dt^2} - \dfrac{g}{l} y = z$ and $f(t) = \tau \dfrac{dz}{dt} + z$.

Our goal is to stabilize this system so the closed-loop feedback control will be defined as shown in the block diagram in Fig. 2P-45(b).

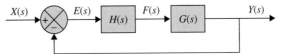

Figure 2P-45b

Assuming $f(t) = k_p e + k_d \dfrac{de}{dt}$.

(a) Find the open-loop transfer function.
(b) Find the closed-loop transfer function.
(c) Find the range of k_p and k_d in which the system is stable.
(d) Suppose $\dfrac{g}{l} = 10$ and $\tau = 0.1$. If $y(0) = 10$ and $\dfrac{dy}{dt} = 0$, then plot the step response of the system with three different values for k_p and k_d. Then show that some values are better than others; however, all values must satisfy the Routh-Hurwitz criterion.

2-46. The block diagram of a motor-control system with tachometer feedback is shown in Fig. 2P-46. Find the range of the tachometer constant K_t so that the system is asymptotically stable.

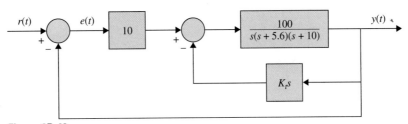

Figure 2P-46

2-47. The block diagram of a control system is shown in Fig. 2P-47. Find the region in the K-versus-α plane for the system to be asymptotically stable. (Use K as the vertical and α as the horizontal axis.)

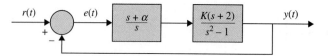

Figure 2P-47

2-48. The conventional Routh-Hurwitz criterion gives information only on the location of the zeros of a polynomial $F(s)$ with respect to the left half and right half of the s-plane. Devise a linear transformation $s = f(p, \alpha)$, where p is a complex variable, so that the Routh-Hurwitz criterion can be applied to determine whether $F(s)$ has zeros to the right of the line $s = -\alpha$, where α is a positive real number. Apply the transformation to the following characteristic equations to determine how many roots are to the right of the line $s = -1$ in the s-plane.

(a) $F(s) = s^2 + 5s + 3 = 0$
(b) $s^3 + 3s^2 + 3s + 1 = 0$
(c) $F(s) = s^3 + 4s^2 + 3s + 10 = 0$
(d) $s^3 + 4s^2 + 4s + 4 = 0$

2-49. The payload of a space-shuttle-pointing control system is modeled as a pure mass M. The payload is suspended by magnetic bearings so that no friction is encountered in the control. The attitude of the payload in the y direction is controlled by magnetic actuators located at the base. The total force produced by the magnetic actuators is $f(t)$. The controls of the other degrees of motion are independent and are not considered here. Because there are experiments located on the payload, electric power must be brought to the payload through cables. The linear spring with spring constant K_s is used to model the cable attachment. The dynamic system model for the

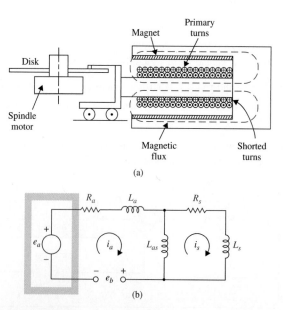

Figure 2P-49

control of the y-axis motion is shown in Figure 2P-49. The force equation of motion in the y-direction is

$$f(t) = K_s y(t) + M \frac{d^2 y(t)}{dt^2}$$

where $K_s = 0.5$ N-m/m and $M = 500$ kg. The magnetic actuators are controlled through state feedback, so that

$$f(t) = -K_P y(t) - K_D \frac{dy(t)}{dt}$$

(a) Draw a functional block diagram for the system.
(b) Find the characteristic equation of the closed-loop system.
(c) Find the region in the K_D-versus-K_P plane in which the system is asymptotically stable.

2-50. An inventory-control system is modeled by the following differential equations:

$$\frac{dx_1(t)}{dt} = -x_2(t) + u(t)$$

$$\frac{dx_2(t)}{dt} = -Ku(t)$$

where $x_1(t)$ is the level of inventory; $x_2(t)$, the rate of sales of product; $u(t)$, the production rate; and K, a real constant. Let the output of the system by $y(t) = x_1(t)$ and $r(t)$ be the reference set point for the desired inventory level. Let $u(t) = r(t) - y(t)$. Determine the constraint on K so that the closed-loop system is asymptotically stable.

2-51. Use MATLAB to solve Problem 2-50.

2-52. Use MATLAB to
(a) Generate symbolically the time function of $f(t)$

$$f(t) = 5 + 2e^{-2t}\sin\left(2t + \frac{\pi}{4}\right) - 4e^{-2t}\cos\left(2t - \frac{\pi}{2}\right) + 3e^{-4t}$$

(b) Generate symbolically $G(s) = \dfrac{(s+1)}{s(s+2)(s^2+2s+2)}$

(c) Find the Laplace transform of $f(t)$ and name it $F(s)$.
(d) Find the inverse Laplace transform of $G(s)$ and name it $g(t)$.
(e) If $G(s)$ is the forward-path transfer function of unity-feedback control systems, find the transfer function of the closed-loop system and apply the Routh-Hurwitz criterion to determine its stability.
(f) If $F(s)$ is the forward-path transfer function of unity-feedback control systems, find the transfer function of the closed-loop system and apply the Routh-Hurwitz criterion to determine its stability.

CHAPTER 3

Block Diagrams and Signal-Flow Graphs

In this chapter, we discuss graphical techniques for modeling control systems and their underlying mathematics. We also utilize the block diagram reduction techniques and the Mason's gain formula to find the transfer function of the overall control system. Later on in Chapters 4 and 5, we use the material presented in this chapter and Chapter 2 to fully model and study the performance of various control systems. The main objectives of this chapter are:

1. To study block diagrams, their components, and their underlying mathematics.
2. To obtain transfer function of systems through block diagram manipulation and reduction.
3. To introduce the signal-flow graphs.
4. To establish a parallel between block diagrams and signal-flow graphs.
5. To use Mason's gain formula for finding transfer function of systems.
6. To introduce state diagrams.
7. To demonstrate the MATLAB tools using case studies.

▶ 3-1 BLOCK DIAGRAMS

The **block diagram** modeling may provide control engineers with a better understanding of the composition and interconnection of the components of a system. Or it can be used, together with transfer functions, to describe the cause-and-effect relationships throughout the system. For example, consider a simplified block diagram representation of the heating system in your lecture room, shown in Fig. 3-1, where by setting a desired temperature, also defined as the **input**, one can set off the furnace to provide heat to the room. The process is relatively straightforward. The actual room temperature is also known as the **output** and is measured by a **sensor** within the thermostat. A simple electronic circuit within the thermostat compares the actual room temperature to the desired room temperature

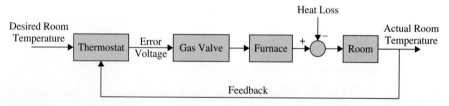

Figure 3-1 A simplified block diagram representation of a heating system.

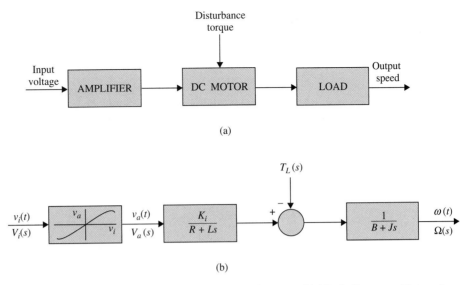

Figure 3-2 (a) Block diagram of a dc-motor control system. (b) Block diagram with transfer functions and amplifier characteristics.

(**comparator**). If the room temperature is below the desired temperature, an **error** voltage will be generated. The error voltage acts as a switch to open the gas valve and turn on the furnace (or the **actuator**). Opening the windows and the door in the classroom would cause heat loss and, naturally, would disturb the heating process (**disturbance**). The room temperature is constantly monitored by the output sensor. The process of sensing the output and comparing it with the input to establish an error signal is known as **feedback**. Note that the error voltage here causes the furnace to turn on, and the furnace would finally shut off when the error reaches zero.

As another example, consider the block diagram of Fig. 3-2 (a), which models an open-loop, dc-motor, speed-control system. The block diagram in this case simply shows how the system components are interconnected, and no mathematical details are given. If the mathematical and functional relationships of all the system elements are known, the block diagram can be used as a tool for the analytic or computer solution of the system. In general, block diagrams can be used to model linear as well as nonlinear systems. For example, the input–output relations of the dc-motor control system may be represented by the block diagram shown in Fig. 3-2 (b). In this figure, the input voltage to the motor is the output of the power amplifier, which, realistically, has a nonlinear characteristic. If the motor is linear, or, more appropriately, if it is operated in the linear region of its characteristics, its dynamics can be represented by transfer functions. The nonlinear amplifier gain can only be described in time domain and between the time variables $v_i(t)$ and $v_a(t)$. Laplace transform variables do not apply to nonlinear systems; hence, in this case, $V_i(s)$ and $V_a(s)$ do not exist. However, if the magnitude of $v_i(t)$ is limited to the linear range of the amplifier, then the amplifier can be regarded as linear, and the amplifier may be described by the transfer function

$$\frac{V_a(s)}{V_i(s)} = K \tag{3-1}$$

where K is a constant, which is the slope of the linear region of the amplifier characteristics.

Alternatively, we can use signal-flow graphs or state diagrams to provide a graphical representation of a control system. These topics are discussed later in this chapter.

3-1-1 Typical Elements of Block Diagrams in Control Systems

We shall now define the block-diagram elements used frequently in control systems and the related algebra. The common elements in block diagrams of most control systems include:

- Comparators
- Blocks representing individual component transfer functions, including:
  - Reference sensor (or input sensor)
  - Output sensor
  - Actuator
  - Controller
  - Plant (the component whose variables are to be controlled)
- Input or reference signals
- Output signals
- Disturbance signal
- Feedback loops

Fig. 3-3 shows one configuration where these elements are interconnected. You may wish to compare Fig. 3-1 or Fig. 3-2 to Fig. 3-3 to find the control terminology for each system. As a rule, each block represents an element in the control system, and each element can be modeled by one or more equations. These equations are normally in the time domain or preferably (because of ease in manipulation) in the Laplace domain. Once the block diagram of a system is fully constructed, one can study individual components or the overall system behavior.

One of the important components of a control system is the sensing and the electronic device that acts as a junction point for signal comparisons—otherwise known as a **comparator**. In general, these devices possess sensors and perform simple mathematical operations such as addition and subtraction (such as the thermostat in Fig. 3-1). Three examples of comparators are illustrated in Fig. 3-4. Note that the addition and subtraction operations in Fig. 3-4 (a) and (b) are linear, so the input and output variables of these block-diagram elements can be time-domain variables or Laplace-transform variables. Thus, in Fig. 3-4 (a), the block diagram implies

$$e(t) = r(t) - y(t) \tag{3-2}$$

or

$$E(s) = R(s) - Y(s) \tag{3-3}$$

As mentioned earlier, **blocks** represent the equations of the system in time domain or the **transfer function** of the system in the Laplace domain, as demonstrated in Fig. 3-5.

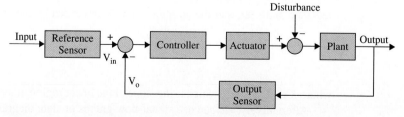

Figure 3-3 Block diagram representation of a general control system.

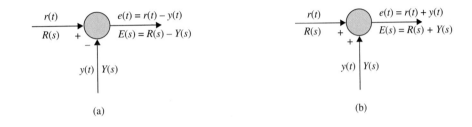

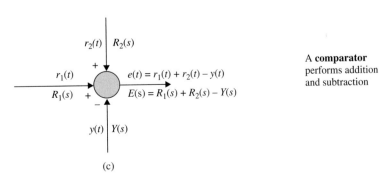

A **comparator** performs addition and subtraction

Figure 3-4 Block-diagram elements of typical sensing devices of control systems. (a) Subtraction. (b) Addition. (c) Addition and subtraction.

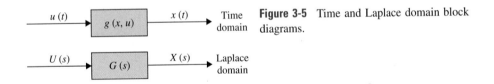

Figure 3-5 Time and Laplace domain block diagrams.

In Laplace domain, the following input–output relationship can be written for the system in Fig. 3-5:

$$X(s) = G(s) U(s) \qquad (3\text{-}4)$$

If signal $X(s)$ is the output and signal $U(s)$ denotes the input, the transfer function of the block in Fig. 3-5 is

$$G(s) = \frac{X(s)}{U(s)} \qquad (3\text{-}5)$$

Typical block elements that appear in the block diagram representation of most control systems include **plant**, **controller**, **actuator**, and **sensor**.

▶ **EXAMPLE 3-1-1** Consider the block diagram of two transfer functions $G_1(s)$ and $G_2(s)$ that are connected in series. Find the transfer function $G(s)$ of the overall system.

SOLUTION The system transfer function can be obtained by combining individual block equations. Hence, for signals $A(s)$ and $X(s)$, we have

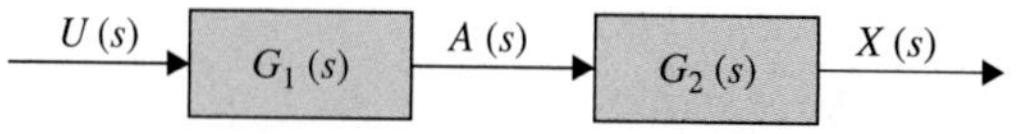

Figure 3-6 Block diagrams $G_1(s)$ and $G_2(s)$ connected in series.

$$X(s) = A(s)G_2(s)$$
$$A(s) = U(s)G_1(s)$$
$$X(s) = G_1(s)G_2(s)U(s)$$
$$G(s) = \frac{X(s)}{U(s)}$$

Hence,
$$G(s) = G_1(s)G_2(s) \tag{3-6}$$

Hence, using Eq. (3-6), the system in Fig. 3-6 can be represented by the system in Fig. 3-5.

▶ **EXAMPLE 3-1-2** Consider a more complicated system of two transfer functions $G_1(s)$ and $G_2(s)$ that are connected in parallel, as shown in Fig. 3-7. Find the transfer function $G(s)$ of the overall system.

SOLUTION The system transfer function can be obtained by combining individual block equations. Note for the two blocks $G_1(s)$ and $G_2(s)$, $A_1(s)$ acts as the input, and $A_2(s)$ and $A_3(s)$ are the outputs, respectively. Further, note that signal $U(s)$ goes through a **branch point P** to become $A_1(s)$. Hence, for the overall system, we combine the equations as follows.

$$A_1(s) = U(s)$$
$$A_2(s) = A_1(s)G_1(s)$$
$$A_3(s) = A_1(s)G_2(s)$$
$$X(s) = A_2(s) + A_3(s)$$
$$X(s) = U(s)(G_1(s) + G_2(s))$$
$$G(s) = \frac{X(s)}{U(s)}$$

Hence,
$$G(s) = G_1(s) + G_2(s) \tag{3-7}$$

For a system to be classified as a **feedback control system**, it is necessary that the controlled variable be fed back and **compared** with the reference input. After the comparison, an **error** signal is generated, which is used to **actuate** the control system. As a result, the actuator is activated in the presence of the error to minimize or eliminate that very error. A necessary component of every feedback control system is an **output sensor**, which is used to convert the output signal to a quantity that has the same units as the reference input. A feedback control system is also known a **closed-loop** system. A system may have multiple feedback loops. Fig. 3-8 shows the block diagram of a linear

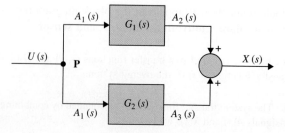

Figure 3-7 Block diagrams $G_1(s)$ and $G_2(s)$ connected in parallel.

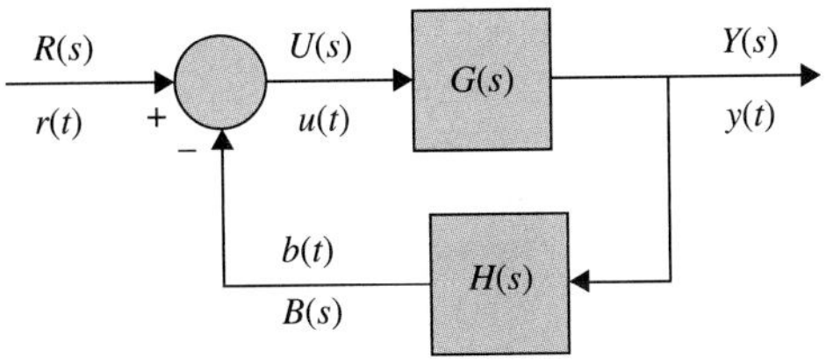

Figure 3-8 Basic block diagram of a feedback control system.

feedback control system with a single feedback loop. The following terminology is defined with reference to the diagram:

$r(t), R(s)$ = reference input(command)
$y(t), Y(s)$ = output (controlled variable)
$b(t), B(s)$ = feedback signal
$u(t), U(s)$ = actuating signal = error signal $e(t), E(s)$, when $H(s) = 1$
$H(s)$ = feedback transfer function
$G(s)H(s) = L(s)$ = loop transfer function
$G(s)$ = forward-path transfer function
$M(s) = Y(s)/R(s)$ = closed-loop transfer function or system transfer function

The closed-loop transfer function $M(s)$ can be expressed as a function of $G(s)$ and $H(s)$. From Fig. 3-8, we write

$$Y(s) = G(s)U(s) \qquad (3\text{-}8)$$

and

$$B(s) = H(s)Y(s) \qquad (3\text{-}9)$$

The actuating signal is written

$$U(s) = R(s) - B(s) \qquad (3\text{-}10)$$

Substituting Eq. (3-10) into Eq. (3-8) yields

$$Y(s) = G(s)R(s) - G(s)H(s) \qquad (3\text{-}11)$$

Substituting Eq. (3-9) into Eq. (3-7) and then solving for $Y(s)/R(s)$ gives the closed-loop transfer function

$$M(s) = \frac{Y(s)}{R(s)} = \frac{G(s)}{1 + G(s)H(s)} \qquad (3\text{-}12)$$

The feedback system in Fig. 3-8 is said to have a **negative feedback loop** because the comparator **subtracts**. When the comparator **adds** the feedback, it is called **positive feedback**, and the transfer function Eq. (3-12) becomes

$$M(s) = \frac{Y(s)}{R(s)} = \frac{G(s)}{1 - G(s)H(s)} \qquad (3\text{-}13)$$

If G and H are constants, they are also called **gains**. If $H = 1$ in Fig. 3-8, the system is said to have a **unity feedback loop**, and if $H = 0$, the system is said to be **open loop**.

3-1-2 Relation between Mathematical Equations and Block Diagrams

Consider the following second-order prototype system:

$$\ddot{x}(t) + 2\zeta\omega_n \dot{x}(t) + \omega_n^2 x(t) = \omega_n^2 u(t) \qquad (3\text{-}14)$$

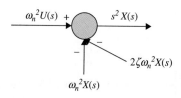

Figure 3-9 Graphical representation of Eq. (3-17) using a comparator.

which has Laplace representation (assuming zero initial conditions $x(0) = \dot{x}(0) = 0$):

$$X(s)s^2 + 2\zeta\omega_n X(s)s + \omega_n^2 X(s) = \omega_n^2 U(s) \tag{3-15}$$

Eq. (3-15) consists of constant damping ratio ζ, constant natural frequency ω_n, input $U(s)$, and output $X(s)$. If we rearrange Eq. (3-15) to

$$\omega_n^2 U(s) - 2\zeta\omega_n X(s)s - \omega_n^2 X(s) = X(s)s^2 \tag{3-16}$$

it can graphically be shown as in Fig. 3-9.

The signals $2\zeta\omega_n sX(s)$ and $\omega_n^2 X(s)$ may be conceived as the signal $X(s)$ going into blocks with transfer functions $2\zeta\omega_n s$ and ω_n^2, respectively, and the signal $X(s)$ may be obtained by integrating $s^2 X(s)$ twice or by post-multiplying by $\frac{1}{s^2}$, as shown in Fig. 3-10.

Because the signals $X(s)$ in the right-hand side of Fig. 3-10 are the same, they can be connected, leading to the block diagram representation of the system Eq. (3-17), as shown in Fig. 3-11. If you wish, you can further dissect the block diagram in Fig. 3-11 by factoring out the term $\frac{1}{s}$ as in Fig. 3-12(a) to obtain Fig. 3-12(b).

If the system studied here corresponds to the spring-mass-damper seen in Fig. 4-5 (see Chapter 4), then we can designate internal variables $A(s)$ and $V(s)$, which represent acceleration and velocity of the system, respectively, as illustrated in Fig. 3-12. The best way to see this is by recalling that $\frac{1}{s}$ is the integration operation in Laplace domain. Hence, if $A(s)$ is integrated once, we get $V(s)$, and after integrating $V(s)$, we get the $X(s)$ signal.

It is evident that there is no unique way of representing a system model with block diagrams. We may use different block diagram forms for different purposes, as long as the

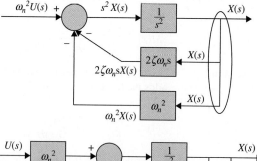

Figure 3-10 Addition of blocks $\frac{1}{s^2}$, $2\zeta\omega_n s$, and ω_n^2 to the graphical representation of Eq. (3-17).

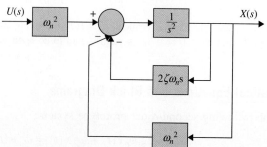

Figure 3-11 Block diagram representation of Eq. (3-17) in Laplace domain.

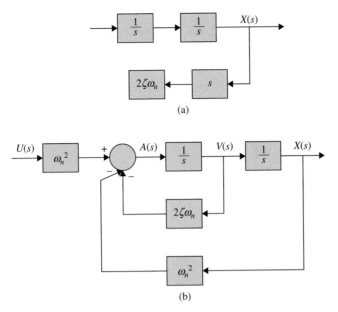

Figure 3-12 (a) Factorization of $\frac{1}{s}$ term in the internal feedback loop of Fig. 3-11. (b) Final block diagram representation of Eq. (3-17) in Laplace domain.

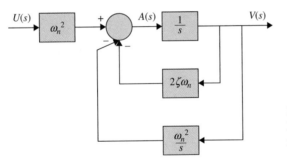

Figure 3-13 Block diagram of Eq. (3-17) in Laplace domain with $V(s)$ represented as the output.

overall transfer function of the system is not altered. For example, to obtain the transfer function $V(s)/U(s)$, we may yet rearrange Fig. 3-12 to get $V(s)$ as the system output, as shown in Fig. 3-13. This enables us to determine the behavior of velocity signal with input $U(s)$.

▶ **EXAMPLE 3-1-3** Find the transfer function of the system in Fig. 3-12 and compare that to the transfer function of system in Eq. (3-15).

SOLUTIONS The ω_n^2 block at the input and feedback signals in Fig. 3-12(b) may be moved to the right-hand side of the comparator. This is the same as factorization of ω_n^2 as shown:

$$\omega_n^2 U(s) - \omega_n^2 X(s) = \omega_n^2 (U(s) - X(s)) \tag{3-17}$$

Fig. 3-14(a) shows the factorization operation of Eq. (3-17), which results in a simpler block diagram representation of the system shown in Fig. 3-14 (b). Note that Fig. 3-12(b) and Fig. 3-14(b) are equivalent systems.

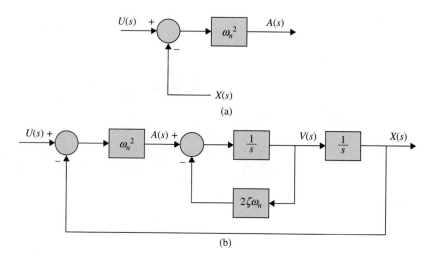

Figure 3-14 (a) Factorization of ω_n^2. (b) Alternative block diagram representation of Eq. (3-17) in Laplace domain.

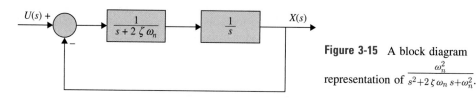

Figure 3-15 A block diagram representation of $\frac{\omega_n^2}{s^2+2\zeta\omega_n s+\omega_n^2}$.

Considering Fig. 3-12(b), it is easy to identify the internal feedback loop, which in turn can be simplified using Eq. (3-12), or

$$\frac{V(s)}{A(s)} = \frac{\frac{1}{s}}{1+\frac{2\zeta\omega_n}{s}} = \frac{s}{s+2\zeta\omega_n} \tag{3-18}$$

After pre- and post-multiplication by ω_2^n and $\frac{1}{s}$, respectively, the block diagram of the system is simplified to what is shown in Fig. 3-15, which ultimately results in

$$\frac{X(s)}{U(s)} = \frac{\frac{\omega_n^2}{s(s+2\zeta\omega_n)}}{1+\frac{\omega_n^2}{s(s+2\zeta\omega_n)}} = \frac{\omega_n^2}{s^2+2\zeta\omega_n s+\omega_n^2} \tag{3-19}$$

Eq. (3-19) is also the transfer function of system Eq. (3-15). ◄

▶ **EXAMPLE 3-1-4** Find the velocity transfer function using Fig. 3-13 and compare that to the derivative of Eq. (3-19).

SOLUTIONS Simplifying the two feedback loops in Fig. 3-13, starting with the internal loop first, we have

$$\frac{V(s)}{U(s)} = \frac{\frac{\frac{1}{s}}{1+\frac{2\zeta\omega_n}{s}}\omega_n^2}{1+\frac{\frac{1}{s}}{1+\frac{2\zeta\omega_n}{s}}\frac{\omega_n^2}{s}}$$

$$\frac{V(s)}{U(s)} = \frac{s\omega_n^2}{s^2+2\zeta\omega_n s+\omega_2^n} \tag{3-20}$$

Eq. (3-20) is the same as the derivative of Eq. (3-19), which is nothing but multiplying Eq. (3-19) by an s term. Try to find the $A(s)/U(s)$ transfer function. Obviously you must get: $s^2 X(s)/U(s)$.

3-1-3 Block Diagram Reduction

As you might have noticed from the examples in the previous section, the transfer function of a control system may be obtained by manipulation of its block diagram and by its ultimate reduction into one block. For complicated block diagrams, it is often necessary to move a **comparator** or a **branch point** to make the block diagram reduction process simpler. The two key operations in this case are:

1. Moving a branch point from **P** to **Q**, as shown in Fig. 3-16(a) and Fig. 3-16(b). This operation must be done such that the signals $Y(s)$ and $B(s)$ are unaltered. In Fig. 3-16(a), we have the following relations:

$$Y(s) = A(s)G_2(s)$$
$$B(s) = Y(s)H_1(s) \quad (3\text{-}21)$$

In Fig. 3-16(b), we have the following relations:

$$Y(s) = A(s)G_2(s)$$
$$B(s) = A(s)\frac{H_1(s)}{G_2(s)} \quad (3\text{-}22)$$

But

$$G_2(s) = \frac{A(s)}{Y(s)} \quad (3\text{-}23)$$
$$\Rightarrow B(s) = Y(s)H_1(s)$$

2. Moving a comparator, as shown in Fig. 3-17(a) and Fig. 3-17(b), should also be done such that the output $Y(s)$ is unaltered. In Fig. 3-17(a), we have the following relations:

$$Y(s) = A(s)G_2(s) + B(s)H_1(s) \quad (3\text{-}24)$$

(a)
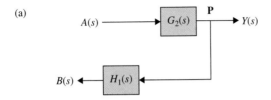

(b)
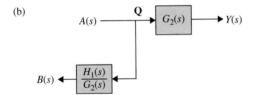

Figure 3-16 (a) Branch point relocation from point **P** to (b) point **Q**.

114 ▶ Chapter 3. Block Diagrams and Signal-Flow Graphs

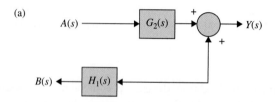

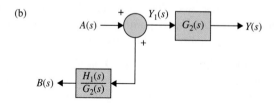

Figure 3-17 (a) Comparator relocation from the right-hand side of block $G_2(s)$ to (b) the left-hand side of block $G_2(s)$.

In Fig. 3-17(b), we have the following relations:

$$Y_1(s) = A(s) + B(s)\frac{H_1(s)}{G_2(s)}$$

$$Y(s) = Y_1(s)G_2(s)$$
(3-25)

So

$$Y(s) = A(s)G_2(s) + B(s)\frac{H_1(s)}{G_2(s)}G_2(s)$$

$$\Rightarrow Y(s) = A(s)G_2(s) + B(s)H_1(s)$$
(3-26)

▶ **EXAMPLE 3-1-5** Find the input–output transfer function of the system shown in Fig. 3-17(a).

SOLUTION To perform the block diagram reduction, one approach is to move the branch point at Y_1 to the left of block G_2, as shown in Fig. 3-18(b). After that, the reduction becomes trivial, first by combining the blocks G_2, G_3, and G_4 as shown in Fig. 3-18(c), and then by eliminating the two

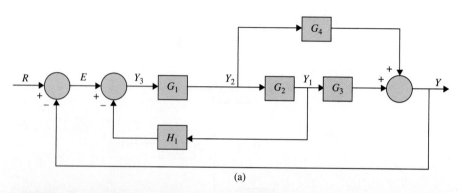

Figure 3-18 (a) Original block diagram. (b) Moving the branch point at Y_1 to the left of block G_2. (c) Combining the blocks G_1, G_2, and G_3. (d) Eliminating the inner feedback loop.

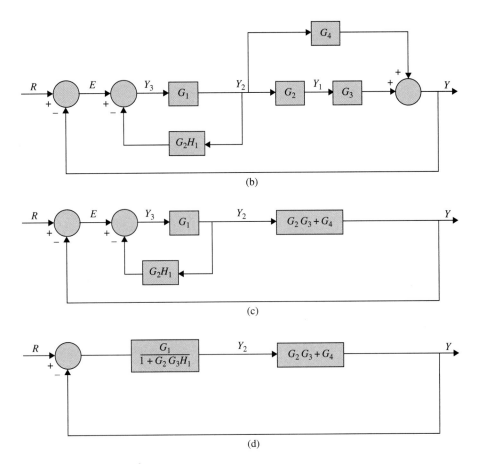

Figure 3-18 (*Continued*)

feedback loops. As a result, the transfer function of the final system after the reduction in Fig. 3-18(d) becomes

$$\frac{Y(s)}{E(s)} = \frac{G_1G_2G_3 + G_1G_4}{1 + G_1G_2H_1 + G_1G_2G_3 + G_1G_4} \quad (3\text{-}27)$$

◀

3-1-4 Block Diagram of Multi-Input Systems—Special Case: Systems with a Disturbance

An important case in the study of control systems is when a disturbance signal is present. Disturbance (such as heat loss in the example in Fig. 3-1) usually adversely affects the performance of the control system by placing a burden on the controller/actuator components. A simple block diagram with two inputs is shown in Fig. 3-19. In this case, one of the inputs, $D(s)$, is known as disturbance, while $R(s)$ is the reference input. Before designing a proper controller for the system, it is always important to learn the effects of $D(s)$ on the system.

We use the method of superposition in modeling a multi-input system.

Super Position: For linear systems, the overall response of the system under multi-inputs is the summation of the responses due to the individual inputs, i.e., in this case,

$$Y_{total} = Y_R|_{D=0} + Y_D|_{R=0} \quad (3\text{-}28)$$

116 ▶ Chapter 3. Block Diagrams and Signal-Flow Graphs

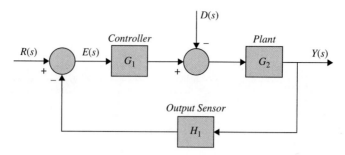

Figure 3-19 Block diagram of a system undergoing disturbance

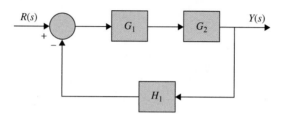

Figure 3-20 Block diagram of the system in Fig. 3-19 when $D(s) = 0$.

When $D(s) = 0$, the block diagram is simplified (Fig. 3-20) to give the transfer function

$$\frac{Y(s)}{R(s)} = \frac{G_1(s)\,G_2(s)}{1 + G_1(s)\,G_2\,H_1(s)} \qquad (3\text{-}29)$$

When $R(s) = 0$, the block diagram is rearranged to give (Fig. 3-21):

$$\frac{Y(s)}{D(s)} = \frac{-G_2(s)}{1 + G_1(s)\,G_2(s)\,H_1(s)} \qquad (3\text{-}30)$$

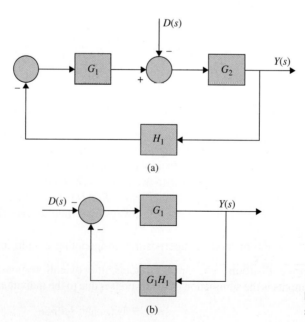

Figure 3-21 Block diagram of the system in Fig. 3-19 when $R(s) = 0$.

As a result, from Eq. (3-28) to Eq. (3-32), we ultimately get

$$Y_{total} = \left.\frac{Y(s)}{R(s)}\right|_{D=0} R(s) + \left.\frac{Y(s)}{D(s)}\right|_{R=0} D(s)$$

$$Y(s) = \frac{G_1 G_2}{1 + G_1 G_2 H_1} R(s) + \frac{-G_2}{1 + G_1 G_2 H_1} D(s)$$
(3-31)

Observations: $\left.\frac{Y}{R}\right|_{D=0}$ and $\left.\frac{Y}{D}\right|_{R=0}$ have the same denominators if the disturbance signal goes to the forward path. The negative sign in the numerator of $\left.\frac{Y}{D}\right|_{R=0}$ shows that the disturbance signal interferes with the controller signal, and, as a result, it adversely affects the performance of the system. Naturally, to compensate, there will be a higher burden on the controller.

3-1-5 Block Diagrams and Transfer Functions of Multivariable Systems

In this section, we shall illustrate the block diagram and matrix representations (see Appendix A) of multivariable systems. Two block-diagram representations of a multivariable system with p inputs and q outputs are shown in Fig. 3-22(a) and (b). In Fig. 3-22 (a), the individual input and output signals are designated, whereas in the block diagram of Fig. 3-22(b), the multiplicity of the inputs and outputs is denoted by vectors. The case of Fig. 3-22(b) is preferable in practice because of its simplicity.

Fig. 3-23 shows the block diagram of a multivariable feedback control system. The transfer function relationships of the system are expressed in vector-matrix form (see Appendix A):

$$\mathbf{Y}(s) = \mathbf{G}(s)\mathbf{U}(s) \quad (3\text{-}32)$$

$$\mathbf{U}(s) = \mathbf{R}(s) - \mathbf{B}(s) \quad (3\text{-}33)$$

$$\mathbf{B}(s) = \mathbf{H}(s)\mathbf{Y}(s) \quad (3\text{-}34)$$

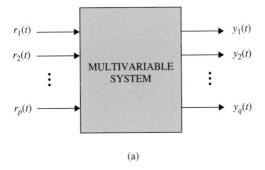

Figure 3-22 Block diagram representations of a multivariable system.

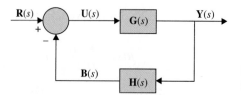

Figure 3-23 Block diagram of a multivariable feedback control system.

where $\mathbf{Y}(s)$ is the $q \times 1$ output vector; $\mathbf{U}(s)$, $\mathbf{R}(s)$, and $\mathbf{B}(s)$ are all $p \times 1$ vectors; and $\mathbf{G}(s)$ and $\mathbf{H}(s)$ are $q \times p$ and $p \times q$ transfer-function matrices, respectively. Substituting Eq. (3-11) into Eq. (3-10) and then from Eq. (3-10) to Eq. (3-9), we get

$$\mathbf{Y}(s) = \mathbf{G}(s)\mathbf{R}(s) - \mathbf{G}(s)\mathbf{H}(s)\mathbf{Y}(s) \tag{3-35}$$

Solving for $\mathbf{Y}(s)$ from Eq. (3-12) gives

$$\mathbf{Y}(s) = [\mathbf{I} + \mathbf{G}(s)\mathbf{H}(s)]^{-1}\mathbf{G}(s)\mathbf{R}(s) \tag{3-36}$$

provided that $\mathbf{I} + \mathbf{G}(s)\mathbf{H}(s)$ is nonsingular. The closed-loop transfer matrix is defined as

$$\mathbf{M}(s) = [\mathbf{I} + \mathbf{G}(s)\mathbf{H}(s)]^{-1}\mathbf{G}(s) \tag{3-37}$$

Then Eq. (3-14) is written

$$\mathbf{Y}(s) = \mathbf{M}(s)\mathbf{R}(s) \tag{3-38}$$

▶ **EXAMPLE 3-1-6** Consider that the forward-path transfer function matrix and the feedback-path transfer function matrix of the system shown in Fig. 3-23 are

$$\mathbf{G}(s) = \begin{bmatrix} \dfrac{1}{s+1} & -\dfrac{1}{s} \\ 2 & \dfrac{1}{s+2} \end{bmatrix} \quad \mathbf{H}(s) = \begin{bmatrix} 1 & 0 \\ 0 & 1 \end{bmatrix} \tag{3-39}$$

respectively. The closed-loop transfer function matrix of the system is given by Eq. (3-15), and is evaluated as follows:

$$\mathbf{I} + \mathbf{G}(s)\mathbf{H}(s) = \begin{bmatrix} 1 + \dfrac{1}{s+1} & -\dfrac{1}{s} \\ 2 & 1 + \dfrac{1}{s+2} \end{bmatrix} = \begin{bmatrix} \dfrac{s+2}{s+1} & -\dfrac{1}{s} \\ 2 & \dfrac{s+3}{s+2} \end{bmatrix} \tag{3-40}$$

The closed-loop transfer function matrix is

$$\mathbf{M}(s) = [\mathbf{I} + \mathbf{G}(s)\mathbf{H}(s)]^{-1}\mathbf{G}(s) = \dfrac{1}{\Delta} \begin{bmatrix} \dfrac{s+3}{s+2} & \dfrac{1}{s} \\ -2 & \dfrac{s+2}{s+1} \end{bmatrix} \begin{bmatrix} \dfrac{1}{s+1} & -\dfrac{1}{s} \\ 2 & \dfrac{1}{s+2} \end{bmatrix} \tag{3-41}$$

where

$$\Delta = \dfrac{s+2}{s+1}\dfrac{s+3}{s+2} + \dfrac{2}{s} = \dfrac{s^2 + 5s + 2}{s(s+1)} \tag{3-42}$$

Thus,

$$\mathbf{M}(s) = \frac{s(s+1)}{s^2+5s+2} \begin{bmatrix} \dfrac{3s^2+9s+4}{s(s+1)(s+2)} & -\dfrac{1}{s} \\ 2 & \dfrac{3s+2}{s(s+1)} \end{bmatrix} \quad (3\text{-}43)$$

3-2 SIGNAL-FLOW GRAPHS (SFGs)

A signal-flow graph (SFG) may be regarded as a simplified version of a block diagram. The SFG was introduced by S. J. Mason [2] for the cause-and-effect representation of linear systems that are modeled by algebraic equations. Besides the differences in the physical appearance of the SFG and the block diagram, the signal-flow graph is constrained by more rigid mathematical rules, whereas the block-diagram notation is more liberal. An SFG may be defined as a graphical means of portraying the input–output relationships among the variables of a set of linear algebraic equations.

Consider a linear system that is described by a set of N algebraic equations:

$$y_j = \sum_{k=1}^{N} a_{kj} y_k \quad j = 1, 2, \ldots, N \quad (3\text{-}34)$$

It should be pointed out that these N equations are written in the form of cause-and-effect relations:

$$j\text{th effect} = \sum_{k=1}^{N} (\text{gain from } k \text{ to } j) \times (k\text{th cause}) \quad (3\text{-}45)$$

or simply

$$\text{Output} = \sum (\text{gain}) \times (\text{input}) \quad (3\text{-}46)$$

This is the single most important axiom in forming the set of algebraic equations for SFGs. When the system is represented by a set of integrodifferential equations, we must first transform these into Laplace-transform equations and then rearrange the latter in the form of Eq. (3-31), or

$$Y_j(s) = \sum_{k=1}^{N} G_{kj}(s) Y_k(s) \quad j = 1, 2, \ldots, N \quad (3\text{-}47)$$

3-2-1 Basic Elements of an SFG

When constructing an SFG, junction points, or nodes, are used to represent variables. The nodes are connected by line segments called branches, according to the cause-and-effect equations. The branches have associated branch gains and directions. *A signal can transmit through a branch only in the direction of the arrow.* In general, given a set of equations such as Eq. (3-31) or Eq. (3-47), the construction of the SFG is basically a matter of

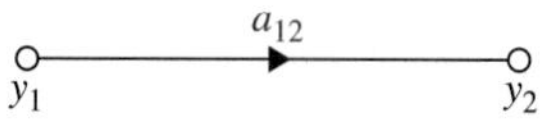

Figure 3-24 Signal flow graph of $y_2 = a_{12}y_1$.

following through the cause-and-effect relations of each variable in terms of itself and the others. For instance, consider that a linear system is represented by the simple algebraic equation

$$y_2 = a_{12}y_1 \tag{3-48}$$

where y_1 is the input, y_2 is the output, and a_{12} is the gain, or transmittance, between the two variables. The SFG representation of Eq. (3-48) is shown in Fig. 3-24. Notice that the branch directing from node y_1 (input) to node y_2 (output) expresses the dependence of y_2 on y_1 but not the reverse. The branch between the input node and the output node should be interpreted as a unilateral amplifier with gain a_{12}, so when a signal of one unit is applied at the input y_1, a signal of strength $a_{12}y_1$ is delivered at node y_2. Although algebraically Eq. (3-48) can be written as

$$y_1 = \frac{1}{a_{12}}y_2 \tag{3-49}$$

the SFG of Fig. 3-24 does not imply this relationship. If Eq. (3-49) is valid as a cause-and-effect equation, a new SFG should be drawn with y_2 as the input and y_1 as the output.

▶ **EXAMPLE 3-2-1** As an example on the construction of an SFG, consider the following set of algebraic equations:

$$\begin{aligned} y_2 &= a_{12}y_1 + a_{32}y_3 \\ y_3 &= a_{23}y_2 + a_{43}y_4 \\ y_4 &= a_{24}y_2 + a_{34}y_3 + a_{44}y_4 \\ y_5 &= a_{25}y_2 + a_{45}y_4 \end{aligned} \tag{3-50}$$

The SFG for these equations is constructed, step by step, in Fig. 3-25. ◀

3-2-2 Summary of the Basic Properties of SFG

The important properties of the SFG that have been covered thus far are summarized as follows.

1. SFG applies only to linear systems.
2. The equations for which an SFG is drawn must be algebraic equations in the form of cause-and-effect.
3. Nodes are used to represent variables. Normally, the nodes are arranged from left to right, from the input to the output, following a succession of cause-and-effect relations through the system.
4. Signals travel along branches only in the direction described by the arrows of the branches.
5. The branch directing from node y_k to y_j represents the dependence of y_j upon y_k but not the reverse.
6. A signal y_k traveling along a branch between y_k and y_j is multiplied by the gain of the branch a_{kj}, so a signal $a_{kj}y_k$ is delivered at y_j.

3-2-3 Definitions of SFG Terms

In addition to the branches and nodes defined earlier for the SFG, the following terms are useful for the purpose of identification and execution of the SFG algebra.

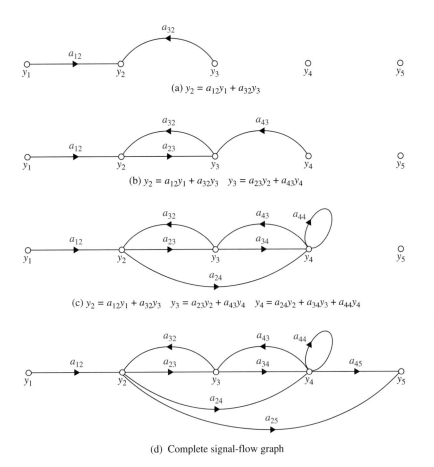

Figure 3-25 Step-by-step construction of the signal-flow graph in Eq. (3-50).

Input Node (Source): *An input node is a node that has only outgoing branches* (example: node y_1 in Fig. 3-24).

Output Node (Sink): *An output node is a node that has only incoming branches:* (example: node y_2 in Fig. 3-24). However, this condition is not always readily met by an output node. For instance, the SFG in Fig. 3-26(a) does not have a node that satisfies the condition of an output node. It may be necessary to regard y_2 and/or y_3 as output nodes to find the effects at these nodes due to the input. To make y_2 an output node, we simply connect a branch with unity gain from the existing node y_2 to a new node also designated as y_2, as shown in Fig. 3-26(b). The same procedure is applied to y_3. Notice that, in the modified SFG of Fig. 3-26(b), the equations $y_2 = y_2$ and $y_3 = y_3$ are added to the original equations. In general, we can make any noninput node of an SFG an output by the procedure just illustrated. However, we **cannot** convert a noninput node into an input node by reversing the branch direction of the procedure described for output nodes. For instance, node y_2 of the SFG in Fig. 3-26(a) is not an input node. If we attempt to convert it into an input node by adding an incoming branch with unity gain from another identical node y_2, the SFG of Fig. 3-27 would result. The equation that portrays the relationship at node y_2 now reads

$$y_2 = y_2 + a_{12}y_1 + a_{32}y_3 \tag{3-51}$$

which is different from the original equation given in Fig. 3-26(a).

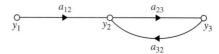

(a) Original signal-flow graph

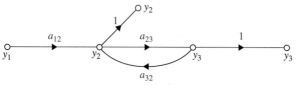

(b) Modified signal-flow graph

Figure 3-26 Modification of a signal-flow graph so that y_2 and y_3 satisfy the condition as output nodes.

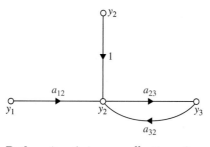

Figure 3-27 Erroneous way to make node y_2 an input node.

Path: *A path is any collection of a continuous succession of branches traversed in the same direction.* The definition of a path is entirely general, since it does not prevent any node from being traversed more than once. Therefore, as simple as the SFG of Fig. 3-26(a) is, it may have numerous paths just by traversing the branches a_{23} and a_{32} continuously.

Forward Path: *A forward path is a path that starts at an input node and ends at an output node and along which no node is traversed more than once.* For example, in the SFG of Fig. 3-25(d), y_1 is the input node, and the rest of the nodes are all possible output nodes. The forward path between y_1 and y_2 is simply the connecting branch between the two nodes. There are two forward paths between y_1 and y_3: One contains the branches from y_1 to y_2 to y_3, and the other one contains the branches from y_1 to y_2 to y_4 (through the branch with gain a_{24}) and then back to y_3 (through the branch with gain a_{43}). The reader should try to determine the two forward paths between y_1 and y_4. Similarly, there are three forward paths between y_1 and y_5.

Path Gain: *The product of the branch gains encountered in traversing a path is called the* path gain. For example, the path gain for the path $y_1 - y_2 - y_3 - y_4$ in Fig. 3-25(d) is $a_{12}a_{23}a_{34}$.

Loop: *A loop is a path that originates and terminates on the same node and along which no other node is encountered more than once.* For example, there are four loops in the SFG of Fig. 3-25(d). These are shown in Fig. 3-28.

Forward-Path Gain: *The* forward-path gain *is the path gain of a forward path.*

Loop Gain: *The* loop gain *is the path gain of a loop.* For example, the loop gain of the loop $y_2 - y_4 - y_3 - y_2$ in Fig. 3-28 is $a_{24}a_{43}a_{32}$.

Nontouching Loops: *Two parts of an SFG are* nontouching *if they do not share a common node.* For example, the loops $y_2 - y_3 - y_2$ and $y_4 - y_4$ of the SFG in Fig. 3-25(d) are nontouching loops.

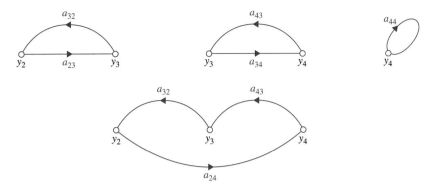

Figure 3-28 Four loops in the signal-flow graph of Fig. 3-25(d).

3-2-4 SFG Algebra

Based on the properties of the SFG, we can outline the following manipulation rules and algebra:

1. The value of the variable represented by a node is equal to the sum of all the signals entering the node. For the SFG of Fig. 3-29, the value of y_1 is equal to the sum of the signals transmitted through all the incoming branches; that is,

$$y_1 = a_{21}y_2 + a_{31}y_3 + a_{41}y_4 + a_{51}y_5 \tag{3-52}$$

2. The value of the variable represented by a node is transmitted through all branches leaving the node. In the SFG of Fig. 3-29, we have

$$\begin{aligned} y_6 &= a_{16}y_1 \\ y_7 &= a_{17}y_1 \\ y_8 &= a_{18}y_1 \end{aligned} \tag{3-53}$$

3. Parallel branches in the same direction connecting two nodes can be replaced by a single branch with gain equal to the sum of the gains of the parallel branches. An example of this case is illustrated in Fig. 3-30.

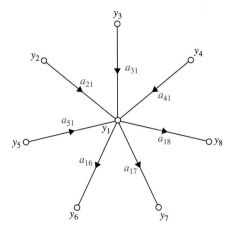

Figure 3-29 Node as a summing point and as a transmitting point.

124 ▶ Chapter 3. Block Diagrams and Signal-Flow Graphs

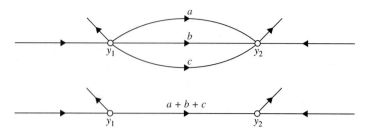

Figure 3-30 Signal-flow graph with parallel paths replaced by one with a single branch.

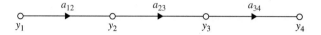

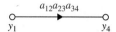

Figure 3-31 Signal-flow graph with cascade unidirectional branches replaced by a single branch.

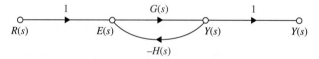

Figure 3-32 Signal-flow graph of the feedback control system shown in Fig. 3-8.

4. A series connection of unidirectional branches, as shown in Fig. 3-31, can be replaced by a single branch with gain equal to the product of the branch gains.

3-2-5 SFG of a Feedback Control System

The SFG of the single-loop feedback control system in Fig. 3-8 is drawn as shown in Fig. 3-32. Using the SFG algebra already outlined, the closed-loop transfer function in Eq. (3-12) can be obtained.

3-2-6 Relation between Block Diagrams and SFGs

The relation between block diagrams and SFGs are tabulated for three important cases, as shown in Table 3-1.

3-2-7 Gain Formula for SFG

Given an SFG or block diagram, the task of solving for the input–output relations by algebraic manipulation could be quite tedious. Fortunately, there is a general gain formula available that allows the determination of the input–output relations of an SFG by inspection.

TABLE 3-1 Block diagrams and their SFG equivalent representations

	Block Diagram	Signal Flow Diagram
Simple Transfer Function $\dfrac{Y(s)}{U(s)} = G(s)$	$U(s) \to \boxed{G(s)} \to Y(s)$	$y_1 \xrightarrow{G(s)} y_2$
Parallel Feedback	$U(s)$ through $G_1(s)$ and $G_2(s)$ summed to $Y(s)$	$U(s) \to y_1 \rightrightarrows^{G_1(s)}_{G_2(s)} y_2 \to Y(s)$
$\dfrac{Y(s)}{R(s)} = \dfrac{G(s)}{1+G(s)H(s)}$	Feedback loop with $G(s)$ forward and $H(s)$ feedback	$R(s) \xrightarrow{1} U(s) \xrightarrow{G(s)} Y(s) \xrightarrow{1} Y(s)$, with $-H(s)$ feedback

Given an SFG with N forward paths and K loops, the gain between the input node y_{in} and output node y_{out} is [3]

$$M = \frac{y_{out}}{y_{in}} = \sum_{k=1}^{N} \frac{M_k \Delta_k}{\Delta} \tag{3-54}$$

where

y_{in} = input-node variable

y_{out} = output-node variable

M = gain between y_{in} and y_{out}

N = total number of forward paths between y_{in} and y_{out}

M_k = gain of the kth forward paths between y_{in} and y_{out}

$$\Delta = 1 - \sum_i L_{i1} + \sum_j L_{j2} - \sum_k L_{k3} + \cdots \tag{3-55}$$

L_{mr} = gain product of the mth ($m = i, j, k, \ldots$) possible combination of r nontouching loops ($1 \le r \le K$).

or

$\Delta = 1 -$ (sum of the gains of **all individual** loops) + (sum of products of gains of all possible combinations of **two** nontouching loops) − (sum of products of gains of all possible combinations of **three** nontouching loops) + $\cdots$

Δ_k = the Δ for that part of the SFG that is nontouching with the kth forward path.

The gain formula in Eq. (3-54) may seem formidable to use at first glance. However, Δ and Δ_k are the only terms in the formula that could be complicated if the SFG has a large number of loops and nontouching loops.

Care must be taken when applying the gain formula to ensure that it is applied between an **input node** and an **output node**.

▶ **EXAMPLE 3-2-2** Consider that the closed-loop transfer function $Y(s)/R(s)$ of the SFG in Fig. 3-32 is to be determined by use of the gain formula, Eq. (3-54). The following results are obtained by inspection of the SFG:

1. There is only one forward path between $R(s)$ and $Y(s)$, and the forward-path gain is

$$M_1 = G(s) \tag{3-56}$$

2. There is only one loop; the loop gain is

$$L_{11} = -G(s)H(s) \tag{3-57}$$

3. There are no nontouching loops since there is only one loop. Furthermore, the forward path is in touch with the only loop. Thus, $\Delta_1 = 1$, and

$$\Delta = 1 - L_{11} = 1 + G(s)H(s) \tag{3-58}$$

Using Eq. (3-54), the closed-loop transfer function is written

$$\frac{Y(s)}{R(s)} = \frac{M_1 \Delta_1}{\Delta} = \frac{G(s)}{1 + G(s)H(s)} \tag{3-59}$$

which agrees with Eq. (3-12). ◀

▶ **EXAMPLE 3-2-3** Consider the SFG shown in Fig. 3-25(d). Let us first determine the gain between y_1 and y_5 using the gain formula.

The three forward paths between y_1 and y_5 and the forward-path gains are

$M_1 = a_{12}a_{23}a_{34}a_{45}$ Forward path: $y_1 - y_2 - y_3 - y_4 - y_5$
$M_2 = a_{12}a_{25}$ Forward path: $y_1 - y_2 - y_5$
$M_3 = a_{12}a_{24}a_{45}$ Forward path: $y_1 - y_2 - y_4 - y_5$

The four loops of the SFG are shown in Fig. 3-28. The loop gains are

$$L_{11} = a_{23}a_{32} \quad L_{21} = a_{34}a_{43} \quad L_{31} = a_{24}a_{43}a_{32} \quad L_{41} = a_{44}$$

There is only one pair of nontouching loops; that is, the two loops are

$$y_2 - y_3 - y_2 \quad \text{and} \quad y_4 - y_4$$

Thus, the product of the gains of the two nontouching loops is

$$L_{12} = a_{23}a_{32}a_{44} \tag{3-60}$$

All the loops are in touch with forward paths M_1 and M_3. Thus, $\Delta_1 = \Delta_3 = 1$. Two of the loops are not in touch with forward path M_2. These loops are $y_3 - y_4 - y_3$ and $y_4 - y_4$. Thus,

$$\Delta_2 = 1 - a_{34}a_{43} - a_{44} \tag{3-61}$$

Substituting these quantities into Eq. (3-54), we have

$$\frac{y_5}{y_1} = \frac{M_1\Delta_1 + M_2\Delta_2 + M_3\Delta_3}{\Delta}$$

$$= \frac{(a_{12}a_{23}a_{34}a_{45}) + (a_{12}a_{25})(1 - a_{34}a_{43} - a_{44}) + a_{12}a_{24}a_{45}}{1 - (a_{23}a_{32} + a_{34}a_{43} + a_{24}a_{32}a_{43} + a_{44}) + a_{23}a_{32}a_{44}} \tag{3-62}$$

where

$$\Delta = 1 - (L_{11} + L_{21} + L_{31} + L_{41}) + L_{12}$$
$$= 1 - (a_{23}a_{32} + a_{34}a_{43} + a_{24}a_{32}a_{43} + a_{44}) + a_{23}a_{32}a_{44} \quad (3\text{-}63)$$

The reader should verify that choosing y_2 as the output,

$$\frac{y_2}{y_1} = \frac{a_{12}(1 - a_{34}a_{43} - a_{44})}{\Delta} \quad (3\text{-}64)$$

where Δ is given in Eq. (3-63).

▶ **EXAMPLE 3-2-4** Consider the SFG in Fig. 3-33. The following input–output relations are obtained by use of the gain formula:

$$\frac{y_2}{y_1} = \frac{1 + G_3 H_2 + H_4 + G_3 H_2 H_4}{\Delta} \quad (3\text{-}65)$$

$$\frac{y_4}{y_1} = \frac{G_1 G_2 (1 + H_4)}{\Delta} \quad (3\text{-}66)$$

$$\frac{y_6}{y_1} = \frac{y_7}{y_1} = \frac{G_1 G_2 G_3 G_4 + G_1 G_5 (1 + G_3 H_2)}{\Delta} \quad (3\text{-}67)$$

where

$$\Delta = 1 + G_1 H_1 + G_3 H_2 + G_1 G_2 G_3 H_3 + H_4 + G_1 G_3 H_1 H_2 \\ + G_1 H_1 H_4 + G_3 H_2 H_4 + G_1 G_2 G_3 H_3 H_4 + G_1 G_3 H_1 H_2 H_4 \quad (3\text{-}68)$$

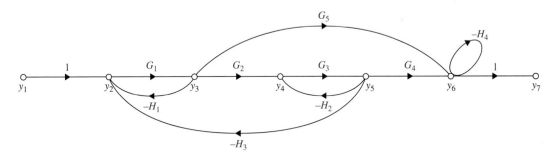

Figure 3-33 Signal-flow graph for Example 3-2-4.

3-2-8 Application of the Gain Formula between Output Nodes and Noninput Nodes

It was pointed out earlier that the gain formula can only be applied between a pair of input and output nodes. Often, it is of interest to find the relation between an output-node variable and a noninput-node variable. For example, in the SFG of Figure 3-33, it may be of interest to find the relation y_7/y_2, which represents the dependence of y_7 upon y_2; the latter is not an input.

We can show that, by including an input node, the gain formula can still be applied to find the gain between a noninput node and an output node. Let y_{in} be an input and y_{out} be an output node of a SFG. The gain, y_{out}/y_2, where y_2 is not an input, may be written as

$$\frac{y_{out}}{y_2} = \frac{\frac{y_{out}}{y_{in}}}{\frac{y_2}{y_{in}}} = \frac{\frac{\Sigma M_k \Delta_k |_{\text{from } y_{in} \text{ to } y_{out}}}{\Delta}}{\frac{\Sigma M_k \Delta_k |_{\text{from } y_{in} \text{ to } y_2}}{\Delta}} \quad (3\text{-}69)$$

Because Δ is independent of the inputs and the outputs, the last equation is written

$$\frac{y_{out}}{y_2} = \frac{\Sigma M_k \Delta_k |_{\text{from } y_{in} \text{ to } y_{out}}}{\Sigma M_k \Delta_k |_{\text{from } y_{in} \text{ to } y_2}} \quad (3\text{-}70)$$

Notice that Δ does not appear in the last equation.

▶ **EXAMPLE 3-2-5** From the SFG in Fig. 3-33, the gain between y_2 and y_7 is written

$$\frac{y_7}{y_2} = \frac{y_7/y_1}{y_2/y_1} = \frac{G_1 G_2 G_3 G_4 + G_1 G_5 (1 + G_3 H_2)}{1 + G_3 H_2 + H_4 + G_3 H_2 H_4} \quad (3\text{-}71)$$

◀

3-2-9 Application of the Gain Formula to Block Diagrams

Because of the similarity between the block diagram and the SFG, the gain formula in Eq. (3-54) can directly be applied to the block diagram to determine the transfer function of the system. However, in complex systems, to be able to identify all the loops and nontouching parts clearly, it may be helpful if an equivalent SFG is drawn for the block diagram first before applying the gain formula.

▶ **EXAMPLE 3-2-6** To illustrate how an equivalent SFG of a block diagram is constructed and how the gain formula is applied to a block diagram, consider the block diagram shown in Fig. 3-34(a). The equivalent SFG of the system is shown in Fig. 3-34(b). Notice that since a node on the SFG is interpreted as the summing point of all incoming signals to the node, the negative feedbacks on the block diagram are represented by assigning negative gains to the feedback paths on the SFG. First we can identify the forward paths and loops in the system and their corresponding gains. That is:

Forward Path Gains: 1. $G_1 G_2 G_3$; 2. $G_1 G_4$

Loop Gains: 1. $-G_1 G_2 H_1$; 2. $-G_2 G_3 H_2$; 3. $-G_1 G_2 G_3$; 4. $-G_4 H_2$; 5. $-G_1 G_4$

The closed-loop transfer function of the system is obtained by applying Eq. (3-54) to either the block diagram or the SFG in Fig. 3-34. That is

$$\frac{Y(s)}{R(s)} = \frac{G_1 G_2 G_3 + G_1 G_4}{\Delta} \quad (3\text{-}72)$$

where

$$\Delta = 1 + G_1 G_2 H_1 + G_2 G_3 H_2 + G_1 G_2 G_3 + G_4 H_2 + G_1 G_4 \quad (3\text{-}73)$$

Similarly,

$$\frac{E(s)}{R(s)} = \frac{1 + G_1 G_2 H_1 + G_2 G_3 H_2 + G_4 H_2}{\Delta} \quad (3\text{-}74)$$

$$\frac{Y(s)}{E(s)} = \frac{G_1 G_2 G_3 + G_1 G_4}{1 + G_1 G_2 H_1 + G_2 G_3 H_2 + G_4 H_2} \quad (3\text{-}75)$$

The last expression is obtained using Eq. (3-70).

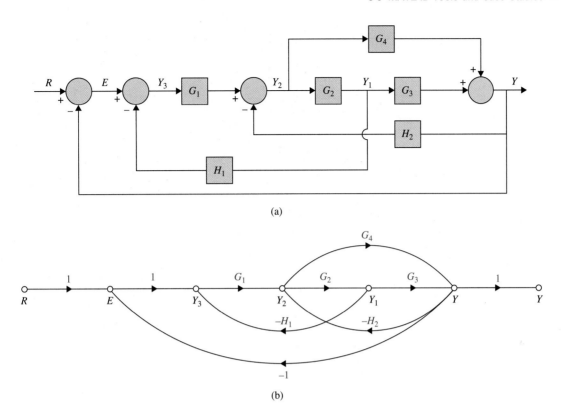

Figure 3-34 (a) Block diagram of a control system. (b) Equivalent signal-flow graph.

3-2-10 Simplified Gain Formula

From Example 3-2-6, we can see that *all loops and forward paths are touching* in this case. As a general rule, if there are no nontouching loops and forward paths (e.g., $y_2 - y_3 - y_2$ and $y_4 - y_4$ in Example 3-2-3) in the block diagram or SFG of the system, then Eq. (3-54) takes a far simpler look, as shown next.

$$M = \frac{y_{\text{out}}}{y_{\text{in}}} = \sum \frac{\text{Forward Path Gains}}{1 - \text{Loop Gains}} \qquad (3\text{-}76)$$

Redo Examples 3-2-2 through 3-2-6 to confirm the validity of Eq. (3-76).

▶ 3-3 MATLAB TOOLS AND CASE STUDIES

There is no specific software developed for this chapter. Although MATLAB Controls Toolbox offers functions for finding the transfer functions from a given block diagram, it was felt that students may master this subject without referring to a computer. For simple operations, however, MATLAB may be used, as shown in the following example.

▶ **EXAMPLE 3-3-1** Consider the following transfer functions, which correspond to the block diagrams shown in Fig. 3-35.

$$G_1(s) = \frac{1}{s+1}, \quad G_2(s) = \frac{s+1}{s+2}, \quad G(s) = \frac{1}{s(s+1)}, \quad H(s) = 10 \qquad (3\text{-}77)$$

Use MATLAB to find the transfer function $Y(s)/R(s)$ for each case. The results are as follows. ◀

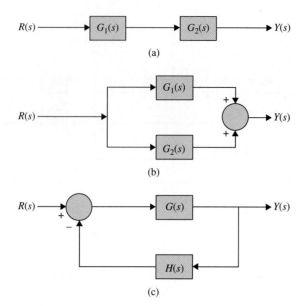

Figure 3-35 Basic block diagrams used for Example 3-3-1.

Toolbox 3-3-1

Case (a) Use MATLAB to find $G_1 * G_2$

$$\frac{Y(s)}{R(s)} = \frac{s+1}{s^2+3s+2} = \frac{1}{(s+2)}$$

Approach 1
```
>> clear all
>> s = tf('s');
>> G1=1/(s+1)

Transfer function:
   1
  ---
  s + 1

>> G2=(s+1)/(s+2)

Transfer function:
  s + 1
  -----
  s + 2
>> YR=G1*G2

Transfer function:
      s + 1
  -------------
   s^2 + 3 s + 2

>> YR_simple=minreal(YR)

Transfer function:
   1
  ---
  s + 2
```

Approach 2
```
>> clear all
>> G1=tf([1],[1 1])

Transfer function:
   1
  ---
  s + 1

>> G2=tf([1 1],[1 2])

Transfer function:
  s + 1
  -----
  s + 2

>> YR=G1*G2

Transfer function:
      s + 1
  -------------
   s^2 + 3 s + 2

>> YR_simple=minreal(YR)

Transfer function:
   1
  ---
  s + 2
```

Use "minreal(YR)" for pole zero cancellation, if necessary
Alternatively use "YR=series(G1,G2)" instead of "YR=G1*G2"

Case (b) Use MATLAB to find $G_1 + G_2$

$$\frac{Y(s)}{R(s)} = \frac{2s+3}{s^2+3s+2} = \frac{2(s+1.5)}{(s+1)(s+2)}$$

Approach 1
```
>> clear all
>> s = tf('s');
>> G=1/(s+1)

Transfer function:
   1
  ---
  s + 1

>> G2 = (s+1)/(s+2)

Transfer function:
  s + 1
  -----
  s + 2

>> YR=G1+G2

Transfer function:
  s^2 + 3 s + 3
  -------------
  s^2 + 3 s + 2

>> YR=parallel(G1,G2)

Transfer function:
  s^2 + 3 s + 3
  -------------
  s^2 + 3 s + 2
```

Approach 2
```
>> clear all
>> G1=tf([1],[1 1])

Transfer function:
   1
  ---
  s + 1

>> G2=tf([1 1],[1 2])

Transfer function:
  s + 1
  -----
  s + 2

>> YR=G1+G2

Transfer function:
  s^2 + 3 s + 3
  -------------
  s^2 + 3 s + 2

>> YR=parallel(G1,G2)

Transfer function:
  s^2 + 3 s + 3
  -------------
  s^2 + 3 s + 2
```

Use "minreal(YR)" for pole zero cancellation, if necessary
Alternatively use "YR=parallel(G1,G2)" instead of "YR=G1+G2"

Use "zpk(YR)" to obtain the real zero/pole/Gain format:
```
>> zpk(YR)

Zero/pole/gain:
 (s^2 + 3s + 3)
 --------------
  (s+2)(s+1)
```

Use "zero(YR)" to obtain transfer function zeros:
```
>> zero(YR)

ans =
  -1.5000 + 0.8660i
  -1.5000 - 0.8660i
```

Use "pole(YR)" to obtain transfer function poles:
```
>> pole(YR)

ans =
  -2
  -1
```

Toolbox 3-3-2

Case (b) Use MATLAB to find the closed-loop feedback function $\dfrac{G}{1+GH}$

Case (c) $\dfrac{Y(s)}{R(s)} = \dfrac{1}{s^2+s+10}$

Approach 1

```
>> clear YR
>> s = tf('s');
>> G=1/(s*(s+1))
```
Transfer function:
```
    1
  -----
  s^2 + s
```
```
>> H=10
```
H =

    10

```
>> YR=G/(1+G*H)
```
Transfer function:
```
          s^2 + s
  -------------------------
  s^4 + 2 s^3 + 11 s^2 + 10 s
```
```
>> YR_simple=minreal(YR)
```
Transfer function:
```
       1
  -----------
  s^2 + s + 10
```

Approach 2

```
>> clear all
>> G=tf([1],[1 1 0])
```
Transfer function:
```
    1
  -----
  s^2 + s
```
```
>> H=10
```
H =

    10

```
>> YR=G/(1+G*H)
```
Transfer function:
```
          s^2 + s
  -------------------------
  s^4 + 2 s^3 + 11 s^2 + 10 s
```
```
>> YR_simple=minreal(YR)
```
Transfer function:
```
       1
  -----------
  s^2 + s + 10
```

Use "minreal(YR)" for pole zero cancellation, if necessary

Alternatively use:

```
>> YR=feedback(G,H)
```
Transfer function:
```
       1
  -----------
  s^2 + s + 10
```

Use "pole(YR)" to obtain transfer function poles:

```
>> pole(YR)
```
ans =

  -0.5000 + 3.1225i
  -0.5000 - 3.1225i

3-4 SUMMARY

This chapter was devoted to the mathematical modeling of physical systems. Transfer functions, block diagrams, and signal-flow graphs were defined. The transfer function of a linear system was defined in terms of impulse response as well as differential equations. Multivariable and single-variable systems were examined.

The block diagram representation was shown to be a versatile method of portraying linear and nonlinear systems. A powerful method of representing the interrelationships between the signals of a linear system is the signal-flow graph, or SFG. When applied properly, an SFG allows the derivation of the transfer functions between input and output variables of a linear system using the gain formula. A state diagram is an SFG that is applied to dynamic systems that are represented by differential equations.

At the end of the chapter, MATLAB was used to calculate transfer functions of simple block diagram systems.

REVIEW QUESTIONS

1. Define the transfer function of a linear time-invariant system in terms of its impulse response.
2. When defining the transfer function, what happens to the initial conditions of the system?
3. Define the characteristic equation of a linear system in terms of the transfer function.
4. What is referred to as a multivariable system?
5. Can signal-flow graphs (SFGs) be applied to nonlinear systems?
6. How can SFGs be applied to systems that are described by differential equations?
7. Define the input node of an SFG.
8. Define the output node of an SFG.
9. State the form to which the equations must first be conditioned before drawing the SFG.
10. What does the arrow on the branch of an SFG represent?
11. Explain how a noninput node of an SFG can be made into an output node.
12. Can the gain formula be applied between any two nodes of an SFG?
13. Explain what the nontouching loops of an SFG are.
14. Does the Δ of an SFG depend on which pair of input and output is selected?
15. List the advantages and utilities of the state diagram.
16. Given the state diagram of a linear dynamic system, how do you define the state variables?
17. Given the state diagram of a linear dynamic system, how do you find the transfer function between a pair of input and output variables?
18. Given the state diagram of a linear dynamic system, how do you write the state equations of the system?
19. The state variables of a dynamic system are not equal to the number of energy-storage elements under what condition?

Answers to these review questions can be found on this book's companion Web site: www.wiley.com/college/golnaraghi.

REFERENCES

Block Diagrams and Signal-Flow Graphs

1. T. D. Graybeal, "Block Diagram Network Transformation," *Elec. Eng.*, Vol. 70, pp. 985–990, 1951.
2. S. J. Mason, "Feedback Theory—Some Properties of Signal Flow Graphs," *Proc. IRE*, Vol. 41, No. 9 pp. 1144–1156, Sept. 1953.

3. S. J. Mason, "Feedback Theory—Further Properties of Signal Flow Graphs," *Proc. IRE*, Vol. 44, No. 7 pp. 920–926, July 1956.
4. L. P. A. Robichaud, M. Boisvert, and J. Robert, *Signal Flow Graphs and Applications*, Prentice Hall, Englewood Cliffs, NJ, 1962.
5. B. C. Kuo, *Linear Networks and Systems*, McGraw-Hill, New York, 1967.

State-Variable Analysis of Electric Networks

6. B. C. Kuo, *Linear Circuits and Systems*, McGraw-Hill, New York, 1967.

▶ PROBLEMS

PROBLEMS FOR SECTION 3-1

3-1. Consider the block diagram shown in Fig. 3P-1.

Find:

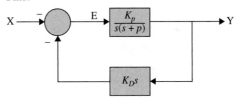

Figure 3P-1

(a) The loop transfer function.
(b) The forward path transfer function.
(c) The error transfer function.
(d) The feedback transfer function.
(e) The closed loop transfer function.

3-2. Reduce the block diagram shown in Fig. 3P-2 to unity feedback form and find the system characteristic equation.

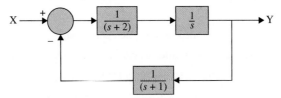

Figure 3P-2

3-3. Reduce the block diagram shown in Fig. 3P-3 and find the Y/X.

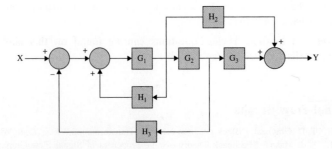

Figure 3P-3

3-4. Reduce the block diagram shown in Fig. 3P-4 to unity feedback form and find the Y/X.

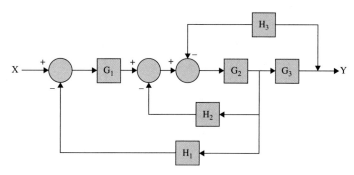

Figure 3P-4

3-5. The aircraft turboprop engine shown in Fig. 3P-5(a) is controlled by a closed-loop system with block diagram shown in Fig. 3P-5(b). The engine is modeled as a multivariable system with input vector $\mathbf{E}(s)$, which contains the fuel rate and propeller blade angle, and output vector $\mathbf{Y}(s)$, consisting of the engine speed and turbine-inlet temperature. The transfer function matrices are given as

$$\mathbf{G}(s) = \begin{bmatrix} \dfrac{2}{s(s+2)} & 10 \\ \dfrac{5}{s} & \dfrac{1}{s+1} \end{bmatrix} \quad \mathbf{H}(s) = \begin{bmatrix} 1 & 0 \\ 0 & 1 \end{bmatrix}$$

Find the closed-loop transfer function matrix $[\mathbf{I} + \mathbf{G}(s)\mathbf{H}(s)]^{-1}\mathbf{G}(s)$.

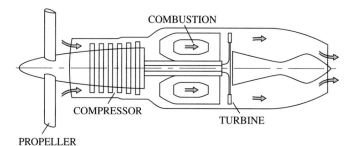

Figure 3P-5(a)

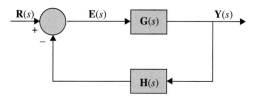

Figure 3P-5(b)

3-6. Use MATLAB to solve Problem 3-5.

3-7. The block diagram of the position-control system of an electronic word processor is shown in Fig. 3P-7.
(a) Find the loop transfer function $\Theta_o(s)/\Theta_e(s)$ (the outer feedback path is open).
(b) Find the closed-loop transfer function $\Theta_o(s)/\Theta_r(s)$.

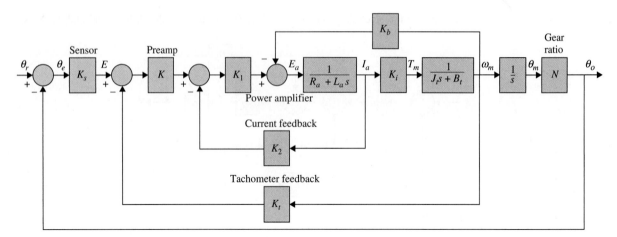

Figure 3P-7

3-8. The block diagram of a feedback control system is shown in Fig. 3P-8. Find the following transfer functions:

(a) $\left.\dfrac{Y(s)}{R(s)}\right|_{N=0}$

(b) $\left.\dfrac{Y(s)}{E(s)}\right|_{N=0}$

(c) $\left.\dfrac{Y(s)}{N(s)}\right|_{R=0}$

(d) Find the output $Y(s)$ when $R(s)$ and $N(s)$ are applied simultaneously.

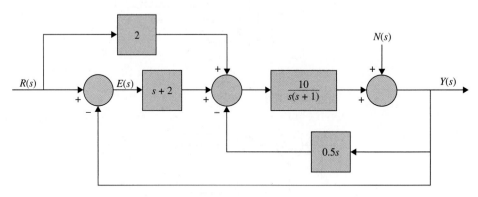

Figure 3P-8

3-9. The block diagram of a feedback control system is shown in Fig. 3P-9.

(a) Apply the SFG gain formula directly to the block diagram to find the transfer functions

$$\left.\dfrac{Y(s)}{R(s)}\right|_{N=0} \qquad \left.\dfrac{Y(s)}{N(s)}\right|_{R=0}$$

Express $Y(s)$ in terms of $R(s)$ and $N(s)$ when both inputs are applied simultaneously.

(b) Find the desired relation among the transfer functions $G_1(s), G_2(s), G_3(s), G_4(s), H_1(s)$, and $H_2(s)$ so that the output $Y(s)$ is not affected by the disturbance signal $N(s)$ at all.

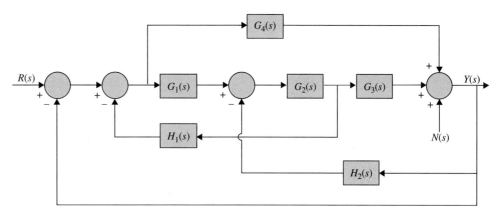

Figure 3P-9

3-10. Fig. 3P-10 shows the block diagram of the antenna control system of the solar-collector field shown in Fig. 1-5. The signal $N(s)$ denotes the wind gust disturbance acted on the antenna. The feedforward transfer function $G_d(s)$ is used to eliminate the effect of $N(s)$ on the output $Y(s)$. Find the transfer function $Y(s)/N(s)|_{R=0}$. Determine the expression of $G_d(s)$ so that the effect of $N(s)$ is entirely eliminated.

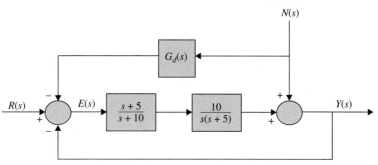

Figure 3P-10

3-11. Fig. 3P-11 shows the block diagram of a dc-motor control system. The signal $N(s)$ denotes the frictional torque at the motor shaft.

(a) Find the transfer function $H(s)$ so that the output $Y(s)$ is not affected by the disturbance torque $N(s)$.

(b) With $H(s)$ as determined in part (a), find the value of K so that the steady-state value of $e(t)$ is equal to 0.1 when the input is a unit-ramp function, $r(t) = tu_s(t), R(s) = 1/s^2$, and $N(s) = 0$. Apply the final-value theorem.

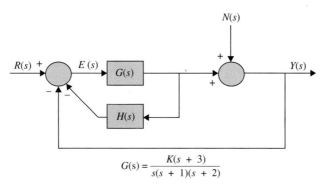

Figure 3P-11

3-12. The block diagram of an electric train control is shown in Fig. 3P-12. The system parameters and variables are

$e_r(t)$ = voltage representing the desired train speed, V
$v(t)$ = speed of train, ft/sec
M = Mass of train = $30,000$ lb/sec^2
K = amplifier gain
K_t = gain of speed indicator = 0.15 V/ft/sec

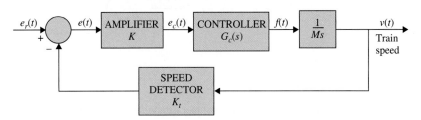

Figure 3P-12

To determine the transfer function of the controller, we apply a step function of 1 volt to the input of the controller, that is, $e_c(t) = u_s(t)$. The output of the controller is measured and described by the following equation:

$$f(t) = 100\left(1 - 0.3e^{-6t} - 0.7e^{-10t}\right)u_s(t)$$

(a) Find the transfer function $G_c(s)$ of the controller.

(b) Derive the forward-path transfer function $V(s)/E(s)$ of the system. The feedback path is opened in this case.

(c) Derive the closed-loop transfer function $V(s)/E_r(s)$ of the system.

(d) Assuming that K is set at a value so that the train will not run away (unstable), find the steady-state speed of the train in feet per second when the input is $e_r(t) = us(t)$V.

3-13. Use MATLAB to solve Problem 3-12.

3-14. Repeat Problem 3-12 when the output of the controller is measured and described by the following expression:

$$f(t) = 100\left(1 - 0.3e^{-6(t-0.5)}\right)u_s(t - 0.5)$$

when a step input of 1 V is applied to the controller.

3-15. Use MATLAB to solve Problem 3-14.

3-16. A linear time-invariant multivariable system with inputs $r_1(t)$ and $r_2(t)$ and outputs $y_1(t)$ and $y_2(t)$ is described by the following set of differential equations.

$$\frac{d^2y_1(t)}{dt^2} + 2\frac{dy_1(t)}{dt} + 3y_2(t) = r_1(t) + r_2(t)$$

$$\frac{d^2y_2(t)}{dt^2} + 3\frac{dy_1(t)}{dt} + y_1(t) - y_2(t) = r_2(t) + \frac{dr_1(t)}{dt}$$

Find the following transfer functions:

$$\left.\frac{Y_1(s)}{R_1(s)}\right|_{R_2=0} \quad \left.\frac{Y_2(s)}{R_1(s)}\right|_{R_2=0} \quad \left.\frac{Y_1(s)}{R_2(s)}\right|_{R_1=0} \quad \left.\frac{Y_2(s)}{R_2(s)}\right|_{R_1=0}$$

PROBLEMS FOR SECTION 3-2

3-17. Find the state-flow diagram for the system shown in Fig. 3P-4.

3-18. Draw a signal-flow diagram for the system with the following state-space representation:

$$\dot{X} = \begin{bmatrix} -5 & -6 & 3 \\ 1 & 0 & -1 \\ -0.5 & 1.5 & 0.5 \end{bmatrix} X + \begin{bmatrix} 0.5 & 0 \\ 0 & 0.5 \\ 0.5 & 0.5 \end{bmatrix} U$$

$$Z = \begin{bmatrix} 0.5 & 0.5 & 0 \\ 0.5 & 0 & 0.5 \end{bmatrix} X$$

3-19. Find the state-space representation of a system with the following transfer function:

$$G(s) = \frac{B_1 s + B_0 s}{s^2 + A_1 s + A_0 s}$$

3-20. Draw signal-flow graphs for the following sets of algebraic equations. These equations should first be arranged in the form of cause-and-effect relations before SFGs can be drawn. Show that there are many possible SFGs for each set of equations.

(a) $x_1 = -x_2 - 3x_3 + 3$

$x_2 = 5x_1 - 2x_2 + x_3$

$x_3 = 4x_1 + x_2 - 5x_3 + 5$

(b) $2x_1 + 3x_2 + x_3 = -1$

$x_1 - 2x_2 - x_3 = 1$

$3x_2 + x_3 = 0$

3-21. The block diagram of a control system is shown in Fig. 3P-21.
(a) Draw an equivalent SFG for the system.
(b) Find the following transfer functions by applying the gain formula of the SFG directly to the block diagram.

$$\left.\frac{Y(s)}{R(s)}\right|_{N=0} \quad \left.\frac{Y(s)}{N(s)}\right|_{R=0} \quad \left.\frac{E(s)}{R(s)}\right|_{N=0} \quad \left.\frac{E(s)}{N(s)}\right|_{R=0}$$

(c) Compare the answers by applying the gain formula to the equivalent SFG.

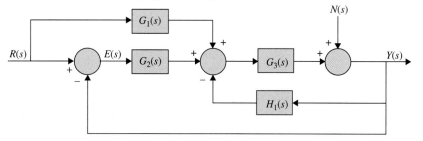

Figure 3P-21

3-22. Apply the gain formula to the SFGs shown in Fig. 3P-22 to find the following transfer functions: $\dfrac{Y_5}{Y_1} \quad \dfrac{Y_4}{Y_1} \quad \dfrac{Y_2}{Y_1} \quad \dfrac{Y_5}{Y_2}$

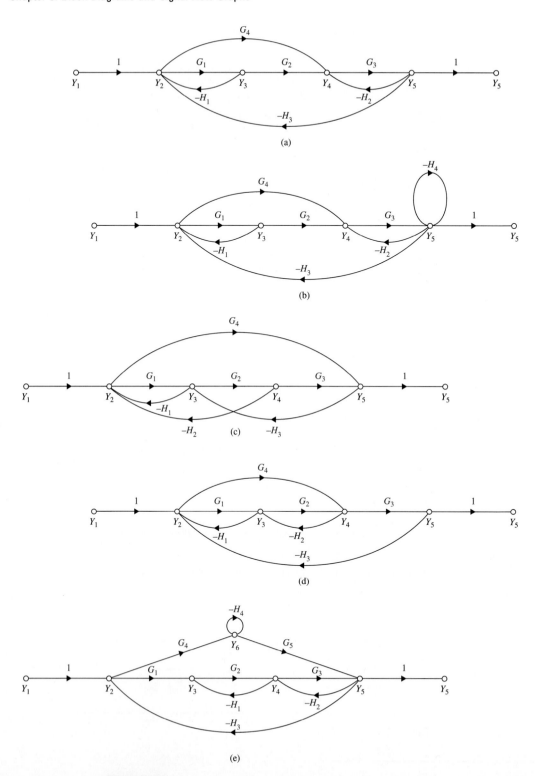

Figure 3P-22

3-23. Find the transfer functions Y_7/Y_1 and Y_2/Y_1 of the SFGs shown in Fig. 3P-23.

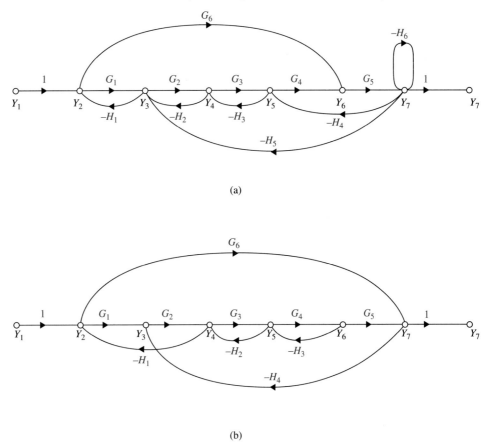

Figure 3P-23

3-24. Signal-flow graphs may be used to solve a variety of electric network problems. Shown in Fig. 3P-24 is the equivalent circuit of an electronic circuit. The voltage source $e_d(t)$ represents a disturbance voltage. The objective is to find the value of the constant k so that the output voltage

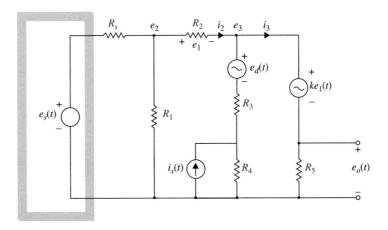

Figure 3P-24

$e_o(t)$ is not affected by $e_d(t)$. To solve the problem, it is best to first write a set of cause-and-effect equations for the network. This involves a combination of node and loop equations. Then construct an SFG using these equations. Find the gain e_o/e_d with all other inputs set to zero. For e_d not to affect e_o, set e_o/e_d to zero.

3-25. Show that the two systems shown in Figs 3P-25(a) and (b) are equivalent.

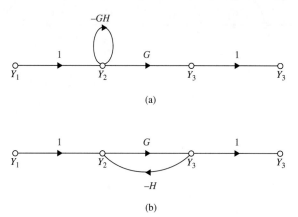

Figure 3P-25

3-26. Show that the two systems shown in Figs. 3P-26(a) and (b) are not equivalent.

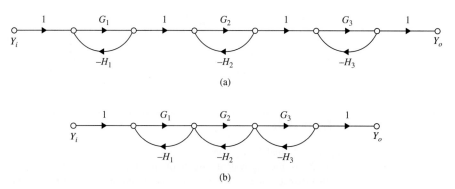

Figure 3P-26

3-27. Find the following transfer functions for the SFG shown in Fig. 3P-27.

$$\left.\frac{Y_6}{Y_1}\right|_{Y_7=0} \qquad \left.\frac{Y_6}{Y_7}\right|_{Y_1=0}$$

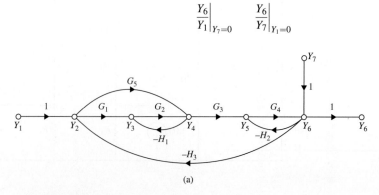

Figure 3P-27(a)

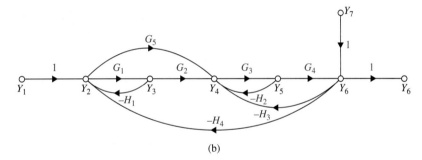

Figure 3P-27(b)

3-28. Find the following transfer functions for the SFG shown in Fig. 3P-28. Comment on why the results for parts (c) and (d) are not the same.

(a) $\left.\dfrac{Y_7}{Y_1}\right|_{Y_8=0}$

(b) $\left.\dfrac{Y_7}{Y_8}\right|_{Y_1=0}$

(c) $\left.\dfrac{Y_7}{Y_4}\right|_{Y_8=0}$

(d) $\left.\dfrac{Y_7}{Y_4}\right|_{Y_1=0}$

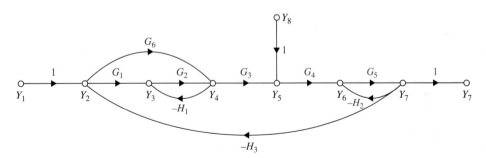

Figure 3P-28

3-29. The coupling between the signals of the turboprop engine shown in Fig. 3P-4(a) is shown in Fig. 3P-29. The signals are defined as

$R_1(s) =$ fuel rate
$R_2(s) =$ propeller blade angle
$Y_1(s) =$ engine speed
$Y_2(s) =$ turbine inlet temperature

(a) Draw an equivalent SFG for the system.
(b) Find the Δ of the system using the SFG gain formula.
(c) Find the following transfer functions:

$$\left.\dfrac{Y_1(s)}{R_1(s)}\right|_{R_2=0} \quad \left.\dfrac{Y_1(s)}{R_2(s)}\right|_{R_1=0} \quad \left.\dfrac{Y_2(s)}{R_1(s)}\right|_{R_2=0} \quad \left.\dfrac{Y_2(s)}{R_2(s)}\right|_{R_1=0}$$

(d) Express the transfer functions in matrix form, $\mathbf{Y}(s) = \mathbf{G}(s)\mathbf{R}(s)$.

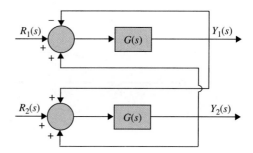

Figure 3P-29

3-30. Figure 3P-30 shows the block diagram of a control system with conditional feedback. The transfer function $G_p(s)$ denotes the controlled process, and $G_c(s)$ and $H(s)$ are the controller transfer functions.

(a) Derive the transfer functions $Y(s)/R(s)|_{N=0}$ and $Y(s)/N(s)|_{R=0}$. Find $Y(s)/R(s)|N=0$ when $G_c(s) = G_p(s)$.

(b) Let

$$G_p(s) = G_c(s) = \frac{100}{(s+1)(s+5)}$$

Find the output response $y(t)$ when $N(s) = 0$ and $r(t) = u_s(t)$.

(c) With $G_p(s)$ and $G_c(s)$ as given in part (b), select $H(s)$ among the following choices such that when $n(t) = u_s(t)$ and $r(t) = 0$, the steady-state value of $y(t)$ is equal to zero. (There may be more than one answer.)

$$H(s) = \frac{10}{s(s+1)} \quad H(s) = \frac{10}{(s+1)(s+2)}$$

$$H(s) = \frac{10(s+1)}{s+2} \quad H(s) = \frac{K}{s^n} \ (n = \text{positive integer}) \text{ Select } n.$$

Keep in mind that the poles of the closed-loop transfer function must all be in the left-half s-plane for the final-value theorem to be valid.

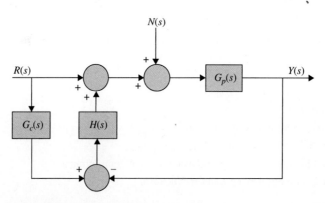

Figure 3P-30

3-31. Use MATLAB to solve Problem 3-30.

3-32. Consider the following differential equations of a system:

$$\frac{dx_1(t)}{dt} = -2x_1(t) + 3x_2(t)$$
$$\frac{dx_2(t)}{dt} = -5x_1(t) - 5x_2(t) + 2r(t)$$

(a) Find the characteristic equation of the system.

(b) Find the transfer functions $X_1(s)/R(s)$ and $X_2(s)/R(s)$.

3-33. The differential equation of a linear system is

$$\frac{d^3y(t)}{dt^3} + 5\frac{d^2y(t)}{dt^2} + 6\frac{dy(t)}{dt} + 10y(t) = r(t)$$

where $y(t)$ is the output, and $r(t)$ is the input.

(a) Write the state equation of the system. Define the state variables from right to left in ascending order.

(b) Find the characteristic equation and its roots. Use MATLAB to find the roots.

(c) Find the transfer function $Y(s)/R(s)$.

(d) Perform a partial-fraction expansion of $Y(s)/R(s)$.

(e) Find the output $y(t)$ for $t \geq 0$ when $r(t) = u_s(t)$.

(f) Find the final value of $y(t)$ by using the final-value theorem.

3-34. Consider the differential equation given in Problem 3-33. Use MATLAB to

(a) Find the partial-fraction expansion of $Y(s)/R(s)$.

(b) Find the Laplace transform of the system.

(c) Find the output $y(t)$ for $t \geq 0$ when $r(t) = u_s(t)$.

(d) Plot the step response of the system.

(e) Verify the final value that you obtained in Problem 3-33 part (f).

3-35. Repeat Problem 3-33 for the following differential equation:

$$\frac{d^4y(t)}{dt^4} + 4\frac{d^3y(t)}{dt^3} + 3\frac{d^2y(t)}{dt^2} + 5\frac{dy(t)}{dt} + y(t) = r(t)$$

3-36. Repeat Problem 3-34 for the differential equation given in Problem 3-35.

3-37. The block diagram of a feedback control system is shown in Fig. 3P-37.

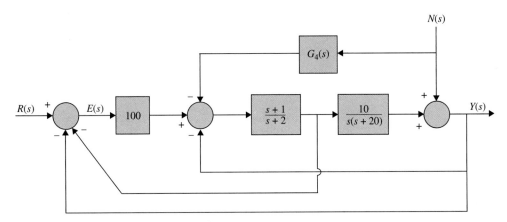

Figure 3P-37

(a) Derive the following transfer functions:

$$\left.\frac{Y(s)}{R(s)}\right|_{N=0} \quad \left.\frac{Y(s)}{N(s)}\right|_{R=0} \quad \left.\frac{E(s)}{R(s)}\right|_{N=0}$$

(b) The controller with the transfer function $G_4(s)$ is for the reduction of the effect of the noise $N(s)$. Find $G_4(s)$ so that the output $Y(s)$ is totally independent of $N(s)$.
(c) Find the characteristic equation and its roots when $G_4(s)$ is as determined in part (b).
(d) Find the steady-state value of $e(t)$ when the input is a unit-step function. Set $N(s) = 0$.
(e) Find $y(t)$ for $t \geq 0$ when the input is a unit-step function. Use $G_4(s)$ as determined in part (b).

3-38. Use MATLAB to solve Problem 3-37.

ADDITIONAL PROBLEMS

3-39. Assuming

$$P_1 = 2s^6 + 9s^5 + 15s^4 + 25s^3 + 25s^2 + 14s + 6$$

$$P_2 = s^6 + 8s^5 + 23s^4 + 36s^3 + 38s^2 + 28s + 16$$

(a) Use MATLAB to find roots of P_1 and P_2.
(b) Use MATLAB to calculate $P_3 = P_2 - P_1$, $P_4 = P_2 + P_1$, and $P_5 = (P_1 - P_2)*P_1$.

3-40. Use MATLAB to calculate the polynomial
(a) $P_6 = (s+1)(s^2+2)(s+3)(2s^2+s+1)$
(b) $P_7 = (s^2+1)(s+2)(s+4)(s^2+2s+1)$

3-41. Use MATLAB to perform partial-fraction expansion to the following functions:

(a) $G_1(s) = \dfrac{(s+1)(s^2+2)(s+4)(s+10)}{s(s+2)(s^2+2s+5)(2s^2+s+4)}$

(b) $G_2(s) = \dfrac{s^3 + 12s^2 + 47s + 60}{4s^6 + 28s^5 + 83s^4 + 135s^3 + 126s^2 + 62s + 12}$

3-42. Use MATLAB to calculate unity feedback control for Problem 3-40.

3-43. Use MATLAB to calculate
(a) $G_3(s) = G_1(s) + G_2(s)$
(b) $G_4(s) = G_1(s) - G_2(s)$
(c) $G_5(s) = \dfrac{G_4(s)}{G_3(s)}$
(d) $G_6(s) = \dfrac{G_4(s)}{G_1(s) * G_2(s)}$

CHAPTER 4

Theoretical Foundation and Background Material: Modeling of Dynamic Systems

One of the most important tasks in the analysis and design of control systems is mathematical modeling of the systems. The two most common methods of modeling linear systems are the transfer function method and the state-variable method. The transfer function is valid only for linear time-invariant systems, whereas the state equations can be applied to linear as well as nonlinear systems.

Although the analysis and design of linear control systems have been well developed, their counterparts for nonlinear systems are usually quite complex. Therefore, the control-systems engineer often has the task of determining not only how to accurately describe a system mathematically but, more importantly, how to make proper assumptions and approximations, whenever necessary, so that the system may be realistically characterized by a linear mathematical model.

A control system may be composed of various components including mechanical, thermal, fluid, pneumatic, and electrical; sensors and actuators; and computers. In this chapter, we review basic properties of these systems, otherwise known as **dynamic systems**. Using the basic modeling principles such as Newton's second law of motion or Kirchoff's law, the models of these dynamic systems are represented by differential equations. It is not difficult to understand that the analytical and computer simulation of any system is only as good as the model used to describe it. It should also be emphasized that the modern control engineer should place special emphasis on the mathematical modeling of systems so that analysis and design problems can be conveniently solved by computers. In this textbook, we consider systems that are modeled by ordinary differential equations. The main objectives of this chapter are:

- To introduce modeling of mechanical systems.
- To introduce modeling of electrical systems.
- To introduce modeling thermal and fluid systems.
- To discuss sensors and actuators.
- To discuss linearization of nonlinear systems.
- To discuss analogies.

Furthermore, the main objectives of the following sections are:

- To demonstrate mathematical modeling of control systems and components.
- To demonstrate how computer solutions are used to obtain the response of these models.
- To provide examples that improve learning.

This chapter represents an introduction to the method of modeling. Because numerous types of control-system components are available, the coverage here is by no means exhaustive. This chapter further is intended to be self-sufficient and will not affect the general flow of the text. In Chapters 5 and 9, through various examples and case studies, the fundamentals discussed here are utilized to model more complex control systems and to establish their behavior.

▶ 4-1 INTRODUCTION TO MODELING OF MECHANICAL SYSTEMS

Mechanical systems may be modeled as systems of lumped masses (rigid bodies) or as distributed mass (continuous) systems. The latter are modeled by partial differential equations, whereas the former are represented by ordinary differential equations. Of course, in reality all systems are continuous, but, in most cases, it is easier and therefore preferred to approximate them with lumped mass models and ordinary differential equations.

Definition: *Mass is considered a property of an element that stores the kinetic energy of translational motion.* Mass is analogous to the inductance of electric networks, as shown in Section 4-10. If W denotes the weight of a body, then M is given by

$$M = \frac{W}{g} \tag{4-1}$$

where g is the acceleration of free fall of the body due to gravity ($g = 32.174$ ft/sec^2 in British units, and $g = 9.8066$ m/sec^2 in SI units).

The equations of a linear mechanical system are written by first constructing a model of the system containing interconnected linear elements and then by applying Newton's law of motion to the **free-body diagram** (FBD). For translational motion, the equation of motion is Eq. (4-2), and for rotational motion, Eq. (4-33) is used.

The motion of mechanical elements can be described in various dimensions as **translational**, **rotational**, or a combination of both. The equations governing the motion of mechanical systems are often directly or indirectly formulated from **Newton's law of motion**.

4-1-1 Translational Motion

The motion of translation is defined as a motion that takes place along a straight or curved path. The variables that are used to describe translational motion are **acceleration**, **velocity**, and **displacement**.

Newton's law of motion states that the algebraic sum of external forces acting on a rigid body in a given direction is equal to the product of the mass of the body and its acceleration in the same direction. The law can be expressed as

$$\sum_{external} \text{forces} = Ma \tag{4-2}$$

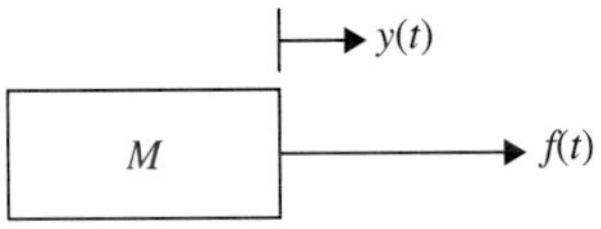

M → y(t), f(t)

Figure 4-1 Force-mass system.

where M denotes the mass, and a is the acceleration in the direction considered. Fig. 4-1 illustrates the situation where a force is acting on a body with mass M. The force equation is written as

$$f(t) = Ma(t) = M\frac{d^2y(t)}{dt^2} = M\frac{dv(t)}{dt} \qquad (4\text{-}3)$$

where $a(t)$ is the acceleration, $v(t)$ denotes linear velocity, and $y(t)$ is the displacement of mass M, respectively.

For linear translational motion, in addition to the mass, the following system elements are also involved.

- **Linear spring.** In practice, a linear spring may be a model of an actual spring or a compliance of a cable or a belt. In general, *a spring is considered to be an element that stores potential energy*.

$$f(t) = Ky(t) \qquad (4\text{-}4)$$

where K is the spring constant, or simply stiffness. Eq. (4-4) implies that the force acting on the spring is directly proportional to the displacement (deformation) of the spring. The model representing a linear spring element is shown in Fig. 4-2. If the spring is preloaded with a preload tension of T, then Eq. (4-4) should be modified to

$$f(t) - T = Ky(t) \qquad (4\text{-}5)$$

- **Friction for translation motion.** Whenever there is motion or tendency of motion between two physical elements, frictional forces exist. The frictional forces encountered in physical systems are usually of a nonlinear nature. The characteristics of the frictional forces between two contacting surfaces often depend on such factors as the composition of the surfaces, the pressure between the surfaces, and their relative velocity among others, so an exact mathematical description of the frictional force is difficult. Three different types of friction are commonly used in practical systems: **viscous friction**, **static friction**, and **Coulomb friction**. These are discussed separately in the following paragraphs.
  - **Viscous friction.** *Viscous friction represents a retarding force that is a linear relationship between the applied force and velocity*. The schematic diagram element for viscous friction is often represented by a dashpot, such as that shown in Fig. 4-3. The mathematical expression of viscous friction is

$$f(t) = B\frac{dy(t)}{dt} \qquad (4\text{-}6)$$

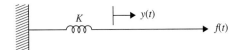

Figure 4-2 Force-spring system.

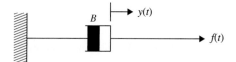

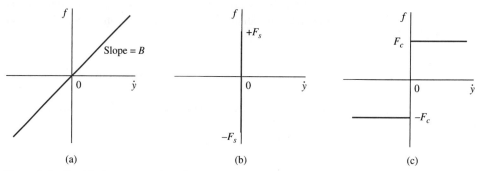

Figure 4-3 Dashpot for viscous friction.

Figure 4-4 Graphical representation of linear and nonlinear frictional forces. (a) Viscous friction. (b) Static friction. (c) Coulomb friction.

where B is the viscous frictional coefficient. Fig. 4-4(a) shows the functional relation between the viscous frictional force and velocity.

- **Static friction.** *Static friction represents a retarding force that tends to prevent motion from beginning.* The static frictional force can be represented by the expression

$$f(t) = \pm(F_s)|_{\dot{y}=0} \tag{4-7}$$

which is defined as a frictional force that exists only when the body is stationary but has a tendency of moving. The sign of the friction depends on the direction of motion or the initial direction of velocity. The force-to-velocity relation of static friction is illustrated in Fig. 4-4(b). Notice that, once motion begins, the static frictional force vanishes and other frictions take over.

- **Coulomb friction.** *Coulomb friction is a retarding force that has constant amplitude with respect to the change of velocity, but the sign of the frictional force changes with the reversal of the direction of velocity.* The mathematical relation for the Coulomb friction is given by

$$f(t) = F_c \frac{\left(\dfrac{dy(t)}{dt}\right)}{\left|\left(\dfrac{dy(t)}{dt}\right)\right|} \tag{4-8}$$

where F_c is the **Coulomb friction coefficient**. The functional description of the friction-to-velocity relation is shown in Fig. 4-4(c).

It should be pointed out that the three types of frictions cited here are merely practical models that have been devised to portray frictional phenomena found in physical systems. They are by no means exhaustive or guaranteed to be accurate. In many unusual situations, we have to use other frictional models to represent the actual phenomenon accurately. One such example is rolling dry friction [3, 4], which is used to model friction in high-precision

TABLE 4-1 Basic Translational Mechanical System Properties and Their Units

Parameter	Symbol Used	SI Units	Other Units	Conversion Factors
Mass	M	kilogram (kg)	slug ft/sec²	1 kg = 1000 g = 2.2046 lb(mass) = 35.274 oz(mass) = 0.06852 slug
Distance	y	meter (m)	ft in	1 m = 3.2808 ft = 39.37 in 1 in. = 25.4 mm 1 ft = 0.3048 m
Velocity	v	m/sec	ft/sec in/sec	
Acceleration	a	m/sec²	ft/sec² in/sec²	
Force	f	Newton (N)	pound (lb force) dyne	1 N = 0.2248 lb(force) = 3.5969 oz(force) 1 N = 1 kg–m/s² 1 dyn = 1 g–cm/s²
Spring Constant	K	N/m	lb/ft	
Viscous Friction Coefficient	B	N/m/sec	lb/ft/sec	

ball bearings used in spacecraft systems. It turns out that rolling dry friction has nonlinear hysteresis properties that make it impossible for use in linear system modeling.

Table 4-1 shows the basic translational mechanical system properties with their corresponding basic SI and other measurement units.

▶ **EXAMPLE 4-1-1** Consider the mass-spring-friction system shown in Fig. 4-5(a). The linear motion concerned is in the horizontal direction. The free-body diagram of the system is shown in Fig. 4-5(b). The force equation of the system is

$$f(t) - B\frac{dy(t)}{dt} - Ky(t) = M\frac{d^2y(t)}{dt^2} \qquad (4\text{-}9)$$

The last equation may be rearranged by equating the highest-order derivative term to the rest of the terms:

$$\frac{d^2y(t)}{dt^2} = -\frac{B}{M}\frac{dy(t)}{dt} - \frac{K}{M}y(t) + \frac{1}{M}f(t) \qquad (4\text{-}10)$$

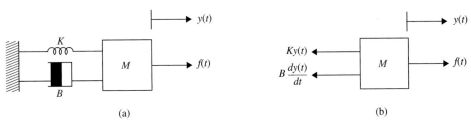

Figure 4-5 (a) Mass-spring-friction system. (b) Free-body diagram.

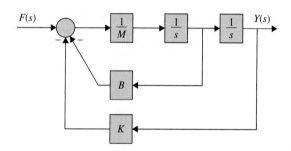

Figure 4-6 The mass-spring-friction system of Eq. (4-11) block diagram representation.

where $\dot{y}(t) = \left(\dfrac{dy(t)}{dt}\right)$ and $\ddot{y}(t) = \left(\dfrac{d^2y(t)}{dt^2}\right)$ represent velocity and acceleration, respectively. Or, alternatively, the former equation may be rewritten into an input–output form as

$$\ddot{y}(t) + \frac{B}{M}\dot{y}(t) + \frac{K}{M}y(t) = \frac{1}{M}f(t) \tag{4-11}$$

where $y(t)$ is the output and $\dfrac{f(t)}{M}$ is considered the input.

For zero initial conditions, the transfer function between $Y(s)$ and $F(s)$ is obtained by taking the Laplace transform on both sides of Eq. (4-11) with zero initial conditions:

$$\frac{Y(s)}{F(s)} = \frac{1}{Ms^2 + Bs + K} \tag{4-12}$$

The same result is obtained by applying the gain formula to the block diagram, which is shown in Fig. 4-6.

Eq. (4-10) may also be represented in the **space state form** using a state vector $\mathbf{x}(t)$ having n rows, where n is the number of state variables, so that

$$\dot{\mathbf{x}} = \mathbf{Ax} + \mathbf{Bu} \tag{4-13}$$

where

$$\mathbf{x}(t) = \begin{bmatrix} x_1(t) \\ x_2(t) \end{bmatrix} \tag{4-14}$$

$$y(t) = x_1(t) \quad \dot{y}(t) = x_2(t) \tag{4-15}$$

and

$$\mathbf{u}(t) = \frac{f(t)}{M} \tag{4-16}$$

So using Eqs. (4-13) through (4-16), Eq. (4-10) is rewritten in vectoral form as

$$\begin{bmatrix} \dot{x}_1 \\ \dot{x}_2 \end{bmatrix} = \begin{pmatrix} 0 & 1 \\ -\dfrac{K}{M} & -\dfrac{B}{M} \end{pmatrix} \begin{bmatrix} x_1 \\ x_2 \end{bmatrix} + \frac{f(t)}{M} \tag{4-17}$$

The state Eq. (4-17) may also be written as a set of first-order differential equations:

$$\begin{aligned}
\frac{dx_1(t)}{dt} &= x_2(t) \\
\frac{dx_2(t)}{dt} &= -\frac{K}{M}x_1(t) - \frac{B}{M}x_2(t) + \frac{1}{M}f(t) \\
y(t) &= x_1(t)
\end{aligned} \tag{4-18}$$

For zero initial conditions, the transfer function between $Y(s)$ and $F(s)$ is obtained by taking the Laplace transform on both sides of Eq. (4-18):

$$\begin{aligned}
sX_1(s) &= X_2(s) \\
sX_2(s) &= -\frac{B}{M}X_2(s) - \frac{K}{M}X_1(s) + \frac{1}{M}F(s) \\
Y(s) &= X_1(s) \\
\frac{Y(s)}{F(s)} &= \frac{1}{Ms^2 + Bs + K}
\end{aligned} \tag{4-19}$$

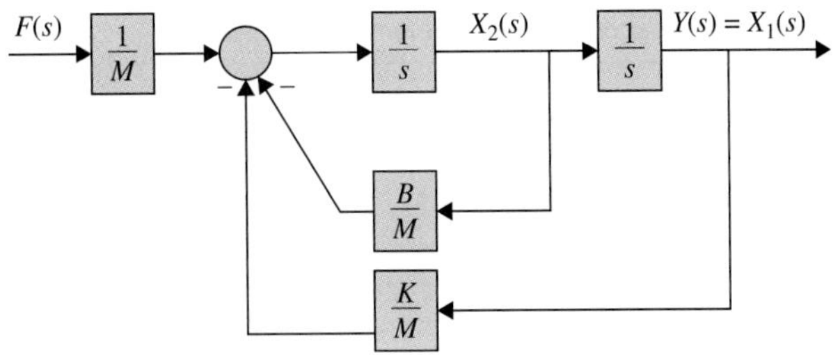

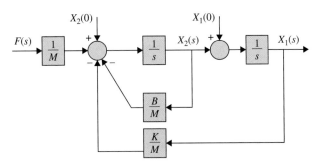

Figure 4-7 Block diagram representation of mass-spring-friction system of Eq. (4-19).

Figure 4-8 Block diagram representation of mass-spring-friction system of Eq. (4-20) with initial conditions $x_1(0)$ and $x_2(0)$.

The same result is obtained by applying the gain formula to the block diagram representation of the system in Eq. (4-19), which is shown in Fig. 4-7.

For nonzero initial conditions, Eq. (4-18) has a different Laplace transform representation that may be written as:

$$sX_1(s) - x_1(0) = X_2(s)$$
$$sX_2(s) - x_2(0) = -\frac{B}{M}X_2(s) - \frac{K}{M}X_1(s) + \frac{1}{M}F(s) \qquad (4\text{-}20)$$
$$Y(s) = X_1(s)$$

Upon simplifying Eq. (4-20) or by applying the gain formula to the block diagram representation of the system, shown in Fig. 4-8, the output becomes

$$Y(s) = \frac{1}{Ms^2 + Bs + K}F(s) + \frac{Ms}{Ms^2 + Bs + K}x_1(0) + \frac{M}{Ms^2 + Bs + K}x_2(0) \qquad (4\text{-}21)$$

Toolbox 4-1-1

Time domain step response for Eq. (4-12) is calculated using MATLAB for K = 1, M = 1, B = 1:

```
K=1; M=1; B=1;
t=0:0.02:30;
num = [1];
den = [M B K];
G = tf(num,den);
y1 = STEP (G,t);
plot(t, y1);
xlabel('Time (Second)');ylabel('Step Response')
title('Response of the system to step input')
```

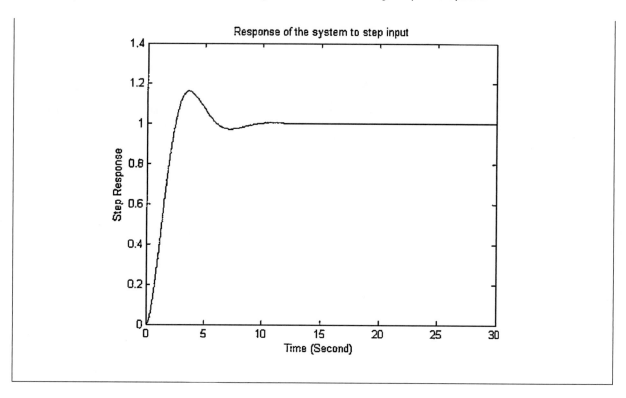

▶ **EXAMPLE 4-1-2** As another example of writing the dynamic equations of a mechanical system with translational motion, consider the system shown in Fig. 4-9(a). Because the spring is deformed when it is subject to a force $f(t)$, two displacements, y_1 and y_2, must be assigned to the end points of the spring. The free-body diagrams of the system are shown in Fig. 4-9(b). The force equations are

$$f(t) = K[y_1(t) - y_2(t)] \tag{4-22}$$

$$-K[y_2(t) - y_1(t)] - B\frac{dy_2(t)}{dt} = M\frac{d^2y_2(t)}{dt^2} \tag{4-23}$$

These equations are rearranged in input–output form as

$$\frac{d^2y_2(t)}{dt^2} + \frac{B}{M}\frac{dy_2(t)}{dt} + \frac{K}{M}y_2(t) = \frac{K}{M}y_1(t) \tag{4-24}$$

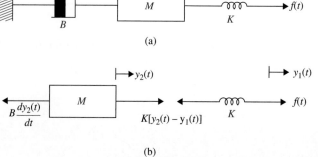

Figure 4-9 Mechanical system for Example 4-1-2. (a) Mass-spring-damper system. (b) Free-body diagram.

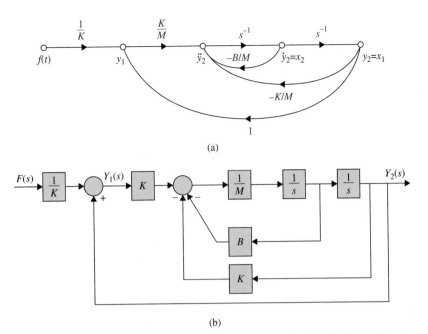

Figure 4-10 Mass-spring-friction system of Eq. (4-25) using Eq. (4-22). (a) The signal-flow graph representation. (b) Block diagram representation.

For zero initial conditions, the transfer function between $Y_1(s)$ and $Y_2(s)$ is obtained by taking the Laplace transform on both sides of Eq. (4-24):

$$\frac{Y_2(s)}{Y_1(s)} = \frac{K}{Ms^2 + Bs + K} \qquad (4\text{-}25)$$

The same result is obtained by applying the gain formula to the block diagram representation of the system, which is shown in Fig. 4-10. Note that in Fig. 4-10, Eq. (4-22) was also used.

For state representation, these equations may be rearranged as

$$y_1(t) = y_2(t) + \frac{1}{K} f(t)$$
$$\frac{d^2 y_2(t)}{dt^2} = -\frac{B}{M} \frac{dy_2(t)}{dt} + \frac{K}{M}[y_1(t) - y_2(t)] \qquad (4\text{-}26)$$

For zero initial conditions, the transfer function of Eq. (4-26) is the same as Eq. (4-25). By using the last two equations, the state variables are defined as $x_1(t) = y_2(t)$ and $x_2(t) = dy_2(t)/dt$. The state equations are therefore written as

$$\frac{dx_1(t)}{dt} = x_2(t)$$
$$\frac{dx_2(t)}{dt} = -\frac{B}{M} x_2(t) + \frac{1}{M} f(t) \qquad (4\text{-}27)$$

The same result is obtained after taking the Laplace transform of Eq. (4-27) and applying the gain formula to the block diagram representation of the system, which is shown in Fig. 4-11. Note that in Fig. 4-11, $F(s)$, $Y_1(s)$, $X_1(s)$, $Y_2(s)$, and $X_2(s)$ are Laplace transforms of $f(t)$, $y_1(t)$, $x_1(t)$, $y_2(t)$, and $x_2(t)$, respectively.

156 ▶ Chapter 4. Theoretical Foundation and Background Material: Modeling of Dynamic Systems

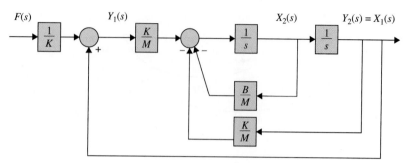

Figure 4-11 Block diagram representation of mass-spring-friction system of Eq. (4-27).

◀

▶ **EXAMPLE 4-1-3** Consider the two degrees of freedom (2-DOF) spring-mass system, with two masses m_1 and m_2, two springs k_1 and k_2, and two forces f_1 and f_2, as shown in Fig. 4-12. Find the equations of motion.

SOLUTION To avoid any confusion, we first draw the free-body diagram (FBD) of the system by assuming the masses are displaced in the positive direction, so that $y_1 > y_2 > 0$ (i.e., springs are both in tension). The FBD of the system is shown in Fig. 4-13. Applying Newton's second law to the masses M_1 and M_2, we have

$$f_1(t) - K_1 y_1 + K_2(y_1 - y_2) = M_1 \ddot{y}_1$$
$$f_2(t) - K_2(y_1 - y_2) = M_2 \ddot{y}_2 \tag{4-28}$$

Rearranging the equations into the standard input–output form, we have

$$M_1 \ddot{y}_1 + (K_1 + K_2) y_1 - K_2 y_2 = f_1(t)$$
$$M_2 \ddot{y}_1 - K_2 y_1 + K_2 y_2 = f_2(t) \tag{4-29}$$

Alternatively, Eq. (4-29) may be represented in the standard second-order matrix form, as

$$\begin{bmatrix} M_1 & 0 \\ 0 & M_2 \end{bmatrix} \begin{bmatrix} \ddot{y}_1 \\ \ddot{y}_2 \end{bmatrix} + \begin{bmatrix} K_1 + K_2 & -K_2 \\ -K_2 & K_2 \end{bmatrix} \begin{bmatrix} y_1 \\ y_2 \end{bmatrix} = \begin{bmatrix} f_1 \\ f_2 \end{bmatrix} \tag{4-30}$$

In state space form, assuming the following state vector $\mathbf{x}(t)$, the inputs $u_1(t)$ and $u_2(t)$, and the output $y(t)$, we get

$$\mathbf{x}(t) = \begin{bmatrix} x_1(t) \\ x_2(t) \\ x_3(t) \\ x_4(t) \end{bmatrix} = \begin{bmatrix} y_1(t) \\ y_2(t) \\ \dot{y}_1(t) \\ \dot{y}_2(t) \end{bmatrix}, \; u_1 = f_1(t), \; u_2 = f_2(t), \; y(t) = x_1(t) \tag{4-31}$$

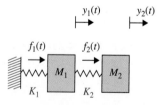

Figure 4-12 A 2-DOF spring-mass system.

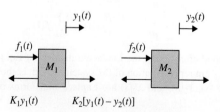

Figure 4-13 FBD of the 2-DOF spring-mass system.

Then, using $\dot{x}_3 = \ddot{y}_1$ and $\dot{x}_4 = \ddot{y}_2$, we get the state-space representation as

$$\begin{bmatrix} \dot{x}_1 \\ \dot{x}_2 \\ \dot{x}_3 \\ \dot{x}_4 \end{bmatrix} = \begin{bmatrix} 0 & 0 & 1 & 0 \\ 0 & 0 & 0 & 1 \\ -K_1/M_1 & K_1/M_1 & 0 & 0 \\ K_2/M_1 & -K_2/M_1 & 0 & 0 \end{bmatrix} \begin{bmatrix} x_1 \\ x_2 \\ x_3 \\ x_4 \end{bmatrix} + \begin{bmatrix} 0 \\ 0 \\ 1/M_1 \\ 0 \end{bmatrix} u_1 + \begin{bmatrix} 0 \\ 0 \\ 0 \\ 1/M_2 \end{bmatrix} u_2 \quad \text{(state equation)}$$

$$y = \begin{bmatrix} 1 & 0 & 0 & 0 \end{bmatrix} \begin{bmatrix} x_1 \\ x_2 \\ x_3 \\ x_4 \end{bmatrix} + 0 \cdot u_1 + 0 \cdot u_2 \quad \text{(output equation)}$$

(4-32)

where the state equation is a set of four first-order differential equations.

4-1-2 Rotational Motion

The rotational motion of a body can be defined as motion about a fixed axis. The extension of Newton's law of motion for rotational motion states that the *algebraic sum of moments or torque about a fixed axis is equal to the product of the inertia and the angular acceleration about the axis*. Or

$$\sum \text{torques} = J\alpha \quad (4\text{-}33)$$

where J denotes the inertia and α is the angular acceleration. The other variables generally used to describe the motion of rotation are **torque** T, **angular velocity** ω, and **angular displacement** θ. The elements involved with the rotational motion are as follows:

- **Inertia.** *Inertia, J, is considered a property of an element that stores the kinetic energy of rotational motion.* The inertia of a given element depends on the geometric composition about the axis of rotation and its density. For instance, the inertia of a circular disk or shaft, of radius r and mass M, about its geometric axis is given by

$$J = \frac{1}{2} M r^2 \quad (4\text{-}34)$$

When a torque is applied to a body with inertia J, as shown in Fig. 4-14, the torque equation is written

$$T(t) = J\alpha(t) = J\frac{d\omega(t)}{dt} = J\frac{d^2\theta(t)}{dt^2} \quad (4\text{-}35)$$

where $\theta(t)$ is the angular displacement; $\omega(t)$, the angular velocity; and $\alpha(t)$, the angular acceleration.

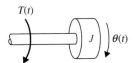

Figure 4-14 Torque-inertia system.

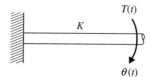

Figure 4-15 Torque torsional spring system.

- **Torsional spring.** As with the linear spring for translational motion, a **torsional spring constant** K, in torque-per-unit angular displacement, can be devised to represent the compliance of a rod or a shaft when it is subject to an applied torque. Fig. 4-15 illustrates a simple torque-spring system that can be represented by the equation

$$T(t) = K\theta(t) \tag{4-36}$$

If the torsional spring is preloaded by a preload torque of TP, Eq. (4-36) is modified to

$$T(t) - TP = K\theta(t) \tag{4-37}$$

- **Friction for rotational motion.** The three types of friction described for translational motion can be carried over to the motion of rotation. Therefore, Eqs. (4-6), (4-7), and (4-8) can be replaced, respectively, by their counterparts:
  - **Viscous friction.**

$$T(t) = B \frac{d\theta(t)}{dt} \tag{4-38}$$

  - **Static friction.**

$$T(t) = \pm (F_s)|_{\dot{\theta}=0} \tag{4-39}$$

  - **Coulomb friction.**

$$T(t) = F_c \frac{\frac{d\theta(t)}{dt}}{\left|\frac{d\theta(t)}{dt}\right|} \tag{4-40}$$

Table 4-2 shows the SI and other measurement units for inertia and the variables in rotational mechanical systems.

▶ **EXAMPLE 4-1-4** The rotational system shown in Fig. 4-16(a) consists of a disk mounted on a shaft that is fixed at one end. The moment of inertia of the disk about the axis of rotation is J. The edge of the disk is riding on the surface, and the viscous friction coefficient between the two surfaces is B. The inertia of the shaft is negligible, but the torsional spring constant is K.

Assume that a torque is applied to the disk, as shown; then the torque or moment equation about the axis of the shaft is written from the free-body diagram of Fig. 4-16(b):

$$T(t) = J \frac{d^2\theta(t)}{dt^2} + B \frac{d\theta(t)}{dt} + K\theta(t) \tag{4-41}$$

Notice that this system is analogous to the translational system in Fig. 4-5. The state equations may be written by defining the state variables as $x_1(t) = \theta(t)$ and $x_2(t) = dx_1(t)/dt$.

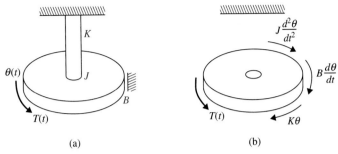

Figure 4-16 Rotational system for Example 4-1-4.

TABLE 4-2 Basic Rotational Mechanical System Properties and Their Units

Parameter	Symbol Used	SI Units	Other Units	Conversion Factors
Inertia	J	kg-m^2	slug-ft^2 lb-ft-sec^2 oz-in.-sec^2	1 g-cm = 1.417 × 10^{-5} oz-in.-sec^2 1 lb-ft-sec^2 = 192 oz-in.-sec^2 = 32.2 lb-ft^2 1 oz-in.-sec^2 = 386 oz-in^2 1 g-cm-sec^2 = 980 g-cm^2
Angular Displacement	T	Radian	Radian	1 rad = $\frac{180}{\pi}$ = 57.3 deg
Angular Velocity	O	radian/sec	radian/sec	1 rpm = $\frac{2\pi}{60}$ = 0.1047 rad/sec 1 rpm = 6 deg/sec
Angular Acceleration	A	radian/sec^2	radian/sec^2	
Torque	T	(N-m) dyne-cm	lb-ft oz-in.	1 g-cm = 0.0139 oz-in. 1 lb-ft = 192 oz-in. 1 oz-in. = 0.00521 lb-ft
Spring Constant	K	N-m/rad	ft-lb/rad	
Viscous Friction Coefficient	B	N-m/rad/sec	ft-lb/rad/sec	
Energy	Q	J (joules)	Btu Calorie	1 J = 1 N-m 1 Btu = 1055 J 1 cal = 4.184 J

▶ **EXAMPLE 4-1-5** Fig. 4-17(a) shows the diagram of a motor coupled to an inertial load through a shaft with a spring constant K. A non-rigid coupling between two mechanical components in a control system often causes torsional resonances that can be transmitted to all parts of the system. The system variables and parameters are defined as follows:

$T_m(t)$ = motor torque

B_m = motor viscous-friction coefficient

K = spring constant of the shaft

$\theta_m(t)$ = motor displacement

$\omega_m(t)$ = motor velocity

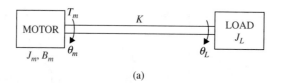

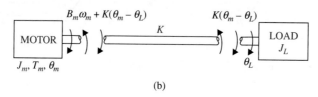

Figure 4-17 (a) Motor–load system. (b) Free-body diagram.

J_m = motor inertia
$\theta_L(t)$ = load displacement
$\omega_L(t)$ = load velocity
J_L = load inertia

The free-body diagrams of the system are shown in Fig. 4-17(b). The torque equations of the system are

$$\frac{d^2\theta_m(t)}{dt^2} = -\frac{B_m}{J_m}\frac{d\theta_m(t)}{dt} - \frac{K}{J_m}[\theta_m(t) - \theta_L(t)] + \frac{1}{J_m}T_m(t) \quad (4\text{-}42)$$

$$K[\theta_m(t) - \theta_L(t)] = J_L\frac{d^2\theta_L(t)}{dt^2} \quad (4\text{-}43)$$

In this case, the system contains three energy-storage elements in J_m, J_L, and K. Thus, there should be three state variables. Care should be taken in constructing the state diagram and assigning the state variables so that a minimum number of the latter are incorporated. Eqs. (4-42) and (4-43) are rearranged as

$$\frac{d^2\theta_m(t)}{dt^2} = -\frac{B_m}{J_m}\frac{d\theta_m(t)}{dt} - \frac{K}{J_m}[\theta_m(t) - \theta_L(t)] + \frac{1}{J_m}T_m(t) \quad (4\text{-}44)$$

$$\frac{d^2\theta_L(t)}{dt^2} = \frac{K}{J_L}[\theta_m(t) - \theta_L(t)] \quad (4\text{-}45)$$

The state variables in this case are defined as $x_1(t) = \theta_m(t) - \theta_L(t)$, $x_2(t) = d\theta_L(t)/dt$, and $x_3(t) = d\theta_m(t)/dt$. The state equations are

$$\begin{aligned}\frac{dx_1(t)}{dt} &= x_3(t) - x_2(t) \\ \frac{dx_2(t)}{dt} &= \frac{K}{J_L}x_1(t) \\ \frac{dx_3(t)}{dt} &= -\frac{K}{J_m}x_1(t) - \frac{B_m}{J_m}x_3(t) + \frac{1}{J_m}T_m(t)\end{aligned} \quad (4\text{-}46)$$

The SFG representation is shown in Fig. 4-18.

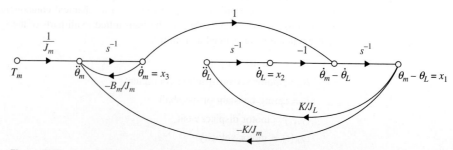

Figure 4-18 Rotational system of Eq. (4-46) signal-flow graph representation.

4-1-3 Conversion between Translational and Rotational Motions

In motion-control systems, it is often necessary to convert rotational motion into translational motion. For instance, a load may be controlled to move along a straight line through a rotary motor-and-lead screw assembly, such as that shown in Fig. 4-19. Fig. 4-20 shows a similar situation in which a rack-and-pinion assembly is used as a mechanical linkage. Another familiar system in motion control is the control of a mass through a pulley by a rotary motor, as shown in Fig. 4-21. The systems shown in Figs. 4-19, 4-20, and 4-21 can all be represented by a simple system with an equivalent inertia connected directly to the drive motor. For instance, the mass in Fig. 4-21 can be regarded as a point mass that moves about the pulley, which has a radius r. By disregarding the inertia of the pulley, the equivalent inertia that the motor sees is

$$J = Mr^2 = \frac{W}{g}r^2 \tag{4-47}$$

If the radius of the pinion in Fig. 4-20 is r, the equivalent inertia that the motor sees is also given by Eq. (4-47).

Now consider the system of Fig. 4-19. The lead of the screw, L, is defined as the linear distance that the mass travels per revolution of the screw. In principle, the two systems in Fig. 4-20 and Fig. 4-21 are equivalent. In Fig. 4-20, the distance traveled by the mass per revolution of the pinion is $2\pi r$. By using Eq. (4-47) as the equivalent inertia for the system of Fig. 4-19, we have

$$J = \frac{W}{g}\left(\frac{L}{2\pi}\right)^2 \tag{4-48}$$

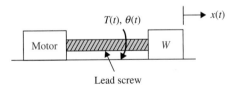

Figure 4-19 Rotary-to-linear motion control system (lead screw).

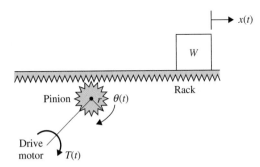

Figure 4-20 Rotary-to-linear motion control system (rack and pinion).

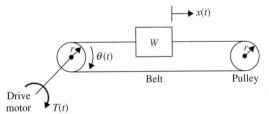

Figure 4-21 Rotary-to-linear motion control system (belt and pulley).

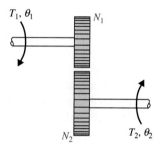

Figure 4-22 Gear train.

4-1-4 Gear Trains

A gear train, lever, or timing belt over a pulley is a mechanical device that transmits energy from one part of the system to another in such a way that force, torque, speed, and displacement may be altered. These devices can also be regarded as matching devices used to attain maximum power transfer. Two gears are shown coupled together in Fig. 4-22. The inertia and friction of the gears are neglected in the ideal case considered.

The relationships between the torques T_1 and T_2, angular displacement θ_1 and θ_2, and the teeth numbers N_1 and N_2 of the gear train are derived from the following facts:

1. The number of teeth on the surface of the gears is proportional to the radii r_1 and r_2 of the gears; that is,

$$r_1 N_2 = r_2 N_1 \tag{4-49}$$

2. The distance traveled along the surface of each gear is the same. Thus,

$$\theta_1 r_1 = \theta_2 r_2 \tag{4-50}$$

3. The work done by one gear is equal to that of the other since there are assumed to be no losses. Thus,

$$T_1 \theta_1 = T_2 \theta_2 \tag{4-51}$$

If the angular velocities of the two gears ω_1 and ω_2 are brought into the picture, Eqs. (4-49) through (4-51) lead to

$$\frac{T_1}{T_2} = \frac{\theta_2}{\theta_1} = \frac{N_1}{N_2} = \frac{\omega_2}{\omega_1} = \frac{r_1}{r_2} \tag{4-52}$$

In practice, gears do have inertia and friction between the coupled gear teeth that often cannot be neglected. An equivalent representation of a gear train with viscous friction, Coulomb friction, and inertia considered as lumped parameters is shown in Fig. 4-23, where T denotes the applied torque, T_1 and T_2 are the transmitted torque, F_{c1} and F_{c2} are the Coulomb friction coefficients, and B_1 and B_2 are the viscous friction coefficients. The torque equation for gear 2 is

$$T_2(t) = J_2 \frac{d^2\theta_2(t)}{dt^2} + B_2 \frac{d\theta_2(t)}{dt} + F_{c2} \frac{\omega_2}{|\omega_2|} \tag{4-53}$$

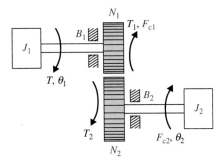

Figure 4-23 Gear train with friction and inertia.

The torque equation on the side of gear 1 is

$$T(t) = J_1 \frac{d^2\theta_1(t)}{dt^2} + B_1 \frac{d\theta_1(t)}{dt} + F_{c1} \frac{\omega_1}{|\omega_1|} + T_1(t) \quad (4\text{-}54)$$

Using Eq. (4-52), Eq. (4-53) is converted to

$$T_1(t) = \frac{N_1}{N_2} T_2(t) = \left(\frac{N_1}{N_2}\right)^2 J_2 \frac{d^2\theta_1(t)}{dt^2} + \left(\frac{N_1}{N_2}\right)^2 B_2 \frac{d\theta_1(t)}{dt} + \frac{N_1}{N_2} F_{c2} \frac{\omega_2}{|\omega_2|} \quad (4\text{-}55)$$

Eq. (4-55) indicates that it is possible to reflect inertia, friction, compliance, torque, speed, and displacement from one side of a gear train to the other. The following quantities are obtained when reflecting from gear 2 to gear 1:

$$\begin{aligned}
&\text{Inertia}: \left(\frac{N_1}{N_2}\right)^2 J_2 \\
&\text{Viscous-friction coefficient}: \left(\frac{N_1}{N_2}\right)^2 B_2 \\
&\text{Torque}: \frac{N_1}{N_2} T_2 \\
&\text{Angular displacement}: \frac{N_1}{N_2} \theta_2 \\
&\text{Angular velocity}: \frac{N_1}{N_2} \omega_2 \\
&\text{Coulomb friction torque}: \frac{N_1}{N_2} F_{c2} \frac{\omega_2}{|\omega_2|}
\end{aligned} \quad (4\text{-}56)$$

Similarly, gear parameters and variables can be reflected from gear 1 to gear 2 by simply interchanging the subscripts in the preceding expressions. If a torsional spring effect is present, the spring constant is also multiplied by $(N_1/N_2)^2$ in reflecting from gear 2 to gear 1. Now substituting Eq. (4-55) into Eq. (4-54), we get

$$T(t) = J_{1e} \frac{d^2\theta_1(t)}{dt^2} + B_{1e} \frac{d\theta_1(t)}{dt} + T_F \quad (4\text{-}57)$$

where

$$J_{1e} = J_1 + \left(\frac{N_1}{N_2}\right)^2 J_2 \quad (4\text{-}58)$$

$$B_{1e} = B_1 + \left(\frac{N_1}{N_2}\right)^2 B_2 \qquad (4\text{-}59)$$

$$T_F = F_{c1}\frac{\omega_1}{|\omega_1|} + \frac{N_1}{N_2}F_{c2}\frac{\omega_2}{|\omega_2|} \qquad (4\text{-}60)$$

▶ **EXAMPLE 4-1-6** Given a load that has inertia of 0.05 oz-in.-sec^2 and a Coulomb friction torque of 2 oz-in., find the inertia and frictional torque reflected through a 1:5 gear train ($N_1/N_2 = 1/5$, with N_2 on the load side). The reflected inertia on the side of N_1 is $(1/5)^2 \times 0.05 = 0.002$ oz-in.-sec^2. The reflected Coulomb friction is $(1/5) \times 2 = 0.4$ oz-in. ◀

4-1-5 Backlash and Dead Zone (Nonlinear Characteristics)

Backlash and dead zone are commonly found in gear trains and similar mechanical linkages where the coupling is not perfect. In a majority of situations, backlash may give rise to undesirable inaccuracy, oscillations, and instability in control systems. In addition, it has a tendency to wear out the mechanical elements. Regardless of the actual mechanical elements, a physical model of backlash or dead zone between an input and an output member is shown in Fig. 4-24. The model can be used for a rotational system as well as for a translational system. The amount of backlash is $b/2$ on either side of the reference position.

In general, the dynamics of the mechanical linkage with backlash depend on the relative inertia-to-friction ratio of the output member. If the inertia of the output member is very small compared with that of the input member, the motion is controlled predominantly by friction. This means that the output member will not coast whenever there is no contact between the two members. When the output is driven by the input, the two members will travel together until the input member reverses its direction; then the output member will be at a standstill until the backlash is taken up on the other side, at which time it is assumed that the output member instantaneously takes on the velocity of the input member. The transfer characteristic between the input and output displacements of a system with backlash with negligible output inertia is shown in Fig. 4-25.

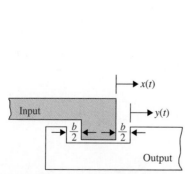

Figure 4-24 Physical model of backlash between two mechanical elements.

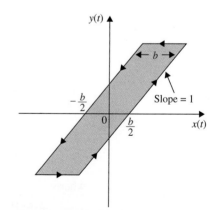

Figure 4-25 Input–output characteristic of backlash.

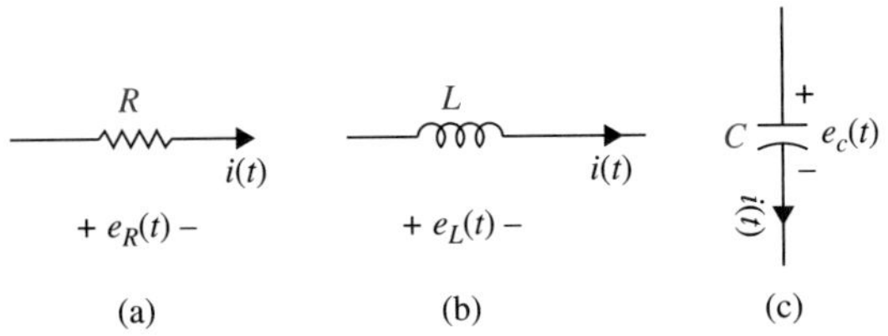

Figure 4-26 Basic passive electrical elements. (a) A resistor. (b) An inductor. (c) A capacitor.

4-2 INTRODUCTION TO MODELING OF SIMPLE ELECTRICAL SYSTEMS

First we address modeling of electrical networks with simple passive elements such as resistors, inductors, and capacitors. Later, in the next section, we address operational amplifiers, which are active electrical elements.

4-2-1 Modeling of Passive Electrical Elements

Consider Fig. 4-26, which shows the basic passive electrical elements: resistors, inductors, and capacitors.

Resistors: Ohm's law states that the voltage drop, $e_R(t)$, across a resistor R is proportional to the current $i(t)$ going through the resistor. Or

$$e_R(t) = i(t)R \tag{4-61}$$

Inductors: The voltage drop, $e_L(t)$, across an inductor L is proportional to the time rate of change of current $i(t)$ going through the inductor. Thus,

$$e_L(t) = L\frac{di(t)}{dt} \tag{4-62}$$

Capacitor: The voltage drop, $e_C(t)$, across a capacitor C is proportional to the integral current $i(t)$ going through the capacitor with respect to time. Therefore,

$$e_c(t) = \int \frac{i(t)}{C} dt \tag{4-63}$$

4-2-2 Modeling of Electrical Networks

The classical way of writing equations of electric networks is based on the loop method or the node method, both of which are formulated from the two laws of Kirchhoff, which state:

Current Law or Loop Method: The algebraic summation of all currents entering a node is zero.
Voltage Law or Node Method: The algebraic sum of all voltage drops around a complete closed loop is zero.

▶ **EXAMPLE 4-2-1** Let us consider the **RLC** network shown in Fig. 4-27. Using the voltage law

$$e(t) = e_R + e_L + e_c \tag{4-64}$$

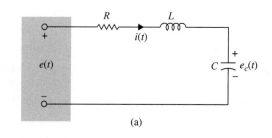
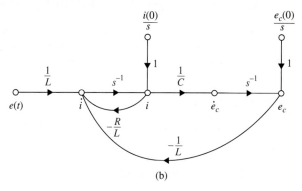
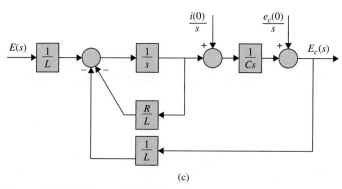

Figure 4-27 RLC network. (a) Electrical schematics. (b) Signal-flow graph representation. (c) Block diagram representation.

where

$e_R =$ Voltage across the resistor R

$e_L =$ Voltage across the inductor L

$e_c =$ Voltage across the capacitor C

Or

$$e(t) = +e_c(t) + Ri(t) + L\frac{di(t)}{dt} \tag{4-65}$$

Using current in C:

$$C\frac{de_c(t)}{dt} = i(t) \tag{4-66}$$

and taking a derivative of Eq. (4-54) with respect to time, we get the equation of the RLC network as

$$L\frac{d^2i(t)}{dt^2} + R\frac{di(t)}{dt} + \frac{i(t)}{C} = \frac{de(t)}{dt} \tag{4-67}$$

A practical approach is to assign the current in the inductor L, $i(t)$, and the voltage across the capacitor C, $e_c(t)$, as the state variables. The reason for this choice is because the state variables are directly related to the energy-storage element of a system. The inductor stores kinetic energy, and the capacitor stores electric potential energy. By assigning $i(t)$ and $e_c(t)$ as state variables, we have a complete description of the past history (via the initial states) and the present and future states of the network.

The state equations for the network in Fig. 4-27 are written by first equating the current in C and the voltage across L in terms of the state variables and the applied voltage $e(t)$. In vector-matrix form, the equations of the system are expressed as

$$\begin{bmatrix} \dfrac{de_c(t)}{dt} \\ \dfrac{di(t)}{dt} \end{bmatrix} = \begin{bmatrix} 0 & \dfrac{1}{C} \\ -\dfrac{1}{L} & -\dfrac{R}{L} \end{bmatrix} \begin{bmatrix} e_c(t) \\ i(t) \end{bmatrix} + \begin{bmatrix} 0 \\ \dfrac{1}{L} \end{bmatrix} e(t) \qquad (4\text{-}68)$$

This format is also known as the state form if we set

$$\begin{bmatrix} x_1(t) \\ x_2(t) \end{bmatrix} = \begin{bmatrix} e_c(t) \\ i(t) \end{bmatrix} \qquad (4\text{-}69)$$

Or

$$\begin{bmatrix} \dot{x}_1 \\ \dot{x}_2 \end{bmatrix} = \begin{bmatrix} 0 & \dfrac{1}{C} \\ -\dfrac{1}{L} & -\dfrac{R}{L} \end{bmatrix} \begin{bmatrix} x_1 \\ x_2 \end{bmatrix} + \begin{bmatrix} 0 \\ \dfrac{1}{L} \end{bmatrix} e(t) \qquad (4\text{-}70)$$

The transfer functions of the system are obtained by applying the gain formula to the SFG or block diagram of the system in Fig. 4-27 when all the initial states are set to zero.

$$\frac{E_c(s)}{E(s)} = \frac{(1/LC)s^{-2}}{1 + (R/L)s^{-1} + (1/LC)s^{-2}} = \frac{1}{1 + RCs + LCs^2} \qquad (4\text{-}71)$$

$$\frac{I(s)}{E(s)} = \frac{(1/L)s^{-1}}{1 + (R/L)s^{-1} + (1/LC)s^{-2}} = \frac{Cs}{1 + RCs + LCs^2} \qquad (4\text{-}72)$$

Toolbox 4-2-1

Time domain step responses for Eqs. (4-71) and (4-72) are shown using MATLAB for $R = 1$, $L = 1$, $C = 1$:

```
R=1; L=1; C=1;
t=0:0.02:30;
num1 = [1];
den1 = [L*C R*C 1];
num2 = [C 0];
den2 = [L*C R*C 1];
G1 = tf(num1,den1);
G2 = tf(num2,den2);
y1 = step (G1,t);
y2 = step (G2,t);
plot(t,y1, 'r');
hold on
plot(t,y2, 'g');
xlabel('Time')
ylabel('Gain')
```

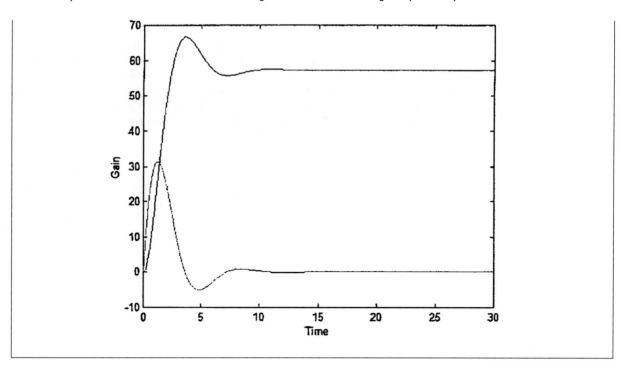

▶ **EXAMPLE 4-2-2** As another example of writing the state equations of an electric network, consider the network shown in Fig. 4-28(a). According to the foregoing discussion, the voltage across the capacitor, $e_c(t)$, and the currents of the inductors, $i_1(t)$ and $i_2(t)$, are assigned as state variables, as shown in Fig. 4-28(a). The state equations of the network are obtained by writing the voltages across the

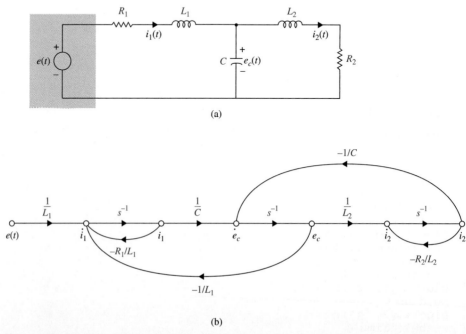

Figure 4-28 Network of Example 4-2-2. (a) Electrical schematic. (b) SFG representation.

inductors and the currents in the capacitor in terms of the three state variables. The state equations are

$$L_1 \frac{di_1(t)}{dt} = -R_1 i_1(t) - e_c(t) + e(t) \quad (4\text{-}73)$$

$$L_2 \frac{di_2(t)}{dt} = -R_2 i_2(t) + e_c(t) \quad (4\text{-}74)$$

$$C \frac{de_c(t)}{dt} = i_1(t) - i_2(t) \quad (4\text{-}75)$$

In vector-matrix form, the state equations are written as

$$\begin{bmatrix} \dot{x}_1 \\ \dot{x}_2 \\ \dot{x}_3 \end{bmatrix} = \begin{bmatrix} -\frac{R_1}{L_1} & 0 & -\frac{1}{L_1} \\ 0 & -\frac{R_2}{L_2} & \frac{1}{L_2} \\ \frac{1}{C} & -\frac{1}{C} & 0 \end{bmatrix} \begin{bmatrix} x_1 \\ x_2 \\ x_3 \end{bmatrix} + \begin{bmatrix} \frac{1}{L_1} \\ 0 \\ 0 \end{bmatrix} e(t) \quad (4\text{-}76)$$

where

$$\begin{bmatrix} x_1 \\ x_2 \\ x_3 \end{bmatrix} = \begin{bmatrix} i_1(t) \\ i_2(t) \\ e_c(t) \end{bmatrix} \quad (4\text{-}77)$$

The signal-flow diagram of the network, without the initial states, is shown in Fig. 4-28(b). The transfer functions between $I_1(s)$ and $E(s)$, $I_2(s)$ and $E(s)$, and $E_c(s)$ and $E(s)$, respectively, are written from the state diagram

$$\frac{I_1(s)}{E(s)} = \frac{L_2 C s^2 + R_2 C s + 1}{\Delta} \quad (4\text{-}78)$$

$$\frac{I_2(s)}{E(s)} = \frac{1}{\Delta} \quad (4\text{-}79)$$

$$\frac{E_c(s)}{E(s)} = \frac{L_2 s + R_2}{\Delta} \quad (4\text{-}80)$$

where

$$\Delta = L_1 L_2 C s^3 + (R_1 L_2 + R_2 L_1) C s^2 + (L_1 + L_2 + R_1 R_2 C) s + R_1 + R_2 \quad (4\text{-}81)$$

◀

Toolbox 4-2-2

Time domain step response for the gain formula explained by step responses of Eqs. 4-78–4-80 are shown using MATLAB as illustrated below (for R1 = 1, R2 = 1, L1 = 1, L2 = 1, C = 1):

```
R1=1; R2=1; L1=1; L2=1; C=1;
t=0:0.02:30;
num1 = [L2*C R2*C 1];
num2 = [1];
num3 = [L2 R2];
den = [L1*L2*C R1*L2*C+R2*L1*C L1+L2+R1*R2*C R1+R2];
G1 = tf(num1,den);
G2 = tf(num2,den);
G3 = tf(num3,den);
y1 = step (G1,t);
y2 = step (G2,t);
y3 = step (G3,t);
```

```
plot(t,y1, 'r');
hold on
plot(t,y2, 'g');
hold on
plot(t,y3, 'b');
xlabel('Time')
ylabel('Gain')
```

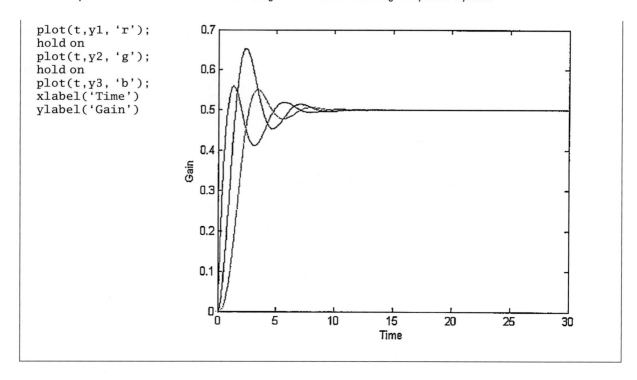

▶ **EXAMPLE 4-2-3** Consider the RC circuit shown in Fig. 4-29. Find the differential equation of the system. Using the voltage law

$$e_{in}(t) = e_R(t) + e_C(t) \tag{4-82}$$

where

$$e_R = iR \tag{4-83}$$

and the voltage across the capacitor $v_c(t)$ is

$$e_C(t) = \frac{1}{C}\int i\,dt \tag{4-84}$$

But from Fig. 4-29

$$e_o(t) = \frac{1}{C}\int i\,dt = e_C(t) \tag{4-85}$$

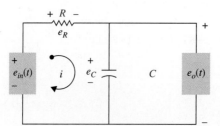

Figure 4-29 Simple electrical RC circuit.

If we differentiate Eq. (4-85) with respect to time, we get

$$\frac{i}{C} = \frac{de_o(t)}{dt} \qquad (4\text{-}86)$$

or

$$C\dot{e}_o(t) = i \qquad (4\text{-}87)$$

This implies that Eq. (4-85) can be written in an input–output form

$$e_{in}(t) = RC\dot{e}_o(t) + e_o(t) \qquad (4\text{-}88)$$

In Laplace domain, we get the system transfer function as

$$\frac{E_0(s)}{E_{in}(s)} = \frac{1}{RCs + 1} \qquad (4\text{-}89)$$

where the $\tau = RC$ is also known as the **time constant** of the system. The significance of this term is discussed earlier in Chapter 2, and the initial conditions are assumed to be $e_{in}(t=0) = \dot{e}_o(t=0) = 0$. ◀

▶ **EXAMPLE 4-2-4** Consider the RC circuit shown in Fig. 4-30. Find the differential equation of the system.

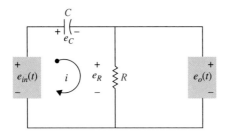

Figure 4-30 Simple electrical RC circuit.

As before, we have

$$e_{in}(t) = e_c(t) + e_R(t) \qquad (4\text{-}90)$$

or

$$e_{in}(t) = \frac{1}{C}\int i\,dt + iR \qquad (4\text{-}91)$$

But $v_o(t) = iR$. So

$$e_{in}(t) = \frac{\int e_o(t)\,dt}{RC} + e_o(t) \qquad (4\text{-}92)$$

is the differential equation of the system. To solve Eq. (4-92), we differentiate once with respect to time:

$$\dot{e}_{in}(t) = \frac{e_o(t)}{RC} + \dot{e}_o(t) \qquad (4\text{-}93)$$

In Laplace domain, we get the system transfer function as

$$\frac{E_0(s)}{E_{in}(s)} = \frac{RCs}{RCs + 1} \qquad (4\text{-}94)$$

where, again, $\tau = RC$ is the **time constant** of the system. ◀

▶ **EXAMPLE 4-2-5** Consider the voltage divider of Fig. 4-31. Given an input voltage $e_0(t)$, find the output voltage $e_1(t)$ in the circuit composed of two resistors R_1 and R_2.

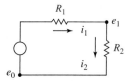

Figure 4-31 A voltage divider.

The currents in the resistors are

$$i_1 = \frac{e_0(t) - e_1(t)}{R_1} \tag{4-95}$$

$$i_2 = \frac{e_1(t)}{R_2} \tag{4-96}$$

The node equation at the $e_1(t)$ node is

$$i_1 - i_2 = 0 \tag{4-97}$$

Substituting Eqs. (4-95) and (4-96) into the previous node equation:

$$\frac{e_0(t) - e_1(t)}{R_1} - \frac{e_1(t)}{R_2} = 0 \tag{4-98}$$

Rearrangement of this equation yields the following equation for the voltage divider:

$$e_1(t) = \frac{R_2}{R_1 + R_2} e_0(t) \tag{4-99}$$

In Laplace domain, we get

$$E_1(s) = \frac{R_2}{R_1 + R_2} E_0(s) \tag{4-100}$$

The SI and most other measurement units for variables in electrical systems are the same, as shown in Table 4-3.

TABLE 4-3 Basic Electrical System Properties and Their Units

Parameter	Notation	Units
Charge	Q	coulomb = newton-meter/volt
Current	i	ampere (A)
Voltage	e	volt (V)
Energy	H	joule = volt × coulomb
Power	P	joule/sec
Resistance	R	ohm (O) = volt/amp
Capacitance	C	farad (F) = coulomb/volt = amp sec/volt = second/ohm
Inductance	L	henry (H) = volt sec/amp = ohm sec

▶ 4-3 MODELING OF ACTIVE ELECTRICAL ELEMENTS: OPERATIONAL AMPLIFIERS

Operational amplifiers, or simply **op-amps**, offer a convenient way to build, implement, or realize continuous-data or s-domain transfer functions. In control systems, op-amps are often used to implement the controllers or compensators that evolve from the control-system design process, so in this section we illustrate common op-amp configurations. An

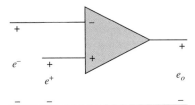

Figure 4-32 Schematic diagram of an op-amp.

in-depth presentation of op-amps is beyond the scope of this text. For those interested, many texts are available that are devoted to all aspects of op-amp circuit design and applications [8, 9].

Our primary goal here is to show how to implement first-order transfer functions with op-amps while keeping in mind that higher-order transfer functions are also important. In fact, simple high-order transfer functions can be implemented by connecting first-order op-amp configurations together. Only a representative sample of the multitude of op-amp configurations will be discussed. Some of the practical issues associated with op-amps are demonstrated in Chapters 5 and 9.

4-3-1 The Ideal Op-Amp

When good engineering practice is used, an op-amp circuit can be accurately analyzed by considering the op-amp to be ideal. The ideal op-amp circuit is shown in Fig. 4-32, and it has the following properties:

1. The voltage between the + and − terminals is zero, that is, $e^+ = e^-$. This property is commonly called the *virtual ground* or *virtual short*.
2. The currents into the + and − input terminals are zero. Thus, the input impedance is infinite.
3. The impedance seen looking into the output terminal is zero. Thus, the output is an ideal voltage source.
4. The input–output relationship is $e_o = A(e^+ - e^-)$, where the gain A approaches infinity.

The input–output relationship for many op-amp configurations can be determined by using these principles. An op-amp cannot be used as shown in Fig. 4-32. Rather, linear operation requires the addition of feedback of the output signal to the − input terminal.

4-3-2 Sums and Differences

As illustrated in Chapter 3, one of the most fundamental elements in a block diagram or an SFG is the addition or subtraction of signals. When these signals are voltages, op-amps provide a simple way to add or subtract signals, as shown in Fig. 4-33, where all the resistors have the same value. Using superposition and the ideal properties given in the preceding section, the input–output relationship in Fig. 4-33(a) is $v_o = -(v_a - v_b)$. Thus, the output is the negative sum of the input voltages. When a positive sum is desired, the circuit shown in Fig. 4-33(b) can be used. Here the output is given by $e_o = e_a + e_b$. Modifying Fig. 4-33(b) slightly gives the differencing circuit shown in Fig. 4-33(c), which has an input–output relationship of $e_o = e_b - e_a$.

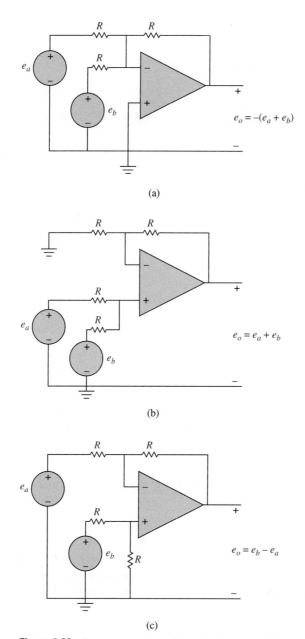

Figure 4-33 Op-amps used to add and subtract signals.

4-3-3 First-Order Op-Amp Configurations

In addition to adding and subtracting signals, op-amps can be used to implement transfer functions of continuous-data systems. While many alternatives are available, we will explore only those that use the inverting op-amp configuration shown in Fig. 4-34. In the figure, $Z_1(s)$ and $Z_2(s)$ are impedances commonly composed of resistors and capacitors. Inductors are not commonly used because they tend to be bulkier and more expensive. Using ideal op-amp properties, the input–output relationship, or

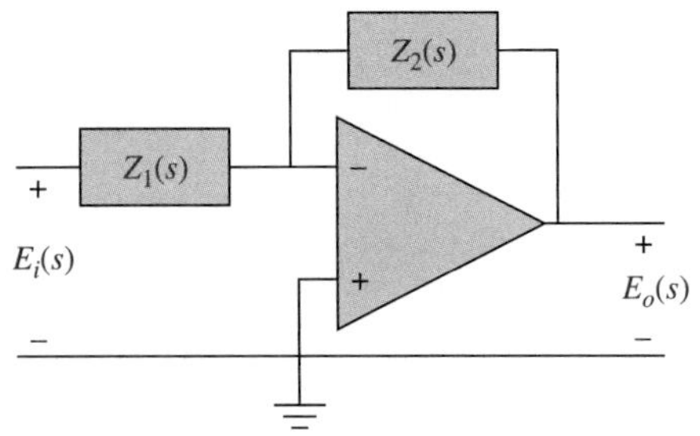

Figure 4-34 Inverting op-amp configuration.

transfer function, of the circuit shown in Fig. 4-34 can be written in a number of ways, such as

$$G(s) = \frac{E_o(s)}{E_i(s)} = -\frac{Z_2(s)}{Z_1(s)} = \frac{-1}{Z_1(s)Y_2(s)}$$
$$= -Z_2(s)Y_1(s) = -\frac{Y_1(s)}{Y_2(s)} \qquad (4\text{-}101)$$

where $Y_1(s) = 1/Z_1(s)$ and $Y_2(s) = 1/Z_2(s)$ are the admittances associated with the circuit impedances. The different transfer function forms given in Eq. (4-101) apply conveniently to the different compositions of the circuit impedances.

Using the inverting op-amp configuration shown in Fig. 4-34 and using resistors and capacitors as elements to compose $Z_1(s)$ and $Z_2(s)$, it is possible to implement poles and zeros along the negative real axis as well as at the origin in the s-plane, as shown in Table 4-4. Because the inverting op-amp configuration has been used, all the transfer functions have negative gains. The negative gain is usually not an issue because it is simple to add a gain of -1 to the input and output signal to make the net gain positive.

TABLE 4-4 Inverting Op-Amp Transfer Functions

	Input Element	Feedback Element	Transfer Function	Comments
(a)	R_1, $Z_1 = R_1$	R_2, $Z_2 = R_2$	$-\dfrac{R_2}{R_1}$	Inverting gain, e.g., if $R_1 = R_2$, $e_o = -e_1$
(b)	R_1, $Z_1 = R_1$	C_2, $Y_2 = sC_2$	$\left(\dfrac{-1}{R_1 C_2}\right)\dfrac{1}{s}$	Pole at the origin, i.e., an integrator
(c)	C_1, $Y_1 = sC_1$	R_2, $Z_2 = R_2$	$(-R_2 C_1)s$	Zero at the origin, i.e., a differentiator
(d)	R_1, $Z_1 = R_1$	$R_2 \parallel C_2$, $Y_2 = \dfrac{1}{R_2} + sC_2$	$-\dfrac{\dfrac{1}{R_1 C_2}}{s + \dfrac{1}{R_2 C_2}}$	Pole at $\dfrac{-1}{R_2 C_2}$ with a dc gain of $-R_2/R_1$

(*Continued*)

TABLE 4-4 (Continued)

	Input Element	Feedback Element	Transfer Function	Comments
(e)	R_1 $Z_1 = R_1$	R_2, C_2 $Z_2 = R_2 + \dfrac{1}{sC_2}$	$\dfrac{-R_2}{R_1}\left(\dfrac{s + 1/R_2C_2}{s}\right)$	Pole at the origin and a zero at $-1/R_2C_2$, i.e., a PI controller
(f)	R_1, C_1 $Y_1 = \dfrac{1}{R_1} + sC_1$	R_2 $Z_2 = R_2$	$-R_2C_1\left(s + \dfrac{1}{R_1C_1}\right)$	Zero at $s = \dfrac{-1}{R_1C_1}$, i.e., a PD controller
(g)	R_1, C_1 $Y_1 = \dfrac{1}{R_1} + sC_1$	R_2, C_2 $Y_2 = \dfrac{1}{R_2} + sC_2$	$\dfrac{\dfrac{-C_1}{C_2}\left(s + \dfrac{1}{R_1C_1}\right)}{s + \dfrac{1}{R_2C_2}}$	Poles at $s = \dfrac{-1}{R_2C_2}$ and a zero at $s = \dfrac{-1}{R_1C_1}$, i.e., a lead or lag controller

▶ **EXAMPLE 4-3-1** As an example of op-amp realization of transfer functions, consider the transfer function

$$G(s) = K_P + \frac{K_I}{s} + K_D s \qquad (4\text{-}102)$$

where K_P, K_D, and K_I are real constants. In Chapters 5 and 9, this transfer function will be called the **PID controller**, since the first term is a **p**roportional gain, the second an **i**ntegral term, and the third a **d**erivative term. Using Table 4-4, the proportional gain can be implemented using line (a), the integral term can be implemented using line (b), and the derivative term can be implemented using line (c). By superposition, the output of $G(s)$ is the sum of the responses due to each term in $G(s)$. This sum can be implemented by adding an additional input resistance to the circuit shown in Fig. 4-33(a). By making the sum negative, the negative gains of the proportional, integral, and derivative term implementations are canceled, giving the desired result shown in Fig. 4-35. The transfer functions of the components of the op-amp circuit in Fig. 4-35 are

Proportional: $\qquad \dfrac{E_P(s)}{E(s)} = -\dfrac{R_2}{R_1} \qquad (4\text{-}103)$

Integral: $\qquad \dfrac{E_I(s)}{E(s)} = -\dfrac{1}{R_i C_i s} \qquad (4\text{-}104)$

Derivative: $\qquad \dfrac{E_D(s)}{E(s)} = -R_d C_d s \qquad (4\text{-}105)$

The output voltage is

$$E_o(s) = -[E_P(s) + E_I(s) + E_D(s)] \qquad (4\text{-}106)$$

Thus, the transfer function of the PID op-amp circuit is

$$G(s) = \frac{E_o(s)}{E(s)} = \frac{R_2}{R_1} + \frac{1}{R_i C_i s} + R_d C_d s \qquad (4\text{-}107)$$

By equating Eqs. (4-102) and (4-107), the design is completed by choosing the values of the resistors and the capacitors of the op-amp circuit so that the desired values of K_P, K_I, and K_D are matched. The design of the controller should be guided by the availability of standard capacitors and resistors.

It is important to note that Fig. 4-35 is just one of many possible implementations of Eq. (4-102). For example, it is possible to implement the PID controller with just three op-amps. Also, it is common to add components to limit the high-frequency gain of the differentiator and to limit the

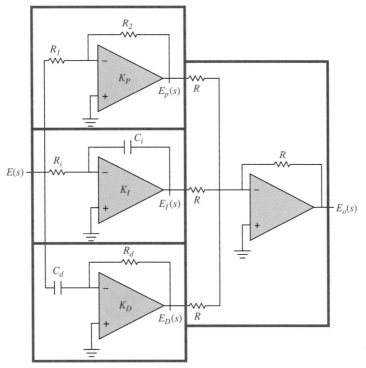

Figure 4-35 Implementation of a PID controller.

integrator output magnitude, which is often referred to as *antiwindup* protection. One advantage of the implementation shown in Fig. 4-35 is that each of the three constants K_P, K_I, and K_D can be adjusted or tuned individually by varying resistor values in its op-amp circuits. Op-amps are also used in control systems for A/D and D/A converters, sampling devices, and realization of nonlinear elements for system compensation.

4-4 INTRODUCTION TO MODELING OF THERMAL SYSTEMS

In this section, we introduce thermal and fluid systems. Because of the complex mathematics associated with these nonlinear systems, we only focus on basic and simplified models.

4-4-1 Elementary Heat Transfer Properties[1]

The two key variables in a thermal process are temperature T and thermal storage or heat stored Q, which has the same units as energy. Heat transfer is related to the heat flow rate q, which has the units of power. That is

$$q = \dot{Q} \tag{4-108}$$

As in the electrical systems, the concept of **capacitance** in a heat transfer problem is related to storage (or discharge) of heat in a body. The capacitance C is related to the change of the body temperature T with respect to time and the rate of heat flow q:

$$q = C\dot{T} \tag{4-109}$$

[1]For more in-depth study of this subject, refer to references [1–7].

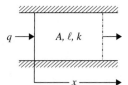

Figure 4-36 One-directional heat conduction flow.

where the thermal capacitance C can be stated as a product of ρ material density, c material specific heat, and volume V:

$$C = \rho c_p V \qquad (4\text{-}110)$$

In a thermal system, there are three different ways that heat is transferred: by conduction, convection, or radiation.

Conduction: Thermal conduction describes how an object conducts heat. In general this type of heat transfer happens in solid materials due to a temperature difference between two surfaces. In this case, heat tends to travel from the hot to the cold region. The transfer of energy in this case takes place by molecule diffusion and in a direction perpendicular to the object surface. Considering one-directional steady state heat conduction along x, as shown in Fig. 4-36, the rate of heat transfer is given by

$$q = \frac{kA}{\ell} \Delta T = D_{1-2}\, \Delta T \qquad (4\text{-}111)$$

where q is the rate of heat transfer (flow), k is the thermal conductivity related to the material used, A is the area normal to the direction of heat flow x, and $\Delta T = T_1 - T_2$ is the difference between the temperatures at $x = 0$ and $x = \ell$, or T_1 and T_2. Note in this case, assuming a perfect insulation, the heat conduction in other directions is zero. Also note that

$$D_{1-2} = \frac{1}{\frac{\ell}{kA}} = \frac{1}{R} \qquad (4\text{-}112)$$

where R is also known as **thermal resistance**. So the rate of heat transfer q may be represented in terms of R as

$$q = \frac{\Delta T}{R} \qquad (4\text{-}113)$$

Convection: This type of heat transfer occurs between a solid surface and a fluid exposed to it, as shown in Fig. 4-37. At the boundary where the fluid and the solid surface meet, the heat transfer process is by conduction. But once the fluid is exposed to the heat, it can be replaced by new fluid. In thermal convection, the heat flow is given by

$$q = hA\, \Delta T = D_0\, \Delta T \qquad (4\text{-}114)$$

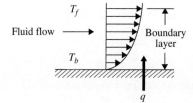

Figure 4-37 Fluid-boundary heat convection.

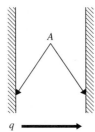

Figure 4-38 A simple heat radiation system with directly opposite ideal radiators.

where q is the rate of heat transfer or heat flow, h is the coefficient of convective heat transfer, A is the area of heat transfer, and $\Delta T = T_b - T_f$ is the difference between the boundary and fluid temperatures. The term hA may be denoted by D_0, where

$$D_0 = hA = \frac{1}{R} \qquad (4\text{-}115)$$

Again, the rate of heat transfer q may be represented in terms of thermal resistance R. Thus

$$q = \frac{\Delta T}{R} \qquad (4\text{-}116)$$

Radiation: The rate of heat transfer through radiation between two separate objects is determined by the Stephan-Boltzmann law,

$$q = \sigma A \left(T_1^4 - T_2^4 \right) \qquad (4\text{-}117)$$

where q is the rate of heat transfer, σ is the Stephan-Boltzmann constant and is equal to 5.667×10^{-8} W/m^2 °K^4, A is the area normal to the heat flow, and T_1 and T_2 are the absolute temperatures of the two bodies. Note that Eq. (4-117) applies to directly opposed ideal radiators of equal surface area A that perfectly absorb all the heat without reflection (see Fig. 4-38).

The SI and other measurement units for variables in thermal systems are shown in Table 4-5.

TABLE 4-5 Basic Thermal System Properties and Their Units

Parameter	Symbol Used	SI Units	Other Units	Conversion Factors
Temperature	T	°C (Celsius) °K (Kelvin)	°F (Fahrenheit)	°C = (°F − 32) × 5/9 °C = °K + 273
Energy (Heat Stored)	Q	J (joule)	Btu calorie	1 J = 1 N-m 1 Btu = 1055 J 1 cal = 4.184 J
Heat Flow Rate	q	J/sec W	Btu/sec	
Resistance	R	°C/W °K/W	°F/(Btu/hr)	
Capacitance	C	J/(kg °C) J/(kg °K)	Btu/°F Btu/°R	

180 ▶ Chapter 4. Theoretical Foundation and Background Material: Modeling of Dynamic Systems

▶ **EXAMPLE 4-4-1** A rectangular object is composed of a material that is in contact with fluid on its top side while being perfectly insulated on three other sides, as shown in Fig. 4-39. Find the equations of the heat transfer process for the following:

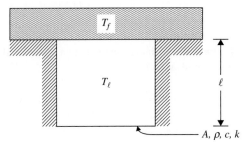

Figure 4-39 Heat transfer problem between a fluid and an insulated solid object.

T_l = Solid object temperature; assume that the temperature distribution is uniform
T_f = Top fluid temperature
ℓ = Length of the object
A = Cross sectional area of the object
ρ = Material density
c = Material specific heat
k = Material thermal conductivity
h = Coefficient of convective heat transfer

SOLUTION The rate of heat storage in the solid from Eq. (4-109) is

$$q = \rho c A \ell \left(\frac{dT_\ell}{dt}\right) \qquad (4\text{-}118)$$

Also, the convection rate of heat transferred from the fluid is

$$q = hA(T_f - T_\ell) \qquad (4\text{-}119)$$

The energy balance equation for the system dictates q to be the same in Eqs. (4-118) and (4-119). Hence, upon introducing thermal capacitance C from Eq. (4-109) and the convective thermal resistance R from Eq. (4-113) and substituting the right-hand sides of Eq. (4-118) into Eq. (4-119), we get

$$RC\dot{T}_\ell = -T_\ell + T_f \qquad (4\text{-}120)$$

In Laplace domain, the transfer function of the system is written as

$$\frac{T_\ell(s)}{T_f(s)} = \frac{1}{RCs + 1} \qquad (4\text{-}121)$$

where the $RC = \tau$ is also known as the **time constant** of the system. The significance of this term is discussed earlier in Chapter 2, and the initial conditions are assumed to be $T_\ell(t=0) = \dot{T}_\ell(t=0) = 0$. ◀

▶ 4-5 INTRODUCTION TO MODELING OF FLUID SYSTEMS

4-5-1 Elementary Fluid and Gas System Properties[2]

In this section, we derive the equations of fluid and pneumatic systems. Understanding the models of these systems will later help in appreciating the models of hydraulic and pneumatic actuators, to be discussed in more detail in Chapter 5. In fluid systems, there are five parameters of importance—pressure, flow mass (and flow rate), temperature, density,

---
[2]For a more in-depth study of this subject, refer to references [1–7].

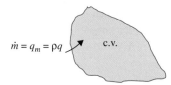

Figure 4-40 Control volume and the net mass flow rate.

and flow volume (and volume rate). **Incompressible fluid systems**, just like electrical systems, can be modeled by passive components including resistance, capacitance, and inductance. In case of incompressible fluids, the fluid volume remains constant.

To understand these concepts better, we must look at the **fluid continuity equation** or **the law of conservation of mass**. For the control volume shown in Fig. 4-40 and the net mass flow rate, we have

$$q_m = \rho q$$
$$m = \int \rho q \, dt \tag{4-122}$$
$$q = q_i - q_o$$

where m is the net mass flow, ρ is fluid density, $q_m = \dot{m}$ is the mass flow rate, and q is the net fluid flow rate (volume flow rate of the ingoing fluid q_i minus volume flow rate of the outgoing fluid q_o). The **conservation of mass** states:

$$\frac{dm}{dt} = \rho q = \frac{d}{dt}(M_{cv}) = \frac{d}{dt}(\rho V) \tag{4-123}$$

$$\frac{dm}{dt} = \rho \dot{V} + V \dot{\rho} \tag{4-124}$$

where $\dot{m}$ is the net mass flow rate, M_{cv} is the mass of the control volume (or for simplicity "the container" fluid), and V is the container volume. Note

$$\frac{dV}{dt} = q_i - q_o \tag{4-125}$$

which is also known as the **conservation of volume** for the fluid. For an incompressible fluid, ρ is constant. Hence,

$$\dot{m} = \rho \dot{V} \tag{4-126}$$

Capacitance—Incompressible Fluids: Similar to the electrical capacitance, fluid capacitance relates to how energy can be stored in a fluid system. The fluid capacitance C is the ratio of the fluid flow rate q to the rate of pressure $\dot{P}$:

$$C = \frac{q}{\dot{P}} \tag{4-127}$$

Or

$$q = C\dot{P} \tag{4-128}$$

EXAMPLE 4-5-1 The pressure in the tank shown in Fig. 4-41, which is filled to height h, is

$$P = \frac{\rho V}{A} = \frac{\rho h g A}{A} = \rho h g \qquad (4\text{-}129)$$

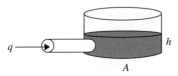

Figure 4-41 Incompressible fluid flow into an open-top cylindrical container.

As a result, noting that $q = \dot{V}$,

$$C = \frac{\dot{V}}{\rho g h} = \frac{A}{\rho g} \qquad (4\text{-}130)$$

For the general case, what is happening in Fig. 4-40, as the fluid flows into the control volume, the fluid mass will change; so does the pressure. Capacitance expresses the rate of change of the fluid mass with respect to pressure. That is

$$C = \frac{dm/dt}{dP/dt} = \frac{dm}{dP} \qquad (4\text{-}131)$$

In general, the fluid density ρ is nonlinear and may depend on temperature and pressure. This nonlinear dependency, $\rho_{nl}(P, T)$, known as **the equation of state**, may be linearized using the first-order Taylor series relating ρ_{nl} to P and T:

$$\rho_{nl} = \rho + \left(\frac{\partial \rho_{nl}}{\partial P}\right)_{P_{ref}, T_{ref}} (P - P_{ref}) + \left(\frac{\partial \rho_{nl}}{\partial T}\right)_{P_{ref}, T_{ref}} (T - T_{ref}) \qquad (4\text{-}132)$$

where ρ, P_{ref}, and T_{ref} are constant reference values of density, pressure, and temperature, respectively. In this case,

$$\beta = \frac{1}{\rho}\left(\frac{\partial \rho_{nl}}{\partial P}\right)_{P_{ref}, T_{ref}} \qquad (4\text{-}133)$$

$$\alpha = -\frac{1}{\rho}\left(\frac{\partial \rho_{nl}}{\partial T}\right)_{P_{ref}, T_{ref}} \qquad (4\text{-}134)$$

are the **bulk modulus** and the **thermal expansion coefficient**, respectively. In most cases of interest, however, the temperatures of the fluid entering and flowing out of the container are almost the same. Further, if the container of volume V is a rigid object, using Eq. (4-133), Eq. (4-124) may be rewritten as

$$\frac{dm}{dt} = \rho q = V \dot{\rho} \Rightarrow q = \frac{V}{\beta}\dot{P} = C\dot{P} \qquad (4\text{-}135)$$

◀

EXAMPLE 4-5-2 In practice, accumulators are fluid capacitors, which may be modeled as a spring-loaded piston system, as shown in Fig. 4-42. In this case, assuming a spring-loaded piston of area A traveling inside a rigid cylindrical container, the pressure rate is shown as

$$\dot{P} = \frac{1}{C}(q - \dot{V}) \qquad (4\text{-}136)$$

where $\dot{V} = A\dot{x}$.

For a **pneumatic system**, the law of conservation of volume does not apply because the volume of a gas varies with pressure or other external effects. In this case, only the **conservation of mass** applies. As a result, it is customary to use the **mass flow rate q_m** as opposed to volume flow rate q in the equations involving pneumatic systems.

Capacitance—Pneumatic Systems: As in the previous case, capacitance relates to how energy can be stored in the system, and it defines the rate of change of gas stored in a control volume, as

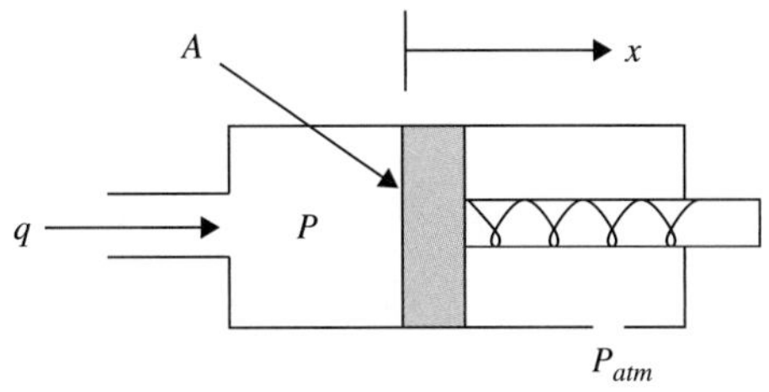

Figure 4-42 A spring-loaded piston system.

shown in Fig. 4-40 with respect to pressure. For a constant volume container, the general gas capacitance relation Eq. (4-131) becomes

$$C = \frac{dm}{dP} = V\frac{d\rho}{dP} \qquad (4\text{-}137)$$

where the container volume V is a constant.

For a perfect gas under normal temperatures and pressures, **the perfect gas law** states:

$$PV = mR_g T \qquad (4\text{-}138)$$

where V is the volume of a gas with absolute pressure P and mass m, T is the absolute temperature of the gas, and R_g is the gas constant, which depends on the type of gas used. Notice in this case four parameters P, V, T, and m are mathematically related. As a result, to solve one, the other three must be somehow related. Using a **polytropic process**, which is a general process for all fluids relating the pressure, volume, and mass, we have:

$$P\left(\frac{V}{m}\right)^n = P\left(\frac{1}{\rho}\right)^n = const. \qquad (4\text{-}139)$$

where n is called the **polytropic exponent** and can vary from 0 to ∞. As a result, the capacitance relation Eq. (4-137) may be restated as

$$C = V\frac{d\rho}{dP} = V\left(\frac{\rho^n}{Pn\rho^{(n-1)}}\right) = V\frac{\rho}{nP} \qquad (4\text{-}140)$$

Or, using Eq. (4-138) and knowing $m = \rho V$,

$$C = \frac{V}{nR_g T} \qquad (4\text{-}141)$$

As a side note, if in a polytropic process the **mass m is constant**, and given a process from state 1 to state 2, the general gas law may also be defined by

$$\frac{P_1 V_1}{T_1} = \frac{P_2 V_2}{T_2} \qquad (4\text{-}142)$$

For a **constant temperature** or an **isothermal process**, the gas temperatures at any two given instants are the same. Or

$$\begin{aligned} T_1 &= T_2 \\ P_1 V_1 &= P_2 V_2 \end{aligned} \qquad (4\text{-}143)$$

In this case, $n = 1$ in the capacitance relation Eq. (4-140).

For a **constant pressure** or an **isobaric process**,

$$\begin{aligned} P_1 &= P_2 \\ \frac{V_1}{T_1} &= \frac{V_2}{T_2} \end{aligned} \qquad (4\text{-}144)$$

In this case, $n = 0$ in the capacitance relation Eq. (4-140).

For a **constant volume** or an **isovolumetric process**, the relation becomes

$$\begin{aligned} V_1 &= V_2 \\ \frac{P_1}{P_2} &= \frac{T_1}{T_2} \end{aligned} \qquad (4\text{-}145)$$

In this case, $n = \infty$ in the capacitance relation Eq. (4-140).

For a **reversible adiabatic** or an **isentropic process**, the relation becomes

$$P_1 V_1^k = P_2 V_2^k \qquad (4\text{-}146)$$

In this case, $n = k$ in the capacitance relation Eq. (4-140), where k is the ratio of specific heats:

$$k = \frac{c_p}{c_v} \qquad (4\text{-}147)$$

where c_p is the specific heat of the gas at constant pressure and c_v is the gas specific heat at constant volume. In pneumatic systems, $k = 1.4$ (for air).

Inductance—Incompressible Fluids: Fluid inductance is also referred to as fluid **inertance** in relation to the inertia of a moving fluid inside a passage (line or a pipe). Inertance occurs mainly in long lines, but it can also occur where an external force (e.g., caused by a pump) causes a significant change in the flow rate. In the case shown in Fig. 4-43, assuming a frictionless pipe with a uniform fluid flow moving at the speed v, in order to accelerate the fluid, an external force F is applied. From Newton's second law,

$$F = A\,\Delta P = M\dot{v} = \rho A \ell \dot{v}$$
$$\Delta P = (P_1 - P_2) \qquad (4\text{-}148)$$

But

$$\dot{V} = Av = \ddot{q} \qquad (4\text{-}149)$$

So

$$(P_1 - P_2) = L\ddot{q} \qquad (4\text{-}150)$$

where

$$L = \frac{\rho \ell}{A} \qquad (4\text{-}151)$$

is known as the **fluid inductance**. The concept of inductance is rarely discussed in the case of compressible fluids and gases and, therefore, is not discussed here.

Resistance—Incompressible Fluids: As in the electrical systems, fluid resistors dissipate energy. For the system shown in Fig. 4-44, the force resisting the fluid passing through a passage like a pipe is

$$F_f = A\Delta P \qquad (4\text{-}152)$$

where $\Delta P = P_1 - P_2$ is the pressure drop and A is the cross-sectional area of the pipe. Depending on the type of flow (i.e., laminar or turbulent) the fluid resistance relationship can be linear or nonlinear and relates the pressure drop to the mass flow rate q_m. For a laminar flow, we define

$$\Delta P = Rq_m = R\rho q \qquad (4\text{-}153)$$

$$R = \frac{\Delta P}{q_m} \qquad (4\text{-}154)$$

where q is the volume flow rate. Table 4-6 shows R for various passage cross sections, assuming a laminar flow.

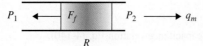

Figure 4-44 Flow of an incompressible fluid through a pipe and a fluid resistor R.

TABLE 4-6   Equations of Resistance R for Laminar Flows

Fluid resistance	$P_1 \longrightarrow \boxed{R} \longrightarrow P_2$ $\quad \xrightarrow{q_m}$
Symbols used	**Fluid volume flow rate:** q **Pressure drop:** $\Delta P = P_{12} = P_1 - P_2$ **Laminar resistance:** R μ: **Fluid viscosity** w = width; h = height; ℓ = length; d = diameter
General case	$R = \dfrac{32\,\mu\ell}{A d_h^2}$ $d_h =$ **hydraulic diameter** $= \dfrac{4A}{perimeter}$
Circular cross section	$R = \dfrac{128\,\mu\ell}{\pi d^4}$
Square cross section	$R = \dfrac{32\,\mu\ell}{w^4}$
Rectangular cross section	$R = \dfrac{8\,\mu\ell}{\dfrac{wh^3}{(1+h/w)^2}}$
Rectangular cross section: Approximation	$R = \dfrac{12\,\mu\ell}{wh^3}$ $w/h =$ small
Annular cross section	$R = \dfrac{8\,\mu\ell}{\pi d_o d_i^3 \left(1 - \dfrac{d_i}{d_o}\right)}$ $d_o =$ outer diameter; $d_i =$ inner diameter
Annular cross section: Approximation	$R = \dfrac{12\,\mu\ell}{\pi d_o d_i^3}$ $d_o/d_i =$ small

When the flow becomes **turbulent**, the pressure drop relation Eq. (4-153) is rewritten as

$$\Delta P = R_T q_m^n \tag{4-155}$$

where R_T is the turbulent resistance and n is a power varying depending on the boundary used—e.g., $n = 7/4$ for a long pipe and, most useful, $n = 2$ for a flow through an orifice or a valve. ◀

▶ **EXAMPLE 4-5-3**
A One-Tank Liquid-Level System

For the liquid-level system shown in Fig. 4-45, water or any incompressible fluid (i.e., fluid density ρ is constant) enters the tank from the top and exits through the valve in the bottom. The volume flow rate at the valve inlet and the volume flow rate at the valve outlet are q_i and q_o, respectively. The fluid height in the tank is h and is variable. The valve resistance is R. Find the system equation for the input, q_i, and output, h.

SOLUTION   The conservation of mass suggests

$$\frac{dm}{dt} = \frac{d(\rho V)}{dt} = \rho q_i - \rho q_o \tag{4-156}$$

where ρq_i and ρq_o are the mass flow rate in and out of the valve, respectively. Because the fluid density ρ is a constant, the conservation of volume also applies, which suggests the time rate of

186 ▶ Chapter 4. Theoretical Foundation and Background Material: Modeling of Dynamic Systems

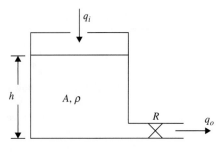

Figure 4-45 A single-tank liquid-level system.

change of the fluid volume inside the tank is equal to the difference of incoming and outgoing flow rates.

$$\frac{d(V)}{dt} = \frac{d(Ah)}{dt} = q_i - q_o \qquad (4\text{-}157)$$

In this case, A is the tank cross-sectional area, and h is the fluid inside the tank height and is a variable. Recall from Eq. (4-131)

$$C = \frac{dm/dt}{dP/dt} \qquad (4\text{-}158)$$

Hence

$$\frac{dm}{dt} = C\frac{dp}{dt} \qquad (4\text{-}159)$$

Or from Eq. (4-158),

$$C\frac{dp}{dt} = \rho q_i - \rho q_o \qquad (4\text{-}160)$$

Using relation Eq. (4-154), the fluid valve resistance R, assuming a laminar flow, is defined as

$$\rho q_o = \frac{\Delta p}{R} = \frac{p_1 - p_2}{R} \qquad (4\text{-}161)$$

where Δp is the pressure drop across the valve. Relating the pressure to fluid height h, which is variable, we get

$$\begin{aligned} p_1 &= p_{atm} + \rho g h \\ p_2 &= p_{atm} \end{aligned} \qquad (4\text{-}162)$$

where p_1 is in the pressure at the valve inlet and p_2 is the outlet pressure, and p_{atm} is the atmospheric pressure. After combining Eqs. (4-157) through (4-162), we get the system equation:

$$\rho A \frac{dh}{dt} = \rho q_i - \frac{\rho g h}{R} \qquad (4\text{-}163)$$

Or

$$RC\dot{h} + h = \frac{R}{g} q_i \qquad (4\text{-}164)$$

where $C = A/g$ is the capacitance and $\rho = R$ is the resistance. As a result, system time constant is $\tau = RC$. ◀

▶ **EXAMPLE 4-5-4**
A Two-Tank Liquid-Level System

Consider a double tank system, as shown in Fig. 4-46, with h_1 and h_2 representing the two tank heights and R_1 and R_2 representing the two valve resistances, respectively. Find the differential equations.

SOLUTION Using the same approach as in Example 4-5-3, it is not difficult to see

$$\rho q_i - \rho q_1 = \rho q_i - \frac{p_1 - p_2}{R_1} = \rho q_i - \frac{(p_{atm} + \rho g h_1) - (p_{atm} + \rho g h_2)}{R_1} \qquad (4\text{-}165)$$

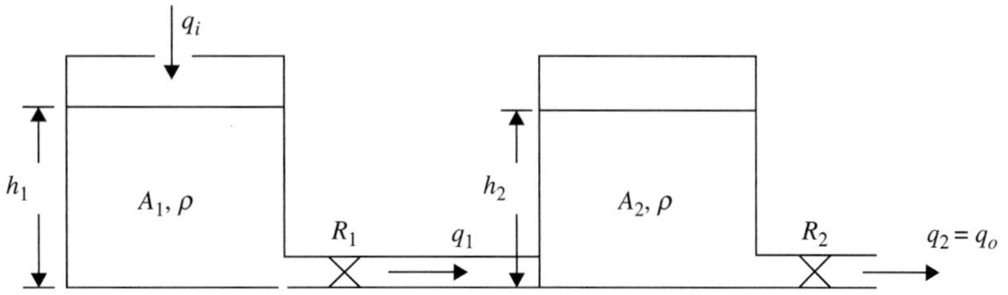

Figure 4-46 Two-tank liquid-level system.

and

$$\rho q_1 - \rho q_2 = \frac{p_1 - p_2}{R_1} - \frac{p_2 - p_3}{R_2}$$

$$= \frac{(p_{atm} + \rho g h_1) - (p_{atm} + \rho g h_2)}{R_1} - \frac{(p_{atm} + \rho g h_2) - p_a}{R_2} \quad (4\text{-}166)$$

Thus, the equations of system are

$$A_1 \dot{h}_1 + \frac{g h_1}{R_1} - \frac{g h_2}{R_1} = q_i \quad (4\text{-}167)$$

$$A_2 \dot{h}_2 - \frac{g h_1}{R_1} + \left(\frac{1}{R_1} + \frac{1}{R_2}\right) g h_2 = 0 \quad (4\text{-}168)$$

Resistance—Pneumatic Systems: The resistance for pneumatic systems is a bit more complicated. For a gas following the perfect gas law Eq. (4-138), the flow through a valve or an orifice of cross-sectional area A, shown in Fig. 4-47, is related to the outlet pressure P. Note that the mass flow rate q_m on both sides of the valve, by the virtue of continuity, is the same. Not considering the theoretical details, for a laminar flow if $\Delta P = P_i - P$ is small, we have

$$R_L = \frac{\Delta P}{q_m} \quad (4\text{-}169)$$

where P_i is the inlet pressure, P is outlet pressure, q is the volumetric flow rate, and R_L is the equivalent resistance, which is obtained experimentally. For a turbulent flow, we get

$$R_T = \frac{\Delta P}{q_m^2} \quad (4\text{-}170)$$

where R_T is the turbulent resistance. We use the next example to better illustrate these concepts.

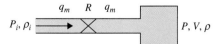

Figure 4-47 Air flow through a pipe with an orifice.

▶ **EXAMPLE 4-5-5** Consider air passing through a valve and entering a rigid container system, as shown in Fig. 4-48. In this case, the valve is modeled as an orifice, inlet pressure is P_i, the mass flow rate is q_m, and the pressure inside the container (or the valve outlet pressure) is P. In this case, it is customary to think of the pressures in both sides of the valve as a constant pressure P_s (or steady state pressure) plus a variation. That is,

$$P_i = P_s + p_i$$
$$P = P_s + p \quad (4\text{-}171)$$

Figure 4-48 Gas flow into a rigid container.

For the rigid container in Fig. 4-48 with constant volume V, the law of conservation of mass dictates that in the container

$$\frac{dm}{dt} = \frac{d(\rho V)}{dt} = V\frac{d\rho}{dt} \tag{4-172}$$

where ρ_i is fluid density before reaching the valve. At the inlet (left side of the valve), we have

$$\frac{dm}{dt} = \rho_i q = q_m \tag{4-173}$$

Recall from Eq. (4-131)

$$C = \frac{dm/dt}{dP/dt} = \frac{dm}{dP} = \frac{dm}{dp} \tag{4-174}$$

But

$$\frac{dm}{dt} = \frac{dm/dp}{dp/dt} \tag{4-175}$$

Hence

$$\frac{dm}{dt} = C\frac{dp}{dt} \tag{4-176}$$

But from Eq. (4-169) we have

$$q_m = \frac{p_i - p}{R_L} \tag{4-177}$$

Thus from Eqs. (4-176) and (4-173), and using $R = R_L$ for simplicity, we get

$$C\frac{dp}{dt} = \frac{p_i - p}{R} \tag{4-178}$$

or

$$C\frac{dp}{dt} + \frac{p}{R} = \frac{p_i}{R} \tag{4-179}$$

The differential equation can be rearranged as

$$RC\dot{p} + p = p_i \tag{4-180}$$

In Laplace domain, the transfer function of the system is written as

$$\frac{P(s)}{P_i(s)} = \frac{1}{RCs + 1} \tag{4-181}$$

where the $RC = \tau$ is also known as the **time constant** of the system. The significance of this term is discussed earlier in Chapter 2, and the initial conditions are assumed to be $p(t=0) = \dot{p}(t=0) = 0$.

Using an isothermal process, where temperature is constant, and taking a derivative of Eq. (4-138) with respect to time, we have

$$\frac{dp}{dt}V = \frac{dm}{dt}R_g T \tag{4-182}$$

From Eq. (4-172) and Eq. (4-182), we get

$$V\frac{d\rho}{dt} = \frac{1}{R_g T}\frac{dp}{dt} \tag{4-183}$$

But from Eqs. (4-169), (4-172), and (4-173), we have

$$V\frac{d\rho}{dt} = \rho_i \frac{p_i - p}{R_L} \tag{4-184}$$

Thus,

$$\frac{V}{R_g T}\frac{dp}{dt} = \rho_i \frac{p_i - p}{R} \tag{4-185}$$

After substituting

$$C = \frac{V}{R_s T} \tag{4-186}$$

the differential Eqs. (4-186) and (4-179) become the same.

TABLE 4-7 Basic Fluid and Pneumatic System Properties and Their Units

Parameter	Symbol Used	SI Units	Other Units	Conversion Factors
Temperature	T	°C (Celsius) °K (Kelvin)	°F (Fahrenheit) °R (Rankin)	°C = (°F − 32) × 5/9 °C =° K + 273
Energy (Heat Stored)	Q	J (joule)	Btu calorie	1 J = 1 N-m 1 Btu = 1055 J 1 cal = 4.184 J
Volume Flow Rate	q	m³/sec	ft³/sec in³/sec	
Mass Flow Rate	q_m	kg/sec	lb/sec	
Resistance (hydraulic)	R	N-sec/m⁵	lb$_f$-sec/in⁵	
Resistance (pneumatic)	R	sec/m²	lb$_f$ hr/(ft² lb$_m$)	
Capacitance (hydraulic)	C	m⁵/N	in⁵/lb	
Capacitance (pneumatic)	C	m²	ft²	
Time Constant	$\tau = RC$	sec		

Using the polytropic process defined in Eq. (4-139), it is easy to see

$$RC\dot{p} + P = p_i \tag{4-187}$$

where

$$C = \frac{V}{nR_g T} \tag{4-188}$$

The SI and other measurement units for variables in electrical systems are tabulated in Table 4-7.

▶ 4-6 SENSORS AND ENCODERS IN CONTROL SYSTEMS

Sensors and encoders are important components used to monitor the performance and for feedback in control systems. In this section, the principle of operation and applications of some of the sensors and encoders that are commonly used in control systems are described.

4-6-1 Potentiometer

A potentiometer is an electromechanical transducer that converts mechanical energy into electrical energy. The input to the device is in the form of a mechanical displacement, either linear or rotational. When a voltage is applied across the fixed terminals of the potentiometer, the output voltage, which is measured across the variable terminal and ground, is proportional to the input displacement, either linearly or according to some nonlinear relation.

Rotary potentiometers are available commercially in single-revolution or multirevolution form, with limited or unlimited rotational motion. The potentiometers are commonly made with wirewound or conductive plastic resistance material. Fig. 4-49 shows a cutaway view of a rotary potentiometer, and Fig. 4-50 shows a linear potentiometer that also contains a built-in operational amplifier. For precision control, the conductive plastic potentiometer is preferable, because it has infinite resolution, long rotational life, good output smoothness, and low static noise.

Fig. 4-51 shows the equivalent circuit representation of a potentiometer, linear or rotary. Because the voltage across the variable terminal and reference is proportional to the shaft displacement of the potentiometer, when a voltage is applied across the fixed terminals, the device can be used to indicate the absolute position of a system or the

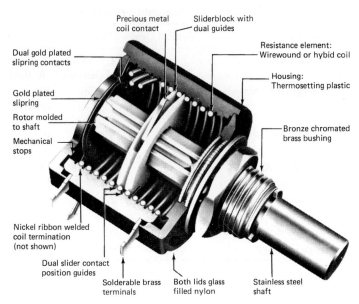

Figure 4-49 Ten-turn rotary potentiometer (courtesy of Helipot Division of Beckman Instruments, Inc.).

Figure 4-50 Linear motion potentiometer with built-in operational amplifier (courtesy of Waters Manufacturing, Inc.).

relative position of two mechanical outputs. Fig. 4-52(a) shows the arrangement when the housing of the potentiometer is fixed at reference; the output voltage $e(t)$ will be proportional to the shaft position $\theta_c(t)$ in the case of a rotary motion. Then

$$e(t) = K_s \theta_c(t) \qquad (4\text{-}189)$$

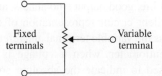

Figure 4-51 Electric circuit representation of a potentiometer.

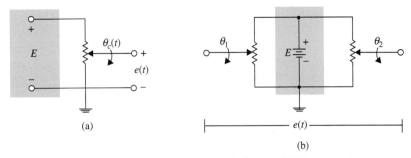

Figure 4-52 Potentiometer used as a position indicator. (b) Two potentiometers used to sense the positions of two shafts.

where K_s is the proportional constant. For an N-turn potentiometer, the total displacement of the variable arm is $2\pi N$ radians. The proportional constant K_s is given by

$$K_s = \frac{E}{2\pi N} \quad \text{V/rad} \tag{4-190}$$

where E is the magnitude of the reference voltage applied to the fixed terminals. A more flexible arrangement is obtained by using two potentiometers connected in parallel, as shown in Fig. 4-52(b). This arrangement allows the comparison of two remotely located shaft positions. The output voltage is taken across the variable terminals of the two potentiometers and is given by

$$e(t) = K_s[\theta_1(t) - \theta_2(t)] \tag{4-191}$$

Fig. 4-53 illustrates the block diagram representation of the setups in Fig. 4-52. In dc-motor control systems, potentiometers are often used for position feedback. Fig. 4-54(a) shows the schematic diagram of a typical dc-motor, position-control system. The potentiometers are used in the feedback path to compare the actual load position with the desired reference position. If there is a discrepancy between the load position and the reference input, an error signal is generated by the potentiometers that will drive the motor in such a way that this error is minimized quickly. As shown in Fig. 4-54 (a), the error signal is amplified by a dc amplifier whose output drives the armature of a permanent-magnet dc motor. Typical waveforms of the signals in the system when the input $\theta_r(t)$ is a step function are shown in Fig. 4-54(b). Note that the electric signals are all unmodulated. *In control-systems terminology, a dc signal usually refers to an unmodulated signal. On the other hand, an ac signal refers to signals that are modulated by a modulation process.* These definitions are different from those commonly used in electrical engineering, where dc simply refers to unidirectional signals and ac indicates alternating signals.

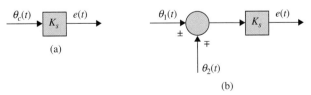

Figure 4-53 Block diagram representation of potentiometer arrangements in Fig. 4-52.

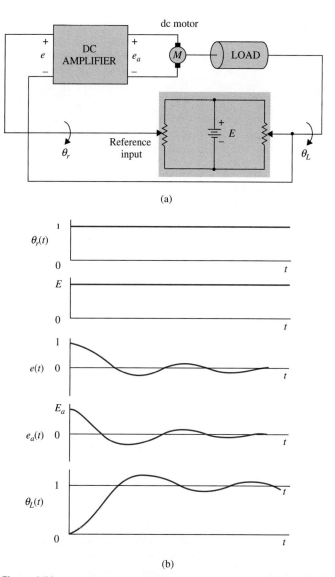

Figure 4-54 (a) A dc-motor, position-control system with potentiometers as error sensors. (b) Typical waveforms of signals in the control system of part (a).

Fig. 4-55(a) illustrates a control system that serves essentially the same purpose as that of the system in Fig. 4-54(a), except that ac signals prevail. In this case, the voltage applied to the error detector is sinusoidal. The frequency of this signal is usually much higher than that of the signal that is being transmitted through the system. Control systems with ac signals are usually found in aerospace systems that are more susceptible to noise.

Typical signals of an ac control system are shown in Fig. 4-55(b). The signal $v(t)$ is referred to as the carrier whose frequency is ω_c, or

$$v(t) = E \sin \omega_c t \tag{4-192}$$

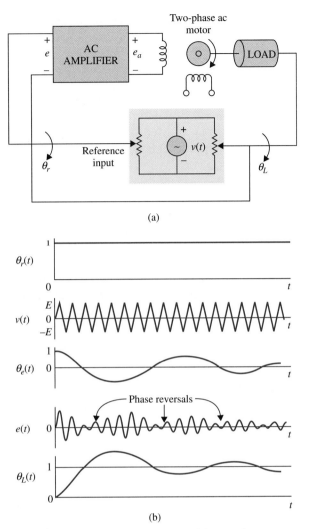

Figure 4-55 (a) An ac control system with potentiometers as error detectors. (b) Typical waveforms of signals in the control system of part (a).

Analytically, the output of the error signal is given by

$$e(t) = K_s \theta_e(t) v(t) \qquad (4\text{-}193)$$

where $\theta_e(t)$ is the difference between the input displacement and the load displacement, or

$$\theta_e(t) = \theta_r(t) - \theta_L(t) \qquad (4\text{-}194)$$

For the $\theta_e(t)$ shown in Fig. 4-55(b), $e(t)$ becomes a **suppressed-carrier-modulated** signal. A reversal in phase of $e(t)$ occurs whenever the signal crosses the zero-magnitude axis. This reversal in phase causes the ac motor to reverse in direction according to the desired sense of correction of the error signal $\theta_e(t)$. The term *suppressed-carrier modulation* stems from the fact that when a signal $\theta_e(t)$ is modulated by a carrier signal $v(t)$ according to Eq.

(4-193), the resultant signal $e(t)$ no longer contains the original carrier frequency ω_c. To illustrate this, let us assume that $\theta_e(t)$ is also a sinusoid given by

$$\theta_e(t) = \sin \omega_s t \qquad (4\text{-}195)$$

where, normally, $\omega_s \ll \omega_c$. Using familiar trigonometric relations and substituting Eqs. (4-192) and (4-195) into Eq. (4-193), we get

$$e(t) = \tfrac{1}{2} K_s E[\cos(\omega_c - \omega_s)t - \cos(\omega_c + \omega_s)t] \qquad (4\text{-}196)$$

Therefore, $e(t)$ no longer contains the carrier frequency ω_c or the signal frequency ω_s but has only the two sidebands $\omega_c + \omega_s$ and $\omega_c - \omega_s$.

When the modulated signal is transmitted through the system, the motor acts as a demodulator, so that the displacement of the load will be of the same form as the dc signal before modulation. This is clearly seen from the waveforms of Fig. 4-55(b). It should be pointed out that a control system need not contain all dc or all ac components. It is quite common to couple a dc component to an ac component through a modulator, or an ac device to a dc device through a demodulator. For instance, the dc amplifier of the system in Fig. 4-55(a) may be replaced by an ac amplifier that is preceded by a modulator and followed by a demodulator.

4-6-2 Tachometers

Tachometers are electromechanical devices that convert mechanical energy into electrical energy. The device works essentially as a voltage generator, with the output voltage proportional to the magnitude of the angular velocity of the input shaft. In control systems, most of the tachometers used are of the dc variety; that is, the output voltage is a dc signal. DC tachometers are used in control systems in many ways; they can be used as velocity indicators to provide shaft-speed readout, velocity feedback, speed control, or stabilization. Fig. 4-56 is a block diagram of a typical velocity-control system in which the tachometer output is compared with the reference voltage, which represents the desired velocity to be achieved. The difference between the two signals, or the error, is amplified and used to drive the motor so that the velocity will eventually reach the desired value. In this type of application, the accuracy of the tachometer is highly critical, as the accuracy of the speed control depends on it.

In a position-control system, velocity feedback is often used to improve the stability or the damping of the closed-loop system. Fig. 4-57 shows the block diagram of such an

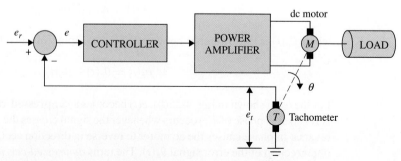

Figure 4-56 Velocity-control system with tachometer feedback.

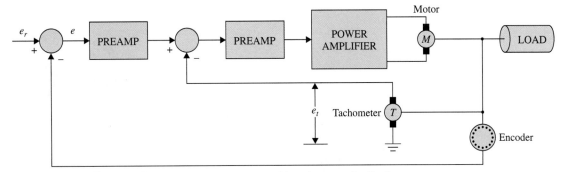

Figure 4-57 Position-control system with tachometer feedback.

application. In this case, the tachometer feedback forms an inner loop to improve the damping characteristics of the system, and the accuracy of the tachometer is not so critical.

The third and most traditional use of a dc tachometer is in providing the visual speed readout of a rotating shaft. Tachometers used in this capacity are generally connected directly to a voltmeter calibrated in revolutions per minute (rpm).

Mathematical Modeling of Tachometers

The dynamics of the tachometer can be represented by the equation

$$e_t(t) = K_t \frac{d\theta(t)}{dt} = K_t \omega(t) \qquad (4\text{-}197)$$

where $e_t(t)$ is the output voltage; $\theta(t)$, the rotor displacement in radians; $\omega(t)$, the rotor velocity in rad/sec; and K_t, the **tachometer constant** in V/rad/sec. The value of K_t is usually given as a catalog parameter in **volts per 1000 rpm** (V/krpm).

The transfer function of a tachometer is obtained by taking the Laplace transform on both sides of Eq. (4-197). The result is

$$\frac{E_t(s)}{\Theta(s)} = K_t s \qquad (4\text{-}198)$$

where $E_t(s)$ and $\Theta(s)$ are the Laplace transforms of $e_t(t)$ and $\theta(t)$, respectively.

4-6-3 Incremental Encoder

Incremental encoders are frequently found in modern control systems for converting linear or rotary displacement into digitally coded or pulse signals. The encoders that output a digital signal are known as absolute encoders. In the simplest terms, absolute encoders provide as output a distinct digital code indicative of each particular least significant increment of resolution. Incremental encoders, on the other hand, provide a pulse for each increment of resolution but do not make distinctions between the increments. In practice, the choice of which type of encoder to use depends on economics and control objectives. For the most part, the need for absolute encoders has much to do with the concern for data loss during power failure or the applications involving periods of mechanical motion without the readout under power. However, the incremental encoder's simplicity in construction, low cost, ease of application, and versatility have made it by far one of the most popular encoders in control systems. Incremental encoders are available in rotary

196 ▶ Chapter 4. Theoretical Foundation and Background Material: Modeling of Dynamic Systems

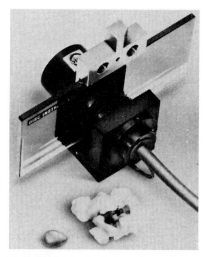

Figure 4-58 Rotary incremental encoder (courtesy of DISC Instruments, Inc.).

Figure 4-59 Linear incremental encoder (courtesy of DISC Instruments, Inc.).

and linear forms. Fig. 4-58 and Fig. 4-59 show typical rotary and linear incremental encoders.

A typical rotary incremental encoder has four basic parts: a light source, a rotary disk, a stationary mask, and a sensor, as shown in Fig. 4-60. The disk has alternate opaque and transparent sectors. Any pair of these sectors represents an incremental period. The mask is used to pass or block a beam of light between the light source and the photosensor located behind the mask. For encoders with relatively low resolution, the mask is not necessary. For fine-resolution encoders (up to thousands of increments per evolution), a multiple-slit mask is often used to maximize reception of the shutter light. The waveforms of the sensor outputs are generally triangular or sinusoidal, depending on the resolution required. Square-wave signals compatible with digital logic are derived by using a linear amplifier followed by a comparator. Fig. 4-61(a) shows a typical rectangular output waveform of a single-channel incremental encoder. In this case, pulses are produced for both directions of shaft rotation. A dual-channel encoder with two sets of output pulses is necessary for direction sensing and other control functions. When the phase of the two-output pulse train is 90° apart electrically, the two signals are said to be in quadrature, as shown in Fig. 4-61 (b). The signals uniquely define 0-to-1 and 1-to-0 logic transitions with respect to the direction of rotation of the encoder disk so that a direction-sending logic circuit can be constructed to decode the signals. Fig. 4-62 shows the single-channel output and the

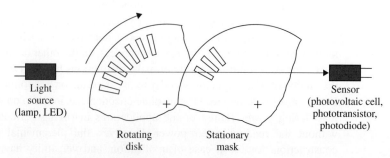

Figure 4-60 Typical incremental optomechanics.

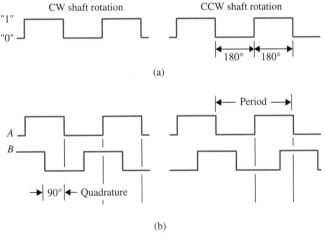

Figure 4-61 (a) Typical rectangular output waveform of a single-channel encoder device (bidirectional). (b) Typical dual-channel encoder signals in quadrature (bidirectional).

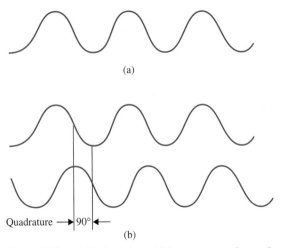

Figure 4-62 (a) Typical sinusoidal output waveform of a single-channel encoder device. (b) Typical dual-channel encoder signals in quadrature.

quadrature outputs with sinusoidal waveforms. The sinusoidal signals from the incremental encoder can be used for fine position control in feedback control systems. The following example illustrates some applications of the incremental encoder in control systems.

▶ **EXAMPLE 4-6** Consider an incremental encoder that generates two sinusoidal signals in quadrature as the encoder disk rotates. The output signals of the two channels are shown in Fig. 4-63 over one cycle. Note that the two encoder signals generate 4 zero crossings per cycle. These zero crossings can be used for position indication, position control, or speed measurements in control systems. Let us assume that the encoder shaft is coupled directly to the rotor shaft of a motor that directly drives the printwheel of an electronic typewriter or word processor. The printwheel has 96 character positions on its periphery, and the encoder has 480 cycles. Thus, there are $480 \times 4 = 1920$ zero crossings per revolution. For the 96-character printwheel, this corresponds to $1920/96 = 20$ zero crossings per character; that is, there are 20 zero crossings between two adjacent characters.

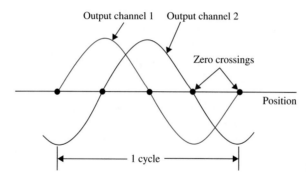

Figure 4-63 One cycle of the output signals of a dual-channel incremental encoder.

One way of measuring the velocity of the printwheel is to count the number of pulses generated by an electronic clock that occur between consecutive zero crossings of the encoder outputs. Let us assume that a 500-kHz clock is used; that is, the clock generates 500,000 pulses/sec. If the counter records, say, 500 clock pulses while the encoder rotates from the zero crossing to the next, the shaft speed is

$$\frac{500,000 \text{ pulses/sec}}{500 \text{ pulses/zero crossing}} = 1000 \text{ zero crossings/sec}$$

$$= \frac{1000 \text{ zero crossings/sec}}{1920 \text{ zero crossings/rev}} = 0.52083 \text{ rev/sec} \quad (4\text{-}199)$$

$$= 31.25 \text{ rpm}$$

The encoder arrangement described can be used for fine position control of the printwheel. Let the zero crossing A of the waveforms in Fig. 4-63 correspond to a character position on the printwheel (the next character position is 20 zero crossings away), and the point corresponds to a stable equilibrium point. The coarse position control of the system must first drive the printwheel position to within 1 zero crossing on either side of position A; then, by using the slope of the sine wave at position A, the control system should null the error quickly.

▶ 4-7 DC MOTORS IN CONTROL SYSTEMS

Direct-current (dc) motors are one of the most widely used prime movers in the industry today. Years ago, the majority of the small servomotors used for control purposes were ac. In reality, ac motors are more difficult to control, especially for position control, and their characteristics are quite nonlinear, which makes the analytical task more difficult. DC motors, on the other hand, are more expensive, because of their brushes and commutators, and variable-flux dc motors are suitable only for certain types of control applications. Before permanent-magnet technology was fully developed, the torque-per-unit volume or weight of a dc motor with a permanent-magnet (PM) field was far from desirable. Today, with the development of the rare-earth magnet, it is possible to achieve very high torque-to-volume PM dc motors at reasonable cost. Furthermore, the advances made in brush-and-commutator technology have made these wearable parts practically maintenance-free. The advancements made in power electronics have made brushless dc motors quite popular in high-performance control systems. Advanced manufacturing techniques have also produced dc motors with ironless rotors that have very low inertia, thus achieving a very high torque-to-inertia ratio. Low-time-constant properties have opened new applications for dc motors in computer peripheral equipment such as tape drives, printers, disk drives, and word processors, as well as in the automation and machine-tool industries.

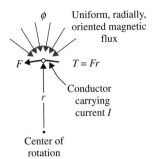

Figure 4-64 Torque production in a dc motor.

4-7-1 Basic Operational Principles of DC Motors

The dc motor is basically a torque transducer that converts electric energy into mechanical energy. The torque developed on the motor shaft is directly proportional to the field flux and the armature current. As shown in Fig. 4-64, a current-carrying conductor is established in a magnetic field with flux ϕ, and the conductor is located at a distance r from the center of rotation. The relationship among the developed torque, flux ϕ, and current i_a is

$$T_m = K_m \phi i_a \tag{4-200}$$

where T_m is the motor torque (in N-m, lb-ft, or oz-in.); ϕ, the magnetic flux (in webers); i_a, the armature current (in amperes); and K_m, a proportional constant.

In addition to the torque developed by the arrangement shown in Fig. 4-64, when the conductor moves in the magnetic field, a voltage is generated across its terminals. This voltage, the **back emf**, which is proportional to the shaft velocity, tends to oppose the current flow. The relationship between the back emf and the shaft velocity is

$$e_b = K_m \phi \omega_m \tag{4-201}$$

where e_b denotes the back emf (volts) and ω_m is the shaft velocity (rad/sec) of the motor. Eqs. (4-200) and (4-201) form the basis of the dc-motor operation.

4-7-2 Basic Classifications of PM DC Motors

In general, the magnetic field of a dc motor can be produced by field windings or permanent magnets. Due to the popularity of PM dc motors in control system applications, we shall concentrate on this type of motor.

PM dc motors can be classified according to commutation scheme and armature design. Conventional dc motors have mechanical brushes and commutators. However, an important type of dc motors in which the commutation is done electronically is called **brushless dc**.

According to the armature construction, the PM dc motor can be broken down into three types of armature design: **iron-core**, **surface-wound**, and **moving-coil** motors.

Iron-Core PM DC Motors

The rotor and stator configuration of an iron-core PM dc motor is shown in Fig. 4-65. The permanent-magnet material can be barium ferrite, Alnico, or a rare-earth compound. The magnetic flux produced by the magnet passes through a laminated rotor structure that contains slots. The armature conductors are placed in the rotor slots. This type of dc motor

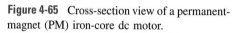 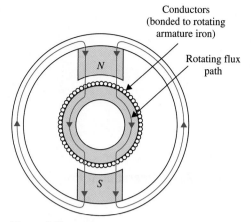

Figure 4-65 Cross-section view of a permanent-magnet (PM) iron-core dc motor.

Figure 4-66 Cross-section view of a surface-wound permanent-magnet (PM) dc motor.

is characterized by relatively high rotor inertia (since the rotating part consists of the armature windings), high inductance, low cost, and high reliability.

Surface-Wound DC Motors

Fig. 4-66 shows the rotor construction of a surface-wound PM dc motor. The armature conductors are bonded to the surface of a cylindrical rotor structure, which is made of laminated disks fastened to the motor shaft. Because no slots are used on the rotor in this design, the armature has no "cogging" effect. The conductors are laid out in the air gap between the rotor and the PM field, so this type of motor has lower inductance than that of the iron-core structure.

Moving-Coil DC Motors

Moving-coil motors are designed to have very low moments of inertia and very low armature inductance. This is achieved by placing the armature conductors in the air gap between a stationary flux return path and the PM structure, as shown in Fig. 4-67. In this case, the conductor structure is supported by nonmagnetic material—usually epoxy resins or fiberglass—to form a hollow cylinder. One end of the cylinder forms a hub, which is attached to the motor shaft. A cross-section view of such a motor is shown in Fig. 4-68. Because all

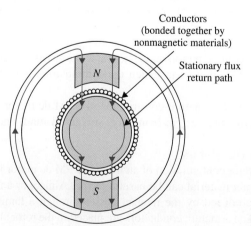

Figure 4-67 Cross-section view of a surface-wound permanent-magnet (PM) dc motor.

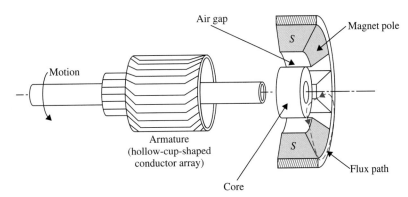

Figure 4-68 Cross-section side view of a moving-coil dc motor.

unnecessary elements have been removed from the armature of the moving-coil motor, its moment of inertia is very low. Because the conductors in the moving-coil armature are not in direct contact with iron, the motor inductance is very low, and values of less than 100 μH are common in this type of motor. Its low-inertia and low-inductance properties make the moving-coil motor one of the best actuator choices for high-performance control systems.

Brushless DC Motors

Brushless dc motors differ from the previously mentioned dc motors in that they employ electrical (rather than mechanical) commutation of the armature current. The most common configuration of brushless dc motors—especially for incremental-motion applications—is one in which the rotor consists of magnets and "back-iron" support and whose commutated windings are located external to the rotating parts, as shown in Fig. 4-69. Compared to the conventional dc motors, such as the one shown in Fig. 4-68, it is an inside-out configuration.

Depending on the specific application, brushless dc motors can be used when a low moment of inertia is needed, such as the spindle drive in high-performance disk drives used in computers.

4-7-3 Mathematical Modeling of PM DC Motors

Dc motors are extensively used in control systems. In this section we establish the mathematical model for dc motors. As it will be demonstrated here, the mathematical model of a dc motor is linear. We use the equivalent circuit diagram in Fig. 4-70 to represent a PM dc

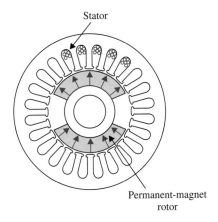

Figure 4-69 Cross-section view of a brushless, permanent-magnet (PM), iron-core dc motor.

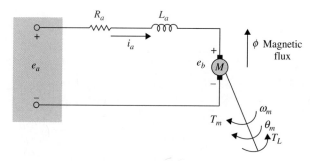

Figure 4-70 Model of a separately excited dc motor.

motor. The armature is modeled as a circuit with resistance R_a connected in series with an inductance L_a, and a voltage source e_b representing the back emf (electromotive force) in the armature when the rotor rotates. The motor variables and parameters are defined as follows:

$i_a(t)$ = armature current L_a = armature inductance
R_a = armature resistance $e_a(t)$ = applied voltage
$e_b(t)$ = back emf K_b = back-emf constant
$T_L(t)$ = load torque ϕ = magnetic flux in the air gap
$T_m(t)$ = motor torque $\omega_m(t)$ = rotor angular velocity
$\theta_m(t)$ = rotor displacement J_m = rotor inertia
K_i = torque constant B_m = viscous-friction coefficient

With reference to the circuit diagram of Fig. 4-70, the control of the dc motor is applied at the armature terminals in the form of the applied voltage $e_a(t)$. For linear analysis, we assume that the torque developed by the motor is proportional to the air-gap flux and the armature current. Thus,

$$T_m(t) = K_m(t)\phi i_a(t) \tag{4-202}$$

Because ϕ is constant, Eq. (4-202) is written

$$T_m(t) = K_i i_a(t) \tag{4-203}$$

where K_i is the **torque constant** in N-m/A, lb-ft/A, or oz-in/A.

Starting with the control input voltage $e_a(t)$, the cause-and-effect equations for the motor circuit in Fig. 4-70 are

$$\frac{di_a(t)}{dt} = \frac{1}{L_a}e_a(t) - \frac{R_a}{L_a}i_a(t) - \frac{1}{L_a}e_b(t) \tag{4-204}$$

$$T_m(t) = K_i i_a(t) \tag{4-205}$$

$$e_b(t) = K_b \frac{d\theta_m(t)}{dt} = K_b \omega_m(t) \tag{4-206}$$

$$\frac{d^2\theta_m(t)}{dt^2} = \frac{1}{J_m}T_m(t) - \frac{1}{J_m}T_L(t) - \frac{B_m}{J_m}\frac{d\theta_m(t)}{dt} \tag{4-207}$$

where $T_L(t)$ represents a load frictional torque such as Coulomb friction.

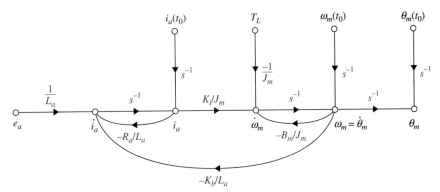

Figure 4-71 Signal-flow graph diagram of a dc-motor system with nonzero initial conditions.

Eqs. (4-204) through (4-207) consider that the applied voltage $e_a(t)$ is the cause; Eq. (4-204) considers that $di_a(t)/dt$ is the immediate effect due to $e_a(t)$; in Eq. (4-205), $i_a(t)$ causes the torque $T_m(t)$; Eq. (4-206) defines the back emf; and, finally, in Eq. (4-207), the torque $T_m(t)$ causes the angular velocity $\omega_m(t)$ and displacement $\theta_m(t)$.

The state variables of the system can be defined as $i_a(t)$, $\omega_m(t)$, and $\theta_m(t)$. By direct substitution and eliminating all the nonstate variables from Eqs. (4-204) through (4-207), the state equations of the dc-motor system are written in vector-matrix form:

$$\begin{bmatrix} \dfrac{di_a(t)}{dt} \\ \dfrac{d\omega_m(t)}{dt} \\ \dfrac{d\theta_m(t)}{dt} \end{bmatrix} = \begin{bmatrix} -\dfrac{R_a}{L_a} & -\dfrac{K_b}{L_a} & 0 \\ \dfrac{K_i}{J_m} & -\dfrac{B_m}{J_m} & 0 \\ 0 & 1 & 0 \end{bmatrix} \begin{bmatrix} i_a(t) \\ \omega_m(t) \\ \theta_m(t) \end{bmatrix} + \begin{bmatrix} \dfrac{1}{L_a} \\ 0 \\ 0 \end{bmatrix} e_a(t) + \begin{bmatrix} 0 \\ -\dfrac{1}{J_m} \\ 0 \end{bmatrix} T_L(t) \quad (4\text{-}208)$$

Notice that, in this case, $T_L(t)$ is treated as a second input in the state equations.

The SFG diagram of the system is drawn as shown in Fig. 4-71, using Eq. (4-208). The transfer function between the motor displacement and the input voltage is obtained from the state diagram as

$$\frac{\Theta_m(s)}{E_a(s)} = \frac{K_i}{L_a J_m s^3 + (R_a J_m + B_m L_a)s^2 + (K_b K_i + R_a B_m)s} \quad (4\text{-}209)$$

where $T_L(t)$ has been set to zero.

Fig. 4-72 shows a block-diagram representation of the dc-motor system. The advantage of using the block diagram is that it gives a clear picture of the transfer function

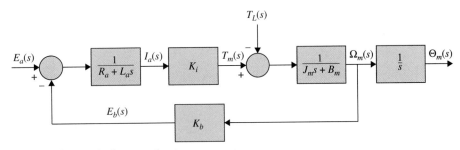

Figure 4-72 Block diagram of a dc-motor system.

relation between each block of the system. Because an s can be factored out of the denominator of Eq. (4-209), *the significance of the transfer function $\Theta_m(s)/E_a(s)$ is that the dc motor is essentially an integrating device between these two variables.* This is expected because, if $e_a(t)$ is a constant input, the output motor displacement will behave as the output of an integrator; that is, it will increase linearly with time.

Although a dc motor by itself is basically an open-loop system, the SFG diagram of Fig. 4-71 and the block diagram of Fig. 4-72 show that the motor has a "built-in" feedback loop caused by the back emf. Physically, the back emf represents the feedback of a signal that is proportional to the negative of the speed of the motor. As seen from Eq. (4-209), the back-emf constant K_b represents an added term to the resistance R_a and the viscous-friction coefficient B_m. Therefore, *the back-emf effect is equivalent to an "electric friction," which tends to improve the stability of the motor and, in general, the stability of the system.*

Relation between K_i and K_b

Although functionally the torque constant K_i and back-emf constant K_b are two separate parameters, for a given motor their values are closely related. To show the relationship, we write the mechanical power developed in the armature as

$$P = e_b(t)i_a(t) \tag{4-210}$$

The mechanical power is also expressed as

$$P = T_m(t)\omega_m(t) \tag{4-211}$$

where, in SI units, $T_m(t)$ is in N-m and $\omega_m(t)$ is in rad/sec. Now, substituting Eqs. (4-205) and (4-206) in Eq. (4-210), we get

$$P = T_m(t)\omega_m(t) = K_b\omega_m(t)\frac{T_m(t)}{K_i} \tag{4-212}$$

from which we get

$$K_b(\text{V/rad/sec}) = K_i(\text{N-m/A}) \tag{4-213}$$

Thus, we see that, in SI units, the values of K_b and K_i are identical if K_b is represented in V/rad/sec and K_i is in N-m/A.

In the British unit system, we convert Eq. (4-210) into horsepower (hp); that is,

$$P = \frac{e_b(t)i_a(t)}{746} \text{ hp} \tag{4-214}$$

In terms of torque and angular velocity, P is

$$P = \frac{T_m(t)\omega_m(t)}{550} \text{ hp} \tag{4-215}$$

where $T_m(t)$ is in ft-lb and $\omega_m(t)$ is in rad/sec. Using Eq. (4-205) and (4-206), and equating Eq. (4-214) to Eq. (4-215), we get

$$\frac{K_b\omega_m(t)T_m(t)}{746K_i} = \frac{T_m(t)\omega_m(t)}{550} \tag{4-216}$$

Thus,

$$K_b = \frac{746}{550}K_i = 1.356K_i \qquad (4\text{-}217)$$

where K_b is in V/rad/sec and K_i is in ft-lb/A.

▶ 4-8 SYSTEMS WITH TRANSPORTATION LAGS (TIME DELAYS)

Thus far, the systems considered all have transfer functions that are quotients of polynomials. In practice, pure time delays may be encountered in various types of systems, especially systems with hydraulic, pneumatic, or mechanical transmissions. Systems with computer control also have time delays, since it takes time for the computer to execute numerical operations. In these systems, the output will not begin to respond to an input until after a given time interval. Fig. 4-73 illustrates systems in which transportation lags or pure time delays are observed. Fig. 4-73(a) outlines an arrangement where two different fluids are to be mixed in appropriate proportions. To assure that a homogeneous solution is measured, the monitoring point is located some distance from the mixing point. A time delay therefore exists between the mixing point and the place where the change in concentration is detected. If the rate of flow of the mixed solution is v inches per second and d is the distance between the mixing and the metering points, the time lag is given by

$$T_d = \frac{d}{v} \text{ seconds} \qquad (4\text{-}218)$$

If it is assumed that the concentration of the mixing point is $y(t)$ and that it is reproduced without change T_d seconds later at the monitoring point, the measured quantity is

$$b(t) = y(t - T_d) \qquad (4\text{-}219)$$

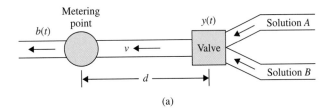

(a)

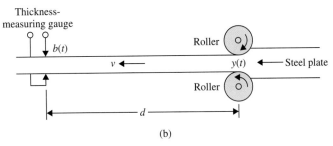

(b)

Figure 4-73 Systems with transportation lag.

The Laplace transform of Eq. (4-219) is

$$B(s) = e^{-T_d s} Y(s) \qquad (4\text{-}220)$$

where $Y(s)$ is the Laplace transform of $y(t)$. The transfer function between $b(t)$ and $y(t)$ is

$$\frac{B(s)}{Y(s)} = e^{-T_d s} \qquad (4\text{-}221)$$

Fig. 4-73(b) illustrates the control of thickness of rolled steel plates. The transfer function between the thickness at the rollers and the measuring point is again given by Eq. (4-221).

4-8-1 Approximation of the Time-Delay Function by Rational Functions

Systems that are described inherently by transcendental transfer functions are more difficult to handle. Many analytical tools such as the Routh-Hurwitz criterion (Chapter 2) are restricted to rational transfer functions. The root-locus technique (Chapter 7) is also more easily applied only to systems with rational transfer functions.

There are many ways of approximating $e^{-T_d s}$ by a rational function. One way is to approximate the exponential function by a Maclaurin series; that is,

$$e^{-T_d s} \cong 1 - T_d s + \frac{T_d^2 s^2}{2} \qquad (4\text{-}222)$$

or

$$e^{-T_d s} \cong \frac{1}{1 + T_d s + T_d^2 s^2/2} \qquad (4\text{-}223)$$

where only three terms of the series are used. Apparently, the approximations are not valid when the magnitude of $T_d s$ is large.

A better approximation is to use the Padé approximation [5, 6], which is given in the following for a two-term approximation:

$$e^{-T_d s} \cong \frac{1 - T_d s/2}{1 + T_d s/2} \qquad (4\text{-}224)$$

The approximation of the transfer function in Eq. (4-224) contains a zero in the right-half s-plane so that the step response of the approximating system may exhibit a small negative undershoot near $t = 0$.

▶ 4-9 LINEARIZATION OF NONLINEAR SYSTEMS

From the discussions given in the preceding sections on basic system modeling, we should realize that most components found in physical systems have nonlinear characteristics. In practice, we may find that some devices have moderate nonlinear characteristics, or nonlinear properties that would occur if they were driven into certain operating regions. For these devices, the modeling by linear-system models may give quite accurate analytical results over a relatively wide range of operating conditions.

However, there are numerous physical devices that possess strong nonlinear characteristics. For these devices, a linearized model is valid only for limited range of operation and often only at the operating point at which the linearization is carried out. More importantly, when a nonlinear system is linearized at an operating point, the linear model may contain time-varying elements.

4-9-1 Linearization Using Taylor Series: Classical Representation

In general, Taylor series may be used to expand a nonlinear function $f(x(t))$ about a reference or operating value $x_0(t)$. An operating value could be the equilibrium position in a spring-mass-damper, a fixed voltage in an electrical system, steady state pressure in a fluid system, and so on. A function $f(x(t))$ can therefore be represented in a form

$$f(x(t)) = \sum_{i=1}^{n} c_i (x(t) - x_0(t))^i \qquad (4\text{-}225)$$

where the constant c_i represents the derivatives of $f(x(t))$ with respect to $x(t)$ and evaluated at the operating point $x_0(t)$. That is

$$c_i = \frac{1}{i!} \frac{d^i f(x_0)}{dx^i} \qquad (4\text{-}226)$$

Or

$$\begin{aligned} f(x(t)) = f(x_0(t)) &+ \frac{d f(x_0(t))}{dt}(x(t) - x_0(t)) + \frac{1}{2}\frac{d^2 f(x_0(t))}{dt^2}(x(t) - x_0(t))^2 + \\ \frac{1}{6}\frac{d^3 f(x_0(t))}{dt^3}(x(t) - x_0(t))^3 &+ \cdots + \frac{1}{n!}\frac{d^n f(x_0(t))}{dt^n}(x(t) - x_0(t))^n \end{aligned} \qquad (4\text{-}227)$$

If $\Delta(x) = x(t) - x_0(t)$ is small, the series Eq. (4-227) converges, and a linearization scheme may be used by replacing $f(x(t))$ with the first two terms in Eq. (4-227). That is,

$$\begin{aligned} f(x(t)) &\approx f(x_0(t)) + \frac{d f(x_0(t))}{dt}(x(t) - x_0(t)) \\ &= c_0 + c_1 \Delta x \end{aligned} \qquad (4\text{-}228)$$

4-9-2 Linearization Using the State Space Approach

Alternatively, let us represent a nonlinear system by the following vector-matrix state equations:

$$\frac{d\mathbf{x}(t)}{dt} = \mathbf{f}[\mathbf{x}(t), \mathbf{r}(t)] \qquad (4\text{-}229)$$

where $\mathbf{x}(t)$ represents the $n \times 1$ state vector; $\mathbf{r}(t)$, the $p \times 1$ input vector; and $\mathbf{f}[\mathbf{x}(t), \mathbf{r}(t)]$, an $n \times 1$ function vector. In general, $\mathbf{f}$ is a function of the state vector and the input vector.

Being able to represent a nonlinear and/or time-varying system by state equations is a distinct advantage of the state-variable approach over the transfer-function method, since the latter is strictly defined only for linear time-invariant systems.

As a simple example, the following nonlinear state equations are given:

$$\frac{dx_1(t)}{dt} = x_1(t) + x_2^2(t) \tag{4-230}$$

$$\frac{dx_2(t)}{dt} = x_1(t) + r(t) \tag{4-231}$$

Because nonlinear systems are usually difficult to analyze and design, it is desirable to perform a linearization whenever the situation justifies it.

A linearization process that depends on expanding the nonlinear state equations into a Taylor series about a nominal operating point or trajectory is now described. All the terms of the Taylor series of order higher than the first are discarded, and the linear approximation of the nonlinear state equations at the nominal point results.

Let the nominal operating trajectory be denoted by $\mathbf{x}_0(t)$, which corresponds to the nominal input $\mathbf{r}_0(t)$ and some fixed initial states. Expanding the nonlinear state equation of Eq. (4-229) into a Taylor series about $\mathbf{x}(t) = \mathbf{x}_0(t)$ and neglecting all the higher-order terms yields

$$\dot{x}_i(t) = f_i(\mathbf{x}_0, \mathbf{r}_0) + \sum_{j=1}^{n} \left. \frac{\partial f_i(\mathbf{x}, \mathbf{r})}{\partial x_j} \right|_{\mathbf{x}_0, \mathbf{r}_0} (x_j - x_{0j}) + \sum_{j=1}^{p} \left. \frac{\partial f_i(\mathbf{x}, \mathbf{r})}{\partial r_j} \right|_{\mathbf{x}_0, \mathbf{r}_0} (r_j - r_{0j}) \tag{4-232}$$

where $i = 1, 2, \ldots, n$. Let

$$\Delta x_i = x_i - x_{0i} \tag{4-233}$$

and

$$\Delta r_j = r_j - r_{0j} \tag{4-234}$$

Then

$$\Delta \dot{x}_i = \dot{x}_i - \dot{x}_{0i} \tag{4-235}$$

Since

$$\dot{x}_{0i} = f_i(\mathbf{x}_0, \mathbf{r}_0) \tag{4-236}$$

Eq. (4-232) is written

$$\Delta \dot{x}_i = \sum_{j=1}^{n} \left. \frac{\partial f_i(\mathbf{x}, \mathbf{r})}{\partial x_j} \right|_{\mathbf{x}_0, \mathbf{r}_0} \Delta x_j + \sum_{j=1}^{p} \left. \frac{\partial f_i(\mathbf{x}, \mathbf{r})}{\partial r_j} \right|_{\mathbf{x}_0, \mathbf{r}_0} \Delta r_j \tag{4-237}$$

Eq. (4-237) may be written in vector-matrix form:

$$\Delta \dot{\mathbf{x}} = \mathbf{A}^* \Delta \mathbf{x} + \mathbf{B}^* \Delta \mathbf{r} \tag{4-238}$$

4-9 Linearization of Nonlinear Systems

where

$$\mathbf{A}^* = \begin{bmatrix} \dfrac{\partial f_1}{\partial x_1} & \dfrac{\partial f_1}{\partial x_2} & \cdots & \dfrac{\partial f_1}{\partial x_n} \\ \dfrac{\partial f_2}{\partial x_1} & \dfrac{\partial f_2}{\partial x_2} & \cdots & \dfrac{\partial f_2}{\partial x_n} \\ \vdots & \vdots & \ddots & \vdots \\ \dfrac{\partial f_n}{\partial x_1} & \dfrac{\partial f_n}{\partial x_2} & \cdots & \dfrac{\partial f_n}{\partial x_n} \end{bmatrix} \qquad (4\text{-}239)$$

$$\mathbf{B}^* = \begin{bmatrix} \dfrac{\partial f_1}{\partial r_1} & \dfrac{\partial f_1}{\partial r_2} & \cdots & \dfrac{\partial f_1}{\partial r_p} \\ \dfrac{\partial f_2}{\partial r_1} & \dfrac{\partial f_2}{\partial r_2} & \cdots & \dfrac{\partial f_2}{\partial r_p} \\ \vdots & \vdots & \ddots & \vdots \\ \dfrac{\partial f_n}{\partial r_1} & \dfrac{\partial f_n}{\partial r_2} & \cdots & \dfrac{\partial f_n}{\partial r_p} \end{bmatrix} \qquad (4\text{-}240)$$

The following examples serve to illustrate the linearization procedure just described.

▶ **EXAMPLE 4-9-1** Find the equation of motion of a pendulum with a mass m and a massless rod of length l, as shown in Fig. 4-74.

SOLUTION Assume the mass is moving in the positive direction as defined by angle θ. Note that θ is measured from the x axis in the counter-clockwise direction. The first step is to draw the free-body diagram of the components of the system, i.e., mass and the rod, as shown in Fig. 4-74(b). For the mass m, the equations of motion are

$$\sum F_x = ma_x \qquad (4\text{-}241)$$

$$\sum F_y = ma_y \qquad (4\text{-}242)$$

where F_x and F_y are the external forces applied to mass m, and a_x and a_y are the components of acceleration of mass m in x and y, respectively. Acceleration of mass m is a vector with tangential and centripetal components. Using the rectangular coordinate frame (x, y) representation, acceleration vector is

$$\mathbf{a} = \left(-l\ddot{\theta}\sin\theta - l\dot{\theta}^2\sin\theta\right)\hat{i} + \left(l\ddot{\theta}\cos\theta - l\dot{\theta}^2\sin\theta\right)\hat{j} \qquad (4\text{-}243)$$

where $\hat{i}$ and $\hat{j}$ are the unit vectors along x and y directions, respectively. As a result,

$$a_x = \left(-l\ddot{\theta}\sin\theta - l\dot{\theta}^2\sin\theta\right) \qquad (4\text{-}244)$$

$$a_y = \left(l\ddot{\theta}\cos\theta - l\dot{\theta}^2\sin\theta\right) \qquad (4\text{-}245)$$

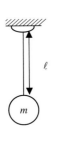
(a)

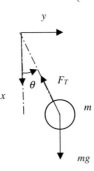
(b)

Figure 4-74 (a) A spring-supported pendulum. (b) Free-body diagram of mass m.

Considering the external forces applied to mass, we have

$$\sum F_x = -F_T \cos\theta + mg \quad (4\text{-}246)$$

$$\sum F_y = -F_T \sin\theta \quad (4\text{-}247)$$

Eqs. (4-241) and (4-242) may therefore be rewritten as

$$-F_T \cos\theta + mg = m\left(-\ell\ddot{\theta}\sin\theta - \ell\dot{\theta}^2 \sin\theta\right) \quad (4\text{-}248)$$

$$-F_T \sin\theta = m\left(\ell\ddot{\theta}\cos\theta - \ell\dot{\theta}^2 \sin\theta\right) \quad (4\text{-}249)$$

Premultiplying Eq. (4-185) by $(-\sin\theta)$ and Eq. (4-186) by $(\cos\theta)$ and adding the two, we get

$$-mg\sin\theta = m\ell\ddot{\theta} \quad (4\text{-}250)$$

where $(\sin^2\theta + \cos^2\theta = 1)$. After rearranging, Eq. (4-250) is rewritten as

$$m\ell\ddot{\theta} + mg\sin\theta = 0 \quad (4\text{-}251)$$

Or

$$\ddot{\theta} + \frac{g}{\ell}\sin\theta = 0 \quad (4\text{-}252)$$

In brief, using static equilibrium position $\theta = 0$ as the operating point, for small motions the linearization of the system implies $\Delta\theta = \theta \approx \sin\theta$. Hence, the linear representation of the system is $\ddot{\theta} + \frac{g}{\ell}\theta = 0$.

Alternatively in the state space form, we define $x_1 = \theta$ and $x_2 = \dot{\theta}$ as state variables, and as a result the state space representation of Eq. (4-252) becomes

$$\dot{x}_1 = x_2$$
$$\dot{x}_2 = -\frac{g}{\ell}\sin x_1 \quad (4\text{-}253)$$

Substituting Eq. (4-253) into (4-173) with $\mathbf{r}(t) = 0$, since there is no input (or external excitations) in this case, we get

$$\Delta\dot{x}_1(t) = \frac{\partial f_1(t)}{\partial x_2}\Delta x_2(t) = \Delta x_2(t) \quad (4\text{-}254)$$

$$\Delta\dot{x}_2(t) = \frac{\partial f_2(t)}{\partial x_1(t)}\Delta x_1(t) = -\frac{g}{\ell}\Delta x_1(t) \quad (4\text{-}255)$$

where $\Delta x_1(t)$ and $\Delta x_2(t)$ denote nominal values of $x_1(t)$ and $x_2(t)$, respectively. Notice that the last two equations are linear and are valid only for small signals. In vector-matrix form, these linearized state equations are written as

$$\begin{bmatrix}\Delta\dot{x}_1(t)\\ \Delta\dot{x}_2(t)\end{bmatrix} = \begin{bmatrix}0 & 1\\ a & 0\end{bmatrix}\begin{bmatrix}\Delta x_1(t)\\ \Delta x_2(t)\end{bmatrix} \quad (4\text{-}256)$$

where

$$a = \frac{g}{\ell} = \text{constant} \quad (4\text{-}257)$$

It is of interest to check the significance of the linearization. If x_{01} is chosen to be at the origin of the nonlinearity, $x_{01} = 0$, then $a = K$; Eq. (4-255) becomes

$$\Delta\dot{x}_2(t) = K\Delta x_1(t) \quad (4\text{-}258)$$

Switching back to classical representation, we get

$$\ddot{\theta} + K\theta = 0 \quad (4\text{-}259)$$

◀

▶ **EXAMPLE 4-9-2** For the pendulum shown in Fig. 4-74, re-derive the differential equation using the moment equation.

SOLUTION The free-body diagram for the moment equation is shown in Fig. 4-74. Applying the moment equation about the fixed point O,

$$\sum M_o = m\ell^2 \alpha$$
$$-\ell\sin\theta \cdot mg = m\ell^2\ddot{\theta} \quad (4\text{-}260)$$

Rearranging the equation in the standard input–output differential equation form,
$$m\ell^2\ddot{\theta} + mg\ell \sin\theta = 0 \tag{4-261}$$
or
$$\ddot{\theta} + \frac{g}{\ell}\sin\theta = 0 \tag{4-262}$$
which is the same result obtained previously. For small motions, as in Example 4-9-1,
$$\sin\theta \approx \theta \tag{4-263}$$
The linearized differential equation is
$$\ddot{\theta} + \omega_n^2 \theta = 0 \tag{4-264}$$
where
$$\omega_n = \sqrt{\frac{g}{\ell}} \tag{4-265}$$

◀

▶ **EXAMPLE 4-9-3** In Example 4-9-1, the linearized system turns out to be time-invariant. As mentioned earlier, linearization of a nonlinear system often results in a linear time-varying system. Consider the following nonlinear system:
$$\dot{x}_1(t) = \frac{-1}{x_2^2(t)} \tag{4-266}$$
$$\dot{x}_2(t) = u(t)x_1(t) \tag{4-267}$$
These equations are to be linearized about the nominal trajectory $[x_{01}(t), x_{02}(t)]$, which is the solution to the equations with initial conditions $x_1(0) = x_2(0) = 1$ and input $u(t) = 0$.

Integrating both sides of Eq. (4-267) with respect to t, we have
$$x_2(t) = x_2(0) = 1 \tag{4-268}$$
Then Eq. (4-266) gives
$$x_1(t) = -t + 1 \tag{4-269}$$
Therefore, the nominal trajectory about which Eqs. (4-266) and (4-267) are to be linearized is described by
$$x_{01}(t) = -t + 1 \tag{4-270}$$
$$x_{02}(t) = 1 \tag{4-271}$$
Now evaluating the coefficients of Eq. (4-237), we get
$$\frac{\partial f_1(t)}{\partial x_1(t)} = 0 \quad \frac{\partial f_1(t)}{\partial x_2(t)} = \frac{2}{x_2^3(t)} \quad \frac{\partial f_2(t)}{\partial x_1(t)} = u(t) \quad \frac{\partial f_2(t)}{\partial u(t)} = x_1(t) \tag{4-272}$$
Eq. (4-237) gives
$$\Delta \dot{x}_1(t) = \frac{2}{x_{02}^3(t)} \Delta x_2(t) \tag{4-273}$$
$$\Delta \dot{x}_2(t) = u_0(t)\Delta x_1(t) + x_{01}(t)\Delta u(t) \tag{4-274}$$
By substituting Eqs. (4-270) and (4-271) into Eqs. (4-273) and (4-274), the linearized equations are
$$\begin{bmatrix} \Delta \dot{x}_1(t) \\ \Delta \dot{x}_2(t) \end{bmatrix} = \begin{bmatrix} 0 & 2 \\ 0 & 0 \end{bmatrix} \begin{bmatrix} \Delta x_1(t) \\ \Delta x_2(t) \end{bmatrix} + \begin{bmatrix} 0 \\ 1-t \end{bmatrix} \Delta u(t) \tag{4-275}$$
which is a set of linear state equations with time-varying coefficients.

Fig. 4-75 shows the diagram of a magnetic-ball-suspension system. The objective of the system is to control the position of the steel ball by adjusting the current in the electromagnet through the input voltage $e(t)$. The differential equations of the system are
$$M\frac{d^2y(t)}{dt^2} = Mg - \frac{i^2(t)}{y(t)} \tag{4-276}$$
$$e(t) = Ri(t) + L\frac{di(t)}{dt} \tag{4-277}$$

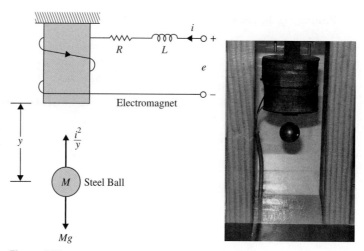

Figure 4-75 Magnetic-ball-suspension system.

where

$e(t)$ = input voltage
$y(t)$ = ball position
$i(t)$ = winding current
R = winding resistance
L = winding inductance
M = mass of ball
g = gravitational acceleration

Let us define the state variables as $x_1(t) = y(t)$, $x_2(t) = dy(t)/dt$, and $x_3(t) = i(t)$. The state equations of the system are

$$\frac{dx_1(t)}{dt} = x_2(t) \tag{4-278}$$

$$\frac{dx_2(t)}{dt} = g - \frac{1}{M}\frac{x_3^2(t)}{x_1(t)} \tag{4-279}$$

$$\frac{dx_3(t)}{dt} = -\frac{R}{L}x_3(t) + \frac{1}{L}e(t) \tag{4-280}$$

Let us linearize the system about the equilibrium point $y_0(t) = x_{01}$ = constant. Then,

$$x_{02}(t) = \frac{dx_{01}(t)}{dt} = 0 \tag{4-281}$$

$$\frac{d^2 y_0(t)}{dt^2} = 0 \tag{4-282}$$

The nominal value of $i(t)$ is determined by substituting Eq. (4-282) into Eq. (4-276)

$$e(t) = Ri(t) + L\frac{di(t)}{dt} \tag{4-283}$$

Thus,

$$i_0(t) = x_{03}(t) = \sqrt{Mgx_{01}} \tag{4-284}$$

The linearized state equation is expressed in the form of Eq. (4-238), with the coefficient matrices $\mathbf{A}^*$ and $\mathbf{B}^*$ evaluated as

$$\mathbf{A}^* = \begin{bmatrix} 0 & 1 & 0 \\ \dfrac{x_{03}^2}{Mx_{01}^2} & 0 & \dfrac{-2x_{03}}{Mx_{01}} \\ 0 & 0 & -\dfrac{R}{L} \end{bmatrix} = \begin{bmatrix} 0 & 1 & 0 \\ \dfrac{g}{x_{01}} & 0 & -2\left(\dfrac{g}{Mx_{01}}\right)^{1/2} \\ 0 & 0 & -\dfrac{R}{L} \end{bmatrix} \qquad (4\text{-}285)$$

$$\mathbf{B}^* = \begin{bmatrix} 0 \\ 0 \\ \dfrac{1}{L} \end{bmatrix} \qquad (4\text{-}286)$$

4-10 ANALOGIES

Comparing Eqs. (4-11), (4-41), and (4-65), it is not difficult to see that the mechanical systems in Eqs. (4-11) and (4-41) are analogous to a series **RLC** electric network shown in Example 4-2-1. As a result, with this analogy, mass M and inertia J are analogous to inductance L, the spring constant K is analogous to the inverse of capacitance $1/C$, and the viscous-friction coefficient B is analogous to resistance R.

▶ **EXAMPLE 4-10-1** It is logical, in Example 4-1-1, to assign $v(t)$, the velocity, and $f_k(t)$, the force acting on the spring, as state variables, since the former is analogous to the current in L and the latter is analogous to the voltage across C. Writing the force on M and the velocity of the spring as functions of the state variables and the input force $f(t)$, we have

Force on mass:
$$M\dfrac{dv(t)}{dt} = -f_k(t) - Bv(t) + f(t) \qquad (4\text{-}287)$$

Velocity of spring:
$$\dfrac{1}{K}\dfrac{df_k(t)}{dt} = v(t) \qquad (4\text{-}288)$$

The final equation of motion Eq. (4-11) may be obtained by dividing both sides of Eq. (4-287) by M and multiplying Eq. (4-288) by K. Hence, in terms of displacement $y(t)$,

$$\dfrac{d^2y(t)}{dt^2} + \dfrac{B}{M}\dfrac{dy(t)}{dt} + \dfrac{K}{M}y(t) = \dfrac{f(t)}{M} \qquad (4\text{-}289)$$

Considering Example 4-2-1, after rewriting Eq. (4-67) as

$$L\dfrac{di(t)}{dt} = -e_c(t) - Ri(t) + e(t) \qquad (4\text{-}290)$$

and using the current relation Eq. (4-66):

$$C\dfrac{de_c(t)}{dt} = i(t) \qquad (4\text{-}291)$$

the comparison of Eq. (4-287) with Eq. (4-290) and Eq. (4-288) with Eq. (4-291) clearly shows the analogies among the mechanical and electrical components. ◀

▶ **EXAMPLE 4-10-2** As another example of writing the dynamic equations of a mechanical system with translational motion, consider the system shown in Fig. 4-9(a). Because the spring is deformed when it is subject to a force $f(t)$, two displacements, y_1 and y_2, must be assigned to the end points of the spring. The

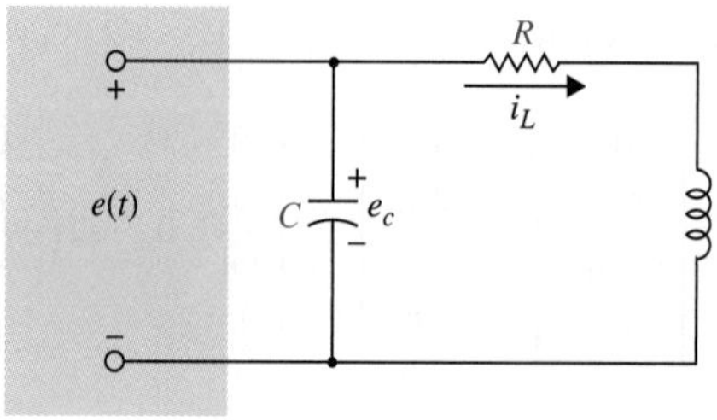

Figure 4-76 Electric network analogous to the mechanical system in Fig. 4-10.

free-body diagram of the system is shown in Fig. 4-9(b). The force equations are

$$f(t) = K[y_1(t) - y_2(t)] \tag{4-292}$$

$$K[y_1(t) - y_2(t)] = M\frac{d^2 y_2(t)}{dt^2} + B\frac{dy_2(t)}{dt} \tag{4-293}$$

These equations are rearranged as

$$y_1(t) = y_2(t) + \frac{1}{K} f(t) \tag{4-294}$$

$$\frac{d^2 y_2(t)}{dt^2} = -\frac{B}{M}\frac{dy_2(t)}{dt} + \frac{K}{M}[y_1(t) - y_2(t)] \tag{4-295}$$

By using the last two equations, the SFG diagram of the system is drawn in Fig. 4-10(a). The state variables are defined as $x_1(t) = y_2(t)$ and $x_2(t) = dy_2(t)/dt$. The state equations are written directly from the state diagram:

$$\frac{dx_1(t)}{dt} = x_2(t) \tag{4-296}$$

$$\frac{dx_2(t)}{dt} = -\frac{B}{M} x_2(t) + \frac{1}{M} f(t) \tag{4-297}$$

As an alternative, we can assign the velocity $v(t)$ of the mass M as one state variable and the force $f_k(t)$ on the spring as the other state variable. We have

$$\frac{dv(t)}{dt} = -\frac{B}{M} v(t) + \frac{1}{M} f_k(t) \tag{4-298}$$

$$f_k(t) = f(t) \tag{4-299}$$

One may wonder why there is only one state equation in Eq. (4-287), whereas there are two state variables in $v(t)$ and $f_k(t)$. The two state equations of Eqs. (4-296) and (4-297) clearly show that the system is of the second order. The situation is better explained by referring to the analogous electric network of the system shown in Fig. 4-76. Although the network has two energy-storage elements in L and C, and thus there should be two state variables, the voltage across the capacitance $e_c(t)$ in this case is redundant, since it is equal to the applied voltage $e(t)$. Eqs. (4-298) and (4-297) can provide only the solutions to the velocity of M, $v(t)$, which is the same as $dy_2(t)/dt$, once $f(t)$ is specified. Then $y_2(t)$ is determined by integrating $v(t)$ with respect to t. The displacement $y_1(t)$ is then found using Eq. (4-292). On the other hand, Eqs. (4-296) and (4-297) give the solutions to $y_2(t)$ and $dy_2(t)/dt$ directly, and $y_1(t)$ is obtained from Eq. (4-292).

The transfer functions of the system are obtained by applying the gain formula to the state diagram.

$$\frac{Y_2(s)}{F(s)} = \frac{1}{s(Ms + B)} \tag{4-300}$$

$$\frac{Y_1(s)}{F(s)} = \frac{Ms^2 + Bs + K}{Ks(Ms + B)} \tag{4-301}$$

◀

▶ **EXAMPLE 4-10-3**
A Pneumatic System
Dry air passes through a valve into a rigid 1 m³ container, as shown in Fig. 4-77, at a constant temperature T = 25°C(= 298°K). The pressure at the left-hand side of the valve is p_i, which is higher than the pressure in the tank p. Assuming a laminar flow, the valve resistance becomes linear, $R = 200 \sec/m^2$. Find the time constant of the system.

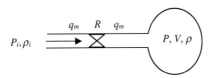

Figure 4-77 A pneumatic system with a valve and a spherical rigid tank.

SOLUTION Assuming air as an ideal gas, isothermal process, and low pressures, from Example 4-5-5, the equation of the system is

$$\frac{RV}{R_{air}T}\dot{p} + p = p_i \qquad (4\text{-}302)$$

where air at standard pressure and temperature is represented as an ideal gas,

$$pv = \frac{p}{\rho} = R_{air}T \quad \rho = \frac{1}{R_{air}T}p \qquad (4\text{-}303)$$

Thus, the time constant is

$$\tau = \frac{RV}{R_{air}T} = \frac{(200)(1)}{88.63(298)} = 7.5(10^{-3})\,\text{sec} \qquad (4\text{-}304)$$

where, from reference [1] at the end of this chapter,

$$R_{air} = 53.35\,\frac{\text{ft lb}_f}{\text{lb}_m\,°R}\,\frac{0.3048\,\text{m}}{\text{ft}}\,\frac{4.45\,\text{N}}{\text{lb}_f}\,\frac{\text{kg m/sec}^2}{\text{N}}\,\frac{\text{lb}_m}{0.4536\,\text{kg}}\,\frac{°R}{°K(9/5)} = 88.63\,\frac{\text{m}^2}{\text{sec}^2\,°K}$$

◀

▶ **EXAMPLE 4-10-4**
A One-Tank Liquid-Level System
For the liquid-level system shown in Fig. 4-45, $C = A/g$ is the capacitance and $\rho = R$ is the resistance. As a result, system time constant is $\tau = RC$. Comparing the thermal, fluid, and electrical systems, similar analogies may be obtained, as shown in Table 4-8.

TABLE 4-8 Mechanical, Thermal, and Fluid Systems and Their Electrical Equivalents

System	R, C, L	Analogy
Mechanical (translation)	$F = Bv(t)$ $R = B$ $F = K\int v(t)dt$ $C = \frac{1}{K}$ $v(t) = \frac{1}{M}\int F\,dt$ $L = M$	$e => F$ $i(t) => v(t)$ where e = voltage $i(t)$ = current F = force $v(t)$ = linear velocity
Mechanical (rotation)	$T = B\omega(t)$ $R = B$ $T = K\int \omega(t)dt$ $C = \frac{1}{K}$ $\omega = \frac{1}{J}\int T\,dt$ $L = J$	$e => T$ $i(t) => \omega(t)$ where e = voltage $i(t)$ = current T = torque $\omega(t)$ = angular velocity
Fluid (incompressible)	$\Delta P = Rq(t)$ (laminar flow) R depends on flow regime $q(t) = C\dot{P}$ C depends on flow regime $L = \frac{\rho \ell}{A}$ (flow in a pipe)	$e => \Delta P$ $i(t) => q(t)$ where e = voltage $i(t)$ = current

(Continued)

TABLE 4-8 (Continued)

System	R, C, L	Analogy
	where A = area of cross section l = length ρ = fluid density	P = pressure $q(t)$ = volume flow rate
Thermal	$R = \dfrac{\Delta T}{q}$ $T = \dfrac{1}{C}\int q\,dt$	$e = > T$ $i(t) = > q(t)$ where e = voltage $i(t)$ = current T = temperature $q(t)$ = heat flow

▶ 4-11 CASE STUDIES

▶ **EXAMPLE 4-11-1** Consider the system in Fig. 4-78. The purpose of the system considered here is to control the positions of the fins of a modern airship. Due to the requirements of improved response and reliability, the surfaces of modern aircraft are controlled by electric actuators with electronic controls. Gone are the days when the ailerons, rudder, and elevators of the aircraft were all linked to the cockpit through mechanical linkages. The so-called fly-by-wire control system used in modern aircraft implies that the attitude of aircraft is no longer controlled entirely be mechanical linkages. Fig. 4-78 illustrates the controlled surfaces and the block diagram of one axis of such a position-control system. Fig. 4-79 shows the analytical block diagram of the system using the dc-motor model given in Fig. 4-72. The system is simplified to the extent that saturation of the amplifier gain and motor torque, gear backlash, and shaft compliances have all been neglected. (When you get into the real world, some of these nonlinear effects should be incorporated into the mathematical model to come up with a better controller design that works in reality. The reader should refer to Chapter 6, where these topics are discussed in more detail.)

The objective of the system is to have the output of the system, $\theta_y(t)$, follow the input, $\theta_r(t)$. The following system parameters are given initially:

Gain of encoder	$K_s = 1$ V/rad
Gain of preamplifier	K = adjustable
Gain of power amplifier	$K_1 = 10$ V/V
Gain of current feedback	$K_2 = 0.5$ V/A
Gain of tachometer feedback	$K_t = 0$ V/rad/sec
Armature resistance of motor	$R_a = 5.0\,\Omega$
Armature inductance of motor	$L_a = 0.003$ H
Torque constant of motor	$K_i = 9.0$ oz-in./A
Back-emf constant of motor	$K_b = 0.0636$ V/rad/sec
Inertia of motor rotor	$J_m = 0.0001$ oz-in.-sec^2
Inertia of load	$J_L = 0.01$ oz-in.-sec^2
Viscous-friction coefficient of motor	$B_m = 0.005$ oz-in.-sec
Viscous-friction coefficient of load	$B_L = 1.0$ oz-in.-sec
Gear-train ratio between motor and load	$N = \theta_y/\theta_m = 1/10$

Because the motor shaft is coupled to the load through a gear train with a gear ratio of N, $\theta_y = N\theta_m$, the total inertia and viscous-friction coefficient seen by the motor are

$$J_t = J_m + N^2 J_L = 0.0001 + 0.01/100 = 0.0002 \text{ oz-in.-sec}^2$$
$$B_t = B_m + N^2 B_L = 0.005 + 1/100 = 0.015 \text{ oz-in.-sec}$$

(4-305)

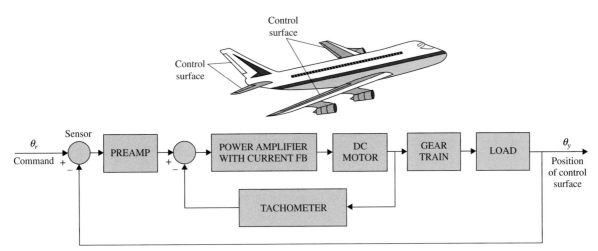

Figure 4-78 Block diagram of an attitude-control system of an aircraft.

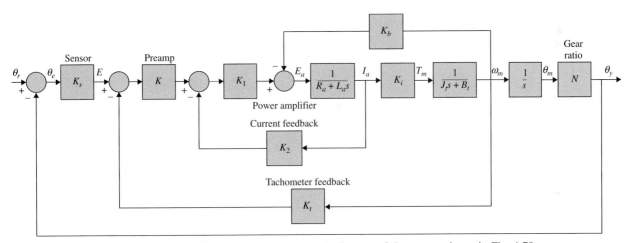

Figure 4-79 Transfer-function block diagram of the system shown in Fig. 4-78.

respectively. The forward-path transfer function of the unity-feedback system is written from Fig. 4-79 by applying the SFG gain formula:

$$G(s) = \frac{\Theta_y(s)}{\Theta_e(s)} = \frac{K_s K_1 K_i K N}{s[L_a J_t s^2 + (R_a J_t + L_a B_t + K_1 K_2 J_t)s + R_a B_t + K_1 K_2 B_t + K_i K_b + K K_1 K_t K_i]} \quad (4\text{-}306)$$

The system is of the third order, since the highest-order term in $G(s)$ is s^3. The electrical time constant of the amplifier-motor system is

$$\tau_a = \frac{L_a}{R_a + K_1 K_2} = \frac{0.003}{5+5} = 0.0003 \text{ sec} \quad (4\text{-}307)$$

The mechanical time constant of the motor-load system is

$$\tau_t = \frac{J_t}{B_t} = \frac{0.0002}{0.015} = 0.01333 \text{ sec} \quad (4\text{-}308)$$

◀

▶ **EXAMPLE 4-11-2** In this case study, we shall model a sun-seeker control system whose purpose is to control the attitude of a space vehicle so that it will track the sun with high accuracy. In the system described here, tracking the sun in only one plane is accomplished. A schematic diagram of the system is shown in

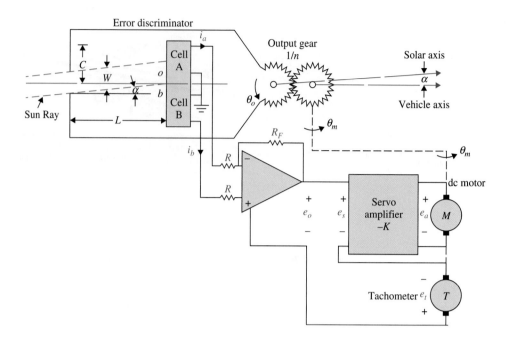

Figure 4-80 Schematic diagram of a sun-seeker system.

Fig. 4-80. The principal elements of the error discriminator are two small rectangular silicon photovoltaic cells mounted behind a rectangular slit in an enclosure. The cells are mounted in such a way that when the sensor is pointed at the sun, a beam of light from the slit overlaps both cells. Silicon cells are used as current sources and connected in opposite polarity to the input of an op-amp. Any difference in the short-circuit current of the two cells is sensed and amplified by the op-amp. Because the current of each cell is proportional to the illumination on the cell, an error signal will be present at the output of the amplifier when the light from the slit is not precisely centered on the cells. This error voltage, when fed to the servoamplifier, will cause the motor to drive the system back into alignment. The description of each part of the system is given in the following sections.

Coordinate System

The center of the coordinate system is considered to be at the output gear of the system. The reference axis is taken to be the fixed frame of the dc motor, and all rotations are measured with respect to this axis. The solar axis, or the line from the output gear to the sun, makes an angle $\theta_r(t)$ with respect to the reference axis, and $\theta_o(t)$ denotes the vehicle axis with respect to the reference axis. The objective of the control system is to maintain the error between $\theta_r(t)$ and $\theta_o(t)$, $\alpha(t)$, near zero:

$$\alpha(t) = \theta_r(t) - \theta_o(t) \tag{4-309}$$

The coordinate system described is illustrated in Fig. 4-81.

Error Discriminator

When the vehicle is aligned perfectly with the sun, $\alpha(t) = 0$, and $i_a(t) = i_b(t) = I$, or $i_a(t) = i_b(t) = 0$. From the geometry of the sun ray and the photovoltaic cells shown in Fig. 4-81, we have

$$oa = \frac{W}{2} + L\tan\alpha(t) \tag{4-310}$$

$$ob = \frac{W}{2} - L\tan\alpha(t) \tag{4-311}$$

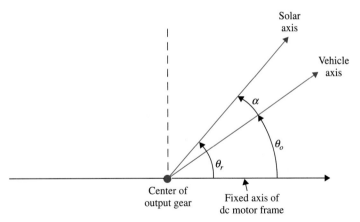

Figure 4-81 Coordinate system of the sun-seeker system.

where oa denotes the width of the sun ray that shines on cell A and ob is the same on cell B, for a given $\alpha(t)$. Because the current $i_a(t)$ is proportional to oa and $i_b(t)$ is proportional to ob, we have

$$i_a(t) = I + \frac{2LI}{W} \tan \alpha(t) \qquad (4\text{-}312)$$

$$i_b(t) = I - \frac{2LI}{W} \tan \alpha(t) \qquad (4\text{-}313)$$

for $0 \leq \tan \alpha(t) \leq W/2L$. For $W/2L \leq \tan \alpha(t) \leq (C - W/2)/L$, the sun ray is completely on cell A, and $i_a(t) = 2I$, $i_b(t) = 0$. For $(C - W/2)L \leq \tan \alpha(t) \leq (C + W/2)L$, $i_a(t)$ decreases linearly from $2I$ to zero. $i_a(t) = i_b(t) = 0$ for $\tan \alpha(t) \geq (C + W/2)/L$. Therefore, the error discriminator may be represented by the nonlinear characteristic of Fig. 4-82, where for small angle $\alpha(t)$, $\tan \alpha(t)$ has been approximated by $\alpha(t)$ on the abscissa.

The relationship between the output of the op-amp and the currents $i_a(t)$ and $i_b(t)$ is

$$e_o(t) = -R_F[i_a(t) - i_b(t)] \qquad (4\text{-}314)$$

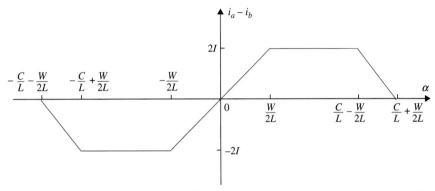

Figure 4-82 Nonlinear characteristic of the error discriminator. The abscissa is $\tan \alpha$, but it is approximated by α for small values of α.

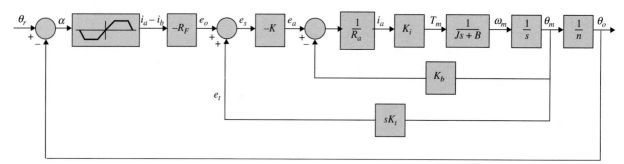

Figure 4-83 Block diagram of the sun-seeker system.

Servoamplifier
The gain of the servoamplifier is $-K$. With reference to Fig. 4-83, the output of the servoamplifier is expressed as

$$e_a(t) = -K[e_o(t) + e_t(t)] = -Ke_s(t) \tag{4-315}$$

Tachometer
The output voltage of the tachometer e_t is related to the angular velocity of the motor through the tachometer constant K_t:

$$e_t(t) = K_t\omega_m(t) \tag{4-316}$$

The angular position of the output gear is related to the motor position through the gear ratio $1/n$. Thus,

$$\theta_o = \frac{1}{n}\theta_m \tag{4-317}$$

DC Motor
The dc motor has been modeled in Section 4-6. The equations are

$$e_a(t) = R_a i_a(t) + e_b(t) \tag{4-318}$$

$$e_b(t) = K_b \omega_m(t) \tag{4-319}$$

$$T_m(t) = K_i i_a(t) \tag{4-320}$$

$$T_m(t) = J\frac{d\omega_m(t)}{dt} + B\omega_m(t) \tag{4-321}$$

where J and B are the inertia and viscous-friction coefficient seen at the motor shaft. The inductance of the motor is neglected in Eq. (4-318). A block diagram that characterizes all the functional relations of the system is shown in Fig. 4-83.

▶ **EXAMPLE 4-11-3** Classically, the quarter-car model is used in the study of vehicle suspension systems and the resulting dynamic response due to various road inputs. Typically, the inertia, stiffness, and damping characteristics of the system as illustrated in Fig. 4-84(a) are modeled in a two degree of freedom (2-DOF) system, as shown in (b). Although a 2-DOF system is a more accurate model, it is sufficient for the following analysis to assume a 1-DOF model, as shown in (c).

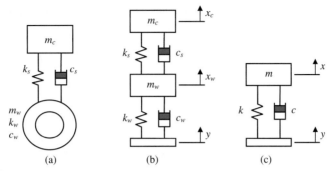

Figure 4-84 Quarter-car model realization. (a) Quarter car. (b) Two degrees of freedom. (c) One degree of freedom.

Open-Loop Base Excitation
Given the system illustrated in Fig. 4-84(c), where

m	Effective ¼ car mass	10 kg
K	Effective stiffness	2.7135 N/m
C	Effective damping	0.9135 N-m/s^{-1}
$x(t)$	Absolute displacement of the mass m	m
$y(t)$	Absolute displacement of the base	m
$z(t)$	Relative displacement $(x(t)-y(t))$	m

the equation of motion of the system is defined as follows:

$$m\ddot{x}(t) + c\dot{x}(t) + kx(t) = c\dot{y}(t) + ky(t) \quad (4\text{-}322)$$

which can be simplified by substituting the relation $z(t) = x(t) - y(t)$ and non-dimensionalizing the coefficients to the form

$$\ddot{z}(t) + 2\zeta\omega_n\dot{z}(t) + \omega_n^2 z(t) = -\ddot{y}(t) = -a(t) \quad (4\text{-}323)$$

The Laplace transform of Eq. (4-323) yields the input–output relationship

$$\frac{Z(s)}{A(s)} = \frac{-1}{s^2 + 2\zeta\omega_n s + \omega_n^2} \quad (4\text{-}324)$$

where the base acceleration $A(s)$ is the Laplace transform of $a(t)$ and is the input, and relative displacement $z(s)$ is the output.

Closed-Loop Position Control
Active control of the suspension system is to be achieved using the same dc motor described in Section 4-7 used in conjunction with a rack as shown in Fig. 4-85.

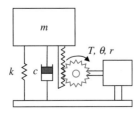

Figure 4-85 Active control of the 1-DOF model via a dc motor and rack.

222 ► Chapter 4. Theoretical Foundation and Background Material: Modeling of Dynamic Systems

In Fig. 4-85, $T(t)$ is the torque produced by the motor with shaft rotation θ, and r is the radius of the motor drive gear. Thus, Eq. (4-322) is rewritten to include the active component, $f(t)$,

$$m\ddot{x} + c\dot{x} + kx = c\dot{y} + ky + f(t) \tag{4-325}$$

where

$$m\ddot{z} + c\dot{z} + kz = f(t) - m\ddot{y} = f(t) - ma(t) \tag{4-326}$$

$$f(t) = \frac{T(t) - (J_m\ddot{\theta} + B_m\dot{\theta})}{r} \tag{4-327}$$

Because $z = \theta r$, we can substitute Eq. (4-327) into Eq. (4-326), rearrange, and take the Laplace transform to get

$$Z(s) = \frac{r}{(mr^2 + J_m)s^2 + (cr^2 + B_m)s + kr^2}[T(s) - mrA(s)] \tag{4-328}$$

Noting that $Z(s)/r = \Theta(s)$, this is analogous to previous input–output relationships where $\Theta(s) = G_{eq}(T(s) - T_d(s))$; hence, the term $mrA(s)$ is interpreted as a disturbance torque. The block diagram in Fig. 4-86 can thus be compared to Fig. 4-85, where $J = mr^2 + J_m$, $B = cr^2 + B_m$, and $K = kr^2$. Using the principle of superposition, this system is rearranged to the following form:

$$Z(s) = \frac{\frac{K_m r}{R_a}}{\left(\frac{L_a}{R_a}s + 1\right)(Js^2 + Bs + K) + \frac{K_m K_b}{R_a}s}V_a(s)$$

$$- \frac{\left(\frac{L_a}{R_a}s + 1\right)r}{\left(\frac{L_a}{R_a}s + 1\right)(Js^2 + Bs + K) + \frac{K_m K_b}{R_a}s}mrA(s) \tag{4-329}$$

▶ 4-12 MATLAB TOOLS

Apart from the MATLAB toolboxes appearing with the chapter, this chapter does not contain any software because of its focus on theoretical development. In Chapters 6 and 9, where we address more complex control-system modeling and analysis, we will introduce the Automatic Control Systems MATLAB and SIMULINK tools. The **Automatic Control**

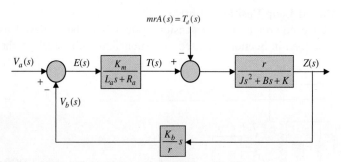

Figure 4-86 Block diagram of an armature-controlled dc motor.

Systems software (**ACSYS**) consists of a number of m-files and GUIs (graphical user interface) for the analysis of simple control engineering transfer functions. It can be invoked from the MATLAB command line by simply typing *Acsys* and then by clicking on the appropriate pushbutton. A specific MATLAB tool has been developed for most chapters of this textbook. Throughout this chapter, we have identified subjects that may be solved using **ACSYS**, with a box in the left margin of the text titled "MATLAB TOOL."

The most relevant components of **ACSYS** to the problems in this chapter are Virtual Lab and SIMLab, which are discussed in detail in Chapter 6. These simulation tools provide the user with virtual experiments and design projects using systems involving dc motors, sensors, electronic components, and mechanical components.

▶ 4-13 SUMMARY

This chapter is devoted to the mathematical modeling of physical systems. The basic mathematical relations of electrical, mechanical, thermal, and fluid systems are described using differential equations, state equations, and transfer functions. Analogies were used to relate the equations of these systems. The operations and mathematical descriptions of some of the commonly used components in control systems, such as error detectors, tachometers, and dc motors, are presented in this chapter.

This chapter includes various examples. However, due to space limitations and the intended scope of this text, only some of the physical devices used in practice are described. The main purpose of this chapter is to illustrate the methods of system modeling, and the coverage is not intended to be exhaustive.

Because nonlinear systems cannot be ignored in the real world, and this book is not devoted to the subject, Section 4-9 introduced the linearization of nonlinear systems at a nominal operating point. Once the linearized model is determined, the performance of the nonlinear system can be investigated under the small-signal conditions at the designated operating point.

Systems with pure time delays are modeled, and methods of approximating the transfer functions by rational ones are described.

In the end, three case study examples were presented that reflect mathematical modeling of practical applications.

▶ REVIEW QUESTIONS

1. Among the three types of friction described, which type is governed by a linear mathematical relation?

2. Given a two-gear system with angular displacement θ_1 and θ_2, numbers of teeth N_1 and N_2, and torques T_1 and T_2, write the mathematical relations between these variables and parameters.

3. How are potentiometers used in control systems?

4. Digital encoders are used in control systems for position and speed detection. Consider that an encoder is set up to output 3600 zero crossings per revolution. What is the angular rotation of the encoder shaft in degrees if 16 zero crossings are detected?

5. The same encoder described in Question 4 and an electronic clock with a frequency of 1 MHz are used for speed measurement. What is the average speed of the encoder shaft in rpm if 500 clock pulses are detected between two consecutive zero crossings of the encoder?

6. Give the advantages of dc motors for control-systems applications.

7. What are the sources of nonlinearities in a dc motor?

8. What are the effects of inductance and inertia in a dc motor?

9. What is back emf in a dc motor, and how does it affect the performance of a control system?

10. What are the electrical and mechanical time constants of an electric motor?

11. Under what condition is the torque constant K_i of a dc motor valid, and how is it related to the back-emf constant K_b?

12. An inertial and frictional load is driven by a dc motor with torque T_m. The dynamic equation of the system is

$$T_m(t) = J_m \frac{d\omega_m(t)}{dt} + B_m \omega_m$$

If the inertia is doubled, how will it affect the steady-state speed of the motor? How will the steady-state speed be affected if, instead, the frictional coefficient B_m is doubled? What is the mechanical constant of the system?

13. What is a tachometer, and how is it used in control systems?

14. Give the transfer function of a pure time delay T_d.

15. Does the linearization technique described in this chapter always result in a linear time-invariant system?

The answers to these review questions can be found on this book's companion Web site: www.wiley.com/college/golnaraghi.

▶ REFERENCES

1. W. J. Palm III, *Modeling, Analysis, and Control of Dynamic Systems*, 2nd Ed., John Wiley & Sons, New York, 1999.
2. K. Ogata, *Modern Control Engineering*, 4th Ed., Prentice Hall, NJ, 2002.
3. I. Cochin and W. Cadwallender, *Analysis and Design of Dynamic Systems*, 3rd Ed., Addison-Wesley, 1997.
4. A. Esposito, *Fluid Power with Applications*, 5th Ed., Prentice Hall, NJ, 2000.
5. H. V. Vu and R. S. Esfandiari, *Dynamic Systems*, Irwin/McGraw-Hill, 1997.
6. J. L. Shearer, B. T. Kulakowski, and J. F. Gardner, *Dynamic Modeling and Control of Engineering Systems*, 2nd Ed., Prentice Hall, NJ, 1997.
7. R. L. Woods and K. L. Lawrence, *Modeling and Simulation of Dynamic Systems*, Prentice Hall, NJ, 1997.
8. E. J. Kennedy, *Operational Amplifier Circuits*, Holt, Rinehart and Winston, Fort Worth, TX, 1988.
9. J. V. Wait, L. P. Huelsman, and G. A. Korn, *Introduction to Operational Amplifier Theory and Applications*, 2nd Ed., McGraw-Hill, New York, 1992.
10. B. C. Kuo, *Automatic Control Systems*, 7th Ed., Prentice Hall, NJ, 1995.
11. B. C. Kuo and F. Golnaraghi, *Automatic Control Systems*, 8th Ed., John Wiley and Sons, NY, 2003.

▶ PROBLEMS

PROBLEMS FOR SECTION 4-1

4-1. Consider the mass-spring system shown in Fig. 4P-1.

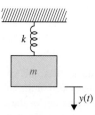

Figure 4P-1

(a) Find the equation of the motion.
(b) Calculate its natural frequency.

4-2. Consider the five-spring one-mass system shown in Fig. 4P-2.
(a) Find its single spring-mass equivalent.
(b) Calculate its natural frequency.

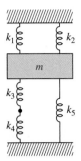

Figure 4P-2

4-3. Fig. 4P-3 shows a simple model of a vehicle suspension system hitting a bump. If the mass of the wheel and its mass moment of inertia are m and J, respectively, then:

(a) Find the equation of the motion.

(b) Determine the transfer function of the system.

(c) Calculate its natural frequency.

(d) Use MATLAB to plot the step response of the system.

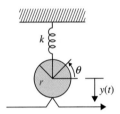

Figure 4P-3

4-4. Write the force equations of the linear translational systems shown in Fig. 4P-4.

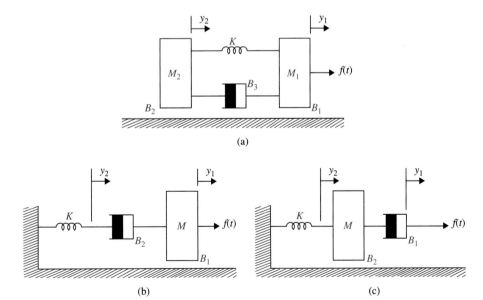

Figure 4P-4

(a) Draw the system block diagrams or SFGs.
(b) Define the state variables as follows:
   (i) $x_1 = y_2$, $x_2 = dy_2/dt$, $x_3 = y_1$, and $x_4 = dy_1/dt$
   (ii) $x_1 = y_2$, $x_2 = y_1$, and $x_3 = dy_1/dt$
   (iii) $x_1 = y_1$, $x_2 = y_2$, and $x_3 = dy_2/dt$
(c) Write the state equations. Find the transfer functions $Y_1(s)/F(s)$ and $Y_2(s)/F(s)$.

4-5. Write the force equations of the linear translational system shown in Fig. 4P-5. Draw system block diagrams. Write the state equations. Find the transfer functions $Y_1(s)/F(s)$ and $Y_2(s)/F(s)$. Set $Mg = 0$ for the transfer functions.

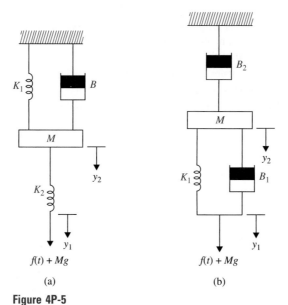

Figure 4P-5

4-6. Consider a train consisting of an engine and a car, as shown in Fig. 4P-6.

Figure 4P-6

A controller is applied to the train so that it has a smooth start and stop, along with a constant-speed ride. The mass of the engine and the car are M and m, respectively. The two are held together by a spring with the stiffness coefficient of K. F represents the force applied by the engine, and μ represents the coefficient of rolling friction. If the train only travels in one direction:
(a) Draw the free-body diagram.
(b) Find the state variables and output equations.
(c) Find the transfer function.
(d) Write the state-space equations of the system.

4-7. A vehicle towing a trailer through a spring-damper coupling hitch is shown in Fig. 4P-7. The following parameters and variables are defined: M is the mass of the trailer; K_h, the spring constant of the hitch; B_h, the viscous-damping coefficient of the hitch; B_t, the viscous-friction coefficient of the trailer; $y_1(t)$, the displacement of the towing vehicle; $y_2(t)$, the displacement of the trailer; and $f(t)$, the force of the towing vehicle.

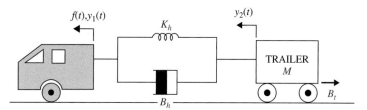

Figure 4P-7

(a) Write the differential equation of the system.

(b) Write the state equations by defining the following state variables: $x_1(t) = y_1(t) - y_2(t)$ and $x_2(t) = dy_2(t)dt$.

4-8. Assume that the displacement angle of the pendulums shown in Fig. 4P-8 are small enough that the spring always remains horizontal. If the rods with the length of L are massless and the spring is attached to the rods $\frac{3}{4}$ from the top, find the state equation of the system.

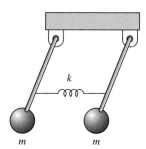

Figure 4P-8

4-9. Fig. 4P-9 shows an inverted pendulum on a cart.

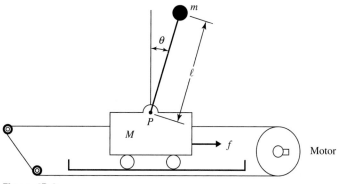

Figure 4P-9

If the mass of the cart is represented by M and the force f is applied to hold the bar at the desired position, then
(a) Draw the free-body diagram.
(b) Determine the dynamic equation of the motion.
(c) Find the transfer function.
(d) Write the state space of the system.

If f is an impulse signal, plot the impulse response of the system by using MATLAB.

4-10. A two-stage inverted pendulum on a cart is shown in Fig. 4P-10.

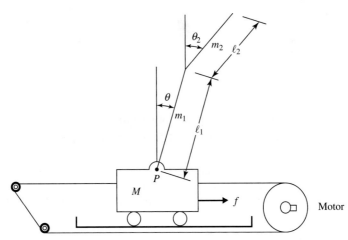

Figure 4P-10

If the mass of the cart is represented by M and the force f is applied to hold the bar at the desired position, then
(a) Draw the free-body diagram of mass M.
(b) Determine the dynamic equation of the motion.
(c) Find the transfer function.
(d) Write the state space equations of the system.

4-11. Fig. 4P-11 shows a well-known "ball and beam" system in control systems. A ball is located on a beam to roll along the length of the beam. A lever arm is attached to the one end of the beam and a servo gear is attached to the other end of the lever arm. As the servo gear turns by an angle θ, the lever arm goes up and down, and then the angle of the beam is changed by α. The change in angle causes the ball to roll along the beam. A controller is desired to manipulate the ball's position.

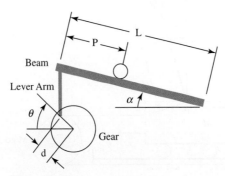

Figure 4P-11

Assuming:

m = mass of the ball

r = radius of the ball

d = lever arm offset

g = gravitational acceleration

L = length of the beam

J = ball's moment of inertia

p = ball position coordinate

α = beam angle coordinate

θ = servo gear angle

(a) Determine the dynamic equation of the motion.

(b) Find the transfer function.

(c) Write the state space equations of the system.

(d) Find the step response of the system by using MATLAB.

4-12. The motion equations of an aircraft are a set of six nonlinear coupled differential equations. Under certain assumptions, they can be decoupled and linearized into the longitudinal and lateral equations. Fig. 4P-12 shows a simple model of airplane during its flight. Pitch control is a longitudinal problem, and an autopilot is designed to control the pitch of the airplane.

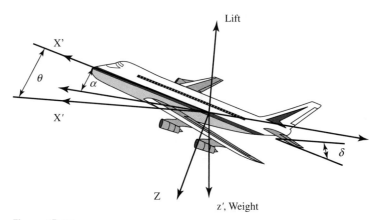

Figure 4P-12

Consider that the airplane is in steady-cruise at constant altitude and velocity, which means the thrust and drag cancel out and the lift and weight balance out each other. To simplify the problem, assume that change in pitch angle does not affect the speed of an aircraft under any circumstance.

(a) Determine the longitudinal equations of motion of the aircraft.

(b) Find the transfer function and state-space variables.

4-13. Write the torque equations of the rotational systems shown in Fig. 4P-13. Write the state equations. Find the transfer function $\Theta(s)/T(s)$ for the system in (a). Find the transfer functions $\Theta_1(s)/T(s)$ and $\Theta_2(s)/T(s)$ for the systems in parts (b), (c), (d), and (e).

4-14. Write the torque equations of the gear-train system shown in Fig. 4P-14. The moments of inertia of gears are lumped as J_1, J_2, and J_3. $T_m(t)$ is the applied torque; N_1, N_2, N_3, and N_4 are the number of gear teeth. Assume rigid shafts

(a) Assume that J_1, J_2, and J_3 are negligible. Write the torque equations of the system. Find the total inertia the motor sees.

(b) Repeat part (a) with the moments of inertia J_1, J_2, and J_3.

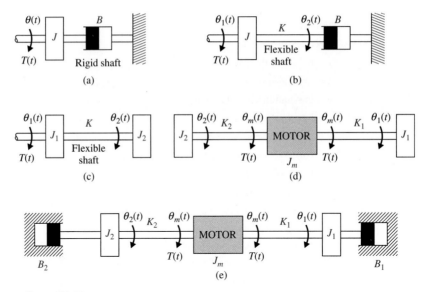

Figure 4P-13

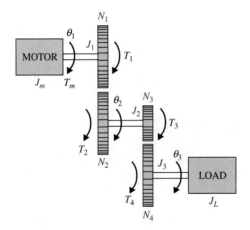

Figure 4P-14

4-15. Fig. 4P-15 shows a motor-load system coupled through a gear train with gear ratio $n = N_1/N_2$. The motor torque is $T_m(t)$, and $T_L(t)$ represents a load torque.
(a) Find the optimum gear ratio n^* such that the load acceleration $\alpha_L = d^2\theta_L/dt^2$ is maximized.
(b) Repeat part (a) when the load torque is zero.

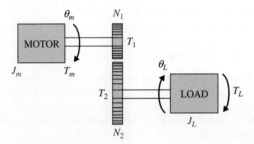

Figure 4P-15

4-16. Fig. 4P-16 shows the simplified diagram of the printwheel control system of an old word processor. The printwheel is controlled by a dc motor through belts and pulleys. Assume that the belts are rigid. The following parameters and variables are defined: $T_m(t)$ is the motor torque; $\theta_m(t)$, the motor displacement; $y(t)$, the linear displacement of the printwheel; J_m, the motor inertia; B_m, the motor viscous-friction coefficient; r, the pulley radius; M, the mass of the printwheel.
(a) Write the differential equation of the system.
(b) Find the transfer function $Y(s)/T_m(s)$.

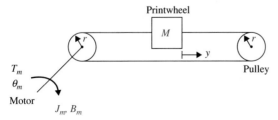

Figure 4P-16

4-17. Fig. 4P-17 shows the diagram of a printwheel system with belts and pulleys. The belts are modeled as linear springs with spring constants K_1 and K_2.
(a) Write the differential equations of the system using θ_m and y as the dependent variables.
(b) Write the state equations using $x_1 = r\theta_m - y$, $x_2 = dy/dt$, and $x_3 = \omega_m = d\theta_m/dt$ as the state variables.
(c) Draw a state-flow diagram for the system.
(d) Find the transfer function $Y(s)/T_m(s)$.
(e) Find the characteristic equation of the system.

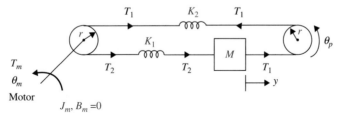

Figure 4P-17

4-18. The schematic diagram of a steel-rolling process is shown in Fig. 4P-18. The steel plate is fed through the rollers at a constant speed of V ft/s. The distance between the rollers and the point where

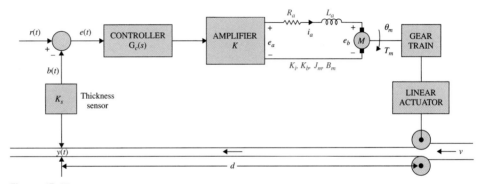

Figure 4P-18

the thickness is measured is d ft. The rotary displacement of the motor, $\theta_m(t)$, is converted to the linear displacement $y(t)$ by the gear box and linear-actuator combination $y(t) = n\theta_m(t)$, where n is a positive constant in ft/rad. The equivalent inertia of the load that is reflected to the motor shaft is J_L.

(a) Draw a functional block diagram for the system.

(b) Derive the forward-path transfer function $Y(s)/E(s)$ and the closed-loop transfer function $Y(s)/R(s)$.

4-19. The schematic diagram of a motor-load system is shown in Fig. 4P-19. The following parameters and variables are defined: $T_m(t)$ is the motor torque; $\omega_m(t)$, the motor velocity; $\theta_m(t)$, the motor displacement; $\omega_L(t)$, the load velocity; $\theta_L(t)$, the load displacement; K, the torsional spring constant; J_m, the motor inertia; B_m, the motor viscous-friction coefficient; and B_L, the load viscous-friction coefficient.

(a) Write the torque equations of the system.

(b) Find the transfer functions $\Theta_L(s)/T_m(s)$ and $\Theta_m(s)/T_m(s)$.

(c) Find the characteristic equation of the system.

(d) Let $T_m(t) = T_m$ be a constant applied torque; show that $\omega_m = \omega_L$ = constant in the steady state. Find the steady-state speeds ω_m and ω_L.

(e) Repeat part (d) when the value of J_L is doubled, but J_m stays the same.

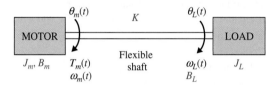

Figure 4P-19

4-20. This problem deals with the attitude control of a guided missile. When traveling through the atmosphere, a missile encounters aerodynamic forces that tend to cause instability in the attitude of the missile. The basic concern from the flight-control standpoint is the lateral force of the air, which tends to rotate the missile about its center of gravity. If the missile centerline is not aligned with the direction in which the center of gravity C is traveling, as shown in Fig. 4P-20, with angle θ, which is also called the angle of attack, a side force is produced by the drag of the air through which the missile travels. The total force F_α may be considered to be applied at the center of pressure P. As shown in Fig. 4P-20, this side force has a tendency to cause the missile to tumble end over end, especially if the point P is in front of the center of gravity C. Let the angular acceleration of the missile about the point C, due to the side force, be denoted by α_F. Normally, α_F is directly proportional to the angle of attack θ and is given by

$$\alpha_F = \frac{K_F d_1}{J}\theta$$

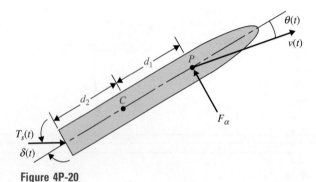

Figure 4P-20

where K_F is a constant that depends on such parameters as dynamic pressure, velocity of the missile, air density, and so on, and

J = missile moment of inertia about C

d_1 = distance between C and P

The main objective of the flight-control system is to provide the stabilizing action to counter the effect of the side force. One of the standard control means is to use gas injection at the tail of the missile to deflect the direction of the rocket engine thrust T_s, as shown in the figure.

(a) Write a torque differential equation to relate among T_s, δ, θ, and the system parameters given. Assume that δ is very small, so that sin $\delta(t)$ is approximated by $\delta(t)$.

(b) Assume that T_s is a constant torque. Find the transfer function $\Theta(s)/\Delta(s)$, where $\Theta(s)$ and $\Delta(s)$ are the Laplace transforms of $\theta(t)$ and $\delta(t)$, respectively. Assume that $\delta(t)$ is very small.

(c) Repeat parts (a) and (b) with points C and P interchanged. The d_1 in the expression of α_F should be changed to d_2.

4-21. Fig. 4P-21(a) shows a well-known "broom-balancing" system in control systems. The objective of the control system is to maintain the broom in the upright position by means of the force $u(t)$ applied to the car as shown. In practical applications, the system is analogous to a one-dimensional control problem of the balancing of a unicycle or a missile immediately after launching. The free-body diagram of the system is shown in Fig. 4P-21(b), where

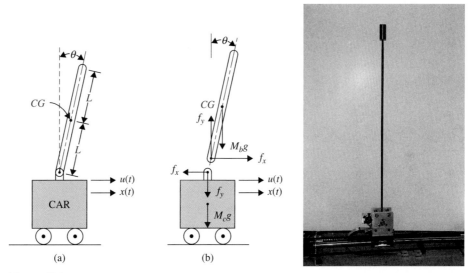

Figure 4P-21

f_x = force at broom base in horizontal direction
f_y = force at broom base in vertical direction
M_b = mass of broom
g = gravitational acceleration
M_c = mass of car
J_b = moment of inertia of broom about center of gravity $CG = M_b L_2/3$

(a) Write the force equations in the x and the y directions at the pivot point of the broom. Write the torque equation about the center of gravity CG of the broom. Write the force equation of the car in the horizontal direction.

(b) Express the equations obtained in part (a) as state equations by assigning the state variables as $x_1 = \theta$, $x_2 = d\theta/dt$, $x_3 = x$, and $x_4 = dx/dt$. Simplify these equations for small θ by making the approximations $\sin\theta \cong \theta$ and $\cos\theta \cong 1$.

(c) Obtain a small-signal linearized state-equation model for the system in the form of
$$\frac{d\Delta \mathbf{x}(t)}{dt} = \mathbf{A}^* \Delta \mathbf{x}(t) + \mathbf{B}^* \Delta \mathbf{r}(t)$$
at the equilibrium point $x_{01}(t) = 1$, $x_{02}(t) = 0$, $x_{03}(t) = 0$, and $x_{04}(t) = 0$.

4-22. Most machines and devices have rotating parts. Even a small irregularity in the mass distribution of rotating components can cause vibration, which is called rotating unbalanced. Fig. 4P-22 represents the schematic of a rotating unbalanced mass of m. Assume that the frequency of rotation of the machine is ω.
(a) Draw the state-flow diagram of the system.
(b) Find the transfer function.
(c) Use MATLAB to obtain the step response of the system.

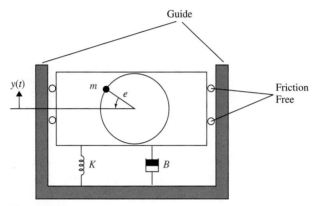

Figure 4P-22

4-23. Vibration absorbers are used to protect machines that work at the constant speed from steady-state harmonic disturbance. Fig. 4P-23 shows a simple vibration absorber.

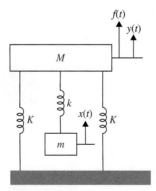

Figure 4P-23

Assuming the harmonic force $F(t) = A\sin(\omega t)$ is the disturbance applied to the mass M:
(a) Derive the state space equations of the system.
(b) Determine the transfer function of the system.

4-24. Fig. 4P-24 represents a damping in the vibration absorption.
Assuming the harmonic force $F(t) = A\sin(\omega t)$ is the disturbance applied to the mass M:
(a) Derive the state space equations of the system.
(b) Determine the transfer function of the system.

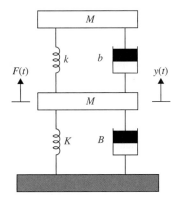

Figure 4P-24

PROBLEMS FOR SECTION 4-2

4-25. Consider the electrical circuits shown in Figs. 4P-25(a) and (b).

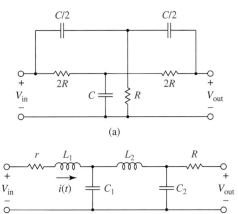

Figure 4P-25

For each circuit:

(a) Find the dynamic equations and state variables.

(b) Determine the transfer function.

(c) Use MATLAB to plot the step response of the system.

4-26. An electromechanical system shown in Fig. 4P-26 represents a moveable-plate capacity.

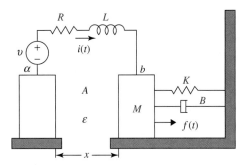

Figure 4P-26

Assume that the plate a of the parallel capacitor is fixed and the plate b with mass M is moved by force f. If $C(x) = \dfrac{\varepsilon A}{d}$, when the ε is the dielectric constant and A is the surface of the plates, then the electric field produces a force opposing the motion of the plates, and it is related to the charge (q) across the plates: $f_c = \dfrac{q^2}{2\varepsilon A}$

(a) Find the differential equations of this system.
(b) Determine $X(s)/C(s)$.

4-27. Consider the electromechanical system shown in Fig. 4P-27.

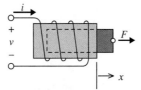

Figure 4P-27

(a) Draw the free-body diagram.
(b) Find the differential equation that describes the operation of the system.
(c) Calculate the transfer function of the system.

4-28. Repeat Problem 4-27 for the electromechanical system shown in Fig. 4P-28.

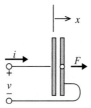

Figure 4P-28

PROBLEMS FOR SECTION 4-3

4-29. Find the transfer function of the circuit for the simple op-amp circuit given in Fig. 4P-29.

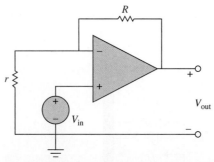

Figure 4P-29

4-30. An op-amp circuit with connection to both terminals is shown in Fig. 4P-30.

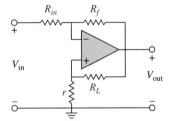

Figure 4P-30

The op-amp can be modeled as

$$V_{out} = \frac{10^7}{s+1}[v_+ - v_-]$$
$$i_+ = i_- = 0$$

when v_+ and v_- represent the voltages of positive and negative terminals, respectively, and i_+ and i_- show the current of these terminals.

(a) Find the positive feedback ratio.
(b) Find the negative feedback ratio.
(c) Determine when the circuit remains stable.

4-31. Find the transfer function for each circuit given in Fig. 4P-31.

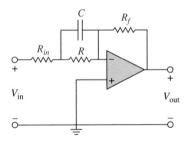

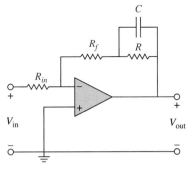

Figure 4P-31

PROBLEMS FOR SECTION 4-4

4-32. A thermal lever is shown in Fig. 4P-32.
As shown, the actuator is a pure electric resistance and the heat flow is generated by the electric power input. The lever (at the top) moves up or down proportionally, depending on the difference between the temperature of the ambient air and the temperature of the actuator. Calculate $V(s)/X(s)$, assuming zero initial conditions.

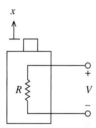

Figure 4P-32

4-33. Hot oil forging in quenching vat with its cross-sectional view is shown in Fig. 4P-33.

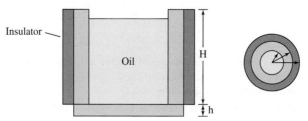

Figure 4P-33

The radii shown in Fig. 4P-33 are r_1, r_2, and r_3 from inside to outside. The heat is transferred to the atmosphere from the sides and bottom of the vat and also the surface of the oil with a convective heat coefficient of k_o. Assuming:

k_v = The thermal conductivity of the vat

k_i = The thermal conductivity of the insulator

c_o = The specific heat of the oil

d_o = The density of the oil

c = The specific heat of the forging

m = Mass of the forging

A = The surface area of the forging

h = The thickness of the bottom of the vat

T_a = The ambient temperature

Determine the system model when the temperature of the oil is desired.

4-34. A power supply within an enclosure is shown in Fig. 4P-34. Because the power supply generates lots of heat, a heat sink is usually attached to dissipate the generated heat. Assuming the rate of heat generation within the power supply is known and constant, Q, the heat transfers from the power supply to the enclosure by radiation and conduction, the frame is an ideal insulator, and the heat sink temperature is constant and equal to the atmospheric temperature, determine the model of the system that can give the temperature of the power supply during its operation. Assign any needed parameters.

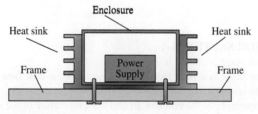

Figure 4P-34

4-35. Fig. 4P-35 shows a heat exchanger system.

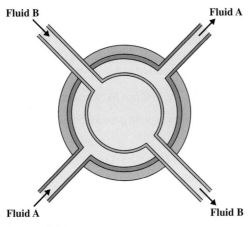

Figure 4P-35

Assuming the simple material transport model represents the rate of heat energy gain for this system, then

$$(\dot{m}c)(T_2 - T_1) = q_{gained}$$

where $\dot{m}$ represents the mass flow, T_1 and T_2 are the entering and leaving fluid temperature, and c shows the specific heat of fluid.

If the length of the heat exchanger cylinder is L, derive a model to give the temperature of Fluid B leaving the heat exchanger. Assign any required parameters, such as radii, thermal conductivity coefficients, and the thickness.

PROBLEMS FOR SECTION 4-5

4-36. The objective of this problem is to develop a linear analytical model of the automobile engine for idle-speed control system shown in Fig. 1-2. The input of the system is the throttle position that controls the rate of air flow into the manifold (see Fig. 4P-36). Engine torque is developed from the buildup of manifold pressure due to air intake and the intake of the air/gas mixture into the cylinder. The engine variations are as follows:

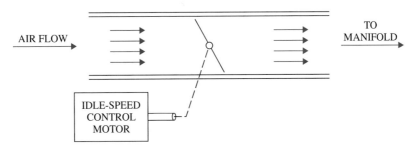

Figure 4P-36

$q_i(t)$ = amount of air flow across throttle into manifold
$dq_i(t)/dt$ = rate of air flow across throttle into manifold
$q_m(t)$ = average air mass in manifold
$q_o(t)$ = amount of air leaving intake manifold through intake valves
$dq_o(t)/dt$ = rate of air leaving intake manifold through intake valves
$T(t)$ = engine torque
T_d = disturbance torque due to application of auto accessories = constant

$\omega(t)$ = engine speed
$\alpha(t)$ = throttle position
τ_D = time delay in engine
J_e = inertia of engine

The following assumptions and mathematical relations between the engine variables are given:

1. The rate of air flow into the manifold is linearly dependent on the throttle position:
$$\frac{dq_i(t)}{dt} = K_1\alpha(t) \quad K_1 = \text{proportional constant}$$

2. The rate of air flow leaving the manifold depends linearly on the air mass in the manifold and the engine speed:
$$\frac{dq_o(t)}{dt} = K_2 q_m(t) + K_3\omega(t) \quad K_2, K_3 = \text{constant}$$

3. A pure time delay of τ_D seconds exists between the change in the manifold air mass and the engine torque:
$$T(t) = K_4 q_m(t - \tau_D) \quad K_4 = \text{constant}$$

4. The engine drag is modeled by a viscous-friction torque $B\omega(t)$, where B is the viscous-friction coefficient.

5. The average air mass $q_m(t)$ is determined from
$$q_m(t) = \int \left(\frac{dq_i(t)}{dt} - \frac{dq_o(t)}{dt}\right) dt$$

6. The equation describing the mechanical components is
$$T(t) = J\frac{d\omega(t)}{dt} + B\omega(t) + T_d$$

(a) Draw a functional block diagram of the system with $\alpha(t)$ as the input, $\omega(t)$ as the output, and T_d as the disturbance input. Show the transfer function of each block.

(b) Find the transfer function $\Omega(s)/\alpha(s)$ of the system.

(c) Find the characteristic equation and show that it is not rational with constant coefficients.

(d) Approximate the engine time delay by
$$e^{-\tau_D s} \cong \frac{1 - \tau_D s/2}{1 + \tau_D s/2}$$
and repeat parts (b) and (c).

4-37. Vibration can also be exhibited in fluid systems. Fig. 4P-37 shows a U tube manometer.

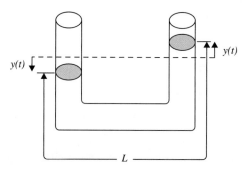

Figure 4P-37

Assume the length of fluid is L, the weight density is μ, and the cross-section area of the tube is A.

(a) Write the state equation of the system.

(b) Calculate the natural frequency of oscillation of the fluid.

4-38. A long pipeline connects a water reservoir to a hydraulic generator system as shown in Fig. 4P-38.

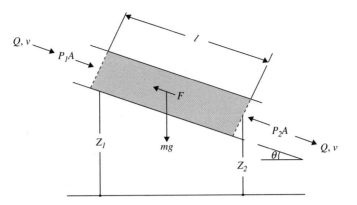

Figure 4P-38

At the end of the pipeline, there is a valve controlled by a speed controller. It may be closed quickly to stop the water flow if the generator loses its load. Determine the dynamic model for the level of the surge tank. Consider the turbine-generator is an energy converter. Assign any required parameters.

4-39. A simplified oil well system is shown in Fig. 4P-39.

In this figure, the drive machinery is replaced by the input torque, $T_{in}(t)$. Assuming the pressure in the surrounding rock is fixed at P and the walking beam moves through small angles, determine a model for this system during the upstroke of the pumping rod.

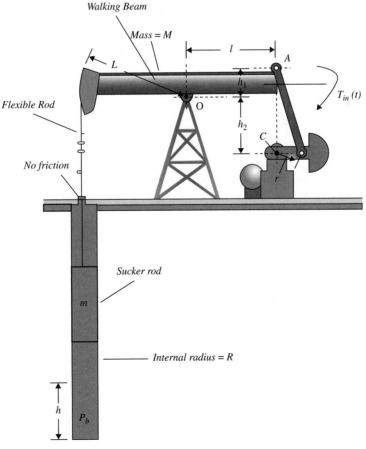

Figure 4P-39

4-40. A hydraulic servomotor usually is used for the speed control of engines. As shown in Fig. 4P-40, the reference speed is controlled by the throttle lever. The flyweight is moved by engine, so then the differential displacement of the spring determines the input to the hydraulic servomotor. Determine the state space model of the system.

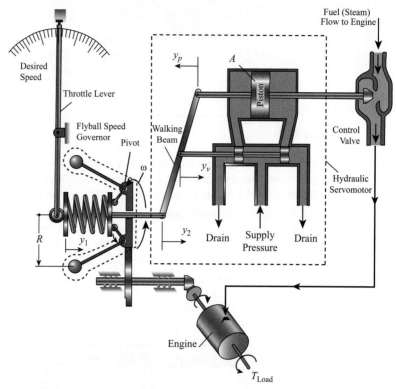

Figure 4P-40

4-41. Fig. 4P-41 shows a two-tank liquid-level system. Assume that Q_1 and Q_2 are the steady-state inflow rates, and H_1 and H_2 are steady-state heads. If the other quantities shown in Fig. 4P-41 are supposed to be small, derive the state-space model of the system when h_1 and h_2 are outputs of the system and q_{i1} and q_{i2} are the inputs.

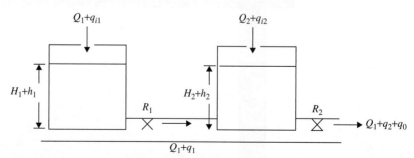

Figure 4P-41

PROBLEMS FOR SECTION 4-6

4-42. An accelerometer is a transducer as shown in Fig. 4P-42.
(a) Find the dynamic equation of motion.
(b) Determine the transfer function.
(c) Use MATLAB to plot its impulse response.

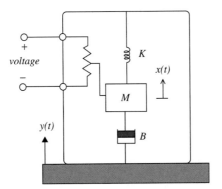

Figure 4P-42

4-43. Fig. 4P-43(a) shows the setup of the temperature control of an air-flow system. The hot-water reservoir supplies the water that flows into the heat exchanger for heating the air. The temperature sensor senses the air temperature T_{AO} and sends it to be compared with the reference temperature T_r.

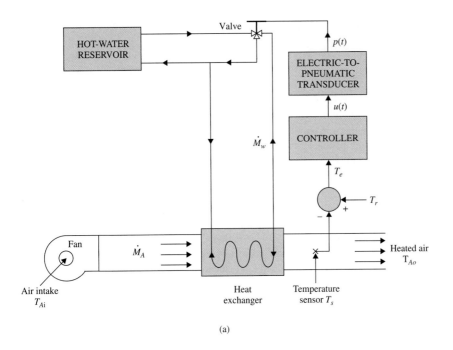

(a)

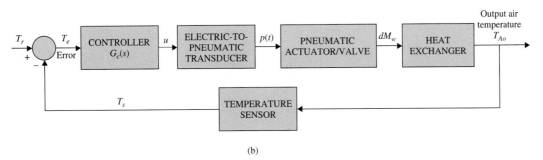

(b)

Figure 4P-43

The temperature error T_e is sent to the controller, which has the transfer function $G_c(s)$. The output of the controller, $u(t)$, which is an electric signal, is converted to a pneumatic signal by a transducer. The output of the actuator controls the water-flow rate through the three-way valve. Fig. 4P-43(b) shows the block diagram of the system.

The following parameters and variables are defined: dM_w is the flow rate of the heating fluid $= K_M u$; $K_M = 0.054$ kg/s/V; T_w, the water temperature $= K_R dM_w$; $K_R = 65°$C/kg/s; and T_{AO}, the output air temperature. Heat-transfer equation between water and air:

$$\tau_c \frac{dT_{AO}}{dt} = T_w - T_{AO} \quad \tau_c = 10 \text{ seconds}$$

Temperature sensor equation:

$$\tau_s \frac{dT_s}{dt} = T_{AO} - T_s \quad \tau_s = 2 \text{ seconds}$$

(a) Draw a functional block diagram that includes all the transfer functions of the system.
(b) Derive the transfer function $T_{AO}(s)/T_r(s)$ when $G_c(s) = 1$.

4-44. An open-loop motor control system is shown in Fig. 4P-44.

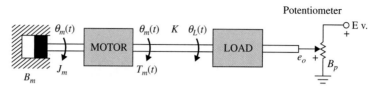

Figure 4P-44

The potentiometer has a maximum range of 10 turns (20π rad). Find the transfer functions $E_o(s)/T_m(s)$. The following parameters and variables are defined: $\theta_m(t)$ is the motor displacement; $\theta_L(t)$, the load displacement; $T_m(t)$, the motor torque; J_m, the motor inertia; B_m, the motor viscous-friction coefficient; B_p, the potentiometer viscous-friction coefficient; $e_o(t)$, the output voltage; and K, the torsional spring constant.

4-45. The schematic diagram of a control system containing a motor coupled to a tachometer and an inertial load is shown in Fig. 4P-45. The following parameters and variables are defined: T_m is the motor torque; J_m, the motor inertia; J_t, the tachometer inertia; J_L, the load inertia; K_1 and K_2, the spring constants of the shafts; θ_t, the tachometer displacement; θ_m, the motor velocity; θ_L, the load displacement; ω_t, the tachometer velocity; ω_L, the load velocity; and B_m, the motor viscous-friction coefficient.

(a) Write the state equations of the system using $\theta_L, \omega_L, \theta_t, \omega_t, \theta_m,$ and ω_m as the state variables (in the listed order). The motor torque T_m is the input.

(b) Draw a signal flow diagram with T_m at the left and ending with θ_L on the far right. The state diagram should have a total of 10 nodes. Leave out the initial states.

(c) Find the following transfer functions: $\dfrac{\Theta_L(s)}{T_m(s)} \quad \dfrac{\Theta_t(s)}{T_m(s)} \quad \dfrac{\Theta_m(s)}{T_m(s)}$

(d) Find the characteristic equation of the system.

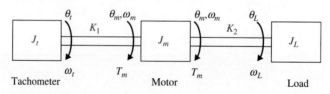

Figure 4P-45

4-46. Phase-locked loops are control systems used for precision motor-speed control. The basic elements of a phase-locked loop system incorporating a dc motor are shown in Fig. 4P-46(a). An input pulse train represents the reference frequency or desired output speed. The digital encoder produces digital pulses that represent motor speed. The phase detector compares the motor speed and the reference frequency and sends an error voltage to the filter (controller) that governs the dynamic response of the system.

Phase detector gain = K_p, encoder gain = K_e, counter gain = $1/N$, and dc-motor torque constant = K_i. Assume zero inductance and zero friction for the motor.

(a) Derive the transfer function $E_c(s)/E(s)$ of the filter shown in Fig. 4P-46(b). Assume that the filter sees infinite impedance at the output and zero impedance at the input.

(b) Draw a functional block diagram of the system with gains or transfer functions in the blocks.

(c) Derive the forward-path transfer function $\Omega_m(s)/E(s)$ when the feedback path is open.

(d) Find the closed-loop transfer function $\Omega_m(s)/F_r(s)$.

(e) Repeat parts (a), (c), and (d) for the filter shown in Fig. 4P-46(c).

(f) The digital encoder has an output of 36 pulses per revolution. The reference frequency f_r is fixed at 120 pulses/s. Find K_e in pulses/rad. The idea of using the counter N is that, with f_r fixed, various desired output speeds can be attained by changing the value of N. Find N if the desired output speed is 200 rpm. Find N if the desired output speed is 1800 rpm.

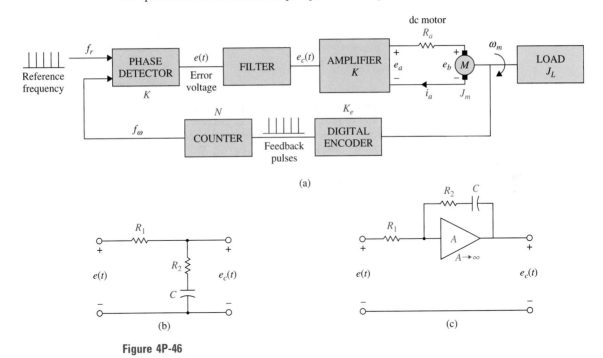

Figure 4P-46

4-47. Describe how an incremental encoder can be used as a frequency divider.

PROBLEMS FOR SECTION 4-7

4-48. The voltage equation of a dc motor is written as

$$e_a(t) = R_a i_a(t) + L_a \frac{di_a(t)}{dt} + K_b \omega_m(t)$$

where $e_a(t)$ is the applied voltage; $i_a(t)$, the armature current; R_a, the armature resistance; L_a, the armature inductance; K_b, the back-emf constant; $\omega_m(t)$, the motor velocity; and $\omega_n(t)$, the reference input voltage. Taking the Laplace transform on both sides of the voltage equation, with zero initial conditions and solving for $\Omega_m(s)$, we get

$$\Omega_m(s) = \frac{E_a(s) - (R_a + L_a s)I_a(s)}{K_b}$$

which shows that the velocity information can be generated by feeding back the armature voltage and current. The block diagram in Fig. 4P-48 shows a dc-motor system, with voltage and current feedbacks, for speed control.

(a) Let K_1 be a very high gain amplifier. Show that when $H_i(s)/H_e(s) = -(R_a + L_a s)$, the motor velocity $\omega_m(t)$ is totally independent of the load-disturbance torque T_L.

(b) Find the transfer function between $\Omega_m(s)$ and $\Omega_r(s)$ ($T_L = 0$) when $H_i(s)$ and $H_e(s)$ are selected as in part (a).

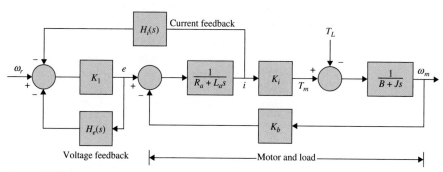

Figure 4P-48

4-49. Fig. 4P-49 shows the schematic diagram of a dc-motor control system for the control of the printwheel of a word processor. The load in this case is the printwheel, which is directly coupled to the motor shaft. The following parameters and variables are defined: K_s is the error-detector gain (V/rad); K_i, the torque constant (oz-in./A); K, the amplifier gain (V/V); K_b, the back-emf constant (V/rad/sec); n, the gear-train ratio $= \theta_2/\theta_m = T_m/T_2$; B_m, the motor viscous-friction coefficient (oz-in.-sec); J_m, the motor inertia (oz-in.-sec^2); K_L, the torsional spring constant of the motor shaft (oz-in./rad); and JL the load inertia (oz-in.-sec^2).

(a) Write the cause-and-effect equations of the system. Rearrange these equations into the form of state equations with $x_1 = \theta_o$, $x_2 = \omega_o$, $x_3 = \theta_m$, $x_4 = \omega_m$, and $x_5 = i_a$.

(b) Draw the signal flow diagram using the nodes shown in Fig. 4P-49(b).

(c) Derive the forward-path transfer function (with the outer feedback path open): $G(s) = \Theta_o(s)/\Theta_e(s)$. Find the closed-loop transfer function $M(s) = \Theta_o(s)/\Theta_r(s)$.

(d) Repeat part (c) when the motor shaft is rigid; i.e., $K_L = \infty$. Show that you can obtain the solutions by taking the limit as K_L approaches infinity in the results in part (c).

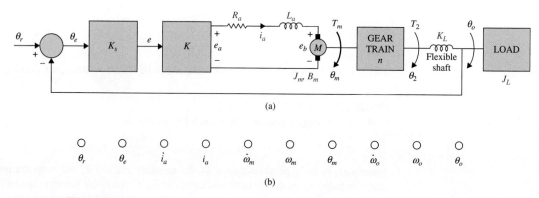

Figure 4P-49

4-50. The schematic diagram of a voice-coil motor (VCM), used as a linear actuator in a disk memory-storage system, is shown in Fig. 4P-50(a). The VCM consists of a cylindrical permanent magnet (PM) and a voice coil. When current is sent through the coil, the magnetic field of the PM interacts with the current-carrying conductor, causing the coil to move linearly. The voice coil of the VCM in Fig. 4P-50(a) consists of a primary coil and a shorted-turn coil. The latter is installed for the purpose of effectively reducing the electric constant of the device. Fig. 4P-50(b) shows the equivalent circuit of the coils. The following parameters and variables are defined: $e_a(t)$ is the applied coil voltage; $i_a(t)$, the primary-coil current; $i_s(t)$, the shorted-turn coil current; R_a, the primary-coil resistance; L_a, the primary-coil inductance; L_{as}, the mutual inductance between the primary and shorted-turn coils; $v(t)$, the velocity of the voice coil; $y(t)$, the displacement of the voice coil; $f(t) = K_i v(t)$, the force of the voice coil; K_i, the force constant; K_b, the back-emf constant; $e_b(t) = K_b v(t)$, the back emf; M_T, the total mass of the voice coil and load; and B_T, the total viscous-friction coefficient of the voice coil and load.

(a) Write the differential equations of the system.

(b) Draw a block diagram of the system with $E_a(s)$, $I_a(s)$, $I_s(s)$, $V(s)$, and $Y(s)$ as variables.

(c) Derive the transfer function $Y(s)/E_a(s)$.

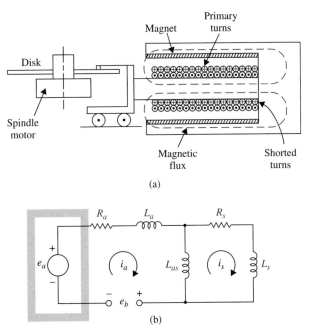

Figure 4P-50

4-51. A dc-motor position-control system is shown in Fig. 4P-51(a). The following parameters and variables are defined: e is the error voltage; e_r, the reference input; θ_L, the load position; K_A, the amplifier gain; e_a, the motor input voltage; e_b, the back emf; i_a, the motor current; T_m, the motor torque; J_m, the motor inertia $= 0.03$ oz-in.-s^2; B_m, the motor viscous-friction coefficient $= 10$ oz-in.-s^2; K_L, the torsional spring constant $= 50,000$ oz-in./rad; J_L, the load inertia $= 0.05$ oz-in.-s^2; K_i, the motor torque constant $= 21$ oz-in./A; K_b, the back-emf constant $= 15.5$ V/1000 rpm; K_s, the error-detector gain $= E/2\pi$; E, the error-detector applied voltage $= 2\pi$ V; R_a, the motor resistance $= 1.15\ \Omega$; and $\theta_e = \theta_r - \theta_L$.

(a) Write the state equations of the system using the following state variables: $x_1 = \theta_L$, $x_2 = d\theta_L/dt = \omega_L$, $x_3 = \theta_3$, and $x_4 = d\theta_m/dt = \omega_m$.

(b) Draw a signal flow diagram using the nodes shown in Fig. 4P-51(b).

(c) Derive the forward-path transfer function $G(s) = \Theta_L(s)/\Theta_e(s)$ when the outer feedback path from θ_L is opened. Find the poles of $G(s)$.

(d) Derive the closed-loop transfer function $M(s) = \Theta_L(s)/\Theta_e(s)$. Find the poles of $M(s)$ when $K_A = 1, 2738,$ and 5476. Locate these poles in the s-plane, and comment on the significance of these values of K_A.

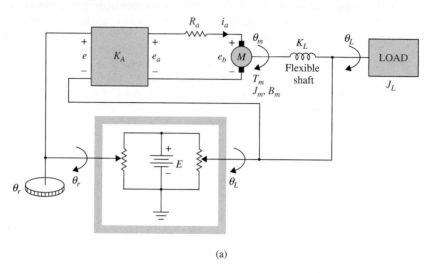

(a)

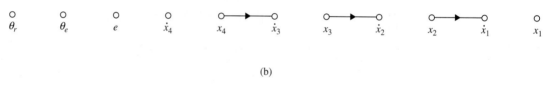

(b)

Figure 4P-51

4-52. The following differential equations describe the motion of an electric train in a traction system:

$$\frac{dx(t)}{dt} = v(t)$$

$$\frac{dv(t)}{dt} = -k(v) - g(x) + f(t)$$

where

$x(t)$ = linear displacement of train
$v(t)$ = linear velocity of train
$k(v)$ = resistance force on train [odd function of v, with the properties $k(0) = 0$ and $dk(v)/dv = 0$]
$g(x)$ = gravitational force for a nonlevel track or due to curvature of track
$f(t)$ = tractive force

The electric motor that provides the tractive force is described by the following equations:

$$e(t) = K_b\phi(t)v(t) + R_a i_a(t)$$
$$f(t) = K_i\phi(t)i_a(t)$$

where $e(t)$ is the applied voltage; $i_a(t)$, the armature current; $i_f(t)$, the field current; R_a, the armature resistance; $\phi(t)$, the magnetic flux from a separately excited field = $K_f i_f(t)$; and K_i, the force constant.

(a) Consider that the motor is a dc series motor with the armature and field windings connected in series, so that $i_a(t) = i_f(t)$, $g(x) = 0$, $k(v) = Bv(t)$, and $R_a = 0$. Show that the system is described by the following nonlinear state equations:

$$\frac{dx(t)}{dt} = v(t)$$

$$\frac{dv(t)}{dt} = -Bv(t) + \frac{K_i}{K_b^2 K_f v^2(t)} e^2(t)$$

(b) Consider that, for the conditions stated in part (a), $i_a(t)$ is the input of the system [instead of $e(t)$]. Derive the state equations of the system.

(c) Consider the same conditions as in part (a) but with $\phi(t)$ as the input. Derive the state equations.

4-53. The linearized model of a robot arm system driven by a dc motor is shown in Fig. 4P-53. The system parameters and variables are given as follows:

DC Motor	Robot Arm
T_m = motor torque = $K_i i_a$	J_L = inertia of arm
K_i = torque constant	T_L = disturbance torque on arm
i_a = armature current of motor	θ_L = arm displacement
J_m = motor inertia	K = torsional spring constant
B_m = motor viscous-friction coefficient	θ_m = motor-shaft displacement
B = viscous-friction coefficient of shaft between the motor and arm	
B_L = viscous-friction coefficient of the robot arm shaft	

(a) Write the differential equations of the system with $i_a(t)$ and $T_L(t)$ as input and $\theta_m(t)$ and $\theta_L(t)$ as outputs.

(b) Draw an SFG using $I_a(s)$, $T_L(s)$, $\Theta_m(s)$, and $\Theta_L(s)$ as node variables.

(c) Express the transfer-function relations as

$$\begin{bmatrix} \Theta_m(s) \\ \Theta_L(s) \end{bmatrix} = \mathbf{G}(s) \begin{bmatrix} I_a(s) \\ -T_L(s) \end{bmatrix}$$

Find $\mathbf{G}(s)$.

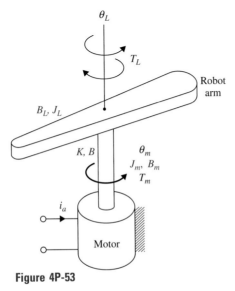

Figure 4P-53

PROBLEMS FOR SECTION 4-8

4-54. The transfer function of the heat exchanger system is given by

$$G(s) = \frac{T(s)}{A(s)} = \frac{Ke^{-T_d s}}{(\tau_1 s + 1)(\tau_2 s + 1)}$$

where T_d is the time delay.
(a) Plot the roots and zeros of the system.
(b) Use MATLAB to verify your answer in part (a).

4-55. Find the polar plot of the following functions by using the approximation of delay function described in Section 2.8.

(a) $G(s) = \dfrac{e^{-sL}}{(Ts + 1)}$

(b) $G(s) = \dfrac{2 + 2se^{-s} + 4e^{-2s}}{s^2 + 3s + 2}$

4-56. Use MATLAB to solve Problem 4-55 and plot the step response of the systems.

PROBLEMS FOR SECTION 4.9

4-57. Fig. 4P-57 shows the schematic diagram of a ball-suspension control system. The steel ball is suspended in the air by the electromagnetic force generated by the electromagnet. The objective of the control is to keep the metal ball suspended at the nominal equilibrium position by controlling the current in the magnet with the voltage $e(t)$. The practical application of this system is the magnetic levitation of trains or magnetic bearings in high-precision control systems. The resistance of the coil is R, and the inductance is $L(y) = L/y(t)$, where L is a constant. The applied voltage $e(t)$ is a constant with amplitude E.

(a) Let E_{eq} be a nominal value of E. Find the nominal values of $y(t)$ and $dy(t)/dt$ at equilibrium.
(b) Define the state variables at $x_1(t) = i(t)$, $x_2(t) = y(t)$, and $x_3(t) = dy(t)/dt$. Find the nonlinear state equations in the form of $\dfrac{d\mathbf{x}(t)}{dt} = \mathbf{f}(\mathbf{x}, e)$.

(c) Linearize the state equations about the equilibrium point and express the linearized state equations as

$$\frac{d\Delta\mathbf{x}(t)}{dt} = \mathbf{A}^* \Delta\mathbf{x}(t) + \mathbf{B}^* \Delta e(t)$$

The force generated by the electromagnet is $Ki^2(t)/y(t)$, where K is a proportional constant, and the gravitational force on the steel ball is Mg.

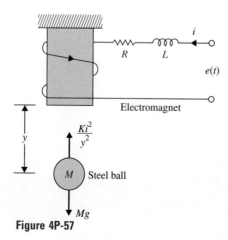

Figure 4P-57

4-58. Fig. 4P-58(a) shows the schematic diagram of a ball-suspension system. The steel ball is suspended in the air by the electromagnetic force generated by the electromagnet. The objective of the control is to keep the metal ball suspended at the nominal position by controlling the current in the electromagnet. When the system is at the stable equilibrium point, any small perturbation of the ball position from its floating equilibrium position will cause the control to return the ball to the equilibrium position. The free-body diagram of the system is shown in Fig. 4P-58(b), where

M_1 = mass of electromagnet = 2.0
M_2 = mass of steel ball = 1.0
B = viscous-friction coefficient of air = 0.1
K = proportional constant of electromagnet = 1.0
g = gravitational acceleration = 32.2

Assume all units are consistent. Let the stable equilibrium values of the variables, $i(t)$, $y_1(t)$, and $y_2(t)$ be I, Y_1, and Y_2, respectively. The state variables are defined as $x_1(t) = y_1(t)$, $x_2(t) = dy_1(t)/dt$, $x_3(t) = y_2(t)$, and $x_4(t) = dy_2(t)/dt$.

(a) Given $Y_1 = 1$, find I and Y_2.

(b) Write the nonlinear state equations of the system in the form of $dx(t)/dt = f(x, i)$.

(c) Find the state equations of the linearized system about the equilibrium state I, Y_1, and Y_2 in the form

$$\frac{d\mathbf{x}(t)}{dt} = \mathbf{A}^* \Delta \mathbf{x}(t) + \mathbf{B}^* \Delta i(t)$$

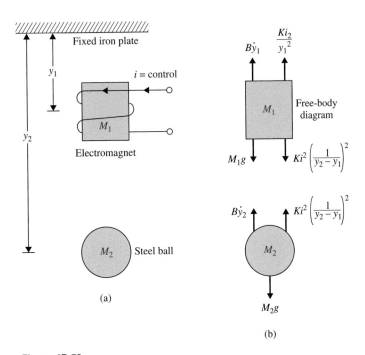

Figure 4P-58

PROBLEMS FOR SECTION 4-10

4-59. Fig. 4P-59 shows a typical grain scale.
Assign any required parameters.

(a) Find the free-body diagram.

(b) Derive a model for the grain scale that determines the waiting time for the reading of the weight of grain after placing on the scale platform.

(c) Develop an analogous electrical circuit for this system.

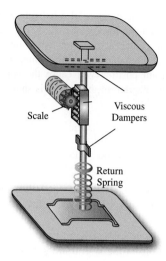

Figure 4P-59

4-60. Develop an analogous electrical circuit for the mechanical system shown in Figure 4P-60.

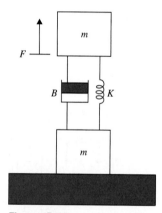

Figure 4P-60

4-61. Develop an analogous electrical circuit for the fluid hydraulic system shown in Fig. 4P-61.

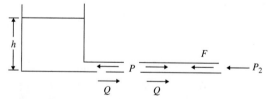

Figure 4P-61

PROBLEMS FOR SECTION 4-11

4-62. The open-loop excitation model of the car suspension system with 1-DOF, illustrated in Fig. 4-84(c), is given in Example 4-11-3. Use MATLAB to find the impulse response of the system.

4-63. An active control designed for the car suspension system with 1-DOF is designed by using a dc motor and rack. Use MATLAB and the transfer function of the system given in Example 4-11-3 to plot the impulse response of the system. Compare your result with the result of Problem 4-62.

CHAPTER 5

Time-Domain Analysis of Control Systems

In this chapter, we depend on the background material discussed in Chapters 1–4 to arrive at the time response of simple control systems. In order to find the time response of a control system, we first need to model the overall system dynamics and find its equation of motion. The system could be composed of mechanical, electrical, or other sub-systems. Each sub-system may have sensors and actuators to sense the environment and to interact with it. Next, using Laplace transforms, we can find the transfer function of all the sub-components and use the block diagram approach or signal flow diagrams to find the interactions among the system components. Depending on our objectives, we can manipulate the system final response by adding feedback or poles and zeros to the system block diagram. Finally, we can find the overall transfer function of the system and, using inverse Laplace transforms, obtain the time response of the system to a test input—normally a step input.

Also in this chapter, we look at more details of the time response analysis, discuss transient and steady state time response of a simple control system, and develop simple design criteria for manipulating the time response. In the end, we look at the effects of adding a simple gain or poles and zeros to the system transfer function and relate them to the concept of control. We finally look at simple proportional, derivative, and integral controller design concepts in time domain. Throughout the chapter, we utilize MATLAB in simple toolboxes to help with our development.

▶ 5-1 TIME RESPONSE OF CONTINUOUS-DATA SYSTEMS: INTRODUCTION

Because time is used as an independent variable in most control systems, it is usually of interest to evaluate the state and output responses with respect to time or, simply, the time response. In the analysis problem, a reference input signal is applied to a system, and the performance of the system is evaluated by studying the system response in the time domain. For instance, if the objective of the control system is to have the output variable track the input signal, starting at some initial time and initial condition, it is necessary to compare the input and output responses as functions of time. Therefore, in most control-system problems, the final evaluation of the performance of the system is based on the time responses.

The time response of a control system is usually divided into two parts: the transient response and the steady-state response. Let $y(t)$ denote the time response of a continuous-data system; then, in general, it can be written as

$$y(t) = y_t(t) + y_{ss}(t) \tag{5-1}$$

where $y_t(t)$ denotes the transient response and $y_{ss}(t)$ denotes the steady-state response. In control systems, *transient response* is defined as the part of the time response that goes to zero as time becomes very large. Thus, $y_t(t)$ has the property

$$\lim_{t \to \infty} y_t(t) = 0 \tag{5-2}$$

The steady-state response is simply the part of the total response that remains after the transient has died out. Thus, the steady-state response can still vary in a fixed pattern, such as a sine wave, or a ramp function that increases with time.

All real, stable control systems exhibit transient phenomena to some extent before the steady state is reached. Because inertia, mass, and inductance are unavoidable in physical systems, the response of a typical control system cannot follow sudden changes in the input instantaneously, and transients are usually observed. Therefore, the control of the transient response is necessarily important, because it is a significant part of the dynamic behavior of the system, and the deviation between the output response and the input or the desired response, before the steady state is reached, must be closely controlled.

The steady-state response of a control system is also very important, because it indicates where the system output ends up when time becomes large. For a position-control system, the steady-state response when compared with the desired reference position gives an indication of the final accuracy of the system. In general, if the steady-state response of the output does not agree with the desired reference exactly, the system is said to have a steady-state error.

The study of a control system in the time domain essentially involves the evaluation of the transient and the steady-state responses of the system. In the design problem, specifications are usually given in terms of the transient and the steady-state performances, and controllers are designed so that the specifications are all met by the designed system.

▶ 5-2 TYPICAL TEST SIGNALS FOR THE TIME RESPONSE OF CONTROL SYSTEMS

Unlike electric networks and communication systems, the inputs to many practical control systems are not exactly known ahead of time. In many cases, the actual inputs of a control system may vary in random fashion with respect to time. For instance, in a radar-tracking system for antiaircraft missiles, the position and speed of the target to be tracked may vary in an unpredictable manner, so that they cannot be predetermined. This poses a problem for the designer, because it is difficult to design a control system so that it will perform satisfactorily to all possible forms of input signals. For the purpose of analysis and design, it is necessary to assume some basic types of test inputs so that the performance of a system can be evaluated. By selecting these basic test signals properly, not only is the mathematical treatment of the problem systematized, but the response due to these inputs allows the prediction of the system's performance to other more complex inputs. In the design problem, performance criteria may be specified with respect to these test signals so that the system may be designed to meet the criteria. This approach is particularly useful for linear systems, since the response to complex signals can be determined by superposing those due to simple test signals.

When the response of a linear time-invariant system is analyzed in the frequency domain, a sinusoidal input with variable frequency is used. When the input frequency is swept from zero to beyond the significant range of the system characteristics, curves in terms of the amplitude ratio and phase between the input and the output are drawn as functions of frequency. It is possible to predict the time-domain behavior of the system from its frequency-domain characteristics.

To facilitate the time-domain analysis, the following deterministic test signals are used.

Step-Function Input: The step-function input represents an instantaneous change in the reference input. For example, if the input is an angular position of a mechanical shaft, a step input represents the sudden rotation of the shaft. The mathematical representation of a step function or magnitude R is

$$r(t) = R \quad t \geq 0 \quad (5\text{-}3)$$
$$= 0 \quad t < 0$$

where R is a real constant. Or

$$r(t) = Ru_s(t) \quad (5\text{-}4)$$

where $u_s(t)$ is the unit-step function. The step function as a function of time is shown in Fig. 5-1(a). The step function is very useful as a test signal because its initial instantaneous jump in amplitude reveals a great deal about a system's quickness in responding to inputs with abrupt changes. Also, because the step function contains, in principle, a wide band of frequencies in its spectrum, as a result of the jump discontinuity, it is equivalent to the application of numerous sinusoidal signals with a wide range of frequencies.

Ramp-Function Input: The ramp function is a signal that changes constantly with time. Mathematically, a ramp function is represented by

$$r(t) = Rtu_s(t) \quad (5\text{-}5)$$

where R is a real constant. The ramp function is shown in Fig. 5-1(b). If the input variable represents the angular displacement of a shaft, the ramp input denotes the constant-speed

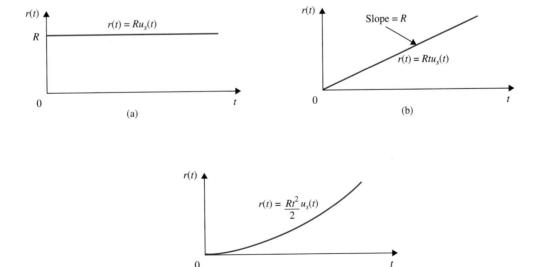

Figure 5-1 Basic time-domain test signals for control systems. (a) Step function. (b) Ramp function. (c) Parabolic function.

rotation of the shaft. The ramp function has the ability to test how the system would respond to a signal that changes linearly with time.

Parabolic-Function Input: The parabolic function represents a signal that is one order faster than the ramp function. Mathematically, it is represented as

$$r(t) = \frac{Rt^2}{2} u_s(t) \quad (5\text{-}6)$$

where R is a real constant and the factor $\frac{1}{2}$ is added for mathematical convenience because the Laplace transform of $r(t)$ is simply R/s^3. The graphical representation of the parabolic function is shown in Fig. 5-1(c).

These signals all have the common feature that they are simple to describe mathematically. From the step function to the parabolic function, the signals become progressively faster with respect to time. In theory, we can define signals with still higher rates, such as t^3, which is called the *jerk function*, and so forth. However, in reality, we seldom find it necessary or feasible to use a test signal faster than a parabolic function. This is because, as we shall see later, in order to track a high-order input accurately, the system must have high-order integrations in the loop, which usually leads to serious stability problems.

▶ 5-3 THE UNIT-STEP RESPONSE AND TIME-DOMAIN SPECIFICATIONS

As defined earlier, the transient portion of the time response is the part that goes to zero as time becomes large. Nevertheless, the transient response of a control system is necessarily important, because both the amplitude and the time duration of the transient response must be kept within tolerable or prescribed limits. For example, in the automobile idle-speed control system described in Chapter 1, in addition to striving for a desirable idle speed in the steady state, the transient drop in engine speed must not be excessive, and the recovery in speed should be made as quickly as possible. For linear control systems, the characterization of the transient response is often done by use of the unit-step function $u_s(t)$ as the input. The response of a control system when the input is a unit-step function is called the unit-step response. Fig. 5-2 illustrates a typical unit-step response of a linear control system. With reference to the unit-step response, performance criteria commonly used for the characterization of linear control systems in the time domain are defined as follows:

1. **Maximum overshoot.** Let $y(t)$ be the unit-step response. Let y_{max} denote the maximum value of $y(t)$; y_{ss}, the steady-state value of $y(t)$; and $y_{max} \geq y_{ss}$. The maximum overshoot of $y(t)$ is defined as

$$\text{maximum overshoot} = y_{max} - y_{ss} \quad (5\text{-}7)$$

The maximum overshoot is often represented as a percentage of the final value of the step response; that is,

$$\text{percent maximum overshoot} = \frac{\text{maximum overshoot}}{y_{ss}} \times 100\% \quad (5\text{-}8)$$

The maximum overshoot is often used to measure the relative stability of a control system. A system with a large overshoot is usually undesirable. For design

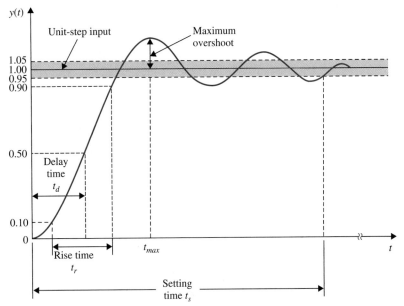

Figure 5-2 Typical unit-step response of a control system illustrating the time-domain specifications.

purposes, the maximum overshoot is often given as a time-domain specification. The unit-step illustrated in Fig. 5-2 shows that the maximum overshoot occurs at the first overshoot. For some systems, the maximum overshoot may occur at a later peak, and, if the system transfer function has an odd number of zeros in the right-half s-plane, a negative undershoot may even occur [4, 5] (Problem 5-23).

2. **Delay time.** The delay time t_d is defined as the time required for the step response to reach 50% of its final value. This is shown in Fig. 5-2.

3. **Rise time.** The rise time t_r is defined as the time required for the step response to rise from 10 to 90% of its final value, as shown in Fig. 5-2. An alternative measure is to represent the rise time as the reciprocal of the slope of the step response at the instant that the response is equal to 50% of its final value.

4. **Settling time.** The settling time t_s is defined as the time required for the step response to decrease and stay within a specified percentage of its final value. A frequently used figure is 5%.

   The four quantities just defined give a direct measure of the transient characteristics of a control system in terms of the unit-step response. These time-domain specifications are relatively easy to measure when the step response is well defined, as shown in Fig. 5-2. Analytically, these quantities are difficult to establish, except for simple systems lower than the third order.

5. **Steady-state error.** The steady-state error of a system response is defined as the discrepancy between the output and the reference input when the steady state $(t \rightarrow \infty)$ is reached.

   It should be pointed out that the steady-state error may be defined for any test signal such as a step-function, ramp-function, parabolic-function, or even a sinusoidal input, although Fig. 5-2 only shows the error for a step input.

5-4 STEADY-STATE ERROR

One of the objectives of most control systems is that the system output response follows a specific reference signal accurately in the steady state. The difference between the output and the reference in the steady state was defined earlier as the steady-state error. In the real world, because of friction and other imperfections and the natural composition of the system, the steady state of the output response seldom agrees exactly with the reference. Therefore, steady-state errors in control systems are almost unavoidable. In a design problem, one of the objectives is to keep the steady-state error to a minimum, or below a certain tolerable value, and at the same time the transient response must satisfy a certain set of specifications.

The accuracy requirement on control systems depends to a great extent on the control objectives of the system. For instance, the final position accuracy of an elevator would be far less stringent than the pointing accuracy on the control of the Large Space Telescope, which is a telescope mounted onboard a space shuttle. The accuracy of position control of such a system is often measured in microradians.

5-4-1 Steady-State Error of Linear Continuous-Data Control Systems

Linear control systems are subject to steady-state errors for somewhat different causes than nonlinear systems, although the reason is still that the system no longer "sees" the error, and no corrective effort is exerted. In general, the steady-state errors of linear control systems depend on the type of the reference signal and the type of the system.

Definition of the Steady-State Error with Respect to System Configuration

Before embarking on the steady-state error analysis, we must first clarify what is meant by system error. In general, we can regard the error as a signal that should be quickly reduced to zero, if possible. Let us refer to the closed-loop system shown in Fig. 5-3, where $r(t)$ is the input; $u(t)$, the actuating signal; $b(t)$, the feedback signal; and $y(t)$, the output. The error of the system may be defined as

$$e(t) = \text{reference signal} - y(t) \qquad (5\text{-}9)$$

where the reference signal is the signal that the output $y(t)$ is to track. When the system has unity feedback, that is, $H(s) = 1$, then the input $r(t)$ is the reference signal, and the error is simply

$$e(t) = r(t) - y(t) \qquad (5\text{-}10)$$

The steady-state error is defined as

$$e_{ss} = \lim_{s \to \infty} e(t) \qquad (5\text{-}11)$$

When $H(s)$ is not unity, the actuating signal $u(t)$ in Fig. 5-2 may or may not be the error, depending on the form and the purpose of $H(s)$. Let us assume that the objective of the

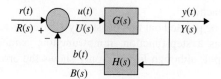

Figure 5-3 Nonunity feedback control system.

system in Fig. 5-3 is to have the output $y(t)$ track the input $r(t)$ as closely as possible, and the system transfer functions are

$$G(s) = \frac{1}{s^2(s+12)} \quad H(s) = \frac{5(s+1)}{(s+5)} \quad (5\text{-}12)$$

We can show that, if $H(s) = 1$, the characteristic equation is

$$s^3 + 12s^2 + 1 = 0 \quad (5\text{-}13)$$

which has roots in the right-half s-plane, and the closed-loop system is unstable. We can show that the $H(s)$ given in Eq. (5-10) stabilizes the system, and the characteristic equation becomes

$$s^4 + 17s^3 + 60s^2 + 5s + 5 = 0 \quad (5\text{-}14)$$

In this case, the system error may still be defined as in Eq. (5-10).

However, consider a velocity control system in which a step input is used to control the system output that contains a ramp in the steady state. The system transfer functions may be of the form

$$G(s) = \frac{1}{s^2(s+12)} \quad H(s) = K_t s \quad (5\text{-}15)$$

where $H(s)$ is the transfer function of an electromechanical or electronic tachometer, and K_t is the tachometer constant. The system error should be defined as in Eq. (5-9), where the reference signal is the *desired velocity* and not $r(t)$. In this case, because $r(t)$ and $y(t)$ are not of the same dimension, it would be meaningless to define the error as in Eq. (5-10). To illustrate the system further, let $K_t = 10$ volts/rad/sec. This means that, for a unit-step input of 1 volt, the desired velocity in the steady state is $1/10$ or 0.1 rad/sec, because when this is achieved, the output voltage of the tachometer would be 1 volt, and the steady-state error would be zero. The closed-loop transfer function of the system is

$$M(s) = \frac{Y(s)}{R(s)} = \frac{G(s)}{1+G(s)H(s)} = \frac{1}{s(s^2+12s+10)} \quad (5\text{-}16)$$

Toolbox 5-4-1

For the system in Eq. 5-15:
$$G(s) = \frac{1}{s^2(s+12)} \quad H(s) = K_t s$$

```
% use Kt=10
%Step input
Kt=10;
Gzpk=zpk([],[0 0 -12],1)
G=tf(Gzpk)
H=zpk(0,[],Kt)
cloop=feedback(G,H)
step(cloop)
xlabel('Time(sec)');
ylabel('Amplitude');
```

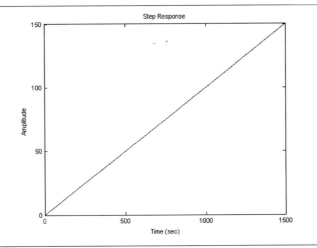

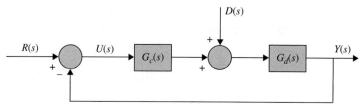

Figure 5-4 System with disturbance input.

For a unit-step function input, $R(s) = 1/s$. The output time response is

$$y(t) = 0.1t - 0.12 - 0.000796e^{-11.1t} + 0.1208e^{-0.901t} \quad t \geq 0 \qquad (5\text{-}17)$$

Because the exponential terms of $y(t)$ in Eq. (5-17) all diminish as $t \to \infty$, the steady-state part of $y(t)$ is $0.1t - 0.12$. Thus, the steady-state error of the system is

$$e_{ss} = \lim_{t \to \infty} [0.1t - y(t)] = 0.12 \qquad (5\text{-}18)$$

More explanations on how to define the reference signal when $H(s) \neq 1$ will be given later when the general discussion on the steady-state error of nonunity feedback systems is given.

Not all system errors are defined with respect to the response due to the input. Fig. 5-4 shows a system with a disturbance $d(t)$, in addition to the input $r(t)$. The output due to $d(t)$ acting alone may also be considered an error.

Because of these reasons, the definition of system error has not been unified in the literature. To establish a systematic study of the steady-state error for linear systems, we shall classify three types of systems and treat these separately.

1. Systems with unity feedback; $H(s) = 1$.
2. Systems with nonunity feedback, but $H(0) = K_H =$ constant.
3. Systems with nonunity feedback, and $H(s)$ has zeros at $s = 0$ of order N.

The objective here is to establish a definition of the error with respect to one basic system configuration so that some fundamental relationships can be determined between the steady-state error and the system parameters.

Type of Control Systems: Unity Feedback Systems

Consider that a control system with unity feedback can be represented by or simplified to the block diagram with $H(s) = 1$ in Fig. 5-3. The steady-state error of the system is written

$$e_{ss} = \lim_{t \to \infty} e(t) = \lim_{s \to 0} sE(s)$$

$$= \lim_{s \to 0} \frac{sR(s)}{1 + G(s)} \qquad (5\text{-}19)$$

Clearly, e_{ss} depends on the characteristics of $G(s)$. More specifically, we can show that e_{ss} depends on the number of poles $G(s)$ has at $s = 0$. This number is known as the type of the control system or, simply, system type.

We can show that the steady-state error e_{ss} depends on the type of the control system. Let us formalize the system type by referring to the form of the forward-path transfer function $G(s)$. In general, $G(s)$ can be expressed for convenience as

$$G(s) = \frac{K(1+T_1 s)(1+T_2 s)\cdots(1+T_{m1}s+T_{m2}s^2)}{s^j(1+T_a s)(1+T_b s)\cdots(1+T_{n1}s+T_{n2}s^2)} e^{-T_d s} \qquad (5\text{-}20)$$

where K and all the T's are real constants. The system type refers to the order of the pole of $G(s)$ at $s=0$. Thus, the closed-loop system having the forward-path transfer function of Eq. (5-20) is type j, where $j = 0, 1, 2, \ldots$. The total number of terms in the numerator and the denominator and the values of the coefficients are not important to the system type, as system type refers only to the number of poles $G(s)$ has at $s=0$. The following example illustrates the system type with reference to the form of $G(s)$.

▶ **EXAMPLE 5-4-1**

$$G(s) = \frac{K(1+0.5s)}{s(1+s)(1+2s)(1+s+s^2)} \quad \text{type 1} \qquad (5\text{-}21)$$

$$G(s) = \frac{K(1+2s)}{s^3} \quad \text{type 3} \qquad (5\text{-}22)$$

◀

Now let us investigate the effects of the types of inputs on the steady-state error. We shall consider only the step, ramp, and parabolic inputs.

Steady-State Error of System with a Step-Function Input
When the input $r(t)$ to the control system with $H(s) = 1$ of Fig. 5-3 is a step function with magnitude R, $R(s) = R/s$, the steady-state error is written from Eq. (5-19),

$$e_{ss} = \lim_{s \to 0} \frac{sR(s)}{1+G(s)} = \lim_{s \to 0} \frac{R}{1+G(s)} = \frac{R}{1+\lim_{s \to 0} G(s)} \qquad (5\text{-}23)$$

For convenience, we define

$$K_p = \lim_{s \to 0} G(s) \qquad (5\text{-}24)$$

as the step-error constant. Then Eq. (5-23) becomes

$$e_{ss} = \frac{R}{1+K_p} \qquad (5\text{-}25)$$

A typical e_{ss} due to a step input when K_p is finite and nonzero is shown in Fig. 5-5. We see from Eq. (5-25) that, for e_{ss} to be zero, when the input is a step function, K_p must be infinite. If $G(s)$ is described by Eq. (5-20), we see that, for K_p to be infinite, j must be at least equal to unity; that is, $G(s)$ must have at least one pole at $s=0$. Therefore, we can summarize the steady-state error due to a step function input as follows:

Type 0 system: $e_{ss} = \dfrac{R}{1+K_p} = $ constant

Type 1 or higher system: $e_{ss} = 0$

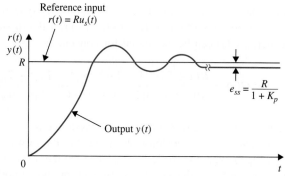

Figure 5-5 Typical steady-state error due to a step input.

Steady-State Error of System with a Ramp-Function Input

When the input to the control system $[H(s) = 1]$ of Fig. 5-3 is a ramp function with magnitude R,

$$r(t) = Rtu_s(t) \tag{5-26}$$

where R is a real constant, the Laplace transform of $r(t)$ is

$$R(s) = \frac{R}{s^2} \tag{5-27}$$

The steady-state error is written using Eq. (5-19),

$$e_{ss} = \lim_{s \to 0} \frac{R}{s + sG(s)} = \frac{R}{\lim_{s \to 0} sG(s)} \tag{5-28}$$

We define the ramp-error constant as

$$K_v = \lim_{s \to 0} sG(s) \tag{5-29}$$

Then, Eq. (5-26) becomes

$$e_{ss} = \frac{R}{K_v} \tag{5-30}$$

which is the steady-state error when the input is a ramp function. A typical e_{ss} due to a ramp input when K_v is finite and nonzero is illustrated in Fig. 5-6.

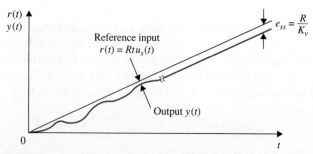

Figure 5-6 Typical steady-state error due to a ramp-function input.

Eq. (5-30) shows that, for e_{ss} to be zero when the input is a ramp function, K_v must be infinite. Using Eqs. (5-20) and (5-29), we obtain

$$K_v = \lim_{s \to 0} sG(s) = \lim_{s \to 0} \frac{K}{s^{j-1}} \quad j = 0, 1, 2, \ldots \quad (5\text{-}31)$$

Thus, for K_v to be infinite, j must be at least equal to 2, or the system must be of type 2 or higher. The following conclusions may be stated with regard to the steady-state error of a system with ramp input:

Type 0 system: $\quad e_{ss} = \infty$

Type 1 system: $\quad e_{ss} = \dfrac{R}{K_v} = $ constant

Type 2 system: $\quad e_{ss} = 0$

Steady-State Error of System with a Parabolic-Function Input

When the input is described by the standard parabolic form

$$r(t) = \frac{Rt^2}{2} u_s(t) \quad (5\text{-}32)$$

the Laplace transform of $r(t)$ is

$$R(s) = \frac{R}{s^3} \quad (5\text{-}33)$$

The steady-state error of the system in Fig. 5-3 with $H(s) = 1$ is

$$e_{ss} = \frac{R}{\lim\limits_{s \to 0} s^2 G(s)} \quad (5\text{-}34)$$

A typical e_{ss} of a system with a nonzero and finite K_a due to a parabolic-function input is shown in Fig. 5-7.

Defining the parabolic-error constant as

$$K_a = \lim_{s \to 0} s^2 G(s) \quad (5\text{-}35)$$

the steady-state error becomes

$$e_{ss} = \frac{R}{K_a} \quad (5\text{-}36)$$

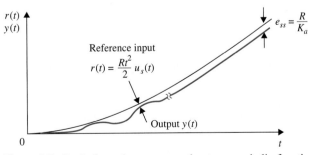

Figure 5-7 Typical steady-state error due to a parabolic-function input.

TABLE 5-1 Summary of the Steady-State Errors Due to Step-, Ramp-, and Parabolic-Function Inputs for Unity-Feedback Systems

Type of System	Error Constants			Steady-State Error e_{ss}		
j	K_p	K_v	K_a	Step Input	Ramp Input	Parabolic
				$\frac{R}{1+K_p}$	$\frac{R}{K_v}$	$\frac{R}{K_a}$
0	K	0	0	$\frac{R}{1+K}$	∞	∞
1	∞	K	0	0	$\frac{R}{K}$	∞
2	∞	∞	K	0	0	$\frac{R}{K}$
3	∞	∞	∞	0	0	0

Following the pattern set with the step and ramp inputs, the steady-state error due to the parabolic input is zero if the system is of type 3 or greater. The following conclusions are made with regard to the steady-state error of a system with parabolic input:

Type 0 system: $\quad e_{ss} = \infty$

Type 1 system: $\quad e_{ss} = \infty$

Type 2 system: $\quad e_{ss} = \dfrac{R}{K_a} = $ constant

Type 3 or higher system: $\quad e_{ss} = 0$

We cannot emphasize often enough that, for these results to be valid, the closed-loop system must be stable.

By using the method described, the steady-state error of any linear closed-loop system subject to an input with order higher than the parabolic function can also be derived if necessary. As a summary of the error analysis, Table 5-1 shows the relations among the error constants, the types of systems with reference to Eq. (5-20), and the input types.

As a summary, the following points should be noted when applying the error-constant analysis just presented.

1. The step-, ramp-, or parabolic-error constants are significant for the error analysis only when the input signal is a step function, ramp function, or parabolic function, respectively.

2. Because the error constants are defined with respect to the forward-path transfer function $G(s)$, the method is applicable to only the system configuration shown in Fig. 5-3 with $H(s) = 1$. Because the error analysis relies on the use of the final-value theorem of the Laplace transform, it is important to check first to see if $sE(s)$ has any poles on the $j\omega$-axis or in the right-half s-plane.

3. The steady-state error properties summarized in Table 5-1 are for systems with unity feedback only.

4. The steady-state error of a system with an input that is a linear combination of the three basic types of inputs can be determined by superimposing the errors due to each input component.

5. When the system configuration differs from that of Fig. 5-3 with $H(s) = 1$, we can either simplify the system to the form of Fig. 5-3 or establish the error signal and apply the final-value theorem. The error constants defined here may or may not apply, depending on the individual situation.

When the steady-state error is infinite, that is, when the error increases continuously with time, the error-constant method does not indicate how the error varies with time. This is one of the disadvantages of the error-constant method. The error-constant method also does not apply to systems with inputs that are sinusoidal, since the final-value theorem cannot be applied. The following examples illustrate the utility of the error constants and their values in the determination of the steady-state errors of linear control systems with unity feedback.

▶ **EXAMPLE 5-4-2** Consider that the system shown in Fig. 5-3 with $H(s) = 1$ has the following transfer functions. The error constants and steady-state errors are calculated for the three basic types of inputs using the error constants.

a. $G(s) = \dfrac{K(s + 3.15)}{s(s + 1.5)(s + 0.5)}$ $H(s) = 1$ Type 1 system

Step input: Step-error constant $K_p = \infty$ $e_{ss} = \dfrac{R}{1 + K_p} = 0$

Ramp input: Ramp-error constant $K_v = 4.2K$ $e_{ss} = \dfrac{R}{K_v} = \dfrac{R}{4.2K}$

Parabolic input: Parabolic-error constant $K_a = 0$ $e_{ss} = \dfrac{R}{K_a} = \infty$

These results are valid only if the value of K stays within the range that corresponds to a stable closed-loop system, which is $0 < K < 1.304$.

b. $G(s) = \dfrac{K}{s^2(s + 12)}$ $H(s) = 1$ Type 2 system

The closed-loop system is unstable for all values of K, and error analysis is meaningless.

c. $G(s) = \dfrac{5(s + 1)}{s^2(s + 12)(s + 5)}$ $H(s) = 1$ Type 2 system

We can show that the closed-loop system is stable. The steady-state errors are calculated for the three basic types of inputs.

Toolbox 5-4-2

For the system in Example 5-4-2:

(a) $G(s) = \dfrac{K(s + 3.15)}{s(s + 1.5)(s + 0.5)}$ $H(s) = 1$ Type 1 system

```
% Step input
K=1; % Use K=1
Gzpk=zpk([-3.15],[0 -1.5 -0.5],1)
G=tf(Gzpk);
H=1;
clooptf=feedback(G,H)
step(clooptf)
xlabel('Time(sec)');
ylabel('Amplitude');
```

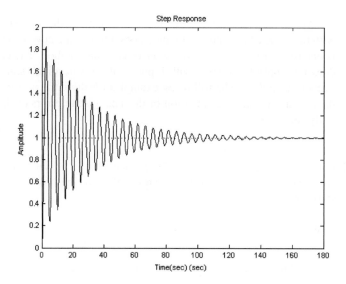

Similarly you may obtain the ramp and parabolic responses

```
%Ramp input
t=0:0.1:50;
u=t;
[y,x]=lsim(clooptf,u,t);
plot(t,y,t,u);
title('Closed-loop response for Ramp Input')
xlabel('Time(sec)')
ylabel('Amplitude')
%Parabolic input
t=0:0.1:50;
u = 0.5*t.*t;
[y,x]=lsim(clooptf,u,t);
plot(t,y,t,u);
title('Closed-loop response for Parabolic Input')
xlabel('Time(sec')
ylabel('Amplitude')
```

Step input: Step-error constant: $K_p = \infty$ $\quad e_{ss} = \dfrac{R}{1+K_p} = 0$

Ramp input: Ramp-error constant: $K_v = \infty$ $\quad e_{ss} = \dfrac{R}{K_v} = 0$

Parabolic input: Parabolic-error constant: $K_a = 1/12$ $\quad e_{ss} = \dfrac{R}{K_a} = 12R$

Relationship between Steady-State Error and Closed-Loop Transfer Function

In the last section, the steady-state error of a closed-loop system was related to the forward-path transfer function $G(s)$ of the system, which is usually known. Often, the closed-loop transfer function is derived in the analysis process, and it would be of interest to establish the relationships between the steady-state error and the coefficients of the closed-loop transfer function. As it turns out, the closed-loop transfer function can be used to find the steady-state error of systems with unity as well as nonunity feedback. For the present discussion, let us impose the following condition:

$$\lim_{s \to 0} H(s) = H(0) = K_H = \text{constant} \qquad (5\text{-}37)$$

which means that $H(s)$ cannot have poles at $s = 0$. Because the signal that is fed back to be compared with the input in the steady state is K_H times the steady-state output, when this feedback signal equals the input, the steady-state error would be zero. Thus, we can define the reference signal as $r(t)/K_H$ and the error signal as

$$e(t) = \frac{1}{K_H} r(t) - y(t) \tag{5-38}$$

or, in the transform domain,

$$E(s) = \frac{1}{K_H} R(s) - Y(s) = \frac{1}{K_H}[1 - K_H M(s)]R(s) \tag{5-39}$$

where $M(s)$ is the closed-loop transfer function, $Y(s)/R(s)$. Notice that the above development includes the unity-feedback case for which $K_H = 1$. Let us assume that $M(s)$ does not have any poles at $s = 0$ and is of the form

$$M(s) = \frac{Y(s)}{R(s)} = \frac{b_m s^m + b_{m-1} s^{m-1} + \cdots + b_1 s + b_0}{s^n + a_{n-1} s^{n-1} + \cdots + a_1 s + a_0} \tag{5-40}$$

where $n > m$. We further require that all the poles of $M(s)$ are in the left-half s-plane, which means that the system is stable. The steady-state error of the system is written

$$e_{ss} = \lim_{s \to 0} sE(s) = \lim_{s \to 0} \frac{1}{K_H}[1 - K_H M(s)]sR(s) \tag{5-41}$$

Substituting Eq. (5-40) into the last equation and simplifying, we get

$$e_{ss} = \frac{1}{K_H} \lim_{s \to 0} \frac{s^n + \cdots + (a_1 - b_1 K_H)s + (a_0 - b_0 K_H)}{s^n + a_{n-1} s^{n-1} + \cdots + a_1 s + a_0} sR(s) \tag{5-42}$$

We consider the three basic types of inputs for $r(t)$.

1. **Step-function input.** $R(s) = R/s$.

    For a step-function input, the steady-state error in Eq. (5-42) becomes

$$e_{ss} = \frac{1}{K_H}\left(\frac{a_0 - b_0 K_H}{a_0}\right) R \tag{5-43}$$

Thus, the steady-state error due to a step input can be zero only if

$$a_0 - b_0 K_H = 0 \tag{5-44}$$

or

$$M(0) = \frac{b_0}{a_0} = \frac{1}{K_H} \tag{5-45}$$

This means that, *for a unity-feedback system $K_H = 1$, the constant terms of the numerator and the denominator of $M(s)$ must be equal, that is, $b_0 = a_0$, for the steady-state error to be zero.*

2. **Ramp-function input.** $R(s) = R/s^2$.

   For a ramp-function input, the steady-state error in Eq. (5-42) becomes

   $$e_{ss} = \frac{1}{K_H} \lim_{s \to 0} \frac{s^n + \cdots + (a_1 - b_1 K_H)s + (a_0 - b_0 K_H)}{s(s^n + a_{n-1}s^{n-1} + \cdots + a_1 s + a_0)} R \quad (5\text{-}46)$$

   The following values of e_{ss} are possible:

   $e_{ss} = 0$ if $a_0 - b_0 K_H = 0$ and $a_1 - b_1 K_H = 0$ (5-47)

   $e_{ss} = \dfrac{a_1 - b_1 K_H}{a_0 K_H} R = $ constant   if $a_0 - b_0 K_H = 0$ and $a_1 - b_1 K_H \neq 0$ (5-48)

   $e_{ss} = \infty$ if $a_0 - b_0 K_H \neq 0$ (5-49)

3. **Parabolic-function input.** $R(s) = R/s^3$.

   For a parabolic input, the steady-state error in Eq. (5-42) becomes

   $$e_{ss} = \frac{1}{K_H} \lim_{s \to 0} \frac{s^n + \cdots + (a_2 - b_2 K_H)s^2 + (a_1 - b_1 K_H)s + (a_0 - b_0 K_H)}{s^2(s^n + a_{n-1}s^{n-1} + \cdots + a_1 s + a_0)} R \quad (5\text{-}50)$$

   The following values of e_{ss} are possible:

   $e_{ss} = 0$ if $a_i - b_i K_H = 0$ for $i = 0, 1,$ and 2   (5-51)

   $e_{ss} = \dfrac{a_2 - b_2 K_H}{a_0 K_H} R = $ constant   if $a_i - b_i K_H = 0$ for $i = 0$ and 1   (5-52)

   $e_{ss} = \infty$ if $a_i - b_i K_H \neq 0$ for $i = 0$ and 1   (5-53)

▶ **EXAMPLE 5-4-3** The forward-path and closed-loop transfer functions of the system shown in Fig. 5-3 are given next. The system is assumed to have unity feedback, so $H(s) = 1$, and thus $K_H = H(0) = 1$.

$$G(s) = \frac{5(s+1)}{s^2(s+12)(s+5)} \quad M(s) = \frac{5(s+1)}{s^4 + 17s^3 + 60s^2 + 5s + 5} \quad (5\text{-}54)$$

The poles of $M(s)$ are all in the left-half s-plane. Thus, the system is stable. The steady-state errors due to the three basic types of inputs are evaluated as follows:

Step input: $e_{ss} = 0$ since $a_0 = b_0 (= 5)$

Ramp input: $e_{ss} = 0$ since $a_0 = b_0 (= 5)$ and $a_1 = b_1 (= 5)$

Parabolic input: $e_{ss} = \dfrac{a_2 - b_2 K_H}{a_0 K_H} R = \dfrac{60}{5} R = 12 R$

Because this is a type 2 system with unity feedback, the same results are obtained with the error-constant method. ◀

▶ **EXAMPLE 5-4-4** Consider the system shown in Fig. 5-3, which has the following transfer functions:

$$G(s) = \frac{1}{s^2(s+12)} \quad H(s) = \frac{5(s+1)}{s+5} \tag{5-55}$$

Then, $K_H = H(0) = 1$. The closed-loop transfer function is

$$M(s) = \frac{Y(s)}{R(s)} = \frac{G(s)}{1+G(s)H(s)} = \frac{s+5}{s^4+17s^3+60s^2+5s+5} \tag{5-56}$$

Comparing the last equation with Eq. (5-40), we have $a_0 = 5$, $a_1 = 5$, $a_2 = 60$, $b_0 = 5$, $b_1 = 1$, and $b_2 = 0$. The steady-state errors of the system are calculated for the three basic types of inputs.

Unit-step input, $r(t) = u_s(t)$: $\quad e_{ss} = \dfrac{a_0 - b_0 K_H}{a_0} = 0$

Unit-ramp input, $r(t) = tu_s(t)$: $\quad e_{ss} = \dfrac{a_1 - b_1 K_H}{a_0 K_H} = \dfrac{5-1}{5} = 0.8$

Unit-parabolic input, $r(t) = tu_s(t)/2$: $\quad e_{ss} = \infty \quad$ since $a_1 - b_1 K_H \neq 0$

It would be illuminating to calculate the steady-state errors of the system from the difference between the input and the output and compare them with the results just obtained.

Applying the unit-step, unit-ramp, and unit-parabolic inputs to the system described by Eq. (5-56), and taking the inverse Laplace transform of $Y(s)$, the outputs are

Unit-step input:

$$\begin{aligned} y(t) = 1 &- 0.00056e^{-12.05t} - 0.00013 81e^{-4.886t} \\ &- 0.9993e^{-0.0302t} \cos 0.2898t - 0.1301e^{-0.0302t} \sin 0.2898t \quad t \geq 0 \end{aligned} \tag{5-57}$$

Thus, the steady-state value of $y(t)$ is unity, and the steady-state error is zero.

Unit-ramp input:

$$\begin{aligned} y(t) = t &- 0.8 + 4.682 \times 10^{-5} e^{-12.05t} + 2.826 \times 10^{-5} e^{-4.886t} \\ &+ 0.8e^{-0.0302t} \cos 0.2898t - 3.365e^{-0.0302t} \sin 0.2898t \quad t \geq 0 \end{aligned} \tag{5-58}$$

Thus, the steady-state portion of $y(t)$ is $t - 0.8$, and the steady-state error to a unit ramp is 0.8.

Unit-parabolic input:

$$\begin{aligned} y(t) = 0.5t^2 &- 0.8t - 11.2 - 3.8842 \times 10^{-6} e^{-12.05t} - 5.784 \times 10^{-6} e^{-4.886t} \\ &+ 11.2 e^{-0.0302t} \cos 0.2898t + 3.9289 e^{-0.0302t} \sin 0.2898t \quad t \geq 0 \end{aligned} \tag{5-59}$$

The steady-state portion of $y(t)$ is $0.5t^2 - 0.8t - 11.2$. Thus, the steady-state error is $0.8t + 11.2$, which becomes infinite as time goes to infinity.

◀

▶ **EXAMPLE 5-4-5** Consider that the system shown in Fig. 5-3 has the following transfer functions:

$$G(s) = \frac{1}{s^2(s+12)} \quad H(s) = \frac{10(s+1)}{s+5} \tag{5-60}$$

Thus,

$$K_H = \lim_{s \to 0} H(s) = 2 \tag{5-61}$$

The closed-loop transfer function is

$$M(s) = \frac{Y(s)}{R(s)} = \frac{G(s)}{1+G(s)H(s)} = \frac{s+5}{s^4+17s^3+60s^2+10s+10} \tag{5-62}$$

The steady-state errors of the system due to the three basic types of inputs are calculated as follows:

Unit-step input $r(t) = u_s(t)$:

$$e_{ss} = \frac{1}{K_H}\left(\frac{a_0 - b_0 K_H}{a_0}\right) = \frac{1}{2}\left(\frac{10 - 5 \times 2}{10}\right) = 0 \qquad (5\text{-}63)$$

Solving for the output using the $M(s)$ in Eq. (5-62), we get

$$y(t) = 0.5 u_s(t) + \text{transient terms} \qquad (5\text{-}64)$$

Thus, the steady-state value of $y(t)$ is 0.5, and because $K_H = 2$, the steady-state error due to a unit-step input is zero.

Unit-ramp input $r(t) = t u_s(t)$:

$$e_{ss} = \frac{1}{K_H}\left(\frac{a_1 - b_1 K_H}{a_0}\right) = \frac{1}{2}\left(\frac{10 - 1 \times 2}{10}\right) = 0.4 \qquad (5\text{-}65)$$

The unit-ramp response of the system is written

$$y(t) = [0.5t - 0.4] u_s(t) + \text{transient terms} \qquad (5\text{-}66)$$

Thus, using Eq. (5-38), the steady-state error is calculated as

$$e(t) = \frac{1}{K_H} r(t) - y(t) = 0.4 u_s(t) - \text{transient terms} \qquad (5\text{-}67)$$

Because the transient terms will die out as t approaches infinity, the steady-state error due to a unit-ramp input is 0.4, as calculated in Eq. (5-66).

Unit-parabolic input $r(t) = t^2 u_s(t)/2$:

$$e_{ss} = \infty \quad \text{since} \quad a_1 - b_1 K_H \neq 0$$

The unit-parabolic input is

$$y(t) = \left[0.25 t^2 - 0.4 t - 2.6\right] u_s(t) + \text{transient terms} \qquad (5\text{-}68)$$

The error due to the unit-parabolic input is

$$e(t) = \frac{1}{K_H} r(t) - y(t) = (0.4t - 2.6) u_s(t) - \text{transient terms} \qquad (5\text{-}69)$$

Thus, the steady-state error is $0.4t + 2.6$, which increases with time. ◀

Steady-State Error of Nonunity Feedback: $H(s)$ Has Nth-Order Zero at $s = 0$

This case corresponds to desired output being proportional to the Nth-order derivative of the input in the steady state. In the real world, this corresponds to applying a tachometer or rate feedback. Thus, for the steady-state error analysis, the reference signal can be defined as $R(s)/K_H s^N$, and the error signal in the transform domain may be defined as

$$E(s) = \frac{1}{K_H s^N} R(s) - Y(s) \qquad (5\text{-}70)$$

where

$$K_H = \lim_{s \to 0} \frac{H(s)}{s^N} \qquad (5\text{-}71)$$

We shall derive only the results for $N = 1$ here. In this case, the transfer function of $M(s)$ in Eq. (5-40) will have a pole at $s = 0$, or $a_0 = 0$. The steady-state error is written from Eq. (5-70) as

$$e_{ss} = \frac{1}{K_H} \lim_{s \to 0} \left[\frac{s^{n-1} + \cdots + (a_2 - b_1 K_H)s + (a_1 - b_0 K_H)}{s^n + a_{n-1} s^{n-1} + \cdots + a_1 s}\right] s R(s) \qquad (5\text{-}72)$$

For a step input of magnitude R, the last equation is written

$$e_{ss} = \frac{1}{K_H} \lim_{s \to 0} \left[\frac{s^{n-1} + \cdots + (a_2 - b_1 K_H)s + (a_1 - b_0 K_H)}{s^n + a_{n-1} s^{n-1} + \cdots + a_1 s} \right] R \quad (5\text{-}73)$$

Thus, the steady-state error is

$$e_{ss} = 0 \quad \text{if} \quad a_2 - b_1 K_H = 0 \quad \text{and} \quad a_1 - b_0 K_H = 0 \quad (5\text{-}74)$$

$$e_{ss} = \frac{a_2 - b_1 K_H}{a_1 K_H} R = \text{constant} \quad \text{if} \quad a_1 - b_0 K_H = 0 \quad \text{but} \quad a_2 - b_1 K_H \neq 0 \quad (5\text{-}75)$$

$$e_{ss} = \infty \quad \text{if} \quad a_1 - b_0 K_H \neq 0 \quad (5\text{-}76)$$

We shall use the following example to illustrate these results.

▶ **EXAMPLE 5-4-6** Consider that the system shown in Fig. 5-3 has the following transfer functions:

$$G(s) = \frac{1}{s^2(s+12)} \quad H(s) = \frac{10s}{s+5} \quad (5\text{-}77)$$

Thus,

$$K_H = \lim_{s \to 0} \frac{H(s)}{s} = 2 \quad (5\text{-}78)$$

The closed-loop transfer function is

$$M(s) = \frac{Y(s)}{R(s)} = \frac{s+5}{s^4 + 17s^3 + 60s^2 + 10s} \quad (5\text{-}79)$$

The velocity control system is stable, although $M(s)$ has a pole at $s = 0$, because the objective is to control velocity with a step input. The coefficients are identified to be $a_0 = 0$, $a_1 = 10$, $a_2 = 60$, $b_0 = 5$, and $b_1 = 1$.

For a unit-step input, the steady-state error, from Eq. (5-75), is

$$e_{ss} = \frac{1}{K_H} \left(\frac{a_2 - b_1 K_H}{a_1} \right) = \frac{1}{2} \left(\frac{60 - 1 \times 2}{10} \right) = 2.9 \quad (5\text{-}80)$$

To verify this result, we find the unit-step response using the closed-loop transfer function in Eq. (5-79). The result is

$$y(t) = (0.5t - 2.9) u_s(t) + \text{transient terms} \quad (5\text{-}81)$$

From the discussion that leads to Eq. (5-70), the reference signal is considered to be $t u_s(t)/K_H = 0.5 t u_s(t)$ in the steady state; thus, the steady-state error is 2.9. Of course, it should be pointed out that if $H(s)$ were a constant for this type 2 system, the closed-loop system would be unstable. So, the derivative control in the feedback path also has a stabilizing effect. ◀

Toolbox 5-4-3

The corresponding responses for Eq. 5-79 are obtained by the following sequence of MATLAB functions

```
t=0:0.1:50;
num = [1 5];
den = [1 17 60 10 0];
sys = tf(num,den);
sys_cl=feedback(sys,1);
[y,t]=step(sys_cl);
```

```
u=ones(size(t));
plot(t,y,'r',t,u,'g')
xlabel('Time(secs)')
ylabel('Amplitude')
title('Input-green, Output-red')
```

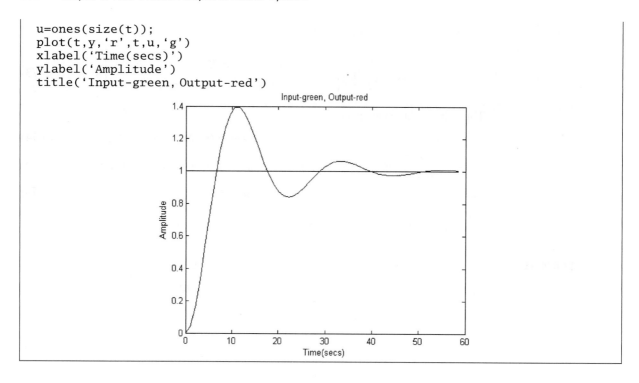

5-4-2 Steady-State Error Caused by Nonlinear System Elements

In many instances, steady-state errors of control systems are attributed to some nonlinear system characteristics such as nonlinear friction or dead zone. For instance, if an amplifier used in a control system has the input–output characteristics shown in Fig. 5-8, then, when the amplitude of the amplifier input signal falls within the dead zone, the output of the amplifier would be zero, and the control would not be able to correct the error if any exists. Dead-zone nonlinearity characteristics shown in Fig. 5-8 are not limited to amplifiers. The flux-to-current relation of the magnetic field of an electric motor may exhibit a similar characteristic. As the current of the motor falls below the dead zone D, no magnetic flux, and, thus, no torque will be produced by the motor to move the load.

The output signals of digital components used in control systems, such as a microprocessor, can take on only discrete or quantized levels. This property is illustrated by the quantization characteristics shown in Fig. 5-9. When the input to the quantizer is within $\pm q/2$, the output is zero, and the system may generate an error in the output whose

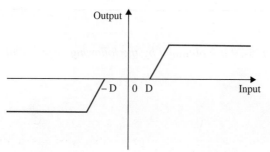

Figure 5-8 Typical input–output characteristics of an amplifier with dead zone and saturation.

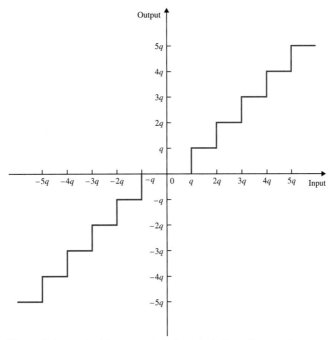

Figure 5-9 Typical input–output characteristics of a quantizer.

magnitude is related to $\pm q/2$. This type of error is also known as the quantization error in digital control systems.

When the control of physical objects is involved, friction is almost always present. Coulomb friction is a common cause of steady-state position errors in control systems. Fig. 5-10 shows a restoring-torque-versus-position curve of a control system. The torque curve typically could be generated by a step motor or a switched-reluctance motor or from a closed-loop system with a position encoder. Point 0 designates a stable equilibrium point on the torque curve, as well as the other periodic intersecting points along the axis where the slope on the torque curve is negative. The torque on either side of point 0 represents a restoring torque that tends to return the output to the equilibrium point when some angular-displacement disturbance takes place. When there is no friction, the position error should be zero, because there is always a restoring torque so long as the position is not at the stable

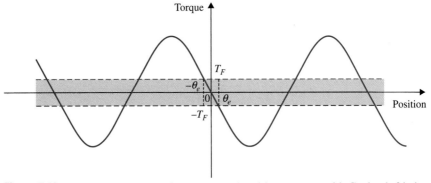

Figure 5-10 Torque-angle curve of a motor or closed-loop system with Coulomb friction.

equilibrium point. If the rotor of the motor sees a Coulomb friction torque T_F, then the motor torque must first overcome this frictional torque before producing any motion. Thus, as the motor torque falls below T_F as the rotor position approaches the stable equilibrium point, it may stop at any position inside the error band bounded by $\pm\theta_e$, as shown in Fig. 5-10.

Although it is relatively simple to comprehend the effects of nonlinearities on errors and to establish maximum upper bounds on the error magnitudes, it is difficult to establish general and closed-form solutions for nonlinear systems. Usually, exact and detailed analysis of errors in nonlinear control systems can be carried out only by computer simulations.

Therefore, we must realize that there are no error-free control systems in the real world, and, because all physical systems have nonlinear characteristics of one form or another, steady-state errors can be reduced but never completely eliminated.

▶ 5-5 TIME RESPONSE OF A PROTOTYPE FIRST-ORDER SYSTEM

Consider the **prototype first-order system** of form

$$\frac{dy(t)}{dt} + \frac{1}{\tau}y(t) = \frac{1}{\tau}f(t) \tag{5-82}$$

where t is known as the **time constant** of the system, which is a measure of how fast the system responds to initial conditions of external excitations. Note that the input in Eq. (5-82) is scaled by $\frac{1}{\tau}$ for cosmetic reasons.

For a unit-step input

$$f(t) = u_s(t) = \begin{cases} 0, & t < 0, \\ 1, & t \geq 0 \end{cases} \tag{5-83}$$

If $y(0) = \dot{y}(0) = 0$, $\mathcal{L}(u_s(t)) = \frac{1}{s}$ and $\mathcal{L}(y(t)) = Y(s)$, then

$$Y(s) = \frac{1}{s}\frac{1/\tau}{s + 1/\tau} \tag{5-84}$$

Applying the inverse Laplace transform to Eq. (5-84), we get the time response of Eq. (5-82):

$$y(t) = 1 - e^{-t/\tau} \tag{5-85}$$

where t is the time for $y(t)$ to reach 63% of its final value of $\lim_{t \to \infty} y(t) = 1$.

Fig. 5-11 shows typical unit-step responses of $y(t)$ for a general value of τ. As the value of time constant τ decreases, the system response approaches faster to the final value. The step response will not have any overshoot for any combination of system parameters.

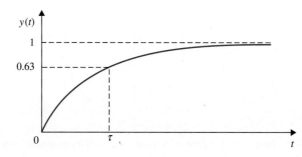

Figure 5-11 Unit-step response of a prototype first-order system.

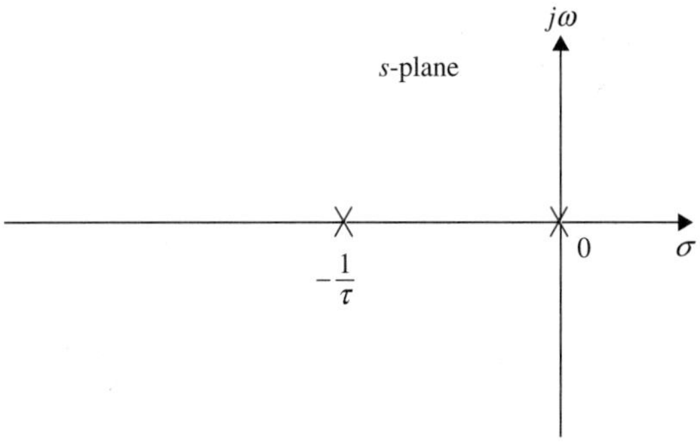

Figure 5-12 Pole configuration of the transfer function of a prototype first-order system.

Fig. 5-12 shows the location of the pole at $s = -\frac{1}{\tau}$ in the s-plane of the system transfer function. For positive τ, the pole at $s = -\frac{1}{\tau}$ will always stay in the left-half s-plane, and the system is always stable.

▶ 5-6 TRANSIENT RESPONSE OF A PROTOTYPE SECOND-ORDER SYSTEM

Although true second-order control systems are rare in practice, their analysis generally helps to form a basis for the understanding of analysis and design of higher-order systems, especially the ones that can be approximated by second-order systems.

Consider that a second-order control system with unity feedback is represented by the block diagram shown in Fig. 5-13. The open-loop transfer function of the system is

$$G(s) = \frac{Y(s)}{E(s)} = \frac{\omega_n^2}{s(s + 2\zeta\omega_n)} \tag{5-86}$$

where ζ and ω_n are real constants. The closed-loop transfer function of the system is

$$\frac{Y(s)}{R(s)} = \frac{\omega_n^2}{s^2 + 2\zeta\omega_n s + \omega_n^2} \tag{5-87}$$

The system in Fig. 5-13 with the transfer functions given by Eqs. (5-86) and (5-87) is defined as the prototype second-order system.

The characteristic equation of the prototype second-order system is obtained by setting the denominator of Eq. (5-87) to zero:

$$\Delta(s) = s^2 + 2\zeta\omega_n s + \omega_n^2 = 0 \tag{5-88}$$

For a unit-step function input, $R(s) = 1/s$, the output response of the system is obtained by taking the inverse Laplace transform of the output transform:

$$Y(s) = \frac{\omega_n^2}{s(s^2 + 2\zeta\omega_n s + \omega_n^2)} \tag{5-89}$$

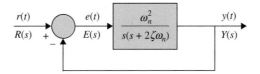

Figure 5-13 Prototype second-order control system.

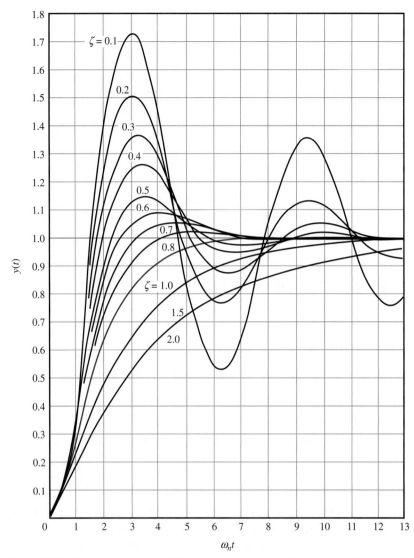

Figure 5-14 Unit-step responses of the prototype second-order system with various damping ratios.

This can be done by referring to the Laplace transform table in Appendix C. The result is

$$y(t) = 1 - \frac{e^{-\zeta\omega_n t}}{\sqrt{1-\zeta^2}} \sin\left(\omega_n \sqrt{1-\zeta^2}\, t + \cos^{-1}\zeta\right) \quad t \geq 0 \quad (5\text{-}90)$$

Fig. 5-14 shows the unit-step responses of Eq. (5-90) plotted as functions of the normalized time $\omega_n t$ for various values of ζ. As seen, the response becomes more oscillatory with larger overshoot as ζ decreases. When $\zeta \geq 1$, the step response does not exhibit any overshoot; that is, $y(t)$ never exceeds its final value during the transient. The responses also show that ω_n has a direct effect on the rise time, delay time, and settling time but does not affect the overshoot. These will be studied in more detail in the following sections.

5-6-1 Damping Ratio and Damping Factor

The effects of the system parameters ζ and ω_n on the step response $y(t)$ of the prototype second-order system can be studied by referring to the roots of the characteristic equation in Eq. (5-88).

Toolbox 5-6-1

The corresponding time responses for Fig. 5-14 are obtained by the following sequence of MATLAB functions

```
clear all
w=10;
for l=[0.2 0.4 0.6 0.8 1 1.2 1.4 1.6 1.8 2]
t=0:0.1:50;
num = [w.^2];
den = [1 2*l*w w.^2];
t=0:0.01:2;
step(num,den,t) hold on;
end
xlabel('Time(secs)')
ylabel('Amplitude')
title('Closed-Loop Step')
```

The two roots can be expressed as

$$s_1, s_2 = -\zeta\omega_n \pm j\omega_n\sqrt{1-\zeta^2}$$
$$= -\alpha \pm j\omega \tag{5-91}$$

where

$$\alpha = \zeta\omega_n \tag{5-92}$$

and

$$\omega = \omega_n\sqrt{1-\zeta^2} \tag{5-93}$$

The physical significance of ζ and α is now investigated. As seen from Eqs. (5-90) and (5-92), α appears as the constant that is multiplied to t in the exponential term of $y(t)$.

Therefore, α controls the rate of rise or decay of the unit-step response $y(t)$. In other words, α controls the "damping" of the system and is called the damping factor, or the damping constant. The inverse of α, $1/\alpha$, is proportional to the time constant of the system.

When the two roots of the characteristic equation are real and equal, we called the system critically damped. From Eq. (5-91), we see that critical damping occurs when $\zeta = 1$. Under this condition, the damping factor is simply $\alpha = \omega_n$. Thus, we can regard ζ as the damping ratio; that is,

$$\zeta = \text{damping ratio} = \frac{\alpha}{\omega_n} = \frac{\text{actual damping factor}}{\text{damping factor at critical damping}} \quad (5\text{-}94)$$

5-6-2 Natural Undamped Frequency

The parameter ω_n is defined as the natural undamped frequency. As seen from Eq. (5-91), when $\zeta = 0$, the damping is zero, the roots of the characteristic equation are imaginary, and Eq. (5-90) shows that the unit-step response is purely sinusoidal. Therefore, ω_n corresponds to the frequency of the undamped sinusoidal response. Eq. (5-91) shows that, when $0 < \zeta < 1$, the imaginary part of the roots has the magnitude of ω. When $\zeta \neq 0$, the response of $y(t)$ is not a periodic function, and ϖ defined in Eq. (5-93) is not a frequency. For the purpose of reference, ω is sometimes defined as the conditional frequency, or the damped frequency.

Fig. 5-15 illustrates the relationships among the location of the characteristic equation roots and α, ζ, ω_n, and ω. For the complex-conjugate roots shown,

- ω_n is the radial distance from the roots to the origin of the s-plane.
- α is the real part of the roots.
- ω is the imaginary part of the roots.
- ζ is the cosine of the angle between the radial line to the roots and the negative axis when the roots are in the left-half s-plane, or

$$\zeta = \cos\theta \quad (5\text{-}95)$$

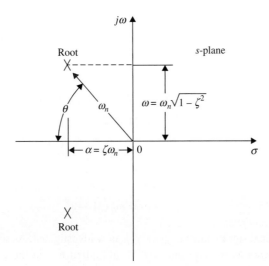

Figure 5-15 Relationships among the characteristic-equation roots of the prototype second-order system and α, ζ, ω_n, and ω.

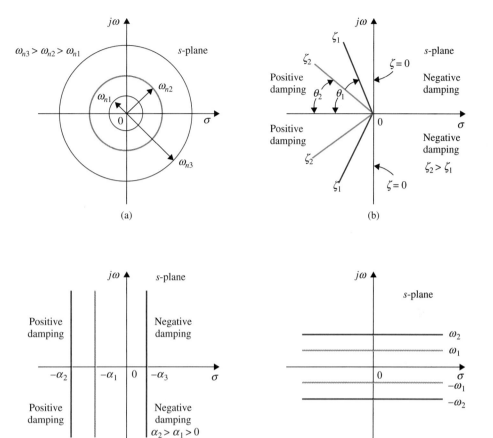

Figure 5-16 (a) Constant-natural-undamped-frequency loci. (b) Constant-damping-ratio loci. (c) Constant-damping-factor loci. (d) Constant-conditional-frequency loci.

Fig. 5-16 shows in the s-plane (a) the constant-ω_n loci, (b) the constant-ζ loci, (c) the constant-α loci, and (d) the constant-ω loci. Regions in the s-plane are identified with the system damping as follows:

- The left-half s-plane corresponds to positive damping; that is, the damping factor or damping ratio is positive. Positive damping causes the unit-step response to settle to a constant final value in the steady state due to the negative exponent of $\exp(-\zeta\omega_n t)$. The system is stable.
- The right-half s-plane corresponds to negative damping. Negative damping gives a response that grows in magnitude without bound, and the system is unstable.
- The imaginary axis corresponds to zero damping ($\alpha = 0$ or $\zeta = 0$). Zero damping results in a sustained oscillation response, and the system is marginally stable or marginally unstable.

Thus, we have demonstrated with the help of the simple prototype second-order system that the location of the characteristic equation roots plays an important role in the transient response of the system.

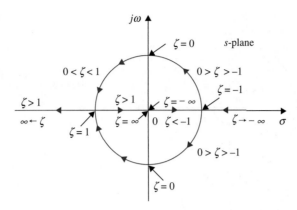

Figure 5-17 Locus of roots of the characteristic equation of the prototype second-order system.

The effect of the characteristic equation roots on the damping of the second-order system is further illustrated by Fig. 5-17 and Fig. 5-18. In Fig. 5-17, ω_n is held constant while the damping ratio ζ is varied from $-\infty$ to $+\infty$. The following classification of the system dynamics with respect to the value of ζ is made:

$0 < \zeta < 1$: $s_1, s_2 = -\zeta\omega_n \pm j\omega_n\sqrt{1-\zeta^2}$ $(-\zeta\omega_n < 0)$ *underdamped*

$\zeta = 1$: $s_1, s_2 = -\omega_n$ *critically damped*

$\zeta > 1$: $s_1, s_2 = -\zeta\omega_n \pm \omega_n\sqrt{\zeta^2-1}$ *overdamped*

$\zeta = 0$: $s_1, s_2 = \pm j\omega_n$ *undamped*

$\zeta < 0$: $s_1, s_2 = -\zeta\omega_n \pm j\omega_n\sqrt{1-\zeta^2}$ $(-\zeta\omega_n > 0)$ *negatively damped*

Fig. 5-18 illustrates typical unit-step responses that correspond to the various root locations already shown.

In practical applications, only stable systems that correspond to $\zeta > 0$ are of interest. Fig. 5-14 gives the unit-step responses of Eq. (5-90) plotted as functions of the normalized time $\omega_n t$ for various values of the damping ratio ζ. As seen, the response becomes more oscillatory as ζ decreases in value. When $\zeta \geq 1$, the step response does not exhibit any overshoot; that is, $y(t)$ never exceeds its final value during the transient.

5-6-3 Maximum Overshoot

The exact relation between the damping ratio and the amount of overshoot can be obtained by taking the derivative of Eq. (5-90) with respect to t and setting the result to zero. Thus,

$$\frac{dy(t)}{dt} = \frac{\omega_n e^{-\zeta\omega_n t}}{\sqrt{1-\zeta^2}}\left[\zeta \sin(\omega t + \theta) - \sqrt{1-\zeta^2}\cos(\omega t + \theta)\right] \quad t \geq 0 \quad (5\text{-}96)$$

where ω and θ are defined in Eqs. (5-93) and (5-95), respectively. We can show that the quantity inside the square bracket in Eq. (5-96) can be reduced to $\sin \omega t$. Thus, Eq. (5-96) is simplified to

$$\frac{dy(t)}{dt} = \frac{\omega_n}{\sqrt{1-\zeta^2}} e^{-\zeta\omega_n t} \sin \omega_n\sqrt{1-\zeta^2}\, t \quad t \geq 0 \quad (5\text{-}97)$$

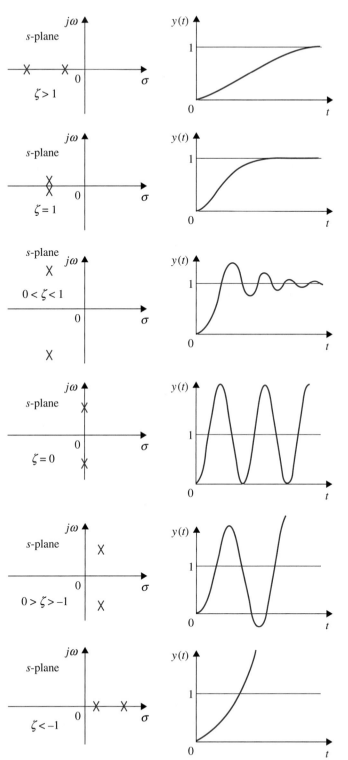

Figure 5-18 Step-response comparison for various characteristic-equation-root locations in the s-plane.

Setting $dy(t)/dt$ to zero, we have the solutions: $t = \infty$ and

$$\omega_n\sqrt{1-\zeta^2}\,t = n\pi \quad n = 0, 1, 2, \ldots \tag{5-98}$$

from which we get

$$t = \frac{n\pi}{\omega_n\sqrt{1-\zeta^2}} \quad n = 0, 1, 2, \ldots \tag{5-99}$$

The solution at $t = \infty$ is the maximum of $y(t)$ only when $\zeta \geq 1$. For the unit-step responses shown in Fig. 5-13, the first overshoot is the maximum overshoot. This corresponds to $n = 1$ in Eq. (5-99). Thus, the time at which the maximum overshoot occurs is

$$t_{\max} = \frac{\pi}{\omega_n\sqrt{1-\zeta^2}} \tag{5-100}$$

With reference to Fig. 5-13, the overshoots occur at odd values of n, that is, $n = 1, 3, 5, \ldots$, and the undershoots occur at even values of n. Whether the extremum is an overshoot or an undershoot, the time at which it occurs is given by Eq. (5-99). It should be noted that, *although the unit-step response for $\zeta \neq 0$ is not periodic, the overshoots and the undershoots of the response do occur at periodic intervals, as shown in Fig. 5-19.*

The magnitudes of the overshoots and the undershoots can be determined by substituting Eq. (5-99) into Eq. (5-90). The result is

$$y(t)|_{\max\, or\, \min} = 1 - \frac{e^{-n\pi\zeta/\sqrt{1-\zeta^2}}}{\sqrt{1-\zeta^2}} \sin(n\pi + \theta) \quad n = 1, 2, \ldots \tag{5-101}$$

or

$$y(t)|_{\max\, or\, \min} = 1 + (-1)^{n-1} e^{-n\pi\zeta/\sqrt{1-\zeta^2}} \quad n = 1, 2, \ldots \tag{5-102}$$

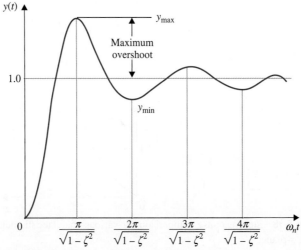

Figure 5-19 Unit-step response illustrating that the maxima and minima occur at periodic intervals.

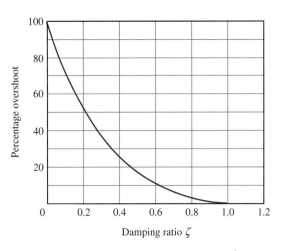

Figure 5-20 Percent overshoot as a function of damping ratio for the step response of the prototype second-order system.

The maximum overshoot is obtained by letting $n = 1$ in Eq. (5-102). Therefore,

$$\text{maximum overshoot} = y_{\max} - 1 = e^{-\pi\zeta/\sqrt{1-\zeta^2}} \quad (5\text{-}103)$$

and

$$\text{percent maximum overshoot} = 100 e^{-\pi\zeta/\sqrt{1-\zeta^2}} \quad (5\text{-}104)$$

Eq. (5-103) shows that the maximum overshoot of the step response of the prototype second-order system is a function of only the damping ratio ζ. The relationship between the percent maximum overshoot and the damping ratio given in Eq. (5-104) is plotted in Fig. 5-20. The time $t_{\max}$ in Eq. (5-100) is a function of both ζ and ω_n.

5-6-4 Delay Time and Rise Time

It is more difficult to determine the exact analytical expressions of the delay time t_d, rise time t_r, and settling time t_s, even for just the simple prototype second-order system. For instance, for the delay time, we would have to set $y(t) = 0.5$ in Eq. (5-90) and solve for t. An easier way would be to plot $\omega_n t_d$ versus ζ, as shown in Fig. 5-21, and then approximate the curve by a straight line or a curve over the range of $0 < \zeta < 1$. From Fig. 5-21, the delay time for the prototype second-order system is approximated as

$$t_d \cong \frac{1 + 0.7\zeta}{\omega_n} \quad 0 < \zeta < 1.0 \quad (5\text{-}105)$$

We can obtain a better approximation by using a second-order equation for t_d:

$$t_d \cong \frac{1.1 + 0.125\zeta + 0.469\zeta^2}{\omega_n} \quad 0 < \zeta < 1.0 \quad (5\text{-}106)$$

For the rise time t_r, which is the time for the step response to reach from 10 to 90% of its final value, the exact value can be determined directly from the responses of Fig. 5-14. The

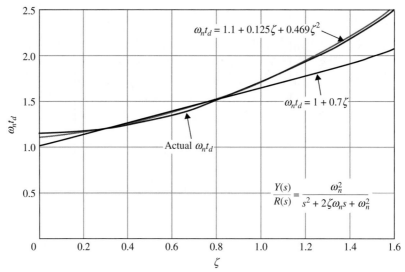

Figure 5-21 Normalized delay time versus ζ for the prototype second-order system.

plot of $\omega_n t_r$ versus ζ is shown in Fig. 5-22. In this case, the relation can again be approximated by a straight line over a limited range of ζ:

$$t_r = \frac{0.8 + 2.5\zeta}{\omega_n} \quad 0 < \zeta < 1 \tag{5-107}$$

A better approximation can be obtained by using a second-order equation:

$$t_r = \frac{1 - 0.4167\zeta + 2.917\zeta^2}{\omega_n} \quad 0 < \zeta < 1 \tag{5-108}$$

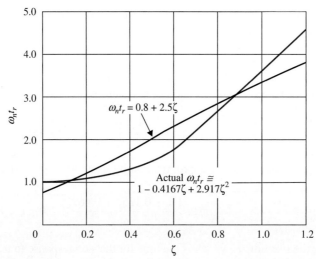

Figure 5-22 Normalized rise time versus ζ for the prototype second-order system.

From this discussion, the following conclusions can be made on the rise time and delay time of the prototype second-order system:

- t_r and t_d are proportional to ζ and inversely proportional to ω_n.
- Increasing (decreasing) the natural undamped frequency ω_n will reduce (increase) t_r and t_d.

5-6-5 Settling Time

From Fig. 5-14, we see that, when $0 < \zeta < 0.69$, the unit-step response has a maximum overshoot greater than 5%, and the response can enter the band between 0.95 and 1.05 for the last time from either the top or the bottom. When ζ is greater than 0.69, the overshoot is less than 5%, and the response can enter the band between 0.95 and 1.05 only from the bottom. Fig. 5-23(a) and (b) show the two different situations. Thus, the settling time has a

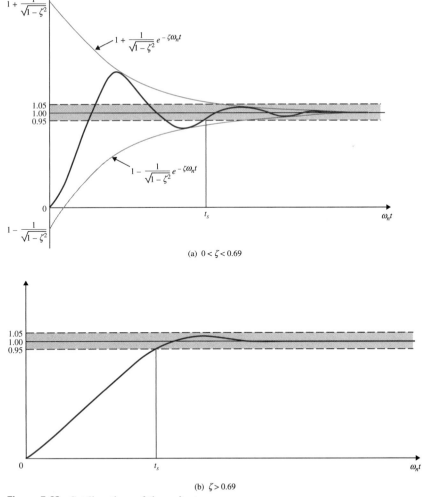

Figure 5-23 Settling time of the unit-step response.

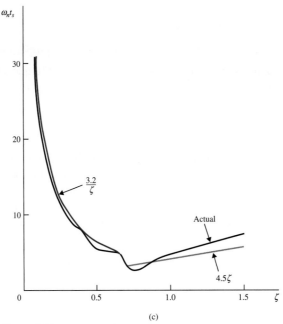

Figure 5-23 (*continued*)

discontinuity at $\zeta = 0.69$. The exact analytical description of the settling time t_s is difficult to obtain. We can obtain an approximation for t_s for $0 < \zeta < 0.69$ by using the envelope of the damped sinusoid of $y(t)$, as shown in Fig. 5-23(a) for a 5% requirement. In general, when the settling time corresponds to an intersection with the upper envelope of $y(t)$, the following relation is obtained:

$$1 + \frac{1}{\sqrt{1-\zeta^2}} e^{-\zeta \omega_n t_s} = \text{upper bound of unit-step response} \qquad (5\text{-}109)$$

When the settling time corresponds to an intersection with the bottom envelope of $y(t)$, t_s must satisfy the following condition:

$$1 - \frac{1}{\sqrt{1-\zeta^2}} e^{-\zeta \omega_n t_s} = \text{lower bound of unit-step response} \qquad (5\text{-}110)$$

For the 5% requirement on settling time, the right-hand side of Eq. (5-109) would be 1.05, and that of Eq. (5-110) would be 0.95. It is easily verified that the same result for t_s is obtained using either Eq. (5-109) or Eq. (5-110).
Solving Eq. (5-109) for $\omega_n t_s$, we have

$$\omega_n t_s = -\frac{1}{\zeta} \ln\left(c_{ts}\sqrt{1-\zeta^2}\right) \qquad (5\text{-}111)$$

where c_{ts} is the percentage set for the settling time. For example, if the threshold is 5 percent, the $c_{ts} = 0.05$. Thus, for a 5-percent settling time, the right-hand side of

Eq. (5-111) varies between 3.0 and 3.32 as ζ varies from 0 to 0.69. We can approximate the settling time for the prototype second-order system as

$$t_s \cong \frac{3.2}{\zeta \omega_n} \quad 0 < \zeta < 0.69 \tag{5-112}$$

The approximation will be poor for small values of $\zeta (< 0.3)$.

When the damping ratio ζ is greater than 0.69, the unit-step response will always enter the band between 0.95 and 1.05 from below. We can show by observing the responses in Fig. 5-14 that the value of $\omega_n t_s$ is almost directly proportional to ζ. The following approximation is used for t_s for $\zeta > 0.69$.

$$t_s = \frac{4.5\zeta}{\omega_n} \quad \zeta > 0.69 \tag{5-113}$$

Fig. 5-23(c) shows the actual values of $\omega_n t_s$ versus ζ for the prototype second-order system described by Eq. (5-87), along with the approximations using Eqs. (5-112) and (5-113) for their respective effective ranges. The numerical values are shown in Table 5-2.

We can summarize the relationships between t_s and the system parameters as follows:

- For $\zeta < 0.69$, the settling time is inversely proportional to ζ and ω_n. A practical way of reducing the settling time is to increase ω_n while holding ζ constant. Although

TABLE 5-2 Comparison of Settling Times of Prototype Second-Order System, $\omega_n t_s$

ζ	Actual	$\frac{3.2}{\zeta}$	4.5ζ
0.10	28.7	30.2	
0.20	13.7	16.0	
0.30	10.0	10.7	
0.40	7.5	8.0	
0.50	5.2	6.4	
0.60	5.2	5.3	
0.62	5.16	5.16	
0.64	5.00	5.00	
0.65	5.03	4.92	
0.68	4.71	4.71	
0.69	4.35	4.64	
0.70	2.86		3.15
0.80	3.33		3.60
0.90	4.00		4.05
1.00	4.73		4.50
1.10	5.50		4.95
1.20	6.21		5.40
1.50	8.20		6.75

the response will be more oscillatory, the maximum overshoot depends only on ζ and can be controlled independently.

- For $\zeta > 0.69$, the settling time is proportional to ζ and inversely proportional to ω_n. Again, t_s can be reduced by increasing ω_n.

Toolbox 5-6-2

To find PO, rise time, and settling time using MATLAB, point at a desired location on the graph and right-click to display the x and y values. For example

```
clear all
w=10;l=0.4;
t=0:0.01:5;
num = [w.^2];
den = [1 2*l*w w.^2];
step(num,den,t)
xlabel('Time(secs)')
ylabel('Amplitude')
title('Closed-Loop Step')
```

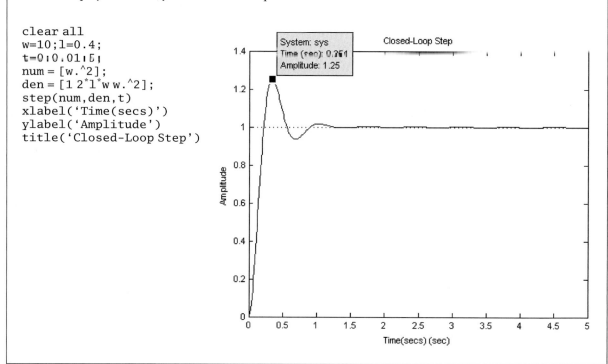

It should be commented that the settling time for $\zeta > 0.69$ is truly a measure of how fast the step response rises to its final value. It seems that, for this case, the rise and delay times should be adequate to describe the response behavior. As the name implies, settling time should be used to measure how fast the step response settles to its final value. It should also be pointed out that the 5% threshold is by no means a number cast in stone. More stringent design problems may require the system response to settle in any number less than 5%.

Keep in mind that, while the definitions on y_{max}, t_{max}, t_d, t_r, and t_s apply to a system of any order, the damping ratio ζ and the natural undamped frequency ω_n strictly apply only to a second-order system whose closed-loop transfer function is given in Eq. (5-87). Naturally, the relationships among t_d, t_r, and t_s and ζ and ω_n are valid only for the same second-order system model. However, these relationships can be used to measure the performance of higher-order systems that can be approximated by second-order ones, under the stipulation that some of the higher-order poles can be neglected.

5-7 SPEED AND POSITION CONTROL OF A DC MOTOR

Servomechanisms are probably the most frequently encountered electromechanical control systems. Applications include robots (each joint in a robot requires a position servo), numerical control (NC) machines, and laser printers, to name but a few. The common characteristic of all such systems is that the variable to be controlled (usually position or velocity) is fed back to modify the command signal. The servomechanism that will be used in the experiments in this chapter comprises a dc motor and amplifier that are fed back the motor speed and position values.

One of the key challenges in the design and implementation of a successful controller is obtaining an accurate model of the system components, particularly the actuator. In Chapter 4, we discussed various issues associated with modeling of dc motors. We will briefly revisit the modeling aspects in this section.

5-7-1 Speed Response and the Effects of Inductance and Disturbance-Open Loop Response

Consider the armature-controlled dc motor shown in Fig. 5-24, where the field current is held constant in this system. The system parameters include

R_a = armature resistance, ohm
L_a = armature inductance, henry
v_a = applied armature voltage, volt
v_b = back emf, volt
θ = angular displacement of the motor shaft, radian
T = torque developed by the motor, N-m
J_L = moment of inertia of the load, kg-m^2
T_L = any external load torque considered as a disturbance, N-m
J_m = moment of inertia of the motor (motor shaft), kg-m^2
J = equivalent moment of inertia of the motor and load connected to the motor-shaft, $J = J_L/n^2 + J_m$, kg-m^2 (refer to Chapter 4 for more details)
n = gear ratio
B = equivalent viscous-friction coefficient of the motor and load referred to the motor shaft, N-m/rad/sec (in the presence of gear ratio, B must be scaled by n; refer to Chapter 4 for more details)
K_t = speed sensor (usually a tachometer) gain

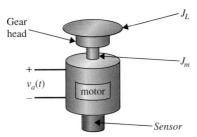

Figure 5-24 An armature-controlled dc motor with a gear head and a load inertia J_L.

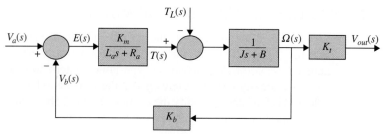

Figure 5-25 Block diagram of an armature-controlled dc motor.

As shown in Fig. 5-25, the armature-controlled dc motor is itself a feedback system, where back-emf voltage is proportional to the speed of the motor. In Fig. 5-25, we have included the effect of any possible external load (e.g., the load applied to a juice machine by the operator pushing in the fruit) as a disturbance torque T_t. The system may be arranged in input–output form such that $V_a(s)$ is the input and $\Omega(s)$ is the output:

$$\Omega(s) = \frac{\dfrac{K_m}{R_a J_m}}{\left(\dfrac{L_a}{R_a}\right)s^2 + \left(1 + \dfrac{BL_a}{R_a J_m}\right)s + \dfrac{K_m K_b + R_a B}{R_a J_m}} V_a(s)$$

$$- \frac{\left\{1 + s\left(\dfrac{L_a}{R_a}\right)\right\}/J_m}{\left(\dfrac{L_a}{R_a}\right)s^2 + \left(1 + \dfrac{BL_a}{R_a J_m}\right)s + \dfrac{K_m K_b + R_a B}{R_a J_m}} T_L(s) \quad (5\text{-}114)$$

The ratio L_a/R_a is called the *motor electric-time constant*, which makes the system speed-response transfer function second order and is denoted by τ_e. Also, it introduces a zero to the disturbance-output transfer function. However, as discussed in Chapter 4, because L_a in the armature circuit is very small, τ_e is neglected, resulting in the simplified transfer functions and the block diagram of the system. Thus, the speed of the motor shaft may be simplified to

$$\Omega(s) = \frac{\dfrac{K_m}{R_a J_m}}{s + \dfrac{K_m K_b + R_a B}{R_a J_m}} V_a(s) - \frac{\dfrac{1}{J_m}}{s + \dfrac{K_m K_b + R_a B}{R_a J_m}} T_L(s) \quad (5\text{-}115)$$

or

$$\Omega(s) = \frac{K_{\text{eff}}}{\tau_m s + 1} V_a(s) - \frac{\dfrac{\tau_m}{J_m}}{\tau_m s + 1} T_L(s) \quad (5\text{-}116)$$

where $K_{\text{eff}} = K_m/(R_a B + K_m K_b)$ is the motor gain constant, and $\tau_m = R_a J_m/(R_a B + K_m K_b)$ is the motor mechanical time constant. If the load inertia and the gear ratio are incorporated into the system model, the inertia J_m in Eqs. (5-114) through (5-116) is replaced with J (total inertia).

Using superposition, we get

$$\Omega(s) = \Omega(s)|_{T_L(s)=0} + \Omega(s)|_{V_a(s)=0} \quad (5\text{-}117)$$

To find the response $\omega(t)$, we use superposition and find the response due to the individual inputs. For $T_L = 0$ (no disturbance and $B = 0$) and an applied voltage $V_a(t) = A$, such that $V_a(s) = A/s$,

$$\omega(t) = \frac{A}{K_b}\left(1 - e^{-t/\tau_m}\right) \quad (5\text{-}118)$$

In this case, note that the motor mechanical time constant τ_m is reflective of how fast the motor is capable of overcoming its own inertia J_m to reach a steady state or constant speed dictated by voltage V_a. From Eq. (5-118), the speed final value is $\omega(t) = A/K_b$. As τ_m increases, the approach to steady state takes longer.

If we apply a constant load torque of magnitude D to the system (i.e., $T_L = D/s$), the speed response from Eq. (5-118) will change to

$$\omega(t) = \frac{1}{K_b}\left(A - \frac{R_a D}{K_m}\right)\left(1 - e^{-t/\tau_m}\right) \quad (5\text{-}119)$$

which clearly indicates that the disturbance T_L affects the final speed of the motor. From Eq. (5-119), at steady state, the speed of the motor is $\omega_{fv} = \frac{1}{K_b}(A - \frac{R_a D}{K_m})$. Here the final value of $\omega(t)$ is reduced by $R_a D/K_m K_b$. A practical note is that the value of $T_L = D$ may never exceed the motor stall torque, and hence for the motor to turn, from Eq. (5-119), $AK_m/R_a > D$, which sets a limit on the magnitude of the torque T_L. For a given motor, the value of the stall torque can be found in the manufacturer's catalog.

If the load inertia is incorporated into the system model, the final speed value becomes $\omega_{fv} = A/K_b$. Does the stall torque of the motor affect the response and the steady-state response? In a realistic scenario, you must measure motor speed using a sensor. How would the sensor affect the equations of the system (see Fig. 5-25)?

5-7-2 Speed Control of DC Motors: Closed-Loop Response

As seen previously, the output speed of the motor is highly dependant on the value of torque T_L. We can improve the speed performance of the motor by using a proportional feedback controller. The controller is composed of a sensor (usually a tachometer for speed applications) to sense the speed and an amplifier with gain K (proportional control) in the configuration shown in Fig. 5-26. The block diagram of the system is also shown in Fig. 5-27.

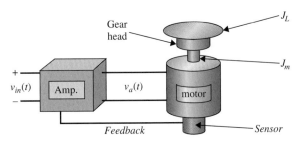

Figure 5-26 Feedback control of an armature-controlled dc motor with a load inertia.

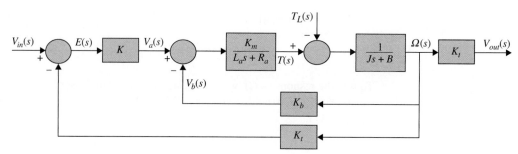

Figure 5-27 Block diagram of a speed-control, armature-controlled dc motor.

Note that the speed at the motor shaft is sensed by the tachometer with a gain K_t. For ease in comparison of input and output, the input to the control system is converted from voltage V_{in} to speed Ω_{in} using the tachometer gain K_t. Hence, assuming $L_a = 0$, we have

$$\Omega(s) = \frac{\frac{K_t K_m K}{R_a J_m}}{s + \left(\frac{K_m K_b + R_a B + K_t K_m K}{R_a J_m}\right)} \Omega_{in}(s)$$

$$- \frac{\frac{1}{J_m}}{s + \left(\frac{K_m K_b + R_a B + K_t K_m K}{R_a J_m}\right)} T_L(s) \qquad (5\text{-}120)$$

For a step input $\Omega_{in} = A/s$ and disturbance torque value $T_L = D/s$, the output becomes

$$\omega(t) = \frac{AKK_m K_t}{R_a J_m} \tau_c (1 - e^{-t/\tau_c}) - \frac{\tau_c D}{J_m}(1 - e^{-t/\tau_c}) \qquad (5\text{-}121)$$

where $\tau_c = \frac{R_a J_m}{K_m K_b + R_a B + K_t K_m K}$ is the system mechanical-time constant. The steady-state response in this case is

$$\omega_{fv} = \left(\frac{AKK_m K_t}{K_m K_b + R_a B + K_t K_m K} - \frac{R_a D}{K_m K_b + R_a B + K_t K_m K}\right) \qquad (5\text{-}122)$$

where $\omega_{fv} \to A$ as $K \to \infty$. So, speed control may reduce the effect of disturbance. As in Section 5-7-1, the reader should investigate what happens if the inertia J_L is included in this model. If the load inertia J_L is too large, will the motor be able to turn? Again, as in Section 5-7-1, you will have to read the speed-sensor voltage to measure speed. How will that affect your equations?

5-7-3 Position Control

The position response in the open-loop case may be obtained by integrating the speed response. Then, considering Fig. 5-25, we have $\Theta(s) = \Omega(s)/s$. The open-loop transfer function is therefore

$$\frac{\Theta(s)}{V_a(s)} = \frac{K_m}{s(L_a J_m s^2 + (L_a B + R_a J_m)s + R_a B + K_m K_b)} \qquad (5\text{-}123)$$

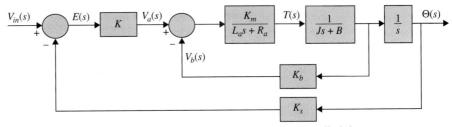

Figure 5-28 Block diagram of a position-control, armature-controlled dc motor.

where we have used the total inertia J. For small L_a, the time response in this case is

$$\theta(t) = \frac{A}{K_b}(t + \tau_m e^{-t/\tau_m} - \tau_m) \tag{5-124}$$

which implies that the motor shaft is turning without stop at a constant steady-state speed. To control the position of the motor shaft, the simplest strategy is to use a proportional controller with gain K. The block diagram of the closed-loop system is shown in Fig. 5-28. The system is composed of an angular position sensor (usually an encoder or a potentiometer for position applications). Note that, for simplicity, the input voltage can be scaled to a position input $\Theta_{in}(s)$ so that the input and output have the same units and scale. Alternatively, the output can be converted into voltage using the sensor gain value. The closed-loop transfer function in this case becomes

$$\frac{\Theta_m(s)}{\Theta_{in}(s)} = \frac{\dfrac{KK_m K_s}{R_a}}{(\tau_e s + 1)\left\{ J_m s^2 + \left(B + \dfrac{K_b K_m}{R_a}\right)s + \dfrac{KK_m K_s}{R_a} \right\}} \tag{5-125}$$

where K_s is the sensor gain, and, as before, $\tau_e = (L_a/R_a)$ may be neglected for small L_a.

$$\frac{\Theta_m(s)}{\Theta_{in}(s)} = \frac{\dfrac{KK_m K_s}{R_a J}}{s^2 + \left(\dfrac{R_a B + K_m K_b}{R_a J_m}\right)s + \dfrac{KK_m K_s}{R_a J_m}} \tag{5-126}$$

Later, in Chapter 6, we set up numerical and experimental case studies to test and verify the preceding concepts and learn more about other practical issues.

▶ 5-8 TIME-DOMAIN ANALYSIS OF A POSITION-CONTROL SYSTEM

In this section, we shall analyze the performance of a system using the time-domain criteria established in the preceding section. The purpose of the system considered here is to control the positions of the fins of an airplane as discussed in Example 4-11-1.

Recall from Chapter 4 that

$$G(s) = \frac{\Theta_y(s)}{\Theta_e(s)} = \frac{K_s K_1 K_i KN}{s[L_a J_t s^2 + (R_a J_t + L_a B_t + K_1 K_2 J_t)s + R_a B_t + K_1 K_2 B_t + K_i K_b + KK_1 K_t K_i]} \tag{5-127}$$

The system is of the third order, since the highest-order term in $G(s)$ is s^3. The electrical time constant of the amplifier-motor system is

$$\tau_a = \frac{L_a}{R_a + K_1 K_2} = \frac{0.003}{5+5} = 0.0003 \text{ sec} \tag{5-128}$$

The mechanical time constant of the motor-load system is

$$\tau_t = \frac{J_t}{B_t} = \frac{0.0002}{0.015} = 0.01333 \text{ sec} \tag{5-129}$$

Because the electrical time constant is much smaller than the mechanical time constant, on account of the low inductance of the motor, we can perform an initial approximation by neglecting the armature inductance L_a. The result is a second-order approximation of the third-order system. Later we will show that this is not the best way of approximating a high-order system by a low-order one. The forward-path transfer function is now

$$G(s) = \frac{K_s K_1 K_i K N}{s[(R_a J_t + K_1 K_2 J_t)s + R_a B_t + K_1 K_2 B_t + K_i K_b + K K_1 K_i K_t]}$$

$$= \frac{\dfrac{K_s K_1 K_i K N}{R_a J_t + K_1 K_2 J_t}}{s\left(s + \dfrac{R_a B_t + K_1 K_2 B_t + K_i K_b + K K_1 K_i K_t}{R_a J_t + K_1 K_2 J_t}\right)} \tag{5-130}$$

Substituting the system parameters in the last equation, we get

$$G(s) = \frac{4500K}{s(s + 361.2)} \tag{5-131}$$

Comparing Eq. (5-131) and (5-132) with the prototype second-order transfer function of Eq. (5-86), we have

$$\text{natural undamped frequency } \omega_n = \pm\sqrt{\frac{K_s K_1 K_i K N}{R_a J_t + K_1 K_2 J_t}} = \pm\sqrt{4500K} \text{ rad/sec} \tag{5-132}$$

$$\text{damping ratio } \zeta = \frac{R_a B_t + K_1 K_2 B_t + K_i K_b + K K_1 K_i K_t}{2\sqrt{K_s K_1 K_i K N (R_a J_t + K_1 K_2 J_t)}} = \frac{2.692}{\sqrt{K}} \tag{5-133}$$

Thus, we see that the natural undamped frequency ω_n is proportional to the square root of the amplifier gain K, whereas the damping ratio ζ is inversely proportional to $\sqrt{K}$.
The closed-loop transfer function of the unity-feedback control system is

$$\frac{\Theta_y(s)}{\Theta_e(s)} = \frac{4500K}{s^2 + 361.2s + 4500K} \tag{5-134}$$

5-8-1 Unit-Step Transient Response

For time-domain analysis, it is informative to analyze the system performance by applying the unit-step input with zero initial conditions. In this way, it is possible to characterize the system performance in terms of the maximum overshoot and some of the other measures, such as rise time, delay time, and settling time, if necessary.

Let the reference input be a unit-step function $\theta_r(t) = u_s(t)$ rad; then $\Theta(s) = 1/s$. The output of the system, with zero initial conditions, is

$$\theta_y(t) = \mathcal{L}^{-1}\left[\frac{4500K}{s(s^2 + 361.2s + 4500K)}\right] \quad (5\text{-}135)$$

The inverse Laplace transform of the right-hand side of the last equation is carried out using the Laplace transform table in Appendix D, or using Eq. (5-90) directly. The following results are obtained for the three values of K indicated.

$K = 7.248 (\zeta \cong 1.0)$:

$$\theta_y(t) = \left(1 - 151e^{-180t} + 150e^{-181.2t}\right)u_s(t) \quad (5\text{-}136)$$

$K = 14.5 (\zeta = 0.707)$:

$$\theta_y(t) = \left(1 - e^{-180.6t}\cos 180.6t - 0.9997e^{-180.6t}\sin 180.6t\right)u_s(t) \quad (5\text{-}137)$$

$K = 181.17 (\zeta = 0.2)$:

$$\theta_y(t) = \left(1 - e^{-180.6t}\cos 884.7t - 0.2041e^{-180.6t}\sin 884.7t\right)u_s(t) \quad (5\text{-}138)$$

The three responses are plotted as shown in Fig. 5-29. Table 5-3 gives the comparison of the characteristics of the three unit-step responses for the three values of K used. When

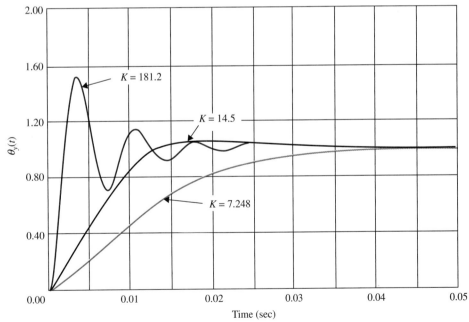

Figure 5-29 Unit-step responses of the attitude-control system in Fig. 4-78; $L_a = 0$.

TABLE 5-3 Comparison of the Performance of the Second-Order Position-Control System with the Gain K Values

Gain K	ζ	ω_n (rad/sec)	% Max overshoot	t_d (sec)	t_r (sec)	t_s (sec)	t_{max} (sec)
7.24808	1.000	180.62	0	0.00929	0.0186	0.0259	—
14.50	0.707	255.44	4.3	0.00560	0.0084	0.0114	0.01735
181.17	0.200	903.00	52.2	0.00125	0.00136	0.0150	0.00369

$K = 181.17$, $\zeta = 0.2$, the system is lightly damped, and the maximum overshoot is 52.7%, which is excessive. When the value of K is set at 7.248, ζ is very close to 1.0, and the system is almost critically damped. The unit-step response does not have any overshoot or oscillation. When K is set at 14.5, the damping ratio is 0.707, and the overshoot is 4.3%. It should be pointed out that, in practice, it would be time consuming, even with the aid of a computer, to compute the time response for each change of a system parameter for either analysis or design purposes. Indeed, one of the main objectives of studying control systems theory, using either the conventional or modern approach, is to establish methods so that the total reliance on computer simulation can be reduced. The motivation behind this discussion is to show that the performance of some control systems can be predicted by investigating the roots of the characteristic equation of the system. For the characteristic equation of Eq. (5-135), the roots are

$$s_1 = -180.6 + \sqrt{32616 - 4500K} \tag{5-139}$$

$$s_2 = -180.6 - \sqrt{32616 - 4500K} \tag{5-140}$$

Toolbox 5-8-1

The Fig. 5-29 responses may be obtained by the following sequence of MATLAB functions.

```
% Equation 5.136
% Unit-Step Transient Response

for k=[7.248 14.5 181.2]
num = [4500*k];
den = [1 361.2 4500*k];
step(num,den)
hold on;
end
xlabel('Time(secs)')
ylabel('Amplitude')
title('Closed-Loop Step')
```

For $K = 7.24808$, 14.5, and 181.2, the roots of the characteristic equation are tabulated as follows:

$K = 7.24808$: $\quad s_1 = s_2 = -180.6$
$K = 14.5$: $\quad s_1 = -180.6 + j180.6 \quad s_2 = -180.6 - j180.6$
$K = 181.2$: $\quad s_1 = -180.6 + j884.7 \quad s_2 = -180.6 + j884.7$

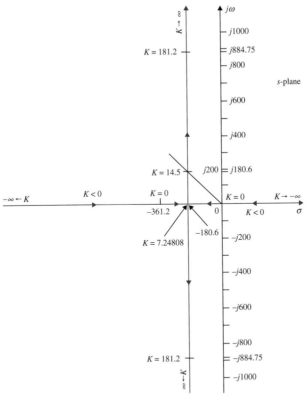

Figure 5-30 Root loci of the characteristic equation in Eq. (5-134) as K varies.

These roots are marked as shown in Fig. 5-30. The trajectories of the two characteristic equation roots when K varies continuously from $-\infty$ to ∞ are also shown in Fig. 5-30. These root trajectories are called the root loci (see Chapter 4) of Eq. (5-135) and are used extensively for the analysis and design of linear control systems.

From Eqs. (5-140) and (5-141), we see that the two roots are real and negative for values of K between 0 and 7.24808. This means that the system is overdamped, and the step response will have no overshoot for this range of K. For values of K greater than 7.24808, the natural undamped frequency will increase with $\sqrt{K}$. When K is negative, one of the roots is positive, which corresponds to a time response that increases monotonically with time, and the system is unstable. The dynamic characteristics of the transient step response as determined from the root loci of Fig. 5-30 are summarized as follows:

Amplifier Gain Dynamics	Characteristic Equation Roots	System
$0 < K < 7.24808$	Two negative distinct real roots	Overdamped ($\zeta > 1$)
$K = 7.24808$	Two negative equal real roots	Critically damped ($\zeta = 1$)
$7.24808 < K < \infty$	Two complex-conjugate roots with negative real parts	Underdamped ($\zeta < 1$)
$-\infty < K < 0$	Two distinct real roots, one positive and one negative	Unstable system ($\zeta < 0$)

5-8-2 The Steady-State Response

Because the forward-path transfer function in Eq. (5-132) has a simple pole at $s = 0$, the system is of type 1. This means that the steady-state error of the system is zero for all positive values of K when the input is a step function. Substituting Eq. (5-132) into Eq. (5-24), the step-error constant is

$$K_p = \lim_{s \to 0} \frac{4500K}{s(s + 361.2)} = \infty \tag{5-141}$$

Thus, the steady-state error of the system due to a step input, as given by Eq. (5-25), is zero. The unit-step responses in Fig. 5-29 verify this result. The zero-steady-state condition is achieved because only viscous friction is considered in the simplified system model. In the practical case, Coulomb friction is almost always present, so the steady-state positioning accuracy of the system can never be perfect.

5-8-3 Time Response to a Unit-Ramp Input

The control of position may be affected by the control of the profile of the output, rather than just by applying a step input. In other words, the system may be designed to follow a reference profile that represents the desired trajectory. It may be necessary to investigate the ability of the position-control system to follow a ramp-function input.

For a unit-ramp input, $\theta_r(t) = tu_s(t)$. The output response of the system in Fig. 4-79 is

$$\theta_y(t) = \mathcal{L}^{-1} \left[\frac{4500K}{s^2(s^2 + 361.2s + 4500K)} \right] \tag{5-142}$$

which can be solved by using the Laplace transform table in Appendix C. The result is

$$\theta_y(t) = \left[t - \frac{2\zeta}{\omega_n} + \frac{e^{-\zeta \omega_n t}}{\omega_n \sqrt{1-\zeta^2}} \sin\left(\omega_n \sqrt{1-\zeta^2} t + \theta\right) \right] u_s(t) \tag{5-143}$$

where

$$\theta = \cos^{-1}(2\zeta^2 - 1) \quad (\zeta < 1) \tag{5-144}$$

The values of ζ and ω_n are given in Eqs. (5-134) and (5-133), respectively. The ramp responses of the system for the three values of K are presented in the following equations.

K = 7.248:

$$\theta_y(t) = \left(t - 0.01107 - 0.8278e^{-181.2t} + 0.8389e^{-180t}\right) u_s(t) \tag{5-145}$$

K = 14.5:

$$\theta_y(t) = (t - 0.005536 + 0.005536e^{-180.6t} \cos 180.6t \\ - 5.467 \times 10^{-7} e^{-180.6t} \sin 180.6t) u_s(t) \tag{5-146}$$

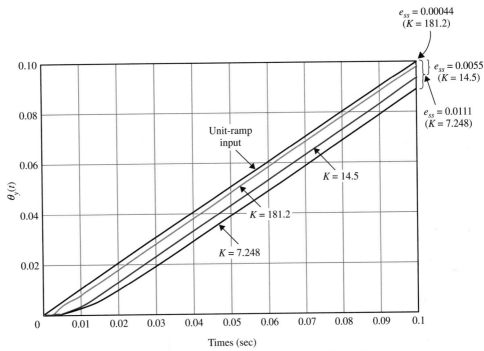

Figure 5-31 Unit-ramp responses of the attitude-control system in Fig. 4-78; $L_a = 0$.

K = 181.2:

$$\theta_y(t) = \left(t - 0.000443 + 0.000443 e^{-180.6t}\cos 884.7t \right. \\ \left. - 0.00104 e^{-180.6t}\sin 884.7t\right) u_s(t) \tag{5-147}$$

These ramp responses are plotted as shown in Fig. 5-31. Notice that the steady-state error of the ramp response is not zero. The last term in Eq. (5-144) is the transient response. The steady-state portion of the unit-ramp response is

$$\lim_{t \to \infty} \theta_y(t) = \lim_{t \to \infty} \left[\left(t - \frac{2\zeta}{\omega_n}\right) u_s(t) \right] \tag{5-148}$$

Thus, the steady-state error of the system due to a unit-ramp input is

$$e_{ss} = \frac{2\zeta}{\omega_n} = \frac{0.0803}{K} \tag{5-149}$$

which is a constant.

A more direct method of determining the steady-state error due to a ramp input is to use the ramp-error constant K_v. From Eq. (5-31),

$$K_v = \lim_{s \to 0} sG(s) = \lim_{s \to 0} \frac{4500K}{s + 361.2} = 12.46K \tag{5-150}$$

Thus, the steady-state error is

$$e_{ss} = \frac{1}{K_v} = \frac{0.0803}{K} \tag{5-151}$$

which agrees with the result in Eq. (5-149).

Toolbox 5-8-2

The Fig. 5-31 responses are obtained by the following sequence of MATLAB functions

```
for k=[7.248 14.5 181.2]
  clnum = [4500*k];
  clden = [1 361.2 4500*k];
  t=0:0.0001:0.3;
  u = t;
  [y,x]=lsim(clnum,clden,u,t);
  plot(t,y,t,u);
  hold on;
end

title('Unit-ramp responses')
xlabel('Time(sec)')
ylabel('Amplitude')
```

The result in Eq. (5-151) shows that the steady-state error is inversely proportional to K. For $K = 14.5$, which corresponds to a damping ratio of 0.707, the steady-state error is 0.0055 rad or, more appropriately, 0.55% of the ramp-input magnitude. Apparently, if we attempt to improve the steady-state accuracy of the system due to ramp inputs by increasing the value of K, the transient step response will become more oscillatory and have a higher overshoot. This phenomenon is rather typical in all control systems. For higher-order systems, if the loop gain of the system is too high, the system can become unstable. Thus, by using the controller in the system loop, the transient and the steady-state error can be improved simultaneously.

5-8-4 Time Response of a Third-Order System

In the preceding section, we have shown that the prototype second-order system, obtained by neglecting the armature inductance, is always stable for all positive values of K. It is not difficult to prove that, in general, all second-order systems with positive coefficients in the characteristic equations are stable.

Let us investigate the performance of the position-control system with the armature inductance $L_a = 0.003$ H. The forward-path transfer function of Eq. (5-128) becomes

$$G(s) = \frac{1.5 \times 10^7 K}{s(s^2 + 3408.3s + 1{,}204{,}000)} = \frac{1.5 \times 10^7 K}{s(s + 400.26)(s + 3008)} \tag{5-152}$$

The closed-loop transfer function is

$$\frac{\Theta_y(s)}{\Theta_r(s)} = \frac{1.5 \times 10^7 K}{s^3 + 3408.3s^2 + 1{,}204{,}000s + 1.5 \times 10^7 K} \tag{5-153}$$

The system is now of the third order, and the characteristic equation is

$$s^3 + 3408.3s^2 + 1{,}204{,}000s + 1.5 \times 10^7 K = 0 \tag{5-154}$$

Transient Response

The roots of the characteristic equation are tabulated for the three values of K used earlier for the second-order system:

$K = 7.248$:	$s_1 = -156.21$	$s_2 = -230.33$	$s_3 = -3021.8$
$K = 14.5$:	$s_1 = -186.53 + j192$	$s_2 = -186.53 - j192$	$s_3 = -3035.2$
$K = 181.2$:	$s_1 = -57.49 + j906.6$	$s_2 = -57.49 - j906.6$	$s_3 = -3293.3$

Comparing these results with those of the approximating second-order system, we see that, when $K = 7.428$, the second-order system is critically damped, whereas the third-order system has three distinct real roots, and the system is slightly overdamped. The root at -3021.8 corresponds to a time constant of 0.33 millisecond, which is more than 13 times faster than the next fastest time constant because of the pole at -230.33. Thus, the transient response due to the pole at -3021.8 decays rapidly, and the pole can be neglected from the transient standpoint. The output transient response is dominated by the two roots at -156.21 and -230.33. This analysis is verified by writing the transformed output response as

$$\Theta_y(s) = \frac{10.87 \times 10^7}{s(s + 156.21)(s + 230.33)(s + 3021.8)} \tag{5-155}$$

Taking the inverse Laplace transform of the last equation, we get

$$\theta_y(t) = \left(1 - 3.28e^{-156.21t} + 2.28e^{-230.33t} - 0.0045e^{-3021.8t}\right)u_s(t) \tag{5-156}$$

The last term in Eq. (5-156), which is due to the root at -3021.8, decays to zero very rapidly. Furthermore, the magnitude of the term at $t = 0$ is very small compared to the other two transient terms. This simply demonstrates that, in general, the contribution of roots that lie relatively far to the left in the s-plane to the transient response will be small. The roots that are closer to the imaginary axis will dominate the transient response, and these are defined as the **dominant roots** of the characteristic equation or of the system.

When $K = 14.5$, the second-order system has a damping ratio of 0.707, because the real and imaginary parts of the two characteristic equation roots are identical. For the third-order system, recall that the damping ratio is strictly not defined. However, because the effect on transient of the root at -3021.8 is negligible, the two roots that dominate the transient response correspond to a damping ratio of 0.697. Thus, for $K = 14.5$, the second-order approximation by setting L_a to zero is not a bad one. It should be noted, however, that the fact that the second-order approximation is justified for $K = 14.5$ does not mean that the approximation is valid for all values of K.

When $K = 181.2$, the two complex-conjugate roots of the third-order system again dominate the transient response, and the equivalent damping ratio due to the two roots is only 0.0633, which is much smaller than the value of 0.2 for the second-order system. Thus, we see that the justification and accuracy of the second-order approximation diminish as the value of K is increased. Fig. 5-32 illustrates the root loci of the third-order characteristic equation of Eq. (5-154) as K varies.

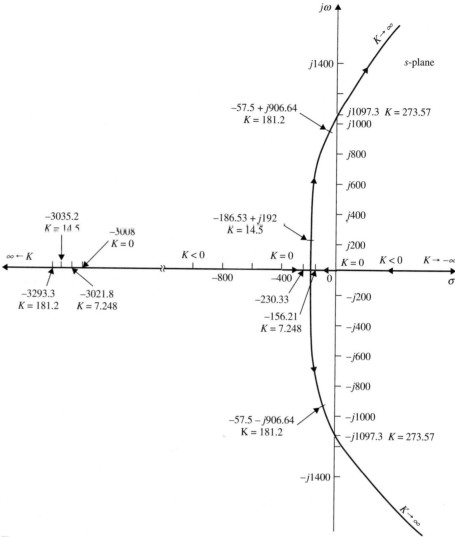

Figure 5-32 Root loci of the third-order attitude-control system.

When $K = 181.2$, the real root at -3293.3 still contributes little to the transient response, but the two complex-conjugate roots at $-57.49 \pm j906.6$ are much closer to the $j\omega$-axis than those of the second-order system for the same K, which are at $-180.6 \pm j884.75$. This explains why the third-order system is a great deal less stable than the second-order system when $K = 181.2$.

By using the Routh-Hurwitz criterion, the marginal value of K for stability is found to be 273.57. With this critical value of K, the closed-loop transfer function becomes

$$\frac{\Theta_y(s)}{\Theta_r(s)} = \frac{1.0872 \times 10^8}{(s + 3408.3)(s^2 + 1.204 \times 10^6)} \tag{5-157}$$

The roots of the characteristic equation are at $s = -3408.3, -j1097.3$, and $j1097.3$. These points are shown on the root loci in Fig. 5-32.

5-8 Time-Domain Analysis of a Position-Control System ◀ 303

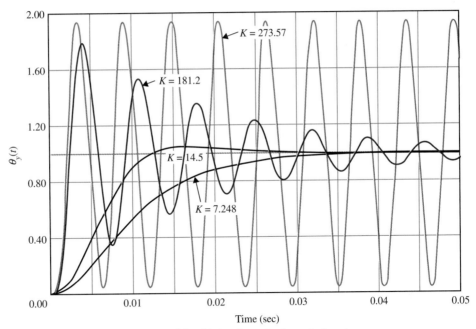

Figure 5-33 Unit-step responses of the third-order attitude-control system.

The unit-step response of the system when $K = 273.57$ is

$$\theta_y(t) = \left[1 - 0.094e^{-3408.3t} - 0.952 \sin(1097.3t + 72.16°)\right] u_s(t) \qquad (5\text{-}158)$$

The steady-state response is an undamped sinusoid with a frequency of 1097.3 rad/sec, and the system is said to be marginally stable. When K is greater than 273.57, the two complex-conjugate roots will have positive real parts, the sinusoidal component of the time response will increase with time, and the system is unstable. Thus, we see that the third-order system is capable of being unstable, whereas the second-order system obtained with $L_a = 0$ is stable for all finite positive values of K.

Fig. 5-33 shows the unit-step responses of the third-order system for the three values of K used. The responses for $K = 7.248$ and $K = 14.5$ are very close to those of the second-order system with the same values of K that are shown in Fig. 5-29. However, the two responses for $K = 181.2$ are quite different.

Toolbox 5-8-3

The root locus plot in Fig. 5-32 is obtained by the following MATLAB commands

```
for k=[7.248 14.5 181.2 273.57]

t=0:0.001:0.05;
num = [1.5*(10^7)*k];
den = [1 3408.3 1204000 1.5*(10^7)*k];
rlocus(num,den)

hold on;
end
```

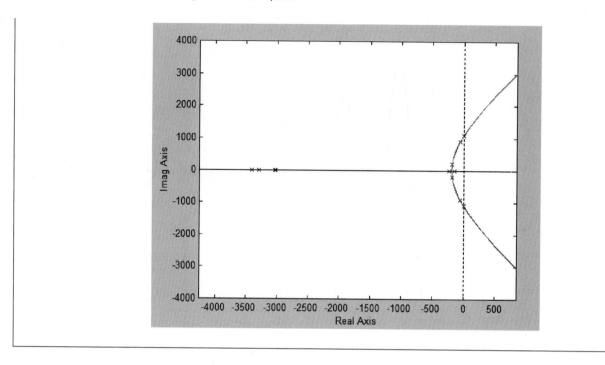

Steady-State Response

From Eq. (5-152), we see that, when the inductance is restored, the third-order system is still of type 1. The value of K_v is still the same as that given in Eq. (5-150). Thus, the inductance of the motor does not affect the steady-state performance of the system, provided that the system is stable. This is expected, since L_a affects only the rate of change and not the final value of the motor current. A good engineer should always try to interpret the analytical results with the physical system.

▶ 5-9 BASIC CONTROL SYSTEMS AND EFFECTS OF ADDING POLES AND ZEROS TO TRANSFER FUNCTIONS

The position-control system discussed in the preceding section reveals important properties of the time responses of typical second- and third-order closed-loop systems. Specifically, the effects on the transient response relative to the location of the roots of the characteristic equation are demonstrated.

In all previous examples of control systems we have discussed thus far, the controller has been typically a simple amplifier with a constant gain K. This type of control action is formally known as **proportional control**, because the control signal at the output of the controller is simply related to the input of the controller by a proportional constant.

Intuitively, one should also be able to use the derivative or integral of the input signal, in addition to the proportional operation. Therefore, we can consider a more general continuous-data controller to be one that contains such components as adders or summers (addition or subtraction), amplifiers, attenuators, differentiators, and integrators — see Section 4-3-3 and Chapter 9 for more details. For example, one of the best-known controllers used in practice is the PID controller, which stands for **proportional**, **integral**, and **derivative**. The integral and derivative components of the PID controller have individual performance implications, and their applications require an understanding of the basics of these elements.

All in all, what these controllers do is *add additional poles and zeros* to the open- or closed-loop transfer function of the overall system. As a result, it is important to appreciate the effects of adding poles and zeros to a transfer function first. We show that—although the roots of the characteristic equation of the system, which are the poles of the closed-loop transfer function, affect the transient response of linear time-invariant control systems, particularly the stability—the zeros of the transfer function are also important. Thus, the addition of poles and zeros and/or cancellation of undesirable poles and zeros of the transfer function often are necessary in achieving satisfactory time-domain performance of control systems.

In this section, we show that the addition of poles and zeros to forward-path and closed-loop transfer functions has varying effects on the transient response of the closed-loop system.

5-9-1 Addition of a Pole to the Forward-Path Transfer Function: Unity-Feedback Systems

For the position-control system described in Section 5-8, when the motor inductance is neglected, the system is of the second order, and the forward-path transfer function is of the prototype given in Eq. (5-131). When the motor inductance is restored, the system is of the third order, and the forward-path transfer function is given in Eq. (5-149). Comparing the two transfer functions of Eqs. (5-131) and (5-149), we see that the effect of the motor inductance is equivalent to adding a pole at $s = -3008$ to the forward-path transfer function of Eq. (5-131) while shifting the pole at -361.2 to -400.26, and the proportional constant is also increased. The apparent effect of adding a pole to the forward-path transfer function is that the third-order system can now become unstable if the value of the amplifier gain K exceeds 273.57. As shown by the root-loci diagrams of Fig. 5-32 and Fig. 5-34, the new pole of $G(s)$ at $s = -3008$ essentially "pushes" and "bends" the complex-conjugate portion of the root loci of the second-order system toward the right-half s-plane. Actually, because of the specific value of the inductance chosen, the additional pole of the third-order system is far to the left of the pole at -400.26, so its effect is small except when the value of K is relatively large.

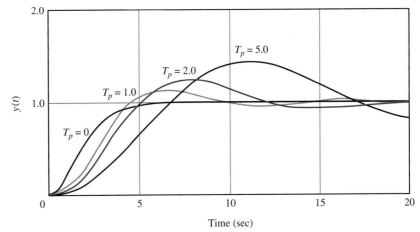

Figure 5-34 Unit-step responses of the system with the closed-loop transfer function in Eq. (5-160): $\zeta = 1$; $\omega_n = 1$; and $T_p = 0, 1, 2,$ and 5.

To study the general effect of the addition of a pole, and its relative location, to a forward-path transfer function of a unity-feedback system, consider the transfer function

$$G(s) = \frac{\omega_n^2}{s(s+2\zeta\omega_n)(1+T_p s)} \tag{5-159}$$

The pole at $s = -1/T_p$ is considered to be added to the prototype second-order transfer function. The closed-loop transfer function is written

$$M(s) = \frac{Y(s)}{R(s)} = \frac{G(s)}{1+G(s)} = \frac{\omega_n^2}{T_p s^3 + (1+2\zeta\omega_n T_p)s^2 + 2\zeta\omega_n s + \omega_n^2} \tag{5-160}$$

Fig. 5-34 illustrates the unit-step responses of the closed-loop system when $\omega_n = 1$; $\zeta = 1$; and $T_p = 0, 1, 2,$ and 5. These responses again show *that the addition of a pole to the forward-path transfer function generally has the effect of increasing the maximum overshoot of the closed-loop system.*

As the value of T_p increases, the pole at $-1/T_p$ moves closer to the origin in the s-plane, and the maximum overshoot increases. These responses also show that the added pole increases the rise time of the step response. This is not surprising, because the additional pole has the effect of reducing the bandwidth (see Chapter 8) of the system, thus cutting out the high-frequency components of the signal transmitted through the system.

Toolbox 5-9-1

The corresponding responses for Fig. 5-34 are obtained by the following sequence of MATLAB functions

```
clear all
w=1; l=1;
for Tp=[0 1 2 5];

t=0:0.001:20;
num = [w];
den = [Tp 1+2*l*w*Tp 2*l*w w^2];

step(num,den,t);
hold on;
end
xlabel('Time(secs)')
ylabel('apos;y(t)')
title('Unit-step responses of the system')
```

The corresponding responses for Fig. 5-37 are obtained by the following sequence of MATLAB functions

```
clear all
w=1;l=0.25;
for Tp=[0 0.2 0.667 1];

t=0:0.001:20;
num = [w];
den = [Tp 1+2*l*w*Tp 2*l*w w^2];
```

```
step(num,den,t);
hold on;
end
xlabel('Time(secs)')
ylabel('y(t)')
title('Unit-step responses of the system')
```

The same conclusion can be drawn from the unit-step responses of Fig. 5-35, which are obtained with $\omega_n = 1$; $\zeta = 0.25$; and $T_p = 0, 0.2, 0.667$, and 1.0. In this case, when T_p is greater than 0.667, the amplitude of the unit-step response increases with time, and the system is unstable.

5-9-2 Addition of a Pole to the Closed-Loop Transfer Function

Because the poles of the closed-loop transfer function are roots of the characteristic equation, they control the transient response of the system directly. Consider the closed-loop transfer function

$$M(s) = \frac{Y(s)}{R(s)} = \frac{\omega_n^2}{\left(s^2 + 2\zeta\omega_n s + \omega_n^2\right)\left(1 + T_p s\right)} \tag{5-161}$$

where the term $(1 + T_p s)$ is added to a prototype second-order transfer function. Fig. 5-36 illustrates the unit-step response of the system with $\omega_n = 1.0$; $\zeta = 0.5$; and $T_p = 0, 0.5, 1.0, 2.0$, and 4.0. As the pole at $s = -1/T_p$ is moved toward the origin

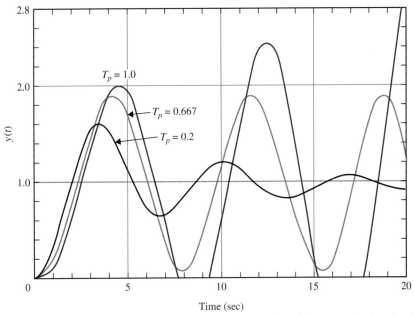

Figure 5-35 Unit-step responses of the system with the closed-loop transfer function in Eq. (5-160): $\zeta = 0.25$; $\omega_n = 1$; and $T_p = 0, 0.2, 0.667$, and 1.0.

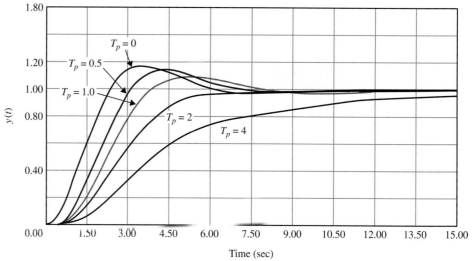

Figure 5-36 Unit-step responses of the system with the closed-loop transfer function in Eq. (5-161): $\zeta = 0.5$; $\omega_n = 1$; and $T_p = 0, 0.5, 1.0, 2.0,$ and 4.0.

in the s-plane, the rise time increases and the maximum overshoot decreases. Thus, as far as the overshoot is concerned, adding a pole to the closed-loop transfer function has just the opposite effect to that of adding a pole to the forward-path transfer function.

Toolbox 5-9-2

The corresponding responses for Fig. 5-36 are obtained by the following sequence of MATLAB functions

```
clear all
w=1;l=0.5;
for Tp=[0 0.5 1 2];

t=0:0.001:15;
num = [w^2];
den = conv([1 2*l*w w^2],[Tp 1]);

step(num,den,t);
hold on;
end
xlabel('Time(secs)')
ylabel('y(t)')
title('Unit-step responses of the system')
```

5-9-3 Addition of a Zero to the Closed-Loop Transfer Function

Fig. 5-37 shows the unit-step responses of the closed-loop system with the transfer function

$$M(s) = \frac{Y(s)}{R(s)} = \frac{\omega_n^2(1 + T_z s)}{\left(s^2 + 2\zeta\omega_n s + \omega_n^2\right)} \tag{5-162}$$

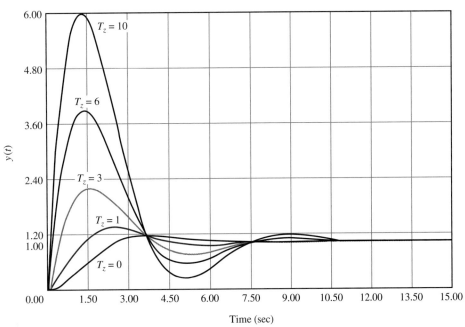

Figure 5-37 Unit-step responses of the system with the closed-loop transfer function in Eq. (5-162): $T_z = 0, 1, 2, 3, 6,$ and 10.

where $\omega_n = 1$; $\zeta = 0.5$; and $T_z = 0, 1, 2, 3, 6,$ and 10. In this case, we see that adding a zero to the closed-loop transfer function decreases the rise time and increases the maximum overshoot of the step response.

We can analyze the general case by writing Eq. (5-162) as

$$M(s) = \frac{Y(s)}{R(s)} = \frac{\omega_n^2}{s^2 + 2\zeta\omega_n s + \omega_n^2} + \frac{T_z \omega_n^2 s}{s^2 + 2\zeta\omega_n s + \omega_n^2} \quad (5\text{-}163)$$

For a unit-step input, let the output response that corresponds to the first term of the right side of Eq. (5-163) be $y_1(t)$. Then, the total unit-step response is

$$y(t) = y_1(t) + T_z \frac{dy_1(t)}{dt} \quad (5\text{-}164)$$

Fig. 5-38 shows why the addition of the zero at $s = -1/T_z$ reduces the rise time and increases the maximum overshoot, according to Eq. (5-164). In fact, as T_z approaches infinity, the maximum overshoot also approaches infinity, and yet the system is still stable as long as the overshoot is finite and ζ is positive.

5-9-4 Addition of a Zero to the Forward-Path Transfer Function: Unity-Feedback Systems

Let us consider that a zero at $-1/T_z$ is added to the forward-path transfer function of a third-order system, so

$$G(s) = \frac{6(1 + T_z s)}{s(s+1)(s+2)} \quad (5\text{-}165)$$

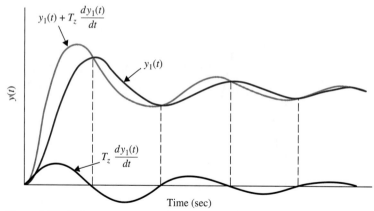

Figure 5-38 Unit-step responses showing the effect of adding a zero to the closed-loop transfer function.

The closed-loop transfer function is

$$M(s) = \frac{Y(s)}{R(s)} = \frac{6(1 + T_z s)}{s^3 + 3s^2 + (2 + 6T_z)s + 6} \qquad (5\text{-}166)$$

The difference between this case and that of adding a zero to the closed-loop transfer function is that, in the present case, not only the term $(1 + T_z s)$ appears in the numerator of $M(s)$, but the denominator of $M(s)$ also contains T_z. The term $(1 + T_z s)$ in the numerator of $M(s)$ increases the maximum overshoot, but T_z appears in the coefficient of the s term in the denominator, which has the effect of improving damping, or reducing the maximum overshoot. Fig. 5-39 illustrates the unit-step responses when $T_z = 0$, 0.2, 0.5, 2.0, 5.0, and 10.0. Notice that, when $T_z = 0$, the closed-loop system is on the verge of becoming unstable. When $T_z = 0.2$ and 0.5, the maximum overshoots are reduced, mainly

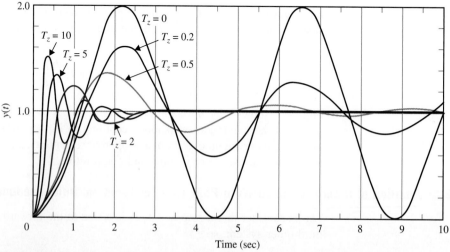

Figure 5-39 Unit-step responses of the system with the closed-loop transfer function in Eq. (5-166): $T_z = 0$, 0.2, 0.5, 2.0, 5.0, and 10.0.

because of the improved damping. As T_z increases beyond 2.0, although the damping is still further improved, the $(1 + T_z s)$ term in the numerator becomes more dominant, so the maximum overshoot actually becomes greater as T_z is increased further.

An important finding from these discussions is that, although the characteristic equation roots are generally used to study the relative damping and relative stability of linear control systems, the zeros of the transfer function should not be overlooked in their effects on the transient performance of the system.

Toolbox 5-9-3

The corresponding responses for Fig. 5-39 are obtained by the following sequence of MATLAB functions

```
clear all
w=1;l=0.5;
for Tz=[0 0.2 0.5 3 5];
t=0:0.001:15;
num = [6*Tz 6];
den = [1 3 2+6*Tz 6];

step(num,den,t);
hold on;
end
xlabel('Time(secs)')
ylabel('y(t)')
title('Unit-step responses of the system')
```

▶ 5-10 DOMINANT POLES AND ZEROS OF TRANSFER FUNCTIONS

From the discussions given in the preceding sections, it becomes apparent that the location of the poles and zeros of a transfer function in the s-plane greatly affects the transient response of the system. For analysis and design purposes, it is important to sort out the poles that have a dominant effect on the transient response and call these the dominant poles.

Because most control systems in practice are of orders higher than two, it would be useful to establish guidelines on the approximation of high-order systems by lower-order ones insofar as the transient response is concerned. In design, we can use the dominant poles to control the dynamic performance of the system, whereas the insignificant poles are used for the purpose of ensuring that the controller transfer function can be realized by physical components.

For all practical purposes, we can divide the s-plane into regions in which the dominant and insignificant poles can lie, as shown in Fig. 5-40. We intentionally do not assign specific values to the coordinates, since these are all relative to a given system.

The poles that are *close* to the imaginary axis in the left-half s-plane give rise to transient responses that will decay relatively slowly, whereas the poles that are *far away* from the axis (relative to the dominant poles) correspond to fast-decaying time responses. The distance D between the dominant region and the least significant region shown in Fig. 5-40 will be subject to discussion. The question is: How large a pole is considered to be really large? It has been recognized in practice and in the literature that if the magnitude of the real part of a pole is at least 5 to 10 times that of a dominant pole or a pair of complex

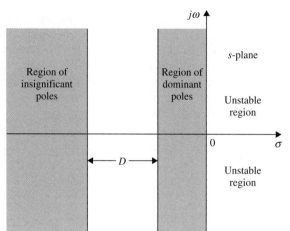

Figure 5-40 Regions of dominant and insignificant poles in the s-plane.

dominant poles, then the pole may be regarded as insignificant insofar as the transient response is concerned. The zeros that are *close* to the imaginary axis in the left-half s-plane affect the transient responses more significantly, whereas the zeros that are *far away* from the axis (relative to the dominant poles) have a smaller effect on the time response.

We must point out that the regions shown in Fig. 5-40 are selected merely for the definitions of dominant and insignificant poles. For design purposes, such as in pole-placement design, the dominant poles and the insignificant poles should most likely be located in the tinted regions in Fig. 5-41. Again, we do not show any absolute coordinates, except that the desired region of the dominant poles is centered around the line that corresponds to $\zeta = 0.707$. It should also be noted that, while designing, we cannot place the insignificant poles arbitrarily far to the left in the s-plane or these may require unrealistic system parameter values when the pencil-and-paper design is implemented by physical components.

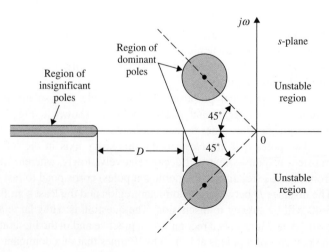

Figure 5-41 Regions of dominant and insignificant poles in the s-plane for design purposes.

5-10-1 Summary of Effects of Poles and Zeros

Based on previous observations, we can summarize the following:

1. Complex-conjugate poles of the closed-loop transfer function lead to a step response that is underdamped. If all system poles are real, the step response is overdamped. However, zeros of the closed-loop transfer function may cause overshoot even if the system is overdamped.
2. The response of a system is dominated by those poles closest to the origin in the s-plane. Transients due to those poles, which are farther to the left, decay faster.
3. The farther to the left in the s-plane the system's dominant poles are, the faster the system will respond and the greater its bandwidth will be.
4. The farther to the left in the s-plane the system's dominant poles are, the more expensive it will be and the larger its internal signals will be. While this can be justified analytically, it is obvious that striking a nail harder with a hammer drives the nail in faster but requires more energy per strike. Similarly, a sports car can accelerate faster, but it uses more fuel than an average car.
5. When a pole and zero of a system transfer function nearly cancel each other, the portion of the system response associated with the pole will have a small magnitude.

5-10-2 The Relative Damping Ratio

When a system is higher than the second order, we can no longer strictly use the damping ratio ζ and the natural undamped frequency ω_n, which are defined for the prototype second-order systems. However, if the system dynamics can be accurately represented by a pair of complex-conjugate dominant poles, then we can still use ζ and ω_n to indicate the dynamics of the transient response, and the damping ratio in this case is referred to as the relative damping ratio of the system. For example, consider the closed-loop transfer function

$$M(s) = \frac{Y(s)}{R(s)} = \frac{20}{(s+10)(s^2+2s+2)} \tag{5-167}$$

The pole at $s = -10$ is 10 times the real part of the complex conjugate poles, which are at $-1 \pm j1$. We can refer to the relative damping ratio of the system as 0.707.

5-10-3 The Proper Way of Neglecting the Insignificant Poles with Consideration of the Steady-State Response

Thus far, we have provided guidelines for neglecting insignificant poles of a transfer function from the standpoint of the transient response. However, going through with the mechanics, the steady-state performance must also be considered. Let us consider the transfer function in Eq. (5-167); the pole at -10 can be neglected from the transient standpoint. To do this, we should first express Eq. (5-167) as

$$M(s) = \frac{20}{10(s/10+1)(s^2+2s+2)} \tag{5-168}$$

Then we reason that $|s/10| \ll 1$ when the absolute value of s is much smaller than 10, because of the dominant nature of the complex poles. The term $s/10$ can be neglected when compared with 1. Then, Eq. (5-168) is approximated by

$$M(s) \cong \frac{20}{10(s^2 + 2s + 2)} \tag{5-169}$$

This way, the steady-state performance of the third-order system will not be affected by the approximation. In other words, the third-order system described by Eq. (5-167) and the second-order system approximated by Eq. (5-169) all have a final value of unity when a unit-step input is applied. On the other hand, if we simply throw away the term $(s + 10)$ in Eq. (5-167), the approximating second-order system will have a steady-state value of 5 when a unit-step input is applied.

▶ 5-11 BASIC CONTROL SYSTEMS UTILIZING ADDITION OF POLES AND ZEROS

In practice we can control the response of a system by adding poles and zeros or a simple amplifier with a constant gain K to its transfer function. So far in this chapter, we have discussed the effect of adding a simple gain in the time response—i.e., proportional control. In this section, we look at controllers that include derivative or integral of the input signal in addition to the proportional operation.

▶ **EXAMPLE 5-11-1** Fig. 5-42 shows the block diagram of a feedback control system that arbitrarily has a second-order prototype process with the transfer function

$$G_p(s) = \frac{\omega_n^2}{s(s + 2\zeta\omega_n)} \tag{5-170}$$

The series controller in this case is a proportional-derivative (PD) type with the transfer function

$$G_c(s) = K_P + K_D s \tag{5-171}$$

In this case, the forward-path transfer function of the compensated system is

$$G(s) = \frac{Y(s)}{E(s)} = G_c(s)G_p(s) = \frac{\omega_n^2(K_P + K_D s)}{s(s + 2\zeta\omega_n)} \tag{5-172}$$

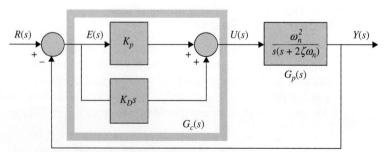

Figure 5-42 Control system with PD controller.

which shows that the PD control is equivalent to adding a simple zero at $s = -K_P/K_D$ to the forward-path transfer function. Consider the second-order model

$$G(s) = \frac{2}{s(s+2)} \tag{5-173}$$

Rewriting the transfer function of the PD controller as

$$G_c(s) = (K_P + K_D s) \tag{5-174}$$

the forward-path transfer function of the system becomes

$$G(s) = \frac{Y(s)}{E(s)} = \frac{2(K_P + K_D s)}{s(s+2)} \tag{5-175}$$

The closed-loop transfer function is

$$\frac{Y(s)}{R(s)} = \frac{2(K_P + K_D s)}{s^2 + (2 + 2K_D)s + 2K_P} \tag{5-176}$$

Eq. (5-176) shows that the effects of the PD controller are the following:

1. Adding a zero at $s = -K_P/K_D$ to the closed-loop transfer function.
2. Increasing the *damping term*, which is the coefficient of the s term in the denominator, from 2 to $2 + 2K_D$.

We should quickly point out that Eq. (5-175) no longer represents a prototype second-order system, since the transient response is also affected by the zero of the transfer function at $s = -K_P/K_D$. It turns out that for this second-order system, as the value of K_D increases, the zero will move very close to the origin and effectively cancel the pole of $G(s)$ at $s = 0$. Thus, as K_D increases, the transfer function in Eq. (5-175) approaches that of a first-order system with the pole at $s = -2$, and the closed-loop system will not have any overshoot. In general, for higher-order systems, however, the zero at $s = -K_P/K_D$ may increase the overshoot when K_D becomes very large.

The characteristic equation is written as

$$s^2 + (2 + 2K_D)s + 2K_P = 0 \tag{5-177}$$

Ignoring the zero of the transfer function in equation (5-177) and comparing (5-177) to prototype second-order system characteristic equation

$$s^2 + 2\zeta\omega_n s + \omega_n^2 = 0 \tag{5-178}$$

we get the damping ratio and natural frequency values of

$$\zeta = \frac{1 + K_D}{\sqrt{2K_P}} \tag{5-179}$$

$$\omega_n = \sqrt{2K_P}$$

which clearly show the positive effect of K_D on damping. For $K_P = 8$, if we wish to have critical damping, $\zeta = 1$, Eq. (5-179) gives $K_D = 3$. Fig. 5-43 shows the unit-step responses of the closed-loop system with $K_P = 8$ and $K_D = 3$. With the PD control, the maximum overshoot is 2%. In the present case, although K_D is chosen for critical damping, the overshoot is due to the zero at $s = -K_P/K_D$ of the closed-loop transfer function. Upon selecting a smaller $K_P = 1$, for $\zeta = 1$, Eq. (5-179) gives $K_D = 0.414$. Fig. 5-43 shows a critically damped unit-step response in this case, which implies the zero at $s = -K_P/K_D$ of the closed-loop transfer function has a smaller impact on the response of the system, and the overall response is similar to that of a prototype second-order system. However, in either case, upon *increasing* K_D, the general conclusion is that the PD controller decreases the maximum overshoot, the rise time, and the settling time.

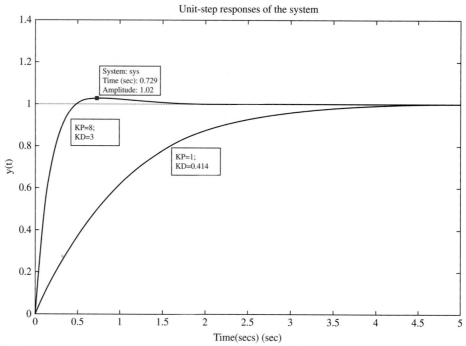

Figure 5-43 Unit-step response of Eq. (5-176) for two sets of K_D and K_P values.

Toolbox 5-11-1

The corresponding responses for Fig. 5-43 are obtained by the following sequence of MATLAB functions

```
clear all
t=0:0.001:5;

num = [2*3 16]; % KP=4 and KD=3
den = [1 2+2*3 16];
step(num,den,t);

hold on;

num = [2*.414 2]; % KP=1 and KD=0.414
den = [1 2+2*.414 2];
step(num,den,t);

xlabel('Time(secs)')
ylabel('y(t)')
title('Unit-step responses of the system')
```

▶ **EXAMPLE 5-11-2** We saw in the previous example that the PD controller can improve the damping and rise time of a control system. Because the PD controller does not change the system type, it may not fulfill the compensation objectives in many situations involving steady-state error. For this purpose, an integral controller may be used. The integral part of the PID controller produces a signal that is proportional to

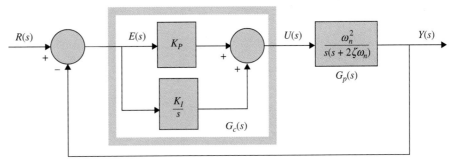

Figure 5-44 Control system with PI controller.

the time integral of the input of the controller. Fig. 5-44 illustrates the block diagram of the prototype second-order system with a series PI controller. The transfer function of the PI controller is

$$G_c(s) = K_P + \frac{K_I}{s} \tag{5-180}$$

Using the circuit elements given in Table 4-4 in Chapter 4, the forward-path transfer function of the compensated system is

$$G(s) = G_c(s)G_p(s) = \frac{\omega_n^2(K_P s + K_I)}{s^2(s + 2\zeta\omega_n)} \tag{5-181}$$

Clearly, the immediate effects of the PI controller are the following:

1. Adding a zero at $s = -K_I/K_P$ to the forward-path transfer function.
2. Adding a pole at $s = 0$ to the forward-path transfer function. This means that the system type is increased by one. Thus, the steady-state error of the original system is improved by one order; that is, if the steady-state error to a given input is constant, the PI control reduces it to zero (provided that the compensated system remains stable).

Consider the second-order model

$$G_p(s) = \frac{2}{(s+1)(s+2)} \tag{5-182}$$

The system in Fig. 5-44, with the forward-path transfer function in Eq. (5-182), will now have a zero steady-state error when the reference input is a step function. However, because the system is now of the third order, *it may be less stable* than the original second-order system or even become *unstable* if the parameters K_P and K_I are not properly chosen. In the case of a type 0 system with a PD control, the magnitude of the steady-state error is inversely proportional to K_P. When a type 0 system is converted to type 1 using a PI controller, the steady-state error due to a step input is always zero if the system is stable. The problem is then to choose the proper combination of K_P and K_I so that the transient response is satisfactory.

The pole-zero configuration of the PI controller in Eq. (5-180) is shown in Fig. 5-45. At first glance, it may seem that PI control will improve the steady-state error at the expense of stability. However, we shall show that, if the location of the zero of $G_c(s)$ is selected properly, both the damping and the steady-state error can be improved. Because the PI controller is essentially a low-pass filter, the compensated system usually will have a slower rise time and longer settling time. *A viable method of designing the PI control is to select the zero at $s = -K_I/K_P$ so that it is relatively close to the origin and away from the most significant poles of the process; the values of K_P and K_I should be relatively small.*

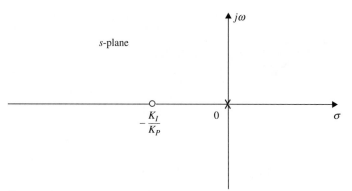

Figure 5-45 Pole-zero configuration of a PI controller.

Applying the PI controller of Eq. (5-180), the forward-path transfer function of the system becomes

$$G(s) = G_c(s)G_P(s) = \frac{2K_P(s + K_I/K_P)}{s(s+1)(s+2)} = \frac{2K_P(s + K_I/K_P)}{s^3 + 3s^2 + 2s} \quad (5\text{-}183)$$

The steady-state error due to a step input $u_s(t)$ is zero. The closed-loop transfer function is

$$\frac{Y(s)}{R(s)} = \frac{2K_P(s + K_I/K_P)}{s^3 + 3s^2 + 2(1 + K_P)s + 2K_I} \quad (5\text{-}184)$$

The characteristic equation of the closed-loop system is

$$s^3 + 3s^2 + 2(1 + K_P)s + 2K_I = 0 \quad (5\text{-}185)$$

Applying Routh's test to Eq. (5-185) yields the result that the system is stable for $0 < K_I/K_P < 13.5$. This means that the zero of $G(s)$ at $s = -K_I/K_P$ cannot be placed too far to the left in the left-half s-plane, or the system will be unstable. Let us place the zero at $-K_I/K_P$ relatively close to the origin. For the present case, the most significant pole of $G_p(s)$ is at -1. Thus, K_I/K_P should be chosen so that the following condition is satisfied:

$$\frac{K_I}{K_P} \ll 1 \quad (5\text{-}186)$$

With the condition in Eq. (5-186) satisfied, Eq. (5-184) can be approximated by

$$G(s) \cong \frac{2K_P}{s^2 + 3s + 2 + 2K_P} \quad (5\text{-}187)$$

where the term K_I/K_P in the numerator and K_I in the denominator are neglected. As a design criterion, we assume a desired percent maximum overshoot value of about 4.3 for a unit-step input, which utilizing expression (5-104) results in a relative damping ratio of 0.707. From the denominator of Eq. (5-187) compared with a prototype second-order system, we get natural frequency value of $\omega_n = 2.1213$ rad/s and the required proportional gain of $K_P = 1.25$. This should also be true for the third-order system with the PI controller if the value of K_I/K_P satisfies Eq. (5-186). Thus, to achieve this, we pick a small K_I. If K_I is too small, however, the system time response is slow and the desired steady-state error requirement is not met fast enough. Upon increasing K_I to 1.125, the desired response is met, as shown in Fig. 5-46.

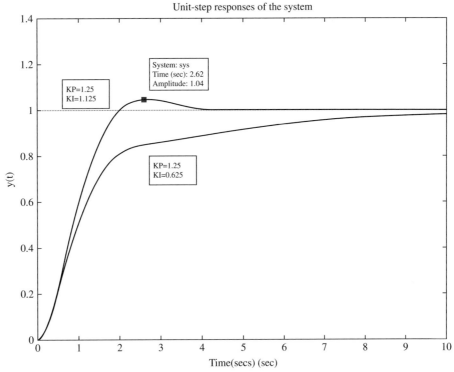

Figure 5-46 Unit-step response of Eq. (5-185) for two sets of K_I and K_P values.

Toolbox 5-11-2

The corresponding responses for Fig. 5-46 are obtained by the following sequence of MATLAB functions

```
clear all
t=0:0.001:10;

num = [2*1.25 1.125]; % KP=1.25 and KI=0.625
den = [1 3 2+2*1.25 1.125];
step(num,den,t);

hold on;

num = [2*1.25 2*1.125]; % KP=1.25 and KI=1.125
den = [1 3 2+2*1.25 2*1.125];
step(num,den,t);

xlabel('Time(secs)')
ylabel('y(t)')
title('Unit-step responses of the system')
```

5-12 MATLAB TOOLS

In this chapter we provided MATLAB toolboxes for finding the time response of simple control systems. We also introduced the concepts of root contours and root locus and included MATLAB codes to draw them for simple control examples. In Chapters 6 and 9, where we address more complex control-system modeling and analysis, we will introduce the **A**utomatic **C**ontrol **Sys**tems software (**ACSYS**) that utilizes MATLAB and SIMULINK m-files and GUIs (graphical user interface) for the analysis of more complex control engineering problems.

The reader is especially encouraged to explore the Control Lab software tools presented in Chapter 6 that simulate dc motor speed and position control topics discussed earlier in this chapter. These simulation tools provide the user with virtual experiments and design projects using systems involving dc motors, sensors, electronic components, and mechanical components.

▶ 5-13 SUMMARY

This chapter was devoted to the time-domain analysis of linear continuous-data control systems. The time response of control systems is divided into the transient and the steady-state responses. The steady-state error is a measure of the accuracy of the system as time approaches infinity. When the system has unity feedback for the step, ramp, and parabolic inputs, the steady-state error is characterized by the error constants K_p, K_v, and K_a, respectively, as well as the system **type**. When applying the steady-state error analysis, the final-value theorem of the Laplace transform is the basis; it should be ascertained that the closed-loop system is stable or the error analysis will be invalid. The error constants are not defined for systems with nonunity feedback. For nonunity-feedback systems, a method of determining the steady-state error was introduced by using the closed-loop transfer function.

The transient response is characterized by such criteria as the **maximum overshoot**, **rise time**, **delay time**, and **settling time**, and such parameters as **damping ratio**, **natural undamped frequency**, and **time constant**. The analytical expressions of these parameters can all be related to the system parameters simply if the transfer function is of the second-order prototype. For second-order systems that are not of the prototype and for higher-order systems, the analytical relationships between the transient parameters and the system constants are difficult to determine. Computer simulations are recommended for these systems.

Time-domain analysis of a position-control system was conducted. The transient and steady-state analyses were carried out first by approximating the system as a second-order system. The effect of varying the amplifier gain K on the transient and steady-state performance was demonstrated. The concept of the root-locus technique was introduced, and the system was then analyzed as a third-order system. It was shown that the second-order approximation was accurate only for low values of K.

The effects of adding poles and zeros to the forward-path and closed-loop transfer functions were demonstrated. The dominant poles of transfer functions were also discussed. This established the significance of the location of the poles of the transfer function in the s-plane and under what conditions the insignificant poles (and zeros) could be neglected with regard to the transient response.

Later in the chapter, simple controllers—namely the PD, PI, and PID—were introduced. Designs were carried out in the time-domain (and s-domain). The time-domain design may be characterized by specifications such as the relative damping ratio, maximum overshoot, rise time, delay time, settling time, or simply the location of the characteristic-equation roots, keeping in mind that the zeros of the system transfer function also affect the transient response. The performance is generally measured by the step response and the steady-state error.

MATLAB toolboxes and the Automatic Control System software tool are good tools to study the time response of control systems. Through the GUI approach provided by **ACSYS**, these programs are intended to create a user-friendly environment to reduce the complexity of control systems design. See Chapters 6 and 9 for more detail.

▶ REVIEW QUESTIONS

1. Give the definitions of the error constants K_p, K_v, and K_a.
2. Specify the type of input to which the error constant K_p is dedicated.

3. Specify the type of input to which the error constant K_v is dedicated.

4. Specify the type of input to which the error constant K_a is dedicated.

5. Define an error constant if the input to a unity-feedback control system is described by $r(t) = t^3 u_s(t)/6$.

6. Give the definition of the system type of a linear time-invariant system.

7. If a unity-feedback control system type is 2, then it is certain that the steady-state error of the system to a step input or a ramp input will be zero. (T) (F)

8. Linear and nonlinear frictions will generally degrade the steady-state error of a control system. (T) (F)

9. The maximum overshoot of a unit-step response of the second-order prototype system will never exceed 100% when the damping ratio ζ and the natural undamped frequency ω_n are all positive. (T) (F)

10. For the second-order prototype system, when the undamped natural frequency ω_n increases, the maximum overshoot of the output stays the same. (T) (F)

11. The maximum overshoot of the following system will never exceed 100% when ζ, ω_n, and T are all positive.

$$\frac{Y(s)}{R(s)} = \frac{\omega_n^2(1+Ts)}{s^2 + 2\zeta\omega_n s + \omega_n^2}$$

(T) (F)

12. Increasing the undamped natural frequency will generally reduce the rise time of the step response. (T) (F)

13. Increasing the undamped natural frequency will generally reduce the settling time of the step response. (T) (F)

14. Adding a zero to the forward-path transfer function will generally improve the system damping and thus will always reduce the maximum overshoot of the system. (T) (F)

15. Given the following characteristic equation of a linear control system, increasing the value of K will increase the frequency of oscillation of the system.

$$s^3 + 3s^2 + 5s + K = 0$$

(T) (F)

16. For the characteristic equation given in question 15, increasing the coefficient of the s^2 term will generally improve the damping of the system. (T) (F)

17. The location of the roots of the characteristic equation in the s-plane will give a definite indication on the maximum overshoot of the transient response of the system. (T) (F)

18. The following transfer function $G(s)$ can be approximated by $G_L(s)$ because the pole at -20 is much larger than the dominant pole at $s = -1$.

$$G(s) = \frac{10}{s(s+1)(s+20)} \quad G_L(s) = \frac{10}{s(s+1)}$$

(T) (F)

19. What is a PD controller? Write its input–output transfer function.

20. A PD controller has the constants K_D and K_P. Give the effects of these constants on the steady-state error of the system. Does the PD control change the type of a system?

21. Give the effects of the PD control on rise time and settling time of a control system.

22. How does the PD controller affect the bandwidth of a control system?

23. Once the value of K_D of a PD controller is fixed, increasing the value of K_P will increase the phase margin monotonically. (T) (F)

24. If a PD controller is designed so that the characteristic-equation roots have better damping than the original system, then the maximum overshoot of the system is always reduced. (T) (F)

25. What does it mean when a control system is described as being robust?

26. A system compensated with a PD controller is usually more robust than the system compensated with a PI controller. (T) (F)

27. What is a PI controller? Write its input–output transfer function.

28. A PI controller has the constants K_P and K_I. Give the effects of the PI controller on the steady-state error of the system. Does the PI control change the system type?

29. Give the effects of the PI control on the rise time and settling time of a control system.

Answers to these review questions can be found on this book's companion Web site: www.wiley.com/college/golnaraghi.

▶ REFERENCES

1. J. C. Willems and S. K. Mitter, "Controllability, Observability, Pole Allocation, and State Reconstruction," *IEEE Trans. Automatic Control*, Vol. AC-16, pp. 582–595, Dec. 1971.
2. H. W. Smith and E. J. Davison, "Design of Industrial Regulators," *Proc. IEE (London)*, Vol. 119, pp. 1210–1216, Aug. 1972.
3. F. N. Bailey and S. Meshkat, "Root Locus Design of a Robust Speed Control," *Proc. Incremental Motion Control Symposium*, pp. 49–54, June 1983.
4. M. Vidyasagar, "On Undershoot and Nonminimum Phase Zeros," *IEEE Trans. Automatic Control*, Vol. AC-31, p. 440, May 1986.
5. T. Norimatsu and M. Ito, "On the Zero Non-Regular Control System," *J. Inst. Elec. Eng. Japan*, Vol. 81, pp. 567–575, 1961.
6. K. Ogata, *Modern Control Engineering*, 4th Ed., Prentice Hall, NJ, 2002.
7. G. F. Franklin and J. D. Powell, *Feedback Control of Dynamic Systems*, 5th Ed., Prentice-Hall, NJ, 2006.
8. J. J. Distefano, III, A. R. Stubberud, and I. J. Williams, *Schaum's Outline of Theory and Problems of Feedback and Control Systems*, 2nd Ed. McGraw-Hill, 1990.

▶ PROBLEMS

In addition to using the conventional approaches, use MATLAB to solve the problems in this chapter.

5-1. A pair of complex-conjugate poles in the *s*-plane is required to meet the various specifications that follow. For each specification, sketch the region in the *s*-plane in which the poles should be located.

(a) $\zeta \geq 0.707$ $\omega_n \geq 2 \, \text{rad/sec}$ (positive damping)
(b) $0 \leq \zeta \leq 0.707$ $\omega_n \leq 2 \, \text{rad/sec}$ (positive damping)
(c) $\zeta \leq 0.5$ $1 \leq \omega_n \leq 5 \, \text{rad/sec}$ (positive damping)
(d) $0.5 \leq \zeta \leq 0.707$ $\omega_n \leq 5 \, \text{rad/sec}$ (positive and negative damping)

5-2. Determine the type of the following unity-feedback systems for which the forward-path transfer functions are given.

(a) $G(s) = \dfrac{K}{(1+s)(1+10s)(1+20s)}$ (b) $G(s) = \dfrac{10e^{-0.2s}}{(1+s)(1+10s)(1+20s)}$

(c) $G(s) = \dfrac{10(s+1)}{s(s+5)(s+6)}$ (d) $G(s) = \dfrac{100(s-1)}{s^2(s+5)(s+6)^2}$

(e) $G(s) = \dfrac{10(s+1)}{s^3(s^2+5s+5)}$ (f) $G(s) = \dfrac{100}{s^3(s+2)^2}$

(g) $G(s) = \dfrac{5(s+2)}{s^2(s+4)}$ (h) $G(s) = \dfrac{8(s+1)}{(s^2+2s+3)(s+1)}$

5-3. Determine the step, ramp, and parabolic error constants of the following unity-feedback control systems. The forward-path transfer functions are given.

(a) $G(s) = \dfrac{1000}{(1+0.1s)(1+10s)}$ (b) $G(s) = \dfrac{100}{s(s^2+10s+100)}$

(c) $G(s) = \dfrac{K}{s(1+0.1s)(1+0.5s)}$ (d) $G(s) = \dfrac{100}{s^2(s^2+10s+100)}$

(e) $G(s) = \dfrac{1000}{s(s+10)(s+100)}$ (f) $G(s) = \dfrac{K(1+2s)(1+4s)}{s^2(s^2+s+1)}$

5-4. For the unity-feedback control systems described in Problem 5-2, determine the steady-state error for a unit-step input, a unit-ramp input, and a parabolic input, $(t^2/2)u_s(t)$. Check the stability of the system before applying the final-value theorem.

5-5. The following transfer functions are given for a single-loop nonunity-feedback control system. Find the steady-state errors due to a unit-step input, a unit-ramp input, and a parabolic input, $(t^2/2)u_s(t)$.

(a) $G(s) = \dfrac{1}{(s^2+s+2)}$ $H(s) = \dfrac{1}{(s+1)}$

(b) $G(s) = \dfrac{1}{s(s+5)}$ $H(s) = 5$

(c) $G(s) = \dfrac{1}{s^2(s+10)}$ $H(s) = \dfrac{s+1}{s+5}$

(d) $G(s) = \dfrac{1}{s^2(s+12)}$ $H(s) = 5(s+2)$

5-6. Find the steady-state errors of the following single-loop control systems for a unit-step input, a unit-ramp input, and a parabolic input, $(t^2/2)u_s(t)$. For systems that include a parameter K, find its value so that the answers are valid.

(a) $M(s) = \dfrac{s+4}{s^4+16s^3+48s^2+4s+4}$, $K_H = 1$

(b) $M(s) = \dfrac{K(s+3)}{s^3+3s^2+(K+2)s+3K}$, $K_H = 1$

(c) $M(s) = \dfrac{s+5}{s^4+15s^3+50s^2+10s}$, $H(s) = \dfrac{10s}{s+5}$

(d) $M(s) = \dfrac{K(s+5)}{s^4+17s^3+60s^2+5Ks+5K}$, $K_H = 1$

5-7. The output of the system shown in Fig. 5P-8 has a transfer function Y/X. Find the poles and zeros of the closed loop system and the system type.

5-8. Find the position, velocity, and acceleration error constants for the system given in Fig. 5P-8.

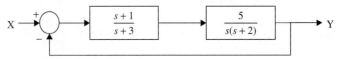

Figure 5P-8

5-9. Find the steady-state error for Problem 5-8 for (a) a unit-step input, (b) a unit-ramp input, and (c) a unit-parabolic input.

5-10. Repeat Problem 5-8 for the system given in Fig. 5P-10.

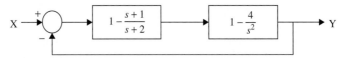

Figure 5P-10

5-11. Find the steady-state error of the system given in Problem 5-10 when the input is

$$X = \frac{5}{2s} - \frac{3}{s^2} + \frac{4}{s^3}$$

5-12. Find the rise time of the following first-order system:

$$G(s) = \frac{1-k}{s-k} \quad \text{with } |k| < 1$$

5-13. The block diagram of a control system is shown in Fig. 5P-13. Find the step-, ramp-, and parabolic-error constants. The error signal is defined to be $e(t)$. Find the steady-state errors in terms of K and K_t when the following inputs are applied. Assume that the system is stable.

(a) $r(t) = u_s(t)$
(b) $r(t) = tu_s(t)$
(c) $r(t) = (t^2/2)u_s(t)$

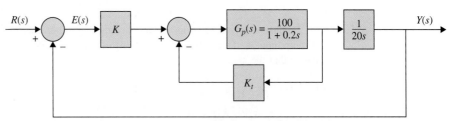

Figure 5P-13

5-14. Repeat Problem 5-13 when the transfer function of the process is, instead,

$$G_p(s) = \frac{100}{(1+0.1s)(1+0.5s)}$$

What constraints must be made, if any, on the values of K and K_t so that the answers are valid? Determine the minimum steady-state error that can be achieved with a unit-ramp input by varying the values of K and K_t.

5-15. For the position-control system shown in Fig. 3P-7, determine the following.
(a) Find the steady-state value of the error signal $\theta_e(t)$ in terms of the system parameters when the input is a unit-step function.
(b) Repeat part (a) when the input is a unit-ramp function. Assume that the system is stable.

5-16. The block diagram of a feedback control system is shown in Fig. 5P-16. The error signal is defined to be $e(t)$.
(a) Find the steady-state error of the system in terms of K and K_t when the input is a unit-ramp function. Give the constraints on the values of K and K_t so that the answer is valid. Let $n(t) = 0$ for this part.
(b) Find the steady-state value of $y(t)$ when $n(t)$ is a unit-step function. Let $r(t) = 0$. Assume that the system is stable.

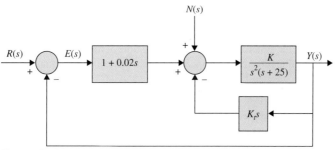

Figure 5P-16

5-17. The block diagram of a linear control system is shown in Fig. 5P-17, where $r(t)$ is the reference input and $n(t)$ is the disturbance.
(a) Find the steady-state value of $e(t)$ when $n(t) = 0$ and $r(t) = tu_s(t)$. Find the conditions on the values of α and K so that the solution is valid.
(b) Find the steady-state value of $y(t)$ when $r(t) = 0$ and $n(t) = u_s(t)$.

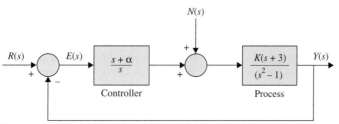

Figure 5P-17

5-18. The unit-step response of a linear control system is shown in Fig. 5P-18. Find the transfer function of a second-order prototype system to model the system.

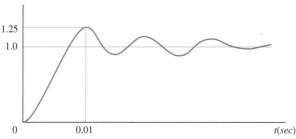

Figure 5P-18

5-19. For the control system shown in Fig. 5P-13, find the values of K and K_t so that the maximum overshoot of the output is approximately 4.3% and the rise time t_r is approximately 0.2 sec. Use Eq. (5-98) for the rise-time relationship. Simulate the system with any time-response simulation program to check the accuracy of your solutions.

5-20. Repeat Problem 5-19 with a maximum overshoot of 10% and a rise time of 0.1 sec.

5-21. Repeat Problem 5-19 with a maximum overshoot of 20% and a rise time of 0.05 sec.

5-22. For the control system shown in Fig. 5P-13, find the values of K and K_t so that the maximum overshoot of the output is approximately 4.3% and the delay time t_d is approximately 0.1 sec. Use Eq. (5-96) for the delay-time relationship. Simulate the system with a computer program to check the accuracy of your solutions.

5-23. Repeat Problem 5-22 with a maximum overshoot of 10% and a delay time of 0.05 sec.

5-24. Repeat Problem 5-22 with a maximum overshoot of 20% and a delay time of 0.01 sec.

5-25. For the control system shown in Fig. 5P-13, find the values of K and K_t so that the damping ratio of the system is 0.6 and the settling time of the unit-step response is 0.1 sec. Use Eq. (5-102) for the settling time relationship. Simulate the system with a computer program to check the accuracy of your results.

5-26. **(a)** Repeat Problem 5-25 with a maximum overshoot of 10% and a settling time of 0.05 sec.
(b) Repeat Problem 5-25 with a maximum overshoot of 20% and a settling time of 0.01 sec.

5-27. Repeat Problem 5-25 with a damping ratio of 0.707 and a settling time of 0.1 sec. Use Eq. (5-103) for the settling time relationship.

5-28. The forward-path transfer function of a control system with unity feedback is

$$G(s) = \frac{K}{s(s+a)(s+30)}$$

where a and K are real constants.
(a) Find the values of a and K so that the relative damping ratio of the complex roots of the characteristic equation is 0.5 and the rise time of the unit-step response is approximately 1 sec. Use Eq. (5-98) as an approximation of the rise time. With the values of a and K found, determine the actual rise time using computer simulation.
(b) With the values of a and K found in part (a), find the steady-state errors of the system when the reference input is (i) a unit-step function and (ii) a unit-ramp function.

5-29. The block diagram of a linear control system is shown in Fig. 5P-29.
(a) By means of trial and error, find the value of K so that the characteristic equation has two equal real roots and the system is stable. You may use any root-finding computer program to solve this problem.
(b) Find the unit-step response of the system when K has the value found in part (a). Use any computer simulation program for this. Set all the initial conditions to zero.
(c) Repeat part (b) when $K = -1$. What is peculiar about the step response for small t, and what may have caused it?

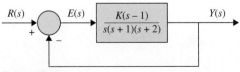

Figure 5P-29

5-30. A controlled process is represented by the following dynamic equations:

$$\frac{dx_1(t)}{dt} = -x_1(t) + 5x_2(t)$$

$$\frac{dx_2(t)}{dt} = -6x_1(t) + u(t)$$

$$y(t) = x_1(t)$$

The control is obtained through state feedback with

$$u(t) = -k_1 x_1(t) - k_2 x_2(t) + r(t)$$

where k_1 and k_2 are real constants, and $r(t)$ is the reference input.

(a) Find the locus in the k_1-versus-k_2 plane (k_1 = vertical axis) on which the overall system has a natural undamped frequency of 10 rad/sec.

(b) Find the locus in the k_1-versus-k_2 plane on which the overall system has a damping ratio of 0.707.

(c) Find the values of k_1 and k_2 such that $\zeta = 0.707$ and $\omega_n = 10$ rad/sec.

(d) Let the error signal be defined as $e(t) = r(t) - y(t)$. Find the steady-state error when $r(t) = u_s(t)$ and k_1 and k_2 are at the values found in part (c).

(e) Find the locus in the k_1-versus-k_2 plane on which the steady-state error due to a unit-step input is zero.

5-31. The block diagram of a linear control system is shown in Fig. 5P-31. Construct a parameter plane of K_p versus K_d (K_p is the vertical axis), and show the following trajectories or regions in the plane.

(a) Unstable and stable regions

(b) Trajectories on which the damping is critical ($\zeta = 1$)

(c) Region in which the system is overdamped ($\zeta > 1$)

(d) Region in which the system is underdamped ($\zeta < 1$)

(e) Trajectory on which the parabolic-error constant K_a is $1000 \sec^{-2}$

(f) Trajectory on which the natural undamped frequency ω_n is 50 rad/sec

(g) Trajectory on which the system is either uncontrollable or unobservable (hint: look for pole-zero cancellation)

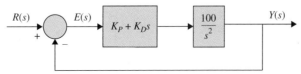

Figure 5P-31

5-32. The block diagram of a linear control system is shown in Fig. 5P-32. The fixed parameters of the system are given as $T = 0.1$, $J = 0.01$, and $K_i = 10$.

(a) When $r(t) = tu_s(t)$ and $T_d(t) = 0$, determine how the values of K and K_t affect the steady-state value of $e(t)$. Find the restrictions on K and K_t so that the system is stable.

(b) Let $r(t) = 0$. Determine how the values of K and K_t affect the steady-state value of $y(t)$ when the disturbance input $T_d(t) = u_s(t)$.

(c) Let $K_t = 0.01$ and $r(t) = 0$. Find the minimum steady-state value of $y(t)$ that can be obtained by varying K, when $T_d(t)$ is a unit-step function. Find the value of this K. From the transient standpoint, would you operate the system at this value of K? Explain.

(d) Assume that it is desired to operate the system with the value of K as selected in part (c). Find the value of K_t so that the complex roots of the characteristic equation will have a real part of -2.5. Find all three roots of the characteristic equation.

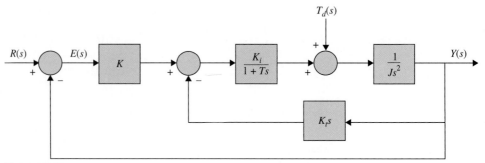

Figure 5P-32

5-33. Consider a second-order unity feedback system with $\zeta = 0.6$ and $\omega_n = 5\,\text{rad/sec}$. Calculate the rise time, peak time, maximum overshoot, and settling time when a unit-step input is applied to the system.

5-34. Fig. 5P-34 shows the block diagram of a servomotor. Assume $J = 1\,\text{kg-m}^2$ and $B = 1\,\text{N-m/rad/sec}$. If the maximum overshoot of the unit-step input and the peak time are 0.2 and 0.1 sec., respectively,

(a) Find its damping ratio and natural frequency.

(b) Find the gain K and velocity feedback K_f. Also, calculate the rise time and settling time.

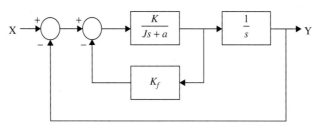

Figure 5P-34

5-35. Find the unit-step response of the following systems assuming zero initial conditions:

(a) $\begin{bmatrix} \dot{x}_1 \\ \dot{x}_2 \end{bmatrix} = \begin{bmatrix} -1 & -1 \\ 6.5 & 0 \end{bmatrix} \begin{bmatrix} x_1 \\ x_2 \end{bmatrix} + \begin{bmatrix} 1 & 1 \\ 1 & 0 \end{bmatrix} \begin{bmatrix} u_1 \\ u_2 \end{bmatrix}$

$\begin{bmatrix} y_1 \\ y_2 \end{bmatrix} = \begin{bmatrix} 1 & 0 \\ 0 & 1 \end{bmatrix} \begin{bmatrix} x_1 \\ x_2 \end{bmatrix} + \begin{bmatrix} 0 & 0 \\ 0 & 0 \end{bmatrix} \begin{bmatrix} u_1 \\ u_2 \end{bmatrix}$

(b) $\begin{bmatrix} \dot{x}_1 \\ \dot{x}_2 \end{bmatrix} = \begin{bmatrix} 0 & 1 \\ -1 & -1 \end{bmatrix} \begin{bmatrix} x_1 \\ x_2 \end{bmatrix} + \begin{bmatrix} 0 \\ 1 \end{bmatrix} u$

$y_1 = \begin{bmatrix} 1 & 0 \end{bmatrix} \begin{bmatrix} x_1 \\ x_2 \end{bmatrix} + [0]u$

(c) $\begin{bmatrix} \dot{x}_1 \\ \dot{x}_2 \\ \dot{x}_3 \end{bmatrix} = \begin{bmatrix} 0 & 1 & 0 \\ -1 & -1 & 0 \\ 1 & 0 & 0 \end{bmatrix} \begin{bmatrix} x_1 \\ x_2 \\ x_3 \end{bmatrix} + \begin{bmatrix} 0 \\ 1 \\ 0 \end{bmatrix} u$

$y = \begin{bmatrix} 0 & 0 & 1 \end{bmatrix} \begin{bmatrix} x_1 \\ x_2 \\ x_3 \end{bmatrix}$

5-36. Use MATLAB to solve Problem 5-35.

5-37. Find the impulse response of the given systems in Problem 5-35.

5-38. Use MATLAB to solve Problem 5-37.

5-39. Fig. 5P-39 shows a mechanical system.
(a) Find the differential equation of the system.
(b) Use MATLAB to find the unit-step input response of the system.

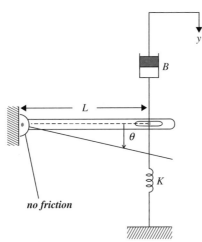

Figure 5P-39

5-40. The dc-motor control system for controlling a printwheel described in Problem 4-49 has the forward-path transfer function

$$G(s) = \frac{\Theta_o(s)}{\Theta_e(s)} = \frac{nK_sK_iK_LK}{\Delta(s)}$$

where $\Delta(s) = s\left[L_aJ_mJ_Ls^4 + J_L(R_aJ_m + B_mL_a)s^3 \right.$
$+ (n^2K_LL_aJ_L + K_LL_aJ_m + K_iK_bJ_L + R_aB_mJ_L)s^2$
$\left. + (n^2R_aK_LJ_L + R_aK_LJ_m + B_mK_LL_a)s + R_aB_mK_L + K_iK_bK_L\right]$

where $K_i = 9$ oz-in./A, $K_b = 0.636$ V/rad/sec, $R_a = 5\Omega$, $L_a = 1$ mH, $K_s = 1$ V/rad, $n = 1/10$, $J_m = J_L = 0.001$ oz-in.-sec^2, and $B_m \cong 0$. The characteristic equation of the closed-loop system is

$$\Delta(s) + nK_sK_iK_LK = 0$$

(a) Let $K_L = 10,000$ oz-in./rad. Write the forward-path transfer function $G(s)$ and find the poles of $G(s)$. Find the critical value of K for the closed-loop system to be stable. Find the roots of the characteristic equation of the closed-loop system when K is at marginal stability.
(b) Repeat part (a) when $K_L = 1000$ oz-in./rad.
(c) Repeat part (a) when $K_L = \infty$; that is, the motor shaft is rigid.
(d) Compare the results of parts (a), (b), and (c), and comment on the effects of the values of K_L on the poles of $G(s)$ and the roots of the characteristic equation.

5-41. The block diagram of the guided-missile attitude-control system described in Problem 4-20 is shown in Fig. 5P-41. The command input is $r(t)$, and $d(t)$ represents disturbance input. The objective of this problem is to study the effect of the controller $G_c(s)$ on the steady-state and transient responses of the system.

(a) Let $G_c(s) = 1$. Find the steady-state error of the system when $r(t)$ is a unit-step function. Set $d(t) = 0$.

(b) Let $G_c(s) = (s + \alpha)/s$. Find the steady-state error when $r(t)$ is a unit-step function.

(c) Obtain the unit-step response of the system for $0 \leq t \leq 0.5$ sec with $G_c(s)$ as given in part (b) and $\alpha = 5$, 50, and 500. Assume zero initial conditions. Record the maximum overshoot of $y(t)$ for each case. Use any available computer simulation program. Comment on the effect of varying the value of α of the controller on the transient response.

(d) Set $r(t) = 0$ and $G_c(s) = 1$. Find the steady-state value of $y(t)$ when $d(t) = u_s(t)$.

(e) Let $G_c(s) = (s + \alpha)/s$. Find the steady-state value of $y(t)$ when $d(t) = u_s(t)$.

(f) Obtain the output response for $0 \leq t \leq 0.5$ sec, with $G_c(s)$ as given in part (e) when $r(t) = 0$ and $d(t) = u_s(t)$; $\alpha = 5$, 50, and 500. Use zero initial conditions.

(g) Comment on the effect of varying the value of α of the controller on the transient response of $y(t)$ and $d(t)$.

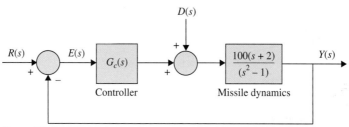

Figure 5P-41

5-42. The block diagram shown in Fig. 5P-42 represents a liquid-level control system. The liquid level is represented by $h(t)$, and N denotes the number of inlets.

(a) Because one of the poles of the open-loop transfer function is relatively far to the left on the real axis of the s-plane at $s = -10$, it is suggested that this pole can be neglected. Approximate the system by a second-order system by neglecting the pole of $G(s)$ at $s = -10$. The approximation should be valid for both the transient and the steady-state responses. Apply the formulas for the maximum overshoot and the peak time t_{max} to the second-order model for $N = 1$ and $N = 10$.

(b) Obtain the unit-step response (with zero initial conditions) of the original third-order system with $N = 1$ and then with $N = 10$. Compare the responses of the original system with those of the second-order approximating system. Comment on the accuracy of the approximation as a function of N.

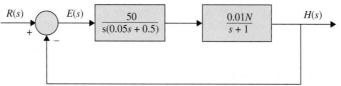

Figure 5P-42

5-43. The forward-path transfer function of a unity-feedback control system is

$$G(s) = \frac{1 + T_z s}{s(s + 1)^2}$$

Compute and plot the unit-step responses of the closed-loop system for $T_z = 0$, 0.5, 1.0, 10.0, and 50.0. Assume zero initial conditions. Use any computer simulation program that is available. Comment on the effects of the various values of T_z on the step response.

5-44. The forward-path transfer function of a unity-feedback control system is

$$G(s) = \frac{1}{s(s+1)^2(1+T_p s)}$$

Compute and plot the unit-step responses of the closed-loop system for $T_p = 0, 0.5,$ and 0.707. Assume zero initial conditions. Use any computer simulation program. Find the critical value of T_p so that the closed-loop system is marginally stable. Comment on the effects of the pole at $s = -1/T_p$ in $G(s)$.

5-45. Compare and plot the unit-step responses of the unity-feedback closed-loop systems with the forward-path transfer functions given. Assume zero initial conditions. Use the timetool program.

(a) $G(s) = \dfrac{1+T_z s}{s(s+0.55)(s+1.5)}$ For $T_z = 0, 1, 5, 20$

(b) $G(s) = \dfrac{1+T_z s}{(s^2+2s+2)}$ For $T_z = 0, 1, 5, 20$

(c) $G(s) = \dfrac{2}{(s^2+2s+2)(1+T_p s)}$ For $T_p = 0, 0.5, 1.0$

(d) $G(s) = \dfrac{10}{s(s+5)(1+T_p s)}$ For $T_p = 0, 0.5, 1.0$

(e) $G(s) = \dfrac{K}{s(s+1.25)(s^2+2.5s+10)}$

     (i) For $K = 5$
     (ii) For $K = 10$
     (iii) For $K = 30$

(f) $G(s) = \dfrac{K(s+2.5)}{s(s+1.25)(s^2+2.5s+10)}$

     (i) For $K = 5$
     (ii) For $K = 10$
     (iii) For $K = 30$

5-46. Fig. 5P-46 shows the block diagram of a servomotor with tachometer feedback.
(a) Find the error signal $E(s)$ in the presence of the reference input $X(s)$ and disturbance input $D(s)$.
(b) Calculate the steady-state error of the system when $X(s)$ is a unit ramp and $D(s)$ is a unit step.
(c) Use MATLAB to plot the response of the system for part (b).
(d) Use MATLAB to plot the response of the system when $X(s)$ is a unit-step input and $D(s)$ is a unit impulse input.

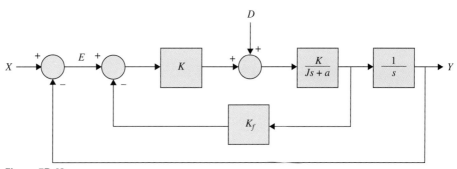

Figure 5P-46

5-47. The feedforward transfer function of a stable unity feedback system is $G(s)$. If the closed-loop transfer function can be rewritten as

$$\frac{Y(s)}{X(s)} = \frac{G(s)}{1+G(s)} = \frac{(A_1 s+1)(A_2 s+1)\ldots(A_n s+1)}{(B_1 s+1)(B_2 s+1)\ldots(B_m s+1)}$$

(a) Find the $\int_0^\infty e(t)$ when $e(t)$ is the error in the unit-step response.

(b) Calculate $\frac{1}{K} = \frac{1}{\lim_{s\to 0} sG(s)}$.

5-48. If the maximum overshoot and 1% settling time of the unit-step response of the closed-loop system shown in Fig. 5P-48 are no more than 25% and 0.1 sec, find the gain K and pole location P of the compensator. Also, use MATLAB to plot the unit-step input response of the system and verify your controller design.

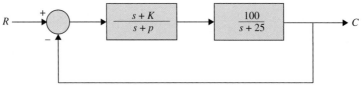

Figure 5P-48

5-49. If a given second-order system is required to have a peak time less than t, find the region in the s-plane corresponding to the poles that meet this specification.

5-50. A unity feedback control system shown in Fig. 5P-50(a) is designed so that its closed-loop poles lie within the region shown in Fig. 5P-50(b).

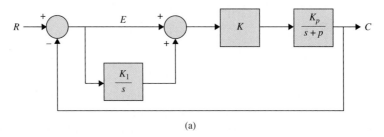

(a)

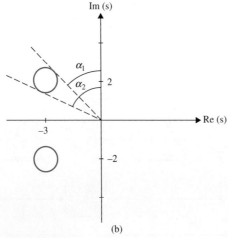

(b)

Figure 5P-50

(a) Find the values for ω_n and ζ.

(b) If $K_p = 2$ and $P = 2$, then find the values for K and K_I.

(c) Show that, regardless of values K_p and P, the controller can be designed to place the poles anywhere in the left side of the s-plane.

5-51. The motion equation of a dc motor is given by

$$J_m \ddot{\theta}_m + \left(B + \frac{K_1 K_2}{R}\right)\dot{\theta}_m = \frac{K_1}{R} v$$

Assuming $J_m = 0.02$ kg-m^2, $B = 0.002$ N-m-sec, $K_1 = 0.04$ N-m/A, $K_2 = 0.04$ V-sec, and $R = 20 \, \Omega$.

(a) Find the transfer function between the applied voltage and the motor speed.

(b) Calculate the steady-state speed of the motor after applying a voltage of 10 V.

(c) Determine the transfer function between the applied voltage and the shaft angle θ_m.

(d) Including a closed-loop feedback to part (c) such that $v = K(\theta_p - \theta_m)$, where K is the feedback gain, obtain the transfer function between θ_p and θ_m.

(e) If the maximum overshoot is less than 25%, determine K.

(f) If the rise time is less than 3 sec, determine K.

(g) Use MATLAB to plot the step response of the position servo system for $K = 0.5, 1.0$, and 2.0. Find the rise time and overshoot.

5-52. In the unity feedback closed-loop system in a configuration similar to that in Fig. 5P-48, the plant transfer function is

$$G(s) = \frac{1}{s(s+3)}$$

and the controller transfer function is

$$H(s) = \frac{k(s+a)}{(s+b)}$$

Design the controller parameters so that the closed-loop system has a 10% overshoot for a unit step input and a 1% settling time of 1.5 sec.

5-53. An autopilot is designed to maintain the pitch attitude α of an airplane. The transfer function between pitch angle α and elevator angle β are given by

$$\frac{\alpha(s)}{\beta(s)} = \frac{60(s+1)(s+2)}{(s^2+6s+40)(s^2+0.04s+0.07)}$$

The autopilot pitch controller uses the pitch error e to adjust the elevator as

$$\frac{\beta_e(s)}{E(s)} = \frac{K(s+3)}{s+10}$$

Use MATLAB to find K with an overshoot of less than 10% and a rise time faster than 0.5 sec for a unit-step input. Explain controller design difficulties for complex systems.

5-54. The block diagram of a control system with a series controller is shown in Fig. 5P-54. Find the transfer function of the controller $G_c(s)$ so that the following specifications are satisfied:

(a) The ramp-error constant K_v is 5.

(b) The closed-loop transfer function is of the form

$$M(s) = \frac{Y(s)}{R(s)} = \frac{K}{(s^2+20s+200)(s+a)}$$

where K and a are real constants. Use MATLAB to find the values of K and a.

The design strategy is to place the closed-loop poles at $-10 + j10$ and $-10 - j10$, and then adjust the values of K and a to satisfy the steady-state requirement. The value of a is large so that it will not affect the transient response appreciably. Find the maximum overshoot of the designed system.

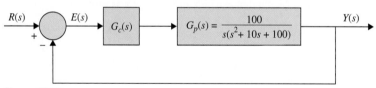

Figure 5P-54

5-55. Repeat Problem 5-54 if the ramp-error constant is to be 9. What is the maximum value of K_v that can be realized? Comment on the difficulties that may arise in attempting to realize a very large K_v.

5-56. A control system with a PD controller is shown in Fig. 5P-56. Use MATLAB to

(a) Find the values of K_P and K_D so that the ramp-error constant K_v is 1000 and the damping ratio is 0.5.

(b) Find the values of K_P and K_D so that the ramp-error constant K_v is 1000 and the damping ratio is 0.707.

(c) Find the values of K_P and K_D so that the ramp-error constant K_v is 1000 and the damping ratio is 1.0.

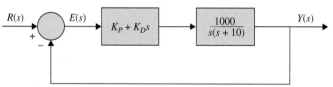

Figure 5P-56

5-57. For the control system shown in Figure 5P-56, set the value of K_P so that the ramp-error constant is 1000. Use MATLAB to

(a) Vary the value of K_D from 0.2 to 1.0 in increments of 0.2 and determine the values of rise time and maximum overshoot of the system.

(b) Vary the value of K_D from 0.2 to 1.0 in increments of 0.2 and find the value of K_D so that the maximum overshoot is minimum.

5-58. Consider the second-order model of the aircraft attitude control system shown in Fig. 5-29. The transfer function of the process is $G_p(s) = \frac{4500 K}{s(s+361.2)}$. Use MATLAB to design a series PD controller with the transfer function $G_c(s) = K_D + K_P s$ so that the following performance specifications are satisfied:

Steady-state error due to a unit-ramp input ≤ 0.001

Maximum overshoot $\leq 5\%$

Rise time $t_r \leq 0.005$ sec

Settling time $t_s \leq 0.005$ sec

5-59. Fig. 5P-59 shows the block diagram of the liquid-level control system described in Problem 5-42. The number of inlets is denoted by N. Set $N = 20$. Use MATLAB to design the PD controller so that with a unit-step input the tank is filled to within 5% of the reference level in less than 3 sec without overshoot.

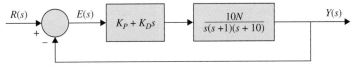

Figure 5P-59

5-60. For the liquid-level control system described in Problem 5-59, set K_P so that the ramp-error constant is 1. Use MATLAB to vary K_D from 0 to 0.5 and determine the values of rise time and maximum overshoot of the system.

5-61. A control system with a type 0 process $G_p(s)$ and a PI controller is shown in Fig. 5P-61. Use MATLAB to

(a) Find the value of K_I so that the ramp-error constant K_v is 10.

(b) Find the value of K_P so that the magnitude of the imaginary parts of the complex roots of the characteristic equation of the system is 15 rad/sec. Find the roots of the characteristic equation.

(c) Sketch the root contours of the characteristic equation with the value of K_I as determined in part (a) and for $0 \leq K_P < \infty$.

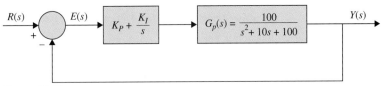

Figure 5P-61

5-62. For the control system described in Problem 5-61, set K_I so that the ramp-error constant is 10. Use MATLAB to vary K_P and determine the values of rise time and maximum overshoot of the system.

5-63. For the control system shown in Fig. 5P-61, use MATLAB to perform the following:

(a) Find the value of K_I so that the ramp-error constant K_v is 100.

(b) With the value of K_I found in part (a), find the critical value of K_P so that the system is stable. Sketch the root contours of the characteristic equation for $0 \leq K_P < \infty$.

(c) Show that the maximum overshoot is high for both large and small values of K_P. Use the value of K_I found in part (a). Find the value of K_P when the maximum overshoot is a minimum. What is the value of this maximum overshoot?

5-64. Repeat Problem 5-63 for $K_v = 10$.

5-65. A control system with a type 0 process and a PID controller is shown in Fig. 5P-65. Use MATLAB to design the controller parameters so that the following specifications are satisfied:

Ramp-error constant $K_v = 100$

Rise time $t_r < 0.01$ sec.

Maximum overshoot $< 2\%$

Plot the unit-step response of the designed system.

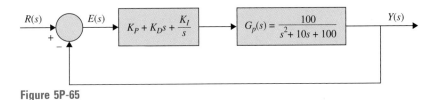

Figure 5P-65

5-66. Consider the quarter-car model of vehicle suspension systems in Example 4-11-3. The Laplace transform between the base acceleration and displacement is given by

$$\frac{Z(s)}{\ddot{Y}(s)} = \frac{-1}{s^2 + 2\zeta\omega_n s + \omega_n^2}$$

(a) It is desired to design a proportional controller. Use MATLAB to design the controller parameters where the rise time is no more than 0.05 sec and the overshoot is no more than 3%. Plot the unit-step response of the designed system.

(b) It is desired to design a PD controller. Use MATLAB to design the controller parameters where the rise time is no more than 0.05 sec and the overshoot is no more than 3%. Plot the unit-step response of the designed system.

(c) It is desired to design a PI controller. Use MATLAB to design the controller parameters where the rise time is no more than 0.05 sec and the overshoot is no more than 3%. Plot the unit-step response of the designed system.

(d) It is desired to design a PID controller. Use MATLAB to design the controller parameters where the rise time is no more than 0.05 sec and the overshoot is no more than 3%. Plot the unit-step response of the designed system.

5-67. Consider the spring-mass system shown in Fig. 5P-67.

Its transfer function is given by $\frac{Y(s)}{F(s)} = \frac{1}{Ms^2+Bs+K}$.

Repeat Problem 5-66 where $M = 1$ kg, $B = 10$ N.s/m, $K = 20$ N/m.

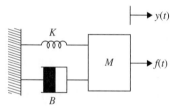

Figure 5P-67

5-68. Consider the vehicle suspension system hitting a bump described in Problem 4-3. Use MATLAB to design a proportional controller where the 1% settling time is less than 0.1 sec and the overshoot is no more than 2%. Assume $m = 25$ kg, $J = 5$ kg-m², $K = 100$ N/m, and $r = 0.35$ m. Plot the impulse response of the system.

5-69. Consider the train system described in Problem 4-6. Use MATLAB to design a proportional controller where the peak time is less than 0.05 sec and the overshoot is no more than 4%. Assume $M = 1$ kg, $m = 0.5$ kg, $k = 1$ N/m, $\mu = 0.002$ sec/m, and $g = 9.8$ m/s².

5-70. Consider the inverted pendulum described in Problem 4-9, where $M = 0.5$ kg, $m = 0.2$ kg, $\mu = 0.1$ N/m/sec (friction of the cart), $I = 0.006$ kg-m², $g = 9.8$ m/s², and $1 = 0.3$ m.
Use MATLAB to design a PD controller where the rise time is less than 0.2 sec and the overshoot is no more than 10%.

CHAPTER 6

The Control Lab

▶ 6-1 INTRODUCTION

The majority of undergraduate courses in control have labs dealing with time response and control of dc motors. The focus of this chapter is therefore on these lab problems—namely, speed response, speed control, position response, and position control of dc motors. In this chapter, using MATLAB and Simulink, we have created a series of virtual lab experiments that are designed to help students understand the concepts discussed in Chapters 4 and 5. This chapter also contains two controller design experiments. There are three classes of simulation experiments designed for this chapter: **SIMLab**, **Virtual Lab**, and **Quarter Car Sim**. There experiments are intended to supplement the experimental exposure of the students in a traditional undergraduate control course.

It is a demanding task to develop software that provides the reader with practical appreciation and understanding of dc motors including modeling uncertainties, nonlinear effects, system identification, and controller design amid these practical challenges. However, through the use of MATLAB and Simulink, we created a virtual dc motor in Virtual Lab, which exhibits many of the same non-idealized behaviors observed in an actual system. All the experiments presented here were compared with real systems in the lab environment, and their accuracy has been verified. These virtual labs include experiments on **speed and position control** of dc motors followed by two **controller design projects**, the first involving control of a simple robotic system and the last one investigating the response of an active suspension system. In this chapter, the focus on dc motors in these experiments is intentional, because of their relative simplicity and wide usage in numerous industrial applications.

The main objectives of this chapter are:

1. To provide an in-depth description of dc motor speed response, speed control, and position control concepts.
2. To provide preliminary instruction on how to identify the parameters of a system.
3. To show how different parameters and nonlinear effects such as friction and saturation affect the response of the motor.
4. To give a better feel for controller design through realistic examples.
5. To get started using the SIMLab and Virtual Lab.
6. To gain practical knowledge of the Quarter Car Sim software.

Before starting the lab, you must have completed the relevant background preparation in Chapters 4 and 5.

6-2 DESCRIPTION OF THE VIRTUAL EXPERIMENTAL SYSTEM

The experiments that you will perform are intended to give you hands-on (virtually!) experience in analyzing the system components and experimenting with various feedback control schemes. To study the speed and position response of a dc motor, a typical experimental test bed is shown in Fig. 6-1.

The setup components are as follows:

- A dc motor with a position sensor (usually an encoder with incremental rotation measurement) or a speed sensor (normally a tachometer or a differencing operation performed on encoder readings)
- A power supply and amplifier to power the motor
- Interface cards to monitor the sensor and provide a command voltage to the amplifier input and a PC running MATLAB and Simulink to control the system and to record the response (alternatively, the controller may be composed of an analog circuit system)

A simple speed control system is composed of a sensor to measure motor shaft speed and an amplifier with gain K (proportional control) in the configuration shown in Fig. 6-1. The block diagram of the system is also shown in Fig. 6-2.

To control the position of the motor shaft, the simplest strategy is to use a proportional controller with gain K. The block diagram of the closed-loop system is shown in Fig. 6-3. The system is composed of an angular position sensor (usually an encoder or a potentiometer for position applications). Note that for simplicity the input voltage can be scaled to a position input $T_{in}(s)$ so that the input and output have the same units and scale.

The components are described in the next sections.

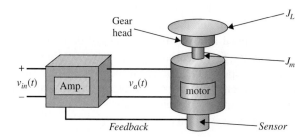

Figure 6-1 Feedback control of an armature-controlled dc motor with load inertia.

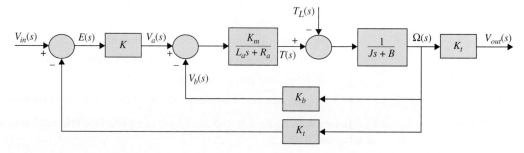

Figure 6-2 Block diagram of a speed-control, armature-controlled dc motor.

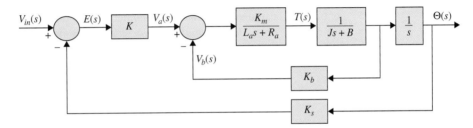

Figure 6-3 Block diagram of a position-control, armature-controlled dc motor.

6-2-1 Motor

Recall from Chapter 5 that for the armature-controlled dc motor shown in Fig. 5-24, the system parameters include

- R_a = armature resistance, ohm
- L_a = armature inductance, henry
- v_a = applied armature voltage, volt
- v_b = back emf, volt
- θ = angular displacement of the motor shaft, radian
- T = torque developed by the motor, N-m
- J_L = moment of inertia of the load, kg-m^2
- T_L = any external load torque considered as a disturbance, N-m
- J_m = moment of inertia of the motor (motor shaft), kg-m^2
- J = equivalent moment of inertia of the motor and load connected to the motor-shaft, $J = J_L/n^2 + J_m$, kg-m^2 (refer to Chapters 4 and 5 for more details)
- n = gear ratio
- B = equivalent viscous-friction coefficient of the motor and load referred to the motor shaft, N-m/rad/sec (in the presence of gear ratio, B must be scaled by n; refer to Chapter 4 for more details)
- K_t = speed sensor (usually a tachometer) gain

The motor used in this experiment is a permanent magnet dc motor with the following parameters (as given by the manufacturer):

- K_m = Motor (torque) constant 0.10 Nm/A
- K_b = Speed Constant 0.10 V/rad/sec
- R_a = Armature resistance 1.35 ohm
- L_a = Armature inductance 0.56 mH
- J_m = Armature moment of inertia 0.0019 kg-m^2
- τ_m = Motor mechanical time constant 2.3172 E-005 sec

A reduction gear head may be attached to the output disk of the motor shaft. If the motor shaft's angular rotation is considered the output, the gear head will scale the inertia of the load by $1/n^2$ in the system model, where n is the gear ratio.

6-2-2 Position Sensor or Speed Sensor

For position-control applications, an incremental encoder or a potentiometer may be attached directly to the motor shaft to measure the rotation of the armature. In speed control, it is customary to connect a tachometer to the motor shaft. *Sensor-shaft inertia and damping are normally too small to be included in the system model.* The output from each sensor is proportional to the variable it is measuring. We will assume a proportionality gain of 1; that is, $K_t = 1$ *(speed control)*, and $K_s = 1$ *(position control)*.

6-2-3 Power Amplifier

The purpose of the amplifier is to increase the current capacity of the voltage signal from the analog output interface card. The output current from the interface should normally be limited, whereas the motor can draw many times this current. The details of the amplifier design are somewhat complex and will not be discussed here. But we should note two important points regarding the amplifier:

1. The maximum voltage that can be output by the amplifier is effectively limited to 20 V.
2. The maximum current that the amplifier can provide to the motor is limited to 8 A. Therefore,
   Amp gain 2 V/V
   Amplifier input saturation limits ± 10 V
   Current saturation limits ± 4 A

6-2-4 Interface

In a real-world scenario, interfacing is an important issue. You would be required to attach all the experimental components and to connect the motor sensor and the amplifier to a computer equipped with MATLAB and Simulink (or some other real-time interface software). Simulink would then provide a voltage output function that would be passed on to the amplifier via a digital-to-analog (D/A) interface card. The sensor output would also have to go through an analog-to-digital (A/D) card to reach the computer. Alternatively, you could avoid using a computer and an A/D or D/A card by using an analog circuit for control.

▶ 6-3 DESCRIPTION OF SIMLAB AND VIRTUAL LAB SOFTWARE

As shown in Fig. 6-4, SIMLab and Virtual Lab are series of MATLAB and Simulink files within the **A**utomatic **C**ontrol **Sys**tems (**ACSYS**) applet that makes up an educational tool for students learning about dc motors and control systems. SIMLab was created to allow students to understand the basic simulation model of a dc motor. The parameters of the motor can be adjusted to see how they affect the system. The Virtual Lab was designed to exhibit some of the key behaviors of real dc motor systems. Real motors have issues such as gear backlash and saturation, which may cause the motor response to deviate from expected behavior. Users should be able to cope with these problems. The motor parameters cannot be modified in the Virtual Lab because, in a realistic scenario, a motor may not be modified but must be replaced by a new one!

In both the SIMLab and the Virtual Lab, there are five experiments. In the first two experiments, feedback speed control and position control are explored. Open-loop step response of the motor appears in the third experiment. In the fourth experiment, the frequency response of the open-loop system can be examined by applying a sinusoidal input. A controller design project is the last experiment.

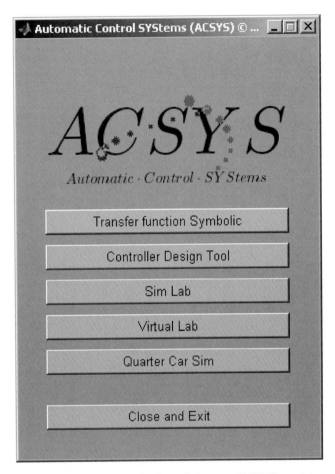

Figure 6-4 The Automatic Control Systems (**ACSYS**) applet.

To launch the **ACSYS** applet, navigate to the appropriate directory in the MATLAB command window, and type *Acsys* at the command prompt. The SIMLab or Virtual Lab experiment windows can be called from the **ACSYS** applet by clicking on the appropriate button. Alternatively, you may directly call SIMLab or Virtual Lab from the MATLAB command window by navigating to the VirtualLab subdirectory and typing *simlab* or *virtuallab*, respectively.

When SIMLab or Virtual Lab is opened, the experiment control window will be displayed. The Experiment menu can be used to switch between different control experiments, as in Fig. 6-5. The grey control panel on the left contains the control buttons for the experiment. Every experiment has a button to enter model parameters, a field to enter simulation time, and additional experiment-specific plot controls. The plots or animations that the experiment supports appear in the display panel on the right. Fig. 6-6 shows a typical experiment control window in SIMLab. The time-response plot is displayed, and plot control buttons appear below the axes. Zoom control buttons allow you to view the response at greater detail, and the data cursor gives precise point values for graphed data. SIMLab allows you to display the motor transfer functions in various formats and to access other custom tools from the **SIMLab Tools** dropdown menu. For step-by-step instructions on using the experiment window, click on the Help Me button in the menu bar. The standard

Figure 6-5 The Experiment menu for SIMLab or Virtual Lab.

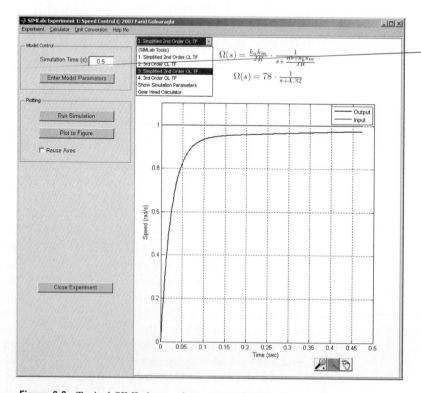

Figure 6-6 Typical SIMLab experiment control window.

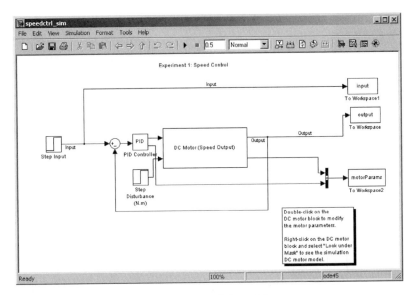

Figure 6-7 SIMLab Speed Control Simulink model.

Microsoft Windows calculator and a unit conversion tool can be accessed from the top menu.

The model parameters must be set first in any experiment. By selecting the Enter Model Parameters button, a Simulink (.mdl) window containing the model for the experiment will be launched. The model, shown in Fig. 6-7, contains a simple closed-loop system using PID speed control, with a reference step input and multiple outputs.

All the simulation parameters for the Simulink model are pre-set. Selecting Simulation from the Simulink menu and next choosing Configuration Parameters allows access to these settings, shown in Fig. 6-8. The Start Time and Stop Time settings in the Solver

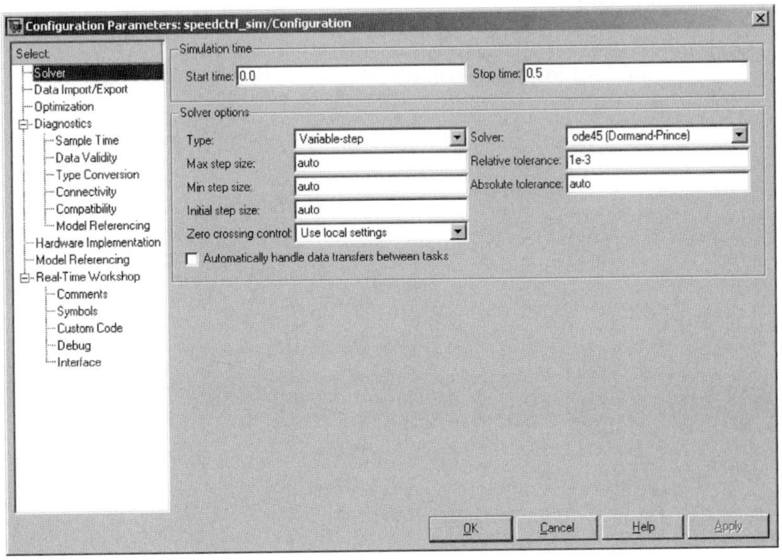

Figure 6-8 Configuration Parameters: Solver properties.

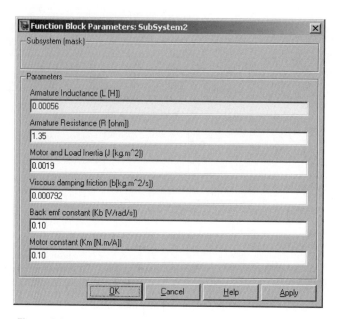

Figure 6-9 Adjustable parameters for the SIMLab motor blocks.

options are most important as far as SIMLab and Virtual Lab examples are concerned, and they can be manipulated in Configuration Parameters, or on the left panel in the SIMLab/Virtual Lab interface. They allow you to modify the simulation running time. Other options in Configuration Parameters should not be modified, as they may cause errors in the SIMLab and Virtual Lab software.

When the Simulink model is opened, double-click on the appropriate model block to modify model parameters such as the PID values. For SIMLab, double-clicking on the motor block brings up a window containing a list of adjustable motor parameters (see Fig. 6-9). All motor parameters, such as the resistance, back-emf constant, load inertia, and damping coefficient, may be modified. Right-clicking on a SIMLab motor block and selecting Look under Mask makes the dc motor model available. However, the Virtual Lab motor blocks are completely opaque to the user since they model actual dc motors. One other feature that SIMLab has, which Virtual Lab does not, is a torque-disturbance input into the motor. This can be used to investigate the stall torque and the effect of an integral controller.

To run the simulation, close the Simulink model and click the button labeled Run Simulation. For more detail, click the zoom button and select the area of the time-response plot to view it closer. The data cursor button allows the graphed values to be displayed as the cursor dot is moved around on the graph using the mouse or arrow keys. The Print to Figure button allows the current response plot to be sent to a separate MATLAB figure. Selecting the Reuse Axes checkbox prints all plots to the same set of axes in an external figure, which is useful for comparing the system response after changing a parameter in the Simulink model. This figure can also be saved as a .fig or image file for future reference and analysis. Again, in the Virtual Lab, you cannot change the system parameters, but PID values are available for modification.

Some of the experiments have additional features, such as animation and calculation tools. These are discussed in the following sections. Selecting Close Experiment in the control window exits the program.

6-4 SIMULATION AND VIRTUAL EXPERIMENTS

It is desired to design and test a controller offline by evaluating the system performance in the safety of the simulation environment. The simulation model can be based on available system parameters, or they may be identified experimentally. Because most of the system parameters are available (see motor specifications in Section 6-2-1), it will be useful to build a model using these values and to simulate the dynamic response for a step input. The response of the actual system (in this case, the virtual system) to the same test input will then verify the validity of the model. Should the actual response to the test input be significantly different from the predicted response, certain model parameter values would have to be revised or the model structure refined to reflect more closely the observed system behavior. Once satisfactory model performance has been achieved, various control schemes can be implemented.

In this chapter, SIMLab represents the simulation model with adjustable parameters, and Virtual Lab represents the actual (virtual) system. Once the model of the Virtual Lab system is identified and confirmed, the controller that was originally designed using SIMLab should be tested on the Virtual Lab model.

6-4-1 Open-Loop Speed

The first step is to model the motor. Using the parameter values in Section 6-2-1 for the model of the motor in Fig. 5-24, simulate the open-loop velocity response of the motor to a step voltage applied to the armature. Start up SIMLab, select 3: Open Loop Speed from the Experiment menu, and perform the following tests:

1. Apply step inputs of +5 V, +15 V, and −10 V. Note that the steady-state speed should be approximately the applied armature voltage divided by K_b as in Eq. (5-118)[1] (try dc motor alone with no gear head or load applied in this case).

2. Study the effect of viscous friction on the steady-state motor speed. First set $B = 0$ in the Simulink motor parameter window. Then gradually increase its value and check the speed response.

3. Repeat Step 2 and connect the gear head with a gear ratio of 5.2:1, using additional load inertia at the output shaft of the gear head of 0.05 kg-m² (requires modification of J in the Simulink motor parameters). Try using the gear head calculator in the SIMLab Tools dropdown menu.

4. Determine the viscous friction required at output shaft to reduce the motor speed by 50% from the speed it would rotate at if there were no viscous friction.

5. Derive and calculate the disturbance torque steady-state gain. Introduce an appropriate step-disturbance input T_L and study its effect on the system in Step 3.

6. Assuming that you do not know the overall inertia J for the system in Step 3, can you use the speed-response plot to estimate its value? If so, confirm the values of motor and load inertia. How about the viscous-damping coefficient? Can you use the time response to find other system parameters?

In this experiment, we use the open-loop model represented in Experiment 3: Open Loop Speed. The Simulink system model is shown in Fig. 6-10, representing a simple open-loop model with a motor speed output.

[1] $\omega(t) = \dfrac{A}{K_b}\left(1 - e^{-t/\tau_m}\right)$ (5-118)

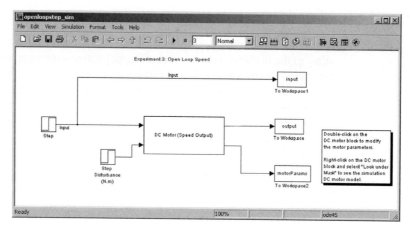

Figure 6-10 SIMLab open-loop speed response of dc motor experiment.

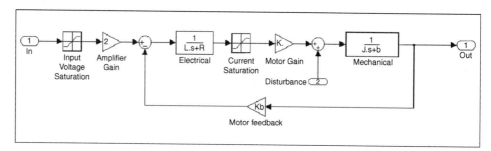

Figure 6-11 DC motor model including voltage and current saturation.

In a realistic scenario, the motor is connected to an amplifier that can output a voltage only in a certain range. Outside of this range, the voltage saturates. The current within the motor may also be saturated. To create these effects in software, right-click the dc motor block and select Look under Mask to obtain the motor model shown in Fig. 6-11. Double-click both the voltage and current blocks and adjust their values (default values of ± 10 volts and ± 4 amps have already been set). If you do not wish to include saturation, you can set the limits very large (or delete these blocks altogether). Run the above experiments again and compare the results.

Assuming a small electric-time constant, we may model the dc motor as a first-order system. As a result, the motor inertia and the viscous-damping friction could be calculated with measurements of the mechanical-time constant using different input magnitudes. For a unit-step input, the open-loop speed response is shown in Fig. 6-12. After measuring the mechanical-time constant of the system τ_m, you can find the inertia J, assuming all other parameters are known. Recall that, for a first-order system, the time constant is the time to reach $(1-e^{-1}) \times 100$, or 63.2% of the final value for a step input [verify using Eq. (5-118) or (5-119)]. A typical open-loop speed response is shown in Fig. 6-12. The steady-state velocity and the time constant τ_m can be found from the time-response plot by using the cursor.

In SIMLab, the disturbance torque default value is set to zero. To change an input value, simply change its final value.

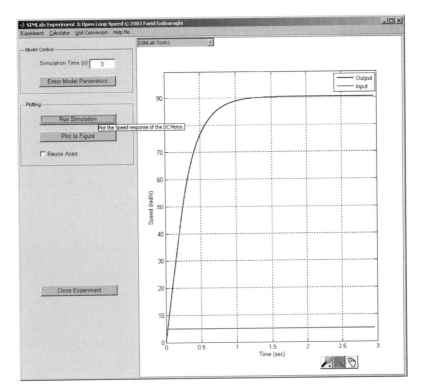

Figure 6-12 Speed response of the open-loop system (SIMLab).

Now that you have gained insight into the motor speed response, it is time to apply your knowledge to test the virtual experiment. Here you have no access to the system parameter values. Use the Virtual Lab to test the following:

7. Apply step inputs of +5 V, +15 V, and −10 V. How different are the results from Step 1?

8. From the transient and steady-state responses, identify the system model as closely as possible.

Recall that the motor and amplifier have built-in nonlinear effects due to noise, friction, and saturation. So, in Step 8, your model may vary for different input values. Distorted values may be obtained if the input to the motor is excessive and saturates. Caution must be taken to ensure that the motor input is low enough such that this does not happen. Use the mechanical time constant and final value of the response in this case to confirm the system parameters defined in Section 6-2-1. These parameters are needed to conduct the speed- and position-control tasks. Fig. 6-13 shows the Virtual Lab motor speed response to a small step input. The friction effect is observed when the motor starts. The noise at steady state may also be observed. For higher input magnitudes, the response will saturate.

6-4-2 Open-Loop Sine Input

The objective of using open-loop sine input is to investigate the frequency response of the motor using both SIMLab and Virtual Lab.

9. For both SIMLab and Virtual Lab, apply a sine wave with a frequency of 1 rad/sec and amplitude of 1 V to the amplifier input, and record both the motor velocity and

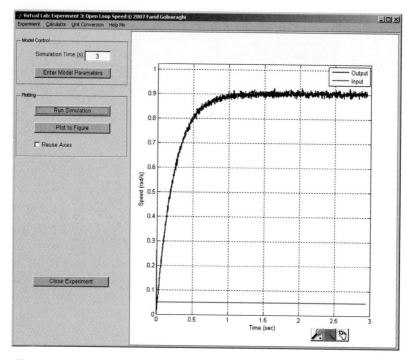

Figure 6-13 Speed response of the open-loop system (Virtual Lab).

sine wave input signals. Repeat this experiment for frequencies of 0.2, 0.5, 2.0, 5.0, 10.0, and 50.0 rad/sec (keeping the sine wave amplitude at 1 V).

10. Change the input magnitude to 20 V and repeat Step 9.

Open Experiment 4: Open Loop Sine Input from the SIMLab or Virtual Lab Experiment menu. The input and disturbance blocks and the motor parameters are adjustable in the SIMLab model. For the Virtual Lab version, the amplitude should be low to avoid amplifier or armature current saturation. The Simulink model is shown in Fig. 6-14. Double-click on the Sine Wave block to modify the properties of the input wave. Amplitude of 1 is a low enough value to avoid saturation in this example. In the SIMLab version, the saturation values are adjustable to allow you practice with their effect. The SIMLab response for sine input with

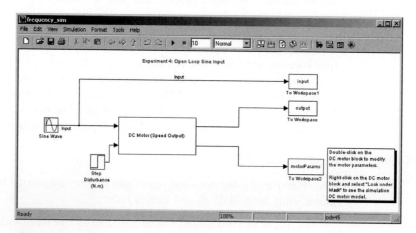

Figure 6-14 Experiment 4: Simulink model.

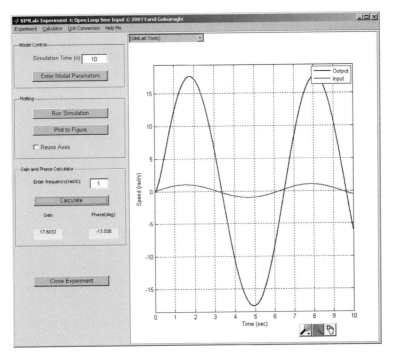

Figure 6-15 SIMLab time response and gain and phase calculation for input = sin(*t*).

magnitude and frequency of 1 is shown in Fig. 6-15. You may also try adding dead zone and backlash to your motor block to test their effects (these functions are available in the Simulink Library Browser, briefly discussed at the end of Section 6-5). For a sine input of magnitude 20 V, the Virtual Lab system exhibits saturation as shown in Fig. 6-16.

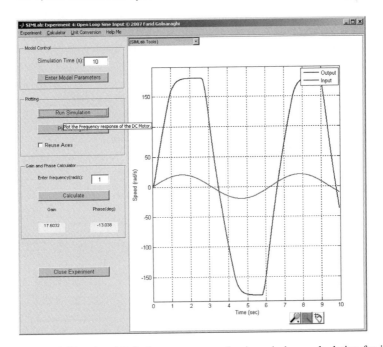

Figure 6-16 Virtual Lab time response and gain and phase calculation for input = 20 sin(*t*).

The frequency of the sine wave will dictate the gain and phase of the response curve. There is a Gain and Phase Calculator in the Experiment 4 control window. To measure the magnitude and phase of the steady-state response, enter a frequency of 1 rad/sec in the edit block. Entering the input frequency and clicking on Calculate displays the gain and phase of the system (see Fig. 6-15). Using the Gain and Phase Calculator, you can record the gain and phase of the response. Repeat with other input frequencies, and discuss any trends.

6-4-3 Speed Control

Having simulated the open-loop motor characteristics in previous sections, we can now extend the model to include velocity feedback from the motor and use a proportional controller. Assume that the motor velocity is measured using a sensor that provides 1 V/rad/sec. The block diagram that you should be modeling is shown in Fig. 6-2. For proportional gains of 1, 10, and 100, perform the following tests using SIMLab:

11. Apply step inputs of +5 V, +15 V, and −10 V (try dc motor alone; no gear head or load applied in this case).

12. Repeat Step 11 using additional load inertia at the output shaft of the gear head (gear ratio 5.2:1) of 0.05 kg-m^2 (requires adjustment of the J value in SIMLab motor parameter block).

13. Apply the same viscous friction to the output shaft as obtained in Step 4 in Section 6-4-1, and observe the effect of the closed-loop control. By how much does the speed change?

14. Repeat Step 5 in Section 6-4-1, and compare the results.

Open Experiment 1: Speed Control from the SIMLab menu window. A screen similar to Fig. 6-5 will be displayed. Next, select the Enter Model Parameters button to get the system Simulink model, as shown in Fig. 6-17. This figure is a simple PID speed-control model. Double-clicking on the PID block displays the editable PID values. The values of

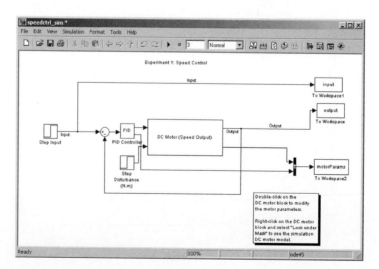

Figure 6-17 Experiment 1: Simulink model.

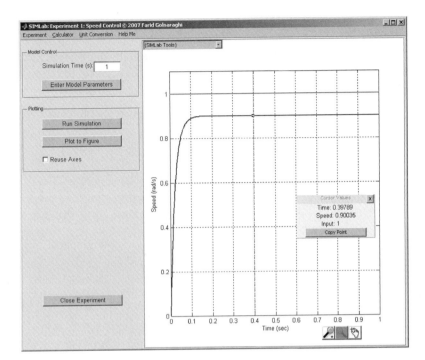

Figure 6-18 Speed-control response in the SIMLab control window.

the step-input and the disturbance-torque blocks may also be adjusted. The disturbance-torque default value is set to zero. To change an input value, simply change the number in the final value field.

Increasing the proportional gain in the PID block will decrease the rise time. For an unsaturated model, the SIMLab version of this experiment could exhibit extremely fast rise times at very high proportional gains, because the dc motor can utilize unlimited voltage and current levels. To create this effect in software, right-click the dc motor block, and select Look under Mask to obtain the motor model similar to Fig. 6-11. Double-click both the voltage and current blocks and adjust their values to very large (or delete their blocks). Recall from Section 6-4-1 that the default saturation limits are ± 10 V and ± 4 A, respectively. Fig. 6-18 displays a typical speed response from the SIMLab.

For a given input to change the proportional gain values, enter the following sets of PID values and print all three plots on the same figure (use the Print to Figure button and the Reuse Axes checkbox in the experiment control panel).

$$P = 1 \qquad I = 0 \qquad D = 0$$
$$P = 10 \qquad I = 0 \qquad D = 0$$
$$P = 100 \qquad I = 0 \qquad D = 0$$

The input units used in these simulations are specified in volts, while the feedback units at the motor are in radians per second. This was done intentionally to illustrate the scaling (or conversion) that is performed by the sensor. Had the velocity been specified in volts per radians per second, a different response would have been obtained. To check the effect of the velocity feedback scaling, repeat the preceding experiments using a proportional gain of 10, but assume that the velocity feedback signal is a voltage generated by a sensor with a conversion factor of 0.2 V/rad/sec. (Note: in commonly used industry standards, the tachometer gain is in volts per RPM.)

Next, for the Virtual Lab, test the following:

15. Apply step inputs of +5 V, +15 V, and −10 V. How different are the results from the SIMLab?

You may again confirm the system parameters obtained in Section 6-4-1.

6-4-4 Position Control

Next, investigate the closed-loop position response; choose Experiment 2: Position Control from the Experiment menu. For proportional gains of 1, 10, and 100 (requires modification of PID block parameters), perform the following tests using SIMLab:

16. For the motor alone, apply a 160° step input. How large is the error when the system reaches steady state?
17. Apply a step disturbance torque (−0.1) and repeat Step 16. Estimate the disturbance-torque gain based on your observations.
18. Eliminate the disturbance torque and repeat Step 16, using additional load inertia at the output shaft of 0.05 kg-m² and the gear ratio 5.2:1 (requires modification of J_m and B in the motor parameters). What can be said about the effect of the increased load on the system performance?
19. Using the disturbance torque in Step 17, examine the effect of integral control by modifying the Simulink PID block. Choose several different integral gain values, and compare the time response for a constant proportional gain. Select the Reuse Axes checkbox, and plot the different simulation results in an external figure for comparison.
20. How does an increase in J affect the system with a PI controller? Compare the transient and steady-state response.
21. Examine the effect of voltage and current saturation blocks (requires modification of the saturation blocks in the motor model).
22. Design a PI controller that will give a 30% overshoot and a rise time of 0.1 seconds. What is the maximum step input amplitude that will meet these calculated requirements (i.e., not cause the amplifier to saturate), given the default current and voltage saturation limits of ±4 A and ±10 V, respectively.
23. In all previous cases, comment on the validity of Eq. (5-126).[2]

Open Experiment 2: Position Control from the SIMLab Experiment menu. A screen similar to Fig. 6-5 will be displayed. Next, select Enter Model Parameters to get the system Simulink model, as shown in Fig. 6-19. This model represents a simple PID position-control system. Double-clicking on the PID block allows you to edit the PID gain values. The Deg to Rad and Rad to Deg gain blocks convert the input and the output

$$2 \; \frac{\Theta(s)}{\Theta_{in}(s)} = \frac{\frac{KK_m K_s}{R_a J}}{s^2 + \left(\frac{R_a B + K_m K_b}{R_a J_m}\right)s + \frac{KK_m K_s}{R_a J_m}} \quad (5\text{-}126)$$

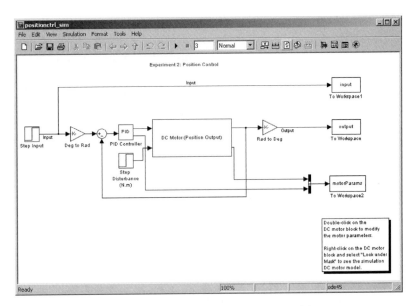

Figure 6-19 Experiment 2: Position Control Simulink model.

such that the user enters inputs and receives outputs in degrees only. The values of the step-input and the disturbance-torque blocks are also adjustable. The disturbance-torque default value is set to zero. To change an input value, double-click on the relevant block and change the number in the final value field. Fig. 6-20 displays a typical position response from the SIMLab.

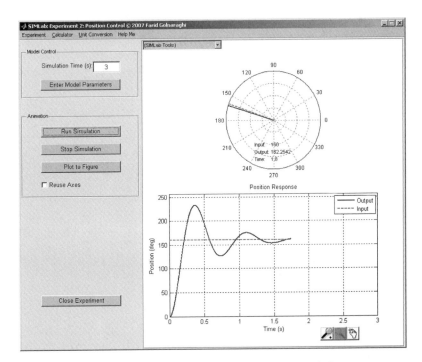

Figure 6-20 Position response in the Experiment 2 control window.

The position time response is also animated when the simulation is run. This is a useful tool that gives the user a physical sense of how a real motor turns. The time, input-angle, and output-angle values are displayed on the animation field, as shown in Fig. 6-20.

The nonlinearities due to voltage and current limits cause the time response to saturate at a high enough proportional gain. The maximum speed and acceleration of the dc motor are dictated by the voltage and current saturation limits.

24. For proportional gains of 1, 10, and 100 (requires modification of PID block parameters), repeat Steps 16 and 19 using Virtual Lab.

▶ 6-5 DESIGN PROJECT 1—ROBOTIC ARM

The primary goal of this section is to help you gain experience in applying your control knowledge to a practical problem. You are encouraged to apply the methods that you have learned throughout this book, particularly in Chapter 5 and later on in Chapter 9, to design a controller for your system. The animation tools provided make this experience more realistic. The project may be more exciting if it is conducted by teams on a competitive basis. The SIMLab and Virtual Lab software are designed to provide enough flexibility to test various scenarios. The SIMLab, in particular, allows introduction of a disturbance function or changes of the system parameters if necessary.

Description of the Project: Consider the system in Fig. 6-21. The system is composed of the dc motor used throughout this chapter. We connect a rigid beam to the motor shaft to create a simple robotic system conducting a pick-and-place operation. A solid disk is attached to the end of the beam through a magnetic device (e.g., a solenoid). If the magnet is on, the disk will stick to the beam, and when the magnet is turned off, the disk is released.

Objective: The objective is to drop the disk into a hole as fast as possible. The hole is 1 in. (25.4 mm) below the disk (see Fig. 6-22).

Design Criteria: The arm is required to move in only one direction from the initial position. The hole location may be anywhere within an angular range of 20° to 180° from the initial position. The arm may not overshoot the desired position by more than 5°.

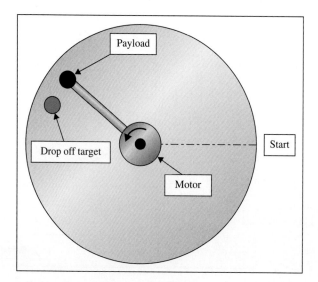

Figure 6-21 Control of a simple robotic arm and a payload.

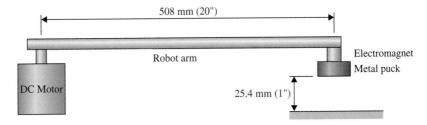

Figure 6-22 Side view of the robot arm.

A tolerance of ±2% is acceptable (settling time). These criteria may easily be altered to create a new scenario.

The objective may be met by looking at the settling time as a key design criterion. However, you may make the design challenge more interesting by introducing other design constraints such as the percent overshoot and rise time. In SIMLab, you can also introduce a disturbance torque to alter the final value properties of the system. The Virtual Lab system contains nonlinear effects that make the controller design more challenging. You may try to confirm the system model parameters first, from earlier experiments. It is highly recommended that you do the design project only after fully appreciating the earlier experiments in this chapter and after understanding Chapter 5. Have fun!

This experiment is similar to the position-control experiment in some respects. The idea of this experiment is to get a metal object attached to a robot arm by an electromagnet from position 0° to a specified angular position with a specified overshoot and minimum overall time.

Select Experiment 5: Control System Design from the SIMLab Experiment menu. A screen similar to Fig. 6-5 will be displayed. Next, select Enter Model Parameters to get the system Simulink model, as shown in Fig. 6-23. As in Section 6-4-1, this figure represents a

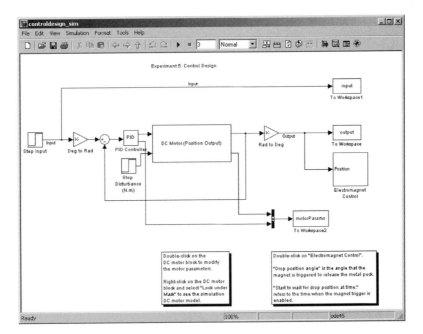

Figure 6-23 Experiment 5: Simulink model.

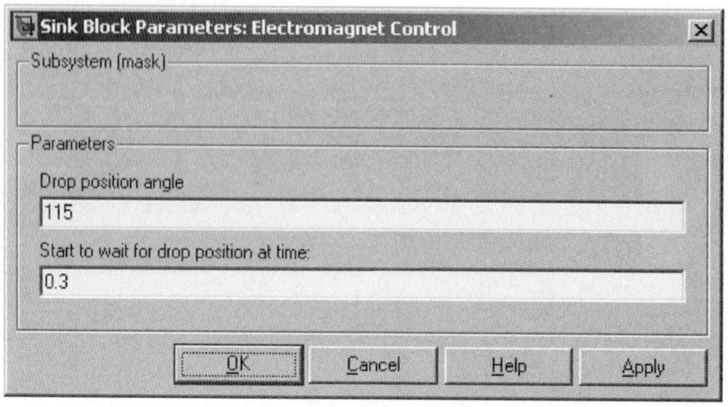

Figure 6-24 Parameter window for the electromagnet control.

simple PID position-control model with the same functionalities. The added feature in this model is the electromagnetic control. By double-clicking the Electromagnet Control block, a parameter window pops up, as in Fig. 6-24, which allows the user to adjust the drop-off payload location and the time delay (in seconds) to turn the magnet off after reaching the target. This feature is particularly useful if the response overshoots and passes through the target more than once. So, in Fig. 6-24, the "Drop position angle" is the angle where the electromagnet turns off, dropping the payload. "Start to wait for drop position at time" refers to the time where the position trigger starts to wait for the position specified by "Drop position angle."

An important note to remember is that in the Virtual Lab the electromagnet will never drop the object exactly where it is specified. Because any electromagnet has residual magnetism even after the current stops flowing, the magnet holds on for a short time after the trigger is tripped. A time response of the system for proportional gain and derivative gain of 3 and 0.05, respectively, is shown in Fig. 6-25.

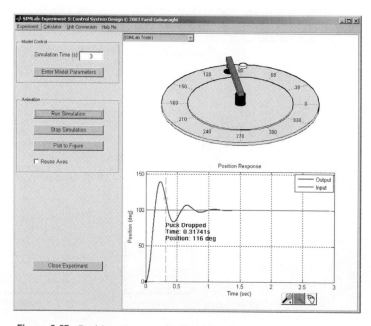

Figure 6-25 Position response for Experiment 5.

The model response is also animated. This feature makes the problem more realistic. The puck has overshot the hole in this case. The drop angle and drop time are displayed on the time-response plot. Note that, in this case, the magnet drop-off takes place prematurely. As a result, the payload has been released earlier and is not on target! In SIMLab, it is possible to change the dimensions of the experiment setup. Choose Modify Puck Drop Setup from the SIMLab Tools menu to adjust the height of the drop and the length of the arm, and change your controller design accordingly.

▶ 6-6 DESIGN PROJECT 2—QUARTER-CAR MODEL

6-6-1 Introduction to the Quarter-Car Model

After studying position and velocity control of the dc motor in the preceding sections of this chapter, you are now well acquainted with the use of the **ACSYS** tools and Simulink and their applications in the study of controls.

In this section a simple one degree of freedom quarter-car model, as shown in Fig. 6-26 (c) is presented for studying base excitation response (i.e., road transmitted effects). The objective here is to control the resulting displacement or acceleration of the mass of the system—which is reflective of the chassis of the car. This study follows the modeling exercise that was discussed in Example 4-11-3.

As discussed in Chapter 4, there are various representations of a quarter-car system, as illustrated in Fig. 6-26, where a two degree of freedom (2-DoF) system in Fig. 6-26(b) takes into account the damping and elastic properties of the tire, shown in Fig. 6-26(a). However, for simplicity, it is customary to ignore tire dynamics and assume a 1-DoF model as shown in Fig. 6-26(c). Hence, for the duration of this design project, we will assume a rigid wheel.

We further assume hereafter the following parameter values for the system illustrated in Fig. 6-26(c):

m	Effective 1/4 car mass	10 kg
k	Effective stiffness	2,7135 N/m
c	Effective damping	0.9135 N-m/s^{-1}
$x(t)$	Absolute displacement of the mass m	m
$y(t)$	Absolute displacement of the base	m
$z(t)$	Relative displacement $(x(t)-y(t))$	m
$a(t)$	Base acceleration $\ddot{y}(t)$	m/s^2

Recall from Eq. (4-324) that the **open loop** response of the system to a base acceleration $a(t)$ has a transfer function:

$$\frac{Z(s)}{A(s)} = \frac{-1}{s^2 + 2\zeta\omega_n s + \omega_n^2} \qquad (6\text{-}1)$$

where the base acceleration $A(s)$ is the input and relative displacement, $Z(s)$, is the output.

Let us next consider the active control suspension system and use the same dc motor described in Section 6-2 used in conjunction with a rack as shown in Fig. 6-27. In this case, T is the torque produced by the motor with shaft position θ, and r is the radius of the motor drive gear.

358 ▶ Chapter 6. The Control Lab

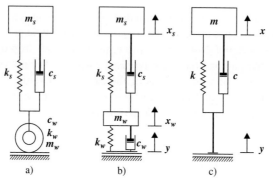

Figure 6-26 Quarter-car model realization. (a) Quarter car. (b) Two degrees of freedom. (c) One degree of freedom.

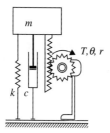

Figure 6-27 Active control of the 1-DOF model via a dc motor and rack.

Recall from Example 4-11-3 that the block diagram in Fig. 6-28 represents the open loop system with no base excitation, where $J = mr^2 + J_m$, $B = cr^2 + B_m$, and $K = kr^2$. Using superposition, this system is rearranged to the following form:

$$Z(s) = \frac{\frac{K_m r}{R_a}}{\left(\frac{L_a}{R_a}s + 1\right)(Js^2 + Bs + K) + \frac{K_m K_b}{R_a}s} V_a(s)$$

$$- \frac{\left(\frac{L_a}{R_a}s + 1\right)r}{\left(\frac{L_a}{R_a}s + 1\right)(Js^2 + Bs + K) + \frac{K_m K_b}{R_a}s} mrA(s)$$

(6-2)

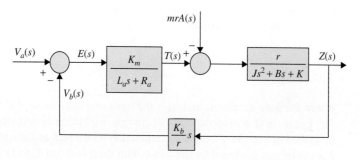

Figure 6-28 Block diagram of an armature-controlled dc motor.

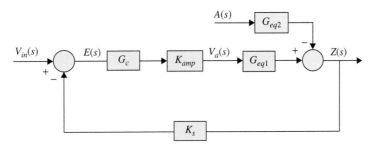

Figure 6-29 Block diagram of a position-control, armature-controlled dc motor.

Again, as in previous sections, we assume the motor electric-time constant is insignificant relative to the mechanical-time constant, Eq. 6-2 and is reduced to

$$Z(s) = \frac{\frac{K_m r}{R_a J}}{s^2 + \frac{BR_a + K_m K_b}{R_a J}s + \frac{K}{J}} V_a(s) - \frac{\frac{mr^2}{J}}{s^2 + \frac{BR_a + K_m K_b}{R_a J}s + \frac{K}{J}} A(s) \quad (6\text{-}3)$$

For simplicity, Eq. (6-3) is written as $Z(s) = G_{eq1}(s)V_a(s) - G_{eq2}(s)A(s)$. The position control block diagram in Fig. 6-29 illustrates the feedback of relative position, $Z(s)$, where K_s is the sensor gain, with units V/m. In this application, the sensor is a linear variable differential transformer (LVDT), which transforms the displacement $z(t)$ between the base $y(t)$ and mass $x(t)$ to voltage. The goal of position control in this scenario is not to create offset as in the previous lab, where a robot arm is given the command signal to displace a metal puck, but rather to reject the so-called disturbance input. Hence the command voltage, or set point, $V_{in}(s) = 0$ V.

Setting $E(s) = 0 - K_s Z(s)$, the block diagram represented in Fig. 6-29 can be reduced to an input–output relation of $\ddot{Y}(s)$ and $Z(s)$, where the simplified **closed-loop** system is represented in Fig. 6-30:

$$\frac{Z(s)}{-A(s)} = \frac{\frac{mr^2}{J}}{s^2 + \frac{BR_a + K_m K_b}{R_a J}s + \frac{K}{J} + \frac{K_m K_{amp} K_s r}{R_a J} G_c} \quad (6\text{-}4)$$

6-6-2 Closed-Loop Acceleration Control

Relative position control is a familiar way to introduce the control of the quarter-car model; however, the vehicle operator cannot really sense displacement except perhaps by comparing their height to fixed objects. If you have ever driven a car too quickly over

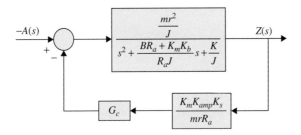

Figure 6-30 Simplified block diagram of the quarter-car dc motor position control.

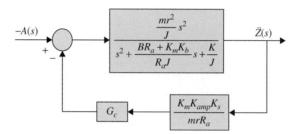

Figure 6-31 Simplified block diagram of the quarter-car relative acceleration control.

a sharp rise and fall of the road, you can feel the effect of acceleration in your stomach. Thus, it is more desirable to control the acceleration because the ultimate goal of the suspension system is to improve ride and driving performance. The block diagram in Fig. 6-30 can be modified to control the relative acceleration of the system.

The second derivative of the forward-path transfer function yields the acceleration control system in Fig. 6-31. The input–output relation is as follows:

$$\frac{\ddot{Z}(s)}{-A(s)} = \frac{\frac{mr^2}{J}s^2}{s^2 + \frac{BR_a + K_m K_b}{R_a J}s + \frac{K}{J} + \frac{K_m K_{amp} K_s r}{R_a J} G_c s^2} \qquad (6\text{-}5)$$

As described above, the position control system used an LVDT to provide the feedback. Just as the LVDT measures relative displacement, two accelerometers can be used to measure both $\ddot{x}(t)$ and $\ddot{y}(t)$, where $\ddot{z}(t) = \ddot{x}(t) - \ddot{y}(t)$. Thus, to control the relative acceleration of the mass, two accelerometers with gain K_s are fixed to the mass and base to provide the relative acceleration feedback.

It is also of interest to control the absolute acceleration of the mass m. The closed-loop system is determined by reconfiguring Fig. 6-29 to yield absolute acceleration from the relation $\ddot{X}(s) = \ddot{Z}(s) + A(s)$ where $\ddot{Z}(s)$ and $\ddot{X}(s)$ are the Laplace transforms of $\ddot{z}(t)$ and $\ddot{x}(t)$, respectively.

The block diagram in Fig. 6-32(a) is simplified to the closed-loop form in Fig. 6-32(b) to obtain the input–output relation

$$\frac{\ddot{X}(s)}{A(s)} = \frac{\left(1 - \frac{mr^2}{J}\right)s^2 + \frac{BR_a + K_m K_b}{R_a J}s + \frac{K}{J}}{s^2 + \frac{BR_a + K_m K_b}{R_a J}s + \frac{K}{J} + \frac{K_m K_{amp} K_s r}{R_a J} G_c s^2} \qquad (6\text{-}6)$$

Note that, in the case of the systems represented by Eq. (6-5) and Eq. (6-6), implementing a compensator will lead to a higher-order transfer function. In this case, designing in the time domain may require that the systems be approximated by lower-order systems, as demonstrated in Chapter 5. Also see Chapter 9, where the controller design topics are studied in more detail.

6-6-3 Description of Quarter Car Modeling Tool

The Quarter Car Modeling Tool allows the students to implement the familiar dc motor and amplifier described in Section 6-2-1 and conduct experiments to observe its effect on a new, slightly more complex system. Designing a controller for a vehicle suspension system requires studying its performance under the influence of different inputs, such as driving

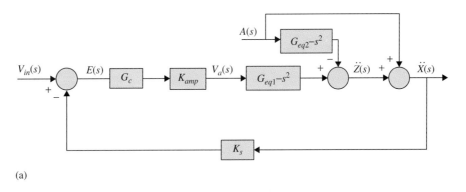

(a)

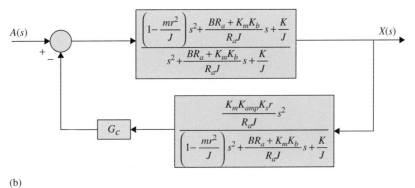

(b)

Figure 6-32 Block diagram of the absolute acceleration control system.

over a curb or speed bump. This tool also incorporates nonlinear effects, such as backlash and saturation in the Virtual Lab component. All of these features are available in one simple window, which automatically controls the Simulink model.

To start the program, click on Quarter Car Sim on the **ACSYS** applet. This launches both a Simulink model file (Fig. 6-33) and MATLAB graphical user interface, to be used as

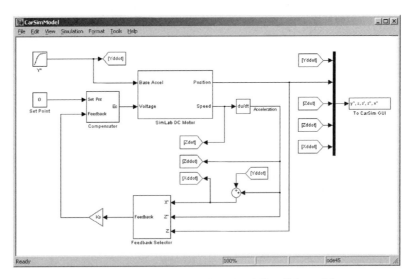

Figure 6-33 Quarter Car Modeling Tool top-level Simulink model.

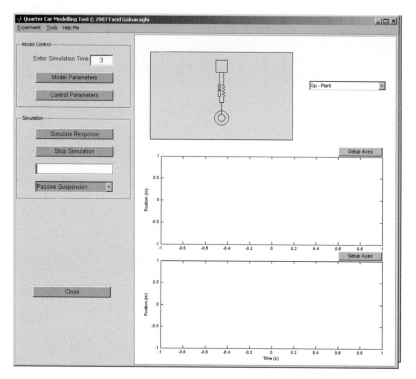

Figure 6-34 Quarter Car Modeling Tool control window.

the control panel (Fig. 6-34). There is very little need to access the Simulink model, other than to reference the model or to modify the simulation parameters.

From the control window, clicking on Model Parameters brings up a window (Fig. 6-35) from which you can modify the parameters of the dc motor, amplifier, sensor gains, and of course the quarter-car model. Parameters from the workspace or a .mat file may be selected in the left IMPORT frame and then assigned to the selected model parameters in the right MODEL frame. Model settings may be saved to, or loaded from, .mat files. Clicking on Defaults assigns the default values to the parameters. Click Apply to implement your changes or Close to cancel.

Selecting Control Parameters calls a window (Fig. 6-36) used to configure the compensator command signal in the left frame and the compensator in the right pane. There are a number of inputs to select from: step, impulse, sin, rounded pulse, rounded step, and random. The compensator frame allows the user to select the sensor output to be used

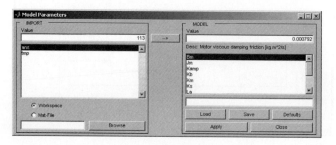

Figure 6-35 Model Parameters window.

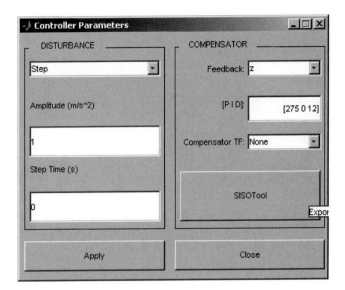

Figure 6-36 Control Parameters window.

as feedback. Also, PID gains are specified in array format, and any existing transfer function object may be selected from the workspace for use as a controller via the dropdown menu. The MATLAB SISO Design Tool may be activated, with the appropriate system transfer functions automatically loaded, using the SISOTool button. Click Apply to implement your changes or Close to cancel.

The closed-loop transfer functions of the system are displayed in the top right corner. The various transfer functions of the form displayed in Fig. 6-37 can be selected from the popup menu.

Once the model and controller parameters are specified, the system is ready for simulation. Click on Simulate Response to begin the simulation. This will start the animation and plot the data on the upper and lower graphs. At the top right corner of both the upper and lower axes, pressing the Setup Axes button will display a small control menu that is used to select which data are to be displayed on the graph. Note that the control menus may be dragged off the axes by clicking and dragging the top bar or closed by clicking the X in the top right corner. Click on Stop Simulation to stop the animation and simulation. Below the progress bar is a popup menu, which allows the user to toggle to different experiment modes. The active suspension system is the dc motor-controlled system from Fig. 6-27. The passive suspension system operates as a spring and damper, without the added control of the dc motor.

To store the input/output plots on a new figure, click Print to Figure. The zoom control and cursor buttons appear at the bottom right corner of the display panel, as seen in Fig. 6-38.

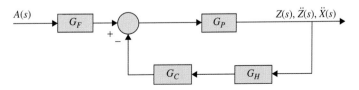

Figure 6-37 Closed-loop transfer functions.

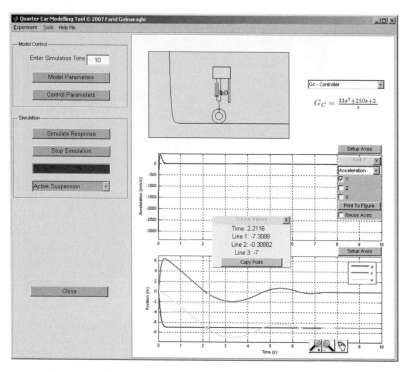

Figure 6-38 A view of the axes controls.

6-6-4 Passive Suspension

The following experiments explore the response of the open-loop quarter-car model. Studying the response of the passive system is essential to understanding the effect of various inputs on the quality of the vehicle's ride and is necessary for appreciating the effect of adding compensators to actively control the suspension system. Run the following tests in order:

1. Set the simulation mode to Passive Suspension and set up the top axes to display $\ddot{y}(t)$. To accomplish this, click on Setup Axes, choose Acceleration from the dropdown menu, and click the checkbox labeled "y." Using the same method, configure the bottom axes to display $z(t)$. Click on Control Parameters, and select a step input with amplitude 0.01 m/s^2 and step time 0 seconds. There is no need to configure the compensator since it is not used in a passive system. Click Simulate Response. When finished, the result will appear similar to the window displayed in Fig. 6-39. Note the shape of the road profile $y(t)$ as well as $\ddot{z}(t)$ and $\ddot{x}(t)$. This data can be accessed in the Setup Axes menu for either axis. Repeat this procedure for 0.1 and 1 m/s^2 input.

2. Experiment with the stiffness and damping of the system by clicking Model Parameters and changing the stiffness, k, from the default 2.17 N/m to 10 N/m. With a step input of 0.01 m/s^2, and the lower axes configured to display $\ddot{z}(t)$, what is the frequency of the oscillatory response? This is the damped frequency of the system using default parameters (ω_d). How does the period of oscillation compare to the value that was observed in Step 1? Open the Setup Axes menu for the lower axes and click the Print to Figure button. This will plot the data on the axes to an

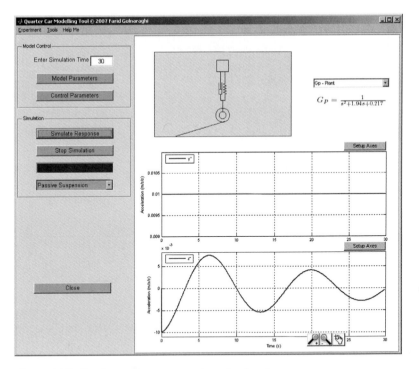

Figure 6-39 Passive system response to a step input.

external figure. Repeat the simulation several more times, gradually reducing the damping (variable c in the Model Parameters control window). The frequency of the oscillation when c is reduced to 0 is the natural frequency of the system (ω_n). Select the checkbox labeled Reuse Axes, and click Print to Figure to plot data from a new simulation in the same external figure. This is useful for comparison.

3. Set the system parameters to the default settings (click Model Parameters, then Defaults, then Apply). Study the effect of a sinusoidal input (washboard bumps) on the response of the system. Select an amplitude of 0.01 m/s² and vary the frequency from 5 rad/s to 0.1 rad/s. What happens to the amplitude of relative displacement at the damped natural frequency, ω_d, measured in Step 2?

4. Now try using the rounded step input with amplitude 0.1 m and duration 0.01 seconds. This input function simulates driving the quarter-car model over a curb. Calculate values of c such that the system is underdamped ($\zeta = 0.1$), critically damped ($\zeta = 1$), and overdamped ($\zeta = 1.5$), and observe the response for each case.

5. Repeat Step 4 using the unidirectional rounded pulse (URP) input (amplitude 0.1 m, duration 0.01 seconds). This emulates driving over a pothole if given amplitude less than zero, or a speed bump if the amplitude is positive.

6-6-5 Closed-Loop Relative Position Control

Now set the simulation mode to active suspension by selecting Active Suspension from the control panel dropdown menu or from the Experiment menu. This activates the feedback control system as defined in the Control Parameters window. Relative position control is the control of $z(t) = x(t) - y(t)$; thus, set the Feedback popup menu to z. The set point is

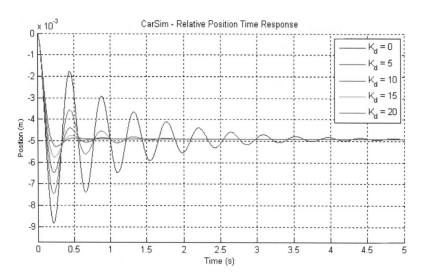

Figure 6-40 Relative Position Time Response plot.

$V_{in}(t) = 0$ V, where feedback gain is $K_s = 1$ V/m. In this section, both a PD and phase-lag controller will be implemented.

6. Click on Control Parameters and select a step disturbance 1 m/s² and step time 0 seconds; feedback z, PID = [1 0 0]. Simulate the response. What is the steady-state error? Apply the final value theorem to $\dfrac{Z(s)}{-A(s)} = \dfrac{\dfrac{mr^2}{J}}{s^2 + \dfrac{BR_a + K_m K_b}{R_a J}s + \dfrac{K}{J} + \dfrac{K_m K_{amp} K_s r}{R_a J} G_c}$.
Do the values correspond? Validate the observed overshoot and rise time using the time-domain analysis techniques introduced in Chapter 5.

7. What value of K_p (PID = [K_p 0 0]) will yield a steady-state error less than 5 mm? This will require a gain much higher than 1. To reduce the need for trial and error, click SISOTool in the Control Parameters window and increase the gain while observing the LTI Viewer step response. What happens to the system overshoot at this gain? Does this match your calculations?

8. Increase the derivative gain in steps from 0 to 22, keeping the proportional gain that was found in Step 7, and observe the effect of adding derivative gain. Again, validate these results using time-domain analysis techniques. Plot successive trials to an external figure for comparison, as in Fig. 6-40.

9. Design four phase-lead compensators with sufficient gain K_p to meet the steady-state error requirement as specified in Step 7 and with phase margins ϕ_m, of 10, 30, 40, and 60°. Compare the optimized response with the PD response.

10. Test your controller's response to the inputs applied in Steps 3, 4, and 5.

6-6-6 Closed-Loop Acceleration Control

As mentioned previously, it is preferable to control the acceleration of the mass m, because it is the acceleration of the vehicle that affects the comfort of the ride. Set the simulation mode to Active Suspension, and set the feedback to $\ddot{z}$ in Control Parameters. This causes

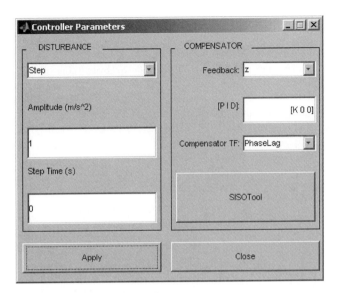

Figure 6-41 The Controller Parameters window.

the relative acceleration, $\ddot{z}(t)$, with gain $K_s = 1$ V/m/s² to be compared with the set point $V_{in}(t) = 0V$.

11. In the Control Parameters window, change the PID gains to PID = [5 0 0], and click SISOTool to apply the changes and launch the SISO Design Tool. Use the LTI Viewer to monitor the step response while tweaking the gain. What gain achieves the smallest rise time? Test this gain on the actual controller, clicking on Print to Figure in the Setup Axes menu to save the results for comparison.

12. Design a phase-lag compensator, with gain $K = 5$ and $\omega'_g = 0.1$ rad/s. Once the compensator is designed, implement it by using the following MATLAB script:

    K = 5, a = 0.1442, T = 231.1996
    PhaseLag = tf([a∗T 1], [T 1])

This creates a transfer function object in the workspace. Now click on Control Parameters and select PhaseLag as the compensator in the Compensator TF dropdown and enter K as the proportional gain (Fig. 6-41). Any transfer function object created in the MATLAB workspace is accessible in this menu and can be used in place of the PID controller $G_c(s)$. Compare the step response to the response in Step 11. Try designing various phase-lag controllers as per Section 9-6-2 and compare the results.

13. Repeat Steps 11 and 12 using absolute acceleration as the feedback (Feedback = $\ddot{x}$). For the phase-lag compensator, try $K = 5$ and $\omega'_g = 0.1$ rad/s.

14. Test the controllers designed in the last few steps with various inputs such as sinusoidal, rounded step, and rounded pulse.

▶ 6-7 SUMMARY

In this chapter, we described the SIMLab and Virtual Lab software to improve your understanding of control and to provide a better practical appreciation of the subject. We discussed that, in a realistic system including an actuator (e.g., a dc motor) and mechanical (gears) and electrical components (amplifiers), issues such as saturation of the amplifier, friction in the motor, or backlash in gears will seriously affect the controller design. This chapter focused on problems involving dc motors including modeling, system identification, and controller design. We presented experiments on speed and position control of dc motors, followed by two controller design projects involving control of a simple robotic system and control of a single degree of freedom quarter-car model. The focus on dc motors in these experiments was intentional, because of their simplicity and wide use in industrial applications. Note that, in the design projects, aside from the speed and position control topics, other controllers such as PID and lead/lag were also discussed. You may wish to visit Chapter 9 to become more acquainted with these topics.

▶ REFERENCE

1. F. Golnaraghi, "ENSC 383 Laboratory Experiment," Simon Fraser University, Mechatronic Systems Engineering Program, British Columbia, Canada, *Lab Manual*, 2008.

▶ PROBLEMS

6-1. Create a model of the motor shown in Fig. 5-25. Use the following parameter values: $J_m = 0.0004$ kg-m^2; $B = 0.001$ Nm/rad/sec, $R_a = 2\,\Omega$, $L_a = 0.008$ H, $K_m = 0.1$ Nm/A, and $K_b = 0.1$ V/rad/sec. Assume that the load torque T_L is zero. Apply a 5-V step input to the motor, and record the motor speed and the current drawn by the motor (requires modification of SIMLab blocks by making current the output) for 10 sec following the step input.
(a) What is the steady-state speed?
(b) How long does it take the motor to reach 63% of its steady-state speed?
(c) How long does it take the motor to reach 75% of its steady-state speed?
(d) What is the maximum current drawn by the motor?

6-2. Set the viscous friction B to zero in Problem 6-1. Apply a 5-V step input to the motor, and record the motor speed and current for 10 sec following the step input. What is the steady-state speed?
(a) How long does it take the motor to reach 63% of its steady-state speed?
(b) How long does it take the motor to reach 75% of its steady-state speed?
(c) What is the maximum current drawn by the motor?
(d) What is the steady-state speed when the applied voltage is 10 V?

6-3. Set the armature inductance L_a to zero in Problem 6-2. Apply a 5-V step input to the motor, and record the motor speed and current drawn by the motor for 10 sec following the step input.
(a) What is the steady-state speed?
(b) How long does it take the motor to reach 63% of its steady-state speed?
(c) How long does it take the motor to reach 75% of its steady-state speed?
(d) What is the maximum current drawn by the motor?
(e) If J_m is increased by a factor of 2, how long does it take the motor to reach 63% of its steady-state speed following a 5-V step voltage input?
(f) If J_m is increased by a factor of 2, how long does it take the motor to reach 75% of its steady-state speed following a 5-V step voltage input?

6-4. Repeat Problems 6-1 through 6-3, and assume the load torque $T_L = -0.1$ N-m (don't forget the minus sign) starting after 0.5 sec (requires change of the disturbance block parameters in SIMLab).
(a) How does the steady-state speed change once T_L is added?
(b) How long does it take the motor to reach 63% of its new steady-state speed?
(c) How long does it take the motor to reach 75% of its new steady-state speed?
(d) What is the maximum current drawn by the motor?
(e) Increase T_L and further discuss its effect on the speed response.

6-5. Repeat Problems 6-1 through 6-3, and assume the load torque $T_L = -0.2$ N-m (don't forget the minus sign) starting after 1 sec (requires change of the disturbance block parameters in SIMLab).
(a) How does the steady-state speed change once T_L is added?
(b) How long does it take the motor to reach 63% of its new steady-state speed?
(c) How long does it take the motor to reach 75% of its new steady-state speed?
(d) What is the maximum current drawn by the motor?
(e) Increase T_L and further discuss its effect on the speed response.

6-6. For the system in Fig. 6-1, use the parameters for Problem 6-1 (but set $L_a = 0$) and an amplifier gain of 2 to drive the motor (ignore the amplifier voltage and current limitations for the time being). What is the steady-state speed when the amplifier input voltage is 5 V?

6-7. Modify the model in Problem 6-6 by adding a proportional controller with a gain of $K_p = 0.1$, apply a 10 rad/sec step input, and record the motor speed and current for 2 sec following the step input.
(a) What is the steady-state speed?
(b) How long does it take the motor to reach 63% of its steady-state speed?
(c) How long does it take the motor to reach 75% of its steady-state speed?
(d) What is the maximum current drawn by the motor?

6-8. Change K_p to 1.0 in Problem 6-7, apply a 10 rad/sec step input, and record the motor speed and current for 2 sec following the step input.
(a) What is the steady-state speed?
(b) How long does it take for the motor to reach 63% of its steady-state speed?
(c) How long does it take for the motor to reach 75% of its steady-state speed?
(d) What is the maximum current drawn by the motor?
(e) How does increasing K_p affect the response (with and without saturation effect in the SIMLab model)?

6-9. Repeat Problem 6-7, and assume the load torque $T_L = -0.1$ N-m starting after 0.5 sec (requires change of the disturbance block parameters in SIMLab).
(a) How does the steady-state speed change once T_L is added?
(b) How long does it take the motor to reach 63% of its new steady-state speed?
(c) How long does it take the motor to reach 75% of its new steady-state speed?

6-10. Repeat Problem 6-7, and assume the load torque $T_L = -0.2$ N-m starting after 1 sec (requires change of the disturbance block parameters in SIMLab).
(a) How does the steady-state speed change once T_L is added?
(b) How long does it take the motor to reach 63% of its new steady-state speed?
(c) How long does it take the motor to reach 75% of its new steady-state speed?

6-11. Insert a velocity sensor transfer function K_s in the feedback loop, where $K_s = 0.2$ V/rad/sec (requires adjustment of the SIMLab model). Apply a 2 rad/sec step input, and record the motor speed and current for 0.5 sec following the step input. Find the value of K_p that gives the same result as in Problem 6-7.

6-12. For the system in Fig. 6-3, select $K_p = 1.0$, apply a 1 rad step input, and record the motor position for 1 sec. Use the same motor parameters as in Problem 6-1.
(a) What is the steady-state position?
(b) What is the maximum rotation?
(c) At what time after the step does the maximum occur?

6-13. Change K_p to 2.0 in Problem 6-12, apply a 1 rad step input, and record the motor position for 1 sec.
(a) At what time after the step does the maximum occur?
(b) What is the maximum rotation?

6-14. Using the SIMLab, investigate the closed-loop position response using a proportional controller. For a position-control case, use proportional controller gains of 0.1, 0.2, 0.5, 1.0, and 2.0; record the step response for a 1 rad change at the output shaft; and estimate what you consider to be the best value for the proportional gain. Use the same motor parameters as in Problem 6-1.

6-15. Using the SIMLab, investigate the closed-loop position response using a PD controller. Modify the controller used in Problem 6-14 by adding derivative action to the proportional controller. Using the best value you obtained for K_p, try various values for K_D, and record the step response in each case.

6-16. Repeat Problem 6-15 and assume a disturbance torque $T_D = -0.1$ N-m in addition to the step input of 1 rad (requires change of the disturbance block parameters in SIMLab).

6-17. Repeat Problem 6-15 and assume a disturbance torque $T_D = -0.2$ N-m in addition to the step input of 1 rad (requires change of the disturbance block parameters in SIMLab).

6-18. Use the SIMLab and parameter values of Problem 6-1 to design a PID controller that eliminates the effect of the disturbance torque, with a percent overshoot of 4.3.

6-19. Use the SIMLab and parameter values of Problem 6-1 to design a PID controller that eliminates the effect of the disturbance torque, with a percent overshoot of 2.8.

6-20. Investigate the frequency response of the motor using the Virtual Lab Tool. Apply a sine wave with a frequency of 0.1 Hz (don't forget: 1 Hz = 2π rad/sec) and amplitude of 1 V the amplifier input, and record both the motor velocity and sine wave input signals. Repeat this experiment for frequencies of 0.2, 0.5, 1.0, 2.0, 5.0, 10.0, and 50.0 Hz (keeping the sine wave amplitude at 1 V).

6-21. Using the Virtual Lab Tool, investigate the closed-loop motor speed response using a proportional controller. Record the closed-loop response of the motor velocity to a step input of 2 rad/sec for proportional gains of 0.1, 0.2, 0.4, and 0.8. What is the effect of the gain on the steady-state velocity?

6-22. Using the Virtual Lab Tool, investigate the closed-loop position response using a proportional controller. For a position-control case, use proportional controller gains of 0.1, 0.2, 0.5, 1.0, and 2.0; record the step response for a 1 rad change at the output shaft; and estimate what you consider to be the best value for the proportional gain.

6-23. Using the Virtual Lab Tool, investigate the closed-loop position response using a PD controller. Modify the controller used in Problem 6-15 by adding derivative action to the proportional controller. Using the best value you obtained for K_p, try various values for K_D, and record the step response in each case.

6-24. In Design Project 2 in Section 6-7, use the CarSim tool to investigate the effects of controlling acceleration $\ddot{X}$ on relative motion (or bounce) Z and vice versa.
(a) Use a PD controller in your investigation.
(b) Use a PI controller in your investigation.
(c) Use a PID controller in your investigation.

6-25. Using the Quarter Car Modeling Tool controlling,
(a) Set the simulation mode to "Passive Suspension" and set up the top axes to display $\ddot{y}(t)$. Select a step input with amplitude 0.02 m/s² and step time 0 seconds. Plot the response. Repeat this procedure for 0.2 and 0.5 m/s² input. Compare the results.
(b) Change the stiffness, k, to 15 N/m. With a step input of 0.02 m/s² and the lower axes configured to display $\ddot{z}(t)$, what is the frequency of the oscillatory response? This is the damped frequency of the system using default parameters (ω_d). How does the period of oscillation compare to the value that was observed in part (a)? Repeat the simulation several more times, gradually reducing the damping (variable c in the Model Parameters control window) to find the natural frequency of the system (ω_n).
(c) Obtain the effect of washboard bumps with an amplitude of 0.02 m/s² on the response of the system. Vary the frequency from 10 rad/s to 0.1 rad/s. What happens to the amplitude of relative displacement at the damped natural frequency, ω_d, measured in part (b)?

(d) Simulate driving the quarter-car model over a curb by using the rounded step input with amplitude 0.2 m and duration 0.02 seconds. Calculate values of c such that the system is underdamped ($\zeta = 0.25$), critically damped ($\zeta = 1$), and overdamped ($\zeta = 2.5$) and observe the response for each case.

(e) Repeat part (d) using the unidirectional rounded pulse (URP) input (amplitude 0.2 m, duration 0.02 seconds).

(f) Add a step disturbance of 2 m/s^2 and step time of 0 seconds; feedback z, PID = [1 0 0]. Simulate the response. Find the steady-state error by simulation and by applying the final-value theorem. Compare the results. Validate the observed overshoot and rise time using the time domain analysis.

(g) What value of K_p (PID = [K_p 0 0]) will yield a steady-state error less than 4 mm? What happens to the system overshoot at this gain? Does this match your calculations?

(h) Increase the derivative gain in steps from 0 to 50, keeping the proportional gain that was found in part (g), and observe the effect of adding derivative gain. Again, validate these results using time domain analysis techniques. Plot successive trials to an external figure for comparison.

(i) Design four phase-lead compensators with sufficient gain K_p to meet the steady-state error requirement as specified in part (g) and with phase margins, ϕ_m, of 15, 20, 25, and 50°. Compare the optimized response with the PD response.

(j) Test your controller's response to the inputs applied in parts (c), (d), and (e).

(k) Change the PID gains to *PID = [5 1 0]* and click SISO Tool to apply the changes and launch the SISO Design Tool. Explain what happens.

(l) What value of K_I (PID = [5 K_I 0]) will yield a steady-state error less than 4 mm? What happens to the system overshoot and rise time at this gain? Does this match your calculations?

(m) Test the controllers designed in the last few parts with sinusoidal, rounded-step, and rounded-pulse input.

CHAPTER 7

Root Locus Analysis

▶ 7-1 INTRODUCTION

In the preceding chapters, we have demonstrated the importance of the poles and zeros of the closed-loop transfer function of a linear control system on the dynamic performance of the system. The roots of the characteristic equation, which are the poles of the closed-loop transfer function, determine the absolute and the relative stability of linear SISO systems. Keep in mind that the transient properties of the system also depend on the zeros of the closed-loop transfer function.

An important study in linear control systems is the investigation of the trajectories of the roots of the characteristic equation—or, simply, the **root loci**—when a certain system parameter varies. In Chapter 5, several examples already illustrated the usefulness of the root loci of the characteristic equation in the study of linear control systems. The basic properties and the systematic construction of the root loci are first due to W. R. Evans [1, 3]. In general, root loci may be sketched by following some simple rules and properties.

For plotting the root loci accurately, the MATLAB root-locus tool in the Control Systems Toolbox component of **ACSYS** can be used. See Chapter 9 for examples. As a design engineer, it may be sufficient for us to learn how to use these computer tools to generate the root loci for design purposes. However, it is important to learn the basics of the root loci and their properties, as well as how to interpret the data provided by the root loci for analysis and design purposes. The material in this text is prepared with these objectives in mind; details on the properties and construction of the root loci are presented in Appendix E.

The root-locus technique is not confined only to the study of control systems. In general, the method can be applied to study the behavior of roots of any algebraic equation with one or more variable parameters. The general root-locus problem can be formulated by referring to the following algebraic equation of the complex variable, say, s:

$$F(s) = P(s) + KQ(s) = 0 \qquad (7\text{-}1)$$

where $P(s)$ is an nth-order polynomial of s,

$$P(s) = s^n + a_{n-1}s^{n-1} + \cdots + a_1 s + a_0 \qquad (7\text{-}2)$$

and $Q(s)$ is an mth-order polynomial of s; n and m are positive integers.

$$Q(s) = s^m + b_{m-1}s^{m-1} + \cdots + b_1 s + b_0 \qquad (7\text{-}3)$$

For the present, we do not place any limitations on the relative magnitudes between n and m. K is a real constant that can vary from $-\infty$ to $+\infty$.

The coefficients $a_1, a_2, \ldots, a_n, b_1, b_2, \ldots, b_m$ are considered to be real and fixed.

Root loci of multiple variable parameters can be treated by varying one parameter at a time. The resultant loci are called the **root contours**, and the subject is treated in Section 7-5. By replacing s with z in Eq. (7-1) through (7-3), the root loci of the characteristic equation of a linear discrete-data system can be constructed in a similar fashion (Appendix E).

For the purpose of identification in this text, we define the following categories of root loci based on the values of K:

1. **Root loci (RL).** Refers to the entire root loci for $-\infty < K < \infty$.
2. **Root contours (RC).** Contour of roots when more than one parameter varies.

In general, for most control-system applications, the values of K are positive. Under unusual conditions, when a system has positive feedback or the loop gain is negative, then we have the situation that K is negative. Although we should be aware of this possibility, we need to place the emphasis only on positive values of K in developing the root-locus techniques.

▶ 7-2 BASIC PROPERTIES OF THE ROOT LOCI (RL)

Because our main interest is control systems, let us consider the closed-loop transfer function of a single-loop control system:

$$\frac{Y(s)}{R(s)} = \frac{G(s)}{1 + G(s)H(s)} \tag{7-4}$$

keeping in mind that the transfer function of multiple-loop SISO systems can also be expressed in a similar form. The characteristic equation of the closed-loop system is obtained by setting the denominator polynomial of $Y(s)/R(s)$ to zero. Thus, the roots of the characteristic equation must satisfy

$$1 + G(s)H(s) = 0 \tag{7-5}$$

Suppose that $G(s)H(s)$ contains a real variable parameter K as a multiplying factor, such that the rational function can be written as

$$G(s)H(s) = \frac{KQ(s)}{P(s)} \tag{7-6}$$

where $P(s)$ and $Q(s)$ are polynomials as defined in Eq. (7-2) and (7-3), respectively. Eq. (7-5) is written

$$1 + \frac{KQ(s)}{P(s)} = \frac{P(s) + KQ(s)}{P(s)} = 0 \tag{7-7}$$

The numerator polynomial of Eq. (7-7) is identical to Eq. (7-1). Thus, by considering that the loop transfer function $G(s)H(s)$ can be written in the form of Eq. (7-6), we have identified the RL of a control system with the general root-locus problem.

When the variable parameter K does not appear as a multiplying factor of $G(s)H(s)$, we can always condition the functions in the form of Eq. (7-1). As an illustrative example, consider that the characteristic equation of a control system is

$$s(s+1)(s+2) + s^2 + (3+2K)s + 5 = 0 \tag{7-8}$$

To express the last equation in the form of Eq. (7-7), we divide both sides of the equation by the terms that do not contain K, and we get

$$1 + \frac{2Ks}{s(s+1)(s+2) + s^2 + 3s + 5} = 0 \qquad (7\text{-}9)$$

Comparing the last equation with Eq. (7-7), we get

$$\frac{Q(s)}{P(s)} = \frac{2s}{s^3 + 4s^2 + 5s + 5} \qquad (7\text{-}10)$$

Now K is isolated as a multiplying factor to the function $Q(s)/P(s)$.

We shall show that the RL of Eq. (7-5) can be constructed based on the properties of $Q(s)/P(s)$. In the case where $G(s)H(s) = KQ(s)/P(s)$, the root-locus problem is another example in which the characteristics of the closed-loop system, in this case represented by the roots of the characteristic equation, are determined from the knowledge of the loop transfer function $G(s)H(s)$.

Now we are ready to investigate the conditions under which Eq. (7-5) or Eq. (7-7) is satisfied.

Let us express $G(s)H(s)$ as

$$G(s)H(s) = KG_1(s)H_1(s) \qquad (7\text{-}11)$$

where $G_1(s)H_1(s)$ does not contain the variable parameter K. Then, Eq. (7-5) is written

$$G_1(s)H_1(s) = -\frac{1}{K} \qquad (7\text{-}12)$$

To satisfy Eq. (7-12), the following conditions must be satisfied simultaneously:
Condition on magnitude

$$|G_1(s)H_1(s)| = \frac{1}{|K|} \quad -\infty < K < \infty \qquad (7\text{-}13)$$

Condition on angles

$$\begin{aligned}\angle G_1(s)H_1(s) &= (2i+1)\pi \quad K \geq 0 \\ &= \text{odd multiples of } \pi \text{ radians or } 180°\end{aligned} \qquad (7\text{-}14)$$

$$\begin{aligned}\angle G_1(s)H_1(s) &= 2i\pi \quad K \leq 0 \\ &= \text{even multiples of } \pi \text{ radians or } 180°\end{aligned} \qquad (7\text{-}15)$$

where $i = 0, \pm 1, \pm 2, \ldots$ (any integer).

In practice, the conditions stated in Eq. (7-13) through (7-15) play different roles in the construction of the root loci.

- The conditions on angles in Eq. (7-14) or Eq. (7-15) are used to determine the trajectories of the root loci in the s-plane.
- Once the root loci are drawn, the values of K on the loci are determined by using the condition on magnitude in Eq. (7-13).

The construction of the root loci is basically a graphical problem, although some of the properties are derived analytically. The graphical construction of the RL is based on the knowledge of the poles and zeros of the function $G(s)H(s)$. In other words, $G(s)H(s)$ must first be written as

$$G(s)H(s) = KG_1(s)H_1(s) = \frac{K(s+z_1)(s+z_2)\cdots(s+z_m)}{(s+p_1)(s+p_2)\cdots(s+p_n)} \quad (7\text{-}16)$$

where the zeros and poles of $G(s)H(s)$ are real or in complex-conjugate pairs.

Applying the conditions in Eqs. (7-13), (7-14), and (7-15) to Eq. (7-16), we have

$$|G_1(s)H_1(s)| = \frac{\prod_{i=1}^{m}|s+z_i|}{\prod_{k=1}^{n}|s+p_k|} = \frac{1}{|K|} \quad -\infty < K < \infty \quad (7\text{-}17)$$

For $0 \leq K < \infty$:

$$\angle G_1(s)H_1(s) = \sum_{k=1}^{m}\angle(s+z_k) - \sum_{j=1}^{n}\angle(s+p_j) = (2i+1) \times 180° \quad (7\text{-}18)$$

For $-\infty < K \leq 0$:

$$\angle G_1(s)H_1(s) = \sum_{k=1}^{m}\angle(s+z_k) - \sum_{j=1}^{n}\angle(s+p_j) = 2i \times 180° \quad (7\text{-}19)$$

where $i = 0, \pm1, \pm2, \ldots$.

The graphical interpretation of Eq. (7-18) is that any point s_1 on the RL that corresponds to a positive value of K must satisfy the following condition:

The difference between the sums of the angles of the vectors drawn from the zeros and those from the poles of $G(s)H(s)$ to s_1 is an odd multiple of 180 degrees.

For negative values of K, any point s_1 on the RL must satisfy the following condition:

The difference between the sums of the angles of the vectors drawn from the zeros and those from the poles of $G(s)H(s)$ to s_1 is an even multiple of 180 degrees, including zero degrees.

Once the root loci are constructed, the values of K along the loci can be determined by writing Eq. (7-17) as

$$|K| = \frac{\prod_{j=1}^{n}|s+p_j|}{\prod_{i=1}^{m}|s+z_i|} \quad (7\text{-}20)$$

The value of K at any point s_1 on the RL is obtained from Eq. (7-20) by substituting the value of s_1 into the equation. Graphically, the numerator of Eq. (7-20) represents the product of the lengths of the vectors drawn from the poles of $G(s)H(s)$ to s_1, and the denominator represents the product of lengths of the vectors drawn from the zeros of $G(s)H(s)$ to s_1.

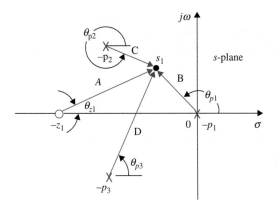

Figure 7-1 Pole-zero configuration of $G(s)H(s) = K(s+z_1)/[s(s+p_2) \times (s+p_3)]$.

To illustrate the use of Eqs. (7-18) to (7-20) for the construction of the root loci, let us consider the function

$$G(s)H(s) = \frac{K(s+z_1)}{s(s+p_2)(s+p_3)} \qquad (7\text{-}21)$$

The location of the poles and zero of $G(s)H(s)$ are arbitrarily assigned, as shown in Fig. 7-1. Let us select an arbitrary trial point s_1 in the s-plane and draw vectors directing from the poles and zeros of $G(s)H(s)$ to the point. If s_1 is indeed a point on the RL for positive K, it must satisfy Eq. (7-18); that is, the angles of the vectors shown in Fig. 7-1 must satisfy

$$\begin{aligned}\angle(s_1+z_1) - \angle s_1 - \angle(s_1+p_2) - \angle(s_1+p_3) \\ = \theta_{z1} - \theta_{p1} - \theta_{p2} - \theta_{p3} = (2i+1) \times 180°\end{aligned} \qquad (7\text{-}22)$$

where $i = 0, \pm 1, \pm 2, \ldots$. As shown in Fig. 7-1, the angles of the vectors are measured with the positive real axis as reference. Similarly, if s_1 is a point on the RL for negative values of K, it must satisfy Eq. (7-19); that is,

$$\begin{aligned}\angle(s_1+z_1) - \angle s_1 - \angle(s_1+p_2) - \angle(s_1+p_3) \\ = \theta_{z1} - \theta_{p1} - \theta_{p2} - \theta_{p3} = 2i \times 180°\end{aligned} \qquad (7\text{-}23)$$

where $i = 0, \pm 1, \pm 2, \ldots$.

If s_1 is found to satisfy either Eq. (7-22) or Eq. (7-23), Eq. (7-20) is used to find the magnitude of K at the point. As shown in Fig. 7-1, the lengths of the vectors are represented by A, B, C, and D. The magnitude of K is

$$|K| = \frac{|s_1||s_1+p_2||s_1+p_3|}{|s_1+z_1|} = \frac{BCD}{A} \qquad (7\text{-}24)$$

The sign of K depends on whether s_1 satisfies Eq. (7-22) ($K \geq 0$) or Eq. (7-23) ($K \leq 0$). Thus, given the function $G(s)H(s)$ with K as a multiplying factor and the poles and zeros are known, the construction of the RL of the zeros of $1 + G(s)H(s)$ involves the following two steps:

1. A search for all the s_1 points in the s-plane that satisfy Eq. (7-18) for positive K. If the RL for negative values of K are desired, then Eq. (7-19) must be satisfied.
2. Use Eq. (7-20) to find the magnitude of K on the RL.

We have established the basic conditions on the construction of the root-locus diagram. However, if we were to use the trial-and-error method just described, the search for all the root-locus points in the s-plane that satisfy Eq. (7-18) or Eq. (7-19) and Eq. (7-20) would be a very tedious task. Years ago, when Evans [1, 2] first invented the root-locus technique, digital computer technology was still at its infancy; he had to devise a special tool, called the **Spirule**, which can be used to assist in adding and subtracting angles of vectors quickly, according to Eq. (7-18) or Eq. (7-19). Even with the Spirule, for the device to be effective, the user still has to first know the general proximity of the roots in the s-plane.

With the availability of digital computers and efficient root-finding subroutines, the Spirule and the trial-and-error method have long become obsolete. Nevertheless, even with a high-speed computer and an effective root-locus program, the analyst should still have an understanding of the properties of the root loci to be able to manually sketch the root loci of simple and moderately complex systems, if necessary, and interpret the computer results correctly, when applying the root loci for analysis and design of control systems.

▶ 7-3 PROPERTIES OF THE ROOT LOCI

The following properties of the root loci are useful for the purpose of constructing the root loci manually and for the understanding of the root loci. The properties are developed based on the relation between the poles and zeros of $G(s)H(s)$ and the zeros of $1 + G(s)H(s)$, which are the roots of the characteristic equation. We shall limit the discussion only to the properties but leave the details of the proofs and the applications of the properties to the construction of the root loci in Appendix E.

7-3-1 $K = 0$ and $K = \pm\infty$ Points

The $K = 0$ points on the root loci are at the poles of $G(s)H(s)$.
The $K = \pm\infty$ points on the root loci are at the zeros of $G(s)H(s)$.

The poles and zeros referred to here include those at infinity, if any. The reason for these properties are seen from the condition of the root loci given by Eq. (7-12), which is

$$G_1(s)H_1(s) = -\frac{1}{K} \tag{7-25}$$

As the magnitude of K approaches zero, $G_1(s)H_1(s)$ approaches infinity, so s must approach the poles of $G_1(s)H_1(s)$ or of $G(s)H(s)$. Similarly, as the magnitude of K approaches infinity, s must approach the zeros of $G(s)H(s)$.

▶ **EXAMPLE 7-3-1** Consider the equation

$$s(s+2)(s+3) + K(s+1) = 0 \tag{7-26}$$

When $K = 0$, the three roots of the equation are at $s = 0, -2,$ and -3. When the magnitude of K is infinite, the three roots of the equation are at $s = -1, \infty,$ and ∞. It is useful to consider that infinity in the s-plane is a point concept. We can visualize that the finite s-plane is only a small portion of a sphere with an infinite radius. Then, infinity in the s-plane is a point on the opposite side of the sphere that we face.

Dividing both sides of Eq. (7-26) by the terms that do not contain K, we get

$$1 + G(s)H(s) = 1 + \frac{K(s+1)}{s(s+2)(s+3)} = 0 \tag{7-27}$$

378 ▶ Chapter 7. Root Locus Analysis

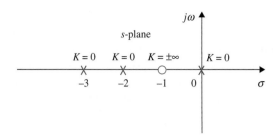

Figure 7-2 Points at which $K = 0$ and $K = \pm\infty$ on the RL of $s(s + 2)(s + 3) + K(s + 1) = 0$.

which gives

$$G(s)H(s) = \frac{K(s+1)}{s(s+2)(s+3)} \quad (7\text{-}28)$$

Thus, the three roots of Eq. (7-26) when $K = 0$ are the same as the poles of the function $G(s)H(s)$. The three roots of Eq. (7-26) when $K = \pm\infty$ are at the three zeros of $G(s)H(s)$, including those at infinity. In this case, one finite zero is at $s = -1$, but there are two zeros at infinity. The three points on the root loci at which $K = 0$ and those at which $K = \pm\infty$ are shown in Fig. 7-2. ◀

7-3-2 Number of Branches on the Root Loci

A branch of the RL is the locus of one root when K varies between $-\infty$ and ∞. The following property of the RL results, since the number of branches of the RL must equal the number of roots of the equation.

The number of branches of the RL of Eq. (7-1) or Eq. (7-5) is equal to the order of the polynomial.

For example, the number of branches of the root loci of Eq. (7-26) when K varies from $-\infty$ to ∞ is three, since the equation has three roots.

Keeping track of the individual branches and the total number of branches of the root-locus diagram is important in making certain that the plot is done correctly. This is particularly true when the root-locus plot is done by a computer, because unless each root locus branch is coded by a different color, it is up to the user to make the distinctions.

7-3-3 Symmetry of the RL

The RL are symmetrical with respect to the real axis of the s-plane. In general, the RL are symmetrical with respect to the axes of symmetry of the pole-zero configuration of $G(s)H(s)$.

The reason behind this property is because for real coefficient, K, in Eq. (7-1), the roots must be real or in complex-conjugate pairs.

7-3-4 Angles of Asymptotes of the RL: Behavior of the RL at $|s| = \infty$

When n, the order of $P(s)$, is not equal to m, the order of $Q(s)$, some of the loci will approach infinity in the s-plane. The properties of the RL near infinity in the s-plane are described by the **asymptotes** of the loci when $|s| \to \infty$. In general when $n \neq m$, there will be $2|n - m|$ asymptotes that describe the behavior of the RL at $|s| = \infty$. The angles of the asymptotes and their intersect with the real axis of the s-plane are described as follows.

For large values of s, the RL for $K \geq 0$ are asymptotic to asymptotes with angles given by

$$\theta_i = \frac{(2i+1)}{|n-m|} \times 180° \quad n \neq m \quad (7\text{-}29)$$

where $i = 0, 1, 2, \ldots, |n-m| - 1$; n and m are the number of finite poles and zeros of $G(s)H(s)$, respectively.

The asymptotes of the root loci for $K \geq 0$ are simply the extensions of the asymptotes for $K \geq 0$.

7-3-5 Intersect of the Asymptotes (Centroid)

The intersect of the $2|n-m|$ asymptotes of the RL lies on the real axis of the s-plane, at

$$\sigma_1 = \frac{\sum \text{finite poles of } G(s)H(s) - \sum \text{finite zeros of } G(s)H(s)}{n-m} \quad (7\text{-}30)$$

where n is the number of finite poles and m is the number of finite zeros of $G(s)H(s)$, respectively. The intersect of the asymptotes σ_1 represents the center of gravity of the root loci and is always a real number, or

$$\sigma_1 = \frac{\sum \text{real parts of poles of } G(s)H(s) - \sum \text{real parts of zeros of } G(s)H(s)}{n-m} \quad (7\text{-}31)$$

The root loci and their asymptotes for Eq. (7-26) for $-\infty \leq K \leq \infty$ are shown in Fig. 7-3. More examples on root-loci asymptotes and constructions are found in Appendix E.

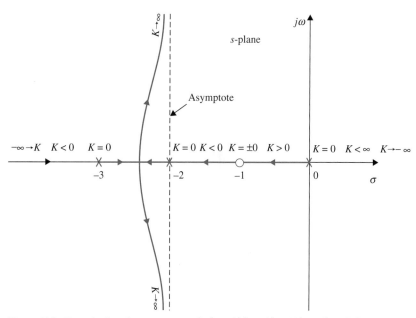

Figure 7-3 Root loci and asymptotes of $s(s+2)(s+3) + K(s+1) = 0$ for $-\infty \leq K \leq \infty$.

7-3-6 Root Loci on the Real Axis

The entire real axis of the s-plane is occupied by the RL for all values K. On a given section of the real axis, RL for $K \geq 0$ are found in the section only if the total number of poles and zeros of $G(s)H(s)$ to the right of the section is odd. Note that the remaining sections of the real axis are occupied by the RL for $K \leq 0$. Complex poles and zeros of $G(s)H(s)$ do not affect the type of RL found on the real axis.

7-3-7 Angles of Departure and Angles of Arrival of the RL

The angle of departure or arrival of a root locus at a pole or zero, respectively, of $G(s)H(s)$ denotes the angle of the tangent to the locus near the point.

7-3-8 Intersection of the RL with the Imaginary Axis

The points where the RL intersect the imaginary axis of the s-plane and the corresponding values of K may be determined by means of the Routh-Hurwitz criterion. For complex situations, when the RL have multiple numbers of intersections on the imaginary axis, the intersects and the critical values of K can be determined with the help of the root-locus computer program. The Bode diagram method in Chapters 2 and 8, associated with the frequency response, can also be used for this purpose.

7-3-9 Breakaway Points (Saddle Points) on the RL

Breakaway points on the RL of an equation correspond to multiple-order roots of the equation.

The breakaway points on the RL of $1 + KG_1(s)H_1(s) = 0$ must satisfy

$$\frac{dG_1(s)H_1(s)}{ds} = 0 \tag{7-32}$$

It is important to point out that the condition for the breakaway point given in Eq. (7-32) is *necessary* but *not sufficient*. In other words, all breakaway points on the root loci must satisfy Eq. (7-32), but not all solutions of Eq. (7-32) are breakaway points. To be a breakaway point, the solution of Eq. (7-32) must also satisfy Eq. (7-5), that is, must also be a point on the root loci for some real K.

Toolbox 7-3-1

MATLAB statements for Fig. 7-3

```
num=[1 1];
den=conv([1 0],[1 2]);
den=conv(den,[1 3]);
mysys=tf(num,den);
rlocus(mysys);
title('Root loci for equation 7.27');
axis([-3 0 -8 8])
[k,poles] = rlocfind(mysys) % rlocfind command in MATLAB can choose the
desired poles on the locus
```

If we take the derivatives on both sides of Eq. (7-12) with respect to s, we get

$$\frac{dK}{ds} = \frac{dG_1(s)H_1(s)/ds}{[G_1(s)H_1(s)]^2} \quad (7\text{-}33)$$

Thus, the condition in Eq. (7-32) is equivalent to

$$\frac{dK}{ds} = 0 \quad (7\text{-}34)$$

In summary, except for extremely complex cases, the properties on the root loci just presented should be adequate for making a reasonably accurate sketch of the root-locus diagram short of plotting it point by point. The computer program can be used to solve for the exact root locations, the breakaway points, and some of the other specific details of the root loci, including the plotting of the final loci. However, one cannot rely on the computer solution completely, since the user still has to decide on the range and resolution of K so that the root-locus plot has a reasonable appearance. For quick reference, the important properties described are summarized in Table 7-1, and the details are given in Appendix E.

TABLE 7-1 Properties of the Root Loci of $1 + KG_1(s) H_1 = 0$

1. $K = 0$ points	The $K = 0$ points are at the poles of $G(s)H(s)$, including those at $s = \infty$.
2. $K = \pm\infty$ points	The $K = \infty$ points are at the zeros of $G(s)H(s)$, including those at $s = \infty$.
3. Number of separate root loci	The total number of root loci is equal to the order of the equation $1 + KG_1(s)H_1(s) = 0$.
4. Symmetry of root loci	The root loci are symmetrical about the axes of symmetry the of pole-zero configuration of $G(s)H(s)$.
5. Asymptotes of root loci as $s \to \infty$	For large values of s, the RL ($K > 0$) are asymptotic to asymptotes with angles given by $$\theta_i = \frac{2i+1}{|n-m|} \times 180°$$ For $K < 0$, the RL are asymptotic to $$\theta_i = \frac{2i}{|n-m|} \times 180°$$ where $i = 0, 1, 2, \ldots, |n-m| - 1$, n = number of finite poles of $G(s)H(s)$, and m = number of finite zeros of $G(s)H(s)$.
6. Intersection of the asymptotes	(a) The intersection of the asymptotes lies only on the real axis in the s-plane. (b) The point of intersection of the asymptotes is given by $$\sigma_1 = \frac{\sum \text{real parts of poles of} G(s)H(s) - \sum \text{real parts of zeros of} G(s)H(s)}{n-m}$$
7. Root loci on the real axis.	RL for $K \geq 0$ are found in a section of the real axis only if the total number of real poles and zeros of $G(s)H(s)$ to the **right** of the section is **odd**. If the total number of real poles and zeros to the right of a given section is **even**, RL for $K \leq 0$ are found.

(Continued)

TABLE 7-1 (Continued)

8. Angles of departure	The angle of departure or arrival of the RL from a pole or a zero of $G(s)H(s)$ can be determined by assuming a point s_1 that is very close to the pole, or zero, and applying the equation $$\angle G(s_1)H(s_1) = \sum_{k=1}^{m} \angle(s_1 - z_k) - \sum_{j=1}^{m} \angle(s_1 - p_j)$$ $$= 2(i+1)180° \quad K \geq 0$$ $$= 2i \times 180° \quad K \leq 0$$ where $i = 0, \pm 1, \pm 2, \ldots$.
9. Intersection of the root loci with the imaginary axis	The crossing points of the root loci on the imaginary axis and the corresponding values of K may be found by use of the Routh-Hurwitz criterion.
10. Breakaway points	The breakaway points on the root loci are determined by finding the roots of $dK/ds = 0$, or $dG(s)H(s)/ds = 0$. These are necessary conditions only.
11. Calculation of the values of K	The absolute value of K at any point s_1 on the root loci is on the root loci determined from the equation $$\|K\| = \frac{1}{\|G_1(s_1)H_1(s_1)\|}$$

7-3-10 The Root Sensitivity

The condition on the breakaway points on the RL in Eq. (7-34) leads to the **root sensitivity** [17, 18, 19] of the characteristic equation. The sensitivity of the roots of the characteristic equation when K varies is defined as the **root sensitivity** and is given by

$$S_K = \frac{ds/s}{dK/K} = \frac{K}{s}\frac{ds}{dK} \qquad (7\text{-}35)$$

Thus, Eq. (7-34) shows that *the root sensitivity at the breakaway points is infinite*. From the root-sensitivity standpoint, we should avoid selecting the value of K to operate at the breakaway points, which correspond to multiple-order roots of the characteristic equation. In the design of control systems, not only it is important to arrive at a system that has the desired characteristics, but, just as important, the system should be insensitive to parameter variations. For instance, a system may perform satisfactorily at a certain K, but if it is very sensitive to the variation of K, it may get into the undesirable performance region or become unstable if K varies by only a small amount. In formal control-system terminology, a system that is insensitive to parameter variations is called a **robust system**. Thus, the root-locus study of control systems must involve not only the shape of the root loci with respect to the variable parameter K but also how the roots along the loci vary with the variation of K.

▶ **EXAMPLE 7-3-2** Fig. 7-4 shows the root locus diagram of

$$s(s+1) + K = 0 \qquad (7\text{-}36)$$

with K incremented uniformly over 100 values from -20 to 20. The RL are computed and plotted digitally. Each dot on the root-locus plot represents one root for a distinct value of K. Thus, we see that the root sensitivity is low when the magnitude of K is large. As the magnitude of K decreases, the movements of the roots become larger for the same incremental change in K. At the breakaway point, $s = -0.5$, the root sensitivity is infinite.

Fig. 7-5 shows the RL of

$$s^2(s+1)^2 + K(s+2) = 0 \qquad (7\text{-}37)$$

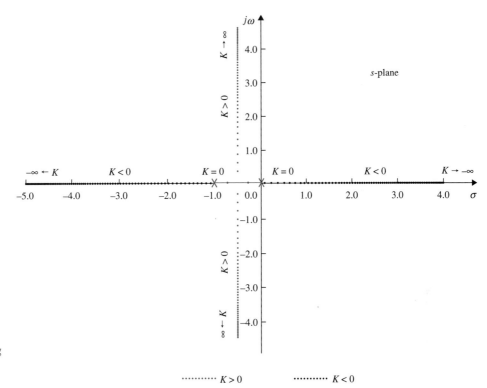

Figure 7-4 RL of $s(s+1)+K=0$ showing the root sensitivity with respect to K.

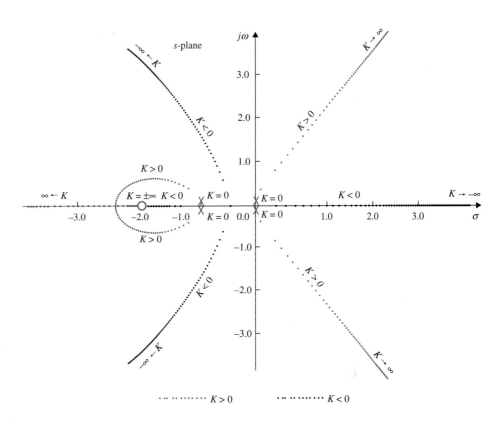

Figure 7-5 RL of $s^2(s+1)^2 + K(s+2) = 0$, showing the root sensitivity with respect to K.

with K incremented uniformly over 200 values from -40 to 50. Again, the loci show that the root sensitivity increases as the roots approach the breakaway points at $s = 0, -0.543, -1.0$, and -2.457. We can investigate the root sensitivity further by using the expression in Eq. (7-34). For the second-order equation in Eq. (7-36),

$$\frac{dK}{ds} = -2s - 1 \tag{7-38}$$

Toolbox 7-3-2

MATLAB statements for Eqs. 7-36 and 7-37

```
num1=[1];
den1=conv([1 0],[1 1]);
mysys1=tf(num1,den1);
subplot(2,1,1);
rlocus(mysys1);
title('Root loci for equation 7.36');
[k,poles] = rlocfind(mysys1) %rlocfind command in MATLAB can choose the
desired poles on the locus.

num2=[1 2];
den2=conv([1 0 0],[1 1]);
den2=conv(den2,[1 1]);
subplot(2,1,2)
mysys2=tf(num2,den2);
rlocus(mysys2);
title('Root loci for equation 7-37');
axis([-3 0 -8 8])
[k,poles] = rlocfind(mysys2)
```

From Eq. (7-36), $K = -s(s + 1)$; the root sensitivity becomes

$$S_K = \frac{ds}{dK}\frac{K}{s} = \frac{s+1}{2s+1} \tag{7-39}$$

where $s = \sigma + j\omega$, and s must take on the values of the roots of Eq. (7-39). For the roots on the real axis, $\omega = 0$. Thus, Eq. (7-39) leads to

$$|S_K|_{\omega=0} = \left|\frac{\sigma+1}{2\sigma+1}\right| \tag{7-40}$$

When the two roots are complex, $s = -0.5$ for all values of ω; Eq. (7-39) gives

$$|S_K|_{\sigma=-0.5} = \left(\frac{0.25 + \omega^2}{4\omega^2}\right)^{1/2} \tag{7-41}$$

From Eq. (7-41), it is apparent that the sensitivities of the pair of complex-conjugate roots are the same, since ω appears only as ω^2 in the equation. Eq. (7-40) indicates that the sensitivities of the two real roots are different for a given value of K. Table 7-2 gives the magnitudes of the sensitivities of the two roots of Eq. (7-36) for several values of K, where $|S_{K1}|$ denotes the root sensitivity of the first root, and $|S_{K2}|$ denotes that of the second root. These values indicate that, although the two real roots reach $\sigma = -0.5$ for the same value of

TABLE 7-2 Root Sensitivity

K	ROOT 1	$|S_{K1}|$	ROOT 2	$|S_{K2}|$
0	0	1.000	−1.000	0
0.04	−0.042	1.045	−0.958	0.454
0.16	−0.200	1.333	−0.800	0.333
0.24	−0.400	3.000	−0.600	2.000
0.25	−0.500	∞	−0.500	∞
0.28	$-0.5 + j0.173$	1.527	$-0.5 - j0.173$	1.527
0.40	$-0.5 + j0.387$	0.817	$-0.5 - j0.387$	0.817
1.20	$-0.5 + j0.975$	0.562	$-0.5 - j0.975$	0.562
4.00	$-0.5 + j1.937$	0.516	$-0.5 - j1.937$	0.516
∞	$-0.5 + j\infty$	0.500	$-0.5 - j\infty$	0.500

$K = 0.25$, and each root travels the same distance from $\sigma = 0$ and $s = -1$, respectively, the sensitivities of the two real roots are not the same.

▶ 7-4 DESIGN ASPECTS OF THE ROOT LOCI

One of the important aspects of the root-locus technique is that, for most control systems with moderate complexity, the analyst or designer can obtain vital information on the performance of the system by making a quick sketch of the RL using some or all of the properties of the root loci. It is of importance to understand all the properties of the RL even when the diagram is to be plotted with the help of a digital computer program. From the design standpoint, it is useful to learn the effects on the RL when poles and zeros of $G(s)H(s)$ are added or moved around in the s-plane. Some of these properties are helpful in the construction of the root-locus diagram. The design of the PI, PID, phase-lead, phase-lag, and the lead-lag controllers discussed in Chapter 9 all have implications of adding poles and zeros to the loop transfer function in the s-plane.

7-4-1 Effects of Adding Poles and Zeros to $G(s)H(s)$

The general problem of controller design in control systems may be treated as an investigation of the effects to the root loci when poles and zeros are added to the loop transfer function $G(s)H(s)$.

Addition of Poles to $G(s)H(s)$

Adding a pole to $G(s)H(s)$ has the effect of pushing the root loci toward the right-half s-plane. The effect of adding a zero to $G(s)H(s)$ can be illustrated with several examples.

▶ **EXAMPLE 7-4-1** Consider the function

$$G(s)H(s) = \frac{K}{s(s+a)} \quad a > 0 \qquad (7\text{-}42)$$

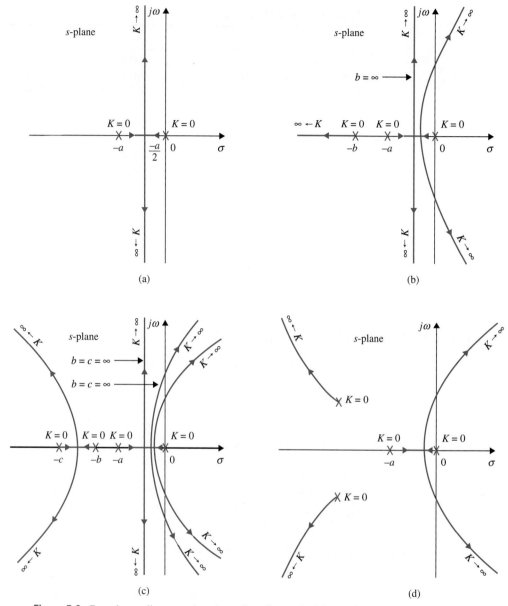

Figure 7-6 Root-locus diagrams that show the effects of adding poles to $G(s)H(s)$.

The RL of $1 + G(s)H(s) = 0$ are shown in Fig. 7-6(a). These RL are constructed based on the poles of $G(s)H(s)$, which are at $s = 0$ and $-a$. Now let us introduce a pole at $s = -b$, with $b > a$. The function $G(s)H(s)$ now becomes

$$G(s)H(s) = \frac{K}{s(s+a)(s+b)} \tag{7-43}$$

Fig. 7-6(b) shows that the pole at $s = -b$ causes the complex part of the root loci to bend toward the right-half s-plane. The angles of the asymptotes for the complex roots are changed from $\pm 90°$ to $\pm 60°$. The intersect of the asymptotes is also moved from $-a/2$ to $-(a+b)/2$ on the real axis.

Toolbox 7-4-1

MATLAB statements for Fig. 7-3

The results for Fig. 7-6 can be obtained by the following Matlab statements:

```
a=2;
b=3;
c=5;

num4=[1];
den4=conv([1 0],[1 a]);
subplot(2,2,1)
mysys4=tf(num4,den4);
rlocus(mysys4);
axis([-3 0 -8 8])

num3=[1];
den3=conv([1 0],conv([1 a],[1 a/2]));
subplot(2,2,2)
mysys3=tf(num3,den3);
rlocus(mysys3);
axis([-3 0 -8 8])

num2=[1];
den2=conv([1 0],conv([1 a],[1 b]));
subplot(2,2,3)
mysys2=tf(num2,den2);
rlocus(mysys2);
axis([-3 0 -8 8])

num1=[1];
den1=conv([1 0],conv([1 a],[1 b]));
den1=conv(den1, [1 c]);
mysys1=tf(num1,den1);
subplot(2,2,4);
rlocus(mysys1);
```

If $G(s)H(s)$ represents the loop transfer function of a control system, the system with the root loci in Fig. 7-6(b) may become unstable if the value of K exceeds the critical value for stability, whereas the system represented by the root loci in Fig. 7-6(a) is always stable for $K > 0$. Fig. 7-6(c) shows the root loci when another pole is added to $G(s)H(s)$ at $s = -c$, $c > b$. The system is now of the fourth order, and the two complex root loci are bent farther to the right. The angles of asymptotes of these two complex loci are now $\pm 45°$. The stability condition of the fourth-order system is even more acute than that of the third-order system. Fig. 7-6(d) illustrates that the addition of a pair of complex-conjugate poles to the transfer function of Eq. (7-42) will result in a similar effect. Therefore, we may draw a general conclusion that the addition of poles to $G(s)H(s)$ has the effect of moving the dominant portion of the root loci toward the right-half s-plane. ◄

Addition of Zeros to $G(s)H(s)$

Adding left-half plane zeros to the function $G(s)H(s)$ generally has the effect of moving and bending the root loci toward the left-half s-plane.

The following example illustrates the effect of adding a zero and zeros to $G(s)H(s)$ on the RL.

▶ **EXAMPLE 7-4-2** Fig. 7-7(a) shows the RL of the $G(s)H(s)$ in Eq. (7-42) with a zero added at $s = -b(b > a)$. The complex-conjugate part of the RL of the original system is bent toward the left and forms a circle. Thus, if $G(s)H(s)$ is the loop transfer function of a control system, the relative stability of the system is improved by the addition of the zero. Fig. 7-7(b) shows that a similar effect will result if a pair of complex-conjugate zeros is added to the function of Eq. (7-42). Fig. 7-7(c) shows the RL when a zero at $s = -c$ is added to the transfer function of Eq. (7-43).

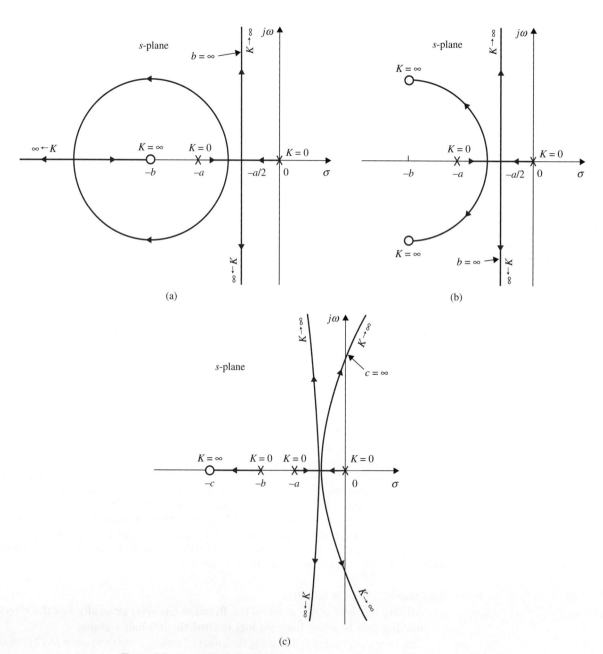

Figure 7-7 Root-locus diagrams that show the effects of adding zeros to $G(s)H(s)$.

Toolbox 7-4-2

MATLAB statements for Fig. 7-7

```
a=2;
b=3;
d=6;
c=20;

num4=[1 d];
den4=conv([1 0],[1 a]);
subplot(2,2,1)
mysys4=tf(num4,den4);
rlocus(mysys4);

num3=[1 c];
den3=conv([1 0],[1 a]);
subplot(2,2,2)
mysys3=tf(num3,den3);
rlocus(mysys3);
axis([-6 0 -8 8])

num2=[1 d];
den2=conv([1 0],conv([1 a],[1 b]));
subplot(2,2,3)
mysys2=tf(num2,den2);
rlocus(mysys2);
axis([-6 0 -8 8])
```

▶ **EXAMPLE 7-4-3** Consider the equation

$$s^2(s+a) + K(s+b) = 0 \tag{7-44}$$

Dividing both sides of Eq. (7-44) by the terms that do not contain K, we have the loop transfer function

$$G(s)H(s) = \frac{K(s+b)}{s^2(s+a)} \tag{7-45}$$

It can be shown that the nonzero breakaway points depend on the value of a and are

$$s = -\frac{a+3}{4} \pm \frac{1}{4}\sqrt{a^2 - 10a + 9} \tag{7-46}$$

Fig. 7-8 shows the RL of Eq. (7-44) with $b = 1$ and several values of a. The results are summarized as follows:

Fig. 7-8(a): $a = 10$. Breakaway points: $s = -2.5$ and -4.0.

Fig. 7-8(b): $a = 9$. The two breakaway points given by Eq. (7-46) converge to one point at $s = -3$. Note the change in the RL when the pole at $-a$ is moved from -10 to -9.

For values of a less than 9, the values of s as given by Eq. (7-46) no longer satisfy Eq. (7-44), which means that there are no finite, nonzero, breakaway points.

Fig. 7-8(c): $a = 8$. No breakaway point on RL.

As the pole at $s = -a$ is moved farther to the right along the real axis, the complex portion of the RL is pushed farther toward the right-half plane.

Fig. 7-8(d): $a = 3$.

Fig. 7-8(e): $a = b = 1$. The pole at $s = -a$ and the zero at $-b$ cancel each other out, and the RL degenerate into a second-order case and lie entirely on the $j\omega$-axis.

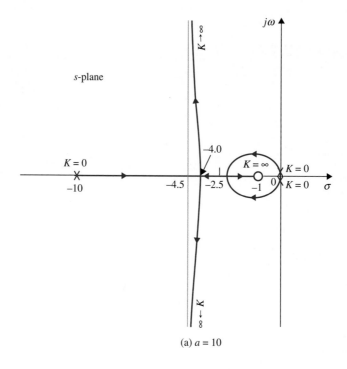

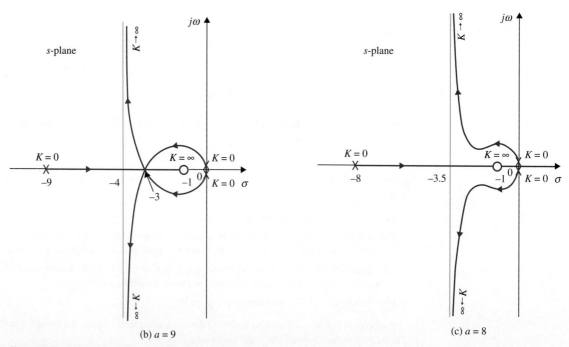

Figure 7-8 Root-locus diagrams that show the effects of moving a pole of $G(s)H(s)$. $G(s)H(s) = K(s+1)/[s2(s+a)]$ (*Continued*).

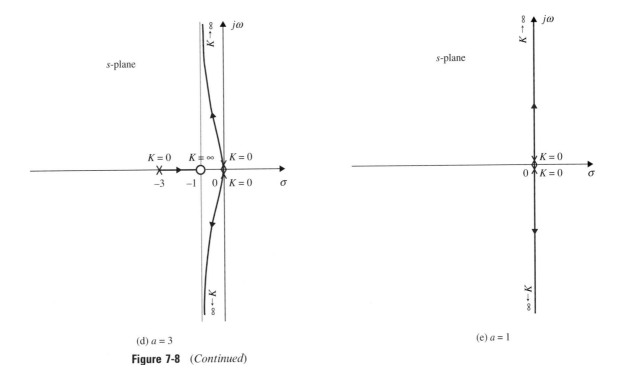

(d) $a = 3$

(e) $a = 1$

Figure 7-8 (*Continued*)

Toolbox 7-4-3

MATLAB statements for Fig. 7-8

```
a1=10;a2=9;a3=8;a4=3;b=1;
  num1=[1 b];
den1=conv([1 0 0],[1 a1]);
subplot(2,2,1)
mysys1=tf(num1,den1);
rlocus(mysys1);

num2=[1 b];
den2=conv([1 0 0],[1 a2]);
subplot(2,2,2)
mysys2=tf(num2,den2);
rlocus(mysys2);

num3=[1 b];
den3=conv([1 0 0],[1 a3]);
subplot(2,2,3)
mysys3=tf(num3,den3);
rlocus(mysys3);

num4=[1 b];
den4=conv([1 0 0],[1 a4]);
subplot(2,2,4)
mysys4=tf(num4,den4);
rlocus(mysys4);
```

EXAMPLE 7-4-4 Consider the equation

$$s(s^2 + 2s + a) + K(s + 2) = 0 \qquad (7\text{-}47)$$

which leads to the equivalent $G(s)H(s)$ as

$$G(s)H(s) = \frac{K(s+2)}{s(s^2 + 2s + a)} \qquad (7\text{-}48)$$

The objective is to study the RL for various values of $a(>0)$. The breakaway point equation of the RL is determined as

$$s^3 + 4s^2 + 4s + a = 0 \qquad (7\text{-}49)$$

Fig. 7-9 shows the RL of Eq. (7-47) under the following conditions.

Fig. 7-9(a): $a = 1$. Breakaway points: $s = -0.38, -1.0$, and -2.618, with the last point being on the RL for $K \geq 0$. As the value of a is increased from unity, the two double poles of $G(s)H(s)$ at $s = -1$ will move vertically up and down with the real parts equal to -1. The breakaway points at $s = -0.38$ and $s = -2.618$ will move to the left, whereas the breakaway point at $s = -1$ will move to the right.

Fig. 7-9(b): $a = 1.12$. Breakaway points: $s = -0.493, -0.857$, and -2.65. Because the real parts of the poles and zeros of $G(s)H(s)$ are not affected by the value of a, the intersect of the asymptotes is always at $s = 0$.

Fig. 7-9(c): $a = 1.185$. Breakaway points: $s = -0.667, -0.667$, and -2.667. The two breakaway points of the RL that lie between $s = 0$ and -1 converge to a point.

Fig. 7-9(d): $a = 3$. Breakaway point: $s = -3$. When a is greater than 1.185, Eq. (7-49) yields only one solution for the breakaway point.

The reader may investigate the difference between the RL in Figs. 7-9(c) and 7-9(d) and fill in the evolution of the loci when the value of a is gradually changed from 1.185 to 3 and beyond.

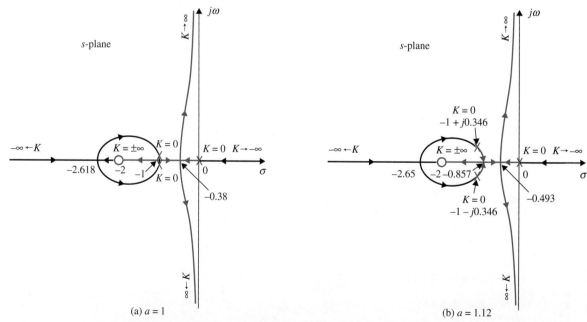

(a) $a = 1$ (b) $a = 1.12$

Figure 7-9 Root-locus diagrams that show the effects of moving a pole of $G(s)H(s) = K(s+2)/[s(s^2 + 2s + a)]$ (*Continued*).

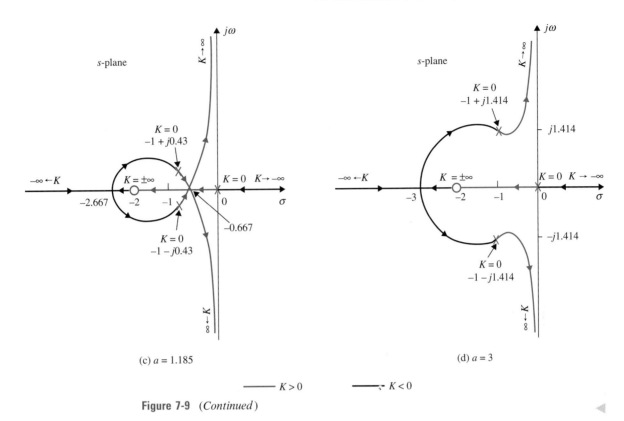

(c) $a = 1.185$

(d) $a = 3$

——— $K > 0$ ---- $K < 0$

Figure 7-9 (*Continued*)

7-5 ROOT CONTOURS (RC): MULTIPLE-PARAMETER VARIATION

The root-locus technique discussed thus far is limited to only one variable parameter in K. In many control-systems problems, the effects of varying several parameters should be investigated. For example, when designing a controller that is represented by a transfer function with poles and zeros, it would be useful to investigate the effects on the characteristic equation roots when these poles and zeros take on various values. In Section 7-4, the root loci of equations with two variable parameters are studied by fixing one parameter and assigning different values to the other. In this section, the multiparameter problem is investigated through a more systematic method of embedding. When more than one parameter varies continuously from $-\infty$ to ∞, the root loci are referred to as the **root contours (RC)**. It will be shown that the root contours still possess the same properties as the single-parameter root loci, so that the methods of construction discussed thus far are all applicable.

The principle of root contour can be described by considering the equation

$$P(s) + K_1 Q_1(s) + K_2 Q_2(s) = 0 \tag{7-50}$$

where K_1 and K_2 are the variable parameters and $P(s)$, $Q_1(s)$, and $Q_2(s)$ are polynomials of s. The first step involves setting the value of one of the parameters to zero. Let us set K_2 to zero. Then, Eq. (7-50) becomes

$$P(s) + K_1 Q_1(s) = 0 \tag{7-51}$$

which now has only one variable parameter in K_1. The root loci of Eq. (7-51) may be determined by dividing both sides of the equation by P(s). Thus,

$$1 + \frac{K_1 Q_1(s)}{P(s)} = 0 \qquad (7-52)$$

Eq. (7-52) is of the form of $1 + K_1 G_1(s)H_1(s) = 0$, so we can construct the RL of the equation based on the pole–zero configuration of $G_1(s)H_1(s)$. Next, we restore the value of K_2, while considering the value of K_1 fixed, and divide both sides of Eq. (7–50) by the terms that do not contain K_2. We have

$$1 + \frac{K_2 Q_2(s)}{P(s) + K_1 Q_1(s)} = 0 \qquad (7-53)$$

which is of the form of $1 + K_2 G_2(s)H_2(s) = 0$. The root contours of Eq. (7-50) when K_2 varies (while K_1 is fixed) are constructed based on the pole–zero configuration of

$$G_2(s)H_2(s) = \frac{Q_2(s)}{P(s) + K_1 Q_1(s)} \qquad (7-54)$$

It is important to note that the poles of $G_2(s)H_2(s)$ are identical to the roots of Eq. (7-51). Thus, the root contours of Eq. (7-50) when K_2 varies must all start ($K_2 = 0$) at the points that lie on the root loci of Eq.(7-51). This is the reason why one root-contour problem is considered to be embedded in another. The same procedure may be extended to more than two variable parameters. The following examples illustrate the construction of RCs when multiparameter-variation situations exist.

▶ **EXAMPLE 7-5-1** Consider the equation

$$s^3 + K_2 s^2 + K_1 s + K_1 = 0 \qquad (7-55)$$

where K_1 and K_2 are the variable parameters, which vary from 0 to ∞.

As the first step, we let $K_2 = 0$, and Eq. (7-55) becomes

$$s^3 + K_1 s + K_1 = 0 \qquad (7-56)$$

Dividing both sides of the last equation by s^3, which is the term that does not contain K_1, we have

$$1 + \frac{K_1(s+1)}{s^3} = 0 \qquad (7-57)$$

The root contours of Eq. (7-56) are drawn based on the pole–zero configuration of

$$G_1(s)H_1(s) = \frac{s+1}{s^3} \qquad (7-58)$$

as shown in Fig. 7-10(a). Next, we let K_2 vary between 0 and ∞ while holding K_1 at a constant nonzero value. Dividing both sides of Eq. (7–55) by the terms that do not contain K_2, we have

$$1 + \frac{K_2 s^2}{s^3 + K_1 s + K_1} = 0 \qquad (7-59)$$

Thus, the root contours of Eq. (7-55) when K_2 varies may be drawn from the pole–zero configuration of

$$G_2(s)H_2(s) = \frac{s^2}{s^3 + K_1 s + K_1} \qquad (7-60)$$

The zeros of $G_2(s)H_2(s)$ are at $s = 0, 0$, but the poles are at the zeros of $1 + K_1 G_1(s)H_1(s)$, which are found on the RL of Fig. 7-10(a). Thus, for fixed K_1, the RC when K_2 varies must all emanate from the root contours of Eq. 7-10(a). Figure 7-10(b) shows the root contours of Eq. (7-55) when K_2 varies from 0 to ∞, for $K_1 = 0.0184$, 0.25, and 2.56.

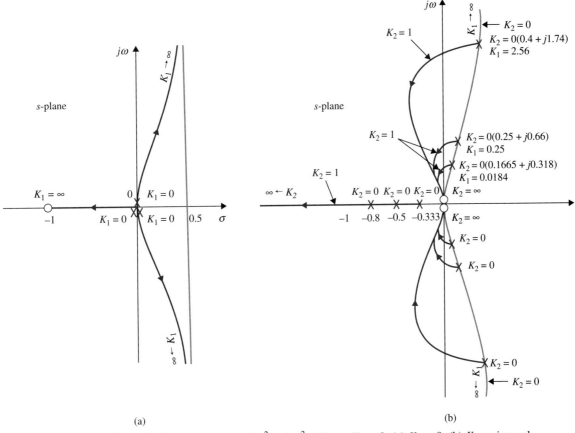

Figure 7-10 Root contours of $s^3 + K_2 s^2 + K_1 s + K_1 = 0$. (a) $K_2 = 0$. (b) K_2 varies and K_1 is a constant.

▶ **EXAMPLE 7-5-2** Consider the loop transfer function

$$G(s)H(s) = \frac{K}{s(1 + Ts)(s^2 + 2s + 2)} \qquad (7\text{-}61)$$

of a closed-loop control system. It is desired to construct the root contours of the characteristic equation with K and T as variable parameters. The characteristic equation of the system is

$$s(1 + Ts)(s^2 + 2s + 2) + K = 0 \qquad (7\text{-}62)$$

First, we set the value of T to zero. The characteristic equation becomes

$$s(s^2 + 2s + 2) + K = 0 \qquad (7\text{-}63)$$

Toolbox 7-5-1

MATLAB statements for Fig. 7-10

```
for k1=[0.0184 0.25 2.56];
num=[1 0 0];
den=[1 0 k1 k1];
mysys=tf(num,den);
rlocus(mysys);
hold on;
end;
```

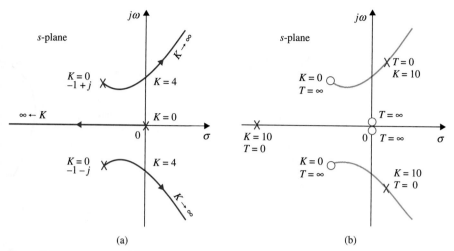

Figure 7-11 (a) RL for $s(s^2 + 2s + 2) + K = 0$. (b) Pole–zero configuration of $G_2(s)H_2(s) = Ts^2(s^2 + 2s + 2)/[s(s^2 + 2s + 2) + K]$.

The root contours of this equation when K varies are drawn based on the pole–zero configuration of

$$G_1(s)H_1(s) = \frac{1}{s(s^2 + 2s + 2)} \tag{7-64}$$

as shown in Fig. 7-11(a). Next, we let K be fixed and consider that T is the variable parameter.

Dividing both sides of Eq. (7-62) by the terms that do not contain T, we get

$$1 + TG_2(s)H_2(s) = 1 + \frac{Ts^2(s^2 + 2s + 2)}{s(s^2 + 2s + 2) + K} = 0 \tag{7-65}$$

The root contours when T varies are constructed based on the pole-zero configuration of $G_2(s)H_2(s)$. When $T = 0$, the points on the root contours are at the poles of $G_2(s)H_2(s)$, which are on the root contours of Eq. (7-63). When $T = \infty$, the roots of Eq. (7-62) are at the zeros of $G_2(s)H_2(s)$, which are at $s = 0, 0, -1 +j$, and $-1 -j$. Figure 7-11(b) shows the pole–zero configuration of $G_2(s)H_2(s)$ for $K = 10$. Notice that $G_2(s)H_2(s)$ has three finite poles and four finite zeros. The root contours for Eq. (7-62) when T varies are shown in Figs. 7-12, 7-13, and 7-14 for three different values of K.

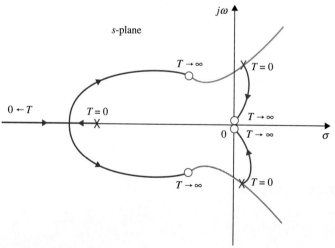

Figure 7-12 Root contours for $s(1 + Ts)(s^2 + 2s + 2) + K = 0$. $K > 4$.

7-5 Root Contours (RC): Multiple-Parameter Variation 397

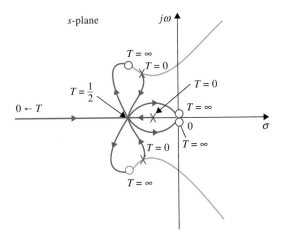

Figure 7-13 Root contours for $s(1 + Ts)(s^2 + 2s + 2) + K = 0$. $K = 0.5$.

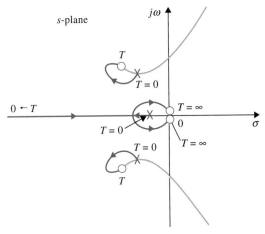

Figure 7-14 Root contours for $s(1 + Ts)(s^2 + 2s + 2) + K = 0$. $K < 0.5$.

The root contours in Fig. 7-13 show that when $K = 0.5$ and $T = 0.5$, the characteristic equation in Eq. (7-62) has a quadruple root at $s = -1$.

Toolbox 7-5-2

MATLAB statements for Example 7-5-2

```
%T = 0
num=[1];den=conv([1 0],conv([0 1],[1 2 2]));
mysys=tf(num,den);
subplot(2,2,1);rlocus(mysys);
  %k>4
for k=4:10;
num=conv([1 0 0],[1 2 2]);den=conv([1 0],[1 2 2]);
den=den+k;
mysys=tf(num,den);
subplot(2,2,2);rlocus(mysys);
end;
k=0.5;
num=conv([1 0 0],[1 2 2]);den = conv([1 0],[1 2 2]);
den=den+k;
mysys=tf(num,den);
```

```
subplot(2,2,3)
rlocus(mysys);
%k<0.5
for k=-100:0.5;
num=conv([1 0 0],[1 2 2]);den=conv([1 0],[1 2 2]);
den=den+k;
mysys=tf(num,den);
subplot(2,2,4);rlocus(mysys);
end;
```

▶ **EXAMPLE 7-5-3** As an example to illustrate the effect of the variation of a zero of $G(s)H(s)$, consider the function

$$G(s)H(s) = \frac{K(1+Ts)}{s(s+1)(s+2)} \qquad (7\text{-}66)$$

The characteristic equation is

$$s(s+1)(s+2) + K(1+Ts) = 0 \qquad (7\text{-}67)$$

Let us first set T to zero and consider the effect of varying K. Eq. (7-67) becomes

$$s(s+1)(s+2) + K = 0 \qquad (7\text{-}68)$$

This leads to

$$G_1(s)H_1(s) = \frac{1}{s(s+1)(s+2)} \qquad (7\text{-}69)$$

The root contours of Eq. (7-68) are drawn based on the pole–zero configuration of Eq. (7-69), and are shown in Fig. 7–15.

When the K is fixed and nonzero, we divide both sides of Eq. (7-67) by the terms that do not contain T, and we get

$$1 + TG_2(s)H_2(s) = 1 + \frac{TKs}{s(s+1)(s+2)+K} = 0 \qquad (7\text{-}70)$$

The points that correspond to $T = 0$ on the root contours are at the poles of $G_2(s)H_2(s)$ or the zeros of $s(s+1)(s+2) + K$, whose root contours are sketched as shown in Fig. 7-15 when K varies. If we choose $K = 20$ just as an illustration, the pole–zero configuration of $G_2(s)H_2(s)$ is shown in Fig. 7-16. The root contours of Eq. (7-67) for $0 \leq T < \infty$ are shown in Fig. 7-17 for three different values of K.

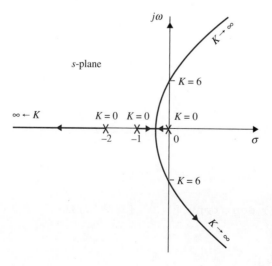

Figure 7-15 Root loci for $s(s+1)(s+2) + K = 0$.

7-5 Root Contours (RC): Multiple-Parameter Variation

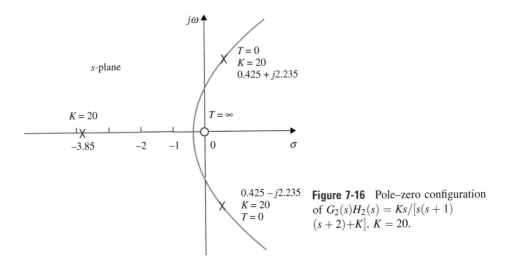

Figure 7-16 Pole–zero configuration of $G_2(s)H_2(s) = Ks/[s(s+1)(s+2)+K]$. $K = 20$.

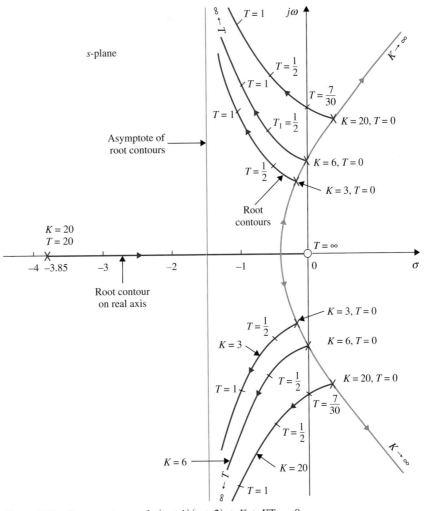

Figure 7-17 Root contours of $s(s+1)(s+2) + K + KTs = 0$.

Toolbox 7-5-3

MATLAB statements for Fig. 7-17

Same results as Fig. 7-17 can be obtained by using the following MATLAB statements:

```
for k= [3 6 20];
num=[k 0];
den=conv([1 0],conv([1 1],[1 2]));
den=den+k;
mysys=tf(num,den);
rlocus(mysys);
axis([-4 4 -10 10]);
hold on
end;
```

Because $G_2(s)H_2(s)$ has three poles and one zero, the angles of the asymptotes of the root contours when T varies are at 90° and –90°. We can show that the intersection of the asymptotes is always at $s = 1.5$. This is because the sum of the poles of $G_2(s)H_2(s)$, which is given by the negative of the coefficient of the s^2 term in the denominator polynomial of Eq. (7-70), is 3; the sum of the zeros of $G_2(s)H_2(s)$ is 0; and $n-m$ in Eq. (7-30) is 2.

The root contours in Fig. 7-17 show that adding a zero to the loop transfer function generally improves the relative stability of the closed-loop system by moving the characteristic equation roots toward the left in the s-plane. As shown in Fig. 7-17, for $K = 20$, the system is stabilized for all values of T greater than 0.2333. However, the largest relative damping ratio that the system can have by increasing T is only approximately 30 percent. ◀

▶ 7-6 MATLAB TOOLS AND CASE STUDIES

Apart from the MATLAB toolboxes appearing in this chapter, this chapter does not contain any software because of its focus on theoretical development. In Chapter 9, when we address more complex control system modeling and analysis, we will introduce the Automatic Control Systems MATLAB tools. The **Automatic Control Systems** software (**ACSYS**) consists of a number of m-files and GUIs (graphical user interface) for the analysis of simple control engineering transfer functions. It can be invoked from the MATLAB command line by simply typing *Acsys* and then by clicking on the appropriate pushbutton. A specific MATLAB tool has been developed for most chapters of this textbook. Throughout this chapter, we have identified subjects that may be solved using **ACSYS**, with a box in the left margin of the text titled "MATLAB TOOL."

▶ 7-7 SUMMARY

In this chapters, we introduced the root-locus technique for linear continuous data control systems. The technique represents a graphical method of investigating the roots of the characteristic equation of a linear time-invariant system when one or more parameters vary. In Chapter 9 the root-locus method will be used heavily for the design of control systems. However, keep in mind that, although the characteristic equation roots give exact indications on the absolute stability of linear SISO systems, they give only qualitative information on the relative stability, since the zeros of the closed-loop transfer function, if any, play an important role in the dynamic performance of the system.

The root-locus technique can also be applied to discrete-data systems with the characteristic equation expressed in the z-transform. As will be shown in Appendix H, the properties and construction of the root loci in the z-plane are essentially the same as those of the continuous-data systems in the s-plane, except that the interpretation of the root location to system performance must be made with respect to the unit circle $|z| = 1$ and the significance of the regions in the z-plane.

The majority of the material in this chapter is designed to provide the basics of constructing the root loci. Computer programs, such as the MATLAB Toolboxes used throughout this chapter, can be used to plot the root loci and provide details of the plot. The final section of Chapter 9 deals with the root-locus tools of MATLAB. However, the authors believe that a computer program can be used only as a tool, and the intelligent investigator should have a thorough understanding of the fundamentals of the subject.

The root-locus technique can also be applied to linear systems with pure time delay in the system loop. The subject is not treated here, since systems with pure time delays are more easily treated with the frequency-domain methods discussed in Chapter 8.

▶ REVIEW QUESTIONS

The following questions and true-and-false problems all refer to the equation $P(s) + KQ(s) = 0$, where $P(s)$ and $Q(s)$ are polynomials of s with constant coefficients.

1. Give the condition from which the root loci are constructed.

2. Determine the points on the complete root loci at which $K = 0$, with reference to the poles and zeros of $Q(s)/P(s)$.

3. Determine the points on the root loci at which $K = \pm\infty$, with reference to the poles and zeros of $Q(s)/P(s)$.

4. Give the significance of the breakaway points with respect to the roots of $P(s) + KQ(s) = 0$.

5. Give the equation of intersect of the asymptotes.

6. The asymptotes of the root loci refer to the angles of the root loci when $K = \pm\infty$. (T) (F)

7. There is only one intersect of the asymptotes of the complete root loci. (T) (F)

8. The intersect of the asymptotes must always be on the real axis. (T) (F)

9. The breakaway points of the root loci must always be on the real axis. (T) (F)

10. Given the equation $1 + KG_1(s)H_1(s) = 0$, where $G_1(s)H_1(s)$ is a rational function of s and does not contain K, the roots of $dG_1(s)H_1(s)/ds$ are all breakaway points on the root loci $(-\infty < K < \infty)$. (T) (F)

11. At the breakaway points on the root loci, the root sensitivity is infinite. (T) (F)

12. Without modification, all the rules and properties for the construction of root loci in the s-plane can be applied to the construction of root loci of discrete-data systems in the z-plane. (T) (F)

13. The determination of the intersections of the root loci in the s-plane with the $j\omega$-axis can be made by solving the auxiliary equation of Routh's tabulation of the equation. (T) (F)

14. Adding a pole to $Q(s)/P(s)$ has the general effect of pushing the root loci to the right, whereas adding a zero pushes the loci to the left. (T) (F)

Answers to these true-and-false questions can be found on this book's companion Web site: www.wiley.com/college/golnaraghi.

REFERENCES

General Subjects

1. W. R. Evans, "Graphical Analysis of Control Systems," *Trans. AIEE*, Vol. 67, pp. 548–551, 1948.
2. W. R. Evans, "Control System Synthesis by Root Locus Method," *Trans. AIEE*, Vol. 69, pp. 66–69, 1950.
3. W. R. Evans, *Control System Dynamics*, McGraw-Hill Book Company, New York, 1954.

Construction and Properties of Root Loci

4. C. C. MacDuff, *Theory of Equations*, pp. 29–104, John Wiley & Sons, New York, 1954.
5. C. S. Lorens and R. C. Titsworth, "Properties of Root Locus Asymptotes," *IRE Trans. Automatic Control*, AC-5, pp. 71–72, Jan. 1960.
6. C. A. Stapleton, "On Root Locus Breakaway Points," *IRE Trans. Automatic Control*, Vol. AC-7, pp. 88–89, April 1962.
7. M. J. Remec, "Saddle-Points of a Complete Root Locus and an Algorithm for Their Easy Location in the Complex Frequency Plane," *Proc. Natl. Electronics Conf.*, Vol. 21, pp. 605–608, 1965.
8. C. F. Chen, "A New Rule for Finding Breaking Points of Root Loci Involving Complex Roots," *IEEE Trans. Automatic Control*, AC-10, pp. 373–374, July 1965.
9. V. Krishran, "Semi-Analytic Approach to Root Locus," *IEEE Trans. Automatic Control*, Vol. AC-11, pp. 102–108, Jan. 1966.
10. R. H. Labounty and C. H. Houpis, "Root Locus Analysis of a High-Grain Linear System with Variable Coefficients: Application of Horowitz's Method," *IEEE Trans. Automatic Control*, Vol. AC-11, pp. 255–263, April 1966.
11. A. Fregosi and J. Feinstein, "Some Exclusive Properties of the Negative Root Locus," *IEEE Trans. Automatic Control*, Vol. AC-14, pp. 304–305, June 1969.

Analytical Representation of Root Loci

12. G. A. Bendrikov and K. F. Teodorchik, "The Analytic Theory of Constructing Root Loci," *Automation and Remote Control*, pp. 340–344, March 1959.
13. K. Steiglitz, "Analytical Approach to Root Loci," *IRE Trans. Automatic Control*, Vol. AC-6, pp. 326–332, Sept. 1961.
14. C. Wojcik, "Analytical Representation of Root Locus," *Trans. ASME*, J. Basic Engineering, Ser. D. Vol. 86, March 1964.
15. C. S. Chang, "An Analytical Method for Obtaining the Root Locus with Positive and Negative Gain," *IEEE Trans. Automatic Control*, Vol. AC-10, pp. 92–94, Jan. 1965.
16. B. P. Bhattacharyya, "Root Locus Equations of the Fourth Degree," *Interna. J. Control*, Vol. 1, No. 6, pp. 533–556, 1965.

Root Sensitivity

17. J. G. Truxal and M. Horowitz, "Sensitivity Consideration in Active Network Synthesis," *Proc. Second Midwest Symposium on Circuit Theory*, East Lansing, MI, 1956.
18. R. Y. Huang, "The Sensitivity of the Poles of Linear Closed-Loop Systems," *IEEE Trans. Appl. Ind.*, Vol. 77, Part 2, pp. 182–187, Sept. 1958.
19. H. Ur, "Root Locus Properties and Sensitivity Relations in Control Systems," *IRE Trans. Automatic Control*, Vol. AC-5, pp. 58–65, Jan. 1960.

PROBLEMS

7-1. Find the angles of the asymptotes and the intersect of the asymptotes of the root loci of the following equations when K varies from $-\infty$ to ∞.

(a) $s^4 + 4s^3 + 4s^2 + (K+8)s + K = 0$

(b) $s^3 + 5s^2 + (K+1)s + K = 0$

(c) $s^2 + K(s^3 + 3s^2 + 2s + 8) = 0$

(d) $s^3 + 2s^2 + 3s + K(s^2 - 1)(s + 3) = 0$

(e) $s^5 + 2s^4 + 3s^3 + K(s^2 + 3s + 5) = 0$

(f) $s^4 + 2s^2 + 10 + K(s + 5) = 0$

7-2. Use MATLAB to solve Problem 7-1.

7-3. Show that the asymptotes angles are

$$\begin{cases} \theta_i = \dfrac{(2i+1)}{|n-m|} \times 180° & K > 0 \\ \theta_i = \dfrac{(2i)}{|n-m|} \times 180° & K < 0 \end{cases}$$

7-4. Prove that the asymptotes center is

$$\sigma_1 = \dfrac{\sum \text{finite poles of } G(s)H(s) - \sum \text{finite zeros of } G(s)H(s)}{n-m}$$

7-5. Plot the asymptotes for $K > 0$ and $K < 0$ for

$$GH = \dfrac{K}{s(s+2)(s^2+2s+2)}$$

7-6. For the loop transfer functions that follow, find the angle of departure or arrival of the root loci at the designated pole or zero.

(a) $G(s)H(s) = \dfrac{Ks}{(s+1)(s^2+1)}$
Angle of arrival ($K < 0$) and angle of departure ($K > 0$) at $s = j$.

(b) $G(s)H(s) = \dfrac{Ks}{(s-1)(s^2+1)}$
Angle of arrival ($K < 0$) and angle of departure ($K > 0$) at $s = j$.

(c) $G(s)H(s) = \dfrac{K}{s(s+2)(s^2+2s+2)}$
Angle of departure ($K > 0$) at $s = -1+j$.

(d) $G(s)H(s) = \dfrac{K}{s^2(s^2+2s+2)}$
Angle of departure ($K > 0$) at $s = -1+j$.

(e) $G(s)H(s) = \dfrac{K(s^2+2s+2)}{s^2(s+2)(s+3)}$
Angle of arrival ($K > 0$) at $s = -1+j$.

7-7. Prove that:
(a) the departure angle of the root locus from a complex pole is $\theta_D = 180° - \arg GH'$, where $\arg GH'$ is the phase angle of GH at the complex pole, ignoring the effect of that pole.
(b) the arrival angle of the root locus at the complex zero is $\theta_D = 180° - \arg GH''$, where $\arg GH''$ is the phase angle of GH at the complex zero, ignoring the contribution of that particular zero.

7-8. Find the angles of departure and arrival for all complex poles and zeros of the open-loop transfer function of

$$G(s)H(s) = \dfrac{K(s^2+2s+2)}{s(s^2+4)} \quad K > 0$$

7-9. Mark the $K = 0$ and $K = \pm\infty$ points and the RL and complementory root loci (CRL) on the real axis for the pole–zero configurations shown in Fig. 7P-9. Add arrows on the root loci on the real axis in the direction of increasing K.

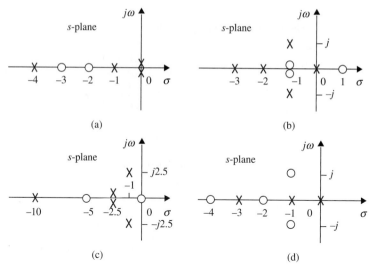

Figure 7P-9

7-10. Prove that a breakaway α satisfies the following:

$$\sum_{i=1}^{n}\frac{1}{\alpha+P_i}=\sum_{i=1}^{m}\frac{1}{\alpha+Z_i}$$

7-11. Find all the breakaway points of the root loci of the systems described by the pole–zero configurations shown in Fig. 7P-9.

7-12. Construct the root-locus diagram for each of the following control systems for which the poles and zeros of $G(s)H(s)$ are given. The characteristic equation is obtained by equating the numerator of $1+G(s)H(s)$ to zero.
(a) Poles at 0, −5, −6; zero at −8
(b) Poles at 0, −1, −3, −4; no finite zeros
(c) Poles at 0, 0, −2, −2; zero at −4
(d) Poles at 0, −1 + j, −1 − j; zero at −2
(e) Poles at 0, −1 + j, −1 − j; zero at −5
(f) Poles at 0, −1 + j, −1 − j; no finite zeros
(g) Poles at 0, 0, −8, −8; zeros at −4, −4
(h) Poles at 0, 0, −8, −8; no finite zeros
(i) Poles at 0, 0, −8, −8; zeros at −4 + j2, −4 − j2
(j) Poles at −2, 2; zeros at 0, 0
(k) Poles at j, −j, j2, −j2; zeros at −2, 2
(l) Poles at j, −j, j2, −j2; zeros at −1, 1
(m) Poles at 0, 0, 0, 1; zeros at −1, −2, −3
(n) Poles at 0, 0, 0, −100, −200; zeros at −5, −40
(o) Poles at 0, −1, −2; zero at 1

7-13. Use MATLAB to solve Problem 7-12.

7-14. The characteristic equations of linear control systems are given as follows. Construct the root loci for $K \geq 0$.
(a) $s^3 + 3s^2 + (K+2)s + 5K = 0$
(b) $s^3 + s^2 + (K+2)s + 3K = 0$

(c) $s^3 + 5Ks^2 + 10 = 0$
(d) $s^4 + (K+3)s^3 + (K+1)s^2 + (2K+5) + 10 = 0$
(e) $s^3 + 2s^2 + 2s + K(s^2-1)(s+2) = 0$
(f) $s^3 - 2s + K(s+4)(s+1) = 0$
(g) $s^4 + 6s^3 + 9s^2 + K(s^2+4s+5) = 0$
(h) $s^3 + 2s^2 + 2s + K(s^2-2)(s+4) = 0$
(i) $s(s^2-1) + K(s+2)(s+0.5) = 0$
(j) $s^4 + 2s^3 + 2s^2 + 2Ks + 5K = 0$
(k) $s^5 + 2s^4 + 3s^3 + 2s^2 + s + K = 0$

7-15. Use MATLAB to solve Problem 7-14.

7-16. The forward-path transfer functions of a unity-feedback control system are given in the following:

(a) $G(s) = \dfrac{K(s+3)}{s(s^2+4s+4)(s+5)(s+6)}$

(b) $G(s) = \dfrac{K}{s(s+2)(s+4)(s+10)}$

(c) $G(s) = \dfrac{K(s^2+2s+8)}{s(s+5)(s+10)}$

(d) $G(s) = \dfrac{K(s^2+4)}{(s+2)^2(s+5)(s+6)}$

(e) $G(s) = \dfrac{K(s+10)}{s^2(s+2.5)(s^2+2s+2)}$

(f) $G(s) = \dfrac{K}{(s+1)(s^2+4s+5)}$

(g) $G(s) = \dfrac{K(s+2)}{(s+1)(s^2+6s+10)}$

(h) $G(s) = \dfrac{K(s+2)(s+3)}{s(s+1)}$

(i) $G(s) = \dfrac{K}{s(s^2+4s+5)}$

Construct the root loci for $K \geq 0$. Find the value of K that makes the relative damping ratio of the closed-loop system (measured by the dominant complex characteristic equation roots) equal to 0.707, if such a solution exists.

7-17. Use MATLAB to verify your answer to Problem 7-16.

7-18. A unity-feedback control system has the forward-path transfer functions given in the following. Construct the root locus diagram for $K \geq 0$. Find the values of K at all the breakaway points.

(a) $G(s) = \dfrac{K}{s(s+10)(s+20)}$

(b) $G(s) = \dfrac{K}{s(s+1)(s+3)(s+5)}$

(c) $G(s) = \dfrac{K(s-0.5)}{(s-1)^2}$

(d) $G(s) = \dfrac{K}{(s+0.5)(s-1.5)}$

(e) $G(s) = \dfrac{K(s+\frac{1}{3})(s+1)}{s(s+\frac{1}{2})(s-1)}$

(f) $G(s) = \dfrac{K}{s(s^2+6s+25)}$

7-19. Use MATLAB to verify your answer to Problem 7-18.

7-20. The forward-path transfer function of a unity-feedback control system is

$$G(s) = \dfrac{K}{(s+4)^n}$$

Construct the root loci of the characteristic equation of the closed-loop system for $K \geq \infty$, with (a) $n=1$, (b) $n=2$, (c) $n=3$, (d) $n=4$, and (e) $n=5$.

7-21. Use MATLAB to solve Problem 7-20.

7-22. The characteristic equation of the control system shown in Fig. 5P-16 when $K=100$ is

$$s^3 + 25s^2 + (100K_t + 2)s + 100 = 0$$

Construct the root loci of the equation for $K_t \geq 0$.

7-23. Use MATLAB to verify your answer to Problem 7-22.

7-24. The block diagram of a control system with tachometer feedback is shown in Fig. 7P-24.
(a) Construct the root loci of the characteristic equation for $K \geq 0$ when $K_t = 0$.
(b) Set $K = 10$. Construct the root loci of the characteristic equation for $K_t \geq 0$.

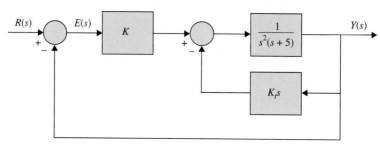

Figure 7P-24

7-25. Use MATLAB to solve Problem 7-24.

7-26. The characteristic equation of the dc-motor control system described in Problems 4-49 and 5-40 can be approximated as

$$2.05 J_L s^3 + (1 + 10.25 J_L) s^2 + 116.84 s + 1843 = 0$$

when $K_L = \infty$ and the load inertia J_L is considered as a variable parameter. Construct the root loci of the characteristic equation for $J_L \geq 0$.

7-27. Use MATLAB to verify your answer to Problem 7-26.

7-28. The forward-path transfer function of the control system shown in Fig. 7P-24 is

$$G(s) = \frac{K(s+\alpha)(s+3)}{s(s^2 - 1)}$$

(a) Construct the root loci for $K \geq 0$ with $\alpha = 5$.
(b) Construct the root loci for $\alpha \geq 0$ with $K = 10$.

7-29. Use MATLAB to solve Problem 7-28.

7-30. The forward-path transfer function of a control system is

$$G(s) = \frac{K(s+0.4)}{s^2(s+3.6)}$$

(a) Construct the root loci for $K \geq 0$.
(b) Use MATLAB to verify your answer to part (a).

7-31. The characteristic equation of the liquid-level control system described in Problem 5-42 is written

$$0.06 s (s + 12.5)(As + K_o) + 250 N = 0$$

(a) For $A = K_o = 50$, construct the root loci of the characteristic equation as N varies from 0 to ∞.
(b) For $N = 10$ and $K_o = 50$, construct the root loci of the characteristic equation for $A \geq 0$.
(c) For $A = 50$ and $N = 20$, construct the root loci for $K_o \geq 0$.

7-32. Use MATLAB to solve Problem 7-31.

7-33. Repeat Problem 7-31 for the following cases.
(a) $A = K_o = 100$ (b) $N = 20$ and $K_o = 50$ (c) $A = 100$ and $N = 20$

7-34. Use MATLAB to verify your answer to Problem 7-33.

7-35. The forward-path transfer function of a unity-feedback system is

$$G(s) = \frac{K(s+2)^2}{(s^2+4)(s+5)^2}$$

(a) Construct the root loci for $K = 25$.
(b) Find the range of K value for which the system is stable.
(c) Use MATLAB to verify your answer to part (a).

7-36. The transfer functions of a single-feedback-loop control system are

$$G(s) = \frac{K}{s^2(s+1)(s+5)} \quad H(s) = 1$$

(a) Construct the loci of the zeros of $1 + G(s)$ for $K \geq 0$.
(b) Repeat part (a) when $H(s) = 1 + 5s$.

7-37. Use MATLAB to solve Problem 7-36.

7-38. The forward-path transfer function of a unity-feedback system is

$$G(s) = \frac{Ke^{-Ts}}{s+1}$$

(a) Construct the root loci for $T = 1$ sec and $K > 0$.
(b) Find the values of K where the system is stable.
(c) Use MATLAB to verify your answer to part (a).

7-39. The transfer functions of a single-feedback-loop control system are

$$G(s) = \frac{10}{s^2(s+1)(s+5)} \quad H(s) = 1 + T_d s$$

(a) Construct the root loci of the characteristic equation for $T_d \geq 0$.
(b) Use MATLAB to verify your answer to part (a).

7-40. For the dc-motor control system described in Problems 4-49 and 5-40, it is of interest to study the effects of the motor-shaft compliance K_L on the system performance.

(a) Let $K = 1$, with the other system parameters as given in Problems 4-49 and 5-40. Find an equivalent $G(s)H(s)$ with K_L as the gain factor. Construct the root loci of the characteristic equation for $K_L \geq 0$. The system can be approximated as a fourth-order system by canceling the large negative pole and zero of $G(s)H(s)$ that are very close to each other.
(b) Repeat part (a) with $K = 1000$.

7-41. Use MATLAB to verify your answer to Problem 7-40.

7-42. The characteristic equation of the dc-motor control system described in Problems 4-49 and 5-40 is given in the following when the motor shaft is considered to be rigid ($K_L = \infty$). Let $K = 1$, $J_m = 0.001$, $L_a = 0.001$, $n = 0.1$, $R_a = 5$, $K_i = 9$, $K_b = 0.0636$, $B_m = 0$, and $K_S = 1$.

$$L_a(J_m + n^2 J_L)s^3 + (R_a J_m + n^2 R_a J_L + B_m L_a)s^2 + (R_a B_m + K_i K_b)s + nK_s K_i K = 0$$

(a) Construct the root loci for $J_L \geq 0$ to show the effects of variation of the load inertia on system performance.
(b) Use MATLAB to verify your answer to part (a).

7-43. Given the equation $s^3 + \alpha s^2 + Ks + K = 0$, it is desired to investigate the root loci of this equation for $-\infty < K < \infty$ and for various values of α.

(a) Construct the root loci for $-\infty < K < \infty$ when $\alpha = 12$.
(b) Repeat part (a) when $\alpha = 4$.
(c) Determine the value of α so that there is only one nonzero breakaway point on the entire root loci for $-\infty < K < \infty$. Construct the root loci.

7-44. Use MATLAB to solve Problem 7-43.

7-45. The forward-path transfer function of a unity-feedback control system is

$$G(s) = \frac{K(s+\alpha)}{s^2(s+3)}$$

Determine the values of α so that the root loci $(-\infty < K < \infty)$ will have zero, one, and two breakaway points, respectively, not including the one at $s = 0$. Construct the root loci for $-\infty < K < \infty$ for all three cases.

7-46. Fig. 7P-46 shows the block diagram of a unity-feedback control system. Design a proper controller $H(s)$ for the system.

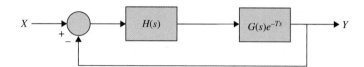

Figure 7P-46

7-47. The pole–zero configuration of $G(s)H(s)$ of a single-feedback-loop control system is shown in Fig. 7P-47(a). Without actually plotting, apply the angle-of-departure (and -arrival) property of the root loci to determine which root-locus diagram shown is the correct one.

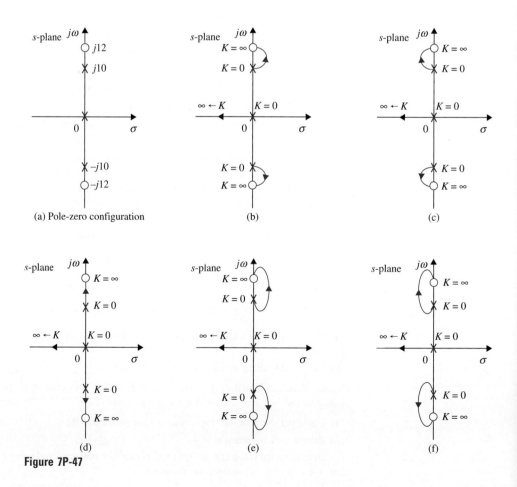

Figure 7P-47

CHAPTER 8

Frequency-Domain Analysis

▶ 8-1 INTRODUCTION

The basic concepts and background material for this subject appear in Chapter 2. In practice, the performance of a control system is more realistically measured by its time-domain characteristics. The reason is that the performance of most control systems is judged based on the time responses due to certain test signals. This is in contrast to the analysis and design of communication systems for which the frequency response is of more importance, since most of the signals to be processed are either sinusoidal or composed of sinusoidal components. We learned in Chapter 5 that the time response of a control system is usually more difficult to determine analytically, especially for high-order systems. In design problems, there are no unified methods of arriving at a designed system that meets the time-domain performance specifications, such as maximum overshoot, rise time, delay time, settling time, and so on. On the other hand, in the frequency domain, there is a wealth of graphical methods available that are not limited to low-order systems. It is important to realize that there are correlating relations between the frequency-domain and the time-domain performances in a linear system, so the time-domain properties of the system can be predicted based on the frequency-domain characteristics. The frequency domain is also more convenient for measurements of system sensitivity to noise and parameter variations. With these concepts in mind, we consider the primary motivation for conducting control systems analysis and design in the frequency domain to be convenience and the availability of the existing analytical tools. Another reason is that it presents an alternative point of view to control-system problems, which often provides valuable or crucial information in the complex analysis and design of control systems. Therefore, to conduct a frequency-domain analysis of a linear control system does not imply that the system will only be subject to a sinusoidal input. It may never be. Rather, from the frequency-response studies, we will be able to project the time-domain performance of the system.

The starting point for frequency-domain analysis of a linear system is its transfer function. It is well known from linear system theory that when the input to a linear time-invariant system is sinusoidal with amplitude R and frequency ω_0,

$$r(t) = R\sin\omega_0 t \tag{8-1}$$

the steady-state output of the system, $y(t)$, will be a sinusoid with the same frequency ω_0 but possibly with different amplitude and phase; that is,

$$y(t) = Y\sin(\omega_0 t + \phi) \tag{8-2}$$

where Y is the amplitude of the output sine wave and ϕ is the phase shift in degrees or radians. Let the transfer function of a linear SISO system be $M(s)$; then the Laplace transforms of the input and the output are related through

$$Y(s) = M(s)R(s) \tag{8-3}$$

For sinusoidal steady-state analysis, we replace s by $j\omega$, and the last equation becomes

$$Y(j\omega) = M(j\omega)R(j\omega) \tag{8-4}$$

By writing the function $Y(j\omega)$ as

$$Y(j\omega) = |Y(j\omega)|\angle Y(j\omega) \tag{8-5}$$

with similar definitions for $M(j\omega)$ and $R(j\omega)$, Eq. (8-4) leads to the magnitude relation between the input and the output:

$$|Y(j\omega)| = |M(j\omega)||R(j\omega)| \tag{8-6}$$

and the phase relation:

$$\angle Y(j\omega) = \angle M(j\omega) + \angle R(j\omega) \tag{8-7}$$

Thus, for the input and output signals described by Eqs. (8-1) and (8-2), respectively, the amplitude of the output sinusoid is

$$Y = R|M(j\omega_0)| \tag{8-8}$$

and the phase of the output is

$$\phi = \angle M(j\omega_0) \tag{8-9}$$

Thus, by knowing the transfer function $M(s)$ of a linear system, the magnitude characteristic, $|M(j\omega)|$, and the phase characteristic, $\angle M(j\omega)$, completely describe the steady-state performance when the input is a sinusoid. The crux of frequency-domain analysis is that the amplitude and phase characteristics of a closed-loop system can be used to predict both time-domain transient and steady-state system performances.

8-1-1 Frequency Response of Closed-Loop Systems

For the single-loop control-system configuration studied in the preceding chapters, the closed-loop transfer function is

$$M(s) = \frac{Y(s)}{R(s)} = \frac{G(s)}{1 + G(s)H(s)} \tag{8-10}$$

Under the sinusoidal steady state, $s = j\omega$, Eq. (8-10) becomes

$$M(j\omega) = \frac{Y(j\omega)}{R(j\omega)} = \frac{G(j\omega)}{1 + G(j\omega)H(j\omega)} \tag{8-11}$$

The sinusoidal steady-state transfer function $M(j\omega)$ may be expressed in terms of its magnitude and phase; that is,

$$M(j\omega) = |M(j\omega)|\angle M(j\omega) \tag{8-12}$$

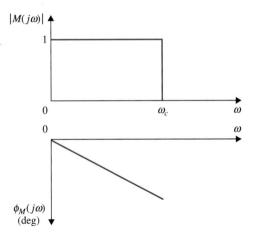

Figure 8-1 Gain-phase characteristics of an ideal low-pass filter.

Or $M(j\omega)$ can be expressed in terms of its real and imaginary parts:

$$M(j\omega) = \text{Re}[M(j\omega)] + j\,\text{Im}[M(j\omega)] \tag{8-13}$$

The magnitude of $M(j\omega)$ is

$$|M(j\omega)| = \left|\frac{G(j\omega)}{1 + G(j\omega)H(j\omega)}\right| = \frac{|G(j\omega)|}{|1 + G(j\omega)H(j\omega)|} \tag{8-14}$$

and the phase of $M(j\omega)$ is

$$\angle M(j\omega) = \phi_M(j\omega) = \angle G(j\omega) - \angle[1 + G(j\omega)H(j\omega)] \tag{8-15}$$

If $M(s)$ represents the input–output transfer function of an electric filter, then the magnitude and phase of $M(j\omega)$ indicate the filtering characteristics on the input signal. Fig. 8-1 shows the gain and phase characteristics of an ideal low-pass filter that has a sharp cutoff frequency at ω_c. It is well known that an ideal filter characteristic is physically unrealizable. In many ways, the design of control systems is quite similar to filter design, and the control system is regarded as a signal processor. In fact, if the ideal low-pass-filter characteristics shown in Fig. 8-1 were physically realizable, they would be highly desirable for a control system, since all signals would be passed without distortion below the frequency ω_c, and completely eliminated at frequencies above ω_c where noise may lie.

If ω_c is increased indefinitely, the output $Y(j\omega)$ would be identical to the input $R(j\omega)$ for all frequencies. Such a system would follow a step-function input in the time domain exactly. From Eq. (8-14), we see that, for $|M(j\omega)|$ to be unity at all frequencies, the magnitude of $G(j\omega)$ must be infinite. An infinite magnitude of $G(j\omega)$ is, of course, impossible to achieve in practice, nor would it be desirable, since most control systems may become unstable when their loop gains become very high. Furthermore, all control systems are subject to noise during operation. Thus, in addition to responding to the input signal, the system should be able to reject and suppress noise and unwanted signals. For control systems with high-frequency noise, such as air-frame vibration of an aircraft, the frequency response should have a finite cutoff frequency ω_c.

The phase characteristics of the frequency response of a control system are also of importance, as we shall see that they affect the stability of the system.

412 ▶ Chapter 8. Frequency-Domain Analysis

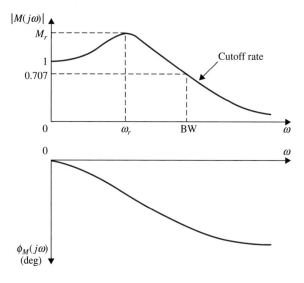

Figure 8-2 Typical gain-phase characteristics of a feedback control system.

Fig. 8-2 illustrates typical gain and phase characteristics of a control system. As shown by Eqs. (8-14) and (8-15), the gain and phase of a closed-loop system can be determined from the forward-path and loop transfer functions. In practice, the frequency responses of $G(s)$ and $H(s)$ can often be determined by applying sine-wave inputs to the system and sweeping the frequency from 0 to a value beyond the frequency range of the system.

8-1-2 Frequency-Domain Specifications

In the design of linear control systems using the frequency-domain methods, it is necessary to define a set of specifications so that the performance of the system can be identified. Specifications such as the maximum overshoot, damping ratio, and the like used in the time domain can no longer be used directly in the frequency domain. The following frequency-domain specifications are often used in practice.

- M_r indicates the relative stability of a stable closed-loop system.

Resonant Peak M_r

The resonant peak M_r is the maximum value of $|M(j\omega)|$.

In general, the magnitude of M_r gives indication on the relative stability of a stable closed-loop system. Normally, a large M_r corresponds to a large maximum overshoot of the step response. For most control systems, it is generally accepted in practice that the desirable value of M_r should be between 1.1 and 1.5.

- BW gives an indication of the transient response properties of a control system.

Resonant Frequency ω_r

The resonant frequency ω_r is the frequency at which the peak resonance M_r occurs.

Bandwidth BW

- BW gives an indication of the noise-filtering characteristics and robustness of the system.

The bandwidth BW is the frequency at which $|M(j\omega)|$ drops to 70.7% of, or 3 dB down from, its zero-frequency value.

In general, the bandwidth of a control system gives indication on the transient-response properties in the time domain. A large bandwidth corresponds to a faster rise time, since higher-frequency signals are more easily passed through the system. Conversely, if the bandwidth is small, only signals of relatively low frequencies are passed, and the time response will be slow and sluggish. Bandwidth also indicates the noise-filtering characteristics and the robustness of the system. The robustness represents a measure of the sensitivity of a system to parameter variations. A robust system is one that is insensitive to parameter variations.

Cutoff Rate
Often, bandwidth alone is inadequate to indicate the ability of a system in distinguishing signals from noise. Sometimes it may be necessary to look at the slope of $|M(j\omega)|$, which is called the cutoff rate of the frequency response, at high frequencies. Apparently, two systems can have the same bandwidth, but the cutoff rates may be different.

The performance criteria for the frequency-domain defined above are illustrated in Fig. 8-2. Other important criteria for the frequency domain will be defined in later sections of this chapter.

▶ 8-2 M_r, ω_r, AND BANDWIDTH OF THE PROTOTYPE SECOND-ORDER SYSTEM

8-2-1 Resonant Peak and Resonant Frequency

For the prototype second-order system defined in Section 5-6, the resonant peak M_r, the resonant frequency ω_r, and the bandwidth BW are all uniquely related to the damping ratio ζ and the natural undamped frequency ω_n of the system.

Consider the closed-loop transfer function of the prototype second-order system

$$M(s) = \frac{Y(s)}{R(s)} = \frac{\omega_n^2}{s^2 + 2\zeta\omega_n s + \omega_n^2} \quad (8\text{-}16)$$

At sinusoidal steady state, $s = j\omega$, Eq. (8-16) becomes

$$M(j\omega) = \frac{Y(j\omega)}{R(j\omega)} = \frac{\omega_n^2}{(j\omega)^2 + 2\zeta\omega_n(j\omega) + \omega_n^2}$$
$$= \frac{1}{1 + j2(\omega/\omega_n)\zeta - (\omega/\omega_n)^2} \quad (8\text{-}17)$$

We can simplify Eq. (8-17) by letting $u = \omega/\omega_n$. Then, Eq. (8-17) becomes

$$M(ju) = \frac{1}{1 + j2u\zeta - u^2} \quad (8\text{-}18)$$

The magnitude and phase of $M(ju)$ are

$$|M(ju)| = \frac{1}{\left[(1 - u^2)^2 + (2\zeta u)^2\right]^{1/2}} \quad (8\text{-}19)$$

and

$$\angle M(ju) = \phi_M(ju) = -\tan^{-1}\frac{2\zeta u}{1 - u^2} \quad (8\text{-}20)$$

respectively. The resonant frequency is determined by setting the derivative of $|M(ju)|$ with respect to u to zero. Thus,

$$\frac{d|M(ju)|}{du} = -\frac{1}{2}\left[(1-u^2)^2 + (2\zeta u)^2\right]^{-3/2}(4u^3 - 4u + 8u\zeta^2) = 0 \qquad (8\text{-}21)$$

from which we get

$$4u^3 - 4u + 8u\zeta^2 = 4u(u^2 - 1 + 2\zeta^2) = 0 \qquad (8\text{-}22)$$

In normalized frequency, the roots of Eq. (8-22) are $u_r = 0$ and

$$u_r = \sqrt{1 - 2\zeta^2} \qquad (8\text{-}23)$$

The solution of $u_r = 0$ merely indicates that the slope of the $|M(ju)|$-versus-ω curve is zero at $\omega = 0$; it is not a true maximum if ζ is less than 0.707. Eq. (8-23) gives the resonant frequency

$$\omega_r = \omega_n\sqrt{1 - 2\zeta^2} \qquad (8\text{-}24)$$

Because frequency is a real quantity, Eq. (8-24) is meaningful only for $2\zeta^2 \leq 1$, or $\zeta \leq 0.707$. This means simply that, for all values of ζ greater than 0.707, the resonant frequency is $\omega_r = 0$ and $M_r = 1$.

Substituting Eq. (8-23) into Eq. (8-20) for u and simplifying, we get

$$M_r = \frac{1}{2\zeta\sqrt{1 - \zeta^2}} \qquad \zeta \leq 0.707 \qquad (8\text{-}25)$$

- For the prototype second-order system, M_R is a function of ζ only.

- For the prototype second-order system, $M_r = 1$ and $\omega_r = 0$ when $\zeta \geq 0.707$.

It is important to note that, for the prototype second-order system, M_r is a function of the damping ratio ζ only, and ω_r is a function of both ζ and ω_n. Furthermore, although taking the derivative of $|M(ju)|$ with respect to u is a valid method of determining M_r and ω_r, for higher-order systems, this analytical method is quite tedious and is not recommended. Graphical methods to be discussed and computer methods are much more efficient for high-order systems.

Toolbox 8-2-1

MATLAB statements for Fig. 8-3

```
i=1;
zeta = [0 0.1 0.2 0.4 0.6 0.707 1 1.5 2.0]
for u=0:0.001:3
    z=1;
    M(z,i)= abs(1/(1+(j*2*zeta(z)*u)-(u^2)));z=z+1;
    M(z,i)= abs(1/(1+(j*2*zeta(z)*u)-(u^2)));z=z+1;
    M(z,i)= abs(1/(1+(j*2*zeta(z)*u)-(u^2)));z=z+1;
    M(z,i)= abs(1/(1+(j*2*zeta(z)*u)-(u^2)));z=z+1;
    M(z,i)= abs(1/(1+(j*2*zeta(z)*u)-(u^2)));z=z+1;
    M(z,i)= abs(1/(1+(j*2*zeta(z)*u)-(u^2)));z=z+1;
    M(z,i)= abs(1/(1+(j*2*zeta(z)*u)-(u^2)));z=z+1;
```

```
    M(z,i)= abs(1/(1+(j*2*zeta(z)*u)-(u^2)));z=z+1;
    M(z,i)= abs(1/(1+(j*2*zeta(z)*u)-(u^2)));z=z+1;
    i=i+1;
end

u=0:0.001:3;

for i = 1:length(zeta)
    plot(u,M(i,:));
    hold on;
end
xlabel('\mu = \omega/\omega_n');
ylabel('|M(j\omega)|');
axis([0 3 0 6]);
grid
```

Fig. 8-3 illustrates the plots of $|M(ju)|$ of Eq. (8-19) versus u for various values of ζ. Notice that, if the frequency scale were unnormalized, the value of $\omega_r = u_r \omega_n$ would increase when ζ decreases, as indicated by Eq. (8-24). When $\zeta = 0$, $\omega_r = \omega_n$. Figs. 8-4 and 8-5 illustrate the relationship between M_r and ζ, and $u_r(=\omega_r/\omega_n)$ and ζ, respectively.

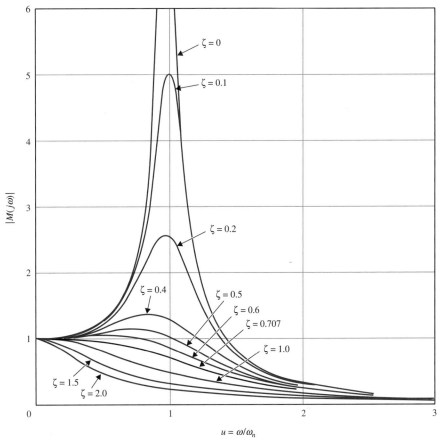

Figure 8-3 Magnification versus normalized frequency of the prototype second-order control system.

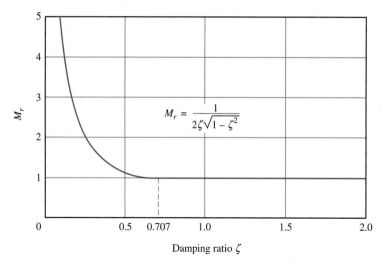

Figure 8-4 M_r versus damping ratio for the prototype second-order system.

8-2-2 Bandwidth

In accordance with the definition of bandwidth, we set the value of $|M(ju)|$ to $1/\sqrt{2} \cong 0.707$.

$$|M(ju)| = \frac{1}{\left[(1-u^2)^2 + (2\zeta u)^2\right]^{1/2}} = \frac{1}{\sqrt{2}} \qquad (8\text{-}26)$$

Thus,

$$\left[(1-u^2)^2 + (2\zeta u)^2\right]^{1/2} = \sqrt{2} \qquad (8\text{-}27)$$

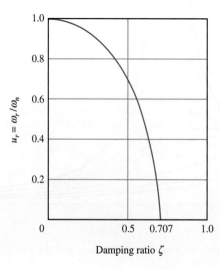

Figure 8-5 Normalized resonant frequency versus damping ratio for the prototype second-order system. $u_r = \sqrt{1-2\zeta^2}$.

- BW/ω_n decreases monotonically as the damping ratio ζ decreases.

which leads to

$$u^2 = (1 - 2\zeta^2) \pm \sqrt{4\zeta^4 - 4\zeta^2 + 2} \qquad (8\text{-}28)$$

Toolbox 8-2-2

MATLAB statements for Fig. 8-6

```
clear all
i=1;
for zetai=0:sqrt(1/2)/100:1.2
    M(i) = sqrt((1-2*zetai.^2)+sqrt(4*zetai.^4-4*zetai.^2+2));
    zeta(i)=zetai
    i=i+1;
end

TMP_COLOR = 1;
plot(zeta,M);
xlabel('\zeta');
ylabel('BW/\omega_n');
axis([0 1.2 0 2]);
grid
```

The plus sign should be chosen in the last equation, since u must be a positive real quantity for any ζ. Therefore, the bandwidth of the prototype second-order system is determined from Eq. (8-28) as

$$\text{BW} = \omega_n \left[(1 - 2\zeta^2) + \sqrt{4\zeta^4 - 4\zeta^2 + 2} \right]^{1/2} \qquad (8\text{-}29)$$

- BW is directly proportional to ω_n.

Fig. 8-6 shows a plot of BW/ω_n as a function of ζ. Notice that, as ζ increases, BW/ω_n decreases monotonically. Even more important, Eq. (8-29) shows that BW is directly proportional to ω_n.

We have established some simple relationships between the time-domain response and the frequency-domain characteristics of the prototype second-order system. The summary of these relationships is as follows.

- When a system is unstable, M_r no longer has any meaning.

1. The resonant peak M_r of the closed-loop frequency response depends on ζ only [Eq. (8-25)]. When ζ is zero, M_r is infinite. When ζ is negative, the system is unstable, and the value of M_r ceases to have any meaning. As ζ increases, M_r decreases.

2. For $\zeta \geq 0.707$, $M_r = 1$ (see Fig. 8-4), and $\omega_r = 0$ (see Fig. 8-5). In comparison with the unit-step time response, the maximum overshoot in Eq. (5-103) also depends only on ζ. However, the maximum overshoot is zero when $\zeta \geq 1$.

3. Bandwidth is directly proportional to ω_n [Eq. (8-29)]; that is, BW increases and decreases linearly with ω_n. BW also decreases with an increase in ζ for a fixed ω_n

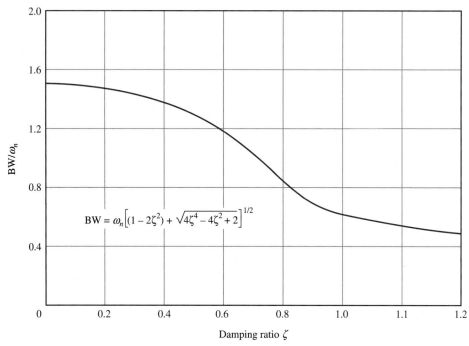

Figure 8-6 Bandwidth/ω_n versus damping ratio for the prototype second-order system.

- Bandwidth and rise time are inversely proportional to each other.

(see Fig. 8-6). For the unit-step response, rise time increases as ω_n decreases, as demonstrated in Eq. (5-108) and Fig. 5-21. Therefore, BW and rise time are inversely proportional to each other.

4. Bandwidth and M_r are proportional to each other for $0 \leq \zeta \leq 0.707$.

The correlations among pole locations, unit-step response, and the magnitude of the frequency response for the prototype second-order system are summarized in Fig. 8-7.

▶ 8-3 EFFECTS OF ADDING A ZERO TO THE FORWARD-PATH TRANSFER FUNCTION

The relationships between the time-domain and the frequency-domain responses arrived at in the preceding section apply only to the prototype second-order system described by Eq. (8-16). When other second-order or higher-order systems are involved, the relationships are different and may be more complex. It is of interest to consider the effects on the frequency-domain response when poles and zeros are added to the prototype second-order transfer function. It is simpler to study the effects of adding poles and zeros to the closed-loop transfer function; however, it is more realistic from a design standpoint to modify the forward-path transfer function.

The closed-loop transfer function of Eq. (8-16) may be considered as that of a unity-feedback control system with the prototype second-order forward-path transfer function

$$G(s) = \frac{\omega_n^2}{s(s + 2\zeta\omega_n)} \qquad (8\text{-}30)$$

8-3 Effects of Adding a Zero to the Forward-Path Transfer Function ◀ 419

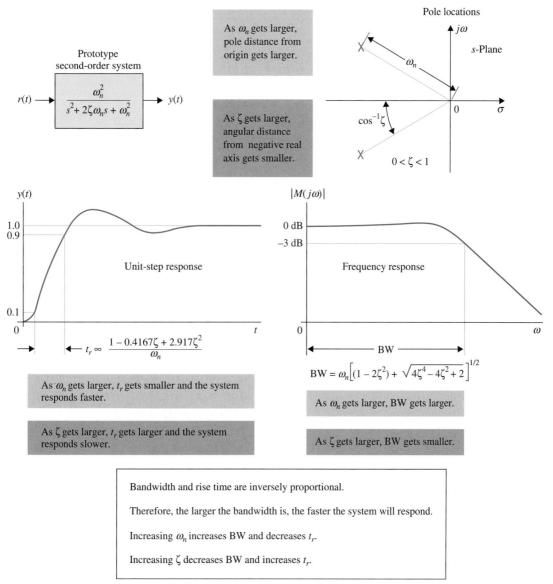

Figure 8-7 Correlation among pole locations, unit-step response, and the magnitude of frequency response of the prototype second-order system.

Let us add a zero at $s = -1/T$ to the transfer function so that Eq. (8-30) becomes

$$G(s) = \frac{\omega_n^2(1 + Ts)}{s(s + 2\zeta\omega_n)} \tag{8-31}$$

The closed-loop transfer function is

$$M(s) = \frac{\omega_n^2(1 + Ts)}{s^2 + (2\zeta\omega_n + T\omega_n^2)s + \omega_n^2} \tag{8-32}$$

- The general effect of adding a zero to the forward-path transfer function is to increase the BW of the closed-loop system.

In principle, M_r, ω_r, and BW of the system can all be derived using the same steps used in the previous section. However, because there are now three parameters in ζ, ω_n, and T, the exact expression for M_r, ω_r, and BW are difficult to obtain analytically even though the system is still second order. After a length derivation, the bandwidth of the system is found to be

$$\text{BW} = \left(-b + \tfrac{1}{2}\sqrt{b^2 + 4\omega_n^4}\right)^{1/2} \tag{8-33}$$

where

$$b = 4\zeta^2\omega_n^2 + 4\zeta\omega_n^3 T - 2\omega_n^2 - \omega_n^4 T^2 \tag{8-34}$$

While it is difficult to see how each of the parameters in Eq. (8-33) affects the bandwidth, Fig. 8-8 shows the relationship between BW and T for $\zeta = 0.707$ and $\omega_n = 1$. Notice that *the general effect of adding a zero to the forward-path transfer function is to increase the bandwidth of the closed-loop system*.

However, as shown in Fig. 8-8, over a range of small values of T, the bandwidth is actually decreased. Figs. 8-9(a) and 8-9(b) give the plots of $|M(j\omega)|$ of the closed-loop system that has the $G(s)$ of Eq. (8-31) as its forward-path transfer function: $\omega_n = 1$; $\zeta = 0.707$ and 0.2, respectively; and T takes on various values. These curves verify that the bandwidth generally increases with the increase of T by the addition of a zero to $G(s)$, except for a range of small values of T, for which BW is actually decreased.

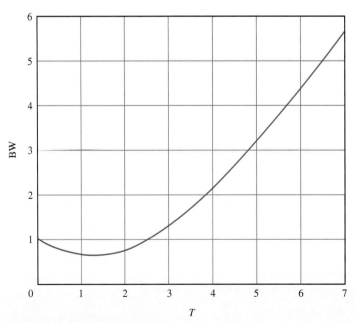

Figure 8-8 Bandwidth of a second-order system with open-loop transfer function $G(s) = (1 + Ts)/[s(s + 1.414)]$.

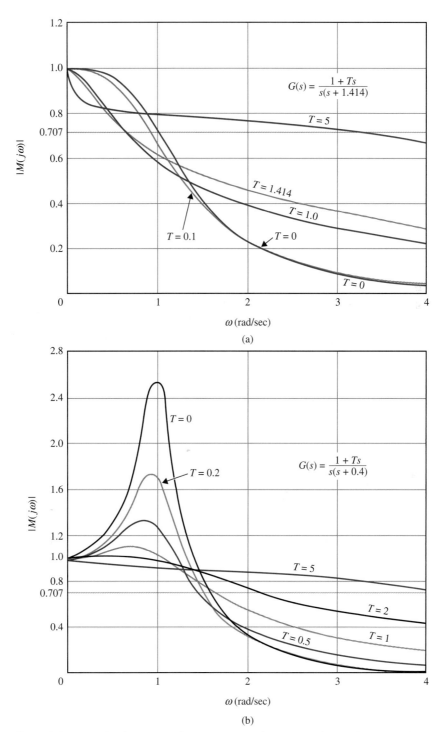

Figure 8-9 Magnification curves for the second-order system with the forward-path transfer function $G(s)$ in Eq. (8-32). (a) $\omega_n = 1$, $\zeta = 0.707$ (b) $\omega_n = 1$, $\zeta = 0.2$.

Toolbox 8-3-1

MATLAB statements for Fig. 8-9(a)

```
clear all
i=1;T=[5 1.414 1 0.1 0];zeta=0.707;
for w=0:0.01:4
    t=1; s=j*w;
    M(t,i) = abs((1+(T(t)*s))/(s^2+(2*zeta+T(t))*s+1));t=t+1;
    M(t,i) = abs((1+(T(t)*s))/(s^2+(2*zeta+T(t))*s+1));t=t+1;
    M(t,i) = abs((1+(T(t)*s))/(s^2+(2*zeta+T(t))*s+1));t=t+1;
    M(t,i) = abs((1+(T(t)*s))/(s^2+(2*zeta+T(t))*s+1));t=t+1;
    M(t,i) = abs((1+(T(t)*s))/(s^2+(2*zeta+T(t))*s+1));t=t+1;
    i=i+1;
end
w=0:0.01:4;
for i = 1:length(T)
    plot(w,M(i,:));
    hold on;
end
xlabel('\omega (rad/sec)');ylabel('|M(j\omega)|');
axis([0 4 0 1.2]);
grid
```

MATLAB statements for Fig. 8-9(b)

```
clear all
i=1;
T=[0 0.2 5 2 1 0.5];
zeta=0.2;
for w=0:0.001:4
    t=1;
    s=j*w;
    M(t,i) = abs((1+(T(t)*s))/(s^2+(2*zeta+T(t))*s+1));t=t+1;
    M(t,i) = abs((1+(T(t)*s))/(s^2+(2*zeta+T(t))*s+1));t=t+1;
    M(t,i) = abs((1+(T(t)*s))/(s^2+(2*zeta+T(t))*s+1));t=t+1;
    M(t,i) = abs((1+(T(t)*s))/(s^2+(2*zeta+T(t))*s+1));t=t+1;
    M(t,i) = abs((1+(T(t)*s))/(s^2+(2*zeta+T(t))*s+1));t=t+1;
    M(t,i) = abs((1+(T(t)*s))/(s^2+(2*zeta+T(t))*s+1));t=t+1;
    i=i+1;
end
w=0:0.001:4; TMP_COLOR = 1;
for i = 1:length(T)
    plot(w,M(i,:));
    hold on;
end
xlabel('\omega (rad/sec)');
ylabel('|M(j\omega)|');
axis([0 4 0 2.8]);
grid
```

Figs. 8-10 and 8-11 show the corresponding unit-step responses of the closed-loop system. These curves show that a high bandwidth corresponds to a faster rise time. However, as T become very large, the zero of the closed-loop transfer function, which is at $s = -1/T$, moves very close to the origin, causing the system to have a large time constant. Thus, Fig. 8-10 illustrates the situation that the rise time is fast, but the large time constant of the zero near the origin of the s-plane causes the time response to drag out in reaching the final steady state (i.e., the settling time will be longer).

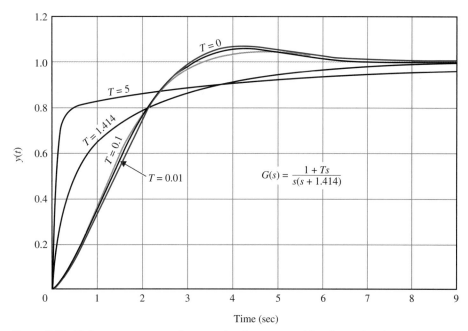

Figure 8-10 Unit-step responses of a second-order system with a forward-path transfer function $G(s)$.

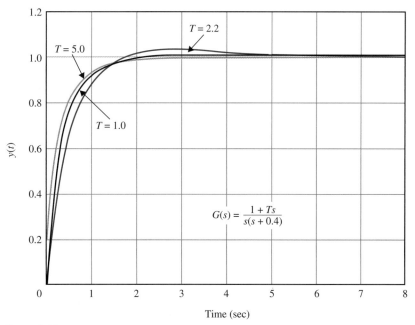

Figure 8-11 Unit-step responses of a second-order system with a forward-path transfer function $G(s)$.

Toolbox 8-3-2

MATLAB statements for Fig. 8-10 – use clear all, close all, and clc if necessary

```
T=[5 1.414 0.1 0.01 0];
t=0:0.01:9;
zeta = 0.707;
for i=1:length(T)
    num=[T(i) 1];
    den = [1 2*zeta+T(i) 1];
    M(i,:)=step(num,den,t);
end

TMP_COLOR = 1;
for i = 1:length(T)
    plot(t,M(i,:));
    hold on;
end
xlabel('Time');
ylabel('y(t)');
grid
```

Toolbox 8-3-3

MATLAB statements for Fig. 8-11 – use clear all, close all, and clc if necessary

```
T=[1 5 0.2];
t=0:0.01:9;
zeta = 0.2;
for i=1:length(T)
    num=[T(i) 1];
    den = [1 2*zeta+T(i) 1];
    M(i,:)=step(num,den,t);
end

for i = 1:length(T)
    plot(t,M(i,:));
    hold on;
end
xlabel('Time');
ylabel('y(t)');
grid
```

▶ 8-4 EFFECTS OF ADDING A POLE TO THE FORWARD-PATH TRANSFER FUNCTION

Adding a pole at $s = -1/T$ to the forward-path transfer function of Eq. (8-30) leads to

$$G(s) = \frac{\omega_n^2}{s(s + 2\zeta\omega_n)(1 + Ts)} \tag{8-35}$$

- Adding a pole to the forward-path transfer function makes the closed-loop system less stable and decreases the bandwidth.

The derivation of the bandwidth of the closed-loop system with $G(s)$ given in Eq. (8-35) is quite tedious. We can obtain a qualitative indication on the bandwidth properties by referring to Fig. 8-12, which shows the plots of $|M(j\omega)|$ versus ω for $\omega_n = 1, \zeta = 0.707$,

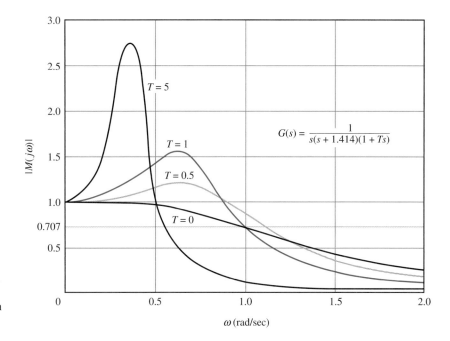

Figure 8-12 Magnification curves for a third-order system with a forward-path transfer function $G(s)$.

and various values of T. Because the system is now of the third order, it can be unstable for a certain set of system parameters. It can be shown that, for $\omega_n = 1$ and $\zeta = 0.707$, the system is stable for all positive values of T. The $|M(j\omega)|$-versus-ω curves of Fig. 8-12 show that, for small values of T, the bandwidth of the system is slightly increased by the addition of the pole, but M_r is also increased. When T becomes large, the pole added to $G(s)$ has the effect of decreasing the bandwidth but increasing M_r. Thus, we can conclude that, in general, *the effect of adding a pole to the forward-path transfer function is to make the closed-loop system less stable while decreasing the bandwidth*.

The unit-step responses of Fig. 8-13 show that, for larger values of T, $T = 1$ and $T = 5$, the following relations are observed:

1. The rise time increases with the decrease of the bandwidth.
2. The larger values of M_r also correspond to a larger maximum overshoot in the unit-step responses.

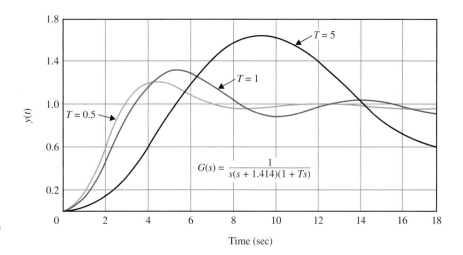

Figure 8-13 Unit-step responses of a third-order system with a forward-path transfer function $G(s)$.

- When $M_r = \infty$, the closed-loop system is marginally stable. When the system is unstable, M_r no longer has any meaning.

The correlation between M_r and the maximum overshoot of the step response is meaningful only when the system is stable. When $G(j\omega) = -1$, $|M(j\omega)|$ is infinite, and the closed-loop system is marginally stable. On the other hand, when the system is unstable, the value of $|M(j\omega)|$ is analytically finite, but it no longer has any significance.

Toolbox 8-4-1

MATLAB statements for Fig. 8-12 – use clear all, close all, and clc if necessary

```
wn=1;
zeta=0.707;
for T=[0 0.5 1 5];
num=[wn^2];
den=conv([1 0],[1 2*zeta*wn]);
den=conv(den,[T 1]);
[nc,dc]=feedback(num,den,1,1,-1);
t=linspace(0,4,1001); %time vector
y=step(nc,dc,t); %step response for basic system
plot(t,y);
hold on
end
xlabel('w (rad/s)');
ylabel('Amplitude');
title('Step Response');
```

The objective of these last two sections is to demonstrate the simple relationships between BW, M_r, and the time-domain response. Typical effects on BW of adding a pole and a zero to the forward-path transfer function are investigated. No attempt is made to include all general cases.

8-5 NYQUIST STABILITY CRITERION: FUNDAMENTALS

- The Nyquist plot of $L(j\omega)$ is done in polar coordinates as ω varies from 0 to ∞.

Thus far we have presented two methods of determining the stability of linear SISO systems: the Routh-Hurwitz criterion and the root-locus method of determining stability by locating the roots of the characteristic equation in the s-plane. Of course, if the coefficients of the characteristic equation are all known, we can solve for the roots of the equation by use of MATLAB.

- The Nyquist criterion also gives indication on relative stability.

The Nyquist criterion is a semigraphical method that determines the stability of a closed-loop system by investigating the properties of the frequency-domain plot, the **Nyquist plot,** of the loop transfer function $G(s)H(s)$, or $L(s)$. Specifically, the Nyquist plot of $L(s)$ is a plot of $L(j\omega)$ in the polar coordinates of $\text{Im}[L(j\omega)]$ versus $\text{Re}[L(j\omega)]$ as ω varies from 0 to ∞. This is another example of using the properties of the loop transfer function to find the performance of the closed-loop system. The Nyquist criterion has the following features that make it an alternative method that is attractive for the analysis and design of control systems.

1. In addition to providing the absolute stability, like the Routh-Hurwitz criterion, the Nyquist criterion also gives information on the relative stability of a stable system and the degree of instability of an unstable system. It also gives an indication of how the system stability may be improved, if needed.

2. The Nyquist plot of $G(s)H(s)$ or of $L(s)$ is very easy to obtain, especially with the aid of a computer.
3. The Nyquist plot of $G(s)H(s)$ gives information on the frequency-domain characteristics such as M_r, ω_r, BW, and others with ease.
4. The Nyquist plot is useful for systems with pure time delay that cannot be treated with the Routh-Hurwitz criterion and are difficult to analyze with the root-locus method.

This subject is also treated in Appendix F for the general case where the loop transfer function is of nonminimum-phase type.

8-5-1 Stability Problem

The Nyquist criterion represents a method of determining the location of the characteristic equation roots with respect to the left half and the right half of the s-plane. Unlike the root-locus method, the Nyquist criterion does not give the exact location of the characteristic equation roots.

Let us consider that the closed-loop transfer function of a SISO system is

$$M(s) = \frac{G(s)}{1 + G(s)H(s)} \tag{8-36}$$

where $G(s)H(s)$ can assume the following form:

$$G(s)H(s) = \frac{K(1 + T_1 s)(1 + T_2 s) \cdots (1 + T_m s)}{s^p (1 + T_a s)(1 + T_b s) \cdots (1 + T_n s)} e^{-T_d s} \tag{8-37}$$

where the T's are real or complex-conjugate coefficients, and T_d is a real time delay.

Because the characteristic equation is obtained by setting the denominator polynomial of $M(s)$ to zero, the roots of the characteristic equation are also the zeros of $1 + G(s)H(s)$. Or, the characteristic equation roots must satisfy

$$\Delta(s) = 1 + G(s)H(s) = 0 \tag{8-38}$$

In general, for a system with multiple number of loops, the denominator of $M(s)$ can be written as

$$\Delta(s) = 1 + L(s) = 0 \tag{8-39}$$

where $L(s)$ is the loop transfer function and is of the form of Eq. (8-37).

Before embarking on the details of the Nyquist criterion, it is useful to summarize the pole–zero relationships of the various system transfer functions.

Identification of Poles and Zeros

Loop transfer function zeros: zeros of $L(s)$

Loop transfer function poles: poles of $L(s)$

Closed-loop transfer function poles: zeros of $1 + L(s)$ = roots of the characteristic equation poles of $1 + L(s)$ = poles of $L(s)$.

Stability Conditions

We define two types of stability with respect to the system configuration.

- **Open-Loop Stability:** A system is said to be **open-loop stable** if the poles of the loop transfer function $L(s)$ are all in the left-half s-plane. For a single-loop system, this is equivalent to the system being stable when the loop is opened at any point.

- **Closed-Loop Stability:** A system is said to be **closed-loop stable,** or simply stable, if the poles of the closed-loop transfer function or the zeros of $1 + L(s)$ are all in the left-half s-plane. Exceptions to the above definitions are systems with poles or zeros intentionally placed at $s = 0$.

8-5-2 Definition of Encircled and Enclosed

Because the Nyquist criterion is a graphical method, we need to establish the concepts of encircled and enclosed, which are used for the interpretation of the Nyquist plots for stability.

Encircled

A point or region in a complex function plane is said to be encircled by a closed path if it is found inside the path.

For example, point A in Fig. 8-14 is encircled by the closed path Γ, because A is *inside* the closed path. Point B is not encircled by the closed path Γ, because it is *outside* the path. Furthermore, when the closed path Γ has a direction assigned to it, the encirclement, if made, can be in the clockwise (CW) or the counterclockwise (CCW) direction. As shown in Fig. 8-14, point A is encircled by Γ in the CCW direction. We can say that the region *inside* the path is encircled in the prescribed direction, and the region *outside* the path is not encircled.

Enclosed

A point or region is said to be enclosed by a closed path if it is encircled in the CCW direction or the point or region lies to the left of the path when the path is traversed in the prescribed direction.

The concept of enclosure is particularly useful if only a portion of the closed path is shown. For example, the shaded regions in Figs. 8-15(a) and (b) are considered to be *enclosed* by the closed path Γ. In other words, point A in Fig. 8-15(a) is *enclosed* by Γ, but point A in Fig. 8-15(b) is not. However, point B and all the points in the shaded region outside Γ in Fig. 8-15(b) are *enclosed*.

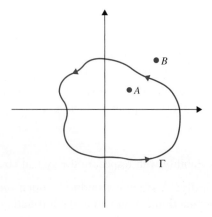

Figure 8-14 Definition of encirclement.

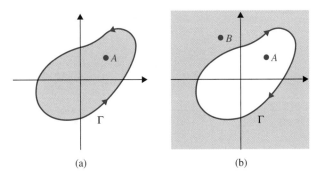

Figure 8-15 Definition of enclosed points and regions. (a) Point A is enclosed by Γ. (b) Point A is not enclosed, but B is enclosed by the locus Γ.

8-5-3 Number of Encirclements and Enclosures

When a point is encircled by a closed path Γ, a number N can be assigned to the number of times it is encircled. The magnitude of N can be determined by drawing an arrow from the point to any arbitrary point s_1 on the closed path Γ and then letting s_1 follow the path in the prescribed direction until it returns to the starting point. The total *net* number of revolutions traversed by this arrow is N, or the net angle is $2\pi N$ radians. For example, point A in Fig. 8-16(a) is *encircled once* or 2π radians by Γ, and point B is *encircled twice* or 4π radians, all in the CW direction. In Fig. 8-16(b), point A is *enclosed once*, and point B is *enclosed twice* by Γ. By definition, N is positive for CCW encirclement and negative for CW encirclement.

8-5-4 Principles of the Argument

The Nyquist criterion was originated as an engineering application of the well-known "principle of the argument" concept in complex-variable theory. The principle is stated in the following in a heuristic manner.

Let $\Delta(s)$ be a single-valued function of the form of the right-hand side of Eq. (8-37), which has a finite number of poles in the s-plane. Single valued means that, for each point in the s-plane, there is one and only one corresponding point, including infinity, in the complex $\Delta(s)$-plane. As defined in Chapter 7, infinity in the complex plane is interpreted as a point.

Suppose that a continuous closed path Γ_s is arbitrarily chosen in the s-plane, as shown in Fig. 8-17(a). If Γ_s does not go through any poles of $\Delta(s)$, then the trajectory Γ_Δ mapped by $\Delta(s)$ into the $\Delta(s)$-plane is also a closed one, as shown in Fig. 8-17(b). Starting from a point s_1, the Γ_s locus is traversed in the arbitrarily chosen direction (CW in the illustrated

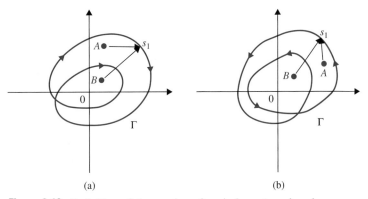

Figure 8-16 Definition of the number of encirclements and enclosures.

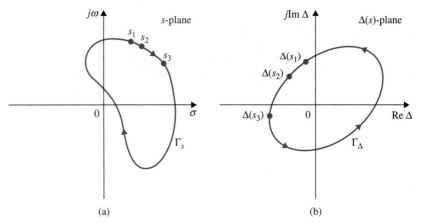

Figure 8-17 (a) Arbitrarily chosen closed path in the s-plane. (b) Corresponding locus Γ_s in the $\Delta(s)$-plane.

case), through the points s_2 and s_3, and then returning to s_1 after going through all the points on the Γ_s locus, as shown in Fig. 8-17(a). The corresponding Γ_Δ locus will start from the point $\Delta(s_1)$ and go through points $\Delta(s_2)$ and $\Delta(s_3)$, corresponding to s_1, s_2, and s_3, respectively, and finally return to the starting point, $\Delta(s_1)$. The direction of traverse of Γ_Δ can be either CW or CCW, that is, in the same direction or the opposite direction as that of Γ_s, depending on the function $\Delta(s)$. In Fig. 8-17(b), the direction of Γ_Δ is arbitrarily assigned, for illustration purposes, to be CCW.

- Do not attempt to relate $\Delta(s)$ with $L(s)$. They are not the same.

Although the mapping from the s-plane to the $\Delta(s)$-plane is single-valued, the reverse process is not a single-valued mapping. For example, consider the function

$$\Delta(s) = \frac{K}{s(s+1)(s+2)} \tag{8-40}$$

which has poles $s = 0$, -1, and -2 in the s-plane. For each point in the s-plane, there is only one corresponding point in the $\Delta(s)$-plane. However, for each point in the $\Delta(s)$-plane, the function maps into three corresponding points in the s-plane. The simplest way to illustrate this is to write Eq. (8-40) as

$$s(s+1)(s+2) - \frac{K}{\Delta(s)} = 0 \tag{8-41}$$

If $\Delta(s)$ is a real constant, which represents a point on the real axis in the $\Delta(s)$-plane, the third-order equation in Eq. (8-41) gives three roots in the s-plane. The reader should recognize the parallel of this situation to the root-locus diagram that essentially represents the mapping of $\Delta(s) = -1 + j0$ onto the loci of roots of the characteristic equation in the s-plane, for a given value of K. Thus, the root loci of Eq. (8-40) have three individual branches in the s-plane.

The principle of the argument can be stated:

Let $\Delta(s)$ be a single-valued function that has a finite number of poles in the s-plane. Suppose that an arbitrary closed path Γ_s is chosen in the s-plane so that the path does not go through any one of the poles or zeros of $\Delta(s)$; the corresponding Γ_Δ locus mapped in the $\Delta(s)$-plane will encircle the origin as many times as the difference between the number of zeros and poles of $\Delta(s)$ that are encircled by the s-plane locus Γ_s.

8-5 Nyquist Stability Criterion: Fundamentals

In equation form, the principle of the argument is stated as

$$N = Z - P \qquad (8\text{-}42)$$

where

N = number of encirclements of the origin made by the $\Delta(s)$-plane locus Γ_Δ.
Z = number of zeros of $\Delta(s)$ encircled by the s-plane locus Γ_s in the s-plane.
P = number of poles of $\Delta(s)$ encircled by the s-plane locus Γ_s in the s-plane.

In general, N can be positive $(Z > P)$, zero $(Z = P)$, or negative $(Z < P)$. These three situations are described in more detail as follows.

1. $N > 0 (Z > P)$. If the s-plane locus encircles more zeros than poles of $\Delta(s)$ in a certain prescribed direction (CW or CCW), N is a positive integer. In this case, the $\Delta(s)$-plane locus Γ_Δ will encircle the origin of the $\Delta(s)$-plane N times in the same direction as that of Γ_s.
2. $N = 0 (Z = P)$. If the s-plane locus encircles as many poles as zeros, or no poles and zeros, of $\Delta(s)$, the $\Delta(s)$-plane locus Γ_Δ will not encircle the origin of the $\Delta(s)$-plane.
3. $N < 0 (Z < P)$. If the s-plane locus encircles more poles than zeros of $\Delta(s)$ in a certain direction, N is a negative integer. In this case, the $\Delta(s)$-plane locus Γ_Δ will encircle the origin N times in the *opposite* direction as that of Γ_s.

A convenient way of determining N with respect to the origin (or any point) of the $\Delta(s)$-plane is to draw a line from the point in any direction to a point as far as necessary; the number of *net* intersections of this line with the $\Delta(s)$ locus gives the magnitude of N. Fig. 8-18 gives

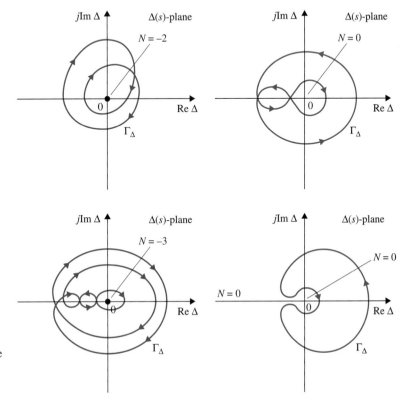

Figure 8-18 Examples of the determination of N in the $\Delta(s)$-plane.

432 ▶ Chapter 8. Frequency-Domain Analysis

several examples of this method of determining N. In these illustrated cases, it is assumed that the Γ_s locus has a CCW sense.

Critical Point

For convenience, we shall designate the origin of the $\Delta(s)$-plane as the **critical point** from which the value of N is determined. Later, we shall designate other points in the complex-function plane as critical points, dependent on the way the Nyquist criterion is applied.

A rigorous proof of the principle of the argument is not given here. The following illustrative example may be considered a heuristic explanation of the principle.

Let us consider the function $\Delta(s)$ is of the form

$$\Delta(s) = \frac{K(s+z_1)}{(s+p_1)(s+p_2)} \tag{8-43}$$

where K is a positive real number. The poles and zeros of $\Delta(s)$ are assumed to be as shown in Fig. 8-19(a). The function $\Delta(s)$ can be written as

$$\begin{aligned}\Delta(s) &= |\Delta(s)|\angle\Delta(s) \\ &= \frac{K|s+z_1|}{|s+p_1||s+p_2|}[\angle(s+z_1) - \angle(s+p_1) - \angle(s+p_2)]\end{aligned} \tag{8-44}$$

Fig. 8-19(a) shows an arbitrarily chosen trajectory Γ_s in the s-plane, with the arbitrary point s_1 on the path, and Γ_s does not pass through any of the poles and the zeros of $\Delta(s)$. The function $\Delta(s)$ evaluated at $s = s_1$ is

$$\Delta(s_1) = \frac{K(s+z_1)}{(s_1+p_1)(s+p_2)} \tag{8-45}$$

- Z and P refer to only the zeros and poles, respectively, of $\Delta(s)$ that are encircled by Γ_s.

The term $(s_1 + z_1)$ can be represented graphically by the vector drawn from $-z_1$ to s_1. Similar vectors can be drawn for $(s_1 + p_1)$ and $(s + p_2)$. Thus, $\Delta(s_1)$ is represented by

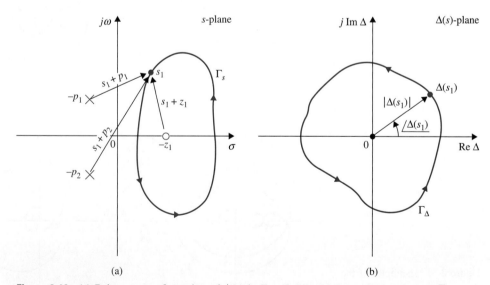

(a)   (b)

Figure 8-19 (a) Pole–zero configuration of $\Delta(s)$ in Eq. (8-44) and the s-plane trajectory Γ_s. (b) $\Delta(s)$-plane locus Γ_Δ, which corresponds to the Γ_s locus of (a) through the mapping of Eq. (8-44).

TABLE 8-1 Summary of All Possible Outcomes of the Principle of the Argument

Direction of $N = Z - P$ Encirclement	Sense of the s-plane Locus	$\Delta(s)$-Plane Locus	
		Number of Encirclements of the Origin	Direction of Encirclement
$N > 0$	CW	N	CW
	CCW	N	CCW
$N < 0$	CW	N	CCW
	CCW	N	CW
$N = 0$	CW	0	No encirclement
	CCW	0	No encirclement

the vectors drawn from the finite poles and zeros of $\Delta(s)$ to the point s_1, as shown in Fig. 8-19(a). Now, if the point s_1 is moved along the locus Γ_s in the prescribed CCW direction until it returns to the starting point, the angles generated by the vectors drawn from the two poles that are not encircled by Γ_s when s_1 completes one roundtrip are zero, whereas the vector $(s_1 + z_1)$ drawn from the zero at $-z_1$, which is encircled by Γ_s, generates a positive angle (CCW) of 2π radians, which means that the corresponding $\Delta(s)$ plot must go around the origin 2π radians, or one revolution, in the CCW direction, as shown in Fig. 8-19(b). This is why only the poles and zeros of $\Delta(s)$ that are inside the Γ_s trajectory in the s-plane will contribute to the value of N of Eq. (8-42). Because the poles of $\Delta(s)$ contribute to a negative phase, and zeros contribute to a positive phase, the value of N depends only on the difference between Z and P. For the case illustrated in Fig. 8-19 (a), $Z = 1$ and $P = 0$.

Thus,

$$N = Z - P = 1 \tag{8-46}$$

which means that the $\Delta(s)$-plane locus Γ_Δ should encircle the origin once in the same direction as that of the s-plane locus Γ_s. It should be kept in mind that Z and P refer only to the zeros and poles, respectively, of $\Delta(s)$ that are encircled by Γ_s and not the total number of zeros and poles of $\Delta(s)$.

In general, the net angle traversed by the $\Delta(s)$-plane locus, as the s-plane locus is traversed once in any direction, is equal to

$$2\pi(Z - P) = 2\pi N \quad \text{radians} \tag{8-47}$$

This equation implies that if there are N more zeros than poles of $\Delta(s)$, which are encircled by the s-plane locus Γ_s, in a prescribed direction, the $\Delta(s)$-plane locus will encircle the origin N times in the *same direction* as that of Γ_s. Conversely, if N more poles than zeros are encircled by Γ_s in a given direction, N in Eq. (8-47) will be negative, and the $\Delta(s)$-plane locus must encircle the origin N times in the *opposite direction* to that of Γ_s.

A summary of all the possible outcomes of the principle of the argument is given in Table 8.1.

8-5-5 Nyquist Path

Years ago when Nyquist was faced with solving the stability problem, which involves determining if the function $\Delta(s) = 1 + L(s)$ has zeros in the right-half s-plane, he

434 ▶ Chapter 8. Frequency-Domain Analysis

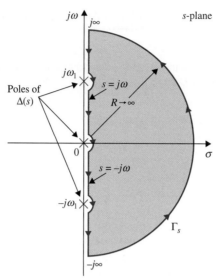

Figure 8-20 Nyquist path.

apparently discovered that the principle of the argument could be applied to solve the stability problem if the s-plane locus Γ_s is taken to be one that encircles the entire right half of the s-plane. Of course, as an alternative, Γ_s can be chosen to encircle the entire left-half s-plane, as the solution is a relative one. Fig. 8-20 illustrates a Γ_s locus with a CCW sense that encircles the entire right half of the s-plane. This path is chosen to be the s-plane trajectory Γ_s for the Nyquist criterion, since in mathematics, CCW is traditionally defined to be the positive sense. The path Γ_s shown in Fig. 8-20 is defined to be the **Nyquist path**. Because the Nyquist path must not pass through any poles and zeros of $\Delta(s)$, the small semicircles shown along the $j\omega$-axis in Fig. 8-20 are used to indicate that the path should go around these poles and zeros if they fall on the $j\omega$-axis. It is apparent that, if any pole or zero of $\Delta(s)$ lies inside the right-half s-plane, it will be encircled by the Nyquist path Γ_s.

- The Nyquist path is defined to encircle the entire right-half s-plane.

8-5-6 Nyquist Criterion and the *L(s)* or the *G(s)H(s)* Plot

The Nyquist criterion is a direct application of the principle of the argument when the s-plane locus is the Nyquist path of Fig. 8-20. In principle, once the Nyquist path is specified, the stability of a closed-loop system can be determined by plotting the $\Delta(s) = 1 + L(s)$ locus when s takes on values along the Nyquist path and investigating the behavior of the $\Delta(s)$ plot with respect to the **critical point,** which in this case is the origin of the $\Delta(s)$-plane.

Because the function $L(s)$ is generally known,

it would be simpler to construct the L(s) plot that corresponds to the Nyquist path, and the same conclusion on the stability of the closed-loop system can be obtained by observing the behavior of the L(s) plot with respect to the (−1, j0) point in the L(s)-plane.

This is because the origin of the $\Delta(s) = 1 + L(s)$ plane corresponds to the $(-1, j0)$ point in the $L(s)$-plane. Thus the $(-1, j0)$ point in the $L(s)$-plane becomes the critical point for the determination of closed-loop stability.

For single-loop systems, $L(s) = G(s)H(s)$, the previous development leads to the determination of the closed-loop stability by investigating the behavior of the $G(s)H(s)$ plot with respect to the $(-1, j0)$ point of the $G(s)H(s)$-plane. Thus, the Nyquist stability criterion is another example of using the loop transfer function properties to find the behavior of closed-loop systems.

Thus, given a control system that has the characteristic equation given by equating the numerator polynomial of $1 + L(s)$ to zero, where $L(s)$ is the loop transfer function, the application of the Nyquist criterion to the stability problem involves the following steps.

1. The Nyquist path Γ_s is defined in the s-plane, as shown in Fig. 8-20.
2. The $L(s)$ plot corresponding to the Nyquist path is constructed in the $L(s)$-plane.
3. The value of N, the number of encirclement of the $(-1, j0)$ point made by the $L(s)$ plot, is observed.
4. The Nyquist criterion follows from Eq. (8-42),

$$N = Z - P \qquad (8\text{-}48)$$

where

$N\;=\;$ number of encirclements of the $(-1, j0)$ point made by the $L(s)$ plot.

$Z\;=\;$ number of zeros of $1 + L(s)$ that are inside the Nyquist path, that is, the right-half s-plane.

$P\;=\;$ number of poles of $1 + L(s)$ that are inside the Nyquist path, that is, the right-half s-plane. Notice that the poles of $1 + L(s)$ are the same as that of $L(s)$.

The stability requirements for the two types of stability defined earlier are interpreted in terms of Z and P.

For closed-loop stability, Z must equal zero.
For open-loop stability, P must equal zero.

Thus, the condition of stability according to the Nyquist criterion is stated as

$$N = -P \qquad (8\text{-}49)$$

That is,

for a closed-loop system to be stable, the $L(s)$ plot must encircle the $(-1, j0)$ point as many times as the number of poles of $L(s)$ that are in the right-half s-plane, and the encirclement, if any, must be made in the clockwise direction (if Γ_s is defined in the CCW sense).

▶ 8-6 NYQUIST CRITERION FOR SYSTEMS WITH MINIMUM-PHASE TRANSFER FUNCTIONS

We shall first apply the Nyquist criterion to systems with $L(s)$ that are **minimum-phase transfer functions.** The properties of the minimum-phase transfer functions are described in Chapter 2 and are summarized as follows:

1. A minimum-phase transfer function does not have poles or zeros in the right-half s-plane or on the $j\omega$-axis, excluding the origin.
2. For a minimum-phase transfer function $L(s)$ with m zeros and n poles, excluding the poles at $s = 0$, when $s = j\omega$ and as ω varies from ∞ to 0, the total phase variation of $L(j\omega)$ is $(n - m)\pi/2$ radians.

3. The value of a minimum-phase transfer function cannot become zero or infinity at any finite nonzero frequency.

4. A nonminimum-phase transfer function will always have a more positive phase shift as ω varies from ∞ to 0. Or, equally true, it will always have a more negative phase shift as ω varies from 0 to ∞.

- A minimum-phase transfer function does not have poles or zeros in the right-half s-plane or on the $j\omega$-axis, except at $s = 0$.

Because a majority of the loop transfer functions encountered in the real world satisfy condition 1 and are of the minimum-phase type, it would be prudent to investigate the application of the Nyquist criterion to this class of systems. As if turns out, this is quite simple.

Because a minimum-phase $L(s)$ does not have any poles or zeros in the right-half s-plane or on the $j\omega$-axis (except at $s = 0$) $P = 0$, and the poles of $\Delta(s) = 1 + L(s)$ also have the same properties. Thus, the Nyquist criterion for a system with $L(s)$ being a minimum-phase transfer function is simplified to

$$N = 0 \qquad (8\text{-}50)$$

Thus, the Nyquist criterion can be stated:

For a closed-loop system with loop transfer function $L(s)$ that is of minimum-phase type, the system is closed-loop stable if the plot of $L(s)$ that corresponds to the Nyquist path does not encircle the critical point $(-1, j0)$ in the $L(s)$-plane.

Furthermore, if the system is unstable, $Z \neq 0$; N in Eq. (8-50) would be a positive integer, which means that the critical point $(-1, j0)$ is **enclosed** N times (corresponding to the direction of the Nyquist path defined here). Thus, the Nyquist criterion of stability for systems with minimum-phase loop transfer functions can be further simplified:

For a closed-loop system with loop transfer function $L(s)$ that is of minimum-phase type, the system is closed-loop stable if the $L(s)$ plot that corresponds to the Nyquist path does not enclose the $(-1, j0)$ point. If the $(-1, j0)$ point is enclosed by the Nyquist plot, the system is unstable.

- For $L(s)$ that is minimum-phase type, Nyquist criterion can be checked by plotting the segment of $L(j\omega)$ from $\omega = \infty$ to 0.

Because the region that is enclosed by a trajectory is defined as the region that lies to the left when the trajectory is traversed in the prescribed direction, the *Nyquist criterion can be checked simply by plotting the segment of $L(j\omega)$ from $\omega = \infty$ to 0, or, points on the positive $j\omega$-axis.* This simplifies the procedure considerably, since the plot can be made easily on a computer. The only drawback to this method is that the Nyquist plot that corresponds to the $j\omega$-axis tells only whether the critical point is enclosed or not and, if it is, not how many times. Thus, if the system is found to be unstable, the enclosure property does not give information on how many roots of the characteristic equation are in the right-half s-plane. However, in practice, this information is not vital. From this point on, we shall define the $L(j\omega)$ plot that corresponds to the positive $j\omega$-axis of the s-plane as the Nyquist plot of $L(s)$.

8-6-1 Application of the Nyquist Criterion to Minimum-Phase Tranfer Functions That Are Not Strictly Proper

Just as in the case of the root locus, it is often necessary in design to create an equivalent loop transfer function $L_{eq}(s)$ so that a variable parameter K will appear as a multiplying factor in $L_{eq}(s)$; that is, $L(s) = KL_{eq}(s)$. Because the equivalent loop transfer function does not correspond to any physical entity, it may not have more poles than zeros, and the transfer function is not strictly proper, as defined in Chapter 2. In principle, there is no difficulty in constructing the Nyquist plot of a transfer function that is not strictly proper, and the Nyquist criterion can be applied for stability studies without any complications. However, some computer programs may not be prepared for handling improper transfer functions, and it may be necessary to reformulate the equation for compatibility with the computer program. To examine this case, consider that the characteristic equation of a system with a variable parameter K is conditioned to

$$1 + KL_{eq}(s) = 0 \tag{8-51}$$

If $L_{eq}(s)$ does not have more poles than zeros, we can rewrite Eq. (8-51) as

$$1 + \frac{1}{KL_{eq}(s)} = 0 \tag{8-52}$$

by dividing both sides of the equation by $KL_{eq}(s)$. Now we can plot the Nyquist plot of $1/L_{eq}(s)$, and the critical point is still $(-1, j0)$ for $K > 0$. The variable parameter on the Nyquist plot is now $1/K$. Thus, with this minor adjustment, the Nyquist criterion can still be applied.

The Nyquist criterion presented here is cumbersome when the loop transfer function is of the nonminimum-phase type, for example, when $L(s)$ has poles or/and zeros in the right-half s-plane. A generalized Nyquist criterion that will take care of transfer functions of all types is presented in Appendix F.

▶ 8-7 RELATION BETWEEN THE ROOT LOCI AND THE NYQUIST PLOT

Because both the root locus analysis and the Nyquist criterion deal with the location of the roots of the characteristic equation of a linear SISO system, the two analyses are closely related. Exploring the relationship between the two methods will enhance the understanding of both methods. Given the characteristic equation

$$1 + L(s) = 1 + KG_1(s)H_1(s) = 0 \tag{8-53}$$

the Nyquist plot of $L(s)$ in the $L(s)$-plane is the mapping of the Nyquist path in the s-plane. Because the root loci of Eq. (8-53) must satisfy the conditions

$$\angle KG_1(s)H_1(s) = (2j+1)\pi \quad K \geq 0 \tag{8-54}$$

$$\angle KG_1(s)H_1(s) = 2j\pi \quad K \leq 0 \tag{8-55}$$

for $j = 0, \pm 1, \pm 2, \ldots$, the root loci simply represent a mapping of the real axis of the $L(s)$-plane or the $G(s)H(s)$-plane onto the s-plane. In fact, for the RL $K \geq 0$, the mapping points are on the negative real axis of the $L(s)$-plane, and, for the RL $K \leq 0$, the mapping points are on the positive real axis of the $L(s)$-plane. It was pointed out earlier that the mapping from the s-plane to the function plane for a rational function is single valued, but the reverse process is multivalued. As a simple illustration, the Nyquist plot of a type-1 third-order transfer function $G(s)H(s)$ that corresponds to points on the $j\omega$-axis of the s-plane is shown in Fig. 8-21. The root loci for the same system are shown in Fig. 8-22 as a

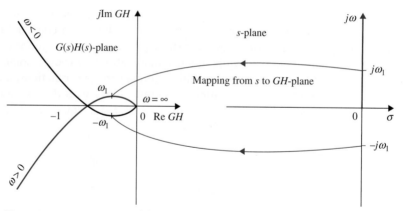

Figure 8-21 Polar plot of $G(s)H(s) = K/[s(s+a)(s+b)]$ interpreted as a mapping of the $j\omega$-axis of the s-plane onto the $G(s)H(s)$-plane.

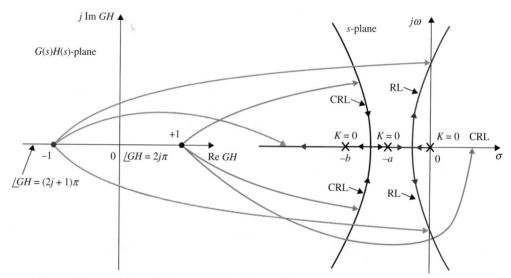

Figure 8-22 Root-locus diagram of $G(s)H(s) = K/[s(s+a)(s+b)]$ interpreted as a mapping of the real axis of the $G(s)H(s)$-plane onto the s-plane.

mapping of the real axis of the $G(s)H(s)$-plane onto the s-plane. Note that, in this case, each point of the $G(s)H(s)$-plane corresponds to three points in the s-plane. The $(-1, j0)$ point of the $G(s)H(s)$-plane corresponds to the two points where the root loci intersect the $j\omega$-axis and a point on the real axis.

The Nyquist plot and the root loci each represent the mapping of only a very limited portion of one domain to the other. In general, it would be useful to consider the mapping of points other than those on the $j\omega$-axis of the s-plane and on the real axis of the $G(s)H(s)$-plane. For instance, we may use the mapping of the constant-damping-ratio lines in the s-plane onto the $G(s)H(s)$-plane for the purpose of determining relative stability of the closed-loop system. Fig. 8-23 illustrates the $G(s)H(s)$ plots that correspond to different constant-damping-ratio lines in the s-plane. As shown by curve (3) in Fig. 8-23, when the $G(s)H(s)$ curve passes through the $(-1, j0)$ point, it means that Eq. (8-52) is satisfied, and the corresponding trajectory in the s-plane passes through the root of the characteristic equation. Similarly, we can construct the root loci that correspond to the straight lines

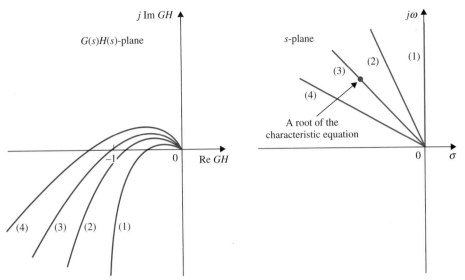

Figure 8-23 $G(s)H(s)$ plots that correspond to constant-damping-ratio lines in the s-plane.

rotated at various angles from the real axis in the $G(s)H(s)$-plane, as shown in Fig. 8-24. Notice that these root loci now satisfy the condition of

$$\angle KG_1(s)H_1(s) = (2j+1)\pi - \theta \quad K \geq 0 \tag{8-56}$$

Or the root loci of Fig. 8-24 must satisfy the equation

$$1 + G(s)H(s)e^{j\theta} = 0 \tag{8-57}$$

for the various values of θ indicated.

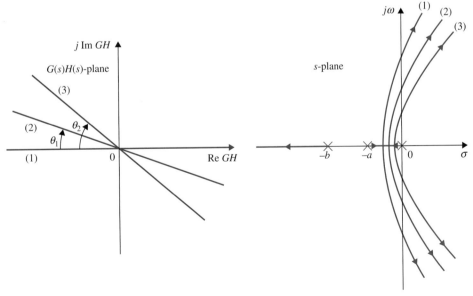

Figure 8-24 Root loci that correspond to different phase-angle loci in the $G(s)H(s)$-plane.

440 ▸ Chapter 8. Frequency-Domain Analysis

▶ 8-8 ILLUSTRATIVE EXAMPLES: NYQUIST CRITERION FOR MINIMUM-PHASE TRANSFER FUNCTIONS

The following examples serve to illustrate the application of the Nyquist criterion to systems with minimum-phase loop transfer functions. All examples in this chapter may also be solved using the **ACSYS** (see Chapter 9) or MATLAB Toolboxes incorporated in this chapter.

▶ **EXAMPLE 8-8-1** Consider that a single-loop feedback control system has the loop transfer function

$$L(s) = G(s)H(s) = \frac{K}{s(s+2)(s+10)} \tag{8-58}$$

which is of minimum-phase type. The stability of the closed-loop system can be conducted by investigating whether the Nyquist plot of $L(j\omega)/K$ for $\omega = \infty$ to 0 encloses the $(-1, j0)$ point. The Nyquist plot of $L(j\omega)/K$ may be plotted using freqtool. Fig. 8-25 shows the Nyquist plot of $L(j\omega)/K$ for $\omega = \infty$ to 0. However, because we are interested only in whether the critical point is enclosed, in general, it is not necessary to produce an accurate Nyquist plot. Because the area that is enclosed by the Nyquist plot is to the left of the curve, traversed in the direction that corresponds to $\omega = \infty$ to 0 on the Nyquist path, all that is necessary to determine stability is to find the point or points at which the Nyquist plot crosses the real axis in the $L(j\omega)/K$-plane. In many cases, information on the intersection on the real axis and the properties of $L(j\omega)/K$ at $\omega = \infty$ and $\omega = 0$ would allow the sketching of the Nyquist plot without actual plotting. We can use the following steps to obtain a sketch of the Nyquist plot of $L(j\omega)/K$.

1. Substitute $s = j\omega$ in $L(s)$.

Setting $s = j\omega$ in Eq. (8-58), we get

$$L(j\omega)/K = \frac{1}{j\omega(j\omega+2)(j\omega+10)} \tag{8-59}$$

2. Substituting $\omega = 0$ in the last equation, we get the zero-frequency property of $L(j\omega)$,

$$L(j0)/K = \infty \angle -90° \tag{8-60}$$

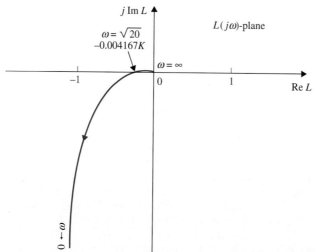

Figure 8-25 Nyquist plot of $L(s)/K = \dfrac{1}{s(s+2)(s+10)}$ for $\omega = \infty$ to $\omega = 0$.

Toolbox 8-8-1

MATLAB statements for Fig. 8-25

```
w=0.1:0.1:1000;
num = [1];
den = conv(conv([1 10],[1,2]),[1 0]);
[re,im,w] = nyquist(num,den,w);
plot(re,im);
axis([-0.1 0.01 -0.6 0.01])
grid
```

3. Substituting $\omega = \infty$ in Eq. (8-59), the property of the Nyquist plot at infinite frequency is established.

$$L(j\infty)/K = 0\angle -270° \tag{8-61}$$

Apparently, these results are verified by the plot shown in Fig. 8-25.

4. To find the intersect(s) of the Nyquist plot with the real axis, if any, we rationalize $L(j\omega)/K$ by multiplying the numerator and the denominator of the equation by the complex conjugate of the denominator. Thus, Eq. (8-59) becomes

$$L(j\omega)/K = \frac{\left[-12\omega^2 - j\omega(20-\omega^2)\right]}{[-12\omega^2 + j\omega(20-\omega^2)][-12\omega^2 - j\omega(20-\omega^2)]}$$

$$= \frac{[-12\omega - j(20-\omega^2)]}{\omega[144\omega^2 + (20-\omega^2)]} \tag{8-62}$$

5. To find the possible intersects on the real axis, we set the imaginary part of $L(j\omega)/K$ to zero. The result is

$$\text{Im}[L(j\omega)/K] = \frac{-(20-\omega^2)}{\omega[144\omega^2 + (20-\omega^2)]} = 0 \tag{8-63}$$

The solutions of the last equation are $\omega = \infty$, which is known to be a solution at $L(j\omega)/K = 0$, and

$$\omega = \pm\sqrt{20} \quad \text{rad/sec} \tag{8-64}$$

Because ω is positive, the correct answer is $\omega = \sqrt{20}$ rad/sec. Substituting this frequency into Eq. (8-62), we have the intersect on the real axis of the $L(j\omega)$-plane at

$$L(j\sqrt{20})/K = -\frac{12}{2880} = -0.004167 \tag{8-65}$$

The last five steps should lead to an adequate sketch of the Nyquist plot of $L(j\omega)/K$ short of plotting it. Thus, we see that, if K is less than 240, the intersect of the $L(j\omega)$ locus on the real axis would be to the right of the critical point $(-1, j0)$; the latter is not enclosed, and the system is stable. If $K = 240$, the Nyquist plot of $L(j\omega)$ would intersect the real axis at the -1 point, and the system would be marginally stable. In this case, the characteristic equation would have two roots on the $j\omega$-axis in the s-plane at $s = \pm j\sqrt{20}$. If the gain is increased to a value beyond 240, the intersect would be to the left of the -1 point on the real axis, and the system would be unstable. When K is negative, we can use the $(+1, j0)$ point in the $L(j\omega)$-plane as the critical point. Fig. 8-25 shows that, under this condition, the $+1$ point on the real axis would be enclosed for all negative

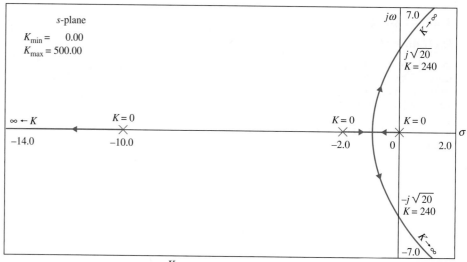

Figure 8-26 RL of $L(s) = \dfrac{K}{s(s+2)(s+10)}$.

values of K, and the system would always be unstable. Thus, the Nyquist criterion leads to the conclusion that the system is stable in the range of $0 < K < 240$. Note that application of the Routh-Hurwitz stability criterion leads to this same result.

Fig. 8-26 shows the root loci of the characteristic equation of the system described by the loop transfer function in Eq. (8-58). The correlation between the Nyquist criterion and the root loci is easily observed.

Toolbox 8-8-2

MATLAB statements for Fig. 8-26

```
den=conv([1 2 0],[1 10]);
mysys=tf(.0001,den);
rlocus(mysys);
title('Root loci of the system');
```

▶ **EXAMPLE 8-8-2** Consider the characteristic equation

$$Ks^3 + (2K+1)s^2 + (2K+5)s + 1 = 0 \qquad (8\text{-}66)$$

Dividing both sides of the last equation by the terms that do not contain K, we have

$$1 + KL_{eq}(s) = 1 + \frac{Ks(s^2 + 2s + 2)}{s^2 + 5s + 1} = 0 \qquad (8\text{-}67)$$

Thus,

$$L_{eq}(s) = \frac{s(s^2 + 2s + 2)}{s^2 + 5s + 1} \qquad (8\text{-}68)$$

which is an improper function. We can obtain the information to manually sketch the Nyquist plot of $L_{eq}(s)$ to determine the stability of the system. Setting $s = j\omega$ in Eq. (8-68), we get

$$\frac{L_{eq}(j\omega)}{K} = \frac{\omega[-2\omega + j(2-\omega^2)]}{(1-\omega^2) + 5j\omega} \qquad (8\text{-}69)$$

From the last equation, we obtain the two end points of the Nyquist plot:

$$L_{eq}(j0) = 0\angle 90° \quad \text{and} \quad L_{eq}(j\infty) = \infty\angle 90° \qquad (8\text{-}70)$$

Rationalizing Eq. (8-69) by multiplying its numerator and denominator by the complex conjugate of the denominator, we get

$$\frac{L_{eq}(j\omega)}{K} = \frac{\omega^2[5(2-\omega^2) - 2(1-\omega^2)] + j\omega[10\omega^2 + (2-\omega^2)(1-\omega^2)]}{(1-\omega^2)^2 + 25\omega^2} \qquad (8\text{-}71)$$

To find the possible intersects of the $L_{eq}(j\omega)/K$ plot on the real axis, we set the imaginary part of Eq. (8-71) to zero. We get $\omega = 0$ and

$$\omega^4 + 7\omega^2 + 2 = 0 \qquad (8\text{-}72)$$

Toolbox 8-8-3

MATLAB statements for Fig. 8-27

```
w=0.1:0.1:1000;
num =[1 2 2 0];
den = [1 5 1];
[re,im,w] = nyquist(num,den,w);
plot(re,im);
axis([-2 1 -1 5]);
grid
```

We can show that all the four roots of Eq. (8-72) are imaginary, which indicates that the $L_{eq}(j\omega)/K$ locus intersects the real axis only at $\omega = 0$. Using the information given by Eq. (8-70) and the fact that there are no other intersections on the real axis than at $\omega = 0$, the Nyquist plot of $L_{eq}(j\omega)/K$ is sketched as shown in Fig. 8-27. Notice that this plot is sketched without any detailed data computed on $L_{eq}(j\omega)/K$ and, in fact, could be grossly inaccurate. However, the sketch is adequate to determine the stability of the system.

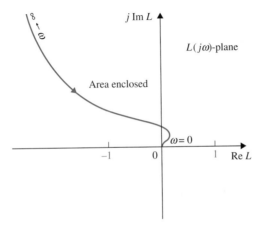

Figure 8-27 Nyquist plot of $\dfrac{L_{eq}(s)}{K} = \dfrac{s(s^2 + 2s + 2)}{s^2 + 5s + 1}$ for $\omega = \infty$ to $\omega = 0$.

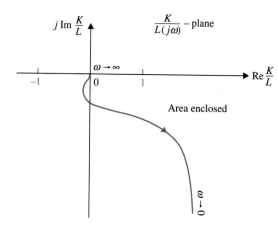

Figure 8-28 Nyquist plot of $K/L_{eq}(j\omega)$ for $\dfrac{L_{eq}(s)}{K} = \dfrac{s(s^2 + 2s + 2)}{s^2 + 5s + 1}$ for $\omega = \infty$ to $\omega = 0$.

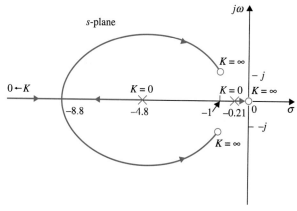

Figure 8-29 RL of $L(s) = \dfrac{Ks(s^2 + 2s + 2)}{s^2 + 5s + 1}$.

Because the Nyquist plot in Fig. 8-27 does not enclose the $(-1, j0)$ point as ω varies from ∞ to 0, the system is stable for all finite positive values of K.

Fig. 8-28 shows the Nyquist plot of Eq. (8-66), based on the poles and zeros of $L_{eq}(s)/K$ in Eq. (8-68). Notice that the RL stays in the left-half s-plane for all positive values of K, and the results confirm the Nyquist criterion results on system stability.

$$\frac{K}{L_{eq}(j\omega)} = \frac{(1 - \omega^2) + 5j\omega}{[-2\omega^2 + j\omega(2 - \omega^2)]} \tag{8-73}$$

for $\omega = \infty$ to 0. The plot again does not enclose the $(-1, j0)$ point, and the system is again stable for all positive values of K by interpreting the Nyquist plot of $K/L_{eq}(j\omega)$.

Fig. 8-29 shows the RL of Eq. (8-67) for $K > 0$, using the pole–zero configuration of $L_{eq}(s)$ of Eq. (8-68). Because the RL stays in the left-half s-plane for all positive values of K, the system is stable for $0 < K < \infty$, which agrees with the conclusion obtained with the Nyquist criterion.

▶ 8-9 EFFECTS OF ADDING POLES AND ZEROS TO L(s) ON THE SHAPE OF THE NYQUIST PLOT

Because the performance of a control system is often affected by adding and moving poles and zeros of the loop transfer function, it is important to investigate how the Nyquist plot is affected when poles and zeros are added to $L(s)$.

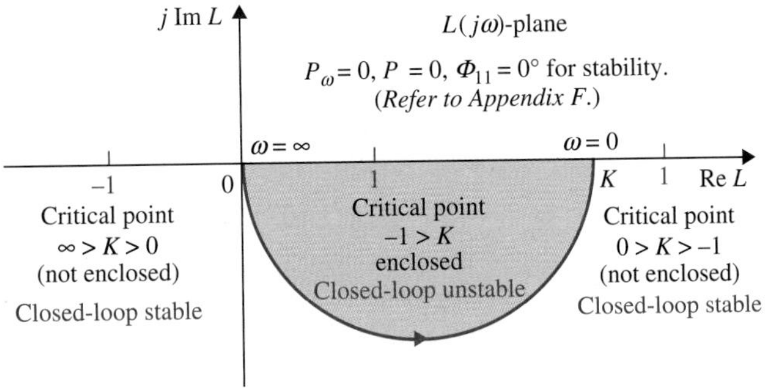

8-9 Effects of Adding Poles and Zeros to L(s) on the Shape of the Nyquist Plot

Figure 8-30 Nyquist plot of $L(s) = \dfrac{K}{(1+T_1 s)}$.

Let us begin with a first-order transfer function

$$L(s) = \frac{K}{1+T_1 s} \tag{8-74}$$

where T_1 is a positive real constant. The Nyquist plot of $L(j\omega)$ for $0 \leq \omega \leq \infty$ is a semicircle, as shown in Fig. 8-30. The figure also shows the interpretation of the closed-loop stability with respect to the critical point for all values of K between $-\infty$ and ∞.

Addition of Poles at $s = 0$
Consider that a pole at $s = 0$ is added to the transfer function of Eq. (8-74); then

$$L(s) = \frac{K}{s(1+T_1 s)} \tag{8-75}$$

Because adding a pole at $s = 0$ is equivalent to dividing $L(s)$ by $j\omega$, the phase of $L(j\omega)$ is reduced by 90° at both zero and infinite frequencies. In addition, the magnitude of $L(j\omega)$ at $\omega = 0$ becomes infinite. Fig. 8-31 illustrates the Nyquist plot of $L(j\omega)$ in Eq. (8-75) and the closed-loop stability interpretations with respect to the critical points for $-\infty < K < \infty$. In

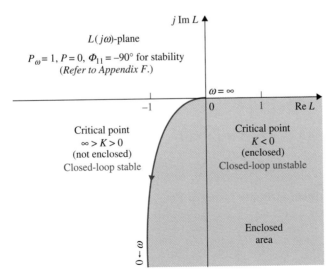

Figure 8-31 Nyquist plot of $L(s) = \dfrac{K}{s(1+T_1 s)}$.

general, adding a pole of multiplicity p at $s = 0$ to the transfer function of Eq. (8-74) will give the following properties to the Nyquist plot of $L(j\omega)$:

$$\lim_{\omega \to \infty} \angle L(j\omega) = -(p+1)90° \quad (8\text{-}76)$$

$$\lim_{\omega \to 0} \angle L(j\omega) = -p \times 90° \quad (8\text{-}77)$$

$$\lim_{\omega \to \infty} |L(j\omega)| = 0 \quad (8\text{-}78)$$

$$\lim_{\omega \to 0} |L(j\omega)| = \infty \quad (8\text{-}79)$$

The following example illustrates the effects of adding multiple-order poles to $L(s)$.

▶ **EXAMPLE 8-9-1** Fig. 8-32 shows the Nyquist plot of

$$L(s) = \frac{K}{s^2(1 + T_1 s)} \quad (8\text{-}80)$$

and the critical points, with stability interpretations. Fig. 8-33 illustrates the same for

$$L(s) = \frac{K}{s^3(1 + T_1 s)} \quad (8\text{-}81)$$

- Adding poles at $s = 0$ to a loop transfer function will reduce stability of the closed-loop system.

The conclusion from these illustrations is that the addition of poles at $s = 0$ to a loop transfer function will affect the stability of the closed-loop system adversely. A system that has a loop transfer function with more than one pole at $s = 0$ (type 2 or higher) is likely to be unstable or difficult to stabilize.

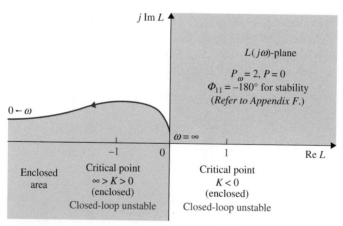

Figure 8-32 Nyquist plot of $L(s) = \dfrac{K}{s^2(1 + T_1 s)}$.

◀

Addition of Finite Nonzero Poles

When a pole at $s = -1/T_2 (T_2 > 0)$ is added to the function $L(s)$ of Eq. (8-74), we have

$$L(s) = \frac{K}{(1 + T_1 s)(1 + T_2 s)} \quad (8\text{-}82)$$

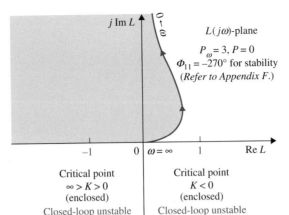

Figure 8-33 Nyquist plot of $L(s) = \dfrac{K}{s^3(1+T_1s)}$.

The Nyquist plot of $L(j\omega)$ at $\omega = 0$ is not affected by the addition of the pole, since

$$\lim_{\omega \to 0} L(j\omega) = K \qquad (8\text{-}83)$$

The value of $L(j\omega)$ at $\omega = \infty$ is

$$\lim_{\omega \to \infty} L(j\omega) = \lim_{\omega \to \infty} \frac{-K}{T_1 T_2 \omega^2} = 0\angle -180° \qquad (8\text{-}84)$$

Thus, the effect of adding a pole at $s = -1/T_2$ to the transfer function of Eq. (8-75) is to shift the phase of the Nyquist plot by $-90°$ at $\omega = \infty$, as shown in Fig. 8-34. The figure also shows the Nyquist plot of

$$L(s) = \frac{K}{(1+T_1s)(1+T_2s)(1+T_3s)} \qquad (8\text{-}85)$$

- Adding nonzero poles to the loop transfer function also reduces stability of the closed-loop system.

where two nonzero poles have been added to the transfer function of Eq. (8-74) ($T_1, T_2, T_3, > 0$). In this case, the Nyquist plot at $\omega = \infty$ is rotated clockwise by another $90°$ from that of Eq. (8-82). *These examples show the adverse effects on closed-loop stability when poles are added to the loop transfer function.* The closed-loop systems with the loop transfer functions of Eqs. (8-74) and (8-82) are all stable as long as K is positive. The system represented by Eq. (8-85) is unstable if the intersect of the Nyquist plot on the negative real axis is to the left of the $(-1, j0)$ point when K is positive.

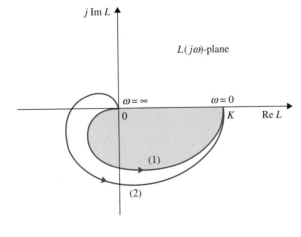

Figure 8-34 Nyquist plots. Curve (1): $L(s) = \dfrac{K}{(1+T_1s)(1+T_2s)}$. Curve (2): $L(s) = \dfrac{K}{(1+T_1s)}(1+T_2s)(1+T_3s)$.

Toolbox 8-9-1

MATLAB statements for Fig. 8-34

```
w=0:0.01:100;
num = [1];

den = conv([1 1],[1 1])
[re,im,w] = nyquist(num,den,w);
plot(re,im,'b');

hold on

den = conv(conv([1 1],[1 1]),[1 1])
[re,im,w] = nyquist(num,den,w);
plot(re,im,'r');

axis([-1 2 -1 1])
grid
```

Addition of Zeros

It was demonstrated in Chapter 5 that adding zeros to the loop transfer function has the effect of reducing the overshoot and the general effect of stabilization. In terms of the Nyquist criterion, this stabilization effect is easily demonstrated, since the multiplication of the term $(1 + T_d s)$ to the loop transfer function increases the phase of $L(s)$ by 90° at $\omega = \infty$. The following example illustrates the effect on stability of adding a zero at $-1/T_d$ to a loop transfer function.

▶ **EXAMPLE 8-9-2** Consider that the loop transfer function of a closed-loop control system is

$$L(s) = \frac{K}{s(1 + T_1 s)(1 + T_2 s)} \tag{8-86}$$

It can be shown that the closed-loop system is stable for

$$0 < K < \frac{T_1 + T_2}{T_1 T_2} \tag{8-87}$$

Suppose that a zero at $s = -1/T_d (T_d > 0)$ is added to the transfer function of Eq. (8-86); then,

- Adding zeros to the loop transfer function has the effect of stabilizing the closed-loop system.

$$L(s) = \frac{K(1 + T_d s)}{s(1 + T_1 s)(1 + T_2 s)} \tag{8-88}$$

The Nyquist plots of the two transfer functions of Eqs. (8-86) and (8-88) are shown in Fig. 8-35. The effect of the zero in Eq. (8-88) is to add 90° to the phase of the $L(j\omega)$ in Eq. (8-86) at $\omega = \infty$ while not

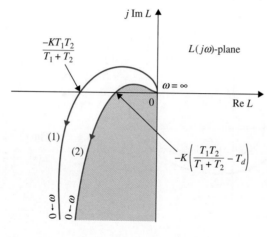

Figure 8-35 Nyquist plots. Curve (1): $L(s) = \dfrac{K}{s(1 + T_1 s)(1 + T_2 s)}$. Curve (2): $L(s) = \dfrac{K(1 + T_d s)}{s(1 + T_1 s)(1 + T_2 s)}$; $T_d < T_1; T_2$.

affecting the value at $\omega = 0$. The intersect on the negative real axis of the $L(j\omega)$-plane is moved from $-KT_1T_2/(T_1+T_2)$ to $-K(T_1T_2 - T_dT_1 - T_dT_2)/(T_1+T_2)$. Thus, the system with the loop transfer function in Eq. (8-88) is stable for

$$0 < K < \frac{T_1 + T_2}{T_1T_2 - T_d(T_1+T_2)} \qquad (8\text{-}89)$$

which, for positive T_d and K, has a higher upper bound than that of Eq. (8-87).

▶ 8-10 RELATIVE STABILITY: GAIN MARGIN AND PHASE MARGIN

We have demonstrated in Sections 8-2 through 8-4 the general relationship between the resonance peak M_p of the frequency response and the maximum overshoot of the time response. Comparisons and correlations between frequency-domain and time-domain parameters such as these are useful in the prediction of the performance of control systems. In general, we are interested not only in the absolute stability of a system but also how stable it is. The latter is often called **relative stability**. In the time domain, relative stability is measured by parameters such as the maximum overshoot and the damping ratio. In the frequency domain, the resonance peak M_p can be used to indicate relative stability. Another way of measuring relative stability in the frequency domain is by how close the Nyquist plot of $L(j\omega)$ is to the $(-1, j0)$ point.

• Relative stability is used to indicate how stable a system is.

Toolbox 8-10-1

MATLAB statements for Fig. 8-35

```
w=0:0.01:100;
num = [1];
den = conv(conv([1 1],[1 1]),[1 0])
[re,im,w] = nyquist(num,den,w);
plot(re,im,'b');
hold on

num = [1 1];
den = conv(conv([1 1],[1 1]),[1 0])
[re,im,w] = nyquist(num,den,w);
plot(re,im,'r');

axis([-2 2 -1 1])
grid
hold on

den = conv(conv([1 1],[1 1]),[1 1])
[re,im,w] = nyquist(num,den,w);
plot(re,im,'r');

axis([-1 2 -1 1])
grid
```

To demonstrate the concept of relative stability in the frequency domain, the Nyquist plots and the corresponding step responses and frequency responses of a typical third-order system are shown in Fig. 8-36 for four different values of loop gain K. It is assumed that the function $L(j\omega)$ is of minimum-phase type, so that the enclosure of the $(-1, j0)$ point is sufficient for stability analysis. The four cases are evaluated as follows.

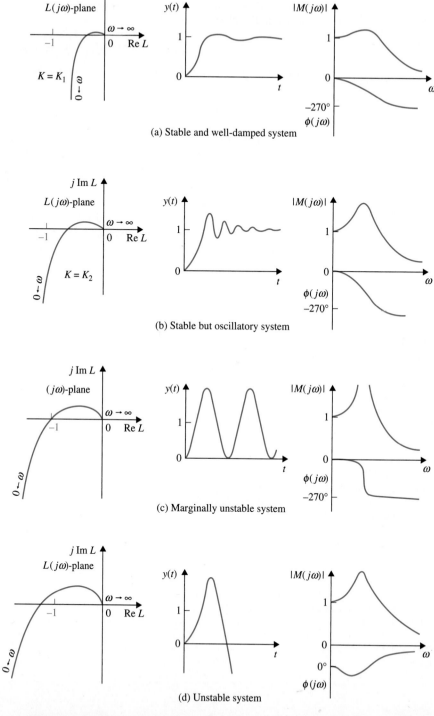

Figure 8-36 Correlation among Nyquist plots, step responses, and frequency responses.

1. **Fig. 8-36(a); the loop gain K is low:** The Nyquist plot of $L(j\omega)$ intersects the negative real axis at a point that is quite far to the right of the $(-1, j0)$ point. The corresponding step response is quite well damped, and the value of M_r of the frequency response is low.

2. *Fig. 8-36(b); K is increased:* The intersect is moved closer to the $(-1, j0)$ point; the system is still stable, because the critical point is not enclosed, but the step response has a larger maximum overshoot, and M_r is also larger.
3. *Fig. 8-36(c); K is increased further:* The Nyquist plot now passes through the $(-1, j0)$ point, and the system is marginally stable. The step response becomes oscillatory with constant amplitude, and M_r becomes infinite.
4. *Fig. 8-36(d); K is relatively very large:* The Nyquist plot now encloses the $(-1, j0)$ point, and the system is unstable. The step response becomes unbounded. The magnitude curve of $|M(j\omega)|$-versus-ω ceases to have any significance. In fact, for the unstable system, the value of M_r is still finite! In all the above analysis, the phase curve $\phi(j\omega)$ of the closed-loop frequency response also gives qualitative information about stability. Notice that the negative slope of the phase curve becomes steeper as the relative stability decreases. When the system is unstable, the slope beyond the resonant frequency becomes positive. In practice, the phase characteristics of the closed-loop system are seldom used for analysis and design purposes.

- M_r ceases to have any meaning when the closed-loop system is unstable.

8-10-1 Gain Margin (GM)

Gain Margin (GM) is one of the most frequently used criteria for measuring relative stability of control systems. In the frequency domain, gain margin is used to indicate the closeness of the intersection of the negative real axis made by the Nyquist plot of $L(j\omega)$ to the $(-1, j0)$ point. Before giving the definition of gain margin, let us first define the **phase crossover** on the Nyquist plot and the **phase-crossover frequency**.

- The definition of gain margin given here is for minimum-phase loop transfer functions.

Phase Crossover: A phase-crossover on the $L(j\omega)$ plot is a point at which the plot intersects the negative real axis.

Phase-Crossover Frequency: The **phase-crossover frequency** ω_p is the frequency at the phase crossover, or where

$$\angle L(j\omega_p) = 180° \qquad (8\text{-}90)$$

- Gain margin is measured at the phase crossover.

The Nyquist plot of a loop transfer function $L(j\omega)$ that is of minimum-phase type is shown in Fig. 8-37. The phase-crossover frequency is denoted as ω_p, and the magnitude of

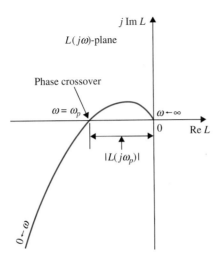

Figure 8-37 Definition of the gain margin in the polar coordinates.

$L(j\omega)$ at $\omega = \omega_p$ is designated as $|L(j\omega_p)|$. Then, the gain margin of the closed-loop system that has $L(s)$ as its loop transfer function is defined as

$$\text{gain margin} = \text{GM} = 20\log_{10}\frac{1}{|L(j\omega_p)|}$$
$$= -20\log_{10}|L(j\omega_p)| \; dB \qquad (8\text{-}91)$$

On the basis of this definition, we can draw the following conclusions about the gain margin of the system shown in Fig. 8-37, depending on the properties of the Nyquist plot.

1. The $L(j\omega)$ plot does not intersect the negative real axis (no finite nonzero phase crossover).

$$|L(j\omega_p)| = 0 \quad \text{GM} = \infty \; dB \qquad (8\text{-}92)$$

2. The $L(j\omega)$ plot intersects the negative real axis between (phase crossover lies between) 0 and the -1 point.

$$0 < |L(j\omega_p)| < 1 \quad \text{GM} > 0 \; dB \qquad (8\text{-}93)$$

3. The $L(j\omega)$ plot passes through (phase crossover is at) the $(-1, j0)$ point.

$$|L(j\omega_p)| = 1 \quad \text{GM} = 0 \; dB \qquad (8\text{-}94)$$

4. The $L(j\omega)$ plot encloses (phase crossover is to the left of) the $(-1, j0)$ point.

$$|L(j\omega_p)| > 1 \quad \text{GM} < 0 \; dB \qquad (8\text{-}95)$$

• Gain margin is the amount of gain in dB that can be added to the loop before the closed-loop system becomes unstable.

Based on the foregoing discussions, the physical significance of gain margin can be summarized as:

Gain margin is the amount of gain in decibels (dB) that can be added to the loop before the closed-loop system becomes unstable.

- When the Nyquist plot does not intersect the negative real axis at any finite nonzero frequency, the gain margin is infinite in dB; this means that, theoretically, the value of the loop gain can be increased to infinity before instability occurs.
- When the Nyquist plot of $L(j\omega)$ passes through the $(-1, j0)$ point, the gain margin is 0 dB, which implies that the loop gain can no longer be increased, as the system is at the margin of instability.
- When the phase-crossover is to the left of the $(-1, j0)$ point, the phase margin is negative in dB, and the loop gain must be reduced by the gain margin to achieve stability.

Gain Margin of Nonminimum-Phase Systems

Care must be taken when attempting to extend gain margin as a measure of relative stability to systems with nonminimum-phase loop transfer functions. For such systems, a system may be stable even when the phase-crossover point is to the left of $(-1, j0)$, and thus a negative gain margin may still correspond to a stable system.

8-10-2 Phase Margin (PM)

The gain margin is only a one-dimensional representation of the relative stability of a closed-loop system. As the name implies, gain margin indicates system stability with respect to the variation in loop gain only. In principle, one would believe a system with a large gain margin should always be relatively more stable than one with a smaller gain margin. Unfortunately, gain margin alone is inadequate to indicate relative stability when system parameters other than the loop gain are subject to variation. For instance, the two systems represented by the $L(j\omega)$ plots in Fig. 8-38 apparently have the same gain margin. However, locus A actually corresponds to a more stable system than locus B, because with any change in the system parameters that affect the phase of $L(j\omega)$, locus B may easily be altered to enclose the $(-1, j0)$ point. Furthermore, we can show that the system B actually has a larger M_r than system A.

To include the effect of phase shift on stability, we introduce the **phase margin**, which requires that we first make the following definitions:

Gain Crossover: The gain crossover is a point on the $L(j\omega)$ plot at which the magnitude of $L(j\omega)$ is equal to 1.

Gain-Crossover Frequency: The gain-crossover frequency, ω_g, is the frequency of $L(j\omega)$ at the gain crossover. Or where

$$|L(j\omega_g)| = 1 \qquad (8\text{-}96)$$

The definition of phase margin is stated as:

Phase margin (PM) is defined as the angle in degrees through which the $L(j\omega)$ plot must be rotated about the origin so that the gain crossover passes through the $(-1, j0)$ point.

Fig. 8-39 shows the Nyquist plot of a typical minimum-phase $L(j\omega)$ plot, and the phase margin is shown as the angle between the line that passes through the gain crossover and the origin. In contrast to the gain margin, which is determined by loop gain, phase margin indicates the effect on system stability due to changes in system parameter, which

- The definition of phase margin given here is for a system with a minimum-phase loop transfer function.

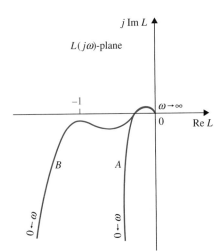

Figure 8-38 Nyquist plots showing systems with the same gain margin but different degrees of relative stability.

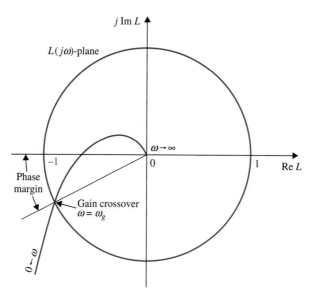

Figure 8-39 Phase margin defined in the $L(j\omega)$-plane.

theoretically alter the phase of $L(j\omega)$ by an equal amount at all frequencies. Phase margin is the amount of pure phase delay that can be added to the loop before the closed-loop system becomes unstable.

When the system is of the minimum-phase type, the analytical expression of the phase margin, as seen from Fig. 8-39, can be expressed as

$$\text{phase margin(PM)} = \angle L(j\omega_g) - 180° \tag{8-97}$$

- Phase margin is measured at the gain crossover.

where ω_g is the gain-crossover frequency.

Care should be taken when interpreting the phase margin from the Nyquist plot of a nonminimum-phase transfer function. When the loop transfer function is of the nonminimum-phase type, the gain crossover can occur in any quadrant of the $L(j\omega)$-plane, and the definition of phase margin given in Eq. (8-97) is no longer valid.

- Phase margin is the amount of pure phase delay that can be added before the system becomes unstable.

▶ **EXAMPLE 8-10-1** As an illustrative example on gain and phase margins, consider that the loop transfer function of a control system is

$$L(s) = \frac{2500}{s(s+5)(s+50)} \tag{8-98}$$

The Nyquist plot of $L(j\omega)$ is shown in Fig. 8-40. The following results are obtained from the Nyquist plot:

Gain crossover $\omega_g = 6.22$ rad/sec

Phase crossover $\omega_p = 15.88$ rad/sec

The gain margin is measured at the phase crossover. The magnitude of $L(j\omega_p)$ is 0.182. Thus, the gain margin is obtained from Eq. (8-91):

$$\text{GM} = 20\log_{10}\frac{1}{|L(j\omega_p)|} = 20\log_{10}\frac{1}{0.182} = 14.80\,\text{dB} \tag{8-99}$$

The phase margin is measured at the gain crossover. The phase of $L(j\omega_g)$ is 211.72°. Thus, the phase margin is obtained from Eq. (8-97):

$$\text{PM} = \angle L(j\omega_g) - 180° = 211.72° - 180° = 31.72° \tag{8-100}$$

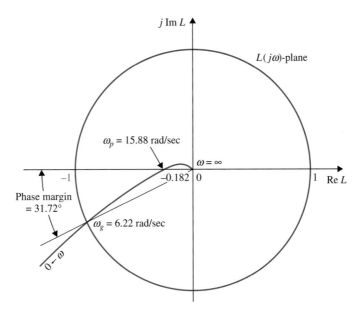

Figure 8-40 Nyquist plot of $L(s) = \dfrac{2500}{s(s+5)(s+50)}$.

Before embarking on the Bode plot technique of stability study, it would be beneficial to summarize advantages and disadvantages of the Nyquist plot.

Advantages of the Nyquist Plot
1. The Nyquist plot can be used for the study of stability of systems with nonminimum-phase transfer functions.
2. The stability analysis of a closed-loop system can be easily investigated by examining the Nyquist plot of the loop transfer function with reference to the $(-1, j0)$ point once the plot is made.

Disadvantage of the Nyquist Plot
1. It's not so easy to carry out the design of the controller by referring to the Nyquist plot.

▶ 8-11 STABILITY ANALYSIS WITH THE BODE PLOT

The Bode plot of a transfer function described in Chapter 2 is a very useful graphical tool for the analysis and design of linear control systems in the frequency domain. Before the inception of computers, Bode plots were often called the "asymptotic plots," because the magnitude and phase curves can be sketched from their asymptotic properties without detailed plotting. Modern applications of the Bode plot for control systems should be identified with the following advantages and disadvantages:

Advantages of the Bode Plot
1. In the absence of a computer, a Bode diagram can be sketched by approximating the magnitude and phase with straightline segments.
2. Gain crossover, phase crossover, gain margin, and phase margin are more easily determined on the Bode plot than from the Nyquist plot.
3. For design purposes, the effects of adding controllers and their parameters are more easily visualized on the Bode plot than on the Nyquist plot.

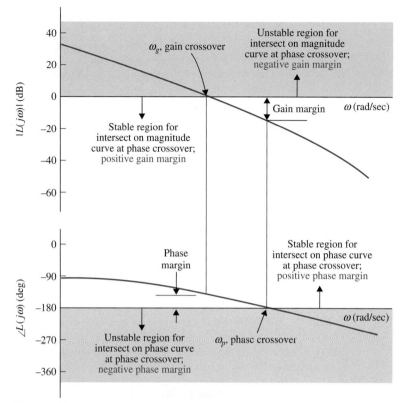

Figure 8-41 Determination of gain margin and phase margin on the Bode plot.

Disadvantage of the Bode Plot

1. Absolute and relative stability of only minimum-phase systems can be determined from the Bode plot. For instance, there is no way of telling what the stability criterion is on the Bode plot.

With reference to the definitions of gain margin and phase margin given in Figs. 8-37 and 8-39, respectively, the interpretation of these parameters from the Bode diagram is illustrated in Fig. 8-41 for a typical minimum-phase loop transfer function. The following observations can be made on system stability with respect to the properties of the Bode plot:

1. The gain margin is positive and the system is stable if the magnitude of $L(j\omega)$ at the phase crossover is negative in dB. That is, the gain margin is measured below the 0-dB-axis. If the gain margin is measured above the 0-dB-axis, the gain margin is negative, and the system is unstable.

- Bode plots are useful only for stability studies of systems with minimum-phase loop transfer functions.

2. The phase margin is positive and the system is stable if the phase of $L(j\omega)$ is greater than $-180°$ at the gain crossover. That is, the phase margin is measured above the $-180°$-axis. If the phase margin is measured below the $-180°$-axis, the phase margin is negative, and the system is unstable.

▶ **EXAMPLE 8-11-1** Consider the loop transfer function given in Eq. (8-98); the Bode plot of the function is drawn as shown in Fig. 8-42. The following results are observed easily from the magnitude and phase plots.

The gain crossover is the point where the magnitude curve intersects the 0-dB axis. The gain-crossover frequency ω_g is 6.22 rad/sec. The phase margin is measured at the gain crossover. The phase

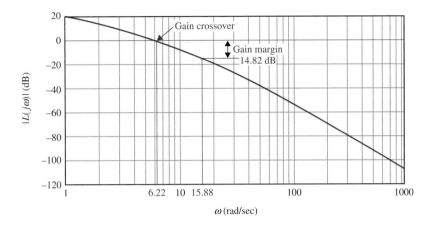

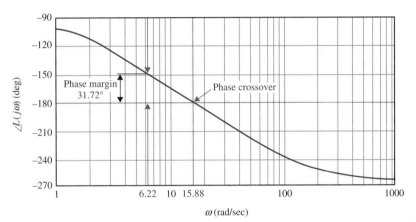

Figure 8-42 Bode plot of $L(s) = \dfrac{2500}{s(s+5)(s+50)}$.

margin is measured from the $-180°$-axis and is $31.72°$. Because the phase margin is measured above the $-180°$-axis, the phase margin is positive, and the system is stable.

The phase crossover is the point where the phase curve intersects the $-180°$-axis. The phase-crossover frequency is $\omega_p = 15.88$ rad/sec. The gain margin is measured at the phase crossover and is 14.8 dB. Because the gain margin is measured below the 0-dB-axis, the gain margin is positive, and the system is stable.

Toolbox 8-11-1

MATLAB statements for Fig. 8-42

```
G = zpk([],[0 -1 -1],2500)
margin(G)

grid
```

The reader should compare the Nyquist plot of Fig. 8-40 with the Bode plot of Fig. 8-42, and the interpretation of ω_g, ω_p, GM, and PM on these plots.

8-11-1 Bode Plots of Systems with Pure Time Delays

The stability analysis of a closed-loop system with a pure time delay in the loop can be conducted easily with the Bode plot. Example 8-11-2 illustrates the standard procedure.

▶ **EXAMPLE 8-11-2** Consider that the loop transfer function of a closed-loop system is

$$L(s) = \frac{Ke^{-T_d s}}{s(s+1)(s+2)} \quad (8\text{-}101)$$

Fig. 8-43 shows the Bode plot of $L(j\omega)$ with $K = 1$ and $T_d = 0$. The following results are obtained:

Gain-crossover frequency = 0.446 rad/sec
Phase margin = 53.4°
Phase-crossover frequency = 1.416 rad/sec
Gain margin = 15.57 dB

Thus, the system with the present parameters is stable.

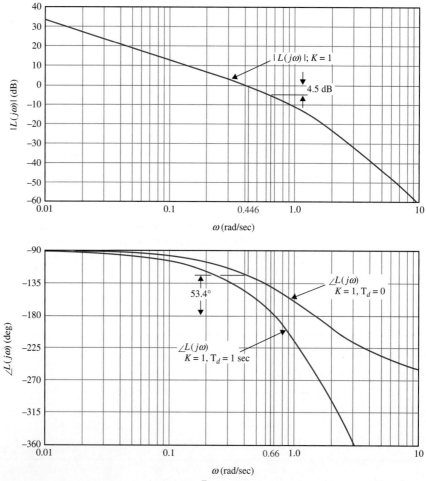

Figure 8-43 Bode plot of $L(s) = \dfrac{Ke^{-T_d s}}{s(s+1)(s+2)}$.

The effect of the pure time delay is to add a phase of $-T_d\omega$ radians to the phase curve while not affecting the magnitude curve. The adverse effect of the time delay on stability is apparent, because the negative phase shift caused by the time delay increases rapidly with the increase in ω. To find the critical value of the time delay for stability, we set

$$T_d\omega_g = 53.4° \frac{\pi}{180°} = 0.932 \quad \text{radians} \tag{8-102}$$

Solving for T_d from the last equation, we get the critical value of T_d to be 2.09 seconds.

Continuing with the example, we set T_d arbitrarily at 1 second and find the critical value of K for stability. Fig. 8-43 shows the Bode plot of $L(j\omega)$ with this new time delay. With K still equal to 1, the magnitude curve is unchanged. The phase curve droops with the increase in ω, and the following results are obtained:

Phase-crossover frequency = 0.66 rad/sec

Gain margin = 4.5 dB

Thus, using the definition of gain margin of Eq. (8-91), the critical value of K for stability is $10^{4.5/20} = 1.68$. ◄

► 8-12 RELATIVE STABILITY RELATED TO THE SLOPE OF THE MAGNITUDE CURVE OF THE BODE PLOT

In addition to GM, PM, and M_p as relative stability measures, the slope of the magnitude curve of the Bode plot of the loop transfer function at the gain crossover also gives a qualitative indication on the relative stability of a closed-loop system. For example, in Fig. 8-42, if the loop gain of the system is decreased from the nominal value, the magnitude curve is shifted downward, while the phase curve is unchanged. This causes the gain-crossover frequency to be lower, and the slope of the magnitude curve at this frequency is less negative; the corresponding phase margin is increased. On the other hand, if the loop gain is increased, the gain-crossover frequency is increased, and the slope of the magnitude curve is more negative. This corresponds to a smaller phase margin, and the system is less stable. The reason behind these stability evaluations is quite simple. For a minimum-phase transfer function, the relation between its magnitude and phase is unique. Because the negative slope of the magnitude curve is a result of having more poles than zeros in the transfer function, the corresponding phase is also negative. In general, the steeper the slope of the magnitude curve, the more negative the phase. Thus, if the gain crossover is at a point where the slope of the magnitude curve is steep, it is likely that the phase margin will be small or negative.

8-12-1 Conditionally Stable System

The illustrative examples given thus far are uncomplicated in the sense that the slopes of the magnitude and phase curves are monotonically decreasing as ω increases. The following example illustrates a **conditionally stable system** that is capable of going through stable/unstable conditions as the loop gain varies.

► **EXAMPLE 8-12-1** Consider that the loop transfer function of a closed-loop system is

$$L(s) = \frac{100K(s+5)(s+40)}{s^3(s+100)(s+200)} \tag{8-103}$$

The Bode plot of $L(j\omega)$ is shown in Fig. 8-44 for $K = 1$. The following results on the system stability are obtained:

Gain-crossover frequency = 1 rad/sec

Phase margin = $-78°$

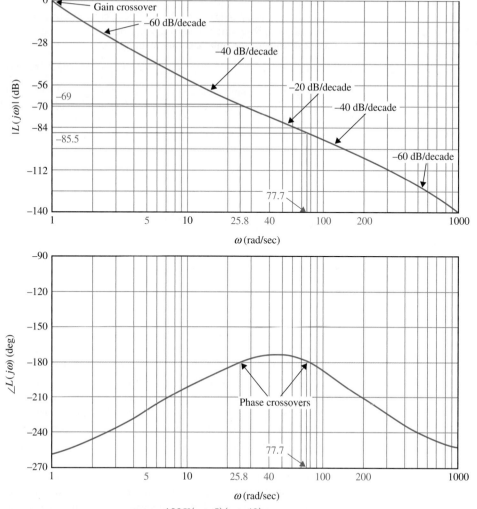

Figure 8-44 Bode plot of $L(s) = \dfrac{100K(s+5)(s+40)}{s^3(s+100)(s+200)}$, $K = 1$.

There are two phase crossovers: one at 25.8 rad/sec and the other at 77.7 rad/sec. The phase characteristics between these two frequencies indicate that, if the gain crossover lies in this range, the system would be stable. From the magnitude curve, the range of K for stable operation is found to be between 69 and 85.5 dB. For values of K above and below this range, the phase of $L(j\omega)$ is less than $-180°$, and the system is unstable. This example serves as a good example of the relation between relative stability and the slope of the magnitude curve at the gain crossover. As observed from Fig. 8-44, at both very low and very high frequencies, the slope of the magnitude curve is -60 dB/decade; if the gain crossover falls in either one of these two regions, the phase margin is negative, and the system is unstable. In the two sections of the magnitude curve that have a slope of -40 dB/decade, the system is stable only if the gain crossover falls in about half of these regions, but even then the phase margin is small. If the gain crossover falls in the region in which the magnitude curve has a slope of -20 dB/decade, the system is stable.

Fig. 8-45 shows the Nyquist plot of $L(j\omega)$. It is of interest to compare the results on stability derived from the Bode plot and the Nyquist plot. The root-locus diagram of the system is shown in Fig. 8-46. The root loci give a clear picture on the stability condition of the system with respect to K. The number of crossings of the root loci on the $j\omega$-axis of the s-plane equals the number of crossings

8-12 Relative Stability Related to the Slope of the Magnitude Curve of the Bode Plot ◄ 461

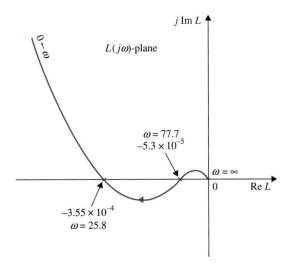

Figure 8-45 Nyquist plot of
$L(s) = \dfrac{100K(s+5)(s+40)}{s^3(s+100)(s+200)}$, $K = 1$.

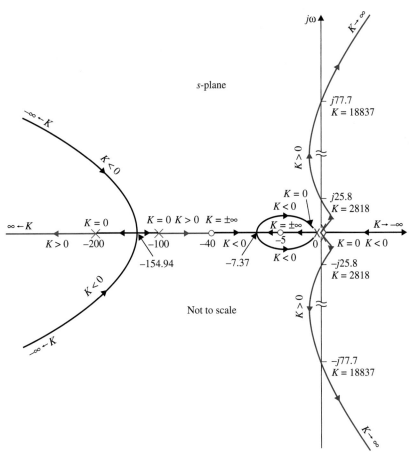

Figure 8-46 Root loci of $G(s) = \dfrac{100K(s+5)(s+40)}{s^3(s+100)(s+200)}$.

462 ▶ Chapter 8. Frequency-Domain Analysis

of the phase curve of $L(j\omega)$ of the $-180°$ axis of the Bode plot and the number of crossings of the Nyquist plot of $L(j\omega)$ with the negative real axis. The reader should check the gain margins obtained from the Bode plot and the coordinates of the crossover points on the negative real axis of the Nyquist plot with the values of K at the $j\omega$-axis crossings on the root loci. ◀

▶ 8-13 STABILITY ANALYSIS WITH THE MAGNITUDE-PHASE PLOT

The magnitude-phase plot described in Chapter 2 is another form of the frequency-domain plot that has certain advantages for analysis and design in the frequency domain. The magnitude-phase plot of a transfer function $L(j\omega)$ is done in $|L(j\omega)|$(dB) versus $\angle L(j\omega)$ (degrees). The magnitude-phase plot of the transfer function in Eq. (8-98) is constructed in Fig. 8-47 by use of the data from the Bode plot of Fig. 8-42. The gain and phase crossovers and the gain and phase margins are clearly indicated on the magnitude-phase plot of $L(j\omega)$.

- The critical point is the intersect of the 0-dB-axis and the $-180°$-axis.
- The phase crossover is where the locus intersects the $-180°$-axis.

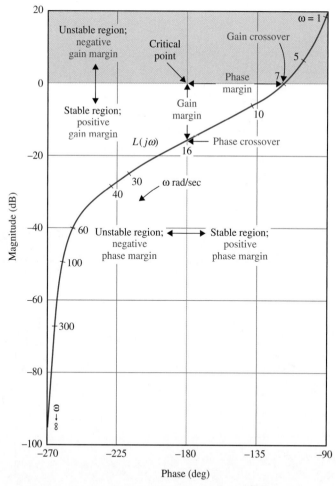

Figure 8-47 Gain-phase plot of $L(s) = \dfrac{10}{s(1+0.2s)(1+0.02s)}$.

- The gain crossover is where the locus intersects the 0-dB-axis.
- The gain margin is the vertical distance in dB measured from the phase crossover to the critical point.
- The phase margin is the horizontal distance measured in degrees from the gain crossover to the critical point.

The regions in which the gain and phase crossovers should be located for stability are also indicated. Because the vertical axis for $|L(j\omega)|$ is in dB, when the loop gain of $L(j\omega)$ changes, the locus is simply shifted up and down along the vertical axis. Similarly, when a constant phase is added to $L(j\omega)$, the locus is shifted horizontally without distortion to the curve. If $L(j\omega)$ contains a pure time delay T_d, the effect of the time delay is to add a phase equal to $-\omega T_d \times 180°/\pi$ along the curve.

Another advantage of using the magnitude-phase plot is that, *for unity-feedback systems*, closed-loop system parameters such as M_r, ω_r, and BW can all be determined from the plot with the help of the constant-M loci. These closed-loop performance parameters are not represented on the Bode plot of the forward-path transfer function of a unity-feedback system.

▶ 8-14 CONSTANT-M LOCI IN THE MAGNITUDE-PHASE PLANE: THE NICHOLS CHART

It was pointed out earlier that, analytically, the resonant peak M_r and bandwidth BW are difficult to obtain for high-order systems, and the Bode plot provides information on the closed-loop system only in the form of gain margin and phase margin. It is necessary to develop a graphical method for the determination of M_r, ω_r, and BW using the forward-path transfer function $G(j\omega)$. As we shall see in the following development, the method is directly applicable only to unity-feedback systems, although with some modification it can also be applied to nonunity-feedback systems.

Consider that $G(s)$ is the forward-path transfer function of a unity-feedback system. The closed-loop transfer function is

$$M(s) = \frac{G(s)}{1 + G(s)} \tag{8-104}$$

For sinusoidal steady state, we replace s with $j\omega$; $G(s)$ becomes

$$\begin{aligned} G(j\omega) &= \text{Re}G(j\omega) + j\text{Im}G(j\omega) \\ &= x + jy \end{aligned} \tag{8-105}$$

where, for simplicity, x denotes $\text{Re}G(j\omega)$ and y denotes $\text{Im}G(j\omega)$. The magnitude of the closed-loop transfer function is written

$$|M(j\omega)| = \left|\frac{G(j\omega)}{1 + G(j\omega)}\right| = \frac{\sqrt{x^2 + y^2}}{\sqrt{(1+x)^2 + y^2}} \tag{8-106}$$

For simplicity of notation, let M denote $|M(j\omega)|$; then Eq. (8-106) leads to

$$M\sqrt{(1+x)^2 + y^2} = \sqrt{x^2 + y^2} \tag{8-107}$$

Squaring both sides of Eq. (8-107) gives

$$M^2\left[(1+x)^2 + y^2\right] = x^2 + y^2 \tag{8-108}$$

Rearranging Eq. (8-108) yields

$$(1 - M^2)x^2 + (1 - M^2)y^2 - 2M^2x = M^2 \quad (8\text{-}109)$$

This equation is conditioned by dividing through by $(1 - M^2)$ and adding the term $[M^2/(1 - M^2)]^2$ on both sides. We have

$$x^2 + y^2 - \frac{2M^2}{1 - M^2}x + \left(\frac{M^2}{1 - M^2}\right)^2 = \frac{M^2}{1 - M^2} + \left(\frac{M^2}{1 - M^2}\right)^2 \quad (8\text{-}110)$$

which is finally simplified to

$$\left(x - \frac{M^2}{1 - M^2}\right)^2 + y^2 = \left(\frac{M}{1 - M^2}\right)^2 \quad M \neq 1 \quad (8\text{-}111)$$

For a given value of M, Eq. (8-111) represents a circle with the center at

$$x = \text{Re}G(j\omega) = \frac{M^2}{1 - M^2} \quad y = 0 \quad (8\text{-}112)$$

The radius of the circle is

$$r = \left|\frac{M}{1 - M^2}\right| \quad (8\text{-}113)$$

When M takes on different values, Eq. (8-111) describes in the $G(j\omega)$-plane a family of circles that are called the **constant-M loci**, or the **constant-M circles**. Fig. 8-48 illustrates a typical set of constant-M circles in the $G(j\omega)$-plane. These circles are symmetrical with

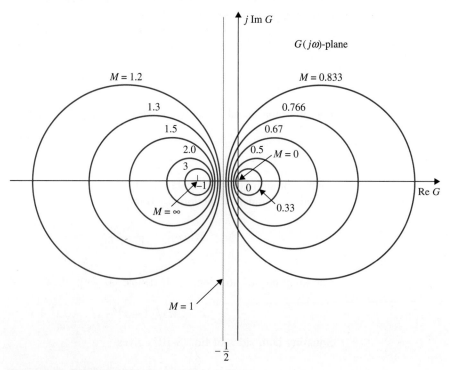

Figure 8-48 Constant-M circles in polar coordinates.

respect to the $M = 1$ line and the real axis. The circles to the left of the $M = 1$ locus correspond to values of M greater than 1, and those to the right of the $M = 1$ line are for M less than 1. Eqs. (8-111) and (8-112) show that, when M becomes infinite, the circle degenerates to a point at $(-1, j0)$. Graphically, the intersection of the $G(j\omega)$ curve and the constant-M circle gives the value of M at the corresponding frequency on the $G(j\omega)$ curve. If we want to keep the value of M_r less than a certain value, the $G(j\omega)$ curve must not intersect the corresponding M circle at any point and at the same time must not enclose the $(-1, j0)$ point. The constant-M circle with the smallest radius that is tangent to the $G(j\omega)$ curve gives the value of M_r, and the resonant frequency ω_r is read off at the tangent point on the $G(j\omega)$ curve.

Fig. 8-49(a) illustrates the Nyquist plot of $G(j\omega)$ for a unity-feedback control system together with several constant-M loci. For a given loop gain $K = K_1$, the intersects between the $G(j\omega)$ curve and the constant-M loci give the points on the $|M(j\omega)|$-versus-ω curve. The resonant peak M_r is found by locating the smallest circle that is tangent to the $G(j\omega)$

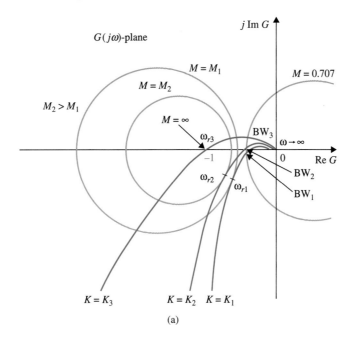

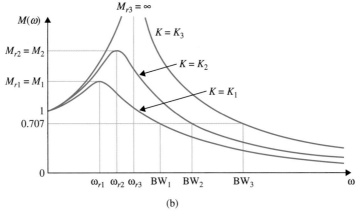

Figure 8-49 (a) Polar plots of $G(s)$ and constant-M loci. (b) Corresponding magnification curves.

466 ▶ Chapter 8. Frequency-Domain Analysis

curve. The resonant frequency is found at the point of tangency and is designated as ω_{r1}. If the loop gain is increased to K_2, and if the system is still stable, a constant-M circle with a smaller radius that corresponds to a larger M is found tangent to the $G(j\omega)$ curve, and thus the resonant peak will be larger. The resonant frequency is shown to be ω_{r2}, which is closer to the phase-crossover frequency ω_p than ω_{r1}. When K is increased to K_3, so that the $G(j\omega)$ curve now passes through the $(-1, j0)$ point, the system is marginally stable, and M_r is infinite; ω_{p3} is now the same as the resonant frequency ω_r.

When enough points of intersection between the $G(j\omega)$ curve and the constant-M loci are obtained, the magnification curves of $|M(j\omega)|$-versus-ω are plotted, as shown in Fig. 8-49(b).

The bandwidth of the closed-loop system is found at the intersect of the $G(j\omega)$ curve and the $M = 0.707$ locus. For values of K beyond K_3, the system is unstable, and the constant-M loci and M_r no longer have any meaning.

- When the system is unstable, the constant-M loci and M_r no longer have any meaning.

A major disadvantage in working in the polar coordinates of the Nyquist plot of $G(j\omega)$ is that the curve no longer retains its original shape when a simple modification such as the change of the loop gain is made to the system. Frequently, in design situations, not only must the loop gain be altered, but a series controller may have to be added to the system. This requires the complete reconstruction of the Nyquist plot of the modified $G(j\omega)$. For design work involving M_r and BW as specifications, it is more convenient to work with the magnitude-phase plot of $G(j\omega)$, because when the loop gain is altered, the entire $G(j\omega)$ curve is shifted up or down vertically without distortion. When the phase properties of $G(j\omega)$ are changed independently, without affecting the gain, the magnitude-phase plot is affected only in the horizontal direction.

- BW is the frequency where the $G(j\omega)$ curve intersects the $M = -3$ dB locus of the Nichols chart.

For that reason, the constant-M loci in the polar coordinates are plotted in magnitude-phase coordinates, and the resulting loci are called the **Nichols chart**. A typical Nichols chart of selected constant-M loci is shown in Fig. 8-50. Once the $G(j\omega)$ curve of the system is constructed in the Nichols chart, the intersects between the constant-M loci and the $G(j\omega)$ trajectory give the value of M at the corresponding frequencies of $G(j\omega)$. The resonant

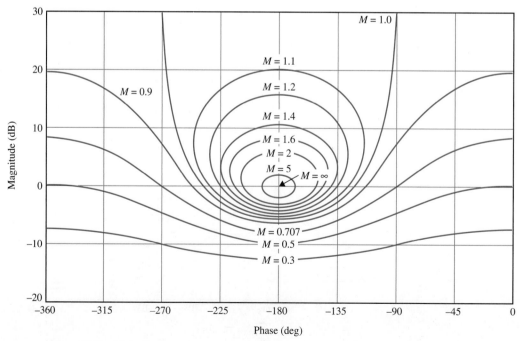

Figure 8-50 Nichols chart.

peak M_r is found by locating the smallest of the constant-M locus ($M \geq 1$) that is tangent to the $G(j\omega)$ curve from above. The resonant frequency is the frequency of $G(j\omega)$ at the point of tangency. *The bandwidth of the closed-loop system is the frequency at which the* $G(j\omega)$ *curve intersects the* $M = 0.707$ *or* $M = -3$ *dB locus.*

The following example illustrates the relationship among the analysis methods using the Bode plot and the Nichols chart.

▶ **EXAMPLE 8-14-1** Consider the position-control system of the control surfaces of the airship analyzed in Section 5-8. The forward-path transfer function of the unity-feedback system is given by Eq. (5-153), and is repeated here:

$$G(s) = \frac{1.5 \times 10^7 K}{s(s + 400.26)(s + 3008)} \qquad (8\text{-}114)$$

The Bode plots for $G(j\omega)$ are shown in Fig. 8-51 for $K = 7.248, 14.5, 181.2,$ and 273.57. The gain and phase margins of the closed-loop system for these values of K are determined and shown on

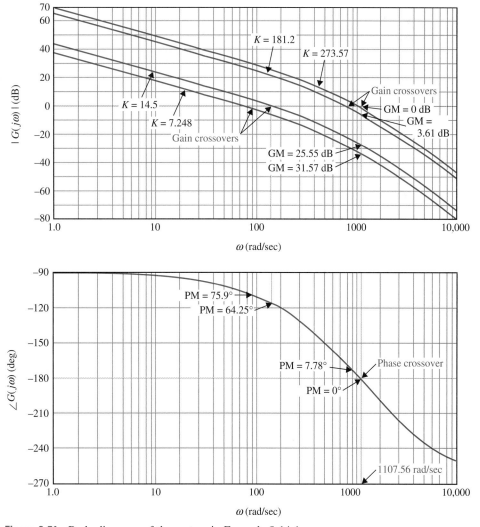

Figure 8-51 Bode diagrams of the system in Example 8-14-1.

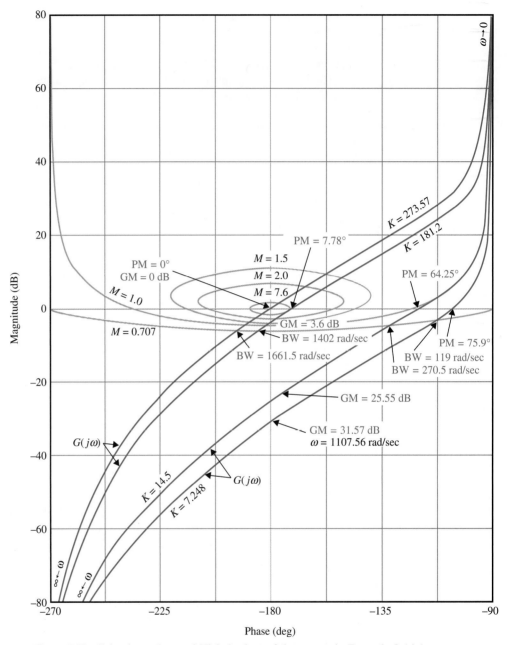

Figure 8-52 Gain-phase plots and Nichols chart of the system in Example 8-14-1.

the Bode plots. The magnitude-phase plots of $G(j\omega)$ corresponding to the Bode plots are shown in Fig. 8-52. These magnitude-phase plots, together with the Nichols chart, give information on the resonant peak M_r, resonant frequency ω_r, and the bandwidth BW. The gain and phase margins are also clearly marked on the magnitude-phase plots. Fig. 8-53 shows the closed-loop frequency responses. Table 8-2 summarizes the results of the frequency-domain analysis for the four different values of K together with the time-domain maximum overshoots determined in Section 5-8.

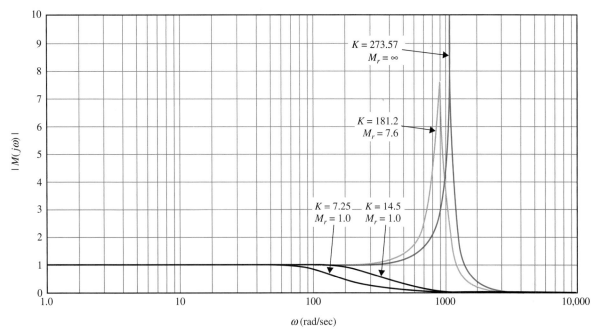

Figure 8-53 Closed-loop frequency response of the system in Example 8-14-1.

TABLE 8-2 Summary of Frequency-Domain Analysis

K	Maximum Overshoot (%)	M_r	ω_r (rad/sec)	Gain Margin (dB)	Phase Margin (deg)	BW (rad/sec)
7.25	0	1.0	1.0	31.57	75.9	119.0
14.5	4.3	1.0	43.33	5.55	64.25	270.5
181.2	15.2	7.6	900.00	3.61	7.78	1402.0
273.57	100.0	∞	1000.00	0	0	1661.5

8-15 NICHOLS CHART APPLIED TO NONUNITY-FEEDBACK SYSTEMS

The constant-M loci and the Nichols chart presented in the preceding sections are limited to closed-loop systems with unity feedback whose transfer function is given by Eq. (8-104). When a system has nonunity feedback, the closed-loop transfer function of the system is expressed as

$$M(s) = \frac{G(s)}{1 + G(s)H(s)} \qquad (8\text{-}115)$$

where $H(s) \neq 1$. The constant-M loci and the Nichols chart cannot be applied directly to obtain the closed-loop frequency response by plotting $G(j\omega)H(j\omega)$, since the numerator of $M(s)$ does not contain $H(j\omega)$.

By proper modification, the constant-M loci and Nichols chart can still be applied to a nonunity-feedback system. Let us consider the function

$$P(s) = H(s)M(s) = \frac{G(s)H(s)}{1 + G(s)H(s)} \qquad (8\text{-}116)$$

Apparently, Eq. (8-116) is of the same form as Eq. (8-104). The frequency response of $P(j\omega)$ can be determined by plotting the function $G(j\omega)H(j\omega)$ in the amplitude-phase coordinates along with the Nichols chart. Once this is done, the frequency-response information for $M(j\omega)$ is obtained as follows.

$$|M(j\omega)| = \frac{|P(j\omega)|}{|H(j\omega)|} \tag{8-117}$$

or, in terms of dB,

$$|M(j\omega)|(dB) = |P(j\omega)|(dB) - |H(j\omega)|(dB) \tag{8-118}$$

$$\phi_m(j\omega) = \angle M(j\omega) = \angle P(j\omega) - \angle H(j\omega) \tag{8-119}$$

▶ 8-16 SENSITIVITY STUDIES IN THE FREQUENCY DOMAIN

• Sensitivity study is easily carried out in the frequency domain.

The advantage of using the frequency domain in linear control systems is that higher-order systems can be handled more easily than in the time domain. Furthermore, the sensitivity of the system with respect to parameter variations can be easily interpreted using frequency-domain plots. We shall show how the Nyquist plot and the Nichols chart can be utilized for the analysis and design of control systems based on sensitivity considerations.

Consider a linear control system with unity feedback described by the transfer function

$$M(s) = \frac{G(s)}{1 + G(s)} \tag{8-120}$$

The sensitivity of $M(s)$ with respect to the loop gain K, which is a multiplying factor in $G(s)$, is defined as

$$S_G^M(s) = \frac{\dfrac{dM(s)}{M(s)}}{\dfrac{dG(s)}{G(s)}} = \frac{dM(s)}{dG(s)} \frac{G(s)}{M(s)} \tag{8-121}$$

Taking the derivative of $M(s)$ with respect to $G(s)$ and substituting the result into Eq. (8-121) and simplifying, we have

$$S_G^M(s) = \frac{1}{1 + G(s)} = \frac{1/G(s)}{1 + 1/G(s)} \tag{8-122}$$

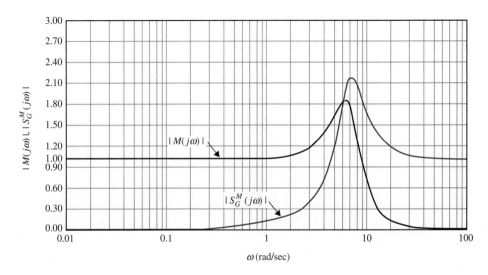

Figure 8-54 $|M(j\omega)|$ and $|S_G^M(j\omega)|$ versus ω for $G(s) = \dfrac{2500}{s(s+5)(s+2500)}$.

Clearly, the sensitivity function $S_G^M(s)$ is a function of the complex variable s. Fig. 8-54 shows the magnitude plot of $S_G^M(s)$ when $G(s)$ is the transfer function given in Eq. (8-98). It is interesting to note that the sensitivity of the closed-loop system is inferior at frequencies greater than 4.8 rad/sec to the open-loop system whose sensitivity to the variation of K is always unity. In general, it is desirable to formulate a design criterion on sensitivity in the following manner:

$$\left|S_G^M(j\omega)\right| = \frac{1}{|1 + G(j\omega)|} = \frac{|1/G(j\omega)|}{|1 + 1/G(j\omega)|} \leq k \tag{8-123}$$

where k is a positive real number. This sensitivity criterion is in addition to the regular performance criteria on the steady-state error and the relative stability.

Eq. (8-123) is analogous to the magnitude of the closed-loop transfer function, $|M(j\omega)|$, given in Eq. (8-106), with $G(j\omega)$ replaced by $1/G(j\omega)$. Thus, the sensitivity function of Eq. (8-123) can be determined by plotting $1/G(j\omega)$ in the magnitude-phase coordinates with the Nichols chart. Fig. 8-55 shows the magnitude-phase plots of $G(j\omega)$ and $1/G(j\omega)$ of Eq. (8-98). Notice that $G(j\omega)$ is tangent to the $M = 1.8$ locus from below,

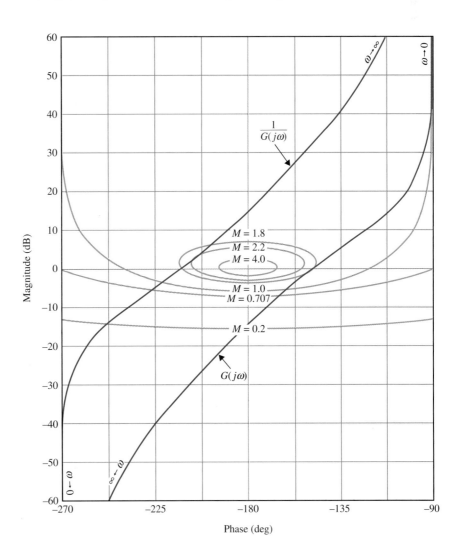

Figure 8-55 Magnitude-phase plots of $G(j\omega)$ and $1/G(j\omega)$ for $G(s) = \dfrac{2500}{s(s+5)(s+50)}$.

which means that M_r of the closed-loop system is 1.8. The $1/G(j\omega)$ curve is tangent to the $M = 2.2$ curve from above and, according to Fig. 8-54, is the maximum value of $\left|S_G^M(s)\right|$.

Eq. (8-123) shows that, for low sensitivity, the loop gain of $G(j\omega)$ must be high, but it is known that, in general, high gain could cause instability. Thus, the designer is again challenged by the task of designing a system with both a high degree of stability and low sensitivity.

The design of robust control systems (low sensitivity) with the frequency-domain methods is discussed in Chapter 9.

▶ 8-17 MATLAB TOOLS AND CASE STUDIES

Apart from the MATLAB toolboxes in this chapter, this chapter does not contain any software. In Chapter 9 we will introduce the Automatic Control Systems MATLAB tools. The **A**utomatic **C**ontrol **Sys**tems software (**ACSYS**) consists of a number of m-files and GUIs (graphical user interface) for the analysis of simple control engineering transfer functions. All the frequency response topics may also be solved utilizing **ACSYS**.

▶ 8-18 SUMMARY

The chapter began by describing typical relationships between the open-loop and closed-loop frequency responses of linear systems. Performance specifications such as the resonance peak M_r, resonant frequency ω_r, and bandwidth BW were defined in the frequency domain. The relationships among these parameters of a second-order prototype system were derived analytically. The effects of adding simple poles and zeros to the loop transfer function on M_r and BW were discussed.

The Nyquist criterion for stability analysis of linear control systems was developed. The stability of a single-loop control system can be investigated by studying the behavior of the Nyquist plot of the loop transfer function $G(s)H(s)$ for $\omega = 0$ to $\omega = \infty$ with respect to the critical point. If $G(s)H(s)$ is a minimum-phase transfer function, the condition of stability is simplified so that the Nyquist plot will not enclose the critical point.

The relationship between the root loci and the Nyquist plot was described in Section 8-7. The discussion should add more perspective to the understanding of both subjects.

Relative stability was defined in terms of gain margin and phase margin. These quantities were defined in the polar coordinates as well as on the Bode diagram. The gain-phase plot allows the Nichols chart to be constructed for closed-loop analysis. The values of M_r and BW can be easily found by plotting the $G(j\omega)$ locus on the Nichols chart.

The stability of systems with pure time delay is analyzed by use of the Bode plot.

Sensitivity function $S_G^M(j\omega)$ was defined as a measure of the variation of $M(j\omega)$ due to variations in $G(j\omega)$. It was shown that the frequency-response plots of $G(j\omega)$ and $1/G(j\omega)$ can be readily used for sensitivity studies.

Finally, using the MATLAB toolboxes developed in this chapter or the **ACSYS** software, described in detail in Chapter 9, the reader may practice all the concepts discussed here.

▶ REVIEW QUESTIONS

1. Explain why it is important to conduct frequency-domain analyses of linear control systems.
2. Define resonance peak M_r of a closed-loop control system.
3. Define bandwidth BW of a closed-loop system.
4. List the advantages and disadvantages of studying stability with the Nyquist plot.
5. List the advantages and disadvantages of carrying out frequency-domain analysis with the Bode plot.

6. List the advantages and disadvantages of carrying out frequency-domain analysis with the magnitude-phase plot.

7. The following quantities are defined:
   Z = number of zeros of $L(s)$ that are in the right-half s-plane
   P = number of poles of $L(s)$ that are in the right-half s-plane
   P_ω = number of poles of $L(s)$ that are on the $j\omega$-axis

   Give the conditions on these parameters for the system to be (a) open-loop stable and (b) closed-loop stable.

8. What condition must be satisfied by the function $L(j\omega)$ so that the Nyquist criterion is simplified to investigating whether the $(-1, j0)$ point is enclosed by the Nyquist plot?

9. Give all the properties of a minimum-phase transfer function.

10. Give the definitions of gain margin and phase margin.

11. By applying a sinusoidal signal of frequency ω_0 to a linear system, the steady-state output of the system will also be of the same frequency. (T) (F)

12. For a prototype second-order system, the value of M_r depends solely on the damping ratio ζ. (T) (F)

13. Adding a zero to the loop transfer function will always increase the bandwidth of the closed-loop system. (T) (F)

14. The general effect of adding a pole to the loop transfer function is to make the closed-loop system less stable while decreasing the bandwidth. (T) (F)

15. For a minimum-phase loop transfer function $L(j\omega)$, if the phase margin is negative, then the closed-loop system is always unstable. (T) (F)

16. Phase-crossover frequency is the frequency at which the phase of $L(j\omega)$ is 0°. (T) (F)

17. Gain-crossover frequency is the frequency at which the gain of $L(j\omega)$ is 0 dB. (T) (F)

18. Gain margin is measured at the phase-crossover frequency. (T) (F)

19. Phase margin is measured at the gain-crossover frequency. (T) (F)

20. A closed-loop system with a pure time delay in the loop is usually less stable than one without a time delay. (T) (F)

21. The slope of the magnitude curve of the Bode plot of $L(j\omega)$ at the gain crossover usually gives indication on the relative stability of the closed-loop system. (T) (F)

22. Nichols chart can be used to find BW and M_r information of a closed-loop system. (T) (F)

23. Bode plot can be used for stability analysis for minimum- as well as nonminimum-phase transfer functions. (T) (F)

Answers to these review questions can be found on this book's companion Web site: www.wiley.com/college/golnaraghi.

▶ REFERENCES

Nyquist Criterion of Continuous-Data Systems

1. H. Nyquist, "Regeneration Theory," *Bell System. Tech. J.*, Vol. 11, pp.126–147, Jan.1932.
2. R. W. Brockett and J. L. Willems, "Frequency Domain Stability Criteria—Part I," *IEEE Trans. Automatic Control*, Vol. AC-10, pp. 255–261, July 1965.
3. R. W. Brockett and J. L. Willems, "Frequency Domain Stability Criteria—Part II," *IEEE Trans. Automatic Control*, Vol. AC-10, pp. 407–413, Oct. 1965.
4. T. R. Natesan, "A Supplement to the Note on the Generalized Nyquist Criterion," *IEEE Trans. Automatic Control*, Vol. AC-12, pp. 215–216, April 1967.
5. K. S. Yeung, "A Reformulation of Nyquist's Criterion," *IEEE Trans. Educ.* Vol. E-28, pp. 59–60, Feb. 1985.

Sensitivity Function

6. A. Gelb, "Graphical Evaluation of the Sensitivity Function Using the Nichols Chart," *IRE Trans. Automatic Control*, Vol. AC-7, pp. 57–58, July 1962.

PROBLEMS

8-1. The forward-path transfer function of a unity-feedback control system is

$$G(s) = \frac{K}{s(s+6.54)}$$

Analytically, find the resonance peak M_r, resonant frequency ω_r, and bandwidth BW of the closed-loop system for the following values of K:
(a) $K = 5$
(b) $K = 21.39$
(c) $K = 100$
Use the formulas for the second-order prototype system given in the text.

8-2. Use MATLAB to verify your answer to Problem 8-1.

8-3. The transfer function of a system is

$$G(s) = \frac{s + \dfrac{1}{A_1}}{s + \dfrac{1}{A_2}}$$

Determine when the system is a lead-network and lag-network.

8-4. Use MATLAB to solve the following problems. Do not attempt to obtain the solutions analytically. The forward-path transfer functions of unity-feedback control systems are given in the following equations. Find the resonance peak M_r, resonant frequency ω_r, and bandwidth BW of the closed-loop systems. (Reminder: Make certain that the system is stable.)

(a) $G(s) = \dfrac{5}{s(1+0.5s)(1+0.1s)}$

(b) $G(s) = \dfrac{10}{s(1+0.5s)(1+0.1s)}$

(c) $G(s) = \dfrac{500}{(s+1.2)(s+4)(s+10)}$

(d) $G(s) = \dfrac{10(s+1)}{s(s+2)(s+10)}$

(e) $G(s) = \dfrac{0.5}{s(s^2+s+1)}$

(f) $G(s) = \dfrac{100e^{-s}}{s(s^2+10s+50)}$

(g) $G(s) = \dfrac{100e^{-s}}{s(s^2+10s+100)}$

(h) $G(s) = \dfrac{10(s+5)}{s(s^2+5s+5)}$

8-5. The specifications on a second-order unity-feedback control system with the closed-loop transfer function

$$M(s) = \frac{Y(s)}{R(s)} = \frac{\omega_n^2}{s^2 + 2\zeta\omega_n s + \omega_n^2}$$

are that the maximum overshoot must not exceed 10% and the rise time must be less than 0.1 sec. Find the corresponding limiting values of M_r and BW analytically.

8-6. Repeat Problem 8-5 for maximum overshoot $\leq$ 20% and $t_r \leq 0.2$ sec.

8-7. Repeat Problem 8-5 for maximum overshoot $\leq$ 30% and $t_r \leq 0.2$ sec.

8-8. Consider the forward-path transfer function of a unity-feedback control system given by

$$G(s) = \frac{0.5K}{s(0.25s^2 + 0.375s + 1)}$$

(a) Analytically find K such that the closed-loop bandwidth is about 1.5 rad/s (0.24 Hz).
(b) Use MATLAB to verify your answer to part (a).

8-9. Repeat Problem 8-8 with a resonance peak of 2.2.

8-10. The closed-loop frequency response $|M(j\omega)|$-versus-frequency of a second-order prototype system is shown in Fig 8P-10. Sketch the corresponding unit-step response of the system; indicate the values of the maximum overshoot, peak time, and the steady-state error due to a unit-step input.

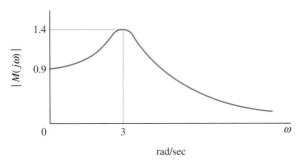

Figure 8P-10

8-11. The forward-path transfer function of a system with an integral control $H(s) = \dfrac{K}{s}$ is

$$G(s) = \dfrac{1}{10s+1}$$

(a) Find K when the closed-loop resonance peak is 1.4.

(b) Determine the frequency at resonance, overshoot for step input, phase margin, and closed-loop BW according to the result of part (a).

8-12. The forward-path transfer function of a unity-feedback control system is

$$G(s) = \dfrac{1+Ts}{2s(s^2+s+1)}$$

Use MATLAB to find the values of BW and M_r of the closed-loop system for $T = 0.05, 1, 2, 3, 4$, and 5.

8-13. The forward-path transfer function of a unity-feedback control system is

$$G(s) = \dfrac{1}{2s(s^2+s+1)(1+Ts)}$$

Use MATLAB to find the values of BW and M_r of the closed-loop system for $T = 0, 0.5, 1, 2, 3, 4$, and 5. Use MATLAB to find the solutions.

8-14. If a loop transfer function of a system is given by

$$G(s)H(s) = \dfrac{0.5K}{0.25s^3 + 0.375s^2 + s + 0.5k}$$

(a) Use the second-order approximation to find the BW and the damping ratio.

(b) If BW $= 1.5$ rad/s, find K and the damping ratio.

(c) Use MATLAB to verify your answer to part (b).

8-15. The loop transfer functions $L(s)$ of single-feedback-loop systems are given below. Sketch the Nyquist plot of $L(j\omega)$ for $\omega = 0$ to $\omega = \infty$. Determine the stability of the closed-loop system. If the system is unstable, find the number of poles of the closed-loop transfer function that are in the right-half s-plane. Solve for the intersect of $L(j\omega)$ on the negative real axis of the $L(j\omega)$-plane analytically. You may construct the Nyquist plot of $L(j\omega)$ using MATLAB.

(a) $L(s) = \dfrac{20}{s(1+0.1s)(1+0.5s)}$

(b) $L(s) = \dfrac{10}{s(1+0.1s)(1+0.5s)}$

(c) $L(s) = \dfrac{100(1+s)}{s(1+0.1s)(1+0.2s)(1+0.5s)}$

(d) $L(s) = \dfrac{10}{s^2(1+0.2s)(1+0.5s)}$

(e) $L(s) = \dfrac{3(s+2)}{s(s^3+3s+1)}$

(f) $L(s) = \dfrac{0.1}{s(s+1)(s^2+s+1)}$

(g) $L(s) = \dfrac{100}{s(s+1)(s^2+2)}$

(h) $L(s) = \dfrac{10(s+10)}{s(s+1)(s+100)}$

8-16. The loop transfer functions of single-feedback-loop control systems are given in the following equations. Apply the Nyquist criterion and determine the values of K for the system to be stable. Sketch the Nyquist plot of $L(j\omega)$ with $K = 1$ for $\omega = 0$ to $\omega = \infty$. You may use a computer program to plot the Nyquist plots.

(a) $L(s) = \dfrac{K}{s(s+2)(s+10)}$

(b) $L(s) = \dfrac{K(s+1)}{s(s+2)(s+5)(s+15)}$

(c) $L(s) = \dfrac{K}{s^2(s+2)(s+10)}$

(d) $L(s) = \dfrac{K}{(s+5)(s+2)^2}$

(e) $L(s) = \dfrac{K(s+5)(s+1)}{(s+50)(s+2)^3}$

8-17. The forward-path transfer function of a unity-feedback control system is

$$G(s) = \dfrac{K}{(s+5)^n}$$

Determine by means of the Nyquist criterion, the range of $K(-\infty < K < \infty)$ for the closed-loop system to be stable. Sketch the Nyquist plot of $G(j\omega)$ for $\omega = 0$ to $\omega = \infty$.

(a) $n = 2$
(b) $n = 3$
(c) $n = 4$

8-18. Sketch the Nyquist plot for the controlled system shown in Fig. 8P-18.

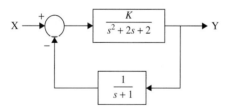

Figure 8P-18

Determine by means of the Nyquist criterion, the range of $K(-\infty < K < \infty)$ for the closed-loop system to be stable.

8-19. The characteristic equation of a linear control system is given in the following equation.

$$s(s^3+2s^2+s+1) + K(s^2+s+1) = 0$$

(a) Apply the Nyquist criterion to determine the values of K for system stability.
(b) Check the answers by means of the Routh-Hurwitz criterion.

8-20. Repeat Problem 8-19 for $s^3 + 3s^2 + 3s + 1 + K = 0$.

8-21. The forward-path transfer function of a unity-feedback control system with a PD (proportional-derivative) controller is

$$G(s) = \dfrac{10(K_P + K_D s)}{s^2}$$

Select the value of K_P so that the parabolic-error constant K_a is 100. Find the equivalent forward-path transfer function $G_{eq}(s)$ for $\omega = 0$ to $\omega = \infty$. Determine the range of K_D for stability by the Nyquist criterion.

8-22. The block diagram of a feedback control system is shown in Fig. 8P-22.
(a) Apply the Nyquist criterion to determine the range of K for stability.
(b) Check the answer obtained in part (a) with the Routh-Hurwitz criterion.

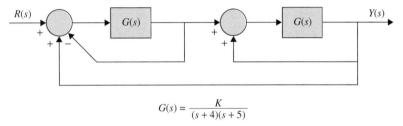

$$G(s) = \frac{K}{(s+4)(s+5)}$$

Figure 8P-22

8-23. The forward-path transfer function of the liquid-level control system in Problem 5-42 is

$$G(s) = \frac{K_a K_i n K_I N}{s(R_a J s + K_i K_b)(A s + K_o)}$$

The following system parameters are given: $K_a = 50$, $K_i = 10$, $K_I = 50$, $J = 0.006$, $K_b = 0.0706$, $n = 0.01$, and $R_a = 10$. The values of A, N, and K_o are variable.

(a) For $A = 50$ and $K_o = 100$, sketch the Nyquist plot of $G(j\omega)$ for $\omega = 0$ to ∞ with N as a variable parameter. Find the maximum integer value of N so that the closed-loop system is stable.

(b) Let $N = 10$ and $K_o = 100$. Sketch the Nyquist plot of an equivalent transfer function $G_{eq}(j\omega)$ that has A as a multiplying factor. Find the critical value of K_o for stability.

(c) For $A = 50$ and $N = 10$, sketch the Nyquist plot of an equivalent transfer function $G_{eq}(j\omega)$ that has K_o as a multiplying factor. Find the critical value of K_o for stability.

8-24. The block diagram of a dc-motor control system is shown in Fig. 8P-24. Determine the range of K for stability using the Nyquist criterion when K_t has the following values:

(a) $K_t = 0$
(b) $K_t = 0.01$
(c) $K_t = 0.1$

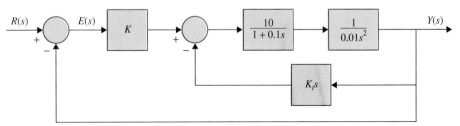

Figure 8P-24

8-25. For the system shown in Fig. 8P-24, let $K = 10$. Find the range of K_t for stability with the Nyquist criterion.

8-26. Fig. 8P-26 shows the block diagram of a servomotor.

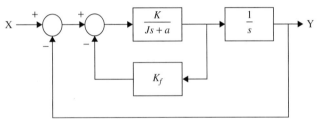

Figure 8P-26

Assume $J = 1$ kg-m² and $B = 1$ N-m/rad/sec. Determine the range of K for stability using the Nyquist criterion when K_f has the following values:
(a) $K_f = 0$
(b) $K_f = 0.1$
(c) $K_f = 0.2$

8-27. For the system shown in Fig. 8P-26, let $K = 10$. Find the range of K_f for stability with the Nyquist criterion.

8-28. For the controlled system shown in Fig. 8P-28, draw the Nyquist plot and apply the Nyquist criterion to determine the range of K for stability and determine the number of roots in the right-half s-plane for the values of K where the system is unstable.

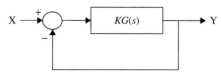

Figure 8P-28

(a) $G(s) = \dfrac{s+1}{(s-1)^2}$

(b) $G(s) = \dfrac{s-1}{(s+1)^2}$

8-29. The steel-rolling control system shown in Fig. 4P-18 has the forward-path transfer function

$$G(s) = \dfrac{100Ke^{-T_d s}}{s(s^2 + 10s + 100)}$$

(a) When $K = 1$, determine the maximum time delay T_d in seconds for the closed-loop system to be stable.

(b) When the time delay T_d is 1 sec, find the maximum value of K for system stability.

8-30. Repeat Problem 8-29 with the following conditions.
(a) When $K = 0.1$, determine the maximum time delay T_d in seconds for the closed-loop system to be stable.
(b) When the time delay T_d is 0.1 sec, find the maximum value of K for system stability.

8-31. The open-loop transfer function of a system is given by

$$G(s)H(s) = \dfrac{K}{s(\tau_1 s + 1)(\tau_2 s + 1)}$$

Study the stability of the system for the following:
(a) K is small.
(b) K is large.

8-32. The system schematic shown in Fig. 8P-32 is devised to control the concentration of a chemical solution by mixing water and concentrated solution in appropriate proportions. The transfer function of the system components between the amplifier output e_a (V) and the valve position x (in.) is

$$\dfrac{X(s)}{E_a(s)} = \dfrac{K}{s^2 + 10s + 100}$$

When the sensor is viewing pure water, the amplifier output voltage e_a is zero; when it is viewing concentrated solution, $e_a = 10$ V; and 0.1 in. of the valve motion changes the output concentration from zero to maximum. The valve ports can be assumed to be shaped so that the output concentration varies linearly with the valve position. The output tube has a cross-sectional area of 0.1 in.^{2}, and the rate of flow is 10^3 in./sec regardless of the valve position. To make sure the sensor views a homogeneous solution, it is desirable to place it at some distance D in. from the valve.

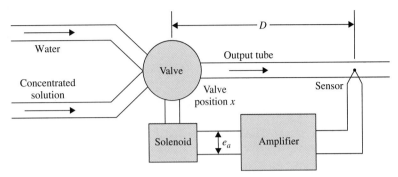

Figure 8P-32

(a) Derive the loop transfer function of the system.

(b) When $K = 10$, find the maximum distance D (in.) so that the system is stable. Use the Nyquist stability criterion.

(c) Let $D = 10$ in. Find the maximum value of K for system stability.

8-33. For the mixing system described in Problem 8-32, the following system parameters are given: When the sensor is viewing pure water, the amplifier output voltage $e_s = 0$ V; when it is viewing concentrated solution, $e_a = 1$ V; and 0.1 in. of the valve motion changes the output concentration from zero to maximum. The rest of the system characteristics are the same as in Problem 8-32. Repeat the three parts of Problem 8-32.

8-34. Figure 8P-34 shows the block diagram of a control system.

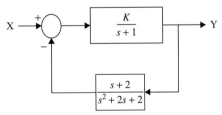

Figure 8P-34

(a) Draw the Nyquist plot and apply the Nyquist criterion to determine the range of K for stability.

(b) Determine the number of roots in the right-half s-plane for the values of K where the system is unstable.

(c) Use Routh's criterion to determine the range of K for stability.

8-35. The forward-path transfer function of a unity-feedback control system is

$$G(s) = \frac{1000}{s(s^2 + 105s + 600)}$$

(a) Find the values of M_r, ω_r, and BW of the closed-loop system.

(b) Find the parameters of the second-order system with the open-loop transfer function

$$G_L(s) = \frac{\omega_n^2}{s(s + 2\zeta\omega_n)}$$

that will give the same values for M_r and ω_r as the third-order system. Compare the values of BW of the two systems.

8-36. Sketch or plot the Bode diagrams of the forward-path transfer functions given in Problem 8-4. Find the gain margin, gain-crossover frequency, phase margin, and the phase-crossover frequency for each system.

8-37. The transfer function of a system is given by

$$G(s)H(s) = \frac{25(s+1)}{s(s+2)(s^2+2s+16)}$$

Use MATLAB to plot the Bode diagrams of the system and find the phase margin and gain margin of the system.

8-38. Use MATLAB to plot the Bode diagrams of the system shown in Fig. 8P-34 when $K = 1$, and determine the stable range of K by using phase margin and gain margin.

8-39. The forward-path transfer functions of unity-feedback control systems are given in the following equations. Plot the Bode diagram of $G(j\omega)/K$, and do the following: **(1)** Find the value of K so that the gain margin of the system is 20 dB. **(2)** Find the value of K so that the phase margin of the system is 45°.

(a) $G(s) = \dfrac{K}{s(1+0.1s)(1+0.5s)}$

(b) $G(s) = \dfrac{K(s+1)}{s(1+0.1s)(1+0.2s)(1+0.5s)}$

(c) $G(s) = \dfrac{K}{(s+3)^3}$

(d) $G(s) = \dfrac{K}{(s+3)^4}$

(e) $G(s) = \dfrac{Ke^{-s}}{s(1+0.1s+0.01s^2)}$

(f) $G(s) = \dfrac{K(1+0.5s)}{s(s^2+s+1)}$

8-40. The forward-path transfer functions of unity-feedback control systems are given in the following equations. Plot $G(j\omega)/K$ in the gain-phase coordinates of the Nichols chart, and do the following: **(1)** Find the value of K so that the gain margin of the system is 10 dB. **(2)** Find the value of K so that the phase margin of the system is 45°. **(3)** Find the value of K so that $M_r = 1.2$.

(a) $G(s) = \dfrac{10K}{s(1+0.1s)(1+0.5s)}$

(b) $G(s) = \dfrac{5K(s+1)}{s(1+0.1s)(1+0.2s)(1+0.5s)}$

(c) $G(s) = \dfrac{10K}{s(1+0.1s+0.01s^2)}$

(d) $G(s) = \dfrac{10Ke^{-s}}{s(1+0.1s+0.01s^2)}$

8-41. The forward-path of a unity-feedback system is given by

$$G(s)H(s) = \frac{K(s+1)(s+2)}{s^2(s+3)(s^2+2s+25)}$$

(a) Plot the Bode diagram.
(b) Plot the root locus.
(c) Find the gain and frequency where instability occurs.
(d) Find the gain at the phase margin of 20°.
(e) Find the gain margin when the phase margin is 20°.

8-42. The Bode diagram of the forward-path transfer function of a unity-feedback control system is obtained experimentally (as shown in Fig. 8P-42) when the forward gain K is set at its nominal value.

(a) Find the gain and phase margins of the system from the diagram as best you can read them. Find the gain- and phase-crossover frequencies.
(b) Repeat part (a) if the gain is doubled from its nominal value.
(c) Repeat part (a) if the gain is 10 times its nominal value.
(d) Find out how much the gain must be changed from its nominal value if the gain margin is 40 dB.
(e) Find out how much the loop gain must be changed from its nominal value if the phase margin is 45°.

(f) Find the steady-state error of the system if the reference input to the system is a unit-step function.
(g) The forward path now has a pure time delay of T_d sec, so that the forward-path transfer function is multiplied by $e^{-T_d s}$. Find the gain margin and the phase margin for $T_d = 0.1$ sec. The gain is set at nominal.
(h) With the gain set at nominal, find the maximum time delay T_d the system can tolerate without going into instability.

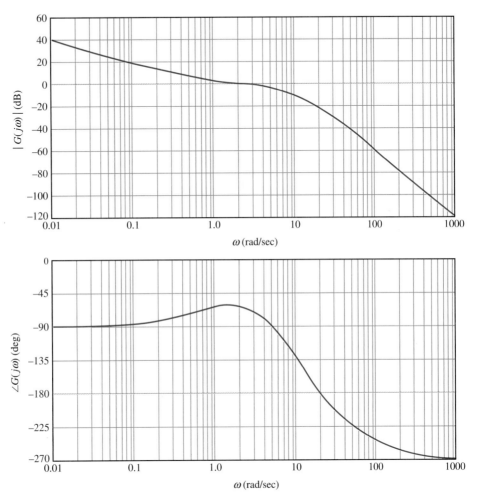

Figure 8P-42

8-43. Repeat Problem 8-42 using Fig. 8P-42 for the following parts.
(a) Find the gain and phase margins if the gain is four times its nominal value. Find the gain- and phase-crossover frequencies.
(b) Find out how much the gain must be changed from its nominal value if the gain margin is 20 dB.
(c) Find the marginal value of the forward-path gain for system stability.
(d) Find out how much the gain must be changed from its nominal value if the phase margin is 60°.
(e) Find the steady-state error of the system if the reference input is a unit-step function and the gain is twice its nominal value.
(f) Find the steady-state error of the system if the reference input is a unit-step function and the gain is 20 times its nominal value.

(g) The system now has a pure time delay so that the forward-path transfer function is multiplied by $e^{-T_d s}$. Find the gain and phase margins when $T_d = 0.1$ sec. The gain is set at its nominal value.

(h) With the gain set at 10 times its nominal, find the maximum time delay T_d the system can tolerate without going into instability.

8-44. The forward-path transfer function of a unity-feedback control system is

$$G(s)H(s) = \frac{80e^{-0.1s}}{s(s+4)(s+10)}$$

(a) Draw the Nyquist plot of the system.
(b) Plot the Bode diagram of the system.
(c) Find the phase margin and gain margin of the system.

8-45. The forward-path transfer function of a unity-feedback control system is

$$G(s) = \frac{K(1+0.2s)(1+0.1s)}{s^2(1+s)(1+0.01s)^2}$$

(a) Construct the Bode and Nyquist plots of $G(j\omega)/K$ and determine the range of K for system stability.

(b) Construct the root loci of the system for $K \geq 0$. Determine the values of K and ω at the points where the root loci cross the $j\omega$-axis, using the information found from the Bode plot.

8-46. Repeat Problem 8-45 for the following transfer function:

$$G(s) = \frac{K(s+1.5)(s+2)}{s^2(s^2+2s+2)}$$

8-47. Repeat Problem 8-45 for the following transfer function:

$$G(s)H(s) = \frac{16000(s+1)(s+5)}{s(s+0.1)(s+8)(s+20)(s+50)}$$

8-48. The forward-path transfer function of the dc-motor control system described in Fig. 3P-11 is

$$G(s) = \frac{6.087 \times 10^8 \, K}{s(s^3 + 423.42s^2 + 2.6667 \times 10^6 s + 4.2342 \times 10^8)}$$

Plot the Bode diagram of $G(j\omega)$ with $K = 1$, and determine the gain margin and phase margin of the system. Find the critical value of K for stability.

8-49. The transfer function between the output position $\Theta_L(s)$ and the motor current $I_a(s)$ of the robot arm modeled in Fig. 4P-53 is

$$G_p(s) = \frac{\Theta_L(s)}{I_a(s)} = \frac{K_i(Bs+K)}{\Delta_o}$$

where

$$\Delta_o(s) = s\{J_L J_m s^3 + [J_L(B_m+B) + J_m(B_L+B)]s^2$$
$$+ [B_L B_m + (B_L+B_m)B + (J_m+J_L)K]s + K(B_L+B_m)\}$$

The arm is controlled by a closed-loop system, as shown in Fig. 8P-49. The system parameters are $K_a = 65, K = 100, K_i = 0.4, B = 0.2, J_m = 0.2, B_L = 0.01, J_L = 0.6,$ and $B_m = 0.25$.

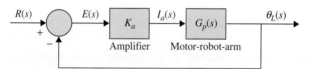

Figure 8P-49

(a) Derive the forward-path transfer function $G(s) = \Theta_L(s)/E(s)$.
(b) Draw the Bode diagram of $G(j\omega)$. Find the gain and phase margins of the system.
(c) Draw $|M(j\omega)|$ versus ω, where $M(s)$ is the closed-loop transfer function. Find M_r, ω_r, and BW.

8-50. For the ball-and-beam system described in Problem 4-11 and shown in Fig. 8P-50, assume the following:

$m = 0.11$ kg	mass of the ball	$I = 9.99 \times 10^{-6}$ kg-m²	ball's moment of inertia
$r = 0.015$	radius of the ball	P	ball position coordinate
$d = 0.03$ m	lever arm offset	α	beam angle coordinate
$g = 9.8$ m/s²	gravitational acceleration	θ	servo gear angle
$L = 1.0$ m	length of the beam		

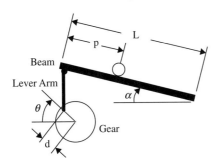

Figure 8P-50

If the system is controlled by a proportional controller in a unity-feedback control system,
(a) Find the transfer function from the gear angle (θ) to the ball position (P).
(b) Find the closed-loop transfer function.
(c) Find the range of K for stability.
(d) Plot the Bode diagram for the system for $K = 1$, and find the gain and phase margins of the system.
(e) Draw $|M(j\omega)|$ versus ω, where $M(s)$ is the closed-loop transfer function. Find M_r, ω_r, and BW.

8-51. The gain-phase plot of the forward-path transfer function of $G(j\omega)/K$ of a unity-feedback control system is shown in Fig. 8P-51. Find the following performance characteristics of the system.
(a) Gain-crossover frequency (rad/sec) when $K = 1$.
(b) Phase-crossover frequency (rad/sec) when $K = 1$.
(c) Gain margin (dB) when $K = 1$.
(d) Phase margin (deg) when $K = 1$.
(e) Resonance peak M_r when $K = 1$.
(f) Resonant frequency ω_r (rad/sec) when $K = 1$.
(g) BW of the closed-loop system when $K = 1$.
(h) The value of K so that the gain margin is 20 dB.
(i) The value of K so that the system is marginally stable. Find the frequency of sustained oscillation in rad/sec.
(j) Steady-state error when the reference input is a unit-step function.

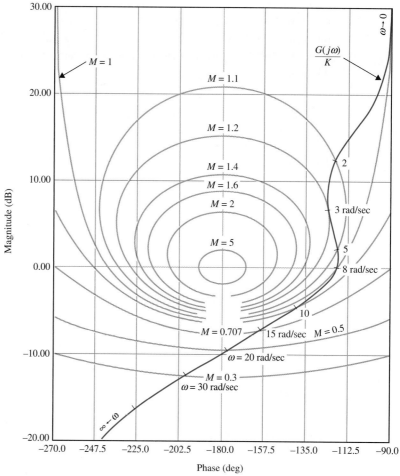

Figure 8P-51

8-52. Repeat parts (a) through (g) of Problem 8-51 when $K = 10$. Repeat part (h) for gain margin = 40 dB.

8-53. For the system in Problem 8-44, plot the Nichols chart and find magnitudes and phase angles of the closed-loop frequency response. Then plot the Bode diagram of the closed-loop system.

8-54. Use **ACSYS** or MATLAB to analyze the frequency response of the following unity-feedback control systems. Plot the Bode diagrams, polar plots, and gain-phase plots, and compute the phase margin, gain margin, M_r, and BW.

(a) $G(s) = \dfrac{1 + 0.1s}{s(s+1)(1+0.01s)}$

(b) $G(s) = \dfrac{0.5(s+1)}{s(1+0.2s)(1+s+0.5s^2)}$

(c) $G(s) = \dfrac{(s+1)}{s(1+0.2s)(1+0.5s)}$

(d) $G(s) = \dfrac{1}{s(1+s)(1+0.5s)}$

(e) $G(s) = \dfrac{50}{s(s+1)(1+0.5s^2)}$

(f) $G(s) = \dfrac{(1+0.1s)e^{-0.1s}}{s(s+1)(1+0.01s)}$

(g) $G(s) = \dfrac{10e^{-0.1s}}{s^2 + 2s + 2}$

8-55. For the gain-phase plot of $G(j\omega)/K$ shown in Fig. 8P-51, the system now has a pure time delay of T_d in the forward path, so that the forward-path transfer function becomes $G(s)e^{-T_d s}$.

(a) With $K = 1$, find T_d so that the phase margin is $40°$.

(b) With $K = 1$, find the maximum value of T_d so that the system will remain stable.

8-56. Repeat Problem 8-55 with $K = 10$.

8-57. Repeat Problem 8-55 so that the gain margin is 5 dB when $K = 1$.

8-58. The block diagram of a furnace-control system is shown in Fig. 8P-58. The transfer function of the process is

$$G_p(s) = \frac{1}{(1 + 10s)(1 + 25s)}$$

The time delay T_d is 2 sec.

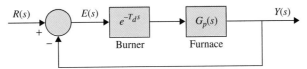

Figure 8P-58

(a) Plot the Bode diagram of $G(s) = Y(s)/E(s)$, and find the gain-crossover and phase-crossover frequencies. Find the gain margin and the phase margin.

(b) Approximate the time delay by [Eq. (4-223)]

$$e^{-T_d s} \cong \frac{1}{1 + T_d s + T_d s^2/2}$$

and repeat part (a). Comment on the accuracy of the approximation. What is the maximum frequency below which the polynomial approximation is accurate?

(c) Repeat part (b) for approximating the time delay term by [Eq. (4-224)]

$$e^{-T_d s} \cong \frac{1 - T_d s/2}{1 + T_d s/2}$$

8-59. Repeat Problem 8-58 with $T_d = 1$ sec.

8-60. Plot the $|S_G^M(j\omega)|$-versus-ω plot for the system described in Problem 8-49 for $K = 1$. Find the frequency at which the sensitivity is maximum and the value of the maximum sensitivity.

8-61. Fig. 8P-61 shows the pitch controller system for an aircraft, as described in Problem 4-12.

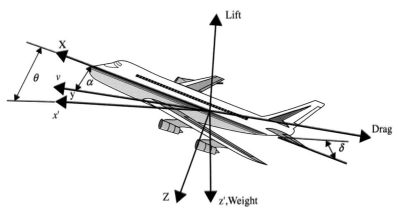

Figure 8P-61

If the system is controlled by a proportional controller in a unity-feedback control system,
(a) Find the transfer function between pitch angle and elevator deflection angle.
(b) Find the closed-loop transfer function.
(c) Find the range of K for stability.
(d) Plot the Bode diagram for the system for $K = 1$, and find the gain and phase margins of the system.
(e) Draw $|M(j\omega)|$ versus ω, where $M(s)$ is the closed-loop transfer function. Find M_r, ω_r, and BW.

CHAPTER 9

Design of Control Systems

9-1 INTRODUCTION

All the foundations of analysis that we have laid in the preceding chapters led to the ultimate goal of design of control systems. Starting with the controlled process such as that shown by the block diagram in Fig. 9-1, control system design involves the following three steps:

1. Determine what the system should do and how to do it (design specifications).
2. Determine the controller or compensator configuration, relative to how it is connected to the controlled process.
3. Determine the parameter values of the controller to achieve the design goals.

These design tasks are explored further in the following sections.

9-1-1 Design Specifications

We often use design specifications to describe what the system should do and how it is done. These specifications are unique to each individual application and often include specifications about **relative stability, steady-state accuracy (error), transient-response characteristics**, and **frequency-response characteristics**. In some applications there may be additional specifications on **sensitivity to parameter variations**, that is, **robustness**, or **disturbance rejection**.

The design of linear control systems can be carried out in either the time domain or the frequency domain. For instance, **steady-state accuracy** is often specified with respect to a step input, a ramp input, or a parabolic input, and the design to meet a certain requirement is more conveniently carried out in the time domain. Other specifications such as **maximum overshoot, rise time**, and **settling time** are all defined for a unit-step input and, therefore, are used specifically for time-domain design. We have learned that relative stability is also measured in terms of **gain margin, phase margin**, and M_r. These are typical frequency-domain specifications, which should be used in conjunction with such tools as the Bode plot, polar plot, gain-phase plot, and Nichols chart.

We have shown that, for a second-order prototype system, there are simple analytical relationships between some of these time-domain and frequency-domain specifications. However, for higher-order systems, correlations between time-domain and frequency-domain specifications are difficult to establish. As pointed out earlier, the analysis and

Figure 9-1 Controlled process.

design of control systems is pretty much an exercise of selecting from several alternative methods for solving the same problem.

Thus, the choice of whether the design should be conducted in the time domain or the frequency domain depends often on the preference of the designer. We should be quick to point out, however, that in most cases, time-domain specifications such as maximum overshoot, rise time, and settling time are usually used as the final measure of system performance. To an inexperienced designer, it is difficult to comprehend the physical connection between frequency-domain specifications such as gain and phase margins and resonance peak to actual system performance. For instance, does a gain margin of 20 dB guarantee a maximum overshoot of less than 10%? To a designer it makes more sense to specify, for example, that the maximum overshoot should be less than 5% and a settling time less than 0.01 sec. It is less obvious what, for example, a phase margin of 60° and an M_r of less than 1.1 may bring in system performance. The following outline will hopefully clarify and explain the choices and reasons for using time-domain versus frequency-domain specifications.

1. Historically, the design of linear control systems was developed with a wealth of graphical tools such as the Bode plot, Nyquist plot, gain-phase plot, and Nichols chart, which are all carried out in the frequency domain. The advantage of these tools is that they can all be sketched by following approximation methods without detailed plotting. Therefore, the designer can carry out designs using frequency-domain specifications such as **gain margin, phase margin, M_r,** and the like. High-order systems do not generally pose any particular problem. For certain types of controllers, design procedures in the frequency domain are available to reduce the trial-and-error effort to a minimum.

2. Design in the time domain using such performance specifications as **rise time, delay time, settling time, maximum overshoot**, and the like is possible *analytically* only for second-order systems or for systems that can be approximated by second-order systems. General design procedures using time-domain specifications are difficult to establish for systems with an order higher than the second.

The development and availability of high-powered and user-friendly computer software, such as MATLAB, is rapidly changing the practice of control system design, which until recently had been dictated by historical development. Now with MATLAB, the designer can go through a large number of design runs using the time-domain specifications within a matter of minutes. This diminishes considerably the historical edge of the frequency-domain design, which is based on the convenience of performing graphical design manually.

Throughout the chapter, we have incorporated small MATLAB toolboxes to help your understanding of the examples, and, at the end of the chapter, we present the **A**utomatic **C**ontrol **Sys**tems software package (**ACSYS**)—it is easy to use and fully graphics based to eliminate the user's need to write code.

Finally, it is generally difficult (except for an experienced designer) to select a meaningful set of frequency-domain specifications that will correspond to the desired time-domain performance requirements. For example, specifying a phase margin of 60° would be meaningless unless we know that it corresponds to a certain maximum overshoot. As it turns out, to control maximum overshoot, usually one has to specify at least phase margin and M_r. Eventually, establishing an intelligent set of frequency-domain specifications becomes a trial-and-error process that precedes the actual design, which often is also a trial-and-error effort. However, frequency-domain methods are still valuable in interpreting noise rejection and sensitivity properties of the system, and, most important, they offer another perspective to the design process. Therefore, in this chapter the design techniques in the time domain and the frequency domain are treated side by side, so that the methods can be easily compared.

9-1-2 Controller Configurations

In general, the dynamics of a linear controlled process can be represented by the block diagram shown in Fig. 9-1. The design objective is to have the controlled variables, represented by the output vector $\mathbf{y}(t)$, behave in certain desirable ways. The problem essentially involves the determination of the control signal $\mathbf{u}(t)$ over the prescribed time interval so that the design objectives are all satisfied.

Most of the conventional design methods in control systems rely on the so-called **fixed-configuration design** in that the designer at the outset decides the basic configuration of the overall designed system and decides where the controller is to be positioned relative to the controlled process. The problem then involves the design of the elements of the controller. Because most control efforts involve the modification or compensation of the system-performance characteristics, the general design using fixed configuration is also called **compensation**.

Fig. 9-2 illustrates several commonly used system configurations with controller compensation. These are described briefly as follows.

- *Series (cascade) compensation:* Fig. 9-2(a) shows the most commonly used system configuration with the controller placed in series with the controlled process, and the configuration is referred to as **series** or **cascade compensation**.
- *Feedback compensation:* In Fig. 9-2(b), the controller is placed in the minor feedback path, and the scheme is called **feedback compensation**.
- *State-feedback compensation:* Fig. 9-2(c) shows a system that generates the control signal by feeding back the state variables through constant real gains, and the scheme is known as **state feedback**. The problem with state-feedback control is that, for high-order systems, the large number of state variables involved would require a large number of transducers to sense the state variables for feedback. Thus, the actual implementation of the state-feedback control scheme may be costly or impractical. Even for low-order systems, often not all the state variables are directly accessible, and an **observer** or **estimator** may be necessary to create the estimated state variables from measurements of the output variables.

The compensation schemes shown in Figs. 9-2(a), (b), and (c) all have one degree of freedom in that there is only one controller in each system, even though the controller may have more than one parameter that can be varied. The disadvantage with a one-degree-of-freedom controller is that the performance criteria that can be realized are limited. For example, if a system is to be designed to achieve a certain amount of relative stability, it may have poor sensitivity to parameter variations. Or if the roots of the characteristic equation are selected to provide a certain amount of relative damping, the maximum overshoot of the step response may still be excessive because of the zeros of the closed-loop transfer function. The compensation schemes shown in Figs. 9-2(d), (e), and (f) all have two degrees of freedom.

- *Series-feedback compensation:* Fig. 9-2(d) shows the series-feedback compensation for which a series controller and a feedback controller are used.

490 ▶ Chapter 9. Design of Control Systems

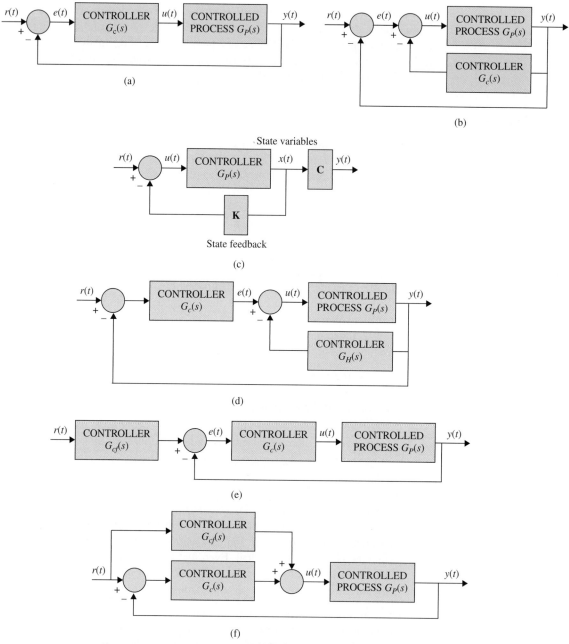

Figure 9-2 Various controller configurations in control-system compensation. (a) Series or cascade compensation. (b) Feedback compensation. (c) State-feedback control. (d) Series-feedback compensation (two degrees of freedom). (e) Forward compensation with series compensation (two degrees of freedom). (f) Feedforward compensation (two degrees of freedom).

- **Feedforward compensation:** Figs. 9-2(e) and (f) show the so-called **feedforward compensation**. In Fig. 9-2(e), the feedforward controller $G_{cf}(s)$ is placed in series with the closed-loop system, which has a controller $G_c(s)$ in the forward path. In Fig. 9-2(f), the feedforward controller $G_{cf}(s)$ is placed in parallel with the forward path. The key to the feedforward compensation is that the controller $G_{cf}(s)$ is not in the loop of the system, so it does not affect the roots of the characteristic equation of the original system. The poles and zeros of $G_{cf}(s)$ may be selected to add or cancel the poles and zeros of the closed-loop system transfer function.

One of the commonly used controllers in the compensation schemes just described is a PID controller, which applies a signal to the process that is proportional to the actuating signal in addition to adding integral and derivative of the actuating signal. Because these signal components are easily realized and visualized in the time domain, PID controllers are commonly designed using time-domain methods. In addition to the PID-type controllers, lead, lag, lead-lag, and notch controllers are also frequently used. The names of these controllers come from properties of their respective frequency-domain characteristics. As a result, these controllers are often designed using frequency-domain concepts. Despite these design tendencies, however, all control system designs will benefit by viewing the resulting design from both time- and frequency-domain viewpoints. Thus, both methods will be used extensively in this chapter.

It should be pointed out that these compensation schemes are by no means exhaustive. The details of these compensation schemes will be discussed in later sections of this chapter. Although the systems illustrated in Fig. 9-2 are all for continuous-data control, the same configurations can be applied to discrete-data control, in which case the controllers are all digital, with the necessary interfacings and signal converters.

9-1-3 Fundamental Principles of Design

After a controller configuration is chosen, the designer must choose a controller type that, with proper selection of its element values, will satisfy all the design specifications. The types of controllers available for control-system design are bounded only by one's imagination. Engineering practice usually dictates that one choose the simplest controller that meets all the design specifications. In most cases, the more complex a controller is, the more it costs, the less reliable it is, and the more difficult it is to design. Choosing a specific controller for a specific application is often based on the designer's past experience and sometimes intuition, and it entails as much *art* as it does *science*. As a novice, you may initially find it difficult to make intelligent choices of controllers with confidence. By understanding that confidence comes only through experience, this chapter provides guided experiences that illustrate the basic elements of control system designs.

After a controller is chosen, the next task is to choose controller parameter values. These parameter values are typically the coefficients of one or more transfer functions making up the controller. The basic design approach is to use the analysis tools discussed in the previous chapters to determine how individual parameter values influence the design specifications and, finally, system performance. Based on this information, controller parameters are selected so that all design specifications are met. While this process is sometimes straightforward, more often than not it involves many design iterations since controller parameters usually interact with each other and influence design specifications in conflicting ways. For example, a particular parameter value may be chosen so that the maximum overshoot is satisfied, but in the process of varying another parameter value in an attempt to meet the rise-time requirement, the maximum overshoot specification may no longer be met! Clearly, the more design specifications there are and the more controller parameters there are, the more complicated the design process becomes.

In carrying out the design either in the time domain or the frequency domain, it is important to establish some basic guidelines or design rules. Keep in mind that time-domain design usually relies heavily on the *s*-plane and the root loci. Frequency-domain design is based on manipulating the gain and phase of the loop transfer function so that the specifications are met.

In general, it is useful to summarize the time-domain and frequency-domain characteristics so that they can be used as guidelines for design purposes.

1. Complex-conjugate poles of the closed-loop transfer function lead to a step response that is underdamped. If all system poles are real, the step response is overdamped. However, zeros of the closed-loop transfer function may cause overshoot even if the system is overdamped.

2. The response of a system is dominated by those poles closest to the origin in the s-plane. Transients due to those poles farther to the left decay faster.

3. The farther to the left in the s-plane the system's dominant poles are, the faster the system will respond and the greater its bandwidth will be.

4. The farther to the left in the s-plane the system's dominant poles are, the more expensive it will be and the larger its internal signals will be. While this can be justified analytically, it is obvious that striking a nail harder with a hammer drives the nail in faster but requires more energy per strike. Similarly, a sports car can accelerate faster, but it uses more fuel than an average car.

5. When a pole and zero of a system transfer function nearly cancel each other, the portion of the system response associated with the pole will have a small magnitude.

6. Time-domain and frequency-domain specifications are loosely associated with each other. Rise time and bandwidth are inversely proportional. Larger phase margin, larger gain margin, and lower M_r will improve damping.

▶ 9-2 DESIGN WITH THE PD CONTROLLER

In all the examples of control systems we have discussed thus far, the controller has been typically a simple amplifier with a constant gain K. This type of control action is formally known as **proportional control**, because the control signal at the output of the controller is simply related to the input of the controller by a proportional constant.

Intuitively, one should also be able to use the derivative or integral of the input signal, in addition to the proportional operation. Therefore, we can consider a more general continuous-data controller to be one that contains such components as adders (addition or subtraction), amplifiers, attenuators, differentiators, and integrators. The designer's task is to determine which of these components should be used, in what proportion, and how they are connected. For example, one of the best-known controllers used in practice is the PID controller, where the letters stand for **proportional, integral**, and **derivative**. The integral and derivative components of the PID controller have individual performance implications, and their applications require an understanding of the basics of these elements. To gain an understanding of this controller, we consider just the PD portion of the controller first.

Fig. 9-3 shows the block diagram of a feedback control system that arbitrarily has a second-order prototype process with the transfer function

$$G_p(s) = \frac{\omega_n^2}{s(s + 2\zeta\omega_n)} \qquad (9\text{-}1)$$

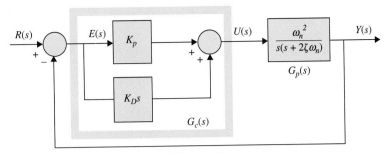

Figure 9-3 Control system with PD controller.

The series controller is a proportional-derivative (PD) type with the transfer function

$$G_c(s) = K_P + K_D s \tag{9-2}$$

Thus, the control signal applied to the process is

$$u(t) = K_P e(t) + K_D \frac{de(t)}{dt} \tag{9-3}$$

where K_P and K_D are the proportional and derivative constants, respectively. Using the components given in Table 4-4, two electronic-circuit realizations of the PD controller are shown in Fig. 9-4. The transfer function of the circuit in Fig. 9-4(a) is

$$\frac{E_o(s)}{E_{in}(s)} = \frac{R_2}{R_1} + R_2 C_1 s \tag{9-4}$$

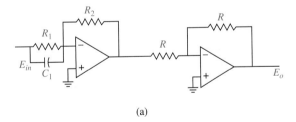

(a)

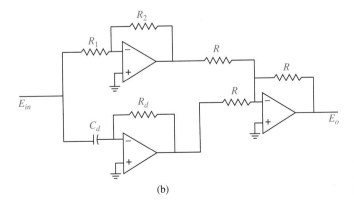

(b)

Figure 9-4 Op-amp circuit realization of the PD controller.

Comparing Eq. (9-2) with Eq. (9-4), we have

$$K_P = R_2/R_1 \quad K_D = R_2 C_1 \tag{9-5}$$

The transfer function of the circuit in Fig. 9-4(b) is

$$\frac{E_o(s)}{E_{in}(s)} = \frac{R_2}{R_1} + R_d C_d s \tag{9-6}$$

Comparing Eq. (9-2) with Eq. (9-6), we have

$$K_P = R_2/R_1 \quad K_D = R_d C_d \tag{9-7}$$

The advantage with the circuit in Fig. 9-4(a) is that only two op-amps are used. However, the circuit does not allow the independent selection of K_P and K_D because they are commonly dependent on R_2. An important concern of the PD controller is that, if the value of K_D is large, a large capacitor C_1 would be required. The circuit in Fig. 9-4(b) allows K_P and K_D to be independently controlled. A large K_D can be compensated by choosing a large value for R_d, thus resulting in a realistic value for C_d. Although the scope of this text does not include all the practical issues involved in controller transfer function implementation, these issues are of the utmost importance in practice.

The forward-path transfer function of the compensated system is

- PD control adds a simple zero at $s = -K_P/K_D$ to the forward-path transfer function.

$$G(s) = \frac{Y(s)}{E(s)} = G_c(s)G_p(s) = \frac{\omega_n^2(K_P + K_D s)}{s(s + 2\zeta\omega_n)} \tag{9-8}$$

which shows that the PD control is equivalent to adding a simple zero at $s = -K_P/K_D$ to the forward-path transfer function.

9-2-1 Time-Domain Interpretation of PD Control

The effect of the PD control on the transient response of a control system can be investigated by referring to the time responses shown in Fig. 9-5. Let us assume that the unit-step response of a stable system with only proportional control is as shown in Fig. 9-5(a), which has a relatively high maximum overshoot and is rather oscillatory. The corresponding error signal, which is the difference between the unit-step input and the output $y(t)$, and its time derivative $de(t)/dt$ are shown in Figs. 9.5(b) and (c), respectively. The overshoot and oscillation characteristics are also reflected in $e(t)$ and $de(t)/dt$. For the sake of illustration, we assume that the system contains a motor of some kind with its torque proportional to $e(t)$. The performance of the system with proportional control is analyzed as follows.

1. During the time interval $0 < t < t_1$: The error signal $e(t)$ is positive. The motor torque is positive and rising rapidly. The large overshoot and subsequent oscillations in the output $y(t)$ are due to the excessive amount of torque developed by the motor and the lack of damping during this time interval.
2. During the time interval $t_1 < t < t_3$: The error signal $e(t)$ is negative, and the corresponding motor torque is negative. This negative torque tends to slow down the output acceleration and eventually causes the direction of the output $y(t)$ to reverse and undershoot.

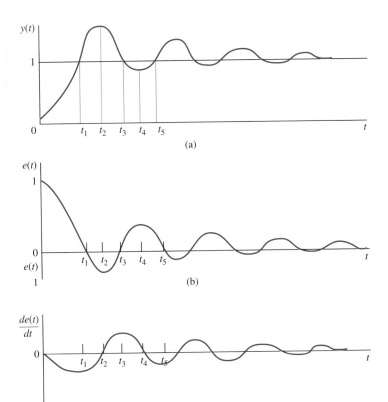

Figure 9-5 Waveforms of $y(t)$, $e(t)$, and $de(t)/dt$, showing the effect of derivative control. (a) Unit-step response. (b) Error signal. (c) Time rate of change of the error signal.

3. During the time interval $t_3 < t < t_5$: The motor torque is again positive, thus tending to reduce the undershoot in the response caused by the negative torque in the previous time interval. Because the system is assumed to be stable, the error amplitude is reduced with each oscillation, and the output eventually settles to its final value.

Considering the above analysis of the system time response, we can say that the contributing factors to the high overshoot are as follows:

1. The positive correcting torque in the interval $0 < t < t_1$ is too large.
2. The retarding torque in the time interval $t_1 < t < t_2$ is inadequate.

Therefore, to reduce the overshoot in the step response, without significantly increasing the rise time, a logical approach would be to

1. Decrease the amount of positive correcting torque during $0 < t < t_1$.
2. Increase the retarding torque during $t_1 < t < t_2$.

Similarly, during the time interval, $t_2 < t < t_4$, the negative corrective torque in $t_2 < t < t_3$ should be reduced, and the retarding torque during $t_3 < t < t_4$, which is now in the positive direction, should be increased to improve the undershoot of $y(t)$.

The PD control described by Eq. (9-2) gives precisely the compensation effect required. Because the control signal of the PD control is given by Eq. (9-3), Fig. 9-5(c) shows the following effects provided by the PD controller:

1. For $0 < t < t_1$, $de(t)/dt$ is negative; this will reduce the original torque developed due to $e(t)$ alone.
2. For $t_1 < t < t_2$, both $e(t)$ and $de(t)/dt$ are negative, which means that the negative retarding torque developed will be greater than that with only proportional control.
3. For $t_2 < t < t_3$, $e(t)$ and $de(t)/dt$ have opposite signs. Thus, the negative torque that originally contributes to the undershoot is reduced also.

- PD is essentially an anticipatory control.

Therefore, all these effects will result in smaller overshoots and undershoots in $y(t)$.

Another way of looking at the derivative control is that since $de(t)/dt$ represents the slope of $e(t)$, the PD control is essentially an *anticipatory* control. That is, by knowing the slope, the controller can anticipate direction of the error and use it to better control the process. Normally, in linear systems, if the slope of $e(t)$ or $y(t)$ due to a step input is large, a high overshoot will subsequently occur. The derivative control measures the instantaneous slope of $e(t)$, predicts the large overshoot ahead of time, and makes a proper corrective effort before the excessive overshoot actually occurs.

- Derivative or PD control will have an effect on a steady-state error only if the error varies with time.

Intuitively, derivative control affects the steady-state error of a system only if the steady-state error varies with time. If the steady-state error of a system is constant with respect to time, the time derivative of this error is zero, and the derivative portion of the controller provides no input to the process. But if the steady-state error increases with time, a torque is again developed in proportion to $de(t)/dt$, which reduces the magnitude of the error. Eq. (9-8) also clearly shows that the PD control does not alter the system type that governs the steady-state error of a unity-feedback system.

9-2-2 Frequency-Domain Interpretation of PD Control

For frequency-domain design, the transfer function of the PD controller is written

$$G_c(s) = K_P + K_D s = K_P \left(1 + \frac{K_D}{K_P} s\right) \tag{9-9}$$

- The PD controller is a high-pass filter.

- The PD controller has the disadvantage that it accentuates high-frequency noise.

- The PD controller will generally increase the BW and reduce the rise time of the step response.

so that it is more easily interpreted on the Bode plot. The Bode plot of Eq. (9-9) is shown in Fig. 9-6 with $K_P = 1$. In general, the proportional-control gain K_P can be combined with a series gain of the system, so that the zero-frequency gain of the PD controller can be regarded as unity. The high-pass filter characteristics of the PD controller are clearly shown by the Bode plot in Fig. 9-6. The phase-lead property may be utilized to improve the phase margin of a control system. Unfortunately, the magnitude characteristics of the PD controller push the gain-crossover frequency to a higher value. Thus, *the design principle of the PD controller involves the placing of the corner frequency of the controller, $\omega = K_P/K_D$, such that an effective improvement of the phase margin is realized at the new gain-crossover frequency*. For a given system, there is a range of values of K_P/K_D that is optimal for improving the damping of the system. Another practical consideration in selecting the values of K_P and K_D is in the physical implementation of the PD controller. Other apparent effects of the PD control in the frequency domain are that, due to its high-pass characteristics, in most cases it will increase the BW of the system and reduce the rise time of the step response. The practical disadvantage of the PD controller is that the differentiator portion is a high-pass filter, which usually accentuates any high-frequency noise that enters at the input.

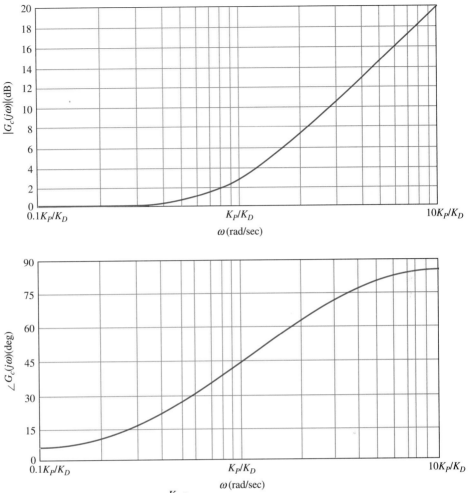

Figure 9-6 Bode diagram of $1 + \dfrac{K_D s}{K_P}, K_P = 1$.

9-2-3 Summary of Effects of PD Control

Though it is not effective with lightly damped or initially unstable systems, a properly designed PD controller can affect the performance of a control system in the following ways:

1. Improving damping and reducing maximum overshoot.
2. Reducing rise time and settling time.
3. Increasing BW.
4. Improving GM, PM, and M_r.
5. Possibly accentuating noise at higher frequencies.
6. Possibly requiring a relatively large capacitor in circuit implementation.

The following example illustrates the effects of the PD controller on the time-domain and frequency-domain responses of a second-order system.

Toolbox 9-2-1

The Bode diagram for Fig. 9-6 is obtained by the following sequence of MATLAB functions

```
%%%%%%%%%%%%%%%%%%%%%%%%%%%%%%%%%%%%%%%%%%%%
close all;
clear all;
clc;
%%%%%%%%%%%%%%%%%%%%%%%%%%%%%%%%%%%%%%%%%%%\
KP = 1;
KD = 1;
num = [KD KP];
den =[1];
bode(tf(num,den))
grid
```

▶ **EXAMPLE 9-2-1** Let us reconsider the second-order model of the aircraft attitude control system shown in Fig. 5-29. The forward-path transfer function of the system is given in Eq. (5-132) and is repeated here:

$$G(s) = \frac{4500K}{s(s+361.2)} \tag{9-10}$$

Let us set the performance specifications as follows:

Steady-state error due to unit-ramp input ≤ 0.000443

Maximum overshoot $\leq 5\%$

Rise time $t_r \leq 0.005$ sec

Settling time $t_s \leq 0.005$ sec

To satisfy the maximum value of the specified steady-state error requirement, K should be set at 181.17. However, with this value of K, the damping ratio of the system is 0.2, and the maximum overshoot is 52.7%, as shown by the unit-step response in Fig. 5-31 and again in Fig. 9-9. Let us consider inserting a PD controller in the forward path of the system so that the damping and the maximum overshoot of the system are improved while maintaining the steady-state error due to the unit-ramp input at 0.000443. ◀

Time-Domain Design

With the PD controller of Eq. (9-9) and $K = 181.17$, the forward-path transfer function of the system becomes

$$G(s) = \frac{\Theta_y(s)}{\Theta_e(s)} = \frac{815,265(K_P + K_D s)}{s(s+361.2)} \tag{9-11}$$

The closed-loop transfer function is

$$\frac{\Theta_y(s)}{\Theta_r(s)} = \frac{815,265(K_P + K_D s)}{s^2 + (361.2 + 815,265 K_D)s + 815,265 K_P} \tag{9-12}$$

The ramp-error constant is

$$K_v = \lim_{s \to 0} sG(s) = \frac{815,265 K_P}{361.2} = 2257.1 K_P \tag{9-13}$$

The steady-state error due to a unit-ramp input is $e_{ss} = 1/K_v = 0.000443/K_P$.

Eq. (9-12) shows that the effects of the PD controller are as follows:

1. Adding a zero at $s = -K_P/K_D$ to the closed-loop transfer function
2. Increasing the *damping term*, which is the coefficient of the s term in the denominator, from 361.2 to $361.2 + 815{,}265K_D$

The characteristic equation is written

$$s^2 + (361.2 + 815{,}265K_D)s + 815{,}265K_P = 0 \tag{9-14}$$

We can set $K_P = 1$, which is acceptable from the steady-state error requirement. The damping ratio of the system is

$$\zeta = \frac{361.2 + 815{,}265K_D}{1805.84} = 0.2 + 451.46K_D \tag{9-15}$$

which clearly shows the positive effect of K_D on damping. If we wish to have critical damping, $\zeta = 1$, Eq. (9-15) gives $K_D = 0.001772$. We should quickly point out that Eq. (9-11) no longer represents a prototype second-order system, since the transient response is also affected by the zero of the transfer function at $s = -K_P/K_D$. It turns out that, for this second-order system, as the value of K_D increases, the zero will move very close to the origin and effectively cancel the pole of $G(s)$ at $s = 0$. Thus, as K_D increases, the transfer function in Eq. (9-11) approaches that of a first-order system with the pole at $s = -361.2$, and the closed-loop system will not have any overshoot. In general, for higher-order systems, however, the zero at $s = -K_P/K_D$ may increase the overshoot when K_D becomes very large.

We can apply the root-contour method to the characteristic equation in Eq. (9-14) to examine the effect of varying K_P and K_D. First, by setting K_D to zero, Eq. (9-14) becomes

$$s^2 + 361.2s + 815{,}265K_P = 0 \tag{9-16}$$

The root loci of the last equation as K_P varies between 0 and ∞ are shown in Fig. 9-7. When $K_D \neq 0$, the characteristic equation in Eq. (9-14) is conditioned as

$$1 + G_{eq}(s) = 1 + \frac{815{,}265K_D s}{s^2 + 361.2s + 815{,}265K_P} = 0 \tag{9-17}$$

The root contours of Eq. (9-14) with K_P = constant and K_D varying are constructed based on the pole–zero configuration of $G_{eq}(s)$ and are shown in Fig. 9-8 for $K_P = 0.25$ and $K_P = 1$. We see that, when $K_P = 1$ and $K_D = 0$, the characteristic equation roots are at $-180.6 + j884.67$ and $-180.6 - j884.67$, and the damping ratio of the closed-loop system is 0.2. When the value of K_D is increased, the two characteristic equation roots move toward the real axis along a circular arc. When K_D is increased to 0.00177, the roots are real and equal at -902.92, and the damping is critical. When K_D is increased beyond 0.00177, the two roots become real and unequal, and the system is overdamped.

When K_P is 0.25 and $K_D = 0$, the two characteristic equation roots are at $-180.6 + j413.76$ and $-180.6 - j413.76$. As K_D increases in value, the root contours again show the improved damping due to the PD controller. Fig. 9-9 shows the unit-step responses of the closed-loop system without PD control and with $K_P = 1$ and $K_D = 0.00177$. With the PD

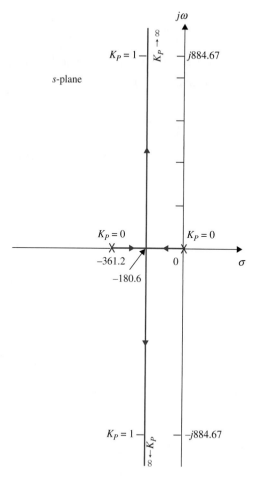

Figure 9-7 Root loci of Eq. (9-16).

control, the maximum overshoot is 4.2%. In the present case, although K_D is chosen for critical damping, the overshoot is due to the zero at $s = -K_P/K_D$ of the closed-loop transfer function. Table 9-1 gives the results on maximum overshoot, rise time, and settling time for $K_P = 1$ and $K_D = 0, 0.0005, 0.00177$, and 0.0025. The results in Table 9-1 show that the performance requirements are all satisfied with $K_D \geq 0.00177$. It should be kept in mind that K_D should only be large enough to satisfy the performance requirements. Large K_D corresponds to large BW, which may cause high-frequency noise problems, and there is also the concern of the capacitor value in the op-amp-circuit implementation.

TABLE 9-1 Attributes of the Unit-Step Responses of the System in Example 9-2-1 with PD Controller

K_D	t_r (sec)	t_s (sec)	Maximum Overshoot (%)
0	0.00125	0.0151	52.2
0.0005	0.0076	0.0076	25.7
0.00177	0.00119	0.0049	4.2
0.0025	0.00103	0.0013	0.7

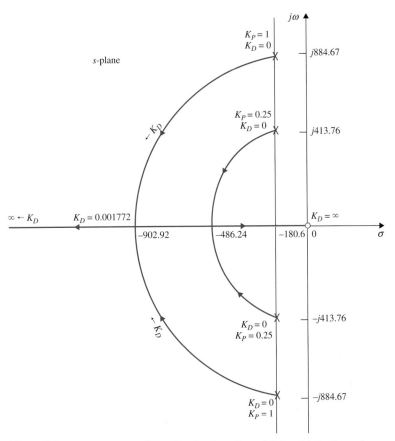

Figure 9-8 Root contours of Eq. (9-14) when $K_P = 0.25$ and 1.0; K_D varies.

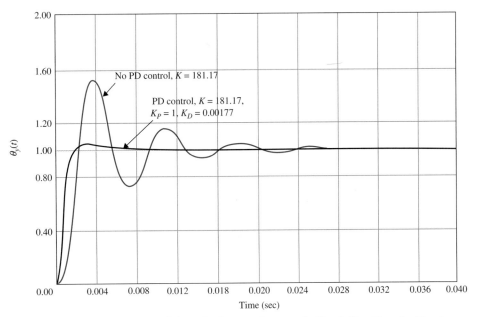

Figure 9-9 Unit-step response of the attitude control system in Fig. 5-29 with and without PD controller.

Toolbox 9-2-2

Root loci of Eq. (9-16) shown in Fig. 9-7 are obtained by the following sequence of MATLAB functions

```
KP = 1;
den = [1 361.2 815265*KP];
num = [1];
rlocus(num,den)
hold on
KP = 0;
den = [1 361.2 815265*KP];
num = [1];
rlocus(num,den)
```

The general conclusion is that the PD controller decreases the maximum overshoot, the rise time, and the settling time.

Another analytic way of studying the effects of the parameters K_P and K_D is to evaluate the performance characteristics in the parameter plane of K_P and K_D. From the characteristic equation of Eq. (9-14), we have

$$\zeta = \frac{0.2 + 451.46 K_D}{\sqrt{K_P}} \tag{9-18}$$

Applying the stability requirement to Eq. (9-14), we find that, for system stability,

$$K_P > 0 \quad \text{and} \quad K_D > -0.000443$$

Toolbox 9-2-3

Root contours of Eq. (9-14) shown in Fig. 9-8 are obtained by the following sequence of MATLAB functions

```
KP = 1; KD = 1;
den = [1 361.2+815265*KD 815265*KP]; num = [1];
rlocus(num,den)
hold on
KD = 1e-6;
for i = 1:1:260
    den = [1 361.2+815265*KD 815265*KP];
    num = [1];
    tf(num,den);
    [numCL,denCL]=cloop(num,den);
    T = tf(numCL,denCL);
    PoleData(:,i)=pole(T);
KD = KD +3e-5;
end
i=60; %%% for continuation of graph
PoleData(1,i+1) = - sqrt((real(PoleData(1,i))^2)+(imag(PoleData(1,i))^2));
PoleData(2,i+1) = - sqrt((real(PoleData(2,i))^2)+(imag(PoleData(2,i))^2));
plot(real(PoleData(1,:)),imag(PoleData(1,:)),real(PoleData(2,:)),imag(PoleData(2,:)));

%%%%%%%%%%%%%%%%%%%%%%%%%%%%%%%%%%%%%%%%%%\
KP = 0.25;
KD = 1;
```

```
den = [1 361.2+815265*KD 815265*KP];
num = [1];
rlocus(num,den)
hold on
KD = 1e-6;
for i = 1:1:260
    den = [1 361.2+815265*KD 815265*KP];
    num = [1];
    tf(num,den);
    [numCL,denCL]=cloop(num,den);
    T = tf(numCL,denCL);
    PoleData(:,i)=pole(T);
    KD = KD +3e-5;
end
i=23; %%% for continuation of graph
PoleData(1,i+1) = - sqrt((real(PoleData(1,i))^2)+(imag(PoleData(1,i))^2));
PoleData(2,i+1) = - sqrt((real(PoleData(2,i))^2)+(imag(PoleData(2,i))^2));
plot(real(PoleData(1,:)),imag(PoleData(1,:)),real(PoleData(2,:)),imag(PoleData(2,:)));
%%%%%%%%%%%%%%%%%%%%%%%%%%%%%%%%%%%%%%%%%%%%%%\
axis([-3000 1000 -1000 1000])
xaxis1 = -181.0076 *ones(1,100);yaxis1 = -1000:20:1000-1
plot(xaxis1,yaxis1);
grid
```

The boundaries of stability in the K_P-versus-K_D parameter plane are shown in Fig. 9-10. The constant-damping-ratio trajectory is described by Eq. (9-18) and is a parabola. Fig. 9-10 illustrates the constant-ζ trajectories for $\zeta = 0.5$, 0.707, and 1.0. The ramp-error constant K_v is given by Eq. (9-13), which describes a horizontal line in the parameter plane, as shown in Fig. 9-10. The figure gives a clear picture as to how the values of K_P and K_D affect the various performance criteria of the system. For instance, if K_v is set at 2257.1, which corresponds to $K_P = 1$, the constant-ζ loci show that the damping is increased monotonically with the increase in K_D. The intersection between the constant-K_v locus and the constant-ζ locus gives the value of K_D for the desired K_v and ζ.

Toolbox 9-2-4

Bode plot of Fig. 9-11 is obtained by the following sequence of MATLAB functions

```
KD = [0 0.0005 0.0025 0.00177];

for i = 1:length(KD)
    num = [815265*KD(i) 815265];
    den =[1 361.2 0];
    bode(tf(num,den));
    hold on;
end
axis([1 10000 -180 -90]);
grid
```

Frequency-Domain Design

Now let us carry out the design of the PD controller in the frequency domain. Fig. 9-11 shows the Bode plot of $G(s)$ in Eq. (9-11) with $K_P = 1$ and $K_D = 0$. The phase margin of

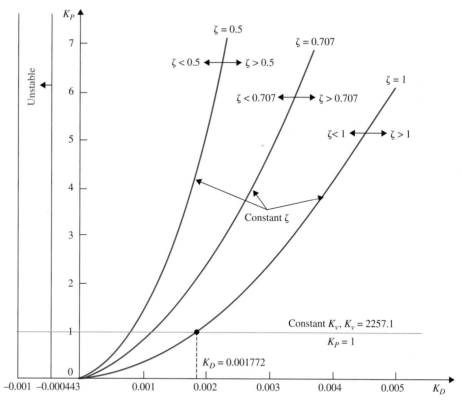

Figure 9-10 K_P-versus-K_D parameter plane for the attitude control system with a PD controller.

the uncompensated system is 22.68°, and the resonant peak M_r is 2.522. These values correspond to a lightly damped system. Let us give the following performance criteria:

Steady-state error due to a unit-ramp input ≤ 0.00443

Phase margin $\geq 80°$

Resonant peak $M_r \leq 1.05$

BW ≤ 2000 rad/sec

The Bode plots of $G(s)$ for $K_P = 1$ and $K_D = 0, 0.005, 0.00177$, and 0.0025 are shown in Fig. 9-11. The performance measures in the frequency domain for the compensated system with these controller parameters are tabulated in Table 9-2, along with the time-domain attributes for comparison. The Bode plots as well as the performance data were easily generated by using MATLAB tools. Use **ACSYS** component controls to reproduce the results in Table 9-2.

The results in Table 9-2 show that the gain margin is always infinite, and thus the relative stability is measured by the phase margin. This is one example where the gain margin is not an effective measure of the relative stability of the system. When $K_D = 0.00177$, which corresponds to critical damping, the phase margin is 82.92°, the resonant peak M_r is 1.025, and BW is 1669 rad/sec. The performance requirements in the frequency domain are all satisfied. Other effects of the PD control are that the BW and the gain-crossover frequency are increased. The phase-crossover frequency is always infinite in this case.

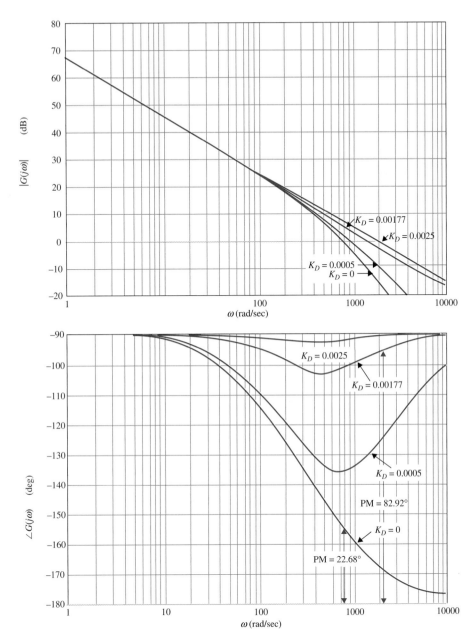

Figure 9-11 Bode plot of $G(s) = \dfrac{815,265(1 + K_D s)}{s(s + 361.2)}$.

TABLE 9-2 Frequency-Domain Characteristics of the System in Example 9-2-1 with PD Controller

K_D	GM (dB)	PM (deg)	Gain CO (rad/sec)	BW (rad/sec)	M_r	t_r (sec)	t_s (sec)	Maximum Overshoot (%)
0	∞	22.68	868	1370	2.522	0.00125	0.0151	52.2
0.0005	∞	46.2	913.5	1326	1.381	0.0076	0.0076	25.7
0.00177	∞	82.92	1502	1669	1.025	0.00119	0.0049	4.2
0.0025	∞	88.95	2046	2083	1.000	0.00103	0.0013	0.7

EXAMPLE 9-2-2

Consider the third-order aircraft attitude control system discussed in Chapter 5 with the forward-path transfer function given in Eq. (5-153),

$$G(s) = \frac{1.5 \times 10^7 K}{s(s^2 + 3408.3s + 1,204,000)} \quad (9\text{-}19)$$

The same set of time-domain specifications given in Example 9-2-1 is to be used. It was shown in Chapter 5 that, when $K = 181.17$, the maximum overshoot of the system is 78.88%.

Let us attempt to meet the transient performance requirements by use of a PD controller with the transfer function given in Eq. (9-2). The forward-path transfer function of the system with the PD controller and $K = 181.17$ is

$$G(s) = \frac{2.718 \times 10^9 (K_P + K_D s)}{s(s^2 + 3408.3s + 1,204,000)} \quad (9\text{-}20)$$

You may also use our MATLAB toolbox **ACSYS** to solve this problem. See Section 9-19.

Time-Domain Design

Setting $K_P = 1$ arbitrarily, the characteristic equation of the closed-loop system is written

$$s^3 + 3408.3s^2 + (1,204,000 + 2.718 \times 10^9 K_D)s + 2.718 \times 10^9 = 0 \quad (9\text{-}21)$$

To apply the root-contour method, we condition Eq. (9-21) as

$$1 + G_{eq}(s) = 1 + \frac{2.718 \times 10^9 K_D s}{s^3 + 3408.2s^2 + 1,204,000s + 2.718 \times 10^9} = 0 \quad (9\text{-}22)$$

where

$$G_{eq}(s) = \frac{2.718 \times 10^9 K_D s}{(s + 3293.3)(s + 57.49 + j906.6)(s + 57.49 - j906.6)} \quad (9\text{-}23)$$

Toolbox 9-2-5

Root contours of Fig. 9-12 are obtained by the following sequence of MATLAB functions

You may wish to use clc, close all, or clear all before running the following

```
% Root contours
kd=0.005;
t=0:0.001:0.05;
num = [2.718*10^9*kd 0];
den = [1 3408.2 1204000 2.718*10^9];
rlocus(num,den)
```

- If a system has very low damping or is unstable, the PD control may not be effective in improving the stability of the system.

The root contours of Eq. (9-21) are plotted as shown in Fig. 9-12, based on the pole–zero configuration of $G_{eq}(s)$. The root contours of Fig. 9-12 reveal the effectiveness of the PD controller for the improvement on the relative stability of the system. Notice that, as the value of K_D increases, one root of the characteristic equation moves from -3293.3 toward the origin, while the two complex roots start out toward the left and eventually approach the vertical asymptotes that intersect at $s = -1704$. The immediate assessment of the situation is that, if the value of K_D is too large, *the two complex roots will actually have reduced*

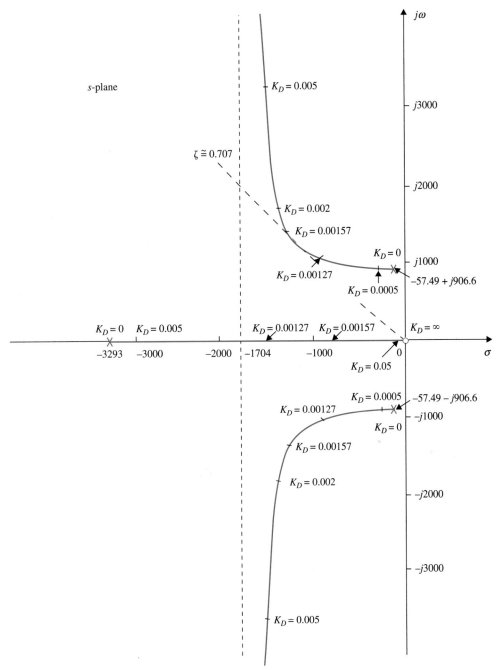

Figure 9-12 Root contours of
$s^3 + 3408.3s^2 + (1,204,000 + 2.718 \times 10^9 K_D)s + 2.718 \times 10^9 = 0$.

damping while increasing the natural frequency of the system. It appears that the ideal location for the two complex characteristic equation roots, from the standpoint of relative stability, is near the bend of the root contour, where the relative damping ratio is approximately 0.707. The root contours of Fig. 9-12 clearly show that, if the original system has low damping or is unstable, the zero introduced by the PD controller may not be able to add sufficient damping or even stabilize the system.

TABLE 9-3 Time-Domain Attributes of the Third-Order System in Example 9-2-2 with PD Controller

K_D	% Maximum Overshoot	t_r (sec)	t_s (sec)	Characteristic Equation Roots	
0	78.88	0.00125	0.0495	−3293.3,	−57.49 ± j906.6
0.0005	41.31	0.00120	0.0106	−2843.07,	−282.62 ± j936.02
0.00127	17.97	0.00100	0.00398	−1523.11,	−942.60 ± j946.58
0.00157	14.05	0.00091	0.00337	−805.33,	−1301.48 ± j1296.59
0.00200	11.37	0.00080	0.00255	−531.89,	−1438.20 ± j1744.00
0.00500	17.97	0.00042	0.00130	−191.71,	−1608.29 ± j3404.52
0.01000	31.14	0.00026	0.00093	−96.85,	−1655.72 ± j5032
0.05000	61.80	0.00010	0.00144	−19.83,	−1694.30 ± j11583

Table 9-3 gives the results of maximum overshoot, rise time, settling time, and the roots of the characteristic equation as functions of the parameter K_D. The following conclusions are drawn on the effects of the PD controller on the third-order system.

1. The minimum value of the maximum overshoot, 11.37%, occurs when K_D is approximately 0.002.
2. Rise time is improved (reduced) with the increase of K_D.
3. Too high a value of K_D will actually increase the maximum overshoot and the settling time substantially. The latter is because the damping is reduced as K_D is increased indefinitely.

Fig. 9-13 shows the unit-step responses of the system with the PD controller for several values of K_D. The conclusion is that, while the PD control does improve the damping of the system, it does not meet the maximum-overshoot requirement.

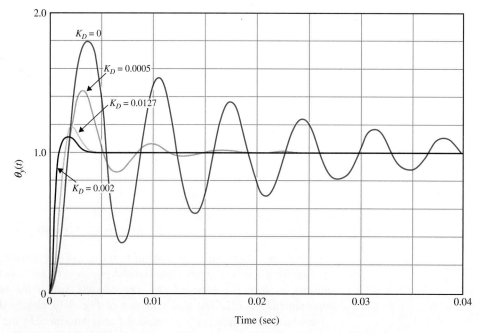

Figure 9-13 Unit-step responses of the system in Example 9-2-2 with PD controller.

Frequency-Domain Design

The Bode plot of Eq. (9-20) is used to conduct the frequency-domain design of the PD controller. Fig. 9-14 shows the Bode plot for $K_P = 1$ and $K_D = 0$. The following performance data are obtained for the uncompensated system:

Gain margin = 3.6 dB

Phase margin = 7.77°

Resonant Peak $M_r = 7.62$

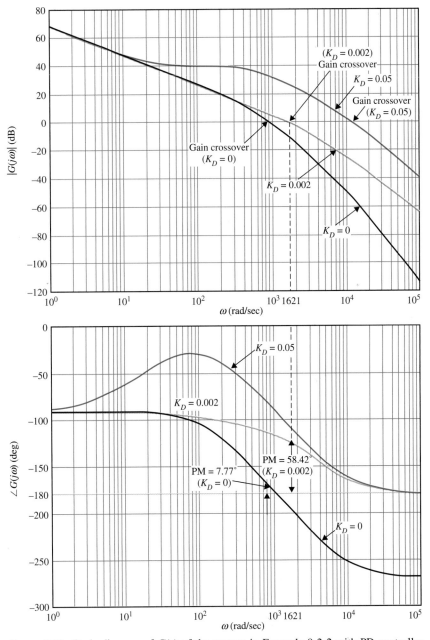

Figure 9-14 Bode diagram of $G(s)$ of the system in Example 9-2-2 with PD controller.

TABLE 9-4 Frequency-Domain Characteristics of the Third-Order System in Example 9-2-2 with PD Controller

K_D	GM (dB)	PM (deg)	M_r	BW (rad/sec)	Gain CO (rad/sec)	Phase CO (rad/sec)
0	3.6	7.77	7.62	1408.83	888.94	1103.69
0.0005	∞	30.94	1.89	1485.98	935.91	∞
0.00127	∞	53.32	1.19	1939.21	1210.74	∞
0.00157	∞	56.83	1.12	2198.83	1372.30	∞
0.00200	∞	58.42	1.07	2604.99	1620.75	∞
0.00500	∞	47.62	1.24	4980.34	3118.83	∞
0.01000	∞	35.71	1.63	7565.89	4789.42	∞
0.0500	∞	16.69	3.34	17989.03	11521.00	∞

Bandwidth BW = 1408.83 rad/sec

Gain crossover (GCO) = 888.94 rad/sec

Phase crossover (PCO) = 1103.69 rad/sec

Let us use the same set of frequency-domain performance requirements listed in Example 9-2-1. The logical way to approach this problem is to first examine how much additional phase is needed to realize a phase margin of 80°. Because the uncompensated system with the gain set to meet the steady-state requirement is only 7.77°, the PD controller must provide an additional phase of 72.23°. This additional phase must be placed at the gain crossover of the compensated system in order to realize a PM of 80°. Referring to the Bode plot of the PD controller in Fig. 9-6, we see that the additional phase is always accompanied by a gain in the magnitude curve. As a result, the gain crossover of the compensated system will be pushed to a higher frequency at which the phase of the uncompensated system would correspond to an even smaller PM. Thus, we may run into the problem of diminishing returns. This symptom is parallel to the situation illustrated by the root-contour plot in Fig. 9-12, in which case the larger K_D would simply push the roots to a higher frequency, and the damping would actually be decreased. The frequency-domain performance data of the compensated system with the values of K_D used in Table 9-3 are obtained from the Bode plots for each case, and the results are shown in Table 9-4. The Bode plots of some of these cases are shown in Fig. 9-14. Notice that the gain margin becomes infinite when the PD controller is added, and the phase margin becomes the dominant measure of relative stability. This is because the phase curve of the PD-compensated system stays above the $-180°$-axis, and the phase crossover is at infinity.

Toolbox 9-2-6

Fig. 9-13 is obtained by the following sequence of MATLAB functions

```
KP = 1;
KD = [ 0.0005 0.0127 0.002];

for i =1:length(KD)
   num =[2.718e9*KD(i) 2.718e9*KP];
   den = [1 3408.3 0 0];
   tf(num,den);
```

```
        [numCL,denCL]=cloop(num,den);
        step(numCL,denCL)
        hold on
end
axis([0 0.04 0 2])
```

When $K_D = 0.002$, the phase margin is at a maximum of 58.42°, and M_r is also minimum at 1.07, which happens to agree with the optimal value obtained in the time-domain design summarized in Table 9-3. When the value of K_D is increased beyond 0.002, the phase margin decreases, which agrees with the findings from the time-domain design that large values of K_D actually decrease damping. However, the BW and the gain crossover increase continuously with the increase in K_D. The frequency-domain design again shows that the PD control falls short in meeting the performance requirements imposed on the system. Just as in the time-domain design, we have demonstrated that if the original system has very low damping, or is unstable, PD control may not be effective in improving the stability of the system. Another situation under which PD control may be ineffective is if the slope of the phase curve near the gain-crossover frequency is steep, in which case the rapid decrease of the phase margin due to the increase of the gain crossover from the added gain of the PD controller may render the additional phase ineffective.

Toolbox 9-2-7

Bode diagram of G(s) in Example 9-2 in Fig. 9-14 is obtained by the following sequence of MATLAB functions

```
KD = [0 0.002 0.05];
KP=1;

for i = 1:length(KD)
    num =[2.718e9*KD(i) 2.718e9*KP];
    den = [1 3408.3 1204000 0];
    bode(num,den);
    hold on;
end
```

▶ 9-3 DESIGN WITH THE PI CONTROLLER

We see from Section 9-2 that the PD controller can improve the damping and rise time of a control system at the expense of higher bandwidth and resonant frequency, and the steady-state error is not affected unless it varies with time, which is typically not the case for step-function inputs. Thus, the PD controller may not fulfill the compensation objectives in many situations.

The integral part of the PID controller produces a signal that is proportional to the time integral of the input of the controller. Fig. 9-15 illustrates the block diagram of a prototype second-order system with a series PI controller. The transfer function of the PI controller is

$$G_c(s) = K_P + \frac{K_I}{s} \tag{9-24}$$

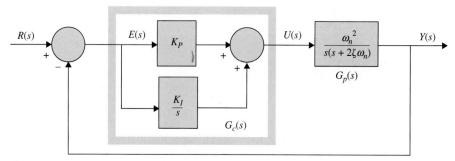

Figure 9-15 Control system with PI controller.

Using the circuit elements given in Table 4-4, two op-amp-circuit realizations of Eq. (9-24) are shown in Fig. 9-16. The transfer function of the two-op-amp circuit in Fig. 9-16(a) is

$$G_c(s) = \frac{E_o(s)}{E_{in}(s)} = \frac{R_2}{R_1} + \frac{R_2}{R_1 C_2 s} \tag{9-25}$$

Comparing Eq. (9-24) with Eq. (9-25), we have

$$K_P = \frac{R_2}{R_1} \quad K_I = \frac{R_2}{R_1 C_2} \tag{9-26}$$

The transfer function of the three-op-amp circuit in Fig. 9-16(b) is

$$G_c(s) = \frac{E_o(s)}{E_{in}(s)} = \frac{R_2}{R_1} + \frac{1}{R_i C_i s} \tag{9-27}$$

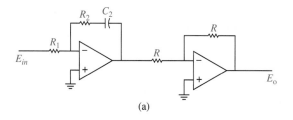

(a)

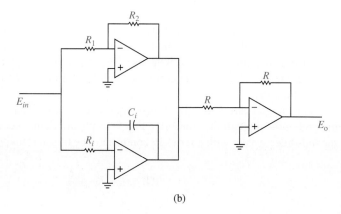

(b)

Figure 9-16 Op-amp-circuit realization of the PI controller, $G_c(s) = K_P + \frac{K_I}{s}$.
(a) Two-op-amp circuit.
(b) Three-op-amp circuit.

Thus, the parameters of the PI controller are related to the circuit parameters as

$$K_P = \frac{R_2}{R_1} \quad K_I = \frac{1}{R_i C_i} \quad (9\text{-}28)$$

The advantage with the circuit in Fig. 9-16(b) is that the values of K_P and K_I are independently related to the circuit parameters. However, in either circuit, K_I is inversely proportional to the value of the capacitor. Unfortunately, effective PI-control designs usually result in small values of K_I, and thus we must again watch out for unrealistically large capacitor values.

The forward-path transfer function of the compensated system is

$$G(s) = G_c(s)G_p(s) = \frac{\omega_n^2(K_P s + K_I)}{s^2(s + 2\zeta\omega_n)} \quad (9\text{-}29)$$

Clearly, the immediate effects of the PI controller are as follows:

1. Adding a zero at $s = -K_I/K_P$ to the forward-path transfer function.
2. Adding a pole at $s = 0$ to the forward-path transfer function. This means that the system type is increased by 1 to a type 2 system. Thus, the steady-state error of the original system is improved by one order; that is, if the steady-state error to a given input is constant, the PI control reduces it to zero (provided that the compensated system remains stable).

The system in Fig. 9-15, with the forward-path transfer function in Eq. (9-29), will now have a zero steady-state error when the reference input is a ramp function. However, because the system is now of the third order, *it may be less stable* than the original second-order system or even become *unstable* if the parameters K_P and K_I are not properly chosen.

In the case of a type 1 system with a PD control, the value of K_P is important because the ramp-error constant K_v is directly proportional to K_P, and thus the magnitude of the steady-state error is inversely proportional to K_P when the input is a ramp. On the other hand, if K_P is too large, the system may become unstable. Similarly, for a type 0 system, the steady-state error due to a step input will be inversely proportional to K_P.

When a type 1 system is converted to type 2 by the PI controller, K_P no longer affects the steady-state error, and the latter is always zero for a stable system with a ramp input. The problem is then to choose the proper combination of K_P and K_I so that the transient response is satisfactory.

9-3-1 Time-Domain Interpretation and Design of PI Control

The pole–zero configuration of the PI controller in Eq. (9-24) is shown in Fig. 9-17. At first glance, it may seem that PI control will improve the steady-state error at the expense of stability. However, we shall show that, if the location of the zero of $G_c(s)$ is selected properly, both the damping and the steady-state error can be improved. Because the PI controller is essentially a low-pass filter, the compensated system usually will have a slower rise time and longer settling time. *A viable method of designing the PI control is to select the zero at* $s = -K_I/K_P$ *so that it is relatively close to the origin and away from the most significant poles of the process; the values of* K_P *and* K_I *should be relatively small.*

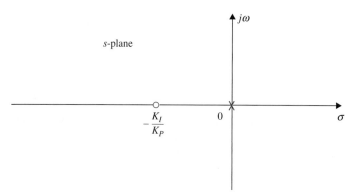

Figure 9-17 Pole–zero configuration of a PI controller.

9-3-2 Frequency-Domain Interpretation and Design of PI Control

For frequency-domain design, the transfer function of the PI controller is written

$$G_c(s) = K_P + \frac{K_I}{s} = \frac{K_I\left(1 + \frac{K_P}{K_I}s\right)}{s} \qquad (9\text{-}30)$$

The Bode plot of $G_c(j\omega)$ is shown in Fig. 9-18. Notice that the magnitude of $G_c(j\omega)$ at $\omega = \infty$ is $20\log_{10}K_P$ dB, which represents an attenuation if the value of K_P is less than 1. This attenuation may be utilized to improve the stability of the system. The phase of $G_c(j\omega)$ is always negative, which is detrimental to stability. Thus, we should place the corner frequency of the controller, $\omega = K_I/K_P$, as far to the left as the bandwidth requirement allows, so the phase-lag properties of $G_c(j\omega)$ do not degrade the achieved phase margin of the system.

The frequency-domain design procedure for the PI control to realize a given phase margin is outlined as follows:

1. The Bode plot of the forward-path transfer function $G_p(s)$ of the uncompensated system is made with the loop gain set according to the steady-state performance requirement.

2. The phase margin and the gain margin of the uncompensated system are determined from the Bode plot. For a specified phase margin requirement, the new gain-crossover frequency ω'_g corresponding to this phase margin is found on the Bode plot. The magnitude plot of the compensated transfer function must pass through the 0-dB-axis at this new gain-crossover frequency in order to realize the desired phase margin.

3. To bring the magnitude curve of the uncompensated transfer function down to 0 dB at the new gain-crossover frequency ω'_g, the PI controller must provide the amount of attenuation equal to the gain of the magnitude curve at the new gain-crossover frequency. In other words, set

$$\left|G_P(j\omega'_g)\right|_{dB} = -20\log_{10}K_P \text{ dB} \qquad K_P < 1 \qquad (9\text{-}31)$$

from which we have

$$K_P = 10^{-|G_P(j\omega'_g)|_{dB}/20} \qquad K_P < 1 \qquad (9\text{-}32)$$

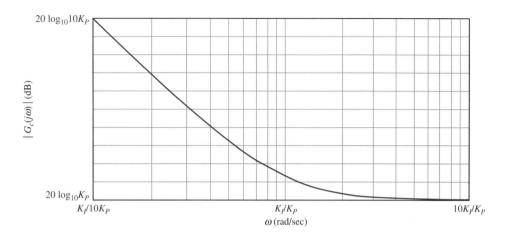

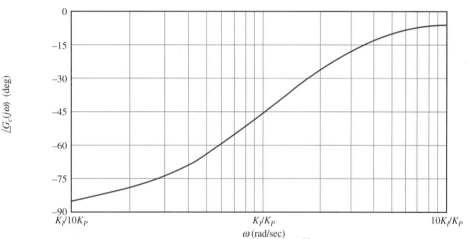

Figure 9-18 Bode diagram of the PI controller. $G_c(s) = K_P + \dfrac{K_I}{s}$.

Once the value of K_P is determined, it is necessary only to select the proper value of K_I to complete the design. Up to this point, we have assumed that, although the gain-crossover frequency is altered by attenuating the magnitude of $G_c(j\omega)$ at ω'_g, the original phase is not affected by the PI controller. This is not possible, however, since, as shown in Fig. 9-18, the attenuation property of the PI controller is accompanied with a phase lag that is detrimental to the phase margin. It is apparent that, if the corner frequency $\omega = K_I/K_P$ is placed far below ω'_g, the phase lag of the PI controller will have a negligible effect on the phase of the compensated system near ω'_g. On the other hand, the value of K_I/K_P should not be too small or the bandwidth of the system will be too low, causing the rise time and settling time to be too long. As a general guideline, K_I/K_P should correspond to a frequency that is at least one decade, sometimes as much as two decades, below ω'_g. That is, we set

$$\frac{K_I}{K_P} = \frac{\omega'_g}{10} \text{ rad/sec} \qquad (9\text{-}33)$$

Within the general guideline, the selection of the value of K_I/K_P is pretty much at the discretion of the designer, who should be mindful of its effect on BW and its practical implementation by an op-amp circuit.

4. The Bode plot of the compensated system is investigated to see if the performance specifications are all met.
5. The values of K_I and K_P are substituted in Eq. (9-30) to give the desired transfer function of the PI controller.

If the controlled process $G_P(s)$ is type 0, the value of K_I may be selected based on the ramp-error-constant requirement, and then there would only be one parameter, K_P, to determine. By computing the phase margin, gain margin, M_r, and BW of the closed-loop system with a range of values of K_P, the best value for K_P can be easily selected.

Based on the preceding discussions, we can summarize the advantages and disadvantages of a properly designed PI controller as the following:

1. Improving damping and reducing maximum overshoot.
2. Increasing rise time.
3. Decreasing BW.
4. Improving gain margin, phase margin, and M_r.
5. Filtering out high-frequency noise.

It should be noted that in the PI controller design process, selection of a proper combination of K_I and K_P, so that the capacitor in the circuit implementation of the controller is not excessively large, is more difficult than in the case of the PD controller.

The following examples will illustrate how the PI control is designed and what its effects are.

▶ **EXAMPLE 9-3-1** Consider the second-order attitude-control system discussed in Example 9-2-1. Applying the PI controller of Eq. (9-24), the forward-path transfer function of the system becomes

$$G(s) = G_c(s)G_P(s) = \frac{4500KK_P(s + K_I/K_P)}{s^2(s + 361.2)} \quad (9\text{-}34)$$

You may use **ACSYS** to solve this problem. ◀

Time-Domain Design
Let the time-domain performance requirements be

Steady-state error due to parabolic input $t^2 u_s(t)/2 \leq 0.2$

Maximum overshoot $\leq 5\%$

Rise time $t_r \leq 0.01$ sec

Settling time $t_s \leq 0.02$ sec

We have to relax the rise time and settling time requirements from those in Example 9-2-1 so that we will have a meaningful design for this system. The significance of the requirement on the steady-state error due to a parabolic input is that it indirectly places a minimum requirement on the speed of the transient response.

9-3 Design with the PI Controller

The parabolic-error constant is

$$K_a = \lim_{s \to 0} s^2 G(s) = \lim_{s \to 0} s^2 \frac{4500KK_P(s + K_I/K_P)}{s^2(s + 361.2)} \quad (9\text{-}35)$$

$$= \frac{4500KK_I}{361.2} = 12.46KK_I$$

The steady-state error due to the parabolic input $t^2 u_s(t)/2$ is

$$e_{ss} = \frac{1}{K_a} = \frac{0.08026}{KK_I} (\leq 0.2) \quad (9\text{-}36)$$

Let us set $K = 181.17$, simply because this was the value used in Example 9-2-1. Apparently, to satisfy a given steady-state error requirement for a parabolic input, the larger the K, the smaller K_I can be. Substituting $K = 181.17$ in Eq. (9-36) and solving K_I for the minimum steady-state error requirement of 0.2, we get the minimum value of K_I to be 0.002215. If necessary, the value of K can be adjusted later.

With $K = 181.17$, the characteristic equation of the closed-loop system is

$$s^3 + 361.2s^2 + 815,265K_P s + 815,265K_I = 0 \quad (9\text{-}37)$$

Applying Routh's test to Eq. (9-37) yields the result that the system is stable for $0 < K_I/K_P < 361.2$. This means that the zero of $G(s)$ at $s = -K_I/K_P$ cannot be placed too far to the left in the left-half s-plane, or the system will be unstable. Let us place the zero at $-K_I/K_P$ relatively close to the origin. For the present case, the most significant pole of $G_P(s)$, besides the pole at $s = 0$, is at -361.2. Thus, K_I/K_P should be chosen so that the following condition is satisfied:

$$\frac{K_I}{K_P} \ll 361.2 \quad (9\text{-}38)$$

The root loci of Eq. (9-37) with $K_I/K_P = 10$ are shown in Fig. 9-19. Notice that, other than the small loop around the zero at $s = -10$, these root loci for the most part are very similar to those shown in Fig. 9-7, which are for Eq. (9-16). With the condition in Eq. (9-38) satisfied, Eq. (9-34) can be approximated by

$$G(s) \cong \frac{815,265K_P}{s(s + 361.2)} \quad (9\text{-}39)$$

where the term K_I/K_P in the numerator is neglected when compared with the magnitude of s, which takes on values along the operating points on the complex portion of the root loci that correspond to, say, a relative damping ratio in the range of $0.7 < \zeta < 1.0$. Let us assume that we wish to have a relative damping ratio of 0.707. From Eq. (9-39), the required value of K_P for this damping ratio is 0.08. This should also be true for the third-order system with the PI controller if the value of K_I/K_P satisfies Eq. (9-38). Thus, with $K_P = 0.08$, $K_I = 0.8$; the root loci in Fig. 9-19 show that the relative damping ratio of the two complex roots is approximately 0.707. In fact, the three characteristic equation roots are at

$$s = -10.605, -175.3 + j175.4, \text{ and } -175.3 - j175.4$$

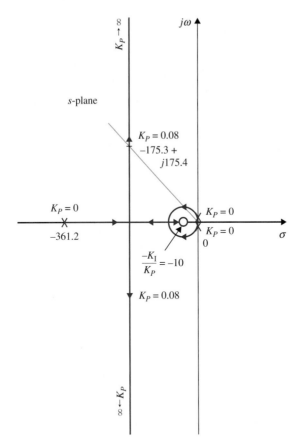

Figure 9-19 Root loci of Eq. (9-37) with $K_I/K_P = 10$; K_P varies.

The reason for this is that when we "stand" at the root at $-175.3 + j175.4$ and "look" toward the neighborhood near the origin, we see that the zero at $s = -10$ is relatively close to the origin and, thus, practically cancels one of the poles at $s = 0$. In fact, we can show that, as long as $K_P = 0.08$ and the value of K_I is chosen such that Eq. (9-38) is satisfied, the relative damping ratio of the complex roots will be very close to 0.707. For example, let us select $K_I/K_P = 5$; the three characteristic equation roots are at

$$s = -5.145, \; -178.03 + j178.03, \text{ and } -178.03 - j178.03$$

and the relative damping ratio is still 0.707. Although the real pole of the closed-loop transfer function is moved, it is very close to the zero at $s = -K_I/K_P$ so that the transient due to the real pole is negligible. For example, when $K_P = 0.08$ and $K_I = 0.4$, the closed-loop transfer function of the compensated system is

$$\frac{\Theta_y(s)}{\Theta_r(s)} = \frac{65,221.2(s+5)}{(s+5.145)(s+178.03+j178.03)(s+178.03-j178.03)} \quad (9\text{-}40)$$

Because the pole at $s = 5.145$ is very close to the zero at $s = -5$, the transient response due to this pole is negligible, and the system dynamics are essentially dominated by the two complex poles.

Toolbox 9-3-1

Root loci of Eq. (9-37) in Fig. 9-19 are obtained by the following sequence of MATLAB functions

```
KP = 0.000001; % start with a very small KP see Figure 9-19
KI=10*KP;
num = [KP KI];
den = [1 361.2 815265*KP 815265*KI];
G=tf(num,den)
rlocus(G)
```

TABLE 9-5 Attributes of the Unit-Step Responses of the System in Example 9-3-1 with PI Controller

K_I/K_P	K_I	K_P	Maximum Overshoot (%)	t_r (sec)	t_s (sec)
0	0	1.00	52.7	0.00135	0.015
20	1.60	0.08	15.16	0.0074	0.049
10	0.80	0.08	9.93	0.0078	0.0294
5	0.40	0.08	7.17	0.0080	0.023
2	0.16	0.08	5.47	0.0083	0.0194
1	0.08	0.08	4.89	0.0084	0.0114
0.5	0.04	0.08	4.61	0.0084	0.0114
0.1	0.008	0.08	4.38	0.0084	0.0115

Table 9-5 gives the attributes of the unit-step responses of the system with PI control for various values of K_I/K_P, with $K_P = 0.08$, which corresponds to a relative damping ratio of 0.707.

The results in Table 9-5 verify the fact that PI control reduces the overshoot but at the expense of longer rise time. For $K_I \leq 1$, the settling times in Table 9-5 actually show a sharp reduction, which is misleading. This is because the settling times for these cases are measured at the points where the response enters the band between 0.95 and 1.00, since the maximum overshoots are less than 5%.

The maximum overshoot of the system can still be reduced further than those shown in Table 9-5 by using smaller values of K_P than 0.08. However, the rise time and settling time will be excessive. For example, with $K_P = 0.04$ and $K_I = 0.04$, the maximum overshoot is 1.1%, but the rise time is increased to 0.0182 seconds, and the settling time is 0.024 seconds.

For the system considered, improvement on the maximum overshoot slows down for K_I less than 0.08, unless K_P is also reduced. As mentioned earlier, the value of the capacitor C_2 is inversely proportional to K_I. Thus, for practical reasons, there is a lower limit on the value of K_I.

Fig. 9-20 shows the unit-step responses of the attitude-control system with PI control, with $K_P = 0.08$ and several values of K_P. The unit-step response of the same system with the PD controller designed in Example 9-2-1, with $K_P = 1$ and $K_D = 0.00177$, is also plotted in the same figure as a comparison.

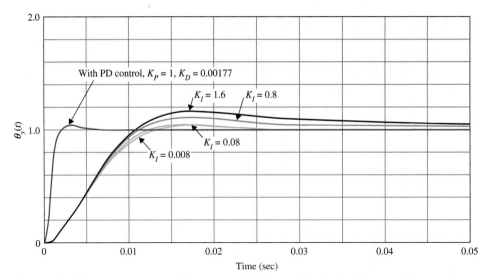

Figure 9-20 Unit-step responses of the system in Example 9-3-1 with PI control. Also, unit-step response of the system in Example 9-2-1 with PD controller.

Frequency-Domain Design

The forward-path transfer function of the uncompensated system is obtained by setting $K_P = 1$ and $K_I = 0$ in the $G(s)$ in Eq. (9-34), and the Bode plot is shown in Fig. 9-21. The phase margin is 22.68°, and the gain-crossover frequency is 868 rad/sec.

Toolbox 9-3-2

Fig. 9-20 is obtained by the following sequence of MATLAB functions

```
K=10;
num =[4500*K];
den = [1 361.2 0];
tf(num,den);
[numCL,denCL]=cloop(num,den);
step(numCL,denCL)
hold on

KI = [ 1.6 0.8 ];
KP=KI/5;
K=100;

for i = 1:length(KI)
    num = [4500*K*KP(i) 4500*K*KI(i)/KP(i)];
    den =[1 361.2 0 0];
    tf(num,den);
    [numCL,denCL]=cloop(num,den);
    step(numCL,denCL)
    hold on
end
axis([0 0.05 0 2])
```

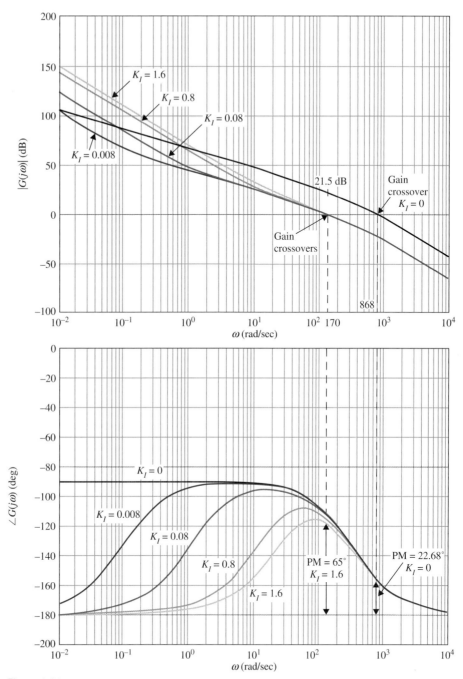

Figure 9-21 Bode plots of the control system in Example 9-3-1 with PI controller.
$$G(s) = \frac{815,265 K_P(s + K_I/K_P)}{s^2(s + 361.2)}.$$

> 522 ▶ Chapter 9. Design of Control Systems

Toolbox 9-3-3

Bode plots of the control system in Example 9-3-1. Fig. 9-21 is obtained by the following sequence of MATLAB functions

```
KI = [ 0 1.6 0.8 0.08 0.008];
KP=0.08;
K=1;

for i = 1:length(KI)
    num = [815265*KP 815265*KI(i)];
    den =[1 361.2 0 0];
    bode(num,den)
    hold on
end
grid
```

Let us specify that the required phase margin should be at least 65°, and this is to be achieved with the PI controller of Eq. (9-30). Following the procedure outlined earlier in Eqs. (9-31) through (9-33) on the design of the PI controller, we conduct the following steps:

1. Look for the new gain-crossover frequency ω'_g at which the phase margin of 65° is realized. From Fig. 9-21, ω'_g is found to be 170 rad/sec. The magnitude of $G(j\omega)$ at this frequency is 21.5 dB. Thus, the PI controller should provide an attenuation of -21.5 dB at $\omega'_g = 170$ rad/sec. Substituting $\left|G\left(j\omega'_g\right)\right| = 21.5$ dB into Eq. (9-32), and solving for K_P, we get

$$K_P = 10^{-|G(j\omega'_g)|_{dB}/20} = 10^{-21.5/20} = 0.084 \qquad (9\text{-}41)$$

Notice that, in the time-domain design conducted earlier, K_P was selected to be 0.08 so that the relative damping ratio of the complex characteristic equation roots will be approximately 0.707. (Perhaps we have cheated a little by selecting the desired phase margin to be 65°. This could not be just a coincidence. Can you believe that we have had no prior knowledge that, in this case, $\zeta = 0.707$ corresponds to PM = 65°?)

2. Let us choose $K_P = 0.08$, so that we can compare the design results of the frequency domain with those of the time-domain design obtained earlier. Eq. (9-33) gives the general guideline of finding K_I once K_P is determined. Thus,

$$K_I = \frac{\omega'_g K_P}{10} = \frac{170 \times 0.08}{10} = 1.36 \qquad (9\text{-}42)$$

As pointed out earlier, the value of K_I is not rigid, as long as the ratio K_I/K_P is sufficiently smaller than the magnitude of the pole of $G(s)$ at -361.2. As it turns out, the value of K_I given by Eq. (9-42) is not sufficiently small for this system.

The Bode plots of the forward-path transfer function with $K_P = 0.08$ and $K_I = 0$, 0.008, 0.08, 0.8, and 1.6 are shown in Fig. 9-21. Table 9-6 shows the frequency-domain

TABLE 9-6 Frequency-Domain Performance Data of the System in Example 9-3-1 with PI Controller

K_I/K_P	K_I	K_P	GM (dB)	PM (deg)	M_r	BW (rad/sec)	Gain CO (rad/sec)	Phase CO (rad/sec)
0	0	1.00	∞	22.6	2.55	1390.87	868	∞
20	1.6	0.08	∞	58.45	1.12	268.92	165.73	∞
10	0.8	0.08	∞	61.98	1.06	262.38	164.96	∞
5	0.4	0.08	∞	63.75	1.03	258.95	164.77	∞
1	0.08	0.08	∞	65.15	1.01	256.13	164.71	∞
0.1	0.008	0.08	∞	65.47	1.00	255.49	164.70	∞

properties of the uncompensated system and the compensated system with various values of K_I. Notice that, for values of K_I/K_P that are sufficiently small, the phase margin, M_r, BW, and gain-crossover frequency all vary little.

It should be noted that the phase margin of the system can be improved further by reducing the value of K_P below 0.08. However, the bandwidth of the system will be further reduced. For example, for $K_P = 0.04$ and $K_I = 0.04$, the phase margin is increased to 75.7°, and $M_r = 1.01$, but BW is reduced to 117.3 rad/sec.

▶ **EXAMPLE 9-3-2** Now let us consider using the PI control for the third-order attitude control system described by Eq. (9-19). First, the time-domain design is carried out as follows. You may use **ACSYS** to solve this problem. ◀

Time-Domain Design

Let the time-domain specifications be as follows:

Steady-state error due to the parabolic input $t^2 u_s(t)/2 \leq 0.2$

Maximum overshoot $\leq 5\%$

Rise time $t_r \leq 0.01$ sec

Settling time $t_s \leq 0.02$ sec

These are identical to the specifications given for the second-order system in Example 9-3-1.

Applying the PI controller of Eq. (9-24), the forward-path transfer function of the system becomes

$$G(s) = G_c(s)G_p(s) = \frac{1.5 \times 10^9 K K_P(s + K_I/K_P)}{s^2(s^2 + 3408.3s + 1,204,000)} \\ = \frac{1.5 \times 10^9 K K_P(s + K_I/K_P)}{s^2(s + 400.26)(s + 3008)} \quad (9\text{-}43)$$

We can show that the steady-state error of the system due to the parabolic input is again given by Eq. (9-36), and, arbitrarily setting $K = 181.17$, the minimum value of K_I is 0.002215.

The characteristic equation of the closed-loop system with $K = 181.17$ is

$$s^4 + 3408.3s^3 + 1,204,000s^2 + 2.718 \times 10^9 K_P s + 2.718 \times 10^9 K_I = 0 \quad (9\text{-}44)$$

The Routh's tabulation of the last equation is performed as follows:

s^4	1	$1,204,000$	$2.718 \times 10^9 K_I$
s^3	3408.3	$2.718 \times 10^9 K_P$	0
s^2	$1,204,000 - 797465 K_P$	$2.718 \times 10^9 K_I$	0
s^1	$\dfrac{1,204,000 K_P - 797465 K_P^2 - 3408.3 K_I}{1,204,000 - 797465 K_P}$	0	
s^0	$2.718 \times 10^9 K_I$	0	

The stability requirements are

$$K_I > 0$$
$$K_P < 1.5098 \qquad (9\text{-}45)$$
$$K_I < 353.255 K_P - 233.98 K_P^2$$

The design of the PI controller calls for the selection of a small value for K_I/K_P, relative to the nearest pole of $G(s)$ to the origin, which is at -400.26. The root loci of Eq. (9-44) are plotted using the pole–zero configuration of Eq. (9-43). Fig. 9-22(a) shows the root loci as K_P

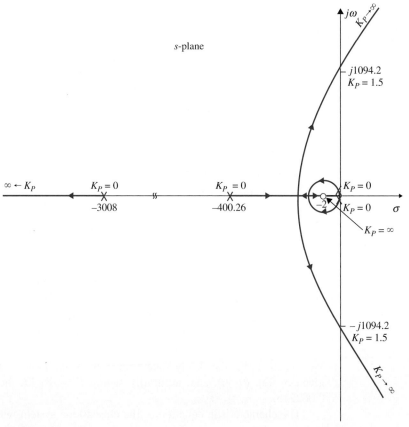

Figure 9-22 (a) Root loci of the control system in Example 9-3-2 with PI controller $K_I/K_P = 2$, $0 \leq K_P < \infty$.

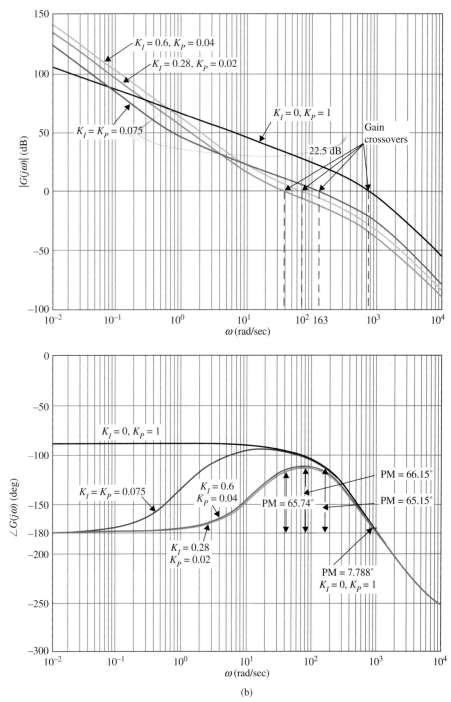

Figure 9-22 (b) Bode plots of the control system in Example 9-3-2 with PI control.

varies for $K_I/K_P = 2$. The root loci near the origin due to the pole and zero of the PI controller again form a small loop, and the root loci at a distance away from the origin will be very similar to those of the uncompensated system, which are shown in Fig. 5-34. By selecting the value of K_P properly along the root loci, it may be possible to satisfy the performance specifications

TABLE 9-7 Attributes of the Unit-Step Responses of the System in Example 9-3-2 with PI Controller

K_I/K_P	K_I	K_P	Maximum Overshoot (%)	t_r (sec)	t_s (sec)	Roots of Characteristic Equation			
0	0	1	76.2	0.00158	0.0487	−3293.3	−57.5	±j906.6	
20	1.6	0.08	15.6	0.0077	0.0471	−3035	−22.7	−175.3	±j180.3
20	0.8	0.04	15.7	0.0134	0.0881	−3021.6	−259	−99	−28
5	0.4	0.08	6.3	0.00883	0.0202	−3035	−5.1	−184	±j189.2
2	0.08	0.04	2.1	0.02202	0.01515	−3021.7	−234.6	−149.9	−2
5	0.2	0.04	4.8	0.01796	0.0202	−3021.7	−240	−141.2	−5.3
2	0.16	0.08	5.8	0.00787	0.01818	−3035.2	−185.5	±j190.8	−2
1	0.08	0.08	5.2	0.00792	0.01616	−3035.2	−186	±j191.4	−1
2	0.15	0.075	4.9	0.0085	0.0101	−3033.5	−187.2	±j178	−1
2	0.14	0.070	4.0	0.00917	0.01212	−3031.8	−187.2	±j164	−1

given above. To minimize the rise time and settling time, we should select K_P so that the dominant roots are complex conjugate. Table 9-7 gives the performance attributes of several combinations of K_I/K_P and K_P. Notice that, although several combinations of these parameters correspond to systems that satisfy the performance specifications, the one with $K_P = 0.075$ and $K_I = 0.15$ gives the best rise and settling times among those shown.

Frequency-Domain Design

The Bode plot of Eq. (9-43) for $K = 181.17$, $K_P = 1$, and $K_I = 0$ is shown in Fig. 9-22(b). The performance data of the uncompensated system are as follows:

Gain margin = 3.578 dB

Phase margin = 7.788°

$M_r = 6.572$

BW = 1378 rad/sec

Let us require that the compensated system has a phase margin of at least 65°, and this is to be achieved with the PI controller of Eq. (9-30). Following the procedure outlined in Eqs. (9-31) through (9-33) on the design of the PI controller, we carry out the following steps.

1. Look for the new gain-crossover frequency ω'_g at which the phase margin of 65° is realized. From Fig. 9-20, ω'_g is found to be 163 rad/sec, and the magnitude of $G(j\omega)$ at this frequency is 22.5 dB. Thus, the PI controller should provide an attenuation of −22.5 dB at $\omega'_g = 163$ rad/sec. Substituting $\left|G(j\omega'_g)\right| = 22.5$ dB into Eq. (9-32), and solving for K_P, we get

$$K_P = 10^{-|G(j\omega'_g)|_{dB}/20} = 10^{-22.5/20} = 0.075 \quad (9-46)$$

This is exactly the same result that was selected for the time-domain design that resulted in a system with a maximum overshoot of 4.9% when $K_I = 0.15$, or $K_I/K_P = 2$.

2. The suggested value of K_I is found from Eq. (9-33):

$$K_I = \frac{\omega'_g K_P}{10} = \frac{163 \times 0.075}{10} = 1.222 \quad (9-47)$$

Toolbox 9-3-4

Bode plots of the control system in Example 9-3-2. Fig. 9-22(b) is obtained by the following sequence of MATLAB functions

```
KI = [ 0.6 0.28 0.075 0];
KP = [0.04 0.02 0.075 1];
K=1;

for i = 1:length(KI)
    num = [1.5e9*KP(i) 1.5e9*KI(i)];
    den =[1 3408.3 1204000 0 0];
    bode(num,den)
    hold on
end
grid
axis([0.01 10000 -270 0]);
```

Thus, $K_I/K_P = 16.3$. However, the phase margin of the system with these design parameters is only 59.52.

To realize the desired PM of 65°, we can reduce the value of K_P or K_I. Table 9-8 gives the results of several designs with various combinations of K_P and K_I. Notice that the last three designs in the table all satisfy the PM requirements. However, the design ramifications show the following:

Reducing K_P would reduce BW and increase M_r.

Reducing K_I would increase the capacitor value in the implementing circuit.

In fact, only the $K_I = K_P = 0.075$ case gives the best all-around performance in both the frequency domain and the time domain. In attempting to increase K_I, the maximum overshoot becomes excessive. This is one example that shows the inadequacy of specifying phase margin only. The purpose of this example is to bring out the properties of the PI controller and the important considerations in its design. No details are explored further.

Fig. 9-23 shows the unit-step responses of the uncompensated system and several systems with PI control.

TABLE 9-8 Performance Summary of the System in Example 9-3-2 with PI Controller

K_I/K_P	K_I	K_P	GM (dB)	PM (deg)	M_r	BW (rad/sec)	Maximum Overshoot (%)	t_r (sec)	t_s (sec)
0	0	1	3.578	7.788	6.572	1378	77.2	0.0015	0.0490
16.3	1.222	0.075	25.67	59.52	1.098	264.4	13.1	0.0086	0.0478
1	0.075	0.075	26.06	65.15	1.006	253.4	4.3	0.0085	0.0116
15	0.600	0.040	31.16	66.15	1.133	134.6	12.4	0.0142	0.0970
14	0.280	0.020	37.20	65.74	1.209	66.34	17.4	0.0268	0.1616

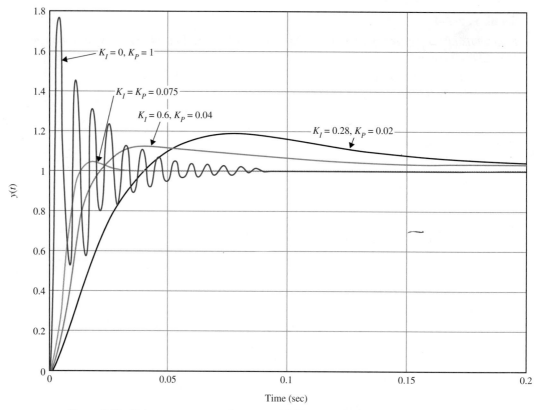

Figure 9-23 Unit-step response of system with PI controller in Example 9-3-2.

▶ 9-4 DESIGN WITH THE PID CONTROLLER

From the preceding discussions, we see that the PD controller could add damping to a system, but the steady-state response is not affected. The PI controller could improve the relative stability and improve the steady-state error at the same time, but the rise time is increased. This leads to the motivation of using a PID controller so that the best features of each of the PI and PD controllers are utilized. We can outline the following procedure for the design of the PID controller.

1. Consider that the PID controller consists of a PI portion connected in cascade with a PD portion. The transfer function of the PID controller is written as

$$G_c(s) = K_P + K_D s + \frac{K_I}{s} = (1 + K_{D1}s)\left(K_{P2} + \frac{K_{I2}}{s}\right) \quad (9\text{-}48)$$

The proportional constant of the PD portion is set to unity, since we need only three parameters in the PID controller. Equating both sides of Eq. (9-48), we have

$$K_P = K_{P2} + K_{D1}K_{I2} \quad (9\text{-}49)$$

$$K_D = K_{D1}K_{P2} \quad (9\text{-}50)$$

$$K_I = K_{I2} \quad (9\text{-}51)$$

2. Consider that the PD portion only is in effect. Select the value of K_{D1} so that a portion of the desired relative stability is achieved. In the time domain, this relative stability may be measured by the maximum overshoot, and in the frequency domain it is the phase margin.

3. Select the parameters K_{I2} and K_{P2} so that the total requirement on relative stability is satisfied.

Toolbox 9-4-1

Fig. 9-23 is obtained by the following sequence of MATLAB functions

```
KI = [0 0.6 0.28 0.075];
KP = [1 0.04 0.02 0.075];
K=1;
t = 0:0.0001:0.2;
for i = 1:length(KI)
    num = [1.5e9*KP(i) 1.5e9*KI(i)];
    den =[1 3408.3 1204000 0 0];
    tf(num,den);
    [numCL,denCL]=cloop(num,den);
    step(numCL,denCL,t)
    hold on
end
grid
axis([0 0.2 0 1.8])
```

As an alternative, the PI portion of the controller can be designed first for a portion of the requirement on relative stability, and, finally, the PD portion is designed.

The following example illustrates how the PID controller is designed in the time domain and the frequency domain.

▶ **EXAMPLE 9-4-1** Consider the third-order attitude control system represented by the forward-path transfer function given in Eq. (9-19). With $K = 181.17$, the transfer function is

$$G_p(s) = \frac{2.718 \times 10^9}{s(s + 400.26)(s + 3008)} \quad (9\text{-}52)$$

You may use **ACSYS** to solve this problem; see Section 9-19. ◀

Time-Domain Design

Let the time-domain performance specifications be as follows:

Steady-state error due to a ramp input $t^2 u_s(t)/2 \leq 0.2$

Maximum overshoot $\leq 5\%$

Rise time $t_r \leq 0.005$ sec

Settling time $t_s \leq 0.005$ sec

We realize from the previous examples that these requirements cannot be fulfilled by either the PI or PD control acting alone. Let us apply the PD control with the transfer function $(1 + K_{D1}s)$. The forward-path transfer function becomes

$$G(s) = \frac{2.718 \times 10^9 (1 + K_{D1}s)}{s(s + 400.26)(s + 3008)} \quad (9\text{-}53)$$

TABLE 9-9 Time-Domain Performance Characteristics of Third-Order Attitude Control System with PID Controller Designed in Example 9-4-1

K_{P2}	Maximum Overshoot (%)	t_r (sec)	t_s (sec)	Roots of Characteristic Equation			
1.0	11.1	0.00088	0.0025	−15.1	−533.2	−1430 ± j	1717.5
0.9	10.8	0.00111	0.00202	−15.1	−538.7	−1427 ± j	1571.8
0.8	9.3	0.00127	0.00303	−15.1	−546.5	−1423 ± j	1385.6
0.7	8.2	0.00130	0.00303	−15.1	−558.4	−1417 ± j	1168.7
0.6	6.9	0.00155	0.00303	−15.2	−579.3	−1406 ± j	897.1
0.5	5.6	0.00172	0.00404	−15.2	−629	−1382 ± j	470.9
0.4	5.1	0.00214	0.00505	−15.3	−1993	−700 ± j	215.4
0.3	4.8	0.00271	0.00303	−15.3	−2355	−519 ± j	263.1
0.2	4.5	0.00400	0.00404	−15.5	−2613	−390 ± j	221.3
0.1	5.6	0.00747	0.00747	−16.1	−284	−284 ± j	94.2
0.08	6.5	0.00895	0.04545	−16.5	−286.3	−266 ± j	4.1

Table 9-3 shows that the best PD controller that can be obtained from the maximum overshoot standpoint is with $K_{D1} = 0.002$, and the maximum overshoot is 11.37%. The rise time and settling time are well within the required values. Next, we add the PI controller, and the forward-path transfer function becomes

$$G(s) = \frac{5.436 \times 10^6 K_{P2}(s + 500)(s + K_{I2}/K_{P2})}{s^2(s + 400.26)(s + 3008)} \quad (9\text{-}54)$$

Following the guideline of choosing a relatively small value for K_{I2}/K_{P2}, we let $K_{I2}/K_{P2} = 15$. Eq. (9-54) becomes

$$G(s) = \frac{5.436 \times 10^6 K_{P2}(s + 500)(s + 15)}{s^2(s + 400.26)(s + 3008)} \quad (9\text{-}55)$$

Table 9-9 gives the time-domain performance characteristics along with the roots of the characteristic equation for various values of K_{P2}. Apparently, the optimal value of K_{P2} is in the neighborhood of between 0.2 and 0.4.

Selecting $K_{P2} = 0.3$, and with $K_{D1} = 0.002$ and $K_{I2} = 15K_{P2} = 4.5$, the following results are obtained for the parameters of the PID controller using Eqs. (9-49) through (9-51):

$$K_I = K_{I2} = 4.5$$
$$K_P = K_{P2} + K_{D1}K_{I2} = 0.3 + 0.002 \times 4.5 = 0.309 \quad (9\text{-}56)$$
$$K_D = K_{D1}K_{P2} = 0.002 \times 0.3 = 0.0006$$

Notice that the PID design resulted in a smaller K_D and a larger K_I, which correspond to smaller capacitors in the implementing circuit.

Fig. 9-24 shows the unit-step responses of the system with the PID controller, as well as those with PD and PI controls designed in Examples 9-2-2 and 9-3-2, respectively. Notice that the PID control, when designed properly, captures the advantages of both the PD and the PI controls.

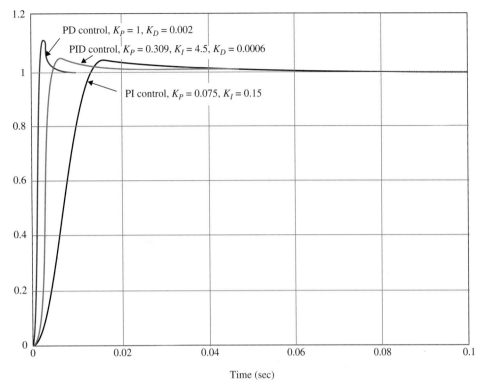

Figure 9-24 Step responses of the system in Example 9-4-1 with PD, PI, and PID controllers.

Frequency-Domain Design

The PD control of the third-order attitude control systems was already carried out in Example 9-2-2, and the results were tabulated in Table 9-3. When $K_P = 1$ and $K_D = 0.002$, the maximum overshoot is 11.37%, but this is the best that the PD control could offer. Using this PD controller, the forward-path transfer function of the system is

$$G(s) = \frac{2.718 \times 10^9 (1 + 0.002s)}{s(s + 400.26)(s + 3008)} \qquad (9\text{-}57)$$

and its Bode plot is shown in Fig. 9-25. Let us estimate that the following set of frequency-domain criteria corresponds to the time-domain specifications given in this problem.

Phase margin $\geq 70°$

$M_r \leq 1.1$

BW ≥ 1.000 rad/sec

From the Bode diagram in Fig. 9-25, we see that, to achieve a phase margin of 70°, the new phase-crossover frequency should be $\omega'_g = 811$ rad/sec, at which the magnitude of $G(j\omega)$ is 7 dB. Thus, using Eq. (9-32), the value of K_{P2} is calculated to be

$$K_{P2} = 10^{-7/20} = 0.45 \qquad (9\text{-}58)$$

Notice that the desirable range of K_{P2} found from the time-domain design with $K_{I2}/K_{P2} = 15$ is from 0.2 to 0.4. The result given in Eq. (9-58) is slightly out of the range. Table 9-10 shows the frequency-domain performance results with $K_D = 0.002$, $K_{I2}/K_{P2} = 15$, and

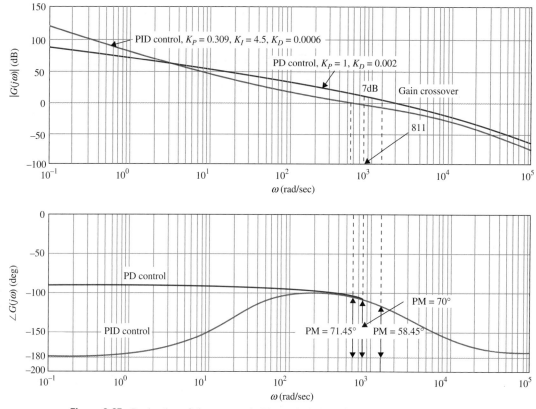

Figure 9-25 Bode plot of the system in Example 9-4-1 with PD and PID controllers.

TABLE 9-10   Frequency-Domain Performance of System in Example 9-4-1 with PID Controller

K_{P2}	K_{I2}	GM (dB)	PM (deg)	M_r	BW (rad/sec)	t_r (sec)	t_s (sec)	Maximum Overshoot (%)
1.00	0	∞	58.45	1.07	2607	0.0008	0.00255	11.37
0.45	6.75	∞	68.5	1.03	1180	0.0019	0.0040	5.6
0.40	6.00	∞	69.3	1.027	1061	0.0021	0.0050	5.0
0.30	4.50	∞	71.45	1.024	1024	0.0027	0.00303	4.8
0.20	3.00	∞	73.88	1.031	528.8	0.0040	0.00404	4.5
0.10	1.5	∞	76.91	1.054	269.5	0.0076	0.0303	5.6
0.08	1.2	∞	77.44	1.065	216.9	0.0092	0.00469	6.5

several values of K_{P2} starting with 0.45. It is interesting to note that, as K_{P2} continues to decrease, the phase margin increases monotonically, but below $K_{P2} = 0.2$, the maximum overshoot actually increases. In this case, the phase margin results are misleading, but the resonant peak M_r is a more accurate indication of this.

▶ 9-5 DESIGN WITH PHASE-LEAD CONTROLLER

The PID controller and its components in the form of PD and PI controls represent simple forms of controllers that utilize derivative and integration operations in the compensation of control systems. In general, we can regard the design of controllers of control systems as

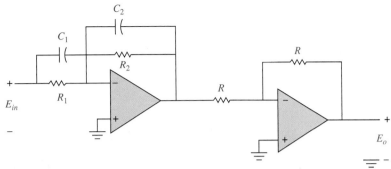

Figure 9-26 Op-amp circuit implementation of $G(s) = K_c \dfrac{s+z_1}{s+p_1}$.

a filter design problem; then there are a large number of possible schemes. From the filtering standpoint, the PD controller is a high-pass filter, the PI controller is a low-pass filter, and the PID controller is a band-pass or band-attenuate filter, depending on the values of the controller parameters. The high-pass filter is often referred to as a **phase-lead controller**, because positive phase is introduced to the system over some frequency range. The low-pass filter is also known as a **phase-lag controller**, because the corresponding phase introduced is negative. These ideas related to filtering and phase shifts are useful if designs are carried out in the frequency domain.

The transfer function of a simple lead or lag controller is expressed as

$$G_c(s) = K_c \frac{s+z_1}{s+p_1} \qquad (9\text{-}59)$$

where the controller is high-pass or phase-lead if $p_1 > z_1$, and low-pass or phase-lag if $p_1 < z_1$.

The op-amp circuit implementation of Eq. (9-59) is given in Table 4-4(g) of Chapter 4 and is repeated in Fig. 9-26 with an inverting amplifier. The transfer function of the circuit is

$$G_c(s) = \frac{E_o(s)}{E_{in}(s)} = \frac{C_1}{C_2} \frac{s + \dfrac{1}{R_1 C_1}}{s + \dfrac{1}{R_2 C_2}} \qquad (9\text{-}60)$$

Comparing the last two equations, we have

$$\begin{aligned} K_c &= C_1/C_2 \\ z_1 &= 1/R_1 C_1 \\ p_1 &= 1/R_2 C_2 \end{aligned} \qquad (9\text{-}61)$$

We can reduce the number of design parameters from four to three by setting $C = C_1 = C_2$. Then Eq. (9-60) is written as

$$\begin{aligned} G_c(s) &= \frac{R_2}{R_1} \left(\frac{1 + R_1 C s}{1 + R_2 C s} \right) \\ &= \frac{1}{a} \left(\frac{1 + aTs}{1 + Ts} \right) \end{aligned} \qquad (9\text{-}62)$$

where

$$a = \frac{R_1}{R_2} \tag{9-63}$$

$$T = R_2 C \tag{9-64}$$

9-5-1 Time-Domain Interpretation and Design of Phase-Lead Control

In this section, we shall first consider that Eqs. (9-60) and (9-62) represent a phase-lead controller ($z_1 < p_1$ or $a > 1$). In order that the phase-lead controller will not degrade the steady-state error, the factor a in Eq. (9-62) should be absorbed by the forward-path gain K. Then, for design purposes, $G_c(s)$ can be written as

$$G_c(s) = \frac{1 + aTs}{1 + Ts} \quad (a > 1) \tag{9-65}$$

The pole–zero configuration of Eq. (9-65) is shown in Fig. 9-27. Based on the discussions given in Chapter 7 on the effects of adding a pole–zero pair (with the zero closer to the origin) to the forward-path transfer function, the phase-lead controller can improve the stability of the closed-loop system if its parameters are chosen properly. The design of phase-lead control is essentially that of placing the pole and zero of $G_c(s)$ so that the design specifications are satisfied. The root-contour method can be used to indicate the proper ranges of the parameters. The **ACSYS** MATLAB tool can be used to speed up the cut-and-try procedure considerably. The following guidelines can be made with regard to the selection of the parameters a and T.

1. Moving the zero $-1/aT$ toward the origin should improve rise time and settling time. If the zero is moved too close to the origin, the maximum overshoot may again increase, because $-1/aT$ also appears as a zero of the closed-loop transfer function.

2. Moving the pole at $-1/T$ farther away from the zero and the origin should reduce the maximum overshoot, but if the value of T is too small, rise time and settling time will again increase.

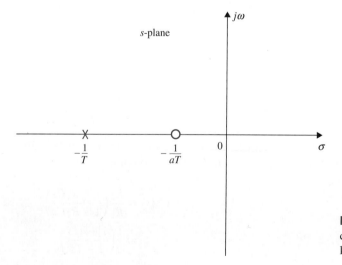

Figure 9-27 Pole–zero configuration of the phase-lead controller.

We can make the following general statements with respect to the effects of phase-lead control on the time-domain performance of a control system:

1. When used properly, it can increase damping of the system.
2. It improves rise time and settling time.
3. In the form of Eq. (9-65), phase-lead control does not affect the steady-state error, because $G_c(0) = 1$.

9-5-2 Frequency-Domain Interpretation and Design of Phase-Lead Control

The Bode plot of the phase-lead controller of Eq. (9-65) is shown in Fig. 9-28. The two corner frequencies are at $\omega = 1/aT$ and $\omega = 1/T$. The maximum value of the phase, ϕ_m, and the frequency at which it occurs, ω_m, are derived as follows. Because ω_m is the geometric mean of the two corner frequencies, we write

$$\log_{10}\omega_m = \frac{1}{2}\left(\log_{10}\frac{1}{aT} + \log_{10}\frac{1}{T}\right) \tag{9-66}$$

Thus,

$$\omega_m = \frac{1}{\sqrt{a}T} \tag{9-67}$$

To determine the maximum phase ϕ_m, the phase of $G_c(j\omega)$ is written

$$\angle G_c(j\omega) = \phi(j\omega) = \tan^{-1}\omega aT - \tan^{-1}\omega T \tag{9-68}$$

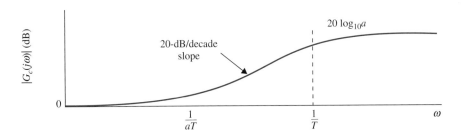

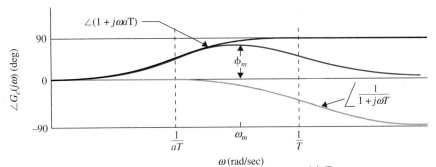

Figure 9-28 Bode plot of phase-lead controller $G_c(s) = a\dfrac{s + 1/aT}{s + 1/T}$ $a > 1$.

from which we get

$$\tan\phi(j\omega) = \frac{\omega aT - \omega T}{1 + (\omega aT)(\omega T)} \qquad (9\text{-}69)$$

Substituting Eq. (9-67) into Eq. (9-69), we have

$$\tan\phi_m = \frac{a - 1}{2\sqrt{a}} \qquad (9\text{-}70)$$

or

$$\sin\phi_m = \frac{a - 1}{a + 1} \qquad (9\text{-}71)$$

Thus, by knowing ϕ_m, the value of a is determined from

$$a = \frac{1 + \sin\phi_m}{1 - \sin\phi_m} \qquad (9\text{-}72)$$

The relationship between the phase ϕ_m and a and the general properties of the Bode plot of the phase-lead controller provide an advantage of designing in the frequency domain. The difficulty is, of course, in the correlation between the time-domain and frequency-domain specifications. The general outline of phase-lead controller design in the frequency domain is given as follows. It is assumed that the design specifications simply include steady-state error and phase-margin requirements.

1. The Bode diagram of the uncompensated process $G_p(j\omega)$ is constructed with the gain constant K set according to the steady-state error requirement. The value of K has to be adjusted upward once the value of a is determined.

2. The phase margin and the gain margin of the uncompensated system are determined, and the additional amount of phase lead needed to realize the phase margin is determined. From the additional phase lead required, the desired value of ϕ_m is estimated accordingly, and the value of a is calculated from Eq. (9-72).

3. Once a is determined, it is necessary only to determine the value of T, and the design is in principle completed. This is accomplished by placing the corner frequencies of the phase-lead controller, $1/aT$ and $1/T$, such that ϕ_m is located at the new gain-crossover frequency ω'_g, so the phase margin of the compensated system is benefited by ϕ_m. It is known that the high-frequency gain of the phase-lead controller is $20 \log_{10} a$ dB. Thus, to have the new gain crossover at ω_m, which is the geometric mean of $1/aT$ and $1/T$, we need to place ω_m at the frequency where the magnitude of the uncompensated $G_p(j\omega)$ is $-10 \log_{10} a$ dB so that adding the controller gain of $10 \log_{10} a$ dB to this makes the magnitude curve go through 0 dB at ω_m.

4. The Bode diagram of the forward-path transfer function of the compensated system is investigated to check that all performance specifications are met; if not, a new value of ϕ_m must be chosen and the steps repeated.

5. If the design specifications are all satisfied, the transfer function of the phase-lead controller is established from the values of a and T.

9-5 Design with Phase-Lead Controller ◀ 537

If the design specifications also include M_r and/or BW, then these must be checked using either the Nichols chart or the output data from a computer program.

We use the following example to illustrate the design of the phase-lead controller in the time domain and frequency domain.

▶ **EXAMPLE 9-5-1** The block diagram of the sun-seeker control system described in Section 4-11 is again shown in Fig. 9-29. The system may be mounted on a space vehicle so that it will track the sun with high accuracy. The variable θ_r represents the reference angle of the solar ray, and θ_o denotes the vehicle axis. The objective of the sun-seeker system is to maintain the error α between θ_r and θ_o near zero. The parameters of the system are as follows:

$$R_F = 10,000\ \Omega \qquad K_b = 0.0125\ \text{V/rad/sec}$$
$$K_i = 0.0125\ \text{N-m/A} \qquad R_a = 6.25\ \Omega$$
$$J = 10^{-6}\ \text{kg-m}^2 \qquad K_s = 0.1\ \text{A/rad}$$
$$K = \text{to be determined} \qquad B = 0$$
$$n = 800$$

The forward-path transfer function of the uncompensated system is

$$G_p(s) = \frac{\Theta_o(s)}{A(s)} = \frac{K_s R_F K K_i / n}{R_a J s^2 + K_i K_b s} \tag{9-73}$$

where $\Theta_o(s)$ and $A(s)$ are the Laplace transforms of $\theta_o(t)$ and $\alpha(t)$, respectively.

Substituting the numerical values of the system parameters in Eq. (9-73), we get

$$G_p(s) = \frac{\Theta_o(s)}{A(s)} = \frac{2500K}{s(s+25)} \tag{9-74}$$

You may use **ACSYS** to solve this problem after reducing the block diagram in Fig. 9-29 to a standard form. See Section 9-15.

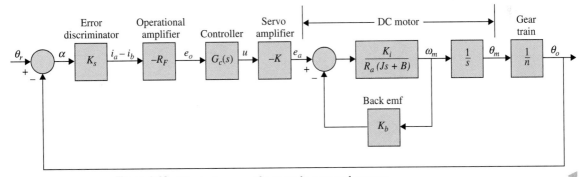

Figure 9-29 Block diagram of sun-seeker control system.

Time-Domain Design
The time-domain specifications of the system are as follows:

1. The steady-state error of $\alpha(t)$ due to a unit-ramp function input for $\theta_r(t)$ should be ≤ 0.01 rad per rad/sec of the final steady-state output velocity. In other words, the steady-state error due to a ramp input should be $\leq 1\%$.
2. The maximum overshoot of the step response should be less than 5% or as small as possible.
3. Rise time $t_r \leq 0.02$ sec.
4. Settling time $t_s \leq 0.02$ sec.

The minimum value of the amplifier gain, K, is determined initially from the steady-state requirement. Applying the final-value theorem to $\alpha(t)$, we have

$$\lim_{t \to \infty} \alpha(t) = \lim_{s \to 0} sA(s) = \lim_{s \to 0} \frac{s\Theta_r(s)}{1 + G_p(s)} \tag{9-75}$$

For a unit-ramp input, $\Theta_r(s) = 1/s^2$. By using Eq. (9-74), Eq. (9-75) leads to

$$\lim_{t \to \infty} \alpha(t) = \frac{0.01}{K} \tag{9-76}$$

Thus, for the steady-state value of $\alpha(t)$ to be ≤ 0.01, K must be ≥ 1. Let us set $K = 1$, the worst case from the steady-state error standpoint. The characteristic equation of the uncompensated system is

$$s^2 + 25s + 2500 = 0 \tag{9-77}$$

We can show that the damping ratio of the uncompensated system with $K = 1$ is only 0.25, which corresponds to a maximum overshoot of 44.4%. Fig. 9-30 shows the unit-step response of the system with $K = 1$.

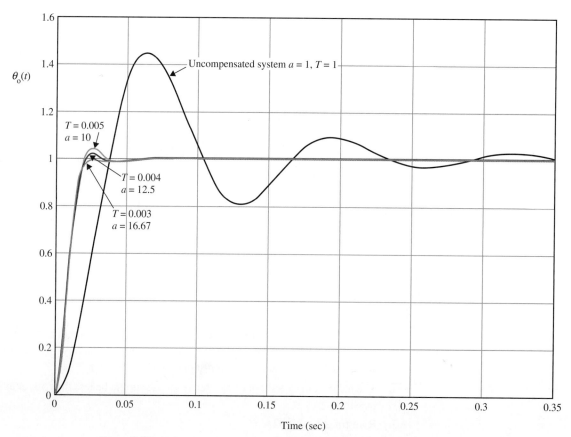

Figure 9-30 Unit-step response of sun-seeker system in Example 9-5-1.

Toolbox 9-5-1

Unit-step response for Example 9-6 in Fig. 9-30 is obtained by the following sequence of MATLAB functions

```
a = [1 10 12.5 16.67];
T = [1 0.005 0.004 0.003];

for i = 1:length(T)
    num = [2500*a(i)*T(i) 2500];
    den =[T(i) 25*T(i)+1 25 0];
    tf(num,den);
    [numCL,denCL]=cloop(num,den);
    step(numCL,denCL)
    hold on
end
grid
axis([0 0.35 0 1.8])
```

A space has been reserved in the forward path of the block diagram of Fig. 9-29 for a controller with transfer function $G_c(s)$. Let us consider using the phase-lead controller of Eq. (9-62), although in the present case, a PD controller or one of the other types of phase-lead controllers may also be effective in satisfying the performance criteria given. The forward-path transfer function of the compensated system is written

$$G(s) = \frac{2500K(1 + aTs)}{as(s + 25)(1 + Ts)} \tag{9-78}$$

For the compensated system to satisfy the steady-state error requirement, K must satisfy

$$K \geq a \tag{9-79}$$

Let us set $K = a$. The characteristic equation of the system is

$$(s^2 + 25s + 2500) + Ts^2(s + 25) + 2500aTs = 0 \tag{9-80}$$

We can use the root-contour method to show the effects of varying a and T of the phase-lead controller. Let us first set $a = 0$. The characteristic equation of Eq. (9-80) becomes

$$s^2 + 25s + 2500 + Ts^2(s + 25) = 0 \tag{9-81}$$

Dividing both sides of the last equation by the terms that do not contain T, we get

$$1 + G_{eq1}(s) = 1 + \frac{Ts^2(s + 25)}{s^2 + 25s + 2500} = 0 \tag{9-82}$$

Thus, the root contours of Eq. (9-81) when T varies are determined using the pole–zero configuration of $G_{eq1}(s)$ in Eq. (9-82). These root contours are drawn as shown in Fig. 9-31. Notice that the poles of $G_{eq1}(s)$ are the roots of the characteristic equation when $a = 0$ and $T = 0$. The root contours in Fig. 9-31 clearly show that adding the factor $(1 + Ts)$ to the denominator of Eq. (9-74) alone would not improve the system performance, since the characteristic equation roots are pushed toward the right-half plane. In fact, the system becomes unstable when T is greater than 0.0133. To achieve the full effect of the phase-lead controller, we must restore the value of a in Eq. (9-80). To prepare for the root contours

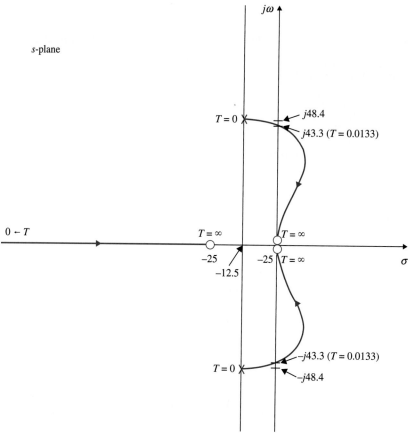

Figure 9-31 Root contours of the sun-seeker system with $a = 0$, and T varies from 0 to ∞.

with a as the variable parameter, we divide both sides of Eq. (9-80) by the terms that do not contain a, and the following equation results:

$$1 + aG_{eq2}(s) = 1 + \frac{2500aTs}{s^2 + 25s + 2500 + Ts^2(s + 25)} = 0 \qquad (9\text{-}83)$$

Toolbox 9-5-2

Root contours for Fig. 9-31 are obtained by the following sequence of MATLAB functions

```
for T = 1:1:260
    num = [T 325*T 0 0];
    den = [1 25 2500];
    tf(num,den);
    [numCL,denCL]=cloop(num,den);
    F = tf(num,den);
    PoleData(:,T)=pole(F);
end
plot(real(PoleData(1,:)),imag(PoleData(1,:)),real(PoleData(2,:)),imag(PoleData(2,:)));
```

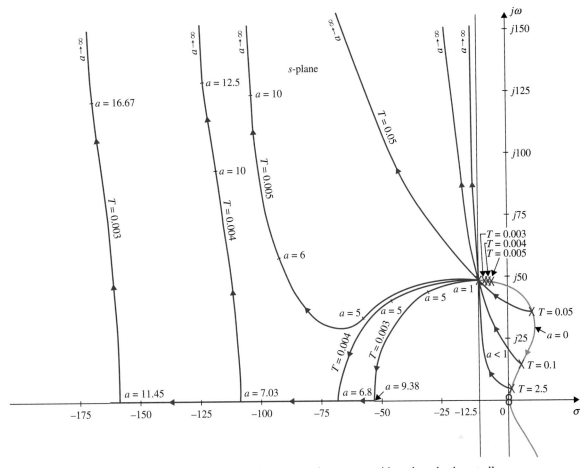

Figure 9-32 Root contours of the sun-seeker system with a phase-lead controller.

For a given T, the root contours of Eq. (9-80) when a varies are obtained based on the poles and zeros of $G_{eq2}(s)$. Notice that the poles of $G_{eq2}(s)$ are the same as the roots of Eq. (9-81). Thus, for a given T, the root contours of Eq. (9-80) when a varies must start $(a = 0)$ at points on the root contours of Fig. 9-31. These root contours end $(a = \infty)$ at $s = 0, \infty, \infty$, which are the zeros of $G_{eq2}(s)$. The complete root contours of Eq. (9-80) are now shown in Fig. 9-32 for several values of T, and a varies from 0 to ∞.

From the root contours of Fig. 9-32, we see that, for effective phase-lead control, the value of T should be small. For large values of T, the natural frequency of the system increases rapidly as a increases, and very little improvement is made on the damping of the system.

Let us choose $T = 0.01$ arbitrarily. Table 9-11 shows the attributes of the unit-step response when the value of aT is varied from 0.02 to 0.1. The **ACSYS** MATLAB tool was used for the calculations of the time responses. The results show that the smallest maximum overshoot is obtained when $aT = 0.05$, although the rise and settling times decrease continuously as aT increases. However, the smallest value of the maximum overshoot is 16.2%, which exceeds the design specification.

TABLE 9-11 Attributes of Unit-Step Response of System with Phase-Lead Controller in Example 9-5-1: $T=0.01$

aT	a	Maximum Overshoot (%)	t_r (sec)	t_s (sec)
0.02	2	26.6	0.0222	0.0830
0.03	3	18.9	0.0191	0.0665
0.04	4	16.3	0.0164	0.0520
0.05	5	16.2	0.0146	0.0415
0.06	6	17.3	0.0129	0.0606
0.08	8	20.5	0.0112	0.0566
0.10	10	23.9	0.0097	0.0485

Toolbox 9-5-3

Root contours for Fig. 9-32 are obtained by the following sequence of MATLAB functions

```
T = [0.003 0.004 0.005 0.05 0.1 2.5];
for j=1:length(T)
   i=1;
   for a = 0:0.005:30
      num = [T(j) 25*T(j)+1 25+2500*a*T(j) 2500];
      den = [T(j) 25*T(j)+1 25 2500]; [numCL,denCL]=cloop(num,den);
      F = tf(numCL,denCL); PoleData(:,i)=pole(F);
      i=i+1;
   end
   Count = i-1; %% for graph continuation
    for i=1:Count
       if imag(PoleData(1,i))~=0
          break;
   end         end
   count = i; %% for graph continuation
   for i=1:count
       PoleData(1,i)=PoleData(1,count);
   end

   for i=1:Count
       if imag(PoleData(2,i))~=0
          break;
   end         end
   count = i; %% for graph continuation
   for i=1:count
       PoleData(2,i)=PoleData(2,count);
   end
   PositivePos = 0;
   for i=1:Count
       if imag(PoleData(1,i)) < 0
          if PositivePos == 0
             PositivePos = i-1;
          end
          PoleData(1,i) = PoleData(1,PositivePos);
   End         end
   PositivePos = 0;
   for i=1:Count
       if imag(PoleData(2,i)) < 0
```

```
            if PositivePos == 0
                PositivePos = i-1;
            end
            PoleData(2,i) = PoleData(2,PositivePos);
    End         end
plot(real(PoleData(1,:)),imag(PoleData(1,:)),real(PoleData(2,:)),imag(PoleData(2,:)));
    hold on
end
axis([-175 0 0 150]); sgrid
```

Next, we set $aT = 0.05$ and vary T from 0.01 to 0.001, as shown in Table 9-12. Table 9-12 shows the attributes of the unit-step responses. As the value of T decreases, the maximum overshoot decreases, but the rise time and settling time increase. The cases that satisfy the design requirements are indicated in Table 9-12 for $aT = 0.05$. Fig. 9-30 shows the unit-step responses of the phase-lead-compensated system with three sets of controller parameters.

Choosing $T = 0.004$, $a = 12.5$, the transfer function of the phase-lead controller is

$$G_c(s) = a\frac{s + 1/aT}{s + 1/T} = 12.5\frac{s + 20}{s + 250} \tag{9-84}$$

The transfer function of the compensated system is

$$G(s) = G_c(s)G_p(s) = \frac{31250(s + 20)}{s(s + 25)(s + 250)} \tag{9-85}$$

To find the op-amp-circuit realization of the phase-lead controller, we arbitrarily set $C = 0.1\ \mu\text{f}$, and the resistors of the circuit are found using Eqs. (9-63) and (9-64) as $R_1 = 500{,}000\ \Omega$ and $R_2 = 40{,}000\ \Omega$.

Frequency-Domain Design

Let us specify that the steady-state error requirement is the same as that given earlier. For frequency-domain design, the phase margin is to be greater than 45°. The following design steps are taken:

1. The Bode diagram of Eq. (9-74) with $K = 1$ is plotted as shown in Fig. 9-33.
2. The phase margin of the uncompensated system, read at the gain-crossover frequency, $\omega_c = 47$ rad/sec, is 28°. Because the minimum desired phase margin

TABLE 9-12 Attributes of Unit-Step Responses of System with Phase-Lead Controller in Example 9-5-1: $aT=0.05$

T	a	Maximum Overshoot (%)	t_r (sec)	t_s (sec)
0.01	5.0	16.2	0.0146	0.0415
0.005	10.0	4.1	0.0133	0.0174
0.004	12.5	1.1	0.0135	0.0174
0.003	16.67	0	0.0141	0.0174
0.002	25.0	0	0.0154	0.0209
0.001	50.0	0	0.0179	0.0244

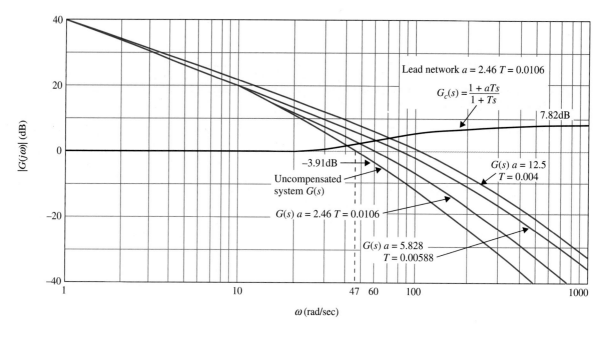

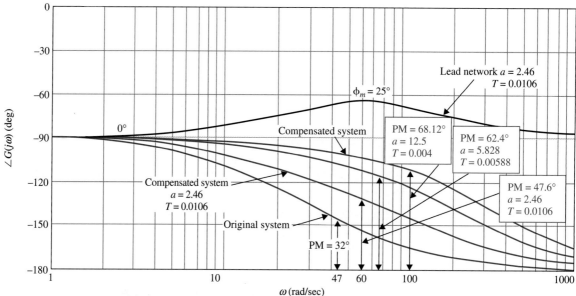

Figure 9-33 Bode diagram of the phase-lead compensation and uncompensated systems in Example 9-5-1, $G(s) = \dfrac{2500(1 + aTs)}{s(s + 25)(s + Ts)}$.

is 45°, at least 17° more phase lead should be added to the loop at the gain-crossover frequency.

3. The phase-lead controller of Eq. (9-65) must provide the additional 17° at the gain-crossover frequency of the compensated system. However, by applying the phase-lead controller, the magnitude curve of the Bode plot is also affected in such a way that the gain-crossover frequency is shifted to a higher frequency. Although it is a

simple matter to adjust the corner frequencies, $1/aT$ and $1/T$, of the controller so that the maximum phase of the controller ϕ_m falls exactly at the new gain-crossover frequency, the original phase curve at this point is no longer 28° (and could be considerably less) because the phase of most control processes decreases with the increase in frequency. In fact, if the phase of the uncompensated process decreases rapidly with increasing frequency near the gain-crossover frequency, the single-stage phase-lead controller will no longer be effective.

In view of the difficulty estimating the necessary amount of phase lead, it is essential to include some safety margin to account for the inevitable phase drop-off. Therefore, in the present case, instead of selecting a ϕ_m of a mere 17°, let ϕ_m be 25°. Using Eq. (9-72), we have

$$a = \frac{1 + \sin 25°}{1 - \sin 25°} = 2.46 \qquad (9\text{-}86)$$

4. To determine the proper location of the two corner frequencies ($1/aT$ and $1/T$) of the controller, it is known from Eq. (9-67) that the maximum phase lead ϕ_m occurs at the geometric mean of the two corner frequencies. To achieve the maximum phase margin with the value of a determined, ϕ_m should occur at the new gain-crossover frequency ω'_g, which is not known. The following steps are taken to ensure that ϕ_m occurs at ω'_g.

   a. The high-frequency gain of the phase-lead controller of Eq. (9-65) is

$$20 \log_{10} a = 20 \log_{10} 2.46 = 7.82 \text{ dB} \qquad (9\text{-}87)$$

   b. The geometric mean ω_m of the two corner frequencies, $1/aT$ and $1/T$, should be located at the frequency at which the magnitude of the uncompensated process transfer function $G_p(j\omega)$ in dB is equal to the negative value in dB of one-half of this gain. This way, the magnitude curve of the compensated transfer function will pass through the 0-dB-axis at $\omega = \omega_m$. Thus, ω_m should be located at the frequency where

$$|G_p(j\omega)|_{\text{dB}} = -10 \log_{10} 2.46 = -3.91 \text{ dB} \qquad (9\text{-}88)$$

Toolbox 9-5-4

Bode diagram for Fig. 9-33 is obtained by the following sequence of MATLAB functions

```
a = [2.46 12.5 5.828];
T = [0.0106 0.004 0.00588];

for i = 1:length(T)
    num = [2500*a(i)*T(i) 2500];
    den =[T(i) 1+25*T(i) 25 0];
    bode(num,den);
    hold on;
end
% axis([1 10000 -180 -90]);
  grid
```

> **Toolbox 9-5-5**
>
> *Plot of G(s) for Fig. 9-34 is obtained by the following sequence of MATLAB functions*
>
> ```
> a = [2.46 12.5 5.828];
> T = [0.0106 0.004 0.00588];
> for i = 1:length(T)
>     num = [2500*a(i)*T(i) 2500];
>     den =[T(i) 1+25*T(i) 25 0];
>     t = tf(num,den)
>     nichols(t); ngrid;
>     hold on;
> end
> ```

From Fig. 9-33, this frequency is found to be $\omega_m = 60$ rad/sec. Now using Eq. (9-67), we have

$$\frac{1}{T} = \sqrt{a}\omega_m = \sqrt{2.46} \times 60 = 94.1 \text{ rad/sec} \tag{9-89}$$

Then, $1/aT = 94.1/2.46 = 38.21$ rad/sec. The transfer function of the phase-lead controller is

$$G_c(s) = a\frac{s + 1/aT}{s + 1/T} = 2.46\frac{s + 38.21}{s + 94.1} \tag{9-90}$$

The forward-path transfer function of the compensated system is

$$G(s) = G_c(s)G_p(s) = \frac{6150(s + 38.21)}{s(s + 25)(s + 94.1)} \tag{9-91}$$

Fig. 9-33 shows that the phase margin of the compensated system is actually 47.6°.

In Fig. 9-34, the magnitude and phase of the original and the compensated systems are plotted on the Nichols chart for display only. These plots can be made by taking the data directly from the Bode plots of Fig. 9-33. The values of M_r, ω_r, and BW can all be determined from the Nichols chart. However, the performance data are more easily obtained with **ACSYS**.

Checking the time-domain performance of the compensated system, we have the following results:

$$\text{Maximum overshoot} = 22.3\% \quad t_r = 0.02045 \text{ sec} \quad t_s = 0.07439 \text{ sec}$$

which fall short of the time-domain specifications listed earlier. Fig. 9-33 also shows the Bode plot of the system compensated with a phase-lead controller with $a = 5.828$ and $T = 0.00588$. The phase margin is improved to 62.4°. Using Eq. (9-71), we can show that the result of $a = 12.5$ obtained in the time-domain design actually corresponds to $\phi_m = 58.41$. Adding this to the original phase of 28°, the corresponding phase margin would be 86.41°. The time-domain and frequency-domain attributes of the system with the three phase-lead controllers are summarized in Table 9-13. The results show that, with

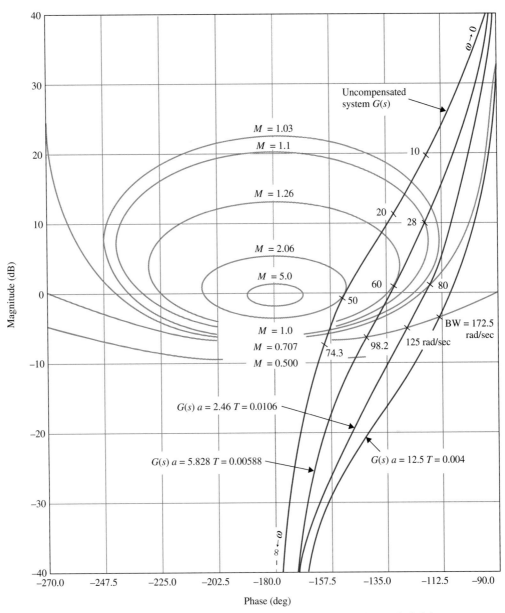

Figure 9-34 Plots of $G(s)$ in the Nichols chart for the system in Example 9-5-1.
$$G(s) = \frac{2500(1 + aTs)}{s(s + 25)(1 + Ts)}.$$

TABLE 9-13 Attributes of System with Phase-Lead Controller in Example 9-5-1

a	T	PM (deg)	M_r	Gain CO (rad/sec)	BW (rad/sec)	Maximum Overshoot (%)	t_r (sec)	t_s (sec)
1	1	28.03	2.06	47.0	74.3	44.4	0.0255	0.2133
2.46	0.0106	47.53	1.26	60.2	98.2	22.3	0.0204	0.0744
5.828	0.00588	62.36	1.03	79.1	124.7	7.7	0.0169	0.0474
12.5	0.0040	68.12	1.00	113.1	172.5	1.1	0.0135	0.0174

$a = 12.5$ and $T = 0.004$, even the projected phase margin is 86.41°; the actual value is 68.12 due to the fall-off of the phase curve at the new gain crossover.

▶ **EXAMPLE 9-5-2** In this example we illustrate the application of a phase-lead controller to a third-order system with relatively high loop gain.

Let us consider that the inductance of the dc motor of the sun-seeker system described in Fig. 9-29 is not zero. The following set of system parameters is given:

$$R_F = 10,000 \, \Omega \qquad K_b = 0.0125 \, \text{V/rad/sec}$$
$$K_i = 0.0125 \, \text{N-m/A} \qquad R_a = 6.25 \, \Omega$$
$$J = 10^{-6} \, \text{kg-m}^2 \qquad K_s = 0.3 \, \text{A/rad}$$
$$K = \text{to be determined} \qquad B = 0$$
$$n = 800 \qquad L_a = 10^{-3} \, \text{H}$$

The transfer function of the dc motor is written

$$\frac{\Omega_m(s)}{E_a(s)} = \frac{K_i}{s(L_a J s^2 + J R_a s + K_i K_b)} \qquad (9\text{-}92)$$

The forward-path transfer function of the system is

$$G_p(s) = \frac{\Theta_o(s)}{A(s)} = \frac{K_s R_F K K_i}{s(L_a J s^2 + J R_a s + K_i K_b)} \qquad (9\text{-}93)$$

Substituting the values of the system parameters in Eq. (9-92), we get

$$G_p(s) = \frac{\Theta_o(s)}{A(s)} = \frac{4.6875 \times 10^7 K}{s(s^2 + 625s + 156,250)} \qquad (9\text{-}94)$$

You may use **ACSYS** to solve this problem. ◀

Time-Domain Design

The time-domain specifications of the system are given as follows:

1. The steady-state error of $\alpha(t)$ due to a unit-ramp function input for $\theta_r(t)$ should be $\leq 1/300$ rad/rad/sec of the final steady-state output velocity.
2. The maximum overshoot of the step response should be less than 5% or as small as possible.
3. Rise time $t_r \leq 0.004$ sec.
4. Settling time $t_s \leq 0.02$ sec.

The minimum value of the amplifier gain K is determined initially from the steady-state requirement. Applying the final-value theorem to $\alpha(t)$, we get

$$\lim_{t \to \infty} \alpha(t) = \lim_{s \to 0} sA(s) = \lim_{s \to 0} \frac{s\Theta_r(s)}{1 + G_p(s)} \qquad (9\text{-}95)$$

Substituting Eq. (9-94) into Eq. (9-95), and $\Theta_r(s) = 1/s^2$, we have

$$\lim_{t \to \infty} \alpha(t) = \frac{1}{300 K} \qquad (9\text{-}96)$$

Thus, for the steady-state value of $\alpha(t)$ to be $\leq 1/300$, K must be ≥ 1. Let us set $K = 1$; the forward-path transfer function in Eq. (9-94) becomes

$$G_p(s) = \frac{4.6875 \times 10^7}{s(s^2 + 625s + 156,250)} \qquad (9\text{-}97)$$

We can show that the closed-loop sun-seeker system with $K = 1$ has the following attributes for the unit-step response.

Maximum overshoot = 43%   Rise time $t_r = 0.004797$ sec   Settling time $t_s = 0.04587$ sec

To improve the system response, let us select the phase-lead controller described by Eq. (9-62). The forward-path transfer function of the compensated system is

$$G(s) = G_c(s)G_p(s) = \frac{4.6875 \times 10^7 K(1 + aTs)}{as(s^2 + 625s + 156,250)(1 + Ts)} \quad (9\text{-}98)$$

Now to satisfy the steady-state requirement, K must be readjusted so that $K \geq a$. Let us set $K = a$. The characteristic equation of the phase-lead compensated system becomes

$$\begin{aligned}(s^3 + 625s^2 + 156,250s + 4.6875 \times 10^7) + Ts^2(s^2 + 625s + 156,250) \\ + 4.6875 \times 10^7 aTs = 0\end{aligned} \quad (9\text{-}99)$$

We can use the root-contour method to examine the effects of varying a and T of the phase-lead controller. Let us first set a to zero. The characteristic equation of Eq. (9-99) becomes

$$(s^3 + 625s^2 + 156,250s + 4.6875 \times 10^7) + Ts^2(s^2 + 625s + 156,250) = 0 \quad (9\text{-}100)$$

Dividing both sides of the last equation by the terms that do not contain T, we get

$$1 + G_{eq1}(s) = 1 + \frac{Ts^2(s^2 + 625s + 156,250)}{s^3 + 625s^2 + 156,250s + 4.6875 \times 10^7} = 0 \quad (9\text{-}101)$$

The root contours of Eq. (9-100) when T varies are determined from the pole–zero configuration of $G_{eq1}(s)$ in Eq. (9-101) and are drawn as shown in Fig. 9-35. When a varies from 0 to ∞, we divide both sides of Eq. (9-99) by the terms that do not contain a, and we have

$$1 + G_{eq2}(s) = 1 + \frac{4.6875 \times 10^7 aTs}{s^3 + 625s^2 + 156,250s + 4.6875 \times 10^7 + Ts^2(s^2 + 625s + 156,250)} = 0 \quad (9\text{-}102)$$

For a given T, the root contours of Eq. (9-99) when a varies are obtained based on the poles and zeros of $G_{eq2}(s)$. The poles of $G_{eq2}(s)$ are the same as the roots of Eq. (9-100). Thus, the root contours when a varies start ($a = 0$) at the root contours for variable T. Fig. 9-34 shows the dominant portions of the root contours when a varies for $T = 0.01, 0.0045, 0.001, 0.0005, 0.0001$, and 0.00001. Notice that, because the uncompensated system is lightly damped, for the phase-lead controller to be effective, the value of T should be very small. Even for very small values of T, there is only a small range of a that could bring increased damping, but the natural frequency of the system increases with the increase in a. The root contours in Fig. 9-35 show the approximate locations of the dominant characteristic equation roots where maximum damping occurs. Table 9-14 gives the roots of the characteristic equation and the unit-step-response attributes for the cases that correspond to near-smallest maximum overshoot for the T selected. Fig. 9-36 shows the unit-step response when $a = 500$ and $T = 0.00001$. Although the maximum overshoot is only 3.8%, the undershoot in this case is greater than the overshoot.

550 ▶ Chapter 9. Design of Control Systems

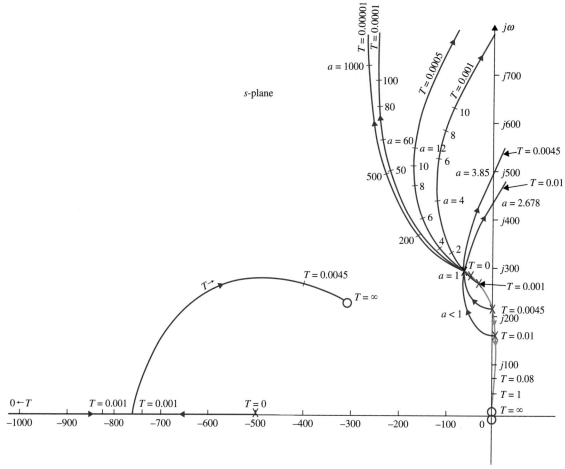

Figure 9-35 Root contours of sun-seeker system in Example 9-5-2 with phase-lead controller. $G_c(s) = \dfrac{1+aTs}{1+Ts}$.

Toolbox 9-5-6

Root contours for Fig. 9-35 are obtained by the following sequence of MATLAB functions

```
a = [50 100 500];
T = [0.0001 0.00005 0.00001];

for i = 1:length(T)
   num = 4.6875e7 * [a(i)*T(i) 1];
   den = conv([1 625 156250 0],[T(i) 1]);
   tf(num,den);
   [numCL,denCL]=cloop(num,den);
   step(numCL,denCL)
   hold on
end
axis([0 0.04 0 1.2])
grid
```

TABLE 9-14 Roots of Characteristic Equation and Time Response Attributes of System with Phase-Lead Controller in Example 9-5-2

T	a	Roots of Characteristic Equation			Maximum Overshoot (%)	t_r (sec)	t_s (sec)
0.001	4	−189.6	−1181.6	−126.9 ± j439.5	21.7	0.0037	0.0184
0.0005	9	−164.6	−2114.2	−173.1 ± j489.3	13.2	0.00345	0.0162
0.0001	50	−147	−10024	−227 ± j517	5.4	0.00348	0.0150
0.00005	100	−147	−20012	−233 ± j515	4.5	0.00353	0.0150
0.00001	500	−146.3	-10^5	−238 ± j513.55	3.8	0.00357	0.0146

Frequency-Domain Design

The Bode plot of $G_p(s)$ in Eq. (9-97) is shown in Fig. 9-37. The performance attributes of the uncompensated system are

PM = 29.74°

$M_r = 2.156$

BW = 426.5 rad/sec

We would like to show that the frequency-domain design procedure outlined earlier does not work effectively here, because the phase curve of $G_p(j\omega)$ shown in Fig. 9-37 has a very steep slope near the gain crossover. For example, if we wish to realize a phase margin

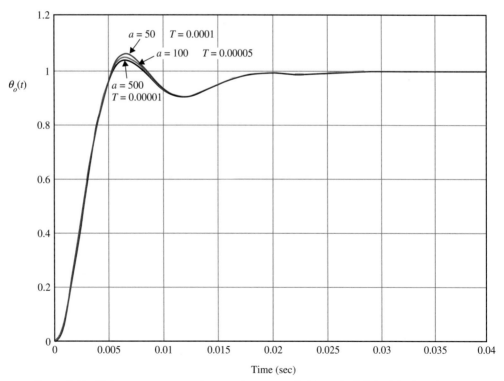

Figure 9-36 Unit-step responses of sun-seeker system in Example 9-5-2 with phase-lead controller. $G_c(s) = \dfrac{1 + aTs}{1 + Ts}$.

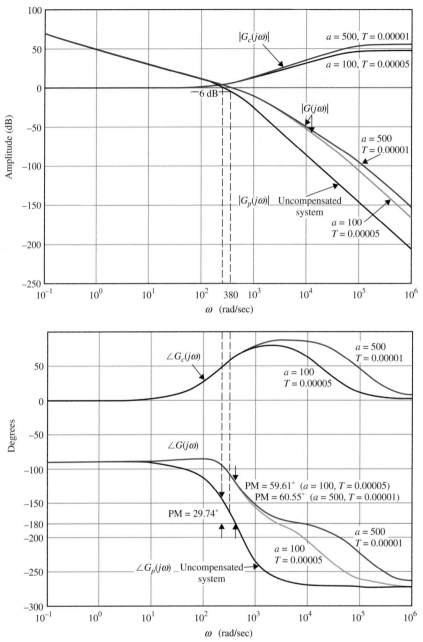

Figure 9-37 Bode plots of phase-lead controller and forward-path transfer function of sun-seeker system in Example 9-5-2. $G_c(s) = \dfrac{1 + aTs}{1 + Ts}$.

of 65°, we need at least $65 - 29.74 = 35.26°$ of phase lead. Or, $\phi_m = 35.26°$. Using Eq. (9-72), the value of a is calculated to be

$$a = \frac{1 + \sin \phi_m}{1 - \sin \phi_m} = \frac{1 + \sin 35.26°}{1 - \sin 35.26°} = 3.732 \qquad (9\text{-}103)$$

Toolbox 9-5-7

Bode plots shown in Fig. 9-37 are obtained by the following sequence of MATLAB functions

```
a = [100 500];
T = [0.00005 0.00001];

for i = 1:length(T)
   num = [a(i)*T(i) 1];
   den =[T(i) 1];
   bode(num,den);
   hold on;
end

for i = 1:length(T)
   num = 4.6875e7 * [a(i)*T(i) 1];
   den = conv([1 625 156250 0],[T(i) 1]);
   bode(num,den);
   hold on;
end
axis([1 1e6 -300 90]);
grid
```

Let us choose $a = 4$. Theoretically, to maximize the utilization of ϕ_m, ω_m should be placed at the new gain crossover, which is located at the frequency where the magnitude of $G_p(j\omega)$ is $-10\log_{10}a$ dB $= -10\log_{10}4 = -6$ dB. From the Bode plot in Fig. 9-37, this frequency is found to be 380 rad/sec. Thus, we let $\omega_m = 380$ rad/sec. The value of T is found by using Eq. (9-67):

$$T = \frac{1}{\omega_m\sqrt{a}} = \frac{1}{380\sqrt{4}} = 0.0013 \tag{9-104}$$

However, checking the frequency response of the phase-lead compensated system with $a = 4$ and $T = 0.0013$, we found that the phase margin is only improved to $38.27°$, and $M_r = 1.69$. The reason is the steep negative slope of the phase curve of $G_p(j\omega)$. The fact is that, at the new gain-crossover frequency of 380 rad/sec, the phase of $G_p(j\omega)$ is $-170°$, as against $-150.26°$ at the original gain crossover—a drop of almost $20°$! From the time-domain design, the first line in Table 9-14 shows that, when $a = 4$ and $T = 0.001$, the maximum overshoot is 21.7%.

Checking the frequency response of the phase-lead compensated system with $a = 500$ and $T = 0.00001$, the following performance data are obtained:

$$\text{PM} = 60.55 \text{ degrees} \quad M_r = 1 \quad \text{BW} = 664.2 \text{ rad/sec}$$

This shows that the value of a has to be increased substantially just to overcome the steep drop of the phase characteristics when the gain crossover is moved upward.

Fig. 9-37 shows the Bode plots of the phase-lead controller and the forward-path transfer functions of the compensated system with $a = 100$, $T = 0.0005$ and $a = 500$, $T = 0.00001$. A summary of performance data is given in Table 9-15.

TABLE 9-15 Attributes of System with Phase-Lead Controller in Example 9-5-2

T	a	PM (deg)	GM (dB)	M_r	BW (rad/sec)	Maximum Overshoot (%)	t_r (sec)	t_r (sec)
1	1	29.74	6.39	2.16	430.4	43.0	0.00478	0.0459
0.00005	100	59.61	31.41	1.009	670.6	4.5	0.00353	0.015
0.00001	500	60.55	45.21	1.000	664.2	3.8	0.00357	0.0146

Selecting $a = 100$ and $T = 0.00005$, the phase-lead controller is described by the transfer function

$$G_c(s) = \frac{1}{a}\frac{1+aTs}{1+Ts} = \frac{1}{100}\frac{1+0.005s}{1+0.00005s} \qquad (9\text{-}105)$$

Using Eqs. (9-63) and (9-64), and letting $C = 0.01\ \mu F$, the circuit parameters of the phase-lead controller are found to be

$$R_2 = \frac{T}{C} = \frac{5\times 10^{-5}}{10^{-8}} = 5000\ \Omega \qquad (9\text{-}106)$$

$$R_1 = aR_2 = 500{,}000\ \Omega \qquad (9\text{-}107)$$

The forward-path transfer function of the compensated system is

$$\frac{\Theta_o(s)}{A(s)} = \frac{4.6875\times 10^7(1+0.005s)}{s(s^2+625s+156{,}250)(1+0.00005s)} \qquad (9\text{-}108)$$

where the amplifier gain K has been set to 100 to satisfy the steady-state requirement.

From the results of the last two illustrative examples, we can summarize the effects and limitations of the single-stage phase-lead controller as follows.

9-5-3 Effects of Phase-Lead Compensation

1. The phase-lead controller adds a zero and a pole, with the zero to the right of the pole, to the forward-path transfer function. The general effect is to add more damping to the closed-loop system. The rise time and settling time are reduced in general.
2. The phase of the forward-path transfer function in the vicinity of the gain-crossover frequency is increased. This improves the phase margin of the closed-loop system.
3. The slope of the magnitude curve of the Bode plot of the forward-path transfer function is reduced at the gain-crossover frequency. This usually corresponds to an improvement in the relative stability of the system in the form of improved gain and phase margins.
4. The bandwidth of the closed-loop system is increased. This corresponds to faster time response.
5. The steady-state error of the system is not affected.

9-5-4 Limitations of Single-Stage Phase-Lead Control

In general, phase-lead control is not suitable for all systems. Successful application of single-stage phase-lead compensation to improve the stability of a control system is hinged on the following conditions:

1. Bandwidth considerations: If the original system is unstable or with a low stability margin, the additional phase lead required to realize a certain desired phase margin may be excessive. This may require a relatively large value of a for the controller, which, as a result, will give rise to a large bandwidth for the compensated system, and the transmission of high-frequency noise entering the system at the input may become objectionable. However, if the noise enters the system near the output, then the increased bandwidth may be beneficial to noise rejection. The larger bandwidth also has the advantage of robustness; that is, the system is insensitive to parameter variations and noise rejection as described before.

2. If the original system is unstable, or with low stability margin, the phase curve of the Bode plot of the forward-path transfer function has a steep negative slope near the gain-crossover frequency. Under this condition, the single-stage phase-lead controller may not be effective because the additional phase lead at the new gain crossover is added to a much smaller phase angle than that at the old gain crossover. The desired phase margin can be realized only by using a very large value of a for the controller. The amplifier gain K must be set to compensate a, so a large value for a requires a high-gain amplifier, which could be costly.

   As shown in Example 9-5-2, the compensated system may have a larger undershoot than overshoot. Often, a portion of the phase curve may still dip below the 180°-axis, resulting in a **conditionally stable** system, even though the desired phase margin is satisfied.

3. The maximum phase lead available from a single-stage phase-lead controller is less than 90°. Thus, if a phase lead of more than 90° is required, a multistage controller should be used.

9-5-5 Multistage Phase-Lead Controller

When the design with a phase-lead controller requires an additional phase of more than 90°, a multistage controller should be used. Fig. 9-38 shows an op-amp-circuit

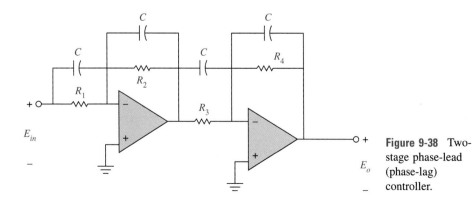

Figure 9-38 Two-stage phase-lead (phase-lag) controller.

realization of a two-stage phase-lead controller. The input–output transfer function of the circuit is

$$G_c(s) = \frac{E_o(s)}{E_{in}(s)} = \left(\frac{s + \frac{1}{R_1 C}}{s + \frac{1}{R_2 C}}\right)\left(\frac{s + \frac{1}{R_3 C}}{s + \frac{1}{R_4 C}}\right) \quad (9\text{-}109)$$

$$= \frac{R_2 R_4}{R_1 R_3}\left(\frac{1 + R_1 C s}{1 + R_2 C s}\right)\left(\frac{1 + R_3 C s}{1 + R_4 C s}\right)$$

or

$$G_c(s) = \frac{1}{a_1 a_2}\left(\frac{1 + a_1 T_1 s}{1 + T_1 s}\right)\left(\frac{1 + a_2 T_2 s}{1 + T_2 s}\right) \quad (9\text{-}110)$$

where $a_1 = R_1/R_2$, $a_2 = R_3/R_4$, $T_1 = R_2 C$, and $T_2 = R_4 C$.

The design of a multistage phase-lead controller in the time domain becomes more cumbersome, since now there are more poles and zeros to be placed. The root-contour method also becomes impractical, since there are more variable parameters. The frequency-domain design in this case does represent a better choice of the design method. For example, for a two-stage controller, we can choose the parameters of the first stage of a two-stage controller so that a portion of the phase margin requirement is satisfied, and then the second stage fulfills the remaining requirement. In general, there is no reason why the two stages cannot be identical. The following example illustrates the design of a system with a two-stage phase-lead controller.

▶ **EXAMPLE 9-5-3** For the sun-seeker system designed in Example 9-5-2, let us alter the rise time and settling time requirements to be

Rise time $t_r \leq 0.001$ sec

Settling time $t_s \leq 0.005$ sec

The other requirements are not altered. One way to meet faster rise time and settling time requirements is to increase the forward-path gain of the system. Let us consider that the forward-path transfer function is

$$G_p(s) = \frac{\Theta_o(s)}{A(s)} = \frac{156,250,000}{s(s^2 + 625s + 156,250)} \quad (9\text{-}111)$$

Another way of interpreting the change in the forward-path gain is that the ramp-error constant is increased to 1000 (up from 300 in Example 9-5-1). The Bode plot of $G_p(s)$ is shown in Fig. 9-39. The closed-loop system is unstable, with a phase margin of $-15.43°$. ◀

Toolbox 9-5-8

Bode plots shown in Fig. 9-39 are obtained by the following sequence of MATLAB functions

```
num = 156250000 * [0.0087 1];
den = conv([0.000087 1],[1 625 156250]);
bode(num,den);
```

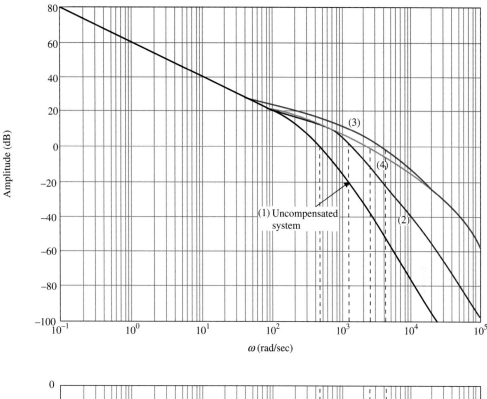
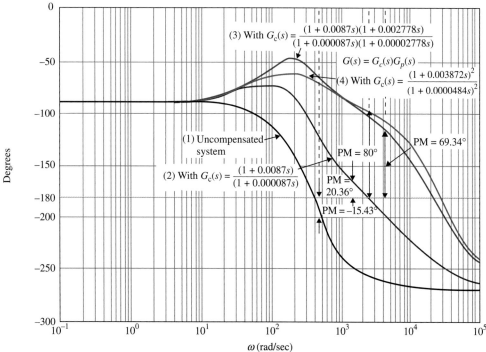

Figure 9-39 Bode plots of uncompensated and compensated sun-seeker systems in Example 9-5-2 with two-state phase-lead controller. $G_p(s) = \dfrac{156{,}250{,}000}{s(s^2 + 625s + 156{,}250)}$.

```
hold on;

num = 156250000 * conv([0.0087 1],[0.002778 1]);
den =conv(conv([0.000087 1],[0.00002778 1]),[1 625 156250 0]);
bode(num,den);
hold on;

num = 156250000 * conv([0.003872 1],[0.003872 1]);
den =conv(conv([0.0000484 1],[0.0000484 1]),[1 625 156250 0]);
bode(num,den);
axis([1 1e5 -300 20]);
grid
```

Because the compensated system in Example 9-5-2 had a phase margin of 60.55°, we would expect that, to satisfy the more stringent time response requirements in this example, the corresponding phase margin would have to be greater. Apparently, this increased phase margin cannot be realized with a single-stage phase-lead controller. It appears that a two-stage controller would be adequate.

The design involves some trial-and-error steps in arriving at a satisfied controller. Because we have two stages of controllers at our disposal, the design has a multitude of flexibility. We can set out by arbitrarily setting $a_1 = 100$ for the first stage of the phase-lead controller. The phase lead provided by the controller is obtained from Eq. (9-71),

$$\phi_m = \sin^{-1}\left(\frac{a_1 - 1}{a_1 + 1}\right) = \sin^{-1}\left(\frac{99}{101}\right) = 78.58° \tag{9-112}$$

To maximize the effect of ϕ_m, the new gain crossover should be at

$$-10 \log_{10} a_1 = -10 \log_{10} 100 = -20 \, \text{dB} \tag{9-113}$$

From Fig. 9-39 the frequency that corresponds to this gain on the amplitude curve is approximately 1150 rad/sec. Substituting $\omega_{m1} = 1150$ rad/sec and $a_1 = 100$ in Eq. (9-67), we get

$$T_1 = \frac{1}{\omega_{m1}\sqrt{a_1}} = \frac{1}{1150\sqrt{100}} = 0.000087 \tag{9-114}$$

The forward-path transfer function with the one-stage phase-lead controller is

$$G(s) = \frac{156,250,000(1 + 0.0087s)}{s(s^2 + 625s + 156,250)(1 + 0.000087s)} \tag{9-115}$$

The Bode plot of the last equation is drawn as curve (2) in Fig. 9-39. We see that the phase margin of the interim design is only 20.36°. Next, we arbitrarily set the value of a_2 of the second stage at 100. From the Bode plot of the transfer function of Eq. (9-115) in Fig. 9-39, we find that the frequency at which the magnitude of $G(j\omega)$ is -20 dB is approximately 3600 rad/sec. Thus,

$$T_2 = \frac{1}{\omega_{m2}\sqrt{a_2}} = \frac{1}{3600\sqrt{100}} = 0.00002778 \tag{9-116}$$

9-5 Design with Phase-Lead Controller

TABLE 9-16 Attributes of Sun-Seeker System in Example 9-5-3 with Two-Stage Phase-Lead Controller

$a_1 = a_2$	T_1	T_2	PM (deg)	M_r	BW (rad/sec)	Maximum Overshoot (%)	t_r (sec)	t_s (sec)
80	0.0000484	0.0000484	80	1	5686	0	0.00095	0.00475
100	0.000087	0.0000278	69.34	1	5686	0	0.000597	0.00404
70	0.0001117	0.000039	66.13	1	5198	0	0.00063	0.00404

The forward-path transfer function of the sun-seeker system with the two-stage phase-lead controller is ($a_1 = a_2 = 100$)

$$G(s) = \frac{156{,}250{,}000(1 + 0.0087s)(1 + 0.002778s)}{s(s^2 + 625s + 156{,}250)(1 + 0.000087s)(1 + 0.00002778s)} \quad (9\text{-}117)$$

Fig. 9-39 shows the Bode plot of the sun-seeker system with the two-stage phase-lead controller designed above [curve (3)]. As seen from Fig. 9-39, the phase margin of the system with $G(s)$ given in Eq. (9-117) is 69.34°. As shown by the system attributes in Table 9-16, the system satisfies all the time-domain specifications. In fact, the selection of $a_1 = a_2 = 100$ appears to be overly stringent. To show that the design is not critical, we can select $a_1 = a_2 = 80$, and then 70 and the time-domain specifications are still satisfied. Following similar design steps, we arrived at $T_1 = 0.0001117$ and $T_2 = 0.000039$ for $a_1 = a_2 = 70$, and $T_1 = T_2 = 0.0000484$ for $a_1 = a_2 = 80$. Curve (4) of Fig. 9-39 shows the Bode plot of the compensated system with $a_1 = a_2 = 80$. Table 9-16 summarizes all the attributes of the system performance with these three controllers.

The unit-step responses of the system with the two-stage phase-lead controller for $a_1 = a_2 = 80$ and 100 are shown in Fig. 9-40.

Toolbox 9-5-9

Fig. 9-40 is obtained by the following sequence of MATLAB functions

```
num = 156250000 * conv([100*0.000087 1],[80*0.00002778 1]);
den =conv(conv([0.000087 1],[0.00002778 1]),[1 625 156250 0]);
[numCL,denCL]=cloop(num,den);
step(numCL,denCL)
hold on

num = 156250000 * conv([80*0.0000484 1],[80*0.0000484 1]);
den =conv(conv([0.0000484 1],[0.0000484 1]),[1 625 156250 0]);
[numCL,denCL]=cloop(num,den);
step(numCL,denCL)
grid
```

9-5-6 Sensitivity Considerations

The sensitivity function defined in Section 8-16, Eq. (8-122), can be used as a design specification to indicate the robustness of the system. In Eq. (9-122), the sensitivity of the

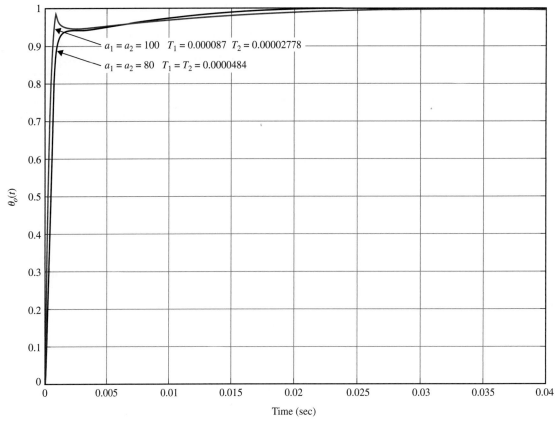

Figure 9-40 Unit-step responses of sun-seeker system in Example 9-5-2 with two-stage phase-lead controller. $G_c(s) = \left(\dfrac{1+a_1T_1s}{1+T_1s}\right)\left(\dfrac{1+a_2T_2s}{1+T_2s}\right) \quad G_p(s) = \dfrac{156{,}250{,}000}{s(s^2+625s+156{,}250)}$.

closed-loop transfer function with respect to the variations of the forward-path transfer function is defined as

$$S_G^M(s) = \frac{\partial M(s)/M(s)}{\partial G(s)/G(s)} = \frac{G^{-1}(s)}{1+G^{-1}(s)} = \frac{1}{1+G(s)} \qquad (9\text{-}118)$$

The plot of $|S_G^M(j\omega)|$ versus frequency gives an indication of the sensitivity of the system as a function of frequency. The ideal robust situation is for $|S_G^M(j\omega)|$ to assume a small value ($\ll 1$) over a wide range of frequencies. As an example, the sensitivity function of the sun-seeker system designed in Example 9-5-2 with the one-stage phase-lead controller with $a = 100$ and $T = 0.00005$ is plotted as shown in Fig. 9-41. Note that the sensitivity function is low at low frequencies and is less than unity for $\omega < 400$ rad/sec. Although the sun-seeker system in Example 9-5-2 does not need a multistage phase-lead controller, we shall show that, if a two-stage phase-lead controller is used, not only the value of a will be substantially reduced, resulting in lower gains for the op-amps, but the system will be more robust. Following the design procedure outlined in Example 9-5-3, a two-stage phase-lead controller is designed for the sun-seeker system with the process transfer function described by Eq. (9-96).

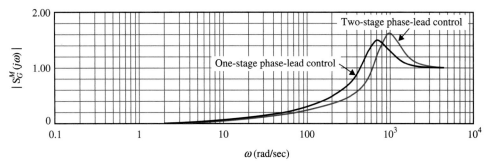

Figure 9-41 Sensitivity functions of sun-seeker system in Example 9-5-2.

The parameters of the controller are $a_1 = a_2 = 5.83$ and $T_1 = T_2 = 0.000673$. The forward-path transfer function of the compensated system is

$$G(s) = \frac{4.6875 \times 10^7 (1 + 0.0039236s)^2}{s(s^2 + 625s + 156{,}250)(1 + 0.000673s)^2} \tag{9-119}$$

Fig. 9-41 shows that the sensitivity function of the system with the two-stage phase-lead controller is less than unity for $\omega < 600$ rad/sec. Thus, the system with the two-stage phase-lead controller is more robust than the system with the single-stage controller. The reason for this is that the more robust system has a higher bandwidth. In general, systems with phase-lead control will be more robust due to the higher bandwidth. However, Fig. 9-41 shows that the system with the two-stage phase-lead controller has a higher sensitivity at high frequencies.

▶ 9-6 DESIGN WITH PHASE-LAG CONTROLLER

The transfer function in Eq. (9-62) represents a phase-lag controller or low-pass filter when $a < 1$. The transfer function is repeated as follows.

$$G_c(s) = \frac{1}{a}\left(\frac{1 + aTs}{1 + Ts}\right) \quad a < 1 \tag{9-120}$$

9-6-1 Time-Domain Interpretation and Design of Phase-Lag Control

The pole–zero configuration of $G_c(s)$ is shown in Fig. 9-42. Unlike the PI controller, which provides a pole at $s = 0$, the phase-lag controller affects the steady-state error only in the

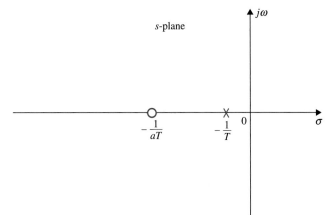

Figure 9-42 Pole–zero configuration of phase-lag controller.

sense that the zero-frequency gain of $G_c(s)$ is greater than unity. Thus, any error constant that is finite and nonzero will be increased by the factor $1/a$ from the phase-lag controller.

Because the pole at $s = -1/T$ is to the right of the zero at $-1/aT$, effective use of the phase-lag controller to improve damping would have to follow the same design principle of the PI control presented in Section 9-3. Thus, *the proper way of applying the phase-lag control is to place the pole and zero close together. For type 0 and type 1 systems, the combination should be located near the origin in the s-plane.* Fig. 9-43 illustrates the design strategies in the *s*-plane for type 0 and type 1 systems. Phase-lag control should not be applied to a type 2 system.

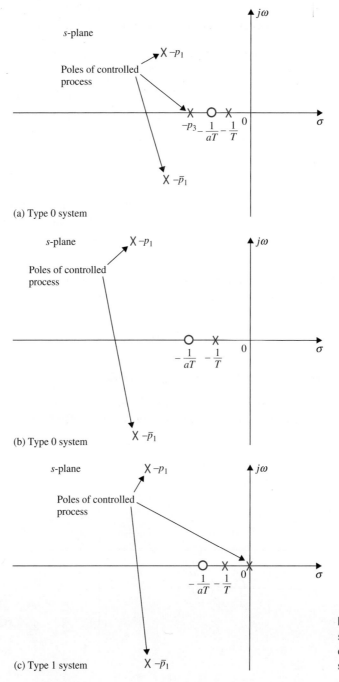

Figure 9-43 Design strategies for phase-lag control for type 0 and type 1 systems.

The design principle described above can be explained by considering that the controlled process of a type 0 control system is

$$G_p(s) = \frac{K}{(s + p_1)(s + \bar{p}_1)(s + p_3)} \tag{9-121}$$

where p_1 and $\bar{p}_1$ are complex-conjugate poles, such as the situation shown in Fig. 9-43.

Just as in the case of the phase-lead controller, we can drop the gain factor $1/a$ in Eq. (9-121), because whatever the value of a is, the value of K can be adjusted to compensate for it. Applying the phase-lag controller of (9-121), without the factor $1/a$, to the system, the forward-path transfer function becomes

$$G(s) = G_c(s)G_p(s) = \frac{K(1 + aTs)}{(s + p_1)(s + \bar{p}_1)(s + p_3)(1 + Ts)} \quad (a < 1) \tag{9-122}$$

Let us assume that the value of K is set to meet the steady-state-error requirement. Also assume that, with the selected value of K, the system damping is low or even unstable. Now let $1/T \cong 1/aT$, and place the pole–zero pair near the pole at $-1/p_3$, as shown in Fig. 9-43. Fig. 9-44 shows the root loci of the system with and without the phase-lag controller. Because the pole–zero combination of the controller is very close to the pole at $-1/p_3$, the shape of the loci of the dominant roots with and without the phase-lag control will be very similar. This is easily explained by writing Eq. (9-122) as

$$\begin{aligned} G(s) &= \frac{Ka(s + 1/aT)}{(s + p_1)(s + \bar{p}_1)(s + p_3)(s + 1/T)} \\ &\cong \frac{Ka}{(s + p_1)(s + \bar{p}_1)(s + p_3)} \end{aligned} \tag{9-123}$$

Because a is less than 1, the application of phase-lag control is equivalent to reducing the forward-path gain from K to Ka, *while not affecting the steady-state performance of the system*. Fig. 9-44 shows that the value of a can be chosen so that the damping of the compensated system is satisfactory. Apparently, the amount of damping that can be added is limited if the poles $-p_1$ and $-\bar{p}_1$ are very close to the imaginary axis. Thus, we can select a using the following equation:

$$a = \frac{K \text{ to realize the desired damping}}{K \text{ to realize the steady-state performance}} \tag{9-124}$$

The value of T should be so chosen that the pole and zero of the controller are very close together and close to $-1/p_3$.

In the time domain, phase-lag control generally has the effect of increasing the rise time and settling time.

9-6-2 Frequency-Domain Interpretation and Design of Phase-Lag Control

The transfer function of the phase-lag controller can again be written as

$$G_c(s) = \frac{1 + aTs}{1 + Ts} \quad (a < 1) \tag{9-125}$$

by assuming that the gain factor $-1/a$ is eventually absorbed by the forward gain K. The Bode diagram of Eq. (9-125) is shown in Fig. 9-45. The magnitude curve has corner frequencies at $\omega = 1/aT$ and $1/T$. Because the transfer functions of the phase-lead and

564 ▶ Chapter 9. Design of Control Systems

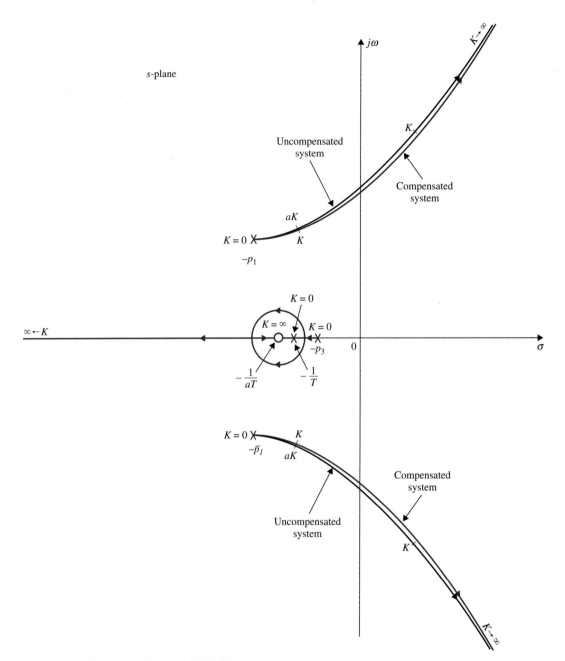

Figure 9-44 Root loci of uncompensated and phase-lag-compensated systems.

phase-lag controllers are identical in form, except for the value of a, the maximum phase lag ϕ_m of the phase curve of Fig. 9-45 is given by

$$\phi_m = \sin^{-1}\left(\frac{a-1}{a+1}\right) \quad (a<1) \tag{9-126}$$

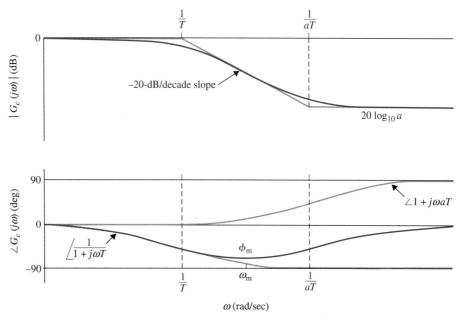

Figure 9-45 Bode diagram of the phase-lag controller. $G_c(s) = \dfrac{(1 + aTs)}{(1 + Ts)} \quad a < 1$.

Fig. 9-45 shows that the phase-lag controller essentially provides an attenuation of $20 \log_{10} a$ at high frequencies. Thus, unlike the phase-lead control that utilizes the maximum phase lead of the controller, phase-lag control utilizes the attenuation of the controller at high frequencies. This is parallel to the situation of introducing an attenuation of a to the forward-path gain in the root-locus design. For phase-lead control, the objective of the controller is to increase the phase of the open-loop system in the vicinity of the gain crossover while attempting to locate the maximum phase lead at the new gain crossover. In phase-lag control, the objective is to move the gain crossover to a lower frequency where the desired phase margin is realized, while keeping the phase curve of the Bode plot relatively unchanged at the new gain crossover.

The design procedure for phase-lag control using the Bode plot is outlined as follows:

1. The Bode plot of the forward-path transfer function of the uncompensated system is drawn. The forward-path gain K is set according to the steady-state performance requirement.

2. The phase and gain margins of the uncompensated system are determined from the Bode plot.

3. Assuming that the phase margin is to be increased, the frequency at which the desired phase margin is obtained is located on the Bode plot. This frequency is also the new gain crossover frequency ω'_g, where the compensated magnitude curve crosses the 0-dB-axis.

4. To bring the magnitude curve down to 0 dB at the new gain-crossover frequency ω'_g, the phase-lag controller must provide the amount of attenuation equal to the value of the magnitude curve at ω'_g. In other words,

$$\left| G_p\left(j\omega'_g \right) \right| = -20 \log_{10} a \text{ dB} \quad (a < 1) \qquad (9\text{-}127)$$

Solving for a from the last equation, we get

$$a = 10^{-|G_p(j\omega'_g)|/20} \quad (a < 1) \tag{9-128}$$

Once the value of a is determined, it is necessary only to select the proper value of T to complete the design. Using the phase characteristics shown in Fig. 9-45, if the corner frequency $1/aT$ is placed far below the new gain-crossover frequency ω'_g, the phase lag of the controller will not appreciably affect the phase of the compensated system near ω'_g. On the other hand, the value of $1/aT$ should not be too small because the bandwidth of the system will be too low, causing the system to be too sluggish and less robust. Usually, as a general guideline, the frequency $1/aT$ should be approximately one decade below ω'_g; that is,

$$\frac{1}{aT} = \frac{\omega'_g}{10} \text{ rad/sec} \tag{9-129}$$

Then,

$$\frac{1}{T} = \frac{a\omega'_g}{10} \text{ rad/sec} \tag{9-130}$$

5. The Bode plot of the compensated system is investigated to see if the phase margin requirement is met; if not, the values of a and T are readjusted, and the procedure is repeated. If design specifications involve gain margin, M_r, or BW, then these values should be checked and satisfied.

Because the phase-lag control brings in more attenuation to a system, then if the design is proper, the stability margins will be improved but at the expense of lower bandwidth. The only benefit of lower bandwidth is reduced sensitivity to high-frequency noise and disturbances.

The following example illustrates the design of the phase-lag controller and all its ramifications.

▶ **EXAMPLE 9-6-1** In this example, we shall use the second-order sun-seeker system described in Example 9-5-1 to illustrate the principle of design of phase-lag control. The forward-path transfer function of the uncompensated system is

$$G_p(s) = \frac{\Theta_o(s)}{A(s)} = \frac{2500K}{s(s+25)} \tag{9-131}$$

You may use **ACSYS** to solve this problem. ◀

Time-Domain Design
The time-domain specifications of the system are as follows:

1. The steady-state error of $\alpha(t)$ due to a unit-ramp function input for $\theta_r(t)$ should be $\leq 1\%$.
2. The maximum overshoot of the step response should be less than 5% or as small as possible.
3. Rise time $t_r \leq 0.5$ sec.

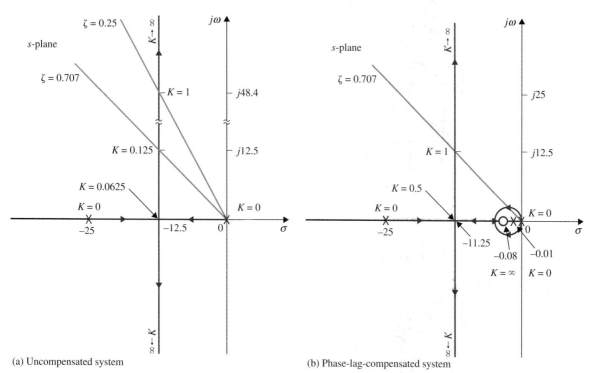

(a) Uncompensated system (b) Phase-lag-compensated system

Figure 9-46 Root loci of sun-seeker system in Example 9-6-1.
$G_p(s) = \dfrac{2500K}{s(s+25)} \quad G_c(s) = \dfrac{1+aTs}{1+Ts} \quad a = 0.125 \quad T = 100$.

4. Settling time $t_s \leq 0.5$ sec.
5. Due to noise problems, the bandwidth of the system must be < 50 rad/sec.

Notice that the rise-time and settling-time requirements have been relaxed considerably from the phase-lead design in Example 9-5-1. The root loci of the uncompensated system are shown in Fig. 9-46(a).

As in Example 9-5-1, we set $K = 1$ initially. The damping ratio of the uncompensated system is 0.25, and the maximum overshoot is 44.4%. Fig. 9-47 shows the unit-step response of the system with $K = 1$.

Let us select the phase-lag controller with the transfer function given in Eq. (9-121). The forward-path transfer function of the compensated system is

$$G(s) = G_c(s)G_p(s) = \frac{2500K(s+1/aT)}{s(s+25)(s+1/T)} \tag{9-132}$$

If the value of K is maintained at 1, the steady-state error will be a percent, which is better than that of the uncompensated system, since $a < 1$. *For effective phase-lag control, the pole and zero of the controller transfer function should be placed close together, and then for the type 1 system, the combination should be located relatively close to the origin of the s-plane.* From the root loci of the uncompensated system in Fig. 9-46(a), we see that, if K could be set to 0.125, the damping ratio would be 0.707, and the maximum overshoot of the system would be 4.32%. By setting the pole and zero of the controller close to $s = 0$, the shape of the loci of the dominant roots of the

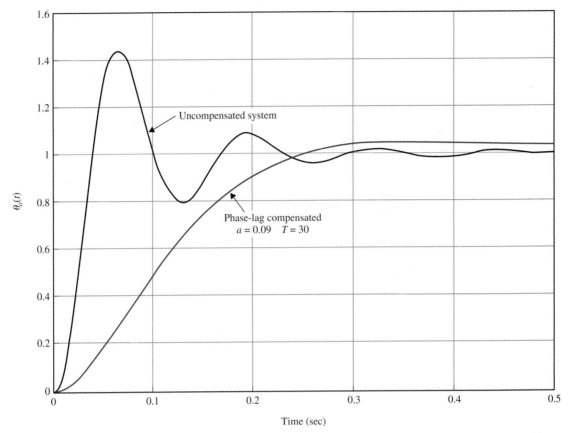

Figure 9-47 Unit-step responses of uncompensated and compensated sun-seeker systems with phase-lag controller in Example 9-6-1. $G_p(s) = \dfrac{2500\,K}{s(s+25)}$ $G_c(s) = \dfrac{1+aTs}{1+Ts}$ $a = 0.09$ $T = 30$.

compensated system will be very similar to those of the uncompensated system. We can find the value of a using Eq. (9-124); that is,

$$a = \frac{K \text{ to realize the desired damping}}{K \text{ to realize the steady-state performance}} = \frac{0.125}{1} = 0.125 \qquad (9\text{-}133)$$

Thus, if the value of T is sufficiently large, when $K = 1$, the dominant roots of the characteristic equation will correspond to a damping ratio of approximately 0.707. Let us arbitrarily select $T = 100$. The root loci of the compensated system are shown in Fig. 9-46(b). The roots of the characteristic equation when $K = 1$, $a = 0.125$, and $T = 100$ are

$$s = -0.0805, \quad -12.465 + j12.465, \quad \text{and} \quad -12.465 - j12.465$$

which corresponds to a damping ratio of exactly 0.707. If we had chosen a smaller value for T, then the damping ratio would be slightly off 0.707. From a practical standpoint, the value of T cannot be too large, since from Eq. (9-64), $T = R_2C$, a large T would correspond to either a large capacitor or an unrealistically large resistor. To reduce the value of T and simultaneously satisfy the maximum overshoot requirement, a should also be reduced. However, a cannot be reduced indefinitely, or the zero of the controller at $-1/aT$ would be too far to the left on the real axis. Table 9-17 gives the attributes of the time-domain

TABLE 9-17 Attributes of Performance of Sun-Seeker System in Example 9-6-1 with Phase-Lag Controller

a	T	Maximum Overshoot (%)	t_r (sec)	t_s (sec)	BW (rad/sec)		Roots of Characteristic Equation	
1.000	1	44.4	0.0255	0.2133	75.00		$-12.500 \pm j48.412$	
0.125	100	4.9	0.1302	0.1515	17.67	-0.0805	$-12.465 \pm j12.465$	
0.100	100	2.5	0.1517	0.2020	13.97	-0.1009	$-12.455 \pm j9.624$	
0.100	50	3.4	0.1618	0.2020	14.06	-0.2037	$-12.408 \pm j9.565$	
0.100	30	4.5	0.1594	0.1515	14.19	-0.3439	$-12.345 \pm j9.484$	
0.100	20	5.9	0.1565	0.4040	14.33	-0.5244	$-12.263 \pm j9.382$	
0.090	50	3.0	0.1746	0.2020	12.53	-0.2274	$-12.396 \pm j8.136$	
0.090	30	4.4	0.1719	0.2020	12.68	-0.3852	$-12.324 \pm j8.029$	
0.090	20	6.1	0.1686	0.5560	12.84	-0.5901	$-12.230 \pm j7.890$	

performance of the phase-lag compensated sun-seeker system with various values for a and T. The ramifications of the various design parameters are clearly displayed.

Thus, a suitable set of controller parameters would be $a = 0.09$ and $T = 30$. With $T = 30$, selecting $C = 1\ \mu F$ would require R_2 to be 30 MΩ. A smaller value for T can be realized by using a two-stage phase-lag controller. The unit-step response of the compensated system with $a = 0.09$ and $T = 30$ is shown in Fig. 9-47. Notice that the maximum overshoot is reduced at the expense of rise time and settling time. Although the settling time of the compensated system is shorter than that of the uncompensated system, it actually takes much longer for the phase-lag-compensated system to reach steady state.

It would be enlightening to explain the design of the phase-lag controller by means of the root contours. The root-contour design conducted earlier in Example 9-5-1 using Eqs. (9-80) through (9-83) for phase-lead control and Figs. 9-31 and 9-32 is still valid for phase-lag control, except that in the present case, $a < 1$. Thus, in Fig. 9-32 only the portions of the root contours that correspond to $a < 1$ are applicable for phase-lag control. These root contours clearly show that, for effective phase-lag control, the value of T should be relatively large. In Fig. 9-48 we illustrate further that the complex poles of the closed-loop transfer function are rather insensitive to the value of T when the latter is relatively large.

Frequency-Domain Design

The Bode plot of $G_p(j\omega)$ of Eq. (9-131) is shown in Fig. 9-49 for $K = 1$. The Bode plot shows that the phase margin of the uncompensated system is only 28°. Not knowing what phase margin will correspond to a maximum overshoot of less than 5%, we conduct the following series of designs using the Bode plot in Fig. 9-49. Starting with a phase margin of 45°, we observe that this phase margin can be realized if the gain-crossover frequency ω'_g is at 25 rad/sec. This means that the phase-lag controller must reduce the magnitude curve of $G_p(j\omega)$ to 0 dB at $\omega = 25$ rad/sec while it does not appreciably affect the phase curve near this frequency. Because the phase-lag controller still contributes a small negative phase when the corner frequency $1/aT$ is placed at $1/10$ of the value of ω'_g, it is a safe measure to choose ω'_g at somewhat less than 25 rad/sec, say, 20 rad/sec.

From the Bode plot, the value of $\left|G_p(j\omega'_g)\right|_{dB}$ at $\omega'_g = 20$ rad/sec is 11.7 dB. Thus, using Eq. (9-128), we have

$$a = 10^{-|G_p(j\omega'_g)|/20} = 10^{-11.7/20} = 0.26 \qquad (9\text{-}134)$$

570 ▶ Chapter 9. Design of Control Systems

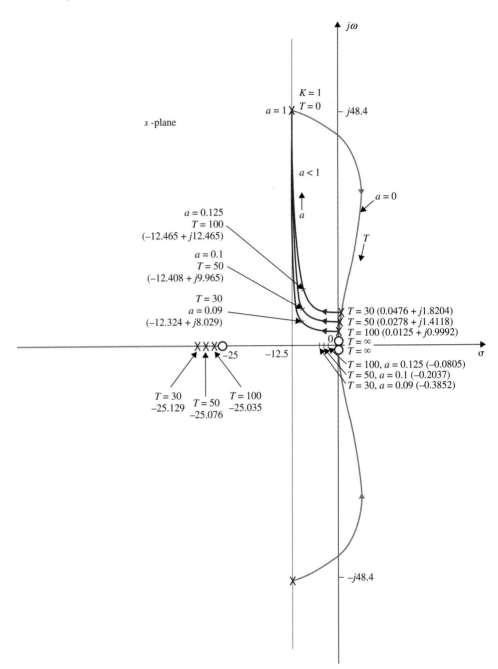

Figure 9-48 Root contours of sun-seeker system in Example 9-6-1 with phase-lag controller.

The value of $1/aT$ is chosen to be at $1/10$ the value of $\omega'_g = 20$ rad/sec. Thus,

$$\frac{1}{aT} = \frac{\omega'_g}{10} = \frac{20}{10} = 2 \text{ rad/sec} \tag{9-135}$$

and

$$T = \frac{1}{2a} = \frac{1}{0.52} = 1.923 \tag{9-136}$$

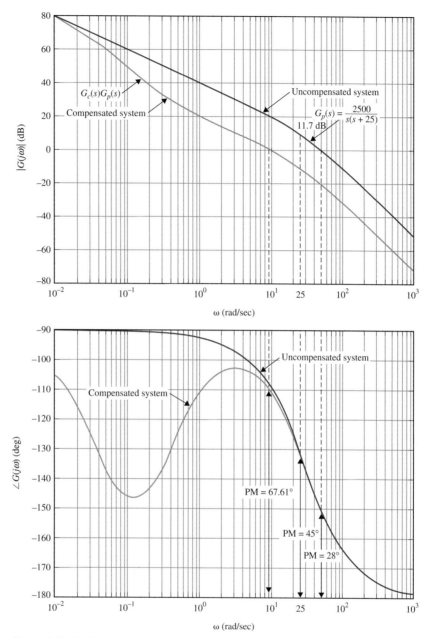

Figure 9-49 Bode plot of uncompensated and compensated systems with phase-lag controller in Example 9-6-1. $G_c(s) = \dfrac{1+3s}{1+30s}$ $G_p(s) = \dfrac{2500}{s(s+25)}$.

Checking out the unit-step response of the system with the designed phase-lag control, we found that the maximum overshoot is 24.5%. The next step is to try aiming at a higher phase margin. Table 9-18 gives the various design results by using various desired phase margins up to 80°.

Examining the results in Table 9-18, we see that none of the cases satisfies the maximum overshoot requirement of $\leq 5\%$. The $a = 0.044$ and $T = 52.5$ case yields the best maximum overshoot, but the value of T is too large to be practical. Thus, we single out the case with $a = 0.1$ and $T = 10$ and refine the design by increasing the value of T. As shown in Table 9-17,

TABLE 9-18 Performance Attributes of Sun-Seeker System in Example 9-6-1 with Phase-Lag Controller

Desired PM (deg)	a	T	Actual PM (deg)	M_r	BW (rad/sec)	Maximum Overshoot (%)	t_r (sec)	t_s (sec)
45	0.26	1.923	46.78	1.27	33.37	24.5	0.0605	0.2222
60	0.178	3.75	54.0	1.19	25.07	17.5	0.0823	0.303
70	0.1	10	63.87	1.08	14.72	10.0	0.1369	0.7778
80	0.044	52.5	74.68	1.07	5.7	7.1	0.3635	1.933

when $a = 0.1$ and $T = 30$, the maximum overshoot is reduced to 4.5%. The Bode plot of the compensated system is shown in Fig. 9-49. The phase margin is $67.61°$.

The unit-step response of the phase-lag-compensated system shown in Fig. 9-47 points out a major disadvantage of the phase-lag control. Because the phase-lag controller is essentially a low-pass filter, the rise time and settling time of the compensated system are usually increased. However, we shall show by the following example that phase-lag control can be more versatile and has a wider range of effectiveness in improving stability than the single-stage phase-lead controller, especially if the system has low or negative damping.

▶ **EXAMPLE 9-6-2** Consider the sun-seeker system designed in Example 9-5-3, with the forward-path transfer function given in Eq. (9-111). Let us restore the gain K, so that a root-locus plot can be made for the system. Then, Eq. (9-111) is written

$$G_p(s) = \frac{156,250,000\,K}{s(s^2 + 625s + 156,250)} \tag{9-137}$$

The root loci of the closed-loop system are shown in Fig. 9-50. When $K = 1$, the system is unstable, and the characteristic equation roots are at -713.14, $44.07 + j466.01$, and $44.07 - j466.01$.

Example 9-5-3 shows that the performance specification on stability cannot be achieved with a single-stage phase-lead controller. Let the performance criteria be as follows:

Maximum overshoot $\leq 5\%$

Rise time $t_r \leq 0.02$ sec

Settling time $t_s \leq 0.02$ sec

Let us assume that the desired relative damping ratio is 0.707. Fig. 9-50 shows that, when $K = 0.10675$, the dominant characteristic equation roots of the uncompensated system are at $-172.77 \pm j172.73$, which correspond to a damping ratio of 0.707. Thus, the value of a is determined from Eq. (9-124),

$$a = \frac{K \text{ to realize the desired damping}}{K \text{ to realize the steady-state performance}} = \frac{0.10675}{1} = 0.10675 \tag{9-138}$$

Let $a = 0.1$. Because the loci of the dominant roots are far away from the origin in the s-plane, the value of T has a wide range of flexibility. Table 9-19 shows the performance results when $a = 0.1$ and for various values of T.

TABLE 9-19 Performance Attributes of Sun-Seeker System in Example 9-6-2 with Phase-Lag Controller

a	T	BW (rad/sec)	PM (deg)	% Max Overshoot	t_r (sec)	t_s (sec)
0.1	20	173.5	66.94	1.2	0.01273	0.01616
0.1	10	174	66.68	1.6	0.01262	0.01616
0.1	5	174.8	66.15	2.5	0.01241	0.01616
0.1	2	177.2	64.56	4.9	0.01601	0.0101

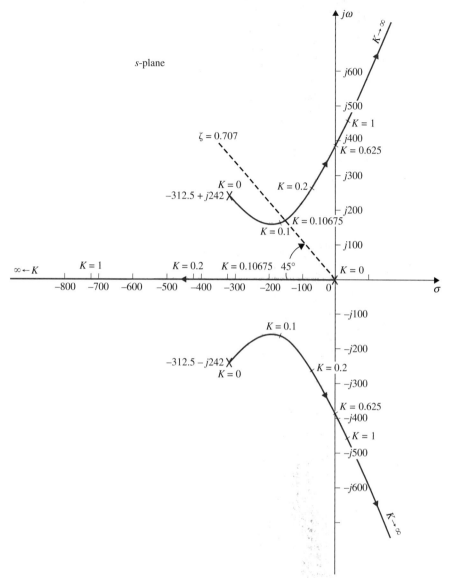

Figure 9-50 Root loci of uncompensated system in Example 9-6-2.
$G_p(s) = \dfrac{156,250,000 K}{s(s^2 + 625s + 156,250)}$.

Therefore, the conclusion is that only one stage of the phase-lag controller is needed to satisfy the stability requirement, whereas two stages of the phase-lead controller are needed, as shown in Example 9-5-3.

Sensitivity Function

The sensitivity function $|S_G^M(j\omega)|$ of the phase-lag compensated system with $a = 0.1$ and $T = 20$ is shown in Fig. 9-51. Notice that the sensitivity function is less than unity for frequencies up to only 102 rad/sec. This is due to the low bandwidth of the system as a result of phase-lag control.

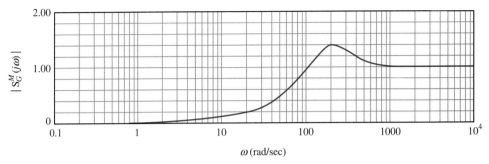

Figure 9-51 Sensitivity function of phase-lag-compensated system in Example 9-6-2.

9-6-3 Effects and Limitations of Phase-Lag Control

From the results of the preceding illustrative examples, the effects and limitations of phase-lag control on the performance of linear control systems can be summarized as follows.

1. For a given forward-path gain K, the magnitude of the forward-path transfer function is attenuated near the gain-crossover frequency, thus improving the relative stability of the system.
2. The gain-crossover frequency is decreased, and thus the bandwidth of the system is reduced.
3. The rise time and settling time of the system are usually longer, because the bandwidth is usually decreased.
4. The system is more sensitive to parameter variations because the sensitivity function is greater than unity for all frequencies approximately greater than the bandwidth of the system.

▶ 9-7 DESIGN WITH LEAD–LAG CONTROLLER

We have learned from preceding sections that phase-lead control generally improves rise time and damping but increases the natural frequency of the closed-loop system. However, phase-lag control when applied properly improves damping but usually results in a longer rise time and settling time. Therefore, each of these control schemes has its advantages, disadvantages, and limitations, and there are many systems that cannot be satisfactorily compensated by either scheme acting alone. It is natural, therefore, whenever necessary, to consider using a combination of the lead and lag controllers, so that the advantages of both schemes are utilized.

The transfer function of a simple lag–lead (or lead–lag) controller can be written

$$G_c(s) = G_{c1}(s)G_{c2}(s) = \left(\frac{1 + a_1 T_1 s}{1 + T_1 s}\right)\left(\frac{1 + a_2 T_2 s}{1 + T_2 s}\right) \quad (a_1 > 1, a_2 < 1) \quad (9\text{-}139)$$
$$|\leftarrow \text{lead} \rightarrow||\leftarrow \text{lag} \rightarrow|$$

The gain factors of the lead and lag controllers are not included because, as shown previously, these gain and attenuation are compensated eventually by the adjustment of the forward gain K.

Because the lead–lag controller transfer function in Eq. (9-139) now has four unknown parameters, its design is not as straightforward as the single-stage phase-lead or phase-lag controller. *In general, the phase-lead portion of the controller is used mainly to achieve a shorter rise time and higher bandwidth, and the phase-lag portion is brought in to provide major damping of the system.* Either the phase-lead or the phase-lag control can be designed first. We shall use Example 9-7-1 to illustrate the design steps.

▶ **EXAMPLE 9-7-1** As an illustrative example of designing a lead–lag controller, let us consider the sun-seeker system of Example 9-5-3. The uncompensated system with $K = 1$ was shown to be unstable. A two-stage phase-lead controller was designed in Example 9-6-1, and a single-stage phase-lag controller was designed in Example 9-6-2.

Based on the design in Example 9-5-3, we can first select a phase-lead control with $a = 70$ and $T_1 = 0.00004$. The remaining phase-lag control can be designed using either the root-locus method or the Bode plot method. Table 9-20 gives the results by letting $T_2 = 2$, which is an insensitive parameter, and various values of a. The results in Table 9-20 show that the optimal value of a_2, from the standpoint of minimizing the maximum overshoot, for $a_1 = 70$ and $T_2 = 0.00004$, is approximately 0.2. Compared with the single-stage phase-lag control designed in Example 9-6-1, the BW is increased to 351.4 rad/sec from 66.94 rad/sec, and the rise time is reduced to 0.00668 sec from 0.01273 sec. The system with the lead–lag controller should be more robust, because the magnitude of the sensitivity function should not increase to unity until near the BW of 351.4 rad/sec. As a comparison, the unit-step responses of the system with the two-stage phase-lead control, the single-stage phase-lag control, and the lead–lag control are shown in Fig. 9-52.

TABLE 9-20 Performance Attributes of Sun-Seeker System in Example 9-7-1 with Lead–Lag Controller: $a_1 = 70$, $T_1 = 0.00004$

a_2	T_2	PM (deg)	M_r	BW (rad/sec)	Maximum Overshoot (%)	t_r (sec)	t_s (sec)
0.1	20	81.81	1.004	122.2	0.4	0.01843	0.02626
0.15	20	76.62	1.002	225.5	0.2	0.00985	0.01515
0.20	20	70.39	1.001	351.4	0.1	0.00668	0.00909
0.25	20	63.87	1.001	443.0	4.9	0.00530	0.00707

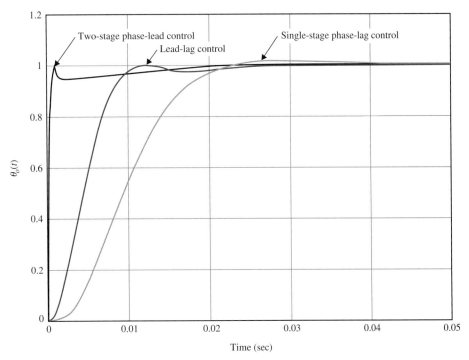

Figure 9-52 Sun-seeker system in Example 9-7-1 with single-stage phase-lag controller, lead–lag controller, and two-stage phase-lead controller.

9-8 POLE-ZERO-CANCELLATION DESIGN: NOTCH FILTER

The transfer functions of many controlled processes contain one or more pairs of complex-conjugate poles that are very close to the imaginary axis of the s-plane. These complex poles usually cause the closed-loop system to be lightly damped or unstable. One immediate solution is to use a controller that has a transfer function with zeros selected, which would cancel the undesirable poles of the controlled process, and to place the poles of the controller at more desirable locations in the s-plane to achieve the desired dynamic performance. For example, if the transfer function of a process is

$$G_p(s) = \frac{K}{s(s^2 + s + 10)} \qquad (9\text{-}140)$$

in which the complex-conjugate poles may cause stability problems in the closed-loop system when the value of K is large, the suggested series controller may be of the form

$$G_c(s) = \frac{s^2 + s + 10}{s^2 + as + b} \qquad (9\text{-}141)$$

The constants a and b may be selected according to the performance specifications of the closed-loop system.

There are practical difficulties with the pole-zero-cancellation design scheme that should prevent the method from being used indiscriminately. The problem is that in practice *exact* cancellation of poles and zeros of transfer functions is rarely possible. In practice, the transfer function of the process, $G_p(s)$, is usually determined through testing and physical modeling; linearization of a nonlinear process and approximation of a complex process are unavoidable. Thus, the *true* poles and zeros of the transfer function of the process may not be accurately modeled. In fact, the true order of the system may even be higher than that represented by the transfer function used for modeling purposes. Another difficulty is that the dynamic properties of the process may vary, even very slowly, due to aging of the system components or changes in the operating environment, so the poles and zeros of the transfer function may move during the operation of the system. The parameters of the controller are constrained by the actual physical components available and cannot be assigned arbitrarily. For these and other reasons, even if we could precisely design the poles and zeros of the transfer function of the controller, exact pole–zero cancellation is almost never possible in practice. We will now show that, in most cases, exact cancellation is *not* really necessary to effectively negate the influence of the undesirable poles using pole-zero-cancellation compensation schemes.

Let us assume that a controlled process is represented by

$$G_p(s) = \frac{K}{s(s + p_1)(s + \bar{p}_1)} \qquad (9\text{-}142)$$

where p_1 and $\bar{p}_1$ are the two complex-conjugate poles that are to be canceled. Let the transfer function of the series controller be

$$G_c(s) = \frac{(s + p_1 + \varepsilon_1)(s + \bar{p}_1 + \bar{\varepsilon}_1)}{s^2 + as + b} \tag{9-143}$$

where ε_1 is a complex number whose magnitude is very small and $\bar{\varepsilon}_1$ is its complex conjugate. The open-loop transfer function of the compensated system is

$$G(s) = G_c(s)G_p(s) = \frac{K(s + p_1 + \varepsilon_1)(s + \bar{p}_1 + \bar{\varepsilon}_1)}{s(s + p_1)(s + \bar{p}_1)(s^2 + as + b)} \tag{9-144}$$

Because of inexact cancellation, we cannot discard the terms $(s + p_1)(s + \bar{p}_1)$ in the denominator of Eq. (9-144). The closed-loop transfer function is

$$\frac{Y(s)}{R(s)} = \frac{K(s + p_1 + \varepsilon_1)(s + \bar{p}_1 + \bar{\varepsilon}_1)}{s(s + p_1)(s + \bar{p}_1)(s^2 + as + b) + K(s + p_1 + \varepsilon_1)(s + \bar{p}_1 + \bar{\varepsilon}_1)} \tag{9-145}$$

The root-locus diagram in Fig. 9-53 explains the effect of inexact pole–zero cancellation. Notice that the two closed-loop poles as a result of inexact cancellation lie between the pairs of poles and zeros at $s = -p_1, -\bar{p}_1$ and $-p_1 - \varepsilon_1, -\bar{p}_1 - \bar{\varepsilon}_1$, respectively. Thus,

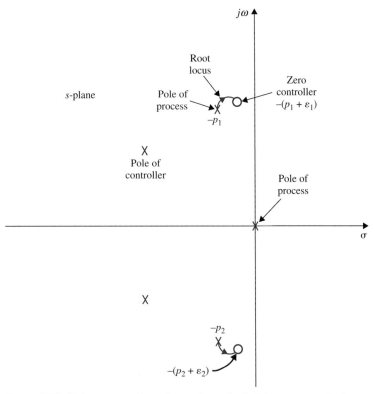

Figure 9-53 Pole–zero configuration and root loci of inexact cancellation.

these closed-loop poles are very close to the open-loop poles and zeros that are meant to be canceled. Eq. (9-145) can be approximated as

$$\frac{Y(s)}{R(s)} \cong \frac{K(s + p_1 + \varepsilon_1)(s + \bar{p}_1 + \bar{\varepsilon}_2)}{(s + p_1 + \delta_1)(s + \bar{p}_1 + \delta_1)(s^3 + as + b + K)} \quad (9\text{-}146)$$

where δ_1 and δ_1 are a pair of very small complex-conjugate numbers that depend on $\varepsilon_1, \bar{\varepsilon}_1$, and all the other parameters. The partial-fraction expansion of Eq. (9-146) is

$$\frac{Y(s)}{R(s)} \cong \frac{K_1}{s + p_1 + \delta_1} + \frac{K_2}{s + \bar{p}_1 + \delta_1} + \text{terms due to the remaining poles} \quad (9\text{-}147)$$

We can show that K_1 is proportional to $\varepsilon_1 - \delta_1$, which is a very small number. Similarly, K_2 is also very small. *This exercise simply shows that, although the poles at $-p_1$ and $-p_2$ cannot be canceled precisely, the resulting transient-response terms due to inexact cancellation will have insignificant amplitudes, so unless the controller zeros earmarked for cancellation are too far off target, the effect can be neglected for all practical purposes.* Another way of viewing this problem is that the zeros of $G(s)$ are retained as the zeros of closed-loop transfer function $Y(s)/R(s)$, so from Eq. (9-146), we see that the two pairs of poles and zeros are close enough to be canceled from the transient-response standpoint.

Keep in mind that we should never attempt to cancel poles that are in the right-half s-plane, because any inexact cancellation will result in an unstable system. Inexact cancellation of poles could cause difficulties if the unwanted poles of the process transfer function are very close to or right on the imaginary axis of the s-plane. In this case, inexact cancellation may also result in an unstable system. Fig. 9-54(a) illustrates a situation in which

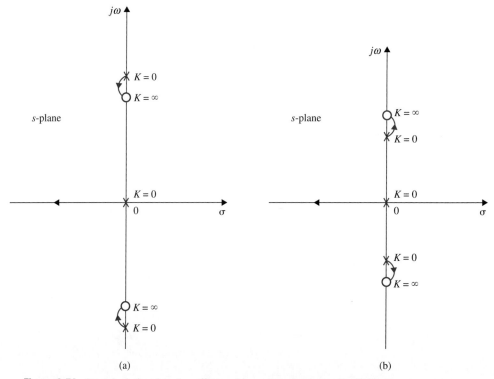

Figure 9-54 Root loci showing the effects of inexact pole–zero cancellations.

the relative positions of the poles and zeros intended for cancellation result in a stable system, whereas in Fig. 9-54(b), the inexact cancellation is unacceptable. The relative distance between the poles and zeros intended for cancellation is small, which results in residual terms in the time response solution. Although these terms have very small amplitudes, they tend to grow without bound as time increases. Hence the system response becomes unstable.

9-8-1 Second-Order Active Filter

Transfer functions with complex poles and/or zeros can be realized by electric circuits with op-amps. Consider the transfer function

$$G_c(s) = \frac{E_2(s)}{E_1(s)} = K\frac{s^2 + b_1 s + b_2}{s^2 + a_1 s + a_2} \qquad (9\text{-}148)$$

where $a_1, a_2, b_1,$ and b_2 are real constants. The active-filter realization of Eq. (9-148) can be accomplished by using the direct decomposition scheme of state variables discussed in Section 10-10. A typical op-amp circuit is shown in Fig. 9-55. The parameters of the transfer function in Eq. (9-148) are related to the circuit parameters as follows:

$$K = -\frac{R_6}{R_7} \qquad (9\text{-}149)$$

$$a_1 = \frac{1}{R_1 C_1} \qquad (9\text{-}150)$$

$$a_2 = \frac{1}{R_2 R_4 C_1 C_2} \qquad (9\text{-}151)$$

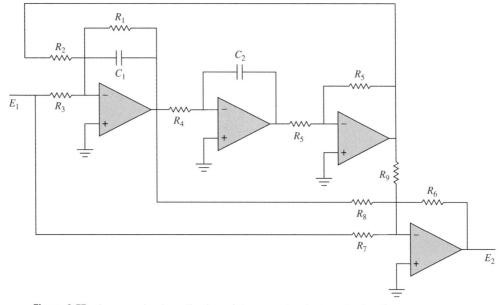

Figure 9-55 Op-amp circuit realization of the second-order transfer function. $\frac{E_2(s)}{E_1(s)} = K\frac{s^2 + b_1 s + b_2}{s^2 + a_1 s + a_2}.$

$$b_1 = \left(1 - \frac{R_1 R_7}{R_3 R_8}\right) a_1 \quad (b_1 < a_1) \tag{9-152}$$

$$b_2 = \left(1 - \frac{R_2 R_7}{R_3 R_9}\right) a_2 \quad (b_2 < a_2) \tag{9-153}$$

Because $b_1 < a_1$, the zeros of $G_c(s)$ in Eq. (9-148) are less damped and are closer to the origin in the s-plane than the poles. By setting various combinations of R_7 and R_8, and R_9 to infinity, a variety of second-order transfer functions can be realized. Note that all the parameters can be adjusted independently of one another. For example, R_1 can be adjusted to set a_1; R_4 can be adjusted to set a_2; and b_1 and b_2 are set by adjusting R_8 and R_9, respectively. The gain factor K is controlled independently by R_6.

9-8-2 Frequency-Domain Interpretation and Design

While it is simple to grasp the idea of pole-zero-cancellation design in the s-domain, the frequency-domain provides added perspective to the design principles. Fig. 9-56

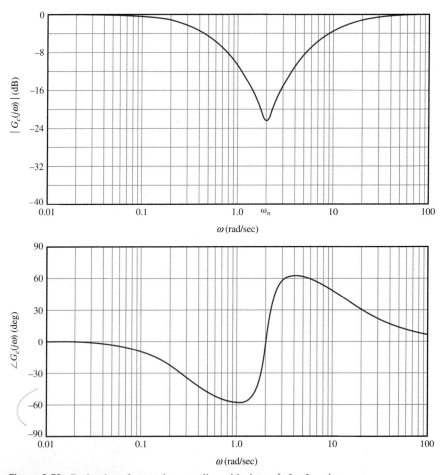

Figure 9-56 Bode plot of a notch controller with the transfer function.
$$G(s) = \frac{(s^2 + 0.8s + 4)}{(s + 0.384)(s + 10.42)}.$$

illustrates the Bode plot of the transfer function of a typical second-order controller with complex zeros. The magnitude plot of the controller typically has a "notch" at the resonant frequency ω_n. The phase plot is negative below and positive above the resonant frequency, while passing through zero degrees at the resonant frequency. The attenuation of the magnitude curve and the positive-phase characteristics can be used effectively to improve the stability of a linear system. Because of the "notch" characteristic in the magnitude curve, the controller is also referred to in the industry as a **notch filter** or **notch controller**.

From the frequency-domain standpoint, the notch controller has advantages over the phase-lead and phase-lag controllers in certain design conditions, because the magnitude and phase characteristics do not affect the high- and low-frequency properties of the system. Without using the pole-zero-cancellation principle, the design of the notch controller for compensation in the frequency domain involves the determination of the amount of attenuation required and the resonant frequency of the controller.

Let us express the transfer function of the notch controller in Eq. (9-148) as

$$G_c(s) = \frac{s^2 + 2\zeta_z\omega_n s + \omega_n^2}{s^2 + 2\zeta_p\omega_n s + \omega_n^2} \tag{9-154}$$

where we have made the simplification by assuming that $a_2 = b_2$.

The attenuation provided by the magnitude of $G_c(j\omega)$ at the resonant frequency ω_n is

$$|G_c(j\omega_n)| = \frac{\zeta_z}{\zeta_p} \tag{9-155}$$

Thus, knowing the maximum attenuation required at ω_n, the ratio of ζ_z/ζ_p is known.

The following example illustrates the design of the notch controller based on pole–zero cancellation and required attenuation at the resonant frequency.

▶ **EXAMPLE 9-8-1** Complex-conjugate poles in system transfer functions are often due to compliances in the coupling between mechanical elements. For instance, if the shaft between the motor and load is nonrigid, the shaft is modeled as a torsional spring, which could lead to complex-conjugate poles in the process transfer function. Fig. 9-57 shows a speed-control system in which the coupling between the motor and the load is modeled as a torsional spring. The system equations are

$$T_m(t) = J_m \frac{d\omega_m(t)}{dt} + B_m\omega_m(t) + J_L \frac{d\omega_L(t)}{dt} \tag{9-156}$$

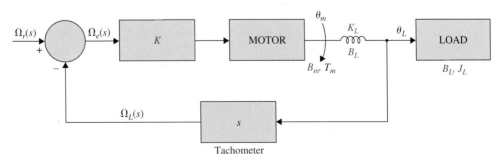

Figure 9-57 Block diagram of speed-control system in Example 9-8-1.

$$K_L[\theta_m(t) - \theta_L(t)] + B_L[\omega_m(t) - \omega_L(t)] = J_L \frac{d\omega_L(t)}{dt} \tag{9-157}$$

$$T_m(t) = K\omega_e(t) \tag{9-158}$$

$$\omega_e(t) = \omega_r(t) - \omega_L(t) \tag{9-159}$$

$T_m(t)$ = motor torque
$\omega_m(t)$ = motor angular velocity
$\omega_L(t)$ = load angular velocity
$\theta_L(t)$ = load angular displacement
$\theta_m(t)$ = motor angular displacement
J_m = motor inertia = 0.0001 oz-in.-sec^2
J_L = load inertia = 0.0005 oz-in.-sec^2
B_m = viscous-friction coefficient of motor = 0.01 oz-in.-sec
B_L = viscous-friction coefficient of shaft = 0.001 oz-in.-sec
K_L = spring constant of shaft = 100 on-in./rad
K = amplifier gain = 1

The loop transfer function of the system is

$$G_p(s) = \frac{\Omega_L(s)}{\Omega_e(s)}$$
$$= \frac{B_L s + K_L}{J_m J_L s^3 + (B_m J_L + B_L J_m + B_L J_L)s^2 + (K_L J_L + B_m B_L + K_L J_m)s + B_m K_L} \tag{9-160}$$

By substituting the system parameters in the last equation, $G_p(s)$ becomes

$$G_p(s) = \frac{20,000(s + 100,000)}{s^3 + 112s^2 + 1,200,200s + 20,000,000}$$
$$= \frac{20,000(s + 100,000)}{(s + 16.69)(s + 47.66 + j1094)(s + 47.66 - j1094)} \tag{9-161}$$

Thus, the shaft compliance between the motor and the load creates two complex-conjugate poles in $G_p(s)$ that are lightly damped. The resonant frequency is approximately 1095 rad/sec, and the closed-loop system is unstable. The complex poles of $G_p(s)$ would cause the speed response to oscillate even if the system were stable.

Pole-Zero-Cancellation Design with Notch Controller
The following are the performance specifications of the system:

The steady-state speed of the load due to a unit-step input should have an error of not more than 1%.

Maximum overshoot of output speed $\leq 5\%$.

Rise time $t_r < 0.5$ sec.

Settling time $t_s < 0.5$ sec.

To compensate the system, we need to get rid, or, perhaps more realistically, minimize the effect, of the complex poles of $G_p(s)$ at $s = -47.66 + j1094$ and $-47.66 - j1094$. Let us select a notch controller with the transfer function given in Eq. (9-154) to improve the performance of the system. The complex-conjugate zeros of the controller should be so placed that they will cancel the undesirable poles of the process. Therefore, the transfer function of the notch controller should be

$$G_c(s) = \frac{s^2 + 95.3s + 1{,}198{,}606.6}{s^2 + 2\zeta_p \omega_n s + \omega_n^2} \qquad (9\text{-}162)$$

The forward-path transfer function of the compensated system is

$$G(s) = G_c(s)G_p(s) = \frac{20{,}000(s + 100{,}000)}{(s + 16.69)\left(s^2 + 2\zeta_p \omega_n s + \omega_n^2\right)} \qquad (9\text{-}163)$$

Because the system is type 0, the step-error constant is

$$K_P = \lim_{s \to 0} G(s) = \frac{2 \times 10^9}{16.69 \times \omega_n^2} = \frac{1.198 \times 10^8}{\omega_n^2} \qquad (9\text{-}164)$$

For a unit-step input, the steady-state error of the system is written

$$e_{ss} = \lim_{t \to \infty} \omega_e(t) = \lim_{s \to 0} s\Omega_e(s) = \frac{1}{1 + K_P} \qquad (9\text{-}165)$$

Thus, for the steady-state error to be less than or equal to 1%, $K_P \geq 99$. The corresponding requirement on ω_n is found from Eq. (9-164),

$$\omega_n \leq 1210 \qquad (9\text{-}166)$$

We can show that, from the stability standpoint, it is better to select a large value for ω_n. Thus, let $\omega_n = 1200$ rad/sec, which is at the high end of the allowable range from the steady-state error standpoint. However, the design specifications given above can only be achieved by using a very large value for ζ_p. For example, when $\zeta_p = 15{,}000$, the time response has the following performance attributes:

Maximum overshoot = 3.7%

Rise time $t_r = 0.1897$ sec

Settling time $t_s = 0.256$ sec

Although the performance requirements are satisfied, the solution is unrealistic, because the extremely large value for ζ_p cannot be realized by physically available controller components.

Let us choose $\zeta_p = 10$ and $\omega_n = 1000$ rad/sec. The forward-path transfer function of the system with the notch controller is

$$G(s) = G_c(s)G_p(s) = \frac{20{,}000(s + 100{,}000)}{(s + 16.69)(s + 50)(s + 19{,}950)} \qquad (9\text{-}167)$$

We can show that the system is stable, but the maximum overshoot is 71.6%. Now we can regard the transfer function in Eq. (9-167) as a new design problem. There are a number of

TABLE 9-21 Time-Domain Performance Attributes of System in Example 9-8-1 with Notch-Phase-Lag Controller

a	T	aT	Maximum Overshoot (%)	t_r (sec)	t_s (sec)
0.001	10	0.01	14.8	0.1244	0.3836
0.002	10	0.02	10.0	0.1290	0.3655
0.004	10	0.04	3.2	0.1348	0.1785
0.005	10	0.05	1.0	0.1375	0.1818
0.0055	10	0.055	0.3	0.1386	0.1889
0.006	10	0.06	0	0.1400	0.1948

possible solutions to the problem of meeting the design specifications given. We can introduce a phase-lag controller or a PI controller, among other possibilities.

Second-Stage Phase-Lag Controller Design

Let us design a phase-lag controller as the second-stage controller for the system. The roots of the characteristic equation of the system with the notch controller are at $s = -19954, -31.328 + j316.36$, and $-31.328 - j316.36$. The transfer function of the phase-lag controller is

$$G_{c1}(s) = \frac{1 + aTs}{1 + Ts} \quad (a < 1) \tag{9-168}$$

where for design purposes we have omitted the gain factor $1/a$ in Eq. (9-168).

Let us select $T = 10$ for the phase-lag controller. Table 9-21 gives time-domain performance attributes for various values of a. The best value of a from the overall performance standpoint appears to be 0.005. Thus, the transfer function of the phase-lag controller is

$$G_{c1}(s) = \frac{1 + aTs}{1 + Ts} = \frac{1 + 0.05s}{1 + 10s} \tag{9-169}$$

The forward-path transfer function of the compensated system with the notch-phase-lag controller is

$$G(s) = G_c(s)G_{c1}(s)G_p(s) = \frac{20,000(s + 100,000)(1 + 0.05s)}{(s + 16.69)(s + 50)(s + 19,950)(1 + 10s)} \tag{9-170}$$

The unit-step response of the system is shown in Fig. 9-58. Because the step-error constant is 120.13, the steady-state speed error due to a step input is 1/120.13, or 0.83%.

Second-Stage PI Controller Design

A PI controller can be applied to the system to improve the steady-state error and the stability simultaneously. The transfer function of the PI control is written

$$G_{c2}(s) = K_P + \frac{K_I}{s} = K_P\left(\frac{s + K_I/K_P}{s}\right) \tag{9-171}$$

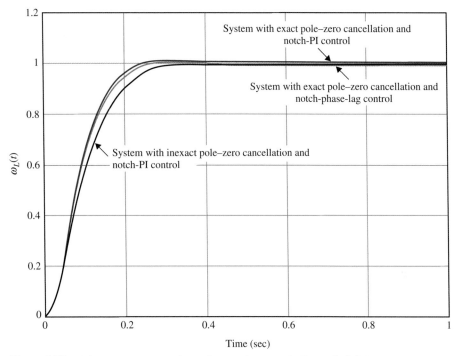

Figure 9-58 Unit-step responses of speed-control system in Example 9-8-1.

We can design the PI controller based on the phase-lag controller by writing Eq. (9-169) as

$$G_{c1}(s) = 0.005\left(\frac{s+20}{s+0.1}\right) \quad (9\text{-}172)$$

Thus, we can set $K_P = 0.005$ and $K_I/K_P = 20$. Then, $K_I = 0.1$. Fig. 9-58 shows the unit-step response of the system with the notch-PI controller. The attributes of the step response are as follows:

% Maximum overshoot = 1%
Rise time $t_r = 0.1380$ sec
Settling time $t_s = 0.1818$ sec

which are extremely close to those with the notch-phase-lag controller, except that in the notch-PI case the steady-state velocity error is zero when the input is a step function.

Sensitivity Due to Imperfect Pole–Zero Cancellation

As mentioned earlier, exact cancellation of poles and zeros is almost never possible in real life. Let us consider that the numerator polynomial of the notch controller in Eq. (9-162) cannot be exactly realized by physical resistor and capacitor components. Rather, the transfer function of the notch controller is more realistically chosen as

$$G_c(s) = \frac{s^2 + 100s + 1{,}000{,}000}{s^2 + 20{,}000s + 1{,}000{,}000} \quad (9\text{-}173)$$

Fig. 9-58 shows the unit-step response of the system with the notch controller in Eq. (9-173). The attributes of the unit-step response are as follows:

% Maximum overshoot = 0.4%

Rise time $t_r = 0.17$ sec

Settling time $t_s = 0.2323$ sec

Frequency-Domain Design

To carry out the design of the notch controller, we refer to the Bode plot of Eq. (9-161) shown in Fig. 9-59. Due to the complex-conjugate poles of $G_p(s)$, the magnitude plot has a peak of 24.86 dB at 1095 rad/sec. From the Bode plot in Fig. 9-59, we see that we may want to bring the magnitude plot down to -20 dB at the resonant frequency of 1095 rad/sec so that the resonance is smoothed out. This requires an attenuation of -44.86 dB. Thus, from Eq. (9-155),

$$|G_c(j\omega_c)| = -44.86 \text{ dB} = \frac{\zeta_z}{\zeta_p} = \frac{0.0435}{\zeta_p} \tag{9-174}$$

where ζ_z is found from Eq. (9-62). Solving for ζ_p from the last equation, we get $\zeta_p = 7.612$. The attenuation should be placed at the resonant frequency of 1095 rad/sec; thus, $\omega_n = 1095$ rad/sec. The notch controller of Eq. (9-162) becomes

$$G_c(s) = \frac{s^2 + 95.3s + 1,198,606.6}{s^2 + 16,670.28s + 1,199,025} \tag{9-175}$$

The Bode plot of the system with the notch controller in Eq. (9-175) is shown in Fig. 9-59. We can see that the system with the notch controller has a phase margin of only 13.7°, and M_r is 3.92.

To complete the design, we can use a PI controller as a second-stage controller. Following the guideline given in Section 9-3 on the design of a PI controller, we assume that the desired phase margin is 80°. From the Bode plot in Fig. 9-59, we see that, to realize a phase margin of 80°, the new gain-crossover frequency should be $\omega_g' = 43$ rad/sec, and the magnitude of $G(j\omega_g')$ is 30 dB. Thus, from Eq. (9-32),

$$K_P = 10^{-|G(j\omega_g')|_{dB}/20} = 10^{-30/20} = 0.0316 \tag{9-176}$$

The value of K_I is determined using the guideline given by Eq. (9-25),

$$K_I = \frac{\omega_g' K_P}{10} = \frac{43 \times 0.0316}{10} = 0.135 \tag{9-177}$$

Because the original system is type 0, the final design needs to be refined by adjusting the value of K_I. Table 9-22 gives the performance attributes when $K_P = 0.0316$ and K_I is varied from 0.135. From the best maximum overshoot, rise time, and settling time measures, the best value of K_I appears to be 0.35. The forward-path transfer function of the compensated system with the notch-PI controller is

$$G(s) = \frac{20,000(s + 100,000)(0.0316s + 0.35)}{s(s + 16.69)(s^2 + 16,670.28s + 1,199,025)} \tag{9-178}$$

9-8 Pole-Zero-Cancellation Design: Notch Filter ◀ 587

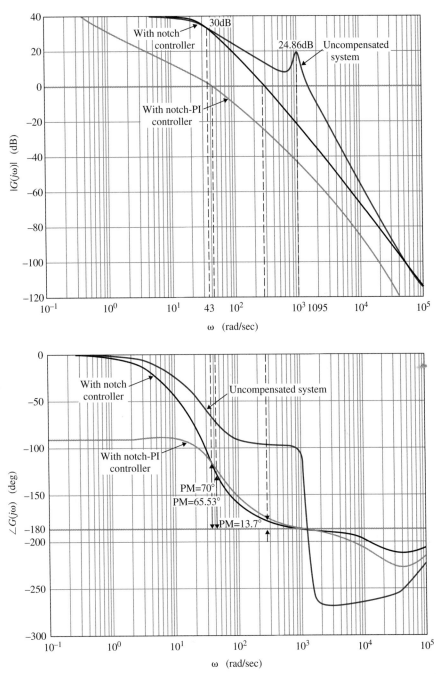

Figure 9-59 Bode plots of the uncompensated speed-control system in Example 9-8-1, with notch controller and with notch-PI controller.

Figure 9-59 shows the Bode plot of the system with the notch-PI controller, with $K_P = 0.0316$ and $K_I = 0.35$. The unit-step responses of the compensated system with $K_P = 0.0316$ and $K_I = 0.135, 0.35$, and 0.40 are shown in Fig. 9-60.

TABLE 9-22 Performance Attributes of System in Example 9-8-1 with Notch-PI Controller Designed in Frequency Domain

K_P	K_I	PM (deg)	M_r	Maximum Overshoot (%)	t_r (sec)	t_s (sec)
0.0316	0.1	76.71	1.00	0	0.2986	0.5758
0.0316	0.135	75.15	1.00	0	0.2036	0.4061
0.0316	0.200	72.22	1.00	0	0.0430	0.2403
0.0316	0.300	67.74	1.00	0	0.0350	0.1361
0.0316	0.350	65.53	1.00	1.6	0.0337	0.0401
0.0316	0.400	63.36	1.00	4.3	0.0323	0.0398

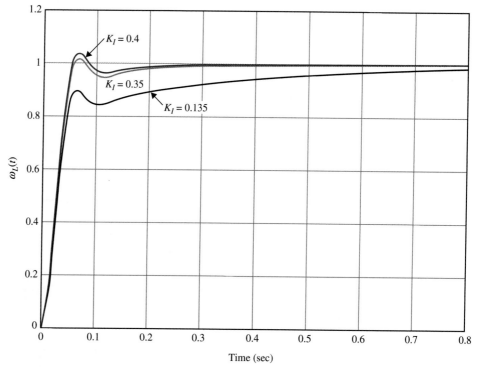

Figure 9-60 Unit-step responses of speed-control system in Example 9-8-1 with notch-PI controller, $G_c(s) = \dfrac{s^2 + 95.3s + 1,198,606.6}{s^2 + 16,670.28s + 1,199,025}$ $G_{c2}(s) = 0.0316 + \dfrac{0.35}{s}$.

▶ 9-9 FORWARD AND FEEDFORWARD CONTROLLERS

The compensation schemes discussed in the preceding sections all have one degree of freedom in that there is essentially one controller in the system, although the controller can contain several stages connected in series or in parallel. The limitations of a one-degree-of-freedom controller were discussed in Section 9-1. The two-degree-of-freedom compensation scheme shown in Fig. 9-2(d) through Fig. 9-2(f) offers design flexibility when a multiple number of design criteria have to be satisfied simultaneously.

From Fig. 9-2(e), the closed-loop transfer function of the system is

$$\frac{Y(s)}{R(s)} = \frac{G_{cf}(s)G_c(s)G_p(s)}{1 + G_c(s)G_p(s)} \qquad (9\text{-}179)$$

and the error transfer function is

$$\frac{E(s)}{R(s)} = \frac{1}{1 + G_c(s)G_p(s)} \qquad (9\text{-}180)$$

Thus, the controller $G_c(s)$ can be designed so that the error transfer function will have certain desirable characteristics, and the controller $G_{cf}(s)$ can be selected to satisfy performance requirements with reference to the input–output relationship. Another way of describing the flexibility of a two-degree-of-freedom design is that the controller $G_c(s)$ is usually designed to provide a certain degree of system stability and performance, but because the zeros of $G_c(s)$ always become the zeros of the closed-loop transfer function, unless some of the zeros are canceled by the poles of the process transfer function, $G_p(s)$, these zeros may cause a large overshoot in the system output even when the relative damping as determined by the characteristic equation is satisfactory. In this case and for other reasons, the transfer function $G_{cf}(s)$ may be used for the control or cancellation of the undesirable zeros of the closed-loop transfer function, while keeping the characteristic equation intact. Of course, we can also introduce zeros in $G_{cf}(s)$ to cancel some of the undesirable poles of the closed-loop transfer function that could not be otherwise affected by the controller $G_c(s)$. The feedforward compensation scheme shown in Fig. 9-2(f) serves the same purpose as the forward compensation, and the difference between the two configurations depends on system and hardware implementation considerations.

It should be kept in mind that, while the forward and feedforward compensations may seem powerful because they can be used directly for the addition or deletion of poles and zeros of the closed-loop transfer function, there is a fundamental question involving the basic characteristics of feedback. If the forward or feedforward controller is so powerful, then why do we need feedback at all? Because $G_{cf}(s)$ in the systems of Figs. 9-2(e) and 9-2(f) are outside the feedback loop, the system is susceptible to parameter variations in $G_{cf}(s)$. Therefore, in reality, these types of compensation cannot be satisfactorily applied to all situations.

▶ **EXAMPLE 9-9-1** As an illustration of the design of the forward and feedforward compensators, consider the second-order sun-seeker system with phase-lag control designed in Example 9-6-1. One of the disadvantages of phase-lag compensation is that the rise time is usually quite long. Let us consider that the phase-lag-compensated sun-seeker system has the forward-path transfer function

$$G(s) = G_c(s)G_p(s) = \frac{2500(1 + 10s)}{s(s + 25)(1 + 100s)} \qquad (9\text{-}181)$$

The time-response attributes are as follows:

Maximum overshoot $= 2.5\%$

$t_r = 0.1637$ sec

$t_s = 0.2020$ sec

We can improve the rise time and the settling time while not appreciably increasing the overshoot by adding a PD controller $G_{cf}(s)$ to the system, as shown in Fig. 9-61(a). This effectively adds a zero to the closed-loop transfer function while not affecting the characteristic equation. Selecting the PD controller as

$$G_{cf}(s) = 1 + 0.05s \qquad (9\text{-}182)$$

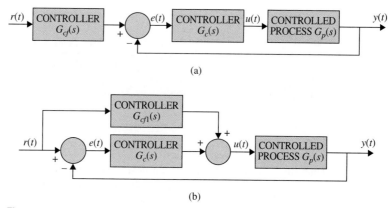

(a)

(b)

Figure 9-61 (a) Forward compensation with series compensation. (b) Feedforward compensation with series compensations.

the time-domain performance attributes are as follows:

Maximum overshoot $= 4.3\%$

$t_r = 0.1069$ sec

$t_s = 0.1313$ sec

If instead the feedforward configuration of Fig. 9-61(b) is chosen, the transfer function of $G_{cf1}(s)$ is directly related to $G_{cf}(s)$; that is, equating the closed-loop transfer functions of the two systems in Figs. 9-61(a) and 9-61(b), we have

$$\frac{[G_{cf1}(s) + G_c(s)]G_p(s)}{1 + G_c(s)G_p(s)} = \frac{G_{cf}G_c(s)G_p(s)}{1 + G_c(s)G_p(s)} \qquad (9\text{-}183)$$

Solving for $G_{cf1}(s)$ from Eq. (9-183) yields

$$G_{cf1}(s) = [G_{cf}(s) - 1]G_c(s) \qquad (9\text{-}184)$$

Thus, with $G_{cf}(s)$ as given in Eq. (9-179), we have the transfer function of the feedforward controller:

$$G_{cf1}(s) = 0.05s\left(\frac{1 + 10s}{1 + 100s}\right) \qquad (9\text{-}185)$$

◀

▶ 9-10 DESIGN OF ROBUST CONTROL SYSTEMS

In many control-system applications, not only must the system satisfy the damping and accuracy specifications, but the control must also yield performance that is **robust** (insensitive) to external disturbance and parameter variations. We have shown that feedback in conventional control systems has the inherent ability to reduce the effects of external disturbance and parameter variations. Unfortunately, robustness with the conventional feedback configuration is achieved only with a high loop gain, which is normally detrimental to stability. Let us consider the control system shown in Fig. 9-62.

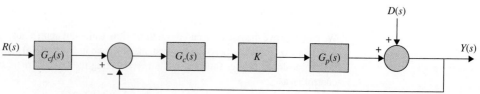

Figure 9-62 Control system with disturbance.

The external disturbance is denoted by the signal $d(t)$, and we assume that the amplifier gain K is subject to variation during operation. The input–output transfer function of the system when $d(t) = 0$ is

$$M(s) = \frac{Y(s)}{R(s)} = \frac{KG_{cf}(s)G_c(s)G_p(s)}{1 + KG_c(s)G_p(s)} \quad (9\text{-}186)$$

and the disturbance-output transfer function when $r(t) = 0$ is

$$T(s) = \frac{Y(s)}{D(s)} = \frac{1}{1 + KG_c(s)G_p(s)} \quad (9\text{-}187)$$

In general, the design strategy is to select the controller $G_c(s)$ so that the output $y(t)$ is insensitive to the disturbance over the frequency range in which the latter is dominant and to design the feedforward controller $G_{cf}(s)$ to achieve the desired transfer function between the input $r(t)$ and the output $y(t)$.

Let us define the sensitivity of $M(s)$ due to the variation of K as

$$S_K^M = \frac{\text{percent change in } M(s)}{\text{percent change in } K} = \frac{dM(s)/M(s)}{dK/K} \quad (9\text{-}188)$$

Then, for the system in Fig. 9-62,

$$S_K^M = \frac{1}{1 + KG_c(s)G_p(s)} \quad (9\text{-}189)$$

which is identical to Eq. (9-187). Thus, the sensitivity function and the disturbance-output transfer function are identical, which means that disturbance suppression and robustness with respect to variations of K can be designed with the same control schemes.

The following example shows how the two-degree-of-freedom control system of Fig. 9-62 can be used to achieve a high-gain system that will satisfy the performance and robustness requirements, as well as noise rejection.

▶ **EXAMPLE 9-10-1** Let us consider the second-order sun-seeker system in Example 9-6-1, which is compensated with phase-lag control. The forward-path transfer function is

$$G_p(s) = \frac{2500K}{s(s + 25)} \quad (9\text{-}190)$$

where $K = 1$. The forward-path transfer function of the phase-lag-compensated system with $a = 0.1$ and $T = 100$ is

$$G(s) = G_c(s)G_p(s) = \frac{2500K(1 + 10s)}{s(s + 25)(1 + 100s)} \quad (K = 1) \quad (9\text{-}191)$$

Because the phase-lag controller is a low-pass filter, the sensitivity of the closed-loop transfer function $M(s)$ with respect to K is poor. The bandwidth of the system is only 13.97 rad/sec, but it is expected that $|S_K^M(j\omega)|$ will be greater than unity at frequencies beyond 13.97 rad/sec. Fig. 9-63 shows the unit-step responses of the system when $K = 1$, the nominal value, and $K = 0.5$ and 2.0. Notice that, if for some reason, the forward gain K is changed from its nominal value, the system response of the phase-lag-compensated system would vary substantially. The attributes of the step responses and the characteristic equation roots are shown in Table 9-23 for the three values of K. Fig. 9-64 shows the root loci of the system with the phase-lag controller. The two complex roots of the characteristic equation vary substantially as K varies from 0.5 to 2.0.

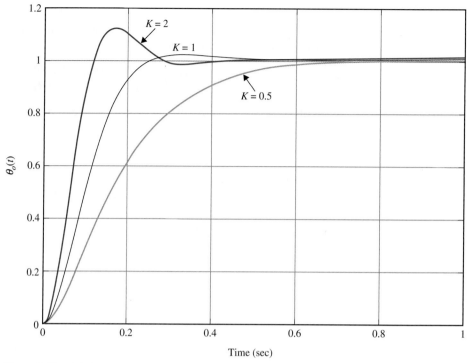

Figure 9-63 Unit-step responses of the second-order sun-seeker system with phase-lag controller, $G(s) = \dfrac{2500(1 + 10s)}{s(s + 25)(1 + 100s)}$.

TABLE 9-23 Attributes of Unit-Step Response of Second-Order Sun-Seeker System with Phase-Lag Controller in Example 9-10-1

K	Maximum Overshoot (%)	t_r (sec)	t_s (sec)	Roots of Characteristic Equation
2.0	12.6	0.07854	0.2323	$-0.1005 - 12.4548 \pm j18.51$
1.0	2.6	0.1519	0.2020	$-0.1009 - 12.4545 \pm j9.624$
0.5	1.5	0.3383	0.4646	$-0.1019 - 6.7628 - 18.1454$

Toolbox 9-10-1

Fig. 9-63 is obtained by the following sequence of MATLAB functions

```
K = 1;
num = K *2500 * [10 1]
den = conv([1 25 0], [100 1]);
[numCL,denCL]=cloop(num,den);
step(numCL,denCL)

hold on;

K = 2;
num = K*2500 * [10 1]
```

```
den = conv([1 25 0], [100 1]);
[numCL,denCL]=cloop(num,den);
step(numCL,denCL)

hold on;

K = 0.5;
num = K*2500 * [10 1]
den = conv([1 25 0], [100 1]);
[numCL,denCL]=cloop(num,den);
step(numCL,denCL)

hold on;
axis([0 1 0 1.2]);
grid
```

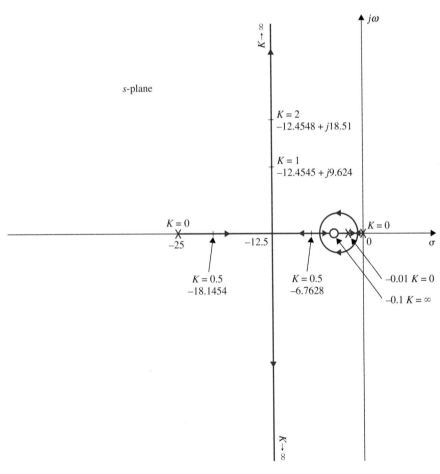

Figure 9-64 Root loci of the second-order sun-seeker system with phase-lag controller, $G(s) = \dfrac{2500(1 + 10s)}{s(s + 25)(1 + 100s)}$.

The design strategy of the robust controller is to place two zeros of the controller near the desired closed-loop poles, which according to the phase-lag-compensated system are at $s = -12.455 \pm j9.624$. Thus, we let the controller transfer function be

$$G_c(s) = \frac{(s + 13 + j10)(s + 13 - j10)}{269} = \frac{(s^2 + 26s + 269)}{269} \qquad (9\text{-}192)$$

The forward-path transfer function of the system with the robust controller is

$$G(s) = \frac{9.2937K(s^2 + 26s + 269)}{s(s + 25)} \qquad (9\text{-}193)$$

Toolbox 9-10-2

Fig. 9-64 is obtained by the following sequence of MATLAB functions

```
K = 1;
num = K *2500 * [10 1]
den = conv([1 25 0], [100 1]);
[numCL, denCL] = cloop(num, den)
rlocus(numCL,denCL);

hold on

K = 2;
num = K *2500 * [10 1]
den = conv([1 25 0], [100 1]);
[numCL, denCL] = cloop(num, den)
rlocus(numCL,denCL);

hold on

K = 1;
num = K *2500 * [10 1]
den = conv([1 25 0], [100 1]);
rlocus(num,den);

hold on

K = 2;
num = K *2500 * [10 1]
den = conv([1 25 0], [100 1]);
rlocus(num,den);

hold on
```

Fig. 9-65 shows the root loci of the system with the robust controller. By placing the two zeros of $G_c(s)$ near the desired characteristic equation roots, the sensitivity of the system is greatly improved. In fact, the root sensitivity near the two complex zeros at which the root loci terminate is very low. Fig. 9-65 shows that, when K approaches infinity, the two characteristic equation roots approach $-13 \pm j10$.

Toolbox 9-10-3

Fig. 9-65 is obtained by the following sequence of MATLAB functions

```
K = 1;
num = 9.2937*K *[1 26 269]
den = [1 25 0];
[numCL, denCL] = cloop(num, den)
rlocus(numCL,denCL);
```

```
hold on

K = 2;
num = 9.2937*K *[1 26 269]
den = [1 25 0];
[numCL, denCL] = cloop(num, den)
rlocus(numCL,denCL);
hold on

K = 1;
num = 9.2937*K *[1 26 269]
den = [1 25 0];
rlocus(num,den);

hold on

K = 2;
num = 9.2937*K *[1 26 269]
den = [1 25 0];
rlocus(num,den);
axis([-35 2 -15 15])
% grid
```

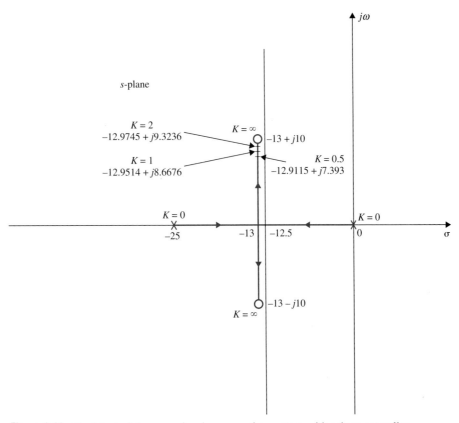

Figure 9-65 Root loci of the second-order sun-seeker system with robust controller, $G(s) = \dfrac{9.2937K\left(s^2 + 26s + 269\right)}{s(s+25)}$.

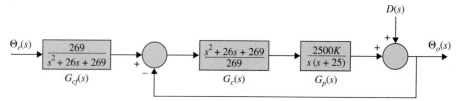

Figure 9-66 Second-order sun-seeker system with robust controller and forward controller.

Because the zeros of the forward-path transfer function are identical to the zeros of the closed-loop transfer function, the design is not complete by using only the series controller $G_c(s)$, because the closed-loop zeros will essentially cancel the closed-loop poles. This means we must add the forward controller, as shown in Fig. 9-62, where $G_{cf}(s)$ should contain poles to cancel the zeros of $s^2 + 26s + 269$ of the closed-loop transfer function. Thus, the transfer function of the forward controller is

$$G_{cf}(s) = \frac{269}{s^2 + 26s + 269} \tag{9-194}$$

The block diagram of the overall system is shown in Fig. 9-66. The closed-loop transfer function of the compensated system with $K = 1$ is

$$\frac{\Theta_o(s)}{\Theta_r(s)} = \frac{242.88}{s^2 + 25.903s + 242.88} \tag{9-195}$$

The unit-step responses of the system for $K = 0.5, 1.0$, and 2.0 are shown in Fig. 9-67, and their attributes are given in Table 9-24. As shown, the system is now very insensitive to the variation of K.

Because the system in Fig. 9-66 is now more robust, it is expected that the disturbance effect will be reduced. However, we cannot evaluate the effect of the controllers in the system of Fig. 9-66 by

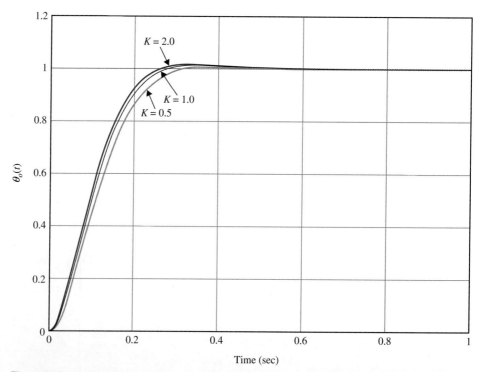

Figure 9-67 Unit-step responses of the second-order sun-seeker system with robust controller and forward controller.

TABLE 9-24 Attributes of Unit-Step Response of Second-Order Sun-Seeker System with Robust Controller in Example 9-10-1

K	Maximum Overshoot (%)	t_r (sec)	t_s (sec)	Roots of Characteristic Equation
2.0	1.3	0.1576	0.2121	$-12.9745 \pm j9.3236$
1.0	0.9	0.1664	0.2222	$-12.9514 \pm j8.6676$
0.5	0.5	0.1846	0.2525	$-12.9115 \pm j7.3930$

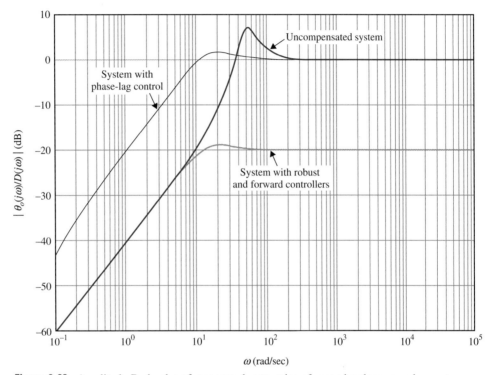

Figure 9-68 Amplitude Bode plot of response due to noise of second-order sun-seeker system.

applying a unit-step function as $d(t)$. The true improvement on the noise rejection properties is more appropriately analyzed by investigating the frequency response of $\Theta_o(s)/D(s)$. The noise-to-output transfer function, written from Fig. 9-66, is

$$\frac{\Theta_o(s)}{D(s)} = \frac{1}{1 + G_c(s)G_p(s)} = \frac{s(s+25)}{10.2937s^2 + 266.636s + 2500} \qquad (9\text{-}196)$$

The amplitude Bode plot of Eq. (9-196) is shown in Fig. 9-68, along with those of the uncompensated system and the system with phase-lag control. Notice that the magnitude of the frequency response between $D(s)$ and $\Theta_o(s)$ is much smaller than those of the system without compensation and with phase-lag control. The phase-lag control also accentuates the noise for frequencies up to approximately 40 rad/sec, adding more stability to the system. ◀

▶ **EXAMPLE 9-10-2** In this example, a robust controller with forward compensation is designed for the third-order sun-seeker system in Example 9-6-2 with phase-lag control. The forward-path transfer function of the uncompensated system is

$$G_p(s) = \frac{156,250,000K}{s(s^2 + 625s + 156,250)} \qquad (9\text{-}197)$$

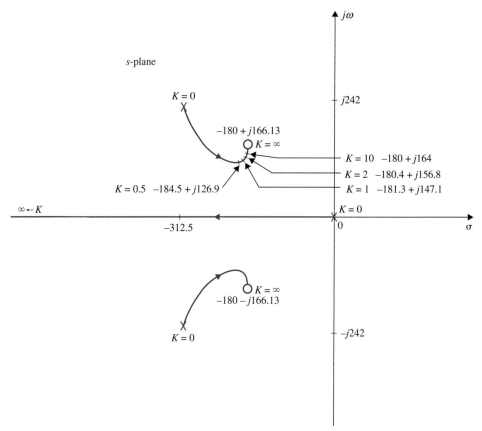

Figure 9-69 Root loci of the third-order sun-seeker system with robust and forward controllers.

where $K = 1$. The root loci of the closed-loop system are shown in Fig. 9-50, which lead to the phase-lag control with results shown in Table 9-19. Let us select the parameters of the phase-lag controller as $a = 0.1$ and $T = 20$. The dominant roots of the characteristic equation are $s = -187.73 \pm j164.93$.

Let us place the two zeros of the second-order robust controller at $-180 \pm j166.13$, so that the controller transfer function is

$$G_c(s) = \frac{s^2 + 360s + 60,000}{60,000} \qquad (9\text{-}198)$$

To ease the high-frequency realization problem of the controller, we may add two nondominant poles to $G_c(s)$. The following analysis is carried out with $G_c(s)$ given in Eq. (9-198), however. The root loci of the compensated system are shown in Fig. 9-69. Thus, by placing the zeros of the controller very

TABLE 9-25 Attributes of Unit-Step Response and Characteristic Equation Roots of Third-Order Sun-Seeker System with Robust and Forward Controllers in Example 9-10-2

K	Maximum Overshoot (%)	t_r (sec)	t_s (sec)	Characteristic Equation Roots
0.5	1.0	0.01115	0.01616	$-1558.1 - 184.5 \pm j126.9$
1.0	2.1	0.01023	0.01414	$-2866.6 - 181.3 \pm j147.1$
2.0	2.7	0.00966	0.01313	$-5472.6 - 180.4 \pm j156.8$
10.0	3.2	0.00924	0.01263	$-26307 - 180.0 \pm j164.0$

close to the desired dominant roots, the system is very insensitive to changing values of K near and beyond the nominal value of K. The forward controller has the transfer function

$$G_{cf}(s) = \frac{60,000}{s^2 + 360s + 60,000} \tag{9-199}$$

The attributes of the unit-step response for $K = 0.5, 1.0, 2.0,$ and 10.0 and the corresponding characteristic equation roots are given in Table 9-25.

▶ **EXAMPLE 9-10-3** In this example, we consider the design of a position-control system that has a variable load inertia. This type of situation is quite common in control systems. For example, the load inertia seen by the motor in an electronic printer will change when different printwheels are used. The system should have satisfactory performance for all the printwheels intended to be used with the system.

To illustrate the design of a robust system that is insensitive to the variation of load inertia, consider that the forward-path transfer function of a unity-feedback control system is

$$G_p(s) = \frac{KK_i}{s[(Js + B)(Ls + B) + K_i K_b]} \tag{9-200}$$

The system parameters are as follows:

K_i = motor torque constant = 1 N-m/A

K_b = motor back-emf constant = 1 V/rad/sec

R = motor resistance = 1 Ω

L = motor inductance = 0.01 H

B = motor and load viscous-friction coefficient $\cong 0$

J = motor and load inertia, varies between 0.01 and 0.02 N-m/rad/sec^2

K = amplifier again

Substituting these system parameters into Eq. (9-200), we get

$$\text{For } J = 0.01 \quad G_p(s) = \frac{10,000K}{s(s^2 + 100s + 10,000)} \tag{9-201}$$

$$\text{For } J = 0.02 \quad G_p(s) = \frac{5000K}{s(s^2 + 100s + 5000)} \tag{9-202}$$

The performance specifications are as follows:

Ramp error constant $K_v \geq 200$

Maximum overshoot $\leq 5\%$ or as small as possible

Rise time $t_r \leq 0.05$ sec

Settling time $t_s \leq 0.05$ sec

These specifications are to be maintained for $0.01 \leq J \leq 0.02$.

To satisfy the ramp-error constant requirement, the value of K must be at least 200. Fig. 9-70 shows the root loci of the uncompensated system for $J = 0.01$ and $J = 0.02$. We see that, regardless of the value of J, the uncompensated system is unstable for $K > 100$.

To achieve robust control, let us choose the system configuration of Fig. 9-61(a). We introduce a second-order series controller with the zeros placed near the desired dominant characteristic equation of the compensated system. The zeros should be so placed that the dominant characteristic equation roots would be insensitive to the variation in J. This is done by placing the two zeros at $-55 \pm j45$, although the exact location is unimportant. By choosing the two controller zeros as designated, the root loci of the compensated system show that the two complex roots of the characteristic equation will be very close to these zeros for various values of J, especially when the value of K is large. The transfer function of the robust controller is

$$G_c(s) = \frac{(s^2 + 110s + 5050)}{5050} \tag{9-203}$$

600 ▸ Chapter 9. Design of Control Systems

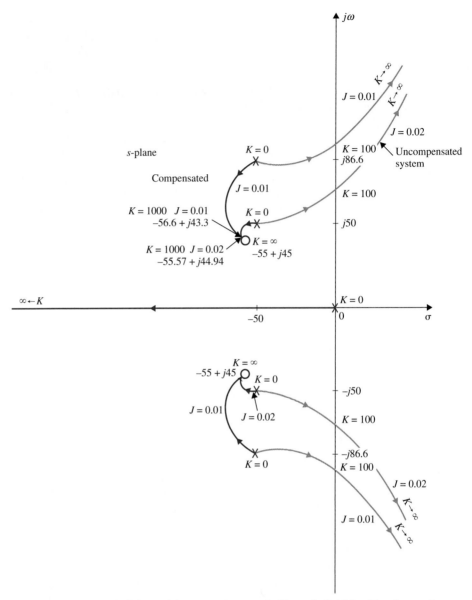

Figure 9-70 Root loci of the position-control system in Example 9-10-3 with robust and forward controllers.

As in the last example, we may add two nondominant poles to $G_c(s)$ to ease the high-frequency realization problem of the controller. The analysis is carried out with $G_c(s)$ given in Eq. (9-203).

Let $K = 1000$, although 200 would have been adequate to satisfy the K_v requirement. Then, for $J = 0.01$, the forward-path transfer function of the compensated system is

$$G(s) = G_c(s)G_p(s) = \frac{1980.198(s^2 + 110s + 5050)}{s(s^2 + 100s + 10,000)} \tag{9-204}$$

and for $J = 0.02$,

$$G(s) = \frac{990.99(s^2 + 110s + 5050)}{s(s^2 + 100s + 5000)} \tag{9-205}$$

TABLE 9-26 Attributes of Unit-Step Response and Characteristic Equation Roots of System with Robust and Forward Controllers in Example 9-10-3

J N-m/rad/sec^2	Maximum Overshoot (%)	t_r (sec)	t_s (sec)	Roots of Characteristic Equation
0.01	1.6	0.03453	0.04444	$-1967 - 56.60 \pm j43.3$
0.02	2.0	0.03357	0.04444	$-978.96 - 55.57 \pm j44.94$

To cancel the two zeros of the closed-loop transfer function, the transfer function of the forward controller is

$$G_{cf}(s) = \frac{5050}{s^2 + 110s + 5050} \tag{9-206}$$

The attributes of the unit-step response and the characteristic equation roots of the compensated system with $K = 1000$, $J = 0.01$, and $J = 0.02$ are given in Table 9-26. ◀

▶ 9-11 MINOR-LOOP FEEDBACK CONTROL

The control schemes discussed in the preceding sections have all utilized series controllers in the forward path of the main loop or feedforward path of the control system. Although series controllers are the most common because of their simplicity in implementation, depending on the nature of the system, sometimes there are advantages in placing the controller in a minor feedback loop, as shown in Fig. 9-2(b). For example, a tachometer may be coupled directly to a dc motor not only for the purpose of speed indication but more often for improving the stability of the closed-loop system by feeding back the output signal of the tachometer. The motor speed can also be generated by processing the back emf of the motor electronically. In principle, the PID controller or phase-lead and phase-lag controllers can all, with varying degrees of effectiveness, be applied as minor-loop feedback controllers. Under certain conditions, minor-loop control can yield systems that are more robust, that is, less sensitive to external disturbance or internal parameter variations.

9-11-1 Rate-Feedback or Tachometer-Feedback Control

The principle of using the derivative of the actuating signal to improve the damping of a closed-loop system can be applied to the output signal to achieve a similar effect. In other words, the derivative of the input signal is fed back and added algebraically to the actuating signal of the system. In practice, if the output variable is mechanical displacement, a tachometer may be used to convert mechanical displacement into an electrical signal that is proportional to the derivative of the displacement. Fig. 9-71 shows the block diagram of a control system with a secondary path that feeds back the derivative of the output. The

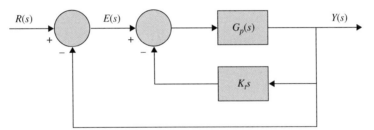

Figure 9-71 Control system with tachometer feedback.

transfer function of the tachometer is denoted by $K_t s$, where K_t is the tachometer constant, usually expressed in volts per radian per second for analytical purposes. Commercially, K_t is given in the data sheet of the tachometer, typically in volts per 1000 rpm. The effects of rate or tachometer feedback can be illustrated by applying it to a second-order prototype system. Consider that the controlled process of the system shown in Fig. 9-71 has the transfer function

$$G_p(s) = \frac{\omega_n^2}{s(s + 2\zeta\omega_n)} \tag{9-207}$$

The closed-loop transfer function of the system is

$$\frac{Y(s)}{R(s)} = \frac{\omega_n^2}{s^2 + (2\zeta\omega_n + K_t\omega_n^2)s + \omega_n^2} \tag{9-208}$$

and the characteristic equation is

$$s^2 + (2\zeta\omega_n + K_t\omega_n^2)s + \omega_n^2 = 0 \tag{9-209}$$

From the characteristic equation, *it is apparent that the effect of the tachometer feedback is the increase of the damping of the system, since* K_t *appears in the same term as the damping ratio* ζ.

In this respect, tachometer-feedback control has exactly the same effect as the PD control. However, the closed-loop transfer function of the system with PD control in Fig. 9-3 is

$$\frac{Y(s)}{R(s)} = \frac{\omega_n^2(K_P + K_D s)}{s^2 + (2\zeta\omega + K_D\omega_n^2)s + \omega_n^2 K_P} \tag{9-210}$$

Comparing the two transfer functions in Eqs. (9-208) and (9-210), we see that the two characteristic equations are identical if $K_P = 1$ and $K_D = K_t$. However, Eq. (9-210) has a zero at $s = -K_P/K_D$, whereas Eq. (9-208) does not. Thus, the response of the system with tachometer feedback is uniquely defined by the characteristic equation, whereas the response of the system with the PD control also depends on the zero at $s = -K_P/K_D$, which could have a significant effect on the overshoot of the step response.

With reference to the steady-state analysis, the forward-path transfer function of the system with tachometer feedback is

$$\frac{Y(s)}{E(s)} = \frac{\omega_n^2}{s(s + 2\zeta\omega_n + K_t\omega_n^2)} \tag{9-211}$$

Because the system is still type 1, the basic characteristics of the steady-state error are not altered by the tachometer feedback; that is, when the input is a step function, the steady-state error is zero. For a unit-ramp function input, the steady-state error of the system is $(2\zeta + K_t\omega_n)/\omega_n$, whereas that of the system with PD control in Fig. 9-3 is $2\zeta/\omega_n$. Thus, *for a type 1 system, tachometer feedback decreases the ramp-error constant* K_v *but does not affect the step-error constant* K_P.

9-11-2 Minor-Loop Feedback Control with Active Filter

Instead of using a tachometer, an active filter with RC elements and op-amps can be used to reduce cost and save space in the minor feedback loop for compensation. We illustrate this approach with the following example.

▶ **EXAMPLE 9-11-1** Consider that, for the second-order sun-seeker system in Example 9-6-1, instead of using a series controller in the forward path, we adopt the minor-loop feedback control, as shown in Fig. 9-72(a), with

$$G_p(s) = \frac{2500}{s(s+25)} \quad (9\text{-}212)$$

and

$$H(s) = \frac{K_t s}{1+Ts} \quad (9\text{-}213)$$

To maintain the system as type 1, it is necessary that $H(s)$ contain a zero at $s = 0$. Eq. (9-213) can be realized by the op-amp circuit shown in Fig. 9-72(b). This circuit cannot be applied as a series controller in the forward path, because it acts as an open circuit in the steady state when the frequency is zero. As a minor-loop controller, the zero-transmission property to dc signals does not pose any problems.

The forward-path transfer function of the system in Fig. 9-72(a) is

$$\frac{\Theta_o(s)}{\Theta_e(s)} = G(s) = \frac{G_p(s)}{1+G_p(s)H(s)}$$

$$= \frac{2500(1+Ts)}{s[(s+25)(1+Ts)+2500K_t]} \quad (9\text{-}214)$$

The characteristic equation of the system is

$$Ts^3 + (25T+1)s^2 + (25+2500T+2500K_t)s + 2500 = 0 \quad (9\text{-}215)$$

To show the effects of the parameters K_t and T, we construct the root contours of Eq. (9-215) by first considering that K_t is fixed and T is variable. Dividing both sides of Eq. (9-215) by the terms that do not contain T, we get

$$1 + \frac{Ts(s^2+25s+2500)}{s^2+(25+2500K_t)s+5000} = 0 \quad (9\text{-}216)$$

When the value of K_t is relatively large, the two poles of the last equation are real with one very close to the origin. It is more effective to choose K_t so that the poles of Eq. (9-216) are complex.

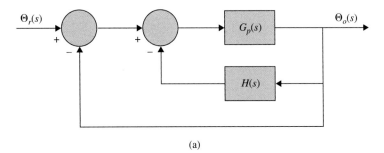

(a)

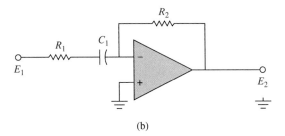

(b)

Figure 9-72 (a) Sun-seeker control system with minor-loop control. (b) Op-amp circuit realization of $\frac{K_t s}{1+Ts}$.

604 ▶ Chapter 9. Design of Control Systems

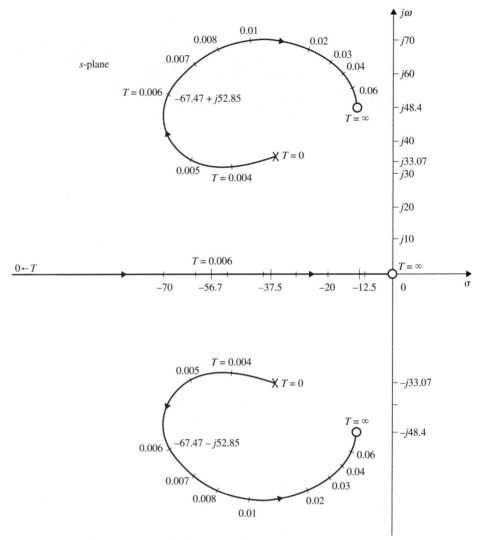

Figure 9-73 Root contours of $Ts^3 + (25T + 1)s^2 + (25 + 2500K_t + 2500T)s + 2500 = 0$, $K_t = 0.02$.

Fig. 9-73 shows the root contours of Eq. (9-215) with $K_t = 0.02$, and T varies from 0 to ∞. When $T = 0.006$, the characteristic equation roots are at -56.72, $-67.47 + j52.85$, and $-67.47 - j52.85$. The attributes of the unit-step response are as follows:

Maximum overshoot $= 0$

$t_r = 0.04485$ sec

$t_s = 0.06061$ sec

$t_{max} = 0.4$ sec

The ramp-error constant of the system is

$$K_v = \lim_{s \to 0} sG(s) = \frac{100}{1 + 100K_t} \tag{9-217}$$

Thus, just as with tachometer feedback, the minor-loop feedback controller of Eq. (9-213) reduces the ramp-error constant K_v, although the system is still type 1. ◀

▶ 9-12 A HYDRAULIC CONTROL SYSTEM

In this case study, we model a four-way electro-hydraulic-valve control of a linear actuator. After deriving the mathematical model for the system, we apply the model to do position control for three different applications: 1) robot arm joint (translational system), 2) dropping a disk into a hole (rotational system), and 3) variable load. At the end, we design P, PD, PI, and PID controllers for the robot-arm-joint system.

A schematic diagram of the system for the first application (robot arm joint) is shown in Fig. 9-74. The diagram is the same for all the applications except for the load. This system consists of a double-acting single rod linear actuator, a two-stage electro-hydraulic valve, controller circuitry, and potentiometer. The input to the system is the voltage corresponding to the desired output position of the load. The output position is fed back through the potentiometer. The output voltage of the potentiometer gets subtracted from the input voltage to produce the error signal. This error signal is used to control the position of the main valve displacement, which controls the pressure level and flow rate entering and leaving the linear actuator. The output force exerted by the linear actuator rod is directly proportional to the pressure difference between two sides of the piston.

9-12-1 Modeling Linear Actuator

A double-acting single rod actuator is shown in Fig. 9-75. We call it double acting since its piston can be forced in both right and left directions. The direction of the piston movement depends on the pressure difference on the two sides of the piston: port A and port B. We call this actuator single rod since the piston is connected to only one rod.

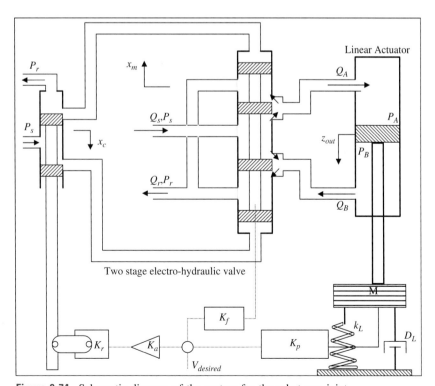

Figure 9-74 Schematic diagram of the system for the robot arm joint.

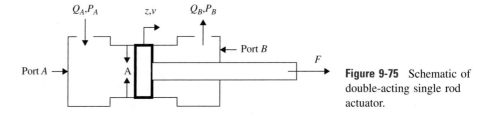

Figure 9-75 Schematic of double-acting single rod actuator.

The hydraulic fluid can enter and exit the actuator from either port A or port B. The magnitude of the flow and pressure of ports A and B are controlled by the four-way valve explained in the following section. For now, we want to develop a relationship between the fluid pressures in port A and port B, the area of the piston (A), and the displacement of the piston rod.

We can take sides A and B of the piston to have the same area A. Ignoring the pressure transient effect of the actuator, we can express the applied force F to the actuator as

$$F = (AP_A - AP_B)\eta_f \tag{9-218}$$

Note that in Eq. (9-218) η_f is the force efficiency of the actuator, which is due to the friction of the piston and the viscous shear of the hydraulic fluid. The force efficiency is always less than one. For ideal actuators, we can consider the force efficiency to be one.

Volumetric efficiency, η_v, is another source of non-ideality in linear actuators. The volumetric efficiency will always be less than one due to hydraulic fluid compression and leakage past the piston. We can express volumetric efficiency as

$$\eta_v = \frac{Av}{Q_A} \tag{9-219}$$

where A is the area of the piston, and v is the velocity of the piston. For an ideal case, we can neglect the leakage and consider the hydraulic fluid to be incompressible. For an ideal case, we can write

$$Q_A = Av = A\dot{z} \tag{9-220}$$

This equation is based on the law of conservation of mass in fluid mechanics shown in Eq. 4-124.

9-12-2 Four-Way Electro-Hydraulic Valve

The four-way electro-hydraulic valve is a two-stage control valve that takes voltage as input and provides the required fluid flow and pressure to the actuator. The first stage of the valve is an electrically actuated hydraulic valve, which controls the displacement of the spool of the second stage of the valve. The second stage is a four-way spool valve, which controls the fluid flow and pressure into and out of ports A and B of the actuator. The first stage itself is not capable of overcoming opposing spring and flow forces; therefore, the second stage is required. Fig. 9-76 shows the schematic of the electro-hydraulic valve.

From Figs. 9-75 and 9-76, we can see that the displacement of the main spool is controlled by the left and right flow rates, which are controlled by the displacement of the

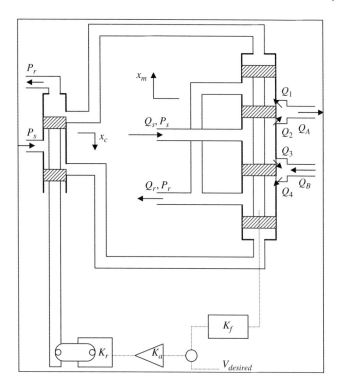

Figure 9-76 Electro-hydraulic two-stage four-way valve.

pilot spool. The control valve consists of a two-way critically centered valve, which has a pilot spool attached to a torque motor. The output from the servo amplifier actuates the torque motor and controls the displacement of the pilot spool.

The displacement of the pilot spool x_c is directly proportional to the output voltage of the servo amplifier. When the main spool is centered, there are no flow rates and both sides of the main spool are zero. We look at the operation of the valve around an operating point to be able to use the linearized flow equations derived in the following sections. At the nominal operating conditions,

$$x_m = x_c = 0, \; P_0 = P_{s0} - P_{B0} = P_{s0} - P_{A0} = \frac{1}{2} P_{s0}, \text{ and } P_r = 0$$

where x_m is the displacement of the main valve, x_c is the displacement of the pilot spool, P_{A0} and P_{B0} are pressure in the lines going from the main valve to the actuator, and P_r is the pressure of the return line to the reservoir. Note that, during the valve operation, the supply pressure coming from the pump is kept constant and equal for both the main valve and the control valve.

Orifice Equation

First, let's see what happens as the fluid passes through an orifice. Consider Fig. 9-77, which shows a flow passage with a sharp-edged orifice separating two sides of the flow passage. The classic orifice equation for the fluid flow can be written as

$$Q = A_0 C_d \sqrt{\frac{2}{\rho}(P_s - P_A)} \tag{9-221}$$

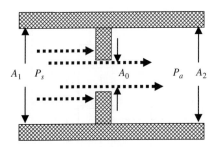

Figure 9-77 Fluid passing through an orifice.

where Q is the volumetric flow rate through the orifice, A_0 is the cross-sectional area of the orifice, ρ is the fluid density, P_s is the fluid pressure before passing through the orifice, P_A is the fluid pressure after passing through the orifice, and C_d is the discharge coefficient and is given by

$$C_d = \frac{\dfrac{A_2}{A_0}}{\sqrt{1 - \left(\dfrac{A_2}{A_1}\right)^2}} \tag{9-222}$$

where A_0, A_1 and A_2 are shown in Fig. 9-77.

Liberalized Flow Equations for the Four-Way Valve

In the orifice equation, we note that A_o and P_a are both variables. A, the opening area of the valve, is a function of the displacement of the spool. Using Taylor series, we can write Eq. (9-221) as

$$Q = Q_0 + \left.\frac{\partial Q}{\partial A_o}\frac{\partial A_o}{\partial x}\right|_{x_0}(x - x_0) + \left.\frac{\partial Q}{\partial P}\right|_{P_0}(P_a - P_0) \tag{9-223}$$

For Eq. (9-223) to be valid, the operating conditions of the valve should not deviate too much from the nominal operating conditions. If we take $x_0 = 0$ and $P_0 = P_{s0} - P_{B0} = P_{s0} - P_{A0} = \frac{1}{2}P_{s0}$, we can write Eq. (9-223) as

$$Q = \frac{1}{2}Q_0 + K_q x + K_c(P_s - P_A) \tag{9-224}$$

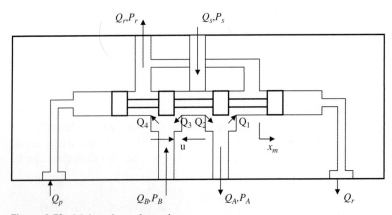

Figure 9-78 Main valve schematic.

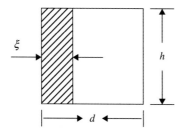

Figure 9-79 Rectangular valve-port geometry.

where K_q is called flow gain and is given by

$$K_q = C_d \sqrt{\frac{2}{\rho}(P_{s0} - P_{A0})} \frac{\partial A}{\partial x}\bigg|_{x_0} \tag{9-225}$$

and K_c is the pressure-flow coefficient and is given by

$$K_c = \frac{AC_d}{\sqrt{2\rho(P_{s0} - P_{A0})}} \tag{9-226}$$

The term $\dfrac{\partial A}{\partial x}\bigg|_{x_0}$ can be evaluated based on the valve-porting geometry. For the rectangular geometry shown in Fig. 9-79, we can write

$$A = h\xi \tag{9-227}$$

where ξ is the open distance of the valve port and h is the width of the valve port. For the open-centered valve shown in Fig. 9-79,

$$\xi = x + u \tag{9-228}$$

where x is the displacement of the spool and u is the fixed underlapped dimension.
Therefore,

$$\frac{\partial A}{\partial x} = h \tag{9-229}$$

Now we can write the following:

$$K_q = hC_d \sqrt{\frac{2}{\rho} P_0} \tag{9-230}$$

$$K_c = \frac{uhC_d}{\sqrt{2\rho P_0}} \tag{9-231}$$

where

$$P_0 = P_{s0} - P_{B0} = P_{s0} - P_{A0} = \frac{1}{2} P_{s0} \tag{9-232}$$

Let's now apply the linearized flow equation to the open-centered four-way valve shown in Fig. 9-76. For the four-way valve, ports A and B are used for directing flow to and from the linear actuator. To direct flow into port A and receive flow from port B, the main spool moves to the right, thus allowing fluid flow from the supply line to port A and, at the same time, allowing fluid flow from port B to the return line. The flow rate and pressure levels at ports A and B are the function of the valve displacement x_m. When modeling the whole system, we can neglect the pressure drop in the pipelines connecting ports A and B of the linear actuator to ports A and B of the four-way valve. Hence, we can assume pressure levels and flow rates at ports A and B of the four-way valve are the same as pressure levels and flow rates at ports A and B of the linear actuator.

Now consider the main spool is moved to the right by a displacement x_m. Using Eq. (9-232), we can write

$$Q_1 = \frac{1}{2}Q_0 - K_q x_m + K_c(P_A - P_r) \tag{9-233}$$

$$Q_2 = \frac{1}{2}Q_0 + K_q x_m + K_c(P_s - P_A) \tag{9-234}$$

$$Q_1 = \frac{1}{2}Q_0 - K_q x_m + K_c(P_s - P_B) \tag{9-235}$$

$$Q_1 = \frac{1}{2}Q_0 + K_q x_m + K_c(P_B - P_r) \tag{9-236}$$

Note that, in the above equations, we can use the same flow gain, K_q, and the same pressure-flow coefficient, K_c, for all the metering lands because the nominal pressure across each metering land is equal to half the supply pressure $P_s/2$ and $P_r = 0$.

The volumetric flow rates into and out of the actuator can now be expressed as

$$Q_A = Q_2 - Q_1 = 2K_q x_m - 2K_c(P_A - P_s/2) \tag{9-237}$$

$$Q_B = Q_4 - Q_3 = 2K_q x_m + 2K_c(P_B - P_s/2) \tag{9-238}$$

Note that the above flow equations are valid for open-centered four-way valves. For other valve designs such as critically centered or closed-centered designs, the flow equations are different. For critically centered valves, such as the control valve in Fig. 9-76, the flow equation can be expressed as

$$Q_A = Q_B = K_q x_c \tag{9-239}$$

The flow-rate equations into and out of the actuator are very important in control analysis, which will follow. The reason is that these equations enable us to relate the flow rate and the pressure level, which delivers power to the actuator. Using these equations, we are able to control the displacement and velocity of the linear actuator (which depends on pressure level and the flow rate delivered to ports A and B of the actuator) based on the main spool displacement x_m. In other words, for this open-centered four-way valve control, we can change flow rate and pressure level simultaneously using a single variable of the main spool displacement x_m.

Relationship between Input Voltage and Main Spool Displacement x_m

Because the control valve is critically centered, the flow equation of the pilot spool can be expressed as

$$Q_P = K_q x_c \tag{9-240}$$

9-12 A Hydraulic Control System

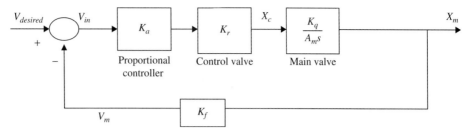

Figure 9-80 Two-stage valve block diagram.

Assuming an incompressible fluid, the interaction between the two stages can be described by

$$Q_p = A_m \frac{dx_m}{dt} \qquad (9\text{-}241)$$

where A_m is the area of the main spool. The displacement of the pilot spool can be expressed as

$$x_c = V_{in} K_a K_r \qquad (9\text{-}242)$$

Equating Eq. (9-241) and (9-242), we get

$$K_a K_r V_{in} = \frac{A_m}{K_q} \frac{dx_m}{dt} \qquad (9\text{-}243)$$

For the two-stage valve alone (ignoring the output position feedback for the moment), using Laplace transforms and assuming zero initial conditions, we have

$$X_m(s) = \frac{K_a K_r K_q}{A_m s} V_{in}(s) \qquad (9\text{-}244)$$

and

$$V_{in}(s) = V_{desired} - K_f X_m(s) \qquad (9\text{-}245)$$

where X_m is the Laplace transform of x_m.

The transfer function of the two-stage valve can be expressed as

$$\frac{X_m(s)}{V_{desired}} = \frac{1}{K_f(1 + T_c s)} \qquad (9\text{-}246)$$

where

$$T_c = \frac{A_m}{K_q K_a K_f K_r} \qquad (9\text{-}247)$$

We can represent the two-stage valve with the block diagrams in Figs. 9-80 and 9-81.

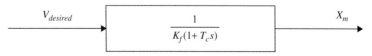

Two-stage valve characteristics

Figure 9-81 Two-stage valve block diagram incorporating Eq. (9-247).

9-12-3 Modeling the Hydraulic System

Fig. 9-82 shows the schematic diagram of the whole system. In the following sections, let us rewrite the mathematical equations we derived for each component.

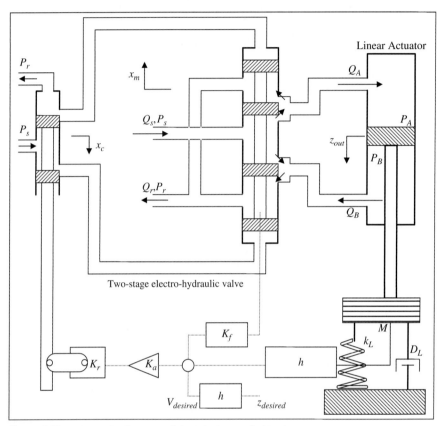

Figure 9-82 Schematic diagram of the whole translational system.

Force Balance Equation for an Ideal Linear Actuator
If the rod of the linear actuator is used for opposing a load force F_L in the z direction, using Eq. (9-247), we can express the force balance equation for the rod as

$$F_L = A(P_A - P_B) \tag{9-248}$$

where we assume the force efficiency to be one.

Expressing the Pressure Level P_A and the Pressure Level P_B
Substituting Eq. (9-220) into Eq. (9-237) and rearranging the terms, we can express P_A as

$$P_A = \frac{K_q}{K_c} x_m - \frac{A}{2K_c} \dot{z}_{out} + P_{s/2} \tag{9-249}$$

Substituting Eq. (9-220) into Eq. (9-239) and rearranging the terms, we can express P_B as

$$P_B = -\frac{K_q}{K_c}x_m + \frac{A}{2K_c}\dot{z}_{out} + P_{s/2} \tag{9-250}$$

Expressing the Pressure Difference Between Two Sides of the Linear Actuator

We can express the pressure difference, $P_A - P_B$, by subtracting Eq. (9-249) from Eq. (9-250):

$$P_A - P_B = 2\frac{K_q}{K_c}x_m - \frac{A}{K_c}\dot{z}_{out} \tag{9-251}$$

General Equation

Assuming an ideal actuator where force and volume efficiencies are one, and by substituting Eq. (9-248) into Eq. (9-251), we get

$$F_L = 2\frac{AK_q}{K_c}x_m - \frac{2A^2}{K_c}\dot{z}_{out} \tag{9-252}$$

Eq. (9-252) is the general equation that will be used to find the transfer functions for different applications in the following section.

9-12-4 Applications

In this section we apply the system model in three different applications: translational motion, rotational system, and variable load.

Position Control of a Hydraulic System (Translational Motion)

In this application, the linear actuator rod is used to move a mass, M; a spring, k_L; and a damper, D_L in the z direction, as shown in Figure 9-82.

For the two-stage valve alone,

$$X_m(s) = \frac{K_a K_r K_q}{A_m s} V_{in}(s) \tag{9-253}$$

$$V_{in}(s) = V_{desired} - K_f x_m(s) \tag{9-254}$$

The transfer function of the two-stage valve can be expressed as

$$\frac{X_m(s)}{V_{desired}} = \frac{1}{K_f(1 + T_c s)} \tag{9-255}$$

where

$$T_c = \frac{A_m}{K_q K_a K_f K_r} \tag{9-256}$$

We can represent the two-stage valve with the block diagram in Fig. 9-81.

Chapter 9. Design of Control Systems

To maintain position control of the actuator, a voltage V_0 is fed back from the actuator displacement z:

$$V_0 = hz_{out} \tag{9-257}$$

where h is the gain of the transducer. The input voltage to the two-stage valve V_{error} can be expressed as

$$V_{error} = V_{desired} - V_0 = V_{desired} - hz_{out} \tag{9-258}$$

It is desired to have position as input instead of voltage. As a result, a potentiometer is used to convert the desired input $z_{desired}$ to the desired voltage $V_{desired}$. The relationship can be expressed as

$$V_{desired} = hz_{desired} \tag{9-259}$$

Using general Eq. (9-252), the equation-relating input (main valve displacement x_m) and output (actuator displacement z_{out}) can be expressed as

$$z_{out}\left(Ms^2 + \left(D + \frac{2A^2}{K_c}\right)s + k_L\right) = 2\frac{AK_q}{K_c}X_m(s) \tag{9-260}$$

From the valve equation derived above, we know that main valve displacement x_m can be expressed as

$$X_m(s) = \left(\frac{1}{K_f(1+T_c s)}\right)(V_{desired} - hz_{out}) \tag{9-261}$$

So the transfer function is

$$\frac{z_{out}}{z_{desired}} = \frac{2\frac{AK_q}{K_c}h}{K_f(1+T_c s)\left(Ms^2 + \left(D + \frac{2A^2}{K_c}\right)s + k_L\right) + 2h\frac{AK_q}{K_c}} \tag{9-262}$$

where $T_c = \frac{A_m}{K_q K_a K_f K_r}$.

In the above equation,

K_q = Flow gain of the control valve
K_c = Pressure-flow coefficient
K_a = Gain of the proportional controller
K_f = Gain of the main spool feedback transducer
K_r = Ratio between the input voltage and displacement of the control spool
A_m = Area of the main spool
M = Mass of the load
D = Damping of the load
k_L = Load spring stiffness
h = Transducer gain
A = Area of the actuator piston

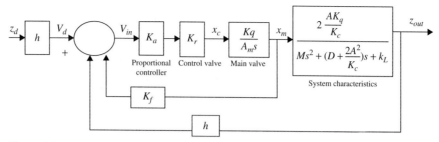

Figure 9-83 Translational system block diagram.

We can represent the system with the block diagrams in Figs. 9-83 and 9-84.

Position Control (Rotational System)
In this application we refer to Fig. 9-85, where the linear actuator rod is used to move a mass, M, θ radians. We start by writing the translational displacement of the rod in terms of the angular displacement θ:

$$z_{out} = \theta_{out} L \tag{9-263}$$

From the valve equation derived in the previous section, we know that main valve displacement x_m can be expressed as

$$X_m(s) = \left(\frac{1}{K_f(1+T_c s)}\right)(V_{desired} - h\theta_{out}L) \tag{9-264}$$

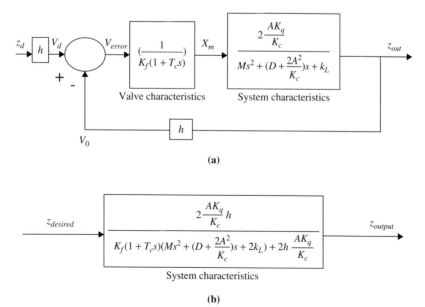

Figure 9-84 Translational reduced system block diagram.

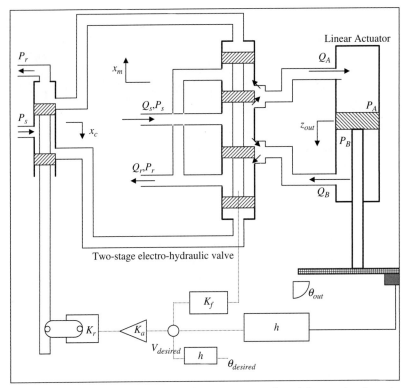

Figure 9-85 Schematic diagram of a linear actuator rod used for rotational motion.

Using Eq. (9-260) and Eq. (9-263) for the system shown, the equation-relating input (main valve displacement x_m) and output (actuator displacement θ_{out}) can be expressed as

$$\theta_{out}\left(Js^2 + \frac{2A^2}{K_c}Ls\right) = 2\frac{AK_q}{K_c}x_m \qquad (9\text{-}265)$$

where J is the moment of inertia. Substituting Eq. (9-264) into Eq. (9-265), we can express the system transfer function as

$$\frac{\theta_{out}}{\theta_{desired}} = \frac{2\frac{AK_q}{K_c}h}{K_f(1+T_c s)\left(Js^2 + \frac{2A^2}{K_c}s\right) + 2hL\frac{AK_q}{K_c}} \qquad (9\text{-}266)$$

Variable Load

In this application, as shown in Figure 9-86, the linear actuator rod is used to exert force on a variable load. From the valve equation derived in the previous section, we know that main valve displacement x_m can be expressed in Laplace domain as

$$X_m(s) = \left(\frac{1}{K_f(1+T_c s)}\right)(V_{desired} - hz_{out}) \qquad (9\text{-}267)$$

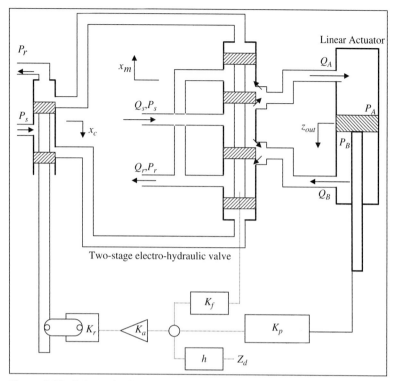

Figure 9-86 Schematic diagram of a variable load application.

Using general Eq. (9-252), the general equation for the system shown in Fig. 9-88 can be expressed as

$$F_L = 2\frac{AK_q}{K_c}\left(\frac{1}{K_f(1+T_c s)}\right)(hz_{desired} - hz_{out}) \qquad (9\text{-}268)$$

9-13 CONTROLLER DESIGN

9-13-1 P Control

Consider the model of the hydraulic system shown in Fig. 9-88, where we added a proportional controller in the forward path. The new system block diagram is shown in Figure 9-87.

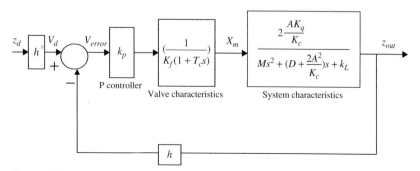

Figure 9-87 Block diagram of the hydraulic system with a proportional controller.

618 ▶ Chapter 9. Design of Control Systems

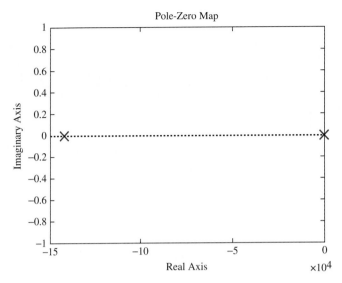

Figure 9-88 Pole–zero map of Eq. (9-269).

This system assumes an ideal actuator where force and volume efficiencies are one.

$K_q = 0.359 \, m^2/s$ Flow gain of the control valve
$K_c = 1.70*10^{-11} \, m^3/Pa/s$ Pressure-flow coefficient
$K_a = 1$ Gain of the proportional controller
$K_f = 1$ Gain of the main spool feedback transducer
$K_r = 1$ Ratio between the input voltage and displacement of the control spool
$A_m = 550 \times 10^{-6} \, m^2$ Area of the main spool
$M = 4 \, kg$ Mass of the load
$D = 1 \, N\text{-}s/m$ Damping of the load
$k_L = 2 \, N/m$ Load spring stiffness
$h = 1$ Transducer gain
$A = 1.1 \times 10^{-3} \, m^2$ Area of the actuator piston

Substituting the preceding values into the hydraulic system transfer function [Eq. (9-262)], we get

$$G(s) = \frac{3.033*10^{10}}{(s + 1.4235*10^5)(s + 653)(s + 7.02*10^{-6})} \quad (9\text{-}269)$$

The poles of Eq. (9-269) are shown in Fig. 9-88. Poles are at −142350, −653, and −7.02∗10^{−6}.

We simplify the forward-path transfer function in Eq. (9-269) by neglecting the pole at −142350. The simplified hydraulic system transfer function is

$$G(s) = \frac{213041}{s^2 + 653s + 0.004584} \quad (9\text{-}270)$$

We note that one of the dominant poles is very close to the origin; therefore, we can write the forward-path transfer function as

$$G(s) = \frac{213041}{s(s + 653)} \quad (9\text{-}271)$$

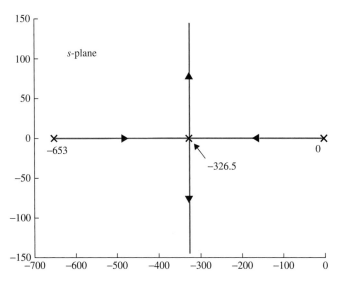

Figure 9-89 Root loci of Eq. (9-274).

Applying a proportional controller, the forward-path transfer function becomes

$$G(s) = \frac{213041(K_P)}{s^2 + 653s} \tag{9-272}$$

The closed-loop transfer function is

$$T(s) = \frac{213041(K_P)}{s^2 + 653s + 213041K_P} \tag{9-273}$$

The characteristic equation is written

$$s^2 + 653s + 213041K_P = 0 \tag{9-274}$$

The root loci of Eq. (9-274) are shown in Fig. 9-89.

By looking at the root loci in Fig. 9-90, we see that, depending on the value of K_p, we can get two real or two complex-conjugate poles. When $K_p = 0.500$, we get a damping

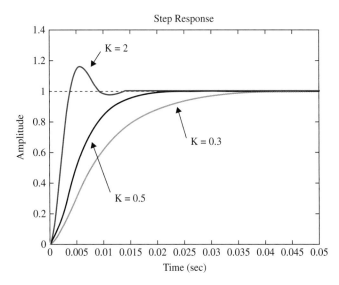

Figure 9-90 Unit-step responses with P control.

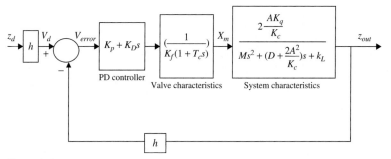

Figure 9-91 System block diagram.

ratio of 1, and the system is critically damped. For values of K_p greater than 0.500, the system is underdamped, and for values of K_p less than 0.500, the system is overdamped. We also note that the system is stable for all the values of $K_p > 0$. The step responses of the system for three different values of K_p are shown in Fig. 9-91.

Toolbox 9-13-1

Fig. 9-89 is obtained by the following sequence of MATLAB functions

```
num = [3.0333e+10];
den = conv(conv([1 1.4235e5],[1 653]),[1 7.02e-6])
pzmap(num,den);
```

Fig. 9-90 is obtained by the following sequence of MATLAB functions

```
KP=1;
KD=0;
num = [1 653+213041*KD 213041*KP];
den = [1 653 213041*KP];
tf(num,den)
rlocus(num,den)
axis([-700 0 -1 1])
```

Fig. 9-91 is obtained by the following sequence of MATLAB functions

```
KP=0.3;
num = [0 213041*KP];
den = [1 653 213041*KP];
step(num,den)
hold on;
KP=0.5;
num = [0 213041*KP];
den = [1 653 213041*KP];
step(num,den)
hold on;
KP=2;
num = [0 213041*KP];
den = [1 653 213041*KP];
step(num,den)
hold on;
```

TABLE 9-27 Attributes of the Unit-Step Response with P control

K_p	Maximum Overshoot (%)	t_s(sec)	t_r(sec)	Steady-State Error Due to Unit Step
2	16.3	0.0124	0.00253	0
0.5	—	0.0179	0.0103	0
0.3	—	0.0348	0.0191	0

Table 9-27 summarizes the attributes of the system's unit-step response for three different values of K_p.

When poles are complex conjugates, as we increase K_p, the overshoot of the system increases, but the settling time and rise time of the system remain unchanged. Also note that steady-state error due to unit-step input decreases as K_p increases.

9-13-2 PD Control

Consider the second-order forward-path transfer function of the hydraulic system discussed in Section 9-13-1. Applying a PD controller, the forward-path transfer function becomes

$$G(s) = \frac{213041(K_P + K_D s)}{s(s + 653)} \tag{9-275}$$

The system block diagram is shown in Fig. 9-91.

Now, let us set the performance specifications as follows:

Settling time $t_s \leq 0.005$ sec

Maximum overshoot $\leq 5\%$

Steady-state error due to unit-ramp input ≤ 0.00061

The closed-loop transfer function is

$$T(s) = \frac{213041(K_P + K_D s)}{s^2 + (653 + 213041 K_D)s + 213041 K_P} \tag{9-276}$$

The characteristic equation is written

$$s^2 + (653 + 213041 K_D)s + 213041 K_P = 0 \tag{9-277}$$

We start by finding the steady-state error for a unit-ramp input:

$$e_{ss}|_{ramp} = \lim_{s \to 0} \frac{1}{1 + sG(s)} = \frac{1}{1 + \frac{213041 K_P}{653}} \tag{9-278}$$

Therefore, for the system to have steady-state error due to unit ramp ≤ 0.00061, we need $K_P \geq 5$. The damping ratio of the system for $K_P = 5$ can be expressed as

$$\zeta = \frac{653 + K_D 213041}{2064.2} = 0.316 + 103.209 K_D \tag{9-279}$$

622 ▶ Chapter 9. Design of Control Systems

If we wish to have critical damping, $\zeta = 1$, the above equation gives $K_D = 0.0066$. One thing we need to check is that this value satisfies the settling-time requirement. Settling time can be expressed as

$$t_s = \frac{8}{(653 + 213041 K_D)} \qquad (9\text{-}280)$$

We see that, for $K_D \geq 0.0044$, we have $t_s \leq 0.005$ sec. Therefore, with $K_D = 0.006624$, we can satisfy the settling-time requirement. We note that Eq. (9-280) is an approximation and, as damping ratio comes closer to one, the actual settling time will be higher; therefore, K_D must be higher than 0.0044 to satisfy the settling time for damping ratio of one. Nevertheless, we can still use the approximation and verify our answer by simulation once a value for K_D is found. Also, we need to make sure that, for the values found for K_P and K_D, the system is stable. Applying the stability requirement, we find that for system stability

$$K_P > 0 \quad \text{and} \quad K_D > -0.00307$$

Alternatively, we can use the system's root contours to find K_P and K_D. We can apply the root-contour method to the characteristic equation in Eq. (9-277) to examine the effect of varying K_P and K_D. First, by setting K_D to zero, Eq. (9-277) becomes

$$s^2 + 653s + 213041 K_P = 0 \qquad (9\text{-}281)$$

The root loci of Eq. (9-281) are shown in Fig. 9-92.

The code used to plot the root locus is given in Toolbox 9-13-2 for $K_D = 0$. Alternatively you can use Toolbox 5-8-1. When $K_D \neq 0$, the characteristic equation in Eq. (9-281) becomes

$$1 + G_{eq}(s) = 1 + \frac{213041 K_D s}{s^2 + 653s + 213041 K_P} = 0 \qquad (9\text{-}282)$$

The root contours for $K_P = 1$ and $K_P = 5$ based on the pole–zero configuration of $G_{eq}(s)$ are shown in Fig. 9-94. Note that we chose $K_P = 5$ to satisfy the steady-state error

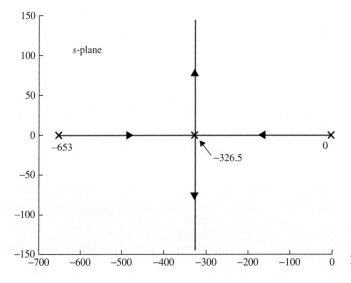

Figure 9-92 Root loci of Eq. (9-281).

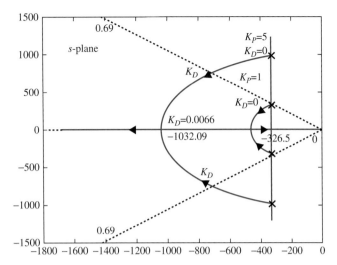

Figure 9-93 The root contours of Eq. (9-281) or Eq. (9-277) when $K_D = 0$.

requirement. For $K_P = 5$ and $K_D = 0$, the characteristic equation roots are at $-326.5 \pm j979.08$ and the damping ratio of the closed-loop system is 0.316. When the value of K_D is increased, the two characteristic equation roots move toward the real axis along a circular arc. The dashed line shows the points on the s-plane with the constant damping ratio of 0.69, which corresponds to 5% overshoot. We see that this line intersects the root contour for $K_P = 5$ at $K_D = 0.00362$. Because this value of K_D is not big enough to satisfy the settling-time requirement of less than 0.002 sec, we need a larger K_D. When K_D is increased to 0.006624, the roots are real and equal at −1032.09, and the damping is critical. At this point $K_D \geq 0.0044$; therefore, our settling-time requirement is met. When $K_p = 1$ and $K_D = 0$, the two characteristic equation roots are at $-326.5 \pm j326.3$. As K_D increases in value, the root contours again show the improved damping due to the PD controller.

Fig. 9-95 shows the unit-step responses of the closed-loop system without PD control and with $K_p = 5$ and $K_D = 0.006624$. With the PD control, although K_D is chosen for critical damping, the maximum overshoot is 0.888%. This is because of the zero at $s = -\dfrac{K_P}{K_D}$ for the

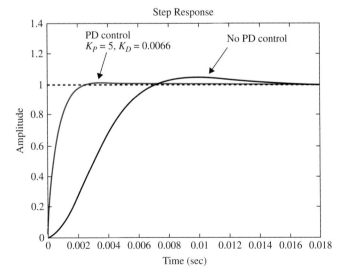

Figure 9-94 Unit-step responses with and without PD control using Toolbox 9-13-3.

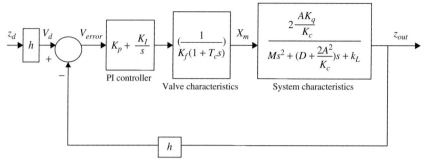

Figure 9-95 System block diagram with PI controller.

closed-loop transfer function. When $K_p = 5$ and $K_D = 0.0066$, the zero is at -754.5, which is close to the dominant poles of the system at -1032.09; therefore, it has a significant effect on the transient response of the system. Note that the zero makes the response faster by lowering the rise time and the settling time of the system.

Toolbox 9-13-2

Fig. 9-94 is obtained by the following sequence of MATLAB functions

```
%Closed loop transfer function for Kp=5
sgrid(0.69);
Kp = 5;Kd = 1e-6;
for i = 1:1:260
    if (i == 222)
        Kd = (((Kp*213041)^(0.5)*2)-653)/213041; % if this statement is not used
then the graph will not continuous
    end
    num =[213041*Kd 213041*Kp];den = [1 653 0];
    tf(num,den);
    [numCL,denCL]=cloop(num,den);
    T = tf(numCL,denCL);
    PoleData(:,i)=pole(T);
    Kd = Kd +3e-5;
end
plot(real(PoleData(1,:)),imag(PoleData(1,:)),real(PoleData(2,:)),imag(PoleData(2,:)));

%Closed loop transfer function for Kp=1
Kp = 1; Kd = 1e-6;
for i = 1:1:260
    if (i == 44)
        Kd = (((Kp*213041)^(0.5)*2)-653)/213041;
    end
    num =[213041*Kd 213041*Kp];den = [1 653 0];
    tf(num,den);
    [numCL,denCL]=cloop(num,den);
    T = tf(numCL,denCL);
    PoleData(:,i)=pole(T);
    Kd = Kd +3e-5;
end
hold on
plot(real(PoleData(1,:)),imag(PoleData(1,:)),real(PoleData(2,:)),imag(PoleData(2,:)));
```

```
%open loop transfer function
Kp = 1; Kd = 0;
num =[213041*Kd 213041*Kp];
den = [1 653 0];
hold on
rlocus(num,den);
axis([-1800 0 -1500 1500])
step(num,den)
hold on;

KP=2;num = [0 213041*KP];den = [1 653 213041*KP];
step(num,den)
hold on;
```

Table 9-28 summarizes the attributes of the unit-step response for $K_p = 5$ and different values of K_D.

TABLE 9-28   Attributes of the Unit-Step Response with PD Control

K_D	Maximum Overshoot (%)	t_s(sec)	t_r(sec)
0.001	23.9	0.00794	0.00141
0.006624	0.888	0.00215	0.00138
0.01	—	0.00363	0.00126

Toolbox 9-13-3

Fig. 9-94 is obtained by the following sequence of MATLAB functions

```
Kp = 5;Kd = 0.0066;
num =[213041*Kd 213041*Kp];
den = [1 653 0];
tf(num,den);
[numCL,denCL]=cloop(num,den);
step(numCL,denCL)
hold on
Kp = 1;Kd = 0;
num =[213041*Kd 213041*Kp];
den = [1 653 0];
tf(num,den);
[numCL,denCL]=cloop(num,den);
step(numCL,denCL)
```

At this point we need to go back to the original hydraulic transfer function expressed in Eq. (9-269), which has three poles. We need to make sure that the third pole is still far away from the origin so that the second-order approximation is valid. In the root locus of the pure proportional control system, the third pole moves to the left as K increases; therefore, the second-order approximation is valid for any value of $K \geq 0$. For the PD control design system, we introduce a zero in the forward path, which changes the behavior of the third pole. In this case, the third pole starts to move toward the origin as we increase K_D. For $K_p = 5$ and $K_D = 0.0066$, the poles of the closed-loop third-order system are at -14093, -1090, and -980.

Because the third pole is still far away from the origin relative to the other poles, it can be safely neglected, and our second-order approximation holds.

9-13-3 PI Control

Consider the second-order forward-path transfer function of the hydraulic system discussed in Section 9-13-1. Applying a PI controller, the forward-path transfer function becomes

$$G(s) = \frac{213041 K_P (s + K_I/K_P)}{s^2(s + 653)} \tag{9-283}$$

The system's block diagram is shown in Fig. 9-95.

Let the time-domain performance requirement be as follows:

Settling time $t_s \leq 0.06$ sec

Rise time $t_r \leq 0.01$ sec

Maximum overshoot $\leq 1.52\%$

Steady-state error due to parabolic input ≤ 0.2

We start by finding the steady-state error for a ramp input:

$$e_{ss}|_{parabolic} = \lim_{s \to 0} \frac{1}{s^2 G(s)} = \frac{1}{213041 K_I / 653} \tag{9-284}$$

Therefore, for the system to have steady-state error due to parabolic input ≤ 0.2, we need $K_I \geq 0.015$. Later we need to make sure that the value of K_I used is above 0.015.

The characteristic equation of the closed-loop system is

$$s^3 + 653s^2 + 213041 K_P s + 213041 K_I = 0 \tag{9-285}$$

Applying Routh's test to Eq. (9-277) yields the result that the system is stable for $0 < \frac{K_I}{K_P} < 653$. This means that, if the zero of $G(s)$ be placed too far to the left in the left-half s-plane, the system will be unstable.

Let us place the zero at $-K_I/K_p$ relatively close to the origin. For this case, the most significant poles of the forward-path transfer function without the PI controller are at -653 and 0. Thus K_I/K_p should be chosen so that the following condition is satisfied

$$\frac{K_I}{K_P} \ll 653 \tag{9-286}$$

The root loci of Eq. (9-283) with $K_I/K_P = 5$ are shown in Fig. 9-96. Notice that one of the poles always has a value close to zero while the other two poles behave the same as those shown in Fig. 9-93, which is for Eq. (9-281).

With the condition in Eq. (9-286) satisfied, the pole near zero is effectively cancelled by the zero at $-K_I/K_p$, and we are left with the poles of Eq. (9-283). Therefore, Eq. (9-284) can be approximated by

$$G(s) \cong \frac{213041 K_P}{(s^2 + 653s + 213041)} \tag{9-287}$$

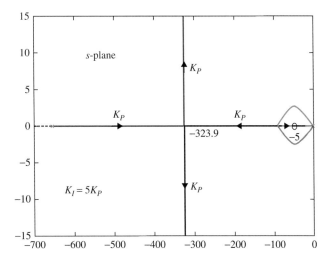

Figure 9-96 Root loci of Eq. (9-283).

Let us assume we wish to have a relative damping ratio of 0.9. From Eq. (9-287), the required value of K_P for this damping ratio is 0.62. Thus with $K_P = 0.62$ and $K_I = 3.09$, we find the roots of the characteristic equation of Eq. (9-285):

$$s = -5.13 \quad -323.94 + j152.8 \quad \text{and} \quad -323.9.3 - j152.8$$

In this case we see that the real pole of the closed-loop system is very close to the zero at $-K_I/K_p$ so that the transient due to the real pole is negligible, and the system dynamics are essentially dominated by the two complex poles; therefore, $K_P = 0.62$ will give us a damping ratio that is close to 0.9. In general, when s takes on values along the operating points on the complex portion of the root loci, we can neglect the effect of the real pole of the closed-loop system and use Eq. (9-287) to find the system characteristics.

Table 9-29 summarizes the attributes of the unit-step response for various values of K_I/K_P with $K_P = 0.62$, which corresponds to a relative damping ratio of 0.9. The results verify that PI control reduces overshoot at the expense of longer rise time.

Fig. 9-97 shows the unit-step response of the hydraulic system with PI control with $K_P = 0.62$ and $K_I = 3.09$. The unit-step response is the same for system with P control designed in Section 9-13-1.

TABLE 9-29 Attributes of the Unit-Step Response with PD Control

K_I/K_P	K_I	K_P	Maximum Overshoot (%)	t_s(sec)	t_r(sec)
0	0	5	35.1	0.0108	0.00131
20	12.4	0.62	8.78	0.0856	0.00688
12	7.44	0.62	5.59	0.099	0.00723
5	3.09	0.62	2.52	0.0555	0.00778
3	1.86	0.62	1.6	0.0121	0.0078
1	0.62	0.62	—	0.0126	0.00787
0.5	0.31	0.62	—	0.0128	0.00791

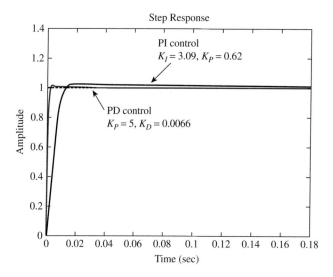

Figure 9-97 Unit-step responses of the system with PD and PI control using Toolbox 9-13-1.

Toolbox 9-13-4

Fig. 9-97 is obtained by the following sequence of MATLAB functions

```
Ki =3.09
Kp = 0.62
num = [213041*Kp 213041*Ki/Kp]
den = [1 653 0 0];
tf(num,den);
[numCL,denCL]=cloop(num,den);
step(numCL,denCL)
hold on
Kp = 5;
Kd = 0.0066;
num =[213041*Kd 213041*Kp];
den = [1 653 0];
tf(num,den);
[numCL,denCL]=cloop(num,den);
step(numCL,denCL)
hold on
```

9-13-4 PID Control

Consider the second-order forward-path transfer function of the hydraulic system discussed previously. Applying a PID control, the forward-path transfer function of the plant is

$$G(s) = \frac{213041(K_{P1} + K_{D1}s)(K_{P2} + K_{I2}/s)}{s(s + 653)} \tag{9-288}$$

The system's block diagram is shown in Fig. 9-98.

Let the time-domain performance requirements be as follows:

Rise time $t_r \leq 0.003$ sec

Settling time $t_s \leq 0.004$ sec

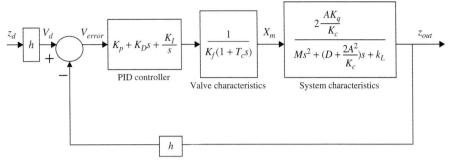

Figure 9-98 System block diagram with PID controller.

Maximum overshoot $\leq 1.2\%$

Steady-state error due to parabolic input ≤ 0.2

We realize that we need a PID controller to fulfill the above requirements. First, we apply the PD control with the transfer function $K_{P1} + k_{D1}s$. The forward-path transfer function becomes

$$G(s) = \frac{213041(K_{P1} + K_{D1}s)}{s(s+653)} \tag{9-289}$$

Table 9-28 shows that the PD controller that can be obtained from the settling time standpoint is with $K_{D1} = 0.0066$ and $K_{P1} = 5$, and the maximum overshoot is 0.888%. The rise time and settling time are well within required values. Next we add the PI controller, and the forward-path transfer function becomes

$$G(s) = \frac{1406.07K_{P2}(s + 757.58)(s + K_{I2}/K_{P2})}{s(s^2 + 653s)} \tag{9-290}$$

Following the guidelines of choosing a relatively small value for K_{I2}/K_{P2}, we let $K_{I2}/K_{P2} = 5$. Eq. (9-290) becomes

$$G(s) = \frac{1406.07K_{P2}(s + 757.58)(s + 5)}{s(s^2 + 653s)} \tag{9-291}$$

Table 9-30 gives the time domain performance characteristics along with the roots of the characteristic equation for various values of K_{P2}.

TABLE 9-30 Attributes of the Unit-Step Response with PID Control

K_{P2}	Maximum Overshoot (%)	t_s(sec)	t_r(sec)	Roots of Characteristic Equation
2	1.23	0.00113	0.000718	−5 −799.5 −2660.6
1.2	1.27	0.00178	0.00116	−5.0 −14464.5 −870.3
0.9	1.2	0.00232	0.00151	−5.02 −956.7 ± j200.2
0.62	1.05	0.00327	0.00213	−5.02 −759.9 ± j282.4
0.40	0.91	0.00505	0.00321	−5.04 −605.9 ± j237.8
0.20	—	0.00983	0.00622	−5.08 −542.6 −386.5

630 ▶ Chapter 9. Design of Control Systems

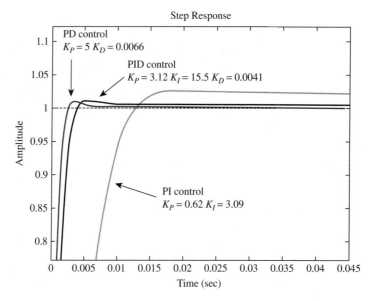

Figure 9-99 Unit-step responses with PD, PI, and PID control using Toolbox 9-13-5.

Setting $K_{P1} = 5$, $K_{D1} = 0.0.0066$, $K_{P2} = 0.62$, and $K_{I2} = 5K_{P2} = 3.09$, the following results are obtained for the parameters of the PID controller.

$K_P = 3.12$

$K_I = 15.5$

$K_D = 0.0041$

Fig. 9-99 shows the unit-step responses of the system with the PID controller, as well as those with PD and PI controls designed before.

Toolbox 9-13-5

Fig. 9-99 is obtained by the following sequence of MATLAB functions

```
Kp1=5;kd1=0.0066;
num = [21304*kd1 21304*Kp1];den = conv([1 0],[1 653]);
tf(num,den)
[numCL,denCL]=cloop(num,den);
step(numCL,denCL)
hold on
Kp2=0.62;KI2=5*Kp2;
num = conv(conv([0 1406.07*Kp2],[1 757.58]),[1 KI2/Kp2])
den = conv([1 0],[1 653 0]);
tf(num,den)
[numCL,denCL]=cloop(num,den);
step(numCL,denCL)
hold on
Kp2=2;num = conv(conv([0 1406.07*Kp2],[1 755.58]),[1 5]);
den = [1 653 0 0];
tf(num,den)
[numCL,denCL]=cloop(num,den);
tf(numCL,denCL)
step(numCL,denCL)
```

9-14 MATLAB TOOLS AND CASE STUDIES

In this section we will go through the steps involved in finding some of the results and displaying many of the graphics from the examples in this chapter using the **ACSYS** software. The tools involved are all contained in the Controller Design Tool. This tool allows the user to conduct the following tasks:

- Enter the transfer function values in polynomial form. (User must use the tftool as discussed in Chapter 2 to convert the transfer function from pole-zero-gain form into polynomial form.)
- Obtain the step, impulse, parabolic, ramp, or other type input time responses.
- Obtain the closed-loop frequency plots.
- Obtain the phase and gain margin Bode plots and the polar plot of the loop transfer functions (in a single feedback loop configuration).
- Understand the effect of adding zeros and poles to the closed-loop or open-loop transfer functions.
- Design and compare various controllers including PID, lead, and lag compensators.

To run the Controller Design Tool, type "Acsys" at the MATLAB command line and click the appropriate button on the menu. Fig. 9-100 will appear on the screen.

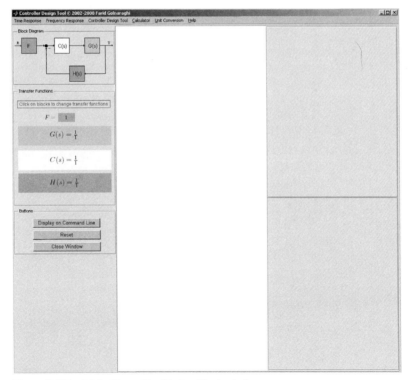

Figure 9-100 Main Controller Design Tool window.

EXAMPLE 9-2-1 Revisited

Recall from Example 9-2-1 the forward-path transfer function for the altitude-control system was

$$G(s) = \frac{4500K}{s(s+361.2)} \quad (9\text{-}292)$$

The design constraints for this problem were as follows:

Steady-state error due to unit-ramp input = 0.000443

Maximum overshoot $\leq 5\%$

Rise time $t_r \leq 0.005$ sec

Settling time $t_s \leq 0.005$ sec

To enter the transfer function into the CONTROLS tool, click on the G(s) box in the transfer function input panel. For the moment we will leave the value of K as 1, so in the numerator text box, enter [4500], and in the denominator text box, enter [1,361.2,0], as is pictured in Fig. 9-101.

Pressing the enter button will display the current closed-loop transfer function. For a more accurate representation of the transfer function, press the "Display on Command Line" button and refer to the MATLAB command window.

Now click on the C(s) box and enter the value [181.2]. Click on the "Step Response" option from the "Time Response" menu to see the step response of this system. The main window should now look like Fig. 9-102.

Press the "Print To Figure" button, and right-click on the resulting plot to see more information about the response.

As in Example 9-2-1, we shall use a PD controller to improve the response of the system. We will use the root contour tool to see the effect that the PD controller has on the poles of the system. From the "Controller Design Tool" menu, choose the "PD design option."

The Root Contour tool plots the poles of a system as functions of certain varying controller parameters. With the PD controller, the closed-loop transfer function is

$$G(s) = \frac{\Theta_y(s)}{\Theta_e(s)} = \frac{815,265(K_P + K_D s)}{s(s+361.2)} \quad (9\text{-}293)$$

K_P is held at 1, and K_D will vary between the limits that we specify. Choose K_D min to be 0.001, choose K_D max to be 0.005, and choose 2000 steps. Press enter and you will see the

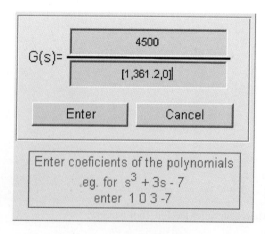

Figure 9-101 Transfer function input module.

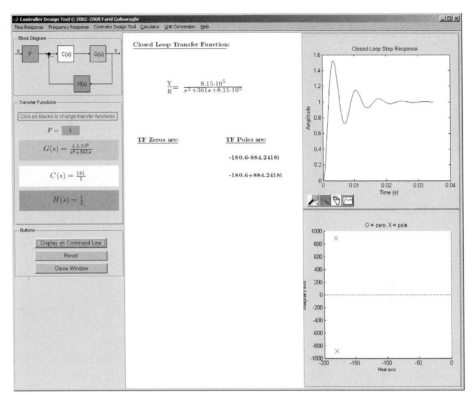

Figure 9-102 Controller Design window showing closed-loop transfer function, step response, and poles of the system.

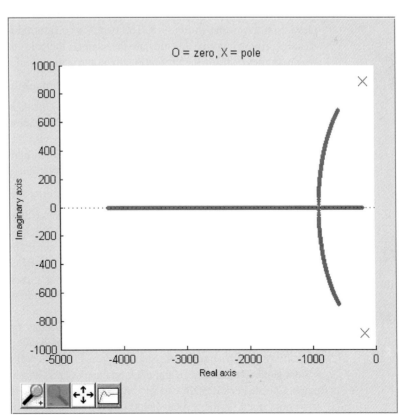

Figure 9-103 Root contours of PD-controlled system from Example 9-2-1.

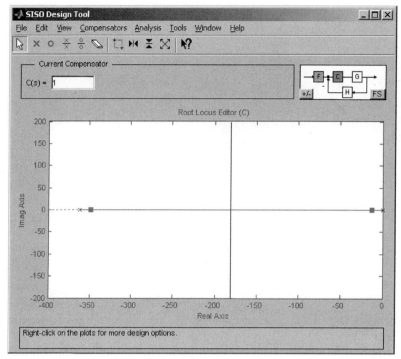

Figure 9-104 Root loci of the system from Example 9-2-1.

contour plot pictured in Fig. 9-104 in the axis in the top right of the window. Notice that this plot is similar to that in Fig. 9-8. The two X's correspond to the poles of the system with $K_D = 0$. You will notice that they are the same as the poles displayed in the lower right axis, which are the poles of the proportionally controlled system we entered earlier. Pressing the

button below the axis will enable you to scroll through the different values of K_D with the arrow keys and see the corresponding poles. As K_D increases, the two poles of the system move together until their imaginary parts disappear. They then move apart on the real axis. We know that, for a second-order system such as this, two real and equal poles is a characteristic of a critically damped system. To find the value of K_D that corresponds to a critically damped system, zoom in on the point where the two root contour lines meet on the real axis. To do this, click on the zoom-in button, , and then click and drag a box on the area you would like to enlarge. By scrolling through the values of K_D, we can see that the two poles are real and equal for some value of K_D between 0.0017712 and 0.0017726. If we choose $K_D = 0.001772$, then we have found exactly what was found analytically in Example 9-2-1.

Now enter the PD controller values in the C(s) box by clicking on the box and entering 181.2*[0.001772,1] in the numerator text box. After pressing enter, the new closed-loop transfer function will be displayed in the middle of the window and the poles of the new system will be plotted in the bottom axis. Notice that while the poles are not equal they are relatively close. To check whether we have met our design constraints, open the SISO tool by clicking on the Root Locus option in the Controller Design menu. Now from the Other Responses option from the Analysis menu, choose Step and press OK. You will now see the step response of the system plotted versus time. Right-click on the plot and choose characteristics from the pop-up menu. Choose Peak Response to see the overshoot of the step response. A dot will appear at the maximum point on the response plot; holding the mouse over it will display the percentage overshoot as 4.17%, which is within our constraint. Checking the other characteristics will show you that all the design constraints have been met.

You will notice that there are many examples in this book where multiple time-responses are plotted on the same axes as in Fig. 9-107. The capability to do this is not built into the **ACSYS** software. However, there is a short tutorial in Section 9-11 that shows you the steps involved in plotting multiple responses on the same axes.

Now let us repeat this same analysis using the SISO design tool, which is built into the Control System Toolbox, which is a component of MATLAB. Click on the C(s) block and change the value in the numerator text box back to [1].

To examine the performance of the proportional controller, we need to find the system root locus. The root locus of the system may be obtained by clicking the "Root Locus" option from the "Controller Design Tool" menu, which will activate the MATLAB SISO design tool. Fig. 9-104 shows the root locus of system. To see the poles and zeros of G and H, go to the View menu and select System Data or, alternatively, double-click the blocks G or H in the top right corner of the block diagram in the top right corner of the figure.

The squares in Fig. 9-104 correspond to the closed-loop system poles for $K = 1$. To see the closed-loop system time response to a unit-step input, select the Closed-Loop Step option from the Other Loop Responses category within the Analysis menu, which is located at the top of the screen shown in Fig. 9-105. Fig. 9-106 shows the unit-step response of the closed-loop system for $K = 1$. You may also obtain the closed-loop system poles by selecting Closed-Loop Poles from the View menu in the SISO Design Tool window. Recall the poles of the closed-loop system are

$$s_{1,2} = -180.6 \pm \sqrt{32616 - 4500K} \tag{9-294}$$

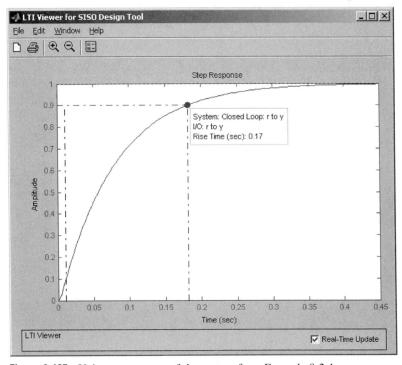

Figure 9-105 Unit-step response of the system from Example 9-2-1.

636 ▶ Chapter 9. Design of Control Systems

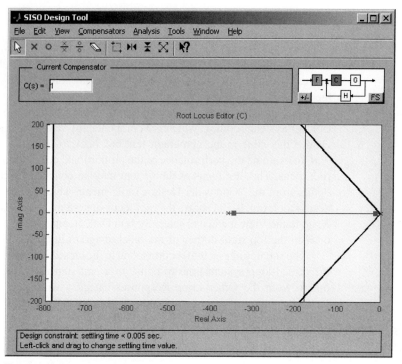

Figure 9-106 The root-locus diagram for Example 9-2-1, after incorporating the percent overshoot and the settling time as design constraints.

Changing K, therefore, affects the pole locations. In the Root Locus windows, "$C(s)$" represents the controller transfer function; in this case, $C(s) = K = 1$. Hence, if $C(s)$ is increased, the effective value of K increases, forcing the closed-loop poles to move together on the real axis and then ultimately to move apart to become complex.

Incorporation of the Design Criteria: As a first step to design a controller, we use the built-in design criteria option within the SISO Design Tool to establish the desired pole regions on the root locus. To add the design constraints, choose the Root Locus option from the Edit menu. "New" must then be selected within the Design Constraints option. The Design Constraints option allows the user to investigate the effect of the following:

- Settling time
- Percent overshoot
- Damping ratio
- Natural frequency

In this case we have included settling time and percent overshoot as design constraints. Enter the two constraints as they are listed at the beginning of this example, one at a time. In order also to enter the rise time as a constraint, the user must first establish a relation between the damping ratio and the natural frequency using an equation for rise time. Recall that the approximate equations for rise time for a second-order prototype system were provided in Chapter 5. Because the settling time and the percent overshoot are more important criteria in this example, we will use them as primary constraints. After designing a controller based on these constraints, we will determine whether the system complies with the rise-time constraint.

Fig. 9-106 shows the desired closed-loop system pole locations on the root locus after inclusion of the design constraints. Obviously, the poles of the system for $K = 1$ are not in

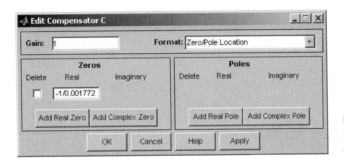

Figure 9-107 Addition of a zero to the controller $C(s)$ to create a PD controller.

the desired area. The desired poles of the system must lie to the left of the boundary imposed by the settling time between -700 and -800 markers in Fig. 9-106. Obviously, it is impossible to use the proportional controller (for any value of K) to move the poles of the closed-loop system farther to the left-half plane. However, a PD controller may be used to accomplish this task. As proposed earlier in the solution to Example 9-2-1, a PD controller with a zero at $z = -1/0.001772$ may be used for this purpose. Recall from the solution of Example 9-2-1 that this number was obtained from the steady-state error criterion by examining Eq. (9-13). In this case, if the proportional component of a PD controller is set to $K_P = 1$, the steady-state error due to a unit-ramp input is $e_{ss} = 0.000443$ for $K_D = 0.001772$.

To enter a zero to the controller, click the C(s) block in the block diagram in the top right corner of Fig. 9-106, or simply follow the instructions on the bottom of the screen shown in Fig. 9-106 and right-click the mouse to edit the controller transfer function. Fig. 9-107 shows the Edit Compensator window and how the PD controller is added.

The new root-locus diagram for the system appears in Fig. 9-108. It is now very easy to establish the required gain to push the poles to the desired region. Using the value of $K = 181.2$, which was proposed in the original solution of Example 9-2-1, would force the

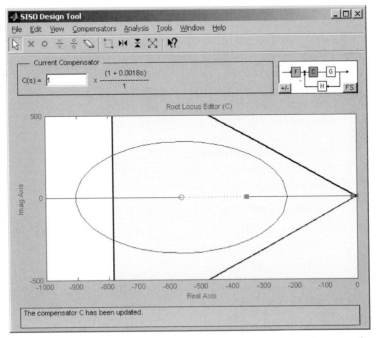

Figure 9-108 The root-locus diagram for Example 9-2-1, after incorporating a zero in the PD controller at $-1/0.001772$.

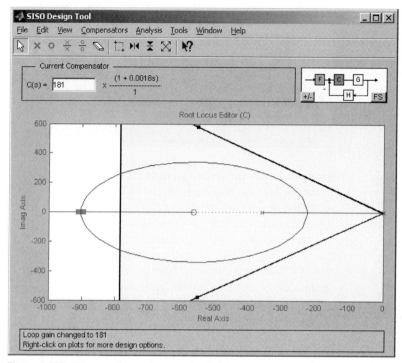

Figure 9-109 The root-locus diagram for Example 9-2-1 after incorporating a zero in the PD controller at $-1/0.001772$ and using a gain of 181.

closed-loop poles to the right region, as shown in Fig. 9-109. After dragging the poles, you can view their current location from the View menu. The step response of the controlled system in Fig. 9-110 shows the system has now complied with all design criteria. The 2% settling time is now 0.0488 sec, while the percent overshoot is 4.17. It is interesting that, although the closed-loop poles are both real, the system has a non-oscillatory response with an overshoot. This is because of the effect of zero on the response. Review the effect of adding a zero to a closed-loop transfer function, which was discussed in Section 5-11, to further appreciate this behavior.

Finally, in practice, always verify that the actuator used has enough torque or load to create such response. The actuator limitations must always be included because they are some of the most important design constraints.

Frequency-Domain Design
Now let us carry out the design of the PD controller in the frequency domain using the following performance criteria:

Steady-state error due to a unit-ramp input ≤ 0.00443

Phase margin $\geq 80°$

Resonant peak $M_r \leq 1.05$

BW ≤ 2000 rad/sec

To start the frequency-domain design process, click the Bode button in the Controller Design main window (Fig. 9-100) to get the MATLAB SISO Design Tool, as shown in Fig. 9-112.

As in the root-locus design approach, a PD controller with a zero at $z = -1/0.001772$ may be used for this purpose. Recall that this number was obtained from the steady-state error criterion. Similar to the root-locus approach, in order to enter zero for the controller,

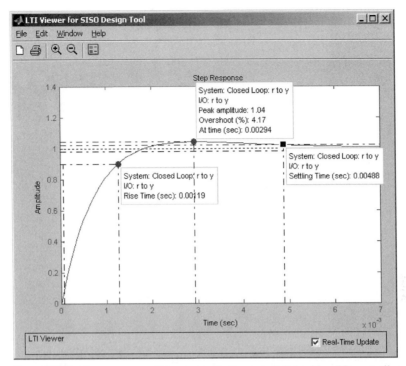

Figure 9-110 Step response of the system in Example 9-2-1 with a PD controller, $C(s) = 181(s + 1/0.001772)$.

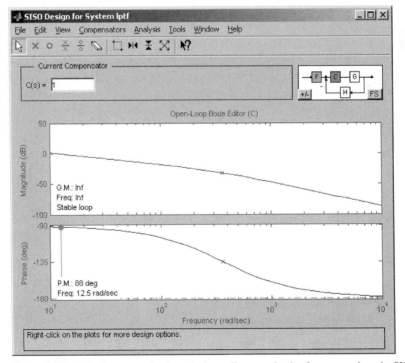

Figure 9-111 The loop magnitude and phase diagrams in the frequency-domain SISO Design Tool for Example 9-2-1.

640 ▶ Chapter 9. Design of Control Systems

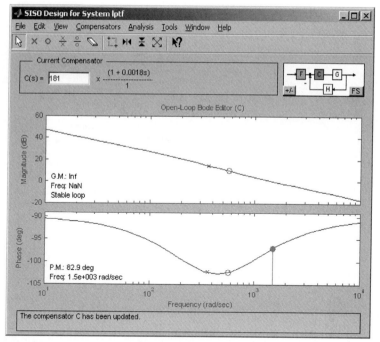

Figure 9-112 The loop magnitude and phase diagrams for Example 9-2-1, after incorporating a zero in the PD controller at $-1/0.001772$ and a gain of $K = 181.2$.

click the C(s) block in the block diagram in the top right corner of the screen shown in Fig. 9-112. The new open-loop Bode diagram of the system appears in Fig. 9-113 for the system with a PD controller $G_c(s) = C(s) = 181.2(1 + 0.001772s)$. The gain margin in this case is infinity, while the phase margin is 83°. As a result, the system has also fully complied with all the design criteria in the frequency domain. Review the effect of adding a zero to a

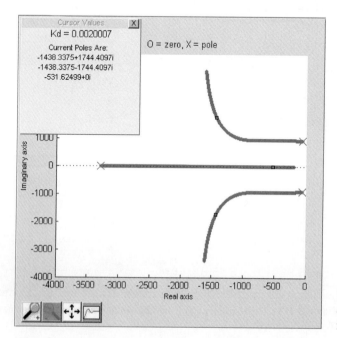

Figure 9-113 Root contours of the PD-controlled systems from Example 9-2-2.

▶ **EXAMPLE 9-2-2** We will now use the **ACSYS** tool to solve the problem from Example 9-2-2. Recall that the forward-
Revisited path transfer function for the system was

$$G(s) = \frac{1.5 \times 10^7 K}{s(s^2 + 3408.3s + 1,204,000)} \tag{9-295}$$

closed-loop transfer function, which was discussed in Section 9-3 (see also Section 9-4 for addition of a pole) to appreciate this behavior. Alternatively, you may select the Open-Loop Bode option from the View menu in the root-locus diagram in Fig. 9-108 to obtain the root-locus and frequency-response representations of the system together on the same figure.

And the design contraints remain the same as in Example 9-2-2.

Click on the G(s) block to enter the plant transfer function. In the numerator text box, enter [1.5e7], and in the denominator text box, enter [1,3408.3,1204000,0]. K will be chosen to be 181.17, which we will enter in the C(s) block. Click on the C(s) block and enter [181.17] in the numerator text box. Let us again use the Root Contour tool. Choose the "PD design: Root Contour" option from the Controller Design menu. Choose K_D min to be 0.00001, choose K_D max to be 0.005, and choose 3000 steps. Pressing enter will display the plot shown in Fig. 9-113, which is similar to Fig. 9-12. Press the "arrow cursor" button and scroll through the values of P_D. You will see the behavior described in Example 9-2-2; as the value of K_D increases, the real pole moves toward the origin and the two complex poles approach asymptotically to the line at $s = -1704$.

In Example 9-2-2 the value of 0.002 is chosen for K_D; let us apply a PD controller with $K_D = 0.002$ to see its effect. K_P will be left at 1. Click on the C(s) block and enter 181.17*[.002,1] in the numerator text box. After pressing enter, click on the Step Response option from the Time Response menu and then press the Print to Figure button. Right-click on the resulting plot and choose "Characteristics" and "Peak Response" from the pop-up menu. Holding the pointer over the marker that appears at the point of maximum amplitude will display the value of the overshoot for this system. Repeating this for the rise time and the settling time will show you all the data listed in Table 9-3.

To see the Bode plots of this PD-controlled system, choose the Bode option from the Controller Design menu and you will see the window shown in Fig. 9-114.

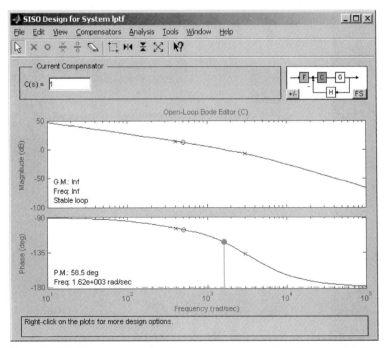

Figure 9-114 Bode plots of the PD-controlled system from Example 9-2-2.

642 ▶ Chapter 9. Design of Control Systems

▶ **EXAMPLE 9-3 Revisited** We will now go through the steps involved in finding some of the results of Example 9-3-1. First input the plant for the system. Click on the G(s) block and enter [4500] in the numerator text box and [1,361.2,0] in the denominator text box. Adding the PI controller is accomplished by setting C(s) equal to a function of the form

$$C(s) = \frac{KK_P\left(s + \dfrac{K_I}{K_P}\right)}{s} \tag{9-296}$$

Adding the PI controller effectively adds a zero at $s = K_I/K_P$ and a pole at $s = 0$. We can use MATLAB's SISO tool to do this. As is done in Example 9-3-1 let's have $K = 181.17$. Click on the C(s) block and enter [181.17] in the numerator text box. From the Controller Design menu, choose the Root Locus option. This will open the SISO Design Tool pictured in Fig. 9-115. The plot shows the effect on the poles of the system of varying the value of a constant multiplied by C(s).

Now to add the PI controller, we need to add in the pole and zero. Click on the C block in the block diagram located in the top right of the SISO window. Now, add a pole at $s = 0$, and as is done in Example 9-3-1, add a zero at $s = -10$. Click OK and you will see the plot pictured in Fig. 9-116.

Notice that this plot is similar to that in Fig. 9-19. As it is now, the value of C(s) is equal to 181.17 times the function pictured just above the Root Locus plot. Let us set the controller parameters equal to those in the third row of Table 9-5. We already have the ratio of K_I/K_P that we want. To change the value of K_P, grab one of the complex poles and drag it toward the real axis. You will see that the value of the gain in front of the controller function starts to decrease. Move the pole until the gain is equal to 0.8. Now from the Analysis menu, choose the Step option from the Other Loop Responses menu. Right-click on the resulting graph and choose "Characteristics" and "Peak Response." Hold the pointer over the dot marking the maximum response to see the percentage overshoot, as pictured in Fig. 9-117. Repeat this for the rise time and the settling time and you will see values close to those listed in Table 9-5. Note that the settling time referred to in Table 9-5 is the 5% settling time, so you may need to change the definition of settling time in the Properties menu, located in the same pop-up menu as the characteristics option. If you put both the main SISO window and the Step Response

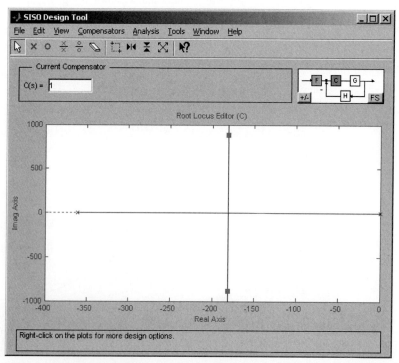

Figure 9-115 Root loci of the system from Example 9-3-1.

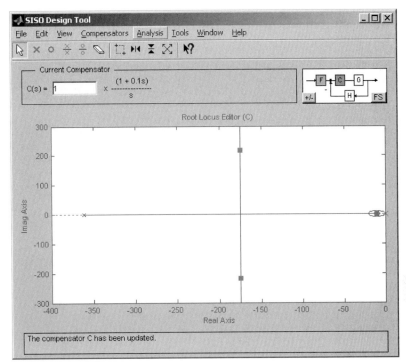

Figure 9-116 Root loci of the PI-controlled system from Example 9-3-1.

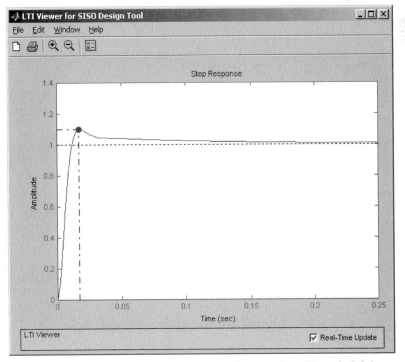

Figure 9-117 Step response of the PI-controlled system from Example 9-3-1.

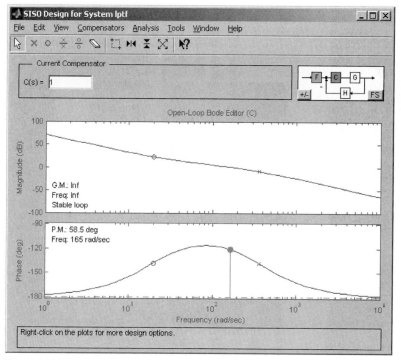

Figure 9-118 Bode plots of the PI-controlled system from Example 9-3-1.

window in view at the same time, you can see the effect that moving the poles has on the response in real time.

Now return to the Controller Design Tool window. To obtain the Bode plots shown at the end of Example 9-3-1, using the parameters from the second row in Table 9-6, enter 181.17*[0.08,1.6] into the C(s) numerator, and enter [1,0] into the C(s) denominator. Then choose the Bode option from the Controller Design Tool menu. You will then see the window pictured in Fig. 9-118.

▶ **EXAMPLE 9-3-2 Revisited** We will now go through the steps involved in finding some of the results of Example 9-3-2 using the **ACSYS** software. First enter the plant transfer function. Click on the C(s) block, and in the numerator text box, enter [1.5e7]. In the denominator text box, enter [1,3408.3,1.204e6,0]. Press enter, and then click the C(s) block, and enter 181.17. From the Time Response menu, choose the Step option. You will obtain one of the plots shown in Fig. 9-23.

To improve the system, we will apply a PI controller. Let us use the parameter values from the fourth row in Table 9-8. Click on the C(s) block. In the numerator text box, enter 181.17*0.075*[1,1], and in the denominator text box, enter [1,0]. Again plot the step response and you will see one of the other plots shown in Fig. 9-23. To see the frequency response, choose the Bode option from the Controller Design Tool menu and you will see some of the information shown in Figure 9-22. ◀

▶ **EXAMPLE 9-4-1 Revisited** We will now go through the steps involved in inputting the parameters that correspond to the PID controller of Example 9-4-1. First enter the plant transfer function. Click on the G(s) block, and enter [2.718e9] in the numerator text box and [1,3408.3,1204000,0] in the denominator text box. To enter the PID controller in the C(s) text boxes, it needs to be written in the form of the left side of Eq. (9-48). We will choose $K_D = 0.0006$, $K_P = 0.309$, and $K_I = 4.5$. In the C(s) numerator text box, enter [0.0006,0.309,4.5], and in the denominator text box, enter [1,0]. Press enter, and then choose the Step Response option from the Time Response menu; you will obtain one of the plots shown in Fig. 9-24.

From the Controller Design Tool menu, choose the Bode option, and you will see the Bode plot shown in Fig. 9-25. ◀

▶ **EXAMPLE 9-5-1** We will now use **ACSYS** to obtain some of the results from Example 9-5-1. First, enter the plant
Revisited of the system. In the C(s) numerator text box, enter [2500], and in the denominator text box,
enter [1,25,0]. We are going to use a phase-lead controller to improve the performance of the
system. Let us use the root contour tool to examine the effect that varying T has on the poles of
the system. From the Controller Design Tool menu, choose the "Lead Lag Design: Root
Contour" option. Set a to be 0. Set $Tmin$ to be 0.0001, set $Tmax$ to be 100, and choose 400 steps.
After pressing enter, you will notice that the x scale on the resulting plot is much wider than that
in Fig. 9-31. To get a better view, click on the zoom-in button below the plot. Then click and drag
the box around the area you would like to be displayed. You should be able to generate
something that looks similar to Fig. 9-119.

Now we will apply the Phase-Lead controller to see its effect on the response. Choosing the
parameter values from the fourth row in Table 9-12, we will have $a = 16.67$ and $T = 0.003$. Click on
the C(s) block, and in the numerator text box, enter [0.05,1]; in the denominator text box, enter
[0.003,1]. Press enter, and then plot the step response from the Time Response menu. Press the Print
to Figure button, and right-click on the plot to see details about the response.

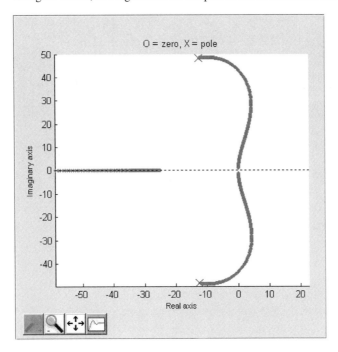

Figure 9-119 Root contours
of the lead–lag controlled
system from Example 9-5-1.

◀

▶ **EXAMPLE 9-5-2** Now we will use some of the **ACSYS** tools to examine the application of a phase-lead controller
Revisited to the third-order system of Example 9-5-2. First input the plant transfer function. In the G(s)
numerator text box, enter [4.6875e7], and in the denominator text box, enter [1,625,156250,0].
As in Example 9-5-2, we will use a phase-lead controller to improve the performance of the
system. First let's use the root contour tool to examine the effect that the controller has on the
poles of the system. From the Controller Design Tool menu, choose the "Lead Lag Design: Root
Contour" option. Choose $a = 0$, $Tmin = 0.0005$, and $Tmax = 1$, and choose 500 steps. After
pressing enter, you will see the plot shown in Fig. 9-121, which is similar to one of the lines
plotted in Fig. 9-35.

Now implement the phase-lead controller with the parameters from the last row in Table 9-14.
In the C(s) numerator text box, enter [0.005,1], and in the denominator text box, enter [0.00001,1].
From the Time Response menu, choose the Step Response option, press the Print to Figure button,
and right-click on the resulting plot to see the characteristics of the system, which are listed in
Table 9-14. From the Controller Design menu, choose the Bode option to see the Bode plots for

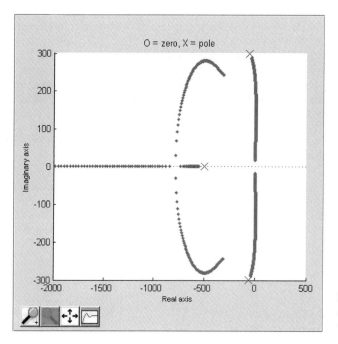

Figure 9-120 Root contours of the system from Example 9-5-2.

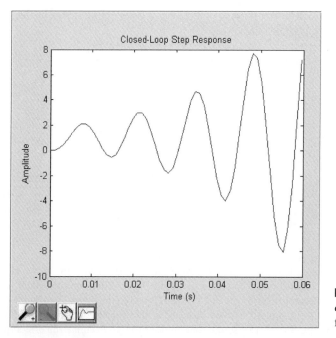

Figure 9-121 Step Response of the uncompensated system from Example 9-5-3.

this system. Right-click on the plot and change the properties to see the full range of frequencies shown in Fig. 9-37. ◀

▶ **EXAMPLE 9-5-3 Revisited** We will now go through some of the steps of Example 9-5-3 using the **ACSYS** tools. First input the plant transfer function. Click on the G(s) block, and in the numerator text box, enter [1.5625e8]. In the denominator text box, enter [1,625,156250,0]. After pressing enter, click the Bode option from the Controller Design Tool menu to see the Bode plot of the uncompensated system, which is shown in Fig. 9-39. From the Time Response menu, choose the Step command and you will see that the system is indeed unstable. The step response of the uncompensated system is shown in Fig. 9-121.

To improve the performance we will use a two-stage phase-lead controller. Let us use the parameter values from the first row of Table 9-16. To input the controller transfer function, we need to rewrite the numerator and denominator in polynomial form. In the C(s) numerator text box, enter [1.499e-5,7.744e-3,1], and in the denominator text box, enter [2.343e-9,9.68e-5,1]. Press enter and plot the step response from the Time Response menu to see one of the plots shown in Fig. 9-40.

▶ **EXAMPLE 9-6-1 Revisited** Now we will go through some of the steps involved in applying the phase-lag controller to a second-order system, as in Example 9-6-1. Input the plant transfer function. Click on the G(s) block. K will be set equal to 1, so in the G(s) denominator text box, enter [2500]; in the numerator text box, enter [1,25,0]. Choose the Step option from the Time Response menu to see the step response of the uncompensated system, which is shown in Fig. 9-47.

Let us use the controller parameters from the second-to-last row in Table 9-17. Click on the C(s) block, and in the numerator text box, enter [2.7,1]. In the denominator text box, enter [30,1]. Press enter, and again plot the step response. Press the Print to Figure button, and right-click on the resulting plot. By clicking the Properties option, you can set the x-limits of the response plot to get a better view of the response. ◀

▶ **EXAMPLE 9-6-2 Revisited** We will now go through some of the steps involved in Example 9-6-2 with the **ACSYS** tools. First input the plant transfer function. Click on the G(s) block, and enter [1.5625e8] in the numerator text box, and enter [1,625,156250,0] in the denominator text box. After pressing enter, select the Root Locus option from the Controller Design menu. You will see the root loci plot pictured in Fig. 9-50.

We will implement a phase-lag controller to improve the performance of the system. Let's use the parameter values from the second row of Table 9-19. Click on the C(s) block, enter [1,1] in the numerator text box, and enter [10,1] in the denominator text box. Click the enter button, and choose the Step option from the Time Response menu. Right-click on the resulting plot to check the characteristics of the current system. ◀

▶ 9-15 PLOTTING TUTORIAL

You will notice that in many of the examples in this book there are figures containing multiple response plots, each corresponding to a certain parameter value. Arranging plots in this way is obviously very helpful in discovering the effect that a certain parameter has on a system. The functionality to overlay the response plots on the same figure is not built into the **ACSYS** software. This short tutorial will go through the steps involved. This tutorial can also serve as an introduction to manipulating graphics and data at the MATLAB command line.

The plots we wish to display are the step responses of the system in Example 9-2-1. One plot will be the response of the system with a proportional controller, and the other will be the system with a PD controller. What we will do is generate the step response plots for each system and then plot the data from each on another figure; this will leave us with a graph similar to that in Fig. 9-110.

Press the G(s) block, enter [4500] in the numerator text box, and enter [1,361.2,0] in the denominator text box. Press the C(s) block, and enter [181.17] in the numerator text box. After pressing enter, choose the Step Response option from the Time Response menu, and press the Print to Figure button under the resulting plot. Now change the value in the C (s) numerator to 181.17*[.00177,1]. Repeat the procedure to generate a second figure. We now have two separate figures, each with its own set of axes and its own set of data plotted.

We now want to extract the plotted data from each figure so we can plot them again on new axes on a new figure. Choose "Data Cursor" from the toolbar on one of the figures, and click somewhere on the line so a marker appears on it. Now, at the MATLAB command line, type the following:

$$t1 = get(gco, \text{'xdata'}); \ y1 = get(gco, \text{'ydata'});$$

Now repeat the procedure for the second figure, this time using the variable names "t2" and "y2."

The line is a graphics object that has a list of properties that define it. GCO stands for "get current object." We make the line the current object by clicking on it with the cursor tool; the "get" command puts the value of the specified property in a variable of our choice. Xdata is a property of the line, which is a row vector of values. We now have two vectors, a time vector $t1$ and an amplitude vector $y1$. When we plot, we will be plotting each $y1$ value versus its corresponding $t1$ value. Repeat the procedure for the second figure and place the xdata and ydata in the variables $t2$ and $y2$, respectively.

Now we are ready to plot the data on a single axis. At the command line, type the following:

$$\text{figure; axes; plot}(t1, y1); \text{hold on; plot}(t2, y2);$$
$$\text{plot}([0, 0.05], [1, 1], \text{'color'}, [0, 0, 0], \text{'linestyle'}, \text{' : '});$$

The "hold on" command stops MATLAB from erasing the first plot when the second plot command is given. The third plot command just draws a black dotted line between the points (0,1) and (.05,1).

There is one final adjustment to make the new graph complete. It would be much better to display the two plots in different colors. Again, choose "Data Cursor" from the toolbar, and click on one of the lines. Now type the following:

$$\text{set(gco, 'color'}, [1, 0, 0]);$$

The "set" command gives the specified value to the specified property of the specified object. You should now have a window that looks like Fig. 9-122.

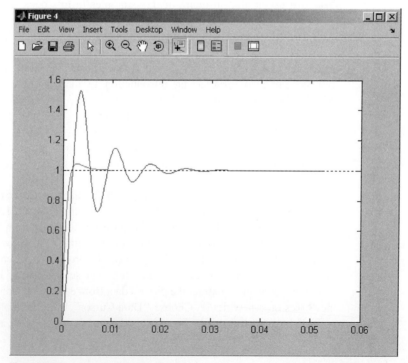

Figure 9-122 Step responses of system.

9-16 SUMMARY

This chapter was devoted to the design of linear continuous-data control systems. It began by giving some fundamental considerations on system design and then reviewed the specifications in the time domain and frequency domain. Fixed configurations of compensation schemes used in practice, such as series, forward and feedforward, feedback and minor loop, and state feedback were illustrated. The types of controllers considered were the PD, PI, PID, phase-lead, phase-lag, lead–lag, pole–zero cancellation, and notch filters. Designs were carried out in the time-domain (s-domain) as well as the frequency domain. The time-domain design was characterized by specifications such as the relative damping ratio, maximum overshoot, rise time, delay time, settling time, or simply the location of the characteristic-equation roots, keeping in mind that the zeros of the system transfer function also affect the transient response. The performance was generally measured by the step response and the steady-state error. Frequency-domain designs were generally carried out using Bode diagrams or gain-phase plots. The performance specifications in the frequency domain were the phase margin, gain margin, M_r, BW, and the like.

The effect of feedforward control on noise and disturbance reduction was demonstrated. A section was devoted to the design of robust control systems.

While the design techniques covered in this chapter were outlined with analytical procedures, the text promotes the use of MATLAB Toolboxes and specifically ACSYS. The Controller Design Tool has been developed by the authors for this purpose. Through the GUI approach, ACSYS creates a user-friendly environment to reduce the complexity of control systems design.

REVIEW QUESTIONS

1. What is a PD controller? Write its input–output transfer function.

2. A PD controller has the constants K_D and K_P. Give the effects of these constants on the steady-state error of the system. Does the PD control change the type of a system?

3. Give the effects of the PD control on rise time and settling time of a control system.

4. How does the PD controller affect the bandwidth of a control system?

5. Once the value of K_D of a PD controller is fixed, increasing the value of K_P will increase the phase margin monotonically. (T) (F)

6. If a PD controller is designed so that the characteristic-equation roots have better damping than the original system, then the maximum overshoot of the system is always reduced. (T) (F)

7. What does it mean when a control system is described as being robust?

8. A system compensated with a PD controller is usually more robust than the system compensated with a PI controller. (T) (F)

9. What is a PI controller? Write its input–output transfer function.

10. A PI controller has the constants K_P and K_I. Give the effects of the PI controller on the steady-state error of the system. Does the PI control change the system type?

11. Give the effects of the PI control on the rise time and settling time of a control system.

12. How does the PI controller affect the bandwidth of a control system?

13. What is a PID controller? Write its input–output transfer function.

14. Give the limitations of the phase-lead controller.

15. How does the phase-lead controller affect the bandwidth of a control system?

16. Give the general effects of the phase-lead controller on rise time and settling time.

17. For the phase-lead controller, $G_c(s) = (1 + aTs)/(1 + Ts)$, $a > 1$, what is the effect of the controller on the steady-state performance of the system?

18. The phase-lead controller is generally less effective if the uncompensated system is very unstable to begin with. (T) (F)

19. The maximum phase that is available from a single-stage phase-lead controller is 90°. **(T) (F)**

20. The design objective of the phase-lead controller is to place the maximum phase lead at the new gain-crossover frequency. **(T) (F)**

21. The design objective of the phase-lead controller is to place the maximum phase lead at the frequency where the magnitude of the uncompensated $G_p(j\omega)$ is $-10 \log_{10} a$, where a is the gain of the phase-lead controller. **(T) (F)**

22. The phase-lead controller may not be effective if the negative slope of the un-compensated process transfer function is too steep near the gain-crossover frequency. **(T) (F)**

23. For the phase-lag controller, $G_c(s) = (1 + aTs)/(1 + Ts)$, $a < 1$, what is the effect of the controller on the steady-state performance of the system?

24. Give the general effects of the phase-lag controller on rise time and settling time.

25. How does the phase-lag controller affect the bandwidth?

26. For a phase-lag controller, if the value of T is large and the value of a is small, it is equivalent to adding a pure attenuation of a to the original uncompensated system at low frequencies. **(T) (F)**

27. The principle of design of the phase-lag controller is to utilize the zero-frequency attenuation property of the controller. **(T) (F)**

28. The corner frequencies of the phase-lag controller should not be too low or else the bandwidth of the system will be too low. **(T) (F)**

29. Give the limitations of the pole-zero-cancellation control scheme.

30. How does the sensitivity function relate to the bandwidth of a system?

Answers to these review questions can be found on this book's companion Web site: www.wiley.com/college/golnaraghi.

▶ REFERENCES

1. N. D. Manring, *Hydraulic Control Systems*, John Wiley & Sons, New York, 2005.
2. D. McCloy and H. R. Martin, *The Control of Fluid Power*, Longman Group Limited, London, 1973.
3. B. C. Kuo and F. Golnaraghi, *Automatic Control Systems*, John Wiley & Sons, New York, 2003.
4. R. L. Woods and K. L. Lawrence, *Modeling and Simulation of Dynamic Systems*, Prentice-Hall, New Jersey, 1997.
5. A. Kleman, *Interfacing Microprocessors in Hydraulic Systems*, Marcel Dekker, New York, 1989.
6. K. Ogata, *Modern Control Engineering*, Prentice-Hall, New Jersey, 1997.
7. W. P. Graebel, *Engineering Fluid Mechanics*, Taylor & Francis, New York, 2001.
8. J. C. Willems and S. K. Mitter, "Controllability, Observability, Pole Allocation, and State Reconstruction," *IEEE Trans. Automatic Control*, Vol. AC-16, pp. 582–595, Dec. 1971.
9. H. W. Smith and E. J. Davison, "Design of Industrial Regulators," *Proc. IEE (London)*, Vol. 119, pp. 1210–1216, Aug. 1972.
10. F. N. Bailey and S. Meshkat, "Root Locus Design of a Robust Speed Control," *Proc. Incremental Motion Control Symposium*, pp. 49–54, June 1983.

▶ PROBLEMS

Most of the following problems can be solved using a computer program. This is highly recommended if the reader has access to such a program.

9-1. The block diagram of a control system with a series controller is shown in Fig. 9P-1. Find the transfer function of the controller $G_c(s)$ so that the following specifications are satisfied:

The ramp-error constant K_v is 5.

The closed-loop transfer function is of the form

$$M(s) = \frac{Y(s)}{R(s)} = \frac{K}{(s^2 + 20s + 200)(s + a)}$$

where K and a are real constants. Find the values of K and a.

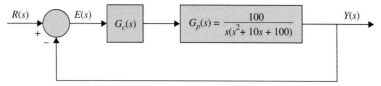

Figure 9P-1

The design strategy is to place the closed-loop poles at $-10 + j10$ and $-10 - j10$, and then adjust the values of K and a to satisfy the steady-state requirement. The value of a is large so that it will not affect the transient response appreciably. Find the maximum overshoot of the designed system.

9-2. Repeat Problem 9-1 if the ramp-error constant is to be 9. What is the maximum value of K_v that can be realized? Comment on the difficulties that may arise in attempting to realize a very large K_v.

9-3. The forward-path transfer function of a unity-feedback control system is

$$G(s) = \frac{K}{s(\tau s + 1)}$$

Find the value of K and τ so that the overshoot of the system is 25.4% at the $\zeta = 0.4$.

9-4. The forward-path transfer function of a system is

$$G(s)H(s) = \frac{24}{s(s+1)(s+6)}$$

Design a PD controller that satisfies the following factors:
(a) The steady-state error is less than $\pi/10$ when the input is a ramp with a slope of 2π rad/sec.
(b) The phase margin is between 40 and 50 degrees.
(c) Gain-crossover frequency is greater than 1 rad/sec.

9-5. A control system with a PD controller is shown in Fig. 9P-5.

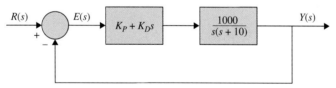

Figure 9P-5

(a) Find the values of K_P and K_D so that the ramp-error constant K_v is 1000 and the damping ratio is 0.5.

(b) Find the values of K_P and K_D so that the ramp-error constant K_v is 1000 and the damping ratio is 0.707.

(c) Find the values of K_P and K_D so that the ramp-error constant K_v is 1000 and the damping ratio is 1.0.

9-6. For the control system shown in Fig. 9P-5, set the value of K_P so that the ramp-error constant is 1000.

(a) Vary the value of K_D from 0.2 to 1.0 in increments of 0.2 and determine the values of phase margin, gain margin, M_r, and BW of the system. Find the value of K_D so that the phase margin is maximum.

(b) Vary the value of K_D from 0.2 to 1.0 in increments of 0.2 and find the value of K_D so that the maximum overshoot is minimum.

9-7. The forward-path transfer function of a system is

$$G(s)H(s) = \frac{1}{(2s+1)(s+1)(0.5s+1)}$$

Design a PD controller such that the $K_P = 9$ and the phase margin is greater than 25 degrees.

9-8. The forward-path transfer function of a system is

$$G(s)H(s) = \frac{60}{s(0.4s+1)(s+1)(s+6)}$$

(a) Design a PD controller to statisfy the following specifications:
  (i) $K_v = 10$.
  (ii) The phase margin is 45 degrees.
(b) Use MATLAB to plot the Bode diagram of the compensated system.

9-9. Consider the second-order model of the aircraft attitude control system shown in Fig. 5-31. The transfer function of the process is

$$G_p(s) = \frac{4500\,K}{s(s+361.2)}$$

(a) Design a series PD controller with the transfer function $G_c(s) = K_D + K_P s$ so that the following performance specifications are satisfied:

Steady-state error due to a unit-ramp input ≤ 0.001
Maximum overshoot $\leq 5\%$
Rise time $t_r \leq 0.005$ sec
Settling time $t_s \leq 0.005$ sec

(b) Repeat part (a) for all the specifications listed, and, in addition, the bandwidth of the system must be less than 850 rad/sec.

9-10. Fig. 9P-10 shows the block diagram of the liquid-level control system described in Problem 5-42. The number of inlets is denoted by N. Set $N = 20$. Design the PD controller so that with a unit-step input the tank is filled to within 5% of the reference level in less than 3 sec without overshoot.

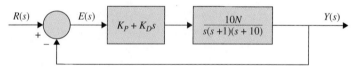

Figure 9P-10

9-11. For the liquid-level control system described in Problem 9-10,
(a) Set K_P so that the ramp-error constant is 1. Vary K_D from 0 to 0.5 and find the value of K_D that gives the maximum phase margin. Record the gain margin, M_r, and BW.
(b) Plot the sensitivity functions $|S_G^M(j\omega)|$ of the uncompensated system and the compensated system with the values of K_D and K_P determined in part (a). How does the PD controller affect the sensitivity?

9-12. The block diagram of a servo system is shown in Fig. 9P-12.

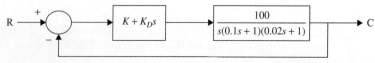

Figure 9P-12

Design the PD controller so that the phase margin is greater than 50 degrees and the BW is greater than 20 rad/sec. Use MATLAB to verify your answer.

9-13. The forward-path transfer function of a unity-feedback system is

$$G(s)H(s) = \frac{1000K}{s(0.2s+1)(0.005s+1)}$$

Design a compensator such that the steady-state error to the unit-step input is less than 0.01 and the closed-loop damping ratio $\zeta > 0.4$.

Use MATLAB to plot the Bode diagram of the compensated system.

9-14. The open-loop transfer function of a dc motor is

$$G(s) = \frac{250}{s(0.2s+1)}$$

Design a PD controller so that the steady-state error to the input ramp is less than 0.005, the maximum overshoot is 20% for the unit-step input, and the BW must be maintained at a value approximately the same as that of the uncompensated system.

9-15. The open-loop plant model of a plastic extrusion is given by

$$G(s) = \frac{40}{(s+1)(0.25s+1)}$$

Design a series of lead compensator, which is described by

$$G_c(s) = \frac{r(\tau s + 1)}{(r\tau s + 1)}$$

so that the phase margin is 45 degrees and the BW must be maintained at a value approximately the same as that of the uncompensated system.

9-16. Repeat Problem 9-15 assuming that the $r < 0.1$.

9-17. The forward-path transfer function of a unity-feedback control system is

$$G(s)H(s) = \frac{1000K}{s(0.2s+1)(0.05s+1)}$$

(a) Design a compensator such that
  (i) The steady-state error is less than 0.01 for a unit-ramp input.
  (ii) The phase margin is greater than 45 degrees.
  (iii) The steady-state error is less than 0.004 for a sinusoidal input with $\omega < 0.2$.
  (iv) The noise for the frequencies greater than 200 rad/sec reduced to 100 at the output.

(b) Use MATLAB to plot the Bode diagram of the compensated system and verify or refine your design in part (a).

9-18. A control system with a type 0 process $G_p(s)$ and a PI controller is shown in Fig. 9P-18.

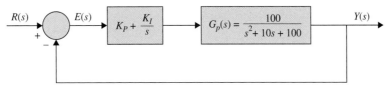

Figure 9P-18

(a) Find the value of K_I so that the ramp-error constant K_v is 10.

(b) Find the value of K_P so that the magnitude of the imaginary parts of the complex roots of the characteristic equation of the system is 15 rad/sec. Find the roots of the characteristic equation.

(c) Sketch the root contours of the characteristic equation with the value of K_I as determined in part (a) and for $0 \leq K_P < \infty$.

9-19. For the control system described in Problem 9-18,

(a) Set K_I so that the ramp-error constant is 10. Find the value of K_P so that the phase margin is minimum. Record the values of the phase margin, gain margin, M_r, and BW.

(b) Plot the sensitivity functions $|S_G^M(j\omega)|$ of the uncompensated and the compensated systems with the values of K_I and K_P selected in part (a). Comment on the effect of the PI control on sensitivity.

9-20. For the control system shown in Fig. 9P-18, perform the following:

(a) Find the value of K_I so that the ramp-error constant K_v is 100.

(b) With the value of K_I found in part (a), find the critical value of K_P so that the system is stable. Sketch the root contours of the characteristic equation for $0 \leq K_P < \infty$.

(c) Show that the maximum overshoot is high for both large and small values of K_P. Use the value of K_I found in part (a). Find the value of K_P when the maximum overshoot is a minimum. What is the value of this maximum overshoot?

9-21. Repeat Problem 9-20 for $K_v = 10$.

9-22. The forward-path transfer function of a system is

$$G(s) = \frac{24}{s(s+1)(s+6)}$$

(a) Design a PI controller that satisfies the following factors:
  (i) The ramp error constant $K_v > 20$.
  (ii) The phase margin is between 40 and 50 degrees.
  (iii) Gain-crossover frequency is greater than 1 rad/sec.

(b) Use MATLAB to plot the Bode diagram of the closed-loop system.

9-23. The forward-path transfer function of a robot arm–positioning system is represented by

$$G(s) = \frac{40}{s(s+2)(s+20)}$$

(a) Design a PI controller such that:
  (i) The steady-state error is less than 5% of the slope for a ramp input.
  (ii) The phase margin is between 32.5 and 37.5 degrees.
  (iii) Gain-crossover frequency is 1 rad/sec.

(b) Use MATLAB to plot the Bode diagram of the closed-loop system and verify your design in part (a).

9-24. The forward-path transfer function of a system is

$$G(s) = \frac{210}{s(5s+7)(s+3)}$$

Design a PI controller with a unity dc gain so that the phase margin of the system is greater than 40 degrees, and then find the BW of the system.

9-25. The transfer function of the steering of a ship is given by

$$G(s) = \frac{2353K(71-500s)}{71s(40s+13)(5000s+181)}$$

Design a PI controller such that:
(a) The ramp error constant $K_v = 2$.
(b) The phase margin is greater than 50 degrees.

(c) For all frequencies greater than crossover frequency, PM > 0. This means the system is always stable without any condition.

(d) Show the closed-loop poles in the root locus with respect to values of K.

9-26. The transfer function of a unity-feedback system is

$$G(s) = \frac{2 \times 10^5}{s(s+20)(s^2+50s+10000)}$$

(a) A PD controller with the transfer function of $H(s) = \frac{(\tau s + 1)}{(r\tau s + 1)}$ is designed with $r = 0.2$ and $\tau = 0.05$. It is desired to find the gain so that the crossover frequency is 31.6 rad/sec.

(b) Find the ramp error constant Kv by applying the controller designed in part (a).

(c) Consider the PD controller designed in part (a) is applied to the system. Find the value of K for a PI controller so that the ramp error constant $K_v = 100$.

(d) If the PI controller pole is at 3.16 rad/sec and the crossover frequency maintains at 31.6 rad/sec, what is the zero of the PI controller? [Consider the transfer function of the PI controller is $H(s) = \frac{r(\tau s + 1)}{(r\tau s + 1)}$.]

(e) Use MATLAB to plot the Bode diagram of the compensated system and find the phase margin.

9-27. A control system with a type 0 process and a PID controller is shown in Fig. 9P-27. Design the controller parameters so that the following specifications are satisfied:

Ramp-error constant $K_v = 100$
Rise time $t_r < 0.01$ sec
Maximum overshoot < 2%

Plot the unit-step response of the designed system.

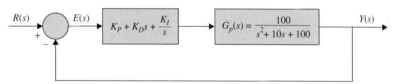

Figure 9P-27

9-28. A considerable amount of effort is being spent by automobile manufacturers to meet the exhaust-emission-performance standards set by the government. Modern automotive-power-plant systems consist of an internal combustion engine that has an internal cleanup device called the catalytic converter. Such a system requires control of the engine air–fuel ratio (A/F), the ignition-spark timing, exhaust-gas recirculation, and injection air. The control system problem considered in this exercise deals with the control of the air–fuel ratio. In general, depending on fuel composition and other factors, a typical stoichiometric A/F is 14.7:1, that is, 14.7 grams of air to each gram of fuel. An A/F greater or less than stoichiometry will cause high hydrocarbons, carbon monoxide, and nitrous oxide in the tailpipe emission. The control system whose block diagram is shown in Fig. 9P-28 is devised to control the air–fuel ratio so that a desired output variable is maintained for a given command signal. Fig. 9P-28 shows that the sensor senses the composition of the exhaust-gas mixture entering the catalytic converter. The electronic controller detects the difference or the error between the command and the sensor signals and computes the control signal necessary to achieve the desired exhaust-gas composition. The output variable $y(t)$ denotes the effective air–fuel ratio. The transfer function of the engine is given by

$$\frac{Y(s)}{U(s)} = G_p(s) = \frac{e^{-T_d s}}{1 + \tau s}$$

where T_d is the time delay and is 0.2 sec. The time constant τ is 0.25 sec. Approximate the time delay by a power series:

$$e^{-T_d s} \cong \frac{1}{1 + T_d s + T_d^2 s^2/2}$$

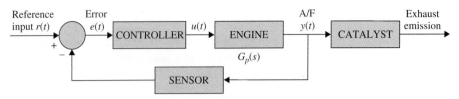

Figure 9P-28

(a) Let the controller be a PI controller so that

$$G_c(s) = \frac{U(s)}{E(s)} = K_P + \frac{K_I}{s}$$

Find the value of K_I so that the ramp-error constant K_v is 2. Determine the value of K_P so that the maximum overshoot of the unit-step response is a minimum and the settling time is a minimum. Give the values of the maximum overshoot and the settling time. Plot the unit-step response of $y(t)$. Find the marginal value of K_P for system stability.

(b) Can the system performance be further improved by using a PID controller?

$$G_c(s) = \frac{U(s)}{E(s)} = K_P + K_D s + \frac{K_I}{s}$$

9-29. One of the advantages of the frequency-domain analysis and design methods is that systems with pure time delays can be treated without approximation. Consider the automobile-engine control system treated in Problem 9-28.

The process has the transfer function

$$G_p(s) = \frac{e^{-0.2s}}{1 + 0.25s}$$

Let the controller be of the PI type so that $G_c(s) = K_P + K_I/s$. Set the value of K_I so that the ramp-error constant K_v is 2. Find the value of K_P so that the phase margin is a maximum. How does this "optimal" K_P compare with the value of K_P found in Problem 9-28(a)? Find the critical value of K_P for system stability. How does this value of K_P compare with the critical value of K_P found in Problem 9-28?

9-30. Fig. 9P-30 shows a simplified design of an airplane attitude controller.

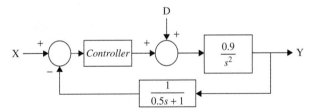

Figure 9P-30

where D is the disturbance torque. Design a PID controller with the following satisfactions:
(a) Zero steady-state error
(b) PM = 65°
(c) High bandwidth (as high as possible)

9-31. Consider the open-loop plant model of a plastic extrusion given in Problem 9-15.
Design a series of lead–lag compensator that is described by

$$H(s) = \frac{(\tau_1 s + 1)(\tau_2 s + 1)}{\tau_1 \tau_1 s^2 + \left(\tau_1 + \dfrac{\tau_2}{r}\right)s + 1}$$

and satisfies the following:
(a) The phase margin is 45 degrees.
(b) The steady-state error of a closed-loop system to the unit-step input is less than 1%.
(c) The gain-crossover frequency is 5 rad/sec.

9-32. A telescope for tracking stars and asteroids on a space shuttle may be modeled as a pure mass M. It is suspended by magnetic bearings so that there is no friction, and its attitude is controlled by magnetic actuators located at the base of the payload. The dynamic model for the control of the z-axis motion is shown in Fig. 9P-32(a). Because there are electrical components on the telescope, electric power must be brought to the telescope through a cable. The spring shown is used to model the wire-cable attachment, which exerts a spring force on the mass. The force produced by the magnetic actuators is denoted by $f(t)$. The force equation of motion in the z direction is

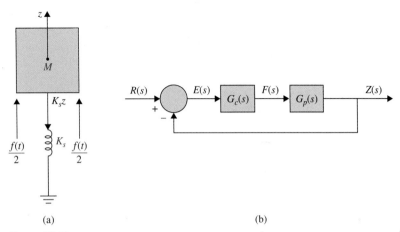

(a)   (b)

Figure 9P-32

$$f(t) - K_s z(t) = M \frac{d^2 z(t)}{dt^2}$$

where $K_s = 1$ lb/ft, and $M = 150$ lb (mass); $f(t)$ is in pounds, and $z(t)$ is measured in feet.

(a) Show that the natural response of the system output $z(t)$ is oscillatory without damping. Find the natural undamped frequency of the open-loop space-shuttle system.

(b) Design the PID controller

$$G_c(s) = K_P + K_D s + \frac{K_I}{s}$$

shown in Fig. 9P-32(b) so that the following performance specifications are satisfied:

Ramp-error constant $K_v = 100$.

The complex charactreristic equation roots correspond to a relative damping ratio of 0.707 and a natural undamped frequency of 1 rad/sec.

Compute and plot the unit-step response of the designed system. Find the maximum overshoot. Comment on the design results.

(c) Design the PID controller so that the following specifications are satisfied:

Ramp-error constant $K_v = 100$

Maximum overshoot $< 5\%$

Compute and plot the unit-step response of the designed system. Find the roots of the characteristic equation of the designed system.

9-33. Repeat Problem 9-32(b) with the following specifications:

Ramp-error constant $K_v = 100$.

The complex charactreristic equation roots correspond to a relative damping ratio of 1.0 and a natural undamped frequency of 1 rad/sec.

9-34. Consider a cruise control system shown in Fig. 9P-34.

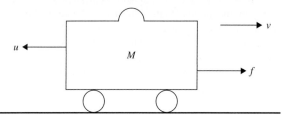

Figure 9P-34

where f is the engine force, v is the velocity, u is the friction force, and $u = \mu v$. Assuming $M = 1000$ kg, $\mu = 50$ Nsec/m, and $f = 500$ N:
(a) Find the transfer function of the system.
(b) Design a PID controller that satisfies the following:
   (i) Rise time is less than 5 sec.
   (ii) Maximum overshoot is less than 10%.
   (iii) Steady-state error is less than 2%.

9-35. An inventory control system is modeled by the following state equations:

$$\frac{dx_1(t)}{dt} = -2x_2(t)$$
$$\frac{dx_2(t)}{dt} = -2u(t)$$

where $x_1(t)$ = level of inventory, $x_2(t)$ = rate of sales of product, and $u(t)$ = production rate. The output equation is $y(t) = x_1(t)$. One unit of time is one day. Fig. 9P-35 shows the block diagram of the closed-loop inventory control system with a series controller. Let the controller be a PD controller, $G_c(s) = K_P + K_D s$.

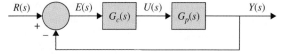

Figure 9P-35

(a) Find the parameters of the PD controller, K_P and K_D, so that the roots of the characteristic equation correspond to a relative damping ratio of 0.707 and $\omega_n = 1$ rad/sec. Plot the unit-step response of $y(t)$ and find the maximum overshoot.
(b) Find the values of K_P and K_D so that the overshoot is zero and the rise time is less than 0.06 sec.
(c) Design the PD controller so that $M_r = 1$ and BW ≤ 40 rad/sec.

9-36. The block diagram of a type 2 control system with a series controller $G_c(s)$ is shown in Fig. 9P-36.

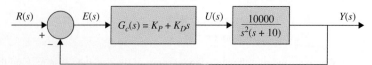

Figure 9P-36

The objective is to design a PD controller so that the following specifications are satisfied:

Maximum overshoot < 10%

Rise time < 0.5 sec

(a) Obtain the characteristic equation of the closed-loop system, and determine the ranges of the values of K_P and K_D for stability. Show the region of stability in the K_D-versus-K_P plane.

(b) Construct the root loci of the characteristic equation with $K_D = 0$ and $0 \leq K_P < \infty$. Then construct the root contours for $0 \leq K_D < \infty$ and several fixed values of K_P ranging from 0.001 to 0.01.

(c) Design the PD controller to satisfy the performance specifications given. Use the information on the root contours to help your design. Plot the unit-step response of $y(t)$.

(d) Check the design results obtained in part (c) in the frequency domain. Determine the phase margin, gain margin, M_r, and BW of the designed system.

9-37. Consider a dc motor shown in Fig. 9P-37 and described in Section 4-7-3.

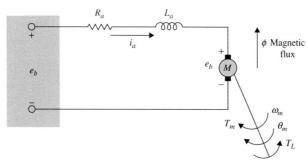

Figure 9P-37

Assuming the following:

The rotor inertia $(J) = 0.01 \text{ kg.m}^2/\text{s}^2$

Damping ratio of the mechanical system $(\zeta) = 0.1 \text{ Nms}$

Back-emf constant $(K_b) = 0.01 \text{ Nm/Amp}$

Torque constant $(K_t) = 0.01 \text{ Nm/Amp}$

Armature resistance $(R_a) = 1 \, \Omega$

Armature inductance $(L_a) = 0.5 \text{ H}$

Design a PID controller that satisfies the following:

(a) Settling time is less than 2 sec.

(b) Maximum overshoot is less than 5%.

(c) Steady-state error is less than 1%.

9-38. For the dc motor described in Problem 9-37, assuming the following:

The rotor inertia $(J) = 3.2284\text{E-6 kg.m}^2/\text{s}^2$

Damping ratio of the mechanical system $(\zeta) = 3.5077\text{E-6 Nms}$

Back-emf constant $(K_b) = 0.0274 \text{ Nm/Amp}$

Torque constant $(K_t) = 0.0274 \text{ Nm/Amp}$

Armature resistance $(R_a) = 4 \, \Omega$

Armature inductance $(L_a) = 2.75\text{E-6 H}$

Design a PID controller that satisfies the following:

(a) Settling time is less than 40 milliseconds.

(b) Maximum overshoot is less than 16%.

(c) Zero steady-state error is less than 1%.

(d) Zero steady-state error due to a disturbance.

9-39. Consider the broom-balancing control system described in Problems 4-21 and 10-51. The $\mathbf{A}_*$ and $\mathbf{B}_*$ matrices are given in Problem 10-51 for the small-signal linearized model.

$$\Delta \dot{\mathbf{x}}(t) = \mathbf{A}^* \Delta \mathbf{x}(t) + \mathbf{B}^* \Delta r(t)$$
$$\Delta y(t) = \mathbf{C} \Delta \mathbf{x}(t)$$
$$\mathbf{D}_* = \begin{bmatrix} 0 & 0 & 1 & 0 \end{bmatrix}$$

Fig. 9P-39 shows the block diagram of the system with a series PD controller. Determine if the PD controller can stabilize the system; if so, find the values of K_P and K_D. If the PD controller cannot stabilize the system, explain why not.

Figure 9P-39

9-40. The process of a unity-feedback control system has the transfer function

$$G_p(s) = \frac{100}{s^2 + 10s + 100}$$

Design a series controller (PD, PI, or PID) so that the following performance specifications are satisfied:

Steady-state error due to a step input $= 0$

Maximum overshoot $< 2\%$

Rise time < 0.02 sec

Carry out the design in the frequency domain and check the design in the time domain.

9-41. The forward path of a unity-feedback control system that includes a disturbance signal $D(s)$ is given by

$$G(s) = \frac{1}{(s^2 + 3.6s + 9)}$$

(a) Design a PID controller with the transfer function of $H(s) = \dfrac{K(\tau_1 s + 1)(\tau_2 s + 1)}{s}$ so that the response to any step disturbance is damped in less than 3 sec at the 2% settling time.

(b) Use MATLAB to plot the response of the closed-loop system to various step disturbance inputs and verify your design in part (a).

9-42. For the inventory control system shown in Fig. 9P-35, let the controller be of the phase-lead type:

$$G_c(s) = \frac{1 + aTs}{1 + Ts} \quad a > 1$$

Determine the values of a and T so that the following performance specifications are satisfied:

Steady-state error due to a step input $= 0$

Maximum overshoot $< 5\%$

(a) Design the controller using the root contours with T and a as variable parameters. Plot the unit-step response of the designed system. Plot the Bode diagram of $G(s) = G_c(s)G_p(s)$, and find PM, GM, M_r, and BW of the designed system.

(b) Design the phase-lead controller so that the following performance specifications are satisfied:

Steady-state error due to a step input $= 0$

Phase margin $> 75°$

$M_r < 1.1$

Construct the Bode diagram of $G(s)$ and carry out the design in the frequency domain. Find the attributes of the time response of the designed system.

9-43. Consider that the process of a unity-feedback control system is

$$G_p(s) = \frac{1000}{s(s+10)}$$

Let the series controller be a single-stage phase-lead controller:

$$G_c(s) = \frac{1+aTs}{1+Ts} \quad a>1$$

(a) Determine the values of a and T so that the zero of $G_c(s)$ cancels the pole of $G_p(s)$ at $s = -10$. The damping ratio of the designed system should be unity. Find the attributes of the unit-step response of the designed system.

(b) Carry out the design in the frequency domain using the Bode plot. The design specifications are as follows:

Phase margin $> 75°$

$M_r < 1.1$

Find the attributes of the unit-step response of the designed system.

9-44. Fig. 9P-44 shows the quarter-car model realization with 2 degrees of freedom.

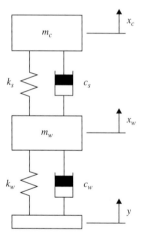

Figure 9P-44

Assuming:
   Body mass $(m_c) = 2500\,\text{kg}$
   Suspension mass $(m_w) = 320\,\text{kg}$
   Spring constant of suspension system $(k_c) = 80,000\,\text{N/m}$
   Spring constant of wheel and tire $(k_w) = 500,000\,\text{N/m}$
   Damping constant of suspension system $(c_s) = 350\,\text{Ns/m}$
   Damping constant of wheel and tire $(c_w) = 15,020\,\text{Ns/m}$

When the vehicle is experiencing any road disturbance, the vehicle body should not have large oscillations, and the oscillations should be damped quickly. If the deformation tire is negligible, and the road disturbance (**D**) is considered a step input,
(a) Design a PID controller that satisfies the following requirements:
   (i) Overshoot less than 5%
   (ii) Settling time shorter than 5 seconds

(b) Use MATLAB to plot the response of the closed-loop system to various step disturbance inputs and verify your design in part (a).

9-45. Consider that the controller in the liquid-level control system shown in Fig. 9P-10 is a phase-lead controller:

$$G_c(s) = \frac{1 + aTs}{1 + Ts} \quad a > 1$$

(a) For $N = 20$, select the values of a and T so that the maximum overshoot is barely 0%. The value of a must not exceed 1000. Find the attributes of the unit-step response of the designed system. Plot the unit-step response.

(b) For $N = 20$, design the phase-lead controller in the frequency domain. Find the values of a and T so that the phase margin is maximized subject to the condition that $BW > 100$. The value of a must not exceed 1000.

9-46. The transfer function of the process of a unity-feedback control system is

$$G_p(s) = \frac{6}{s(1 + 0.2s)(1 + 0.5s)}$$

(a) Construct the Bode diagram of $G_p(j\omega)$ and determine the PM, GM, M_r, and BW of the system.

(b) Design a series single-stage series phase-lead controller with the transfer function

$$G_c(s) = \left(\frac{1 + aTs}{1 + Ts}\right) \quad a > 1$$

so that the phase margin is maximum. The value of a must not be greater than 1000. Determine PM and M_r of the designed system. Determine the attributes of the unit-step response.

(c) Using the system designed in part (b) as a basis, design a two-stage phase-lead controller so that the phase margin is at least 85°. The transfer function of the two-stage phase-lead controller is

$$G_c(s) = \left(\frac{1 + aT_1s}{1 + T_1s}\right)\left(\frac{1 + bT_2s}{1 + T_2s}\right) \quad a > 1 \quad b > 1$$

where a and T_1 are determined in part (b). The value of T_2 should not exceed 1000. Find the values of PM and M_r of the designed system. Find the attributes of the unit-step response.

(d) Plot the unit-step responses of the output in parts (a), (b), and (c).

9-47. Fig. 9P-47 shows an inverted pendulum on a cart.

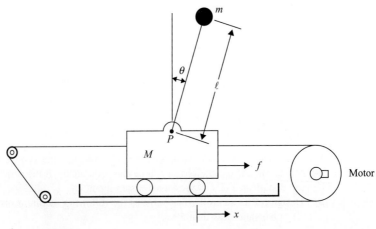

Figure 9P-47

Assuming:

M	mass of the cart	0.5 kg
m	mass of the pendulum	0.2 kg
μ	friction of the cart	0.1 N/m/sec
l	length to pendulum center of mass	0.3 m
I	inertia of the pendulum	0.006 kg*m²

(a) Design a PID controller so that the settling time is less than 5 seconds and the pendulum angle is never more than 0.05 radians from the vertical position.

(b) If the step input is applied to the cart, design a PID controller so that the settling time for x and θ is less than 5 seconds, the rise time for x is less than 0.5 seconds, and the overshoot of theta is less than 20 degrees (0.35 radians).

9-48. A phase-lock-loop, dc-motor-speed-control system is described in Problem 4-46. The block diagram of the system is shown in Fig. 9P-48. The system parameters and transfer functions are given as follows:

Reference speed command, $\omega_r = 120$ pulse/sec
Phase-detector gain, $K_p = 0.06$ V/pulse/sec
Amplifier gain, $K_a = 20$
Encoder gain, $K_e = 5.73$ pulse/rad
Counter gain, $N = 1$
Motor transfer function,

$$\frac{\Omega_m(s)}{E_a(s)} = \frac{10}{s(1+0.05s)}$$

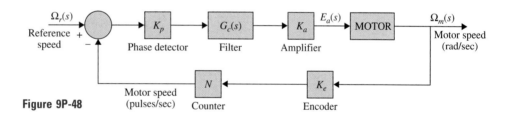

Figure 9P-48

(a) Let the filter (controller) transfer function be of the form

$$G_c(s) = \frac{E_o(s)}{E_i(s)} = \frac{1+R_2Cs}{R_1Cs}$$

where $R_1 = 2 \times 10^6 \, \Omega$ and $C = 1 \, \mu F$. Determine the value of R_2 so that the complex roots of the closed-loop characteristic equation have a maximum relative damping ratio. Sketch the root loci of the characteristic equation for $0 \leq R_2 < \infty$. Compute and plot the unit-step responses of the motor speed $f_\omega(t)$ (pulse/sec) with the values of R_2 found, when the input is 120 pulse/sec. Convert the speed in pulse/sec to rpm.

(b) Let the filter transfer function be

$$G_c(s) = \frac{1+aTs}{1+Ts} \quad a > 1$$

where $T = 0.01$. Find a so that the complex roots of the characteristic equation have a maximum relative damping ratio. Compute and plot the unit-step response of the motor speed $f_\omega(t)$ (pulse/sec) when the input is 120 pulse/sec.

(c) Design the phase-lead controller in the frequency domain so that the phase margin is at least 60°.

9-49. Consider that the controller in the liquid-level control system shown in Fig. 9P-10 is a single-stage phase-lag controller:

$$G_c(s) = \frac{1 + aTs}{1 + Ts} \quad a < 1$$

(a) For $N = 20$, select the values of a and T so that the two complex roots of the characteristic equation correspond to a relative damping ratio of approximately 0.707. Plot the unit-step response of the output $y(t)$. Find the attributes of the unit-step response. Plot the Bode plot of $G_c(s)G_p(s)$ and determine the phase margin of the designed system.

(b) For $N = 20$, design the phase-lag controller in the frequency domain so that the phase margin is approximately 60°. Plot the unit-step response of the output $y(t)$, and find the attributes of the unit-step response.

9-50. The controlled process of a unity-feedback control system is

$$G_p(s) = \frac{K}{s(s+5)^2}$$

The series controller has the transfer function

$$G_c(s) = \frac{1 + aTs}{1 + Ts}$$

(a) Design a phase-lead controller $(a > 1)$ so that the following performance specifications are satisfied:

Ramp-error constant $K_v = 10$

Maximum overshoot is near minimum

The value of a must not exceed 1000. Plot the unit-step response and give its attributes.

(b) Design a phase-lead controller in the frequency domain so that the following performance specifications are satisfied:

Ramp-error constant $K_v = 10$

Phase margin is near maximum

The value of a must not exceed 1000

(c) Design a phase-lag controller $(a < 1)$ so that the following performance specifications are satisfied:

Ramp-error constant $K_v = 10$

Maximum overshoot $< 1\%$

Rise time $t_r < 2$ sec

Settling time $t_s < 2.5$ sec

Find the PM, GM, M_r, and BW of the designed system.

(d) Design the phase-lag controller in the frequency domain so that the following performance specifications are satisfied:

Ramp-error constant $K_v = 10$

Phase margin $\geq 70°$

Check the unit-step response attributes of the designed system and compare with those obtained in part (c).

9-51. Fig. 9P-51 shows the "beam and ball" system that is described in Problem 4-11.

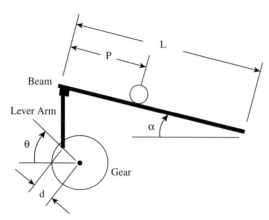

Figure 9P-51

Assuming:

$m = 0.11$ kg	mass of the ball	$I = 9.99\text{e-}6$ kg m^2	ball's moment of inertia
$r = 0.015$	radius of the ball	P	ball position coordinate
$d = 0.03$ m	lever arm offset	α	beam angle coordinate
$g = 9.8$ m/s^2	gravitational acceleration	θ	servo gear angle
$L = 1.0$ m	length of the beam		

Design a PID controller so that the settling time is less than 3 seconds and the maximum overshoot is no more than 5%.

9-52. The controlled process of a dc-motor control system with unity feedback has the transfer function

$$G_p(s) = \frac{6.087 \times 10^{10}}{s(s^3 + 423.42s^2 + 2.6667 \times 10^6 s + 4.2342 \times 10^8)}$$

Due to the compliance in the motor shaft, the process transfer function contains two lightly damped poles, which will cause oscillations in the output response. The following performance criteria are to be satisfied:

Maximum overshoot < 1%

Rise time $t_r < 0.15$ sec

Settling time $t_s < 0.15$ sec

Output response should not have oscillations

Ramp-error constant is not affected

(a) Design a series phase-lead controller,

$$G_c(s) = \frac{1 + aTs}{1 + Ts} \quad a > 1$$

so that all the step-response attributes (except for the oscillations) are satisfied.

(b) To eliminate the oscillations due to the motor shaft compliance, add another stage to the controller with the transfer function

$$G_{c1}(s) = \frac{s^2 + 2\zeta_z\omega_n s + \omega_n^2}{s^2 + 2\zeta_p\omega_n s + \omega_n^2}$$

so that the zeros of $G_{c1}(s)$ will cancel the two complex poles of $G_p(s)$. Set the value of ζ_p so that the two poles of $G_{c1}(s)$ will not have an appreciable affect on the system response. Determine the attributes of the unit-step response to see if all the requirements are satisfied. Plot the unit-step responses of the uncompensated system and the compensated system with the phase-lead controller designed in part (b).

9-53. A computer-tape-drive system utilizing a permanent-magnet dc-motor is shown in Fig. 9P-53(a). The closed-loop system is modeled by the block diagram in Fig. 9P-53(b). The constant K_L represents the spring constant of the elastic tape, and B_L denotes the viscous-friction coefficient between the tape and the capstans. The system parameters are as follows:

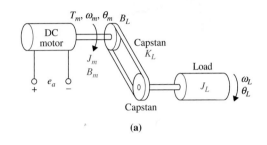

(a)

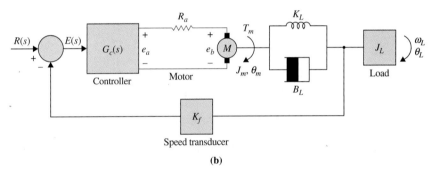

(b)

Figure 9P-53

K_i = motor torque constant = 10 oz-in./A
K_b = motor back-emf constant = 0.0706 V/rad/sec
B_m = motor friction coefficent = 3 oz-in./rad/sec
$R_a = 0.25\,\Omega \quad L_a \cong 0\,H$
$K_L = 3000$ oz-in./rad $\quad B_L = 10$ oz-in./rad/sec
$J_L = 6$ oz-in./rad/sec$^2 K_f = 1$ V/rad/sec
$J_m = 0.05$ oz-in./rad/sec^2

(a) Write the state equations of the system between e_a and θ_L using θ_L, ω_L, θ_m, and ω_m as state variables and e_a as input. Draw a state diagram using the state equations. Derive the transfer functions:

$$\frac{\Omega_m(s)}{E_a(s)} \quad \text{and} \quad \frac{\Omega_L(s)}{E_a(s)}$$

(b) The objective of the system is to control the speed of the tape, ω_L, accurately. Consider that a PI controller with the transfer function $G_c(s) = K_P + K_I/s$ is to be used. Find the values of K_P and K_I so that the following specifications are satisfied:

Ramp-error constant $K_v = 100$
Rise time < 0.02 sec
Settling time < 0.02 sec
Maximum overshoot < 1% or at minimum

Plot the unit-step response of $\omega_L(t)$ of the system.

(c) Design the PI controller in the frequency domain. The value of K_I is to be selected as in part (b). Vary the value of K_P and compute the values of PM, GM, M_r, and BW. Find the value of K_P so that PM is maximum. How does this value of K_P compare with the result obtained in part (b)?

9-54. Fig. 9P-54 shows the block diagram of a motor-control system that has a flexible shaft between the motor and the load. The transfer function between the motor torque and motor displacement is

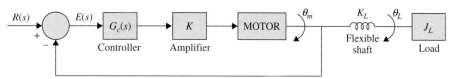

Figure 9P-54

$$G_p(s) = \frac{\Theta_m(s)}{T_m(s)} = \frac{J_L s^2 + B_L s + K_L}{s[J_m J_L s^3 + (B_m J_L + B_L J_m)s^2 + (K_L J_m + K_L J_L + B_m B_L)s + B_m K_L]}$$

where $J_L = 0.01$, $B_L = 0.1$, $K_L = 10$, $J_m = 0.01$, $B_m = 0.1$, and $K = 100$.

(a) Compute and plot the unit-step response of $\theta_m(t)$. Find the attributes of the unit-step response.

(b) Design a second-order notch controller with the transfer function

$$G_c(s) = \frac{s^2 + 2\zeta_z \omega_n s + \omega_n^2}{s^2 + 2\zeta_p \omega_n s + \omega_n^2}$$

so that its zeros cancel the complex poles of $G_p(s)$. The two poles of $G_c(s)$ should be selected so that they do not affect the steady-state response of the system, and the maximum overshoot is a minimum. Compute the attributes of the unit-step response and plot the response.

(c) Carry out design of the second-order controller in the frequency domain. Plot the Bode diagram of the uncompensated $G_p(s)$, and find the values of PM, GM, M_r, and BW. Set the two zeros of $G_c(s)$ to cancel the two complex poles of $G_p(s)$. Determine the value of ζ_p by determining the amount of attenuation required from the second-order notch controller and using Eq. (9-155). Find the PM, GM, M_r, and BW of the compensated system. How do the frequency-domain design results compare with the results in part (b)?

9-55. The transfer function of the process of a unity-feedback control system is

$$G_p(s) = \frac{500(s+10)}{s(s^2 + 10s + 1000)}$$

(a) Plot the Bode diagram of $G_p(s)$ and determine the PM, GM, M_r, and BW of the uncompensated system. Compute and plot the unit-step response of the system.

(b) Design a series second-order notch controller with the transfer function

$$G_c(s) = \frac{s^2 + 2\zeta_z \omega_n s + \omega_n^2}{s^2 + 2\zeta_p \omega_n s + \omega_n^2}$$

so that its zeros cancel the complex poles of $G_p(s)$. Determine the value of ζ_p using the method outlined in Section 9-8-2. Find the PM, GM, M_r, and BW of the designed system. Compute and plot the unit-step response.

(c) Design the series second-order notch controller so that its zeros cancel the complex poles of $G_p(s)$. Determine the value of ζ_p so that the following specifications are satisfied:

Maximum overshoot < 1%
Rise time < 0.4 sec
Settling time < 0.5 sec

9-56. Design the controllers $G_{cf}(s)$ and $G_c(s)$ for the system shown in Fig. 9P-56 so that the following specifications are satisfied:

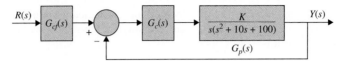

Figure 9P-56

Ramp-error constant $K_v = 50$
Dominant roots of the characteristic equation at $-5 \pm j5$ approximately
Rise time < 0.1 sec
System must be robust when K varies $\pm 20\%$ from the nominal value, with the rise time and overshoot staying within specifications

Compute and plot the unit-step responses to check the design.

9-57. Fig. 9P-57 shows the block diagram of a motor-control system. The transfer function of the controlled process is

$$G_p(s) = \frac{1000K}{s(s+a)}$$

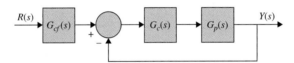

Figure 9P-57

where K denotes the aggregate of the amplifier gain and motor torque constant, and a is the inverse of the motor time constant. Design the controllers $G_{cf}(s)$ and $G_c(s)$ so that the following performance specifications are satisfied.

Ramp-error constant $K_v = 100$ when $a = 10$
Rise time < 0.3 sec
Maximum overshoot $< 8\%$
Dominant characteristic equation roots $= -5 \pm j5$
System must be robust when a varies between 8 and 12

Compute and plot the unit-step responses to verify the design.

9-58. Fig. 9P-58 shows the block diagram of a dc-motor control system with tachometer feedback. Find the values of K and K_t so that the following specifications are satisfied:
Ramp-error constant $K_v = 1$
Dominant characteristic equation roots correspond to a damping ratio of approximately 0.707; if there are two solutions, select the larger value of K

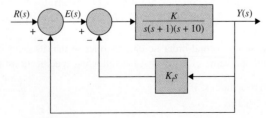

Figure 9P-58

9-59. Carry out the design with the specifications given in Problem 9-58 for the system shown in Fig. 9P-59.

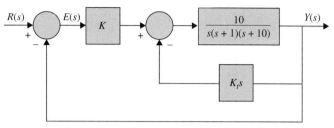

Figure 9P-59

9-60. The block diagram of a control system with a type 2 process is shown in Fig. 9P-60. The system is to be compensated by tachometer feedback and a series controller. Find the values of a, T, K, and K_t so that the following performance specifications are satisfied:

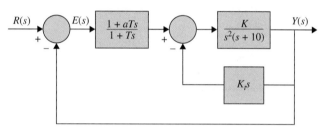

Figure 9P-60

Ramp-error constant $K_v = 100$

Dominant characteristic equation roots correspond to a damping ratio of 0.707

9-61. The aircraft-attitude control system described in Section 5-8 is modeled by the block diagram shown in Fig. 9P-61. The system parameters are as follows:

K = variable	$K_s = 1$	$K_1 = 10$	$K_2 = 0.5$	K_t = variable	$R_a = 5$
$L_a = 0.003$	$K_i = 9.0$	$K_b = 0.0636$	$J_m = 0.0001$	$J_L = 0.01$	
$B_m = 0.005$	$B_L = 1.0$	$N = 0.1$			

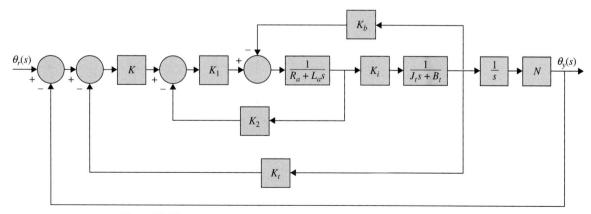

Figure 9P-61

Find the values of K and K_t so that the following specifications are satisfied:
Ramp-error constant $K_v = 100$
Relative damping ratio of the complex roots of the characteristic equation is approximately 0.707

Plot the unit-step response of the designed system. Show that the system performance is extremely insensitive to the value of K. Explain why this is so.

9-62. Fig. 9P-62 shows the block diagram of a position-control system with a series controller $G_c(s)$.

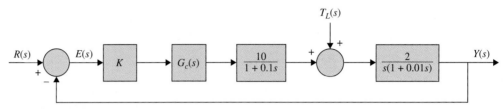

Figure 9P-62

(a) Determine the minimum value of the amplifier gain K so that the steady-state value of the output $y(t)$ due to a unit-step torque disturbance is ≤ 0.01.

(b) Show that the uncompensated system is unstable with the minimum value of K determined in part (a). Construct the Bode diagram for the open-loop transfer function $G(s) = Y(s)/E(s)$, and find the values of PM and GM.

(c) Design a single-stage phase-lead controller with the transfer function

$$G_c(s) = \frac{1 + aTs}{1 + Ts} \quad a > 1$$

so that the phase margin is 30°. Show that this is nearly the highest phase margin that can be achieved with a single-stage phase-lead controller. Find GM, M_r, and BW of the compensated system.

(d) Design a two-stage phase-lead controller using the system arrived at in part (c) as a basis so that the phase margin is 55°. Show that this is the best PM that can be obtained for this system with a two-stage phase-lead controller. Find GM, M_r, and BW of the compensated system.

9-63. The transfer function of the process of a unity-feedback control system is

$$G_c(s) = \frac{60}{s(1 + 0.2s)(1 + 0.5s)}$$

Show that, due to the relative high gain, the uncompensated system is unstable.

(a) Design a two-stage phase-lead controller with

$$G_c(s) = \left(\frac{1 + aT_1s}{1 + T_1s}\right)\left(\frac{1 + bT_2s}{1 + T_2s}\right) \quad a > 1, \quad b > 1$$

so that the phase margin is greater than 60°. Conduct the design by first determining the values of a and T_1 to realize a maximum phase margin that can be achieved with a single-stage phase-lead controller. The second stage of the controller is then designed to realize the balance of the 60° phase margin. Determine GM, M_r, and BW of the compensated system. Compute and plot the unit-step response of the compensated system.

(b) Design a single-stage phase-lag controller with

$$G_c(s) = \frac{1 + aTs}{1 + Ts} \quad a < 1$$

so that the phase margin of the compensated system is greater than 60°. Determine GM, M_r, and BW of the compensated system. Compute and plot the unit-step response of the compensated system.

(c) Design a lag–lead controller with $G_c(s)$ as in the equation in part (a). Design the phase-lag portion first by setting the phase margin at 40°. The resulting system is then compensated by the phase-lead portion to achieve a total of 60° of phase margin. Determine GM, M_r, and BW of the compensated system. Compute and plot the unit-step response of the compensated system.

9-64. The block diagram of the steel-rolling system described in Problem 4-18 is shown in Fig. 9P-64. The transfer function of the process is

$$G_p(s) = \frac{Y(s)}{E(s)} = \frac{5e^{-0.1s}}{s(1+0.1s)(1+0.5s)} \quad K_s = 1$$

Figure 9P-64

(a) Approximate the time delay by

$$e^{-0.1s} \cong \frac{1-0.05s}{1+0.05s}$$

Design a series controller of your choice so that the phase margin of the compensated system is at least 60°. Determine GM, M_r, and BW of the compensated system. Compute and plot the unit-step responses of the compensated and the uncompensated systems.

(b) Repeat part (a) without using the approximation of the time delay.

9-65. Human beings breathe in order to provide for gas exchange for the entire body. A respiratory control system is needed to ensure that the body's needs for this gas exchange are adequately met. The criterion of control is adequate ventilation, which ensures satisfactory levels of both oxygen and carbon dioxide in the arterial blood. Respiration is controlled by neural impulses that originate within the lower brain and are transmitted to the chest cavity and diaphragm to govern the rate and tidal volume. One source of signals consists of the chemoreceptors located near the respiratory center, which are sensitive to carbon dioxide and oxygen concentrations. Fig. 9P-65 shows the block diagram of a simplified model of the human respiratory control system. The objective is to control the effective ventilation of the lungs so that a satisfactory balance of concentrations of carbon dioxide and oxygen is maintained in the blood circulated at the chemoreceptor.

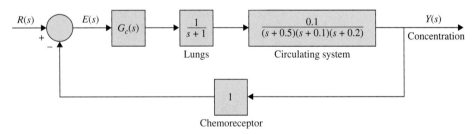

Figure 9P-65

(a) Plot the Bode diagram of the transfer function $G(s) = Y(s)/E(s)$ when $G_c(s) = 1$. Find the PM and GM. Determine the stability of the system.

(b) Design a PI controller, $G_c(s) = K_P + K_I/s$, so that the following specifications are satisfied:

Ramp-error constant $K_v = 1$

Phase margin is maximized

Plot the unit-step response of the system. Find the attributes of the unit-step response.

(c) Design a PI controller so that the following specifications are satisfied:

Ramp-error constant $K_v = 1$

Maximum overshoot is minimized

Plot the unit-step response of the system. Find the attributes of the unit-step response. Compare the design results in parts (b) and (c).

9-66. The block diagram of a control system with state feedback is shown in Fig. 9P-66. Find the real feedback gains k_1, k_2, and k_3 so that:

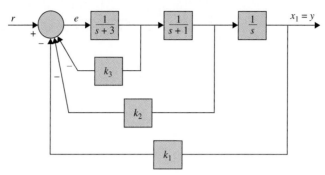

Figure 9P-66

The steady-state error e_{ss} [$e(t)$ is the error signal] due to a step input is zero.

The complex roots of the characteristic equation are at $-1+j$ and $-1-j$.

Find the third root. Can all three roots be arbitrarily assigned while still meeting the steady-state requirement?

9-67. The block diagram of a control system with state feedback is shown in Fig. 9P-67(a). The feedback gains k_1, k_2, and k_3 are real constants.

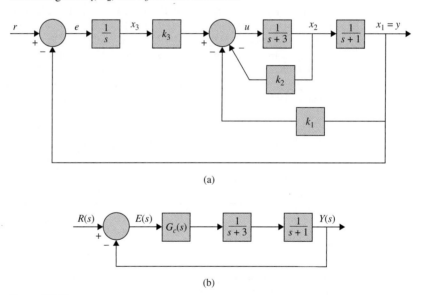

Figure 9P-67

(a) Find the values of the feedback gains so that:

The steady-state error e_{ss} [$e(t)$ is the error signal] due to a step input is zero.

The characteristic equation roots are at $-1+j$, $-1-j$, and -10.

(b) Instead of using state feedback, a series controller is implemented, as shown in Fig. 9P-67(b). Find the transfer function of the controller $G_c(s)$ in terms of k_1, k_2, and k_3 found in part (a) and the other system parameters.

CHAPTER 10

State Variable Analysis

▶ 10-1 INTRODUCTION

In Chapter 2 we presented the concept and definition of state variables and state equations for linear continuous-data and discrete-data dynamic systems. In Chapter 3 we used block-diagram and signal-flow-graph (SFG) methods to obtain the transfer function of linear systems. In this chapter, the SFG concept is extended to the modeling of the state equations, and the result is the **state diagram**. In contrast to the transfer-function approach to the analysis and design of linear control systems, the state-variable method is regarded as modern, since it uses underlying force for optimal control. The basic characteristic of the state-variable formulation is that linear and nonlinear systems, time-invariant and time-varying systems, and single-variable and multivariable systems can all be modeled in a unified manner. Transfer functions, on the other hand, are defined only for linear time-invariant systems.

The objective of this chapter is to introduce the basic methods of state variables and state equations so that the reader can gain a working knowledge of the subject for further studies when the state-space approach is used for modern and optimal control design. Specifically, the closed-form solutions of linear time-invariant state equations are presented. Various transformations that may be used to facilitate the analysis and design of linear control systems in the state-variable domain are introduced. The relationship between the conventional transfer-function approach and the state-variable approach is established so that the analyst will be able to investigate a system problem with various alternative methods. Finally, the controllability and observability of linear systems are defined and their applications investigated. Some state-space controller design problems appear in the end. At the end of the chapter, we also present MATLAB tools to solve most state-space problems.

▶ 10-2 BLOCK DIAGRAMS, TRANSFER FUNCTIONS, AND STATE DIAGRAMS

10-2-1 Transfer Functions (Multivariable Systems)

The definition of a transfer function is easily extended to a system with multiple inputs and outputs. A system of this type is often referred to as a multivariable system. In a multivariable system, a differential equation of the form of Eq. (2-217) may be used to describe the relationship between a pair of input and output variables, when all other inputs are set to zero. This equation is restated as

$$\frac{d^n y(t)}{dt^n} + a_{n-1}\frac{d^{n-1} y(t)}{dt^{n-1}} + \cdots + a_1 \frac{dy(t)}{dt} + a_0\, y(t)$$
$$= b_m \frac{d^m u(t)}{dt^m} + b_{m-1}\frac{d^{m-1} u(t)}{dt^{m-1}} + \cdots + b_1 \frac{du(t)}{dt} + b_0 u(t) \tag{10-1}$$

The coefficients $a_0, a_1, \ldots, a_{n-1}$ and $b_0, b_1, \ldots, b_m$ are real constants. Because the principle of superposition is valid for linear systems, the total effect on any output due to all the inputs acting simultaneously is obtained by adding up the outputs due to each input acting alone.

In general, if a linear system has p inputs and q outputs, the transfer function between the jth input and the ith output is defined as

$$G_{ij}(s) = \frac{Y_i(s)}{R_j(s)} \tag{10-2}$$

with $R_k(s) = 0, k = 1, 2, \ldots, p, k \neq j$. Note that Eq. (10-2) is defined with only the jth input in effect, whereas the other inputs are set to zero. When all the p inputs are in action, the ith output transform is written

$$Y_i(s) = G_{i1}(s)R_1(s) + G_{i2}(s)R_2(s) + \cdots + G_{ip}(s)R_p(s) \tag{10-3}$$

It is convenient to express Eq. (10-3) in matrix-vector form:

$$\mathbf{Y}(s) = \mathbf{G}(s)\mathbf{R}(s) \tag{10-4}$$

where

$$\mathbf{Y}(s) = \begin{bmatrix} Y_1(s) \\ Y_2(s) \\ \vdots \\ Y_q(s) \end{bmatrix} \tag{10-5}$$

is the $q \times 1$ transformed output vector,

$$\mathbf{R}(s) = \begin{bmatrix} R_1(s) \\ R_2(s) \\ \vdots \\ R_p(s) \end{bmatrix} \tag{10-6}$$

is the $p \times 1$ transformed input vector, and

$$\mathbf{G}(s) = \begin{bmatrix} G_{11}(s) & G_{12}(s) & \cdots & G_{1p}(s) \\ G_{21}(s) & G_{22}(s) & \cdots & G_{2p}(s) \\ \cdot & \cdot & \cdots & \cdot \\ G_{q1}(s) & G_{q2}(s) & \cdots & G_{qp}(s) \end{bmatrix} \tag{10-7}$$

is the $q \times p$ transfer-function matrix.

10-2-2 Block Diagrams and Transfer Functions of Multivariable Systems

In this section, we shall illustrate the block diagram and matrix representations of multivariable systems. Two block-diagram representations of a multivariable system with p inputs and q outputs are shown in Fig. 10-1(a) and (b). In Fig. 10-1(a), the

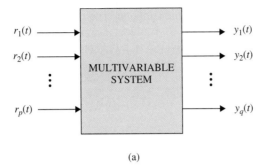

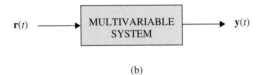

Figure 10-1 Block diagram representations of a multivariable system.

individual input and output signals are designated, whereas in the block diagram of Fig. 10-1(b), the multiplicity of the inputs and outputs is denoted by vectors. The case of Fig. 10-1(b) is preferable in practice because of its simplicity.

Fig. 10-2 shows the block diagram of a multivariable feedback control system. The transfer function relationships of the system are expressed in vector-matrix form (see Section 10-3 for more detail):

$$\mathbf{Y}(s) = \mathbf{G}(s)\mathbf{U}(s) \quad (10\text{-}8)$$
$$\mathbf{U}(s) = \mathbf{R}(s) - \mathbf{B}(s) \quad (10\text{-}9)$$
$$\mathbf{B}(s) = \mathbf{H}(s)\mathbf{Y}(s) \quad (10\text{-}10)$$

where $\mathbf{Y}(s)$ is the $q \times 1$ output vector; $\mathbf{U}(s)$, $\mathbf{R}(s)$, and $\mathbf{B}(s)$ are all $p \times 1$ vectors; and $\mathbf{G}(s)$ and $\mathbf{H}(s)$ are $q \times p$ and $p \times q$ transfer-function matrices, respectively. Substituting Eq. (10-9) into Eq. (10-8) and then from Eq. (10-8) to Eq. (10-10), we get

$$\mathbf{Y}(s) = \mathbf{G}(s)\mathbf{R}(s) - \mathbf{G}(s)\mathbf{H}(s)\mathbf{Y}(s) \quad (10\text{-}11)$$

Solving for $\mathbf{Y}(s)$ from Eq. (10-11) gives

$$\mathbf{Y}(s) = [\mathbf{I} + \mathbf{G}(s)\mathbf{H}(s)]^{-1}\mathbf{G}(s)\mathbf{R}(s) \quad (10\text{-}12)$$

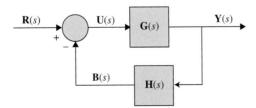

Figure 10-2 Block diagram of a multivariable feedback control system.

provided that $\mathbf{I} + \mathbf{G}(s)\mathbf{H}(s)$ is nonsingular. The closed-loop transfer matrix is defined as

$$\mathbf{M}(s) = [\mathbf{I} + \mathbf{G}(s)\mathbf{H}(s)]^{-1}\mathbf{G}(s) \qquad (10\text{-}13)$$

Then Eq. (10-12) is written

$$\mathbf{Y}(s) = \mathbf{M}(s)\mathbf{R}(s) \qquad (10\text{-}14)$$

▶ **EXAMPLE 10-2-1** Consider that the forward-path transfer function matrix and the feedback-path transfer function matrix of the system shown in Fig. 10-2 are

$$\mathbf{G}(s) = \begin{bmatrix} \dfrac{1}{s+1} & -\dfrac{1}{s} \\ 2 & \dfrac{1}{s+2} \end{bmatrix} \qquad \mathbf{H}(s) = \begin{bmatrix} 1 & 0 \\ 0 & 1 \end{bmatrix} \qquad (10\text{-}15)$$

respectively. The closed-loop transfer function matrix of the system is given by Eq. (10-14) and is evaluated as follows:

$$\mathbf{I} + \mathbf{G}(s)\mathbf{H}(s) = \begin{bmatrix} 1 + \dfrac{1}{s+1} & -\dfrac{1}{s} \\ 2 & 1 + \dfrac{1}{s+2} \end{bmatrix} = \begin{bmatrix} \dfrac{s+2}{s+1} & -\dfrac{1}{s} \\ 2 & \dfrac{s+3}{s+2} \end{bmatrix} \qquad (10\text{-}16)$$

The closed-loop transfer function matrix is

$$\mathbf{M}(s) = [\mathbf{I} + \mathbf{G}(s)\mathbf{H}(s)]^{-1}\mathbf{G}(s) = \dfrac{1}{\Delta}\begin{bmatrix} \dfrac{s+3}{s+2} & \dfrac{1}{s} \\ -2 & \dfrac{s+2}{s+1} \end{bmatrix}\begin{bmatrix} \dfrac{1}{s+1} & -\dfrac{1}{s} \\ 2 & \dfrac{1}{s+2} \end{bmatrix} \qquad (10\text{-}17)$$

where

$$\Delta = \dfrac{s+2}{s+1}\dfrac{s+3}{s+2} + \dfrac{2}{s} = \dfrac{s^2 + 5s + 2}{s(s+1)} \qquad (10\text{-}18)$$

Thus,

$$\mathbf{M}(s) = \dfrac{s(s+1)}{s^2 + 5s + 2}\begin{bmatrix} \dfrac{3s^2 + 9s + 4}{s(s+1)(s+2)} & -\dfrac{1}{s} \\ 2 & \dfrac{3s+2}{s(s+1)} \end{bmatrix} \qquad (10\text{-}19)$$

◀

10-2-3 State Diagram

In this section, we introduce the state diagram, which is an extension of the SFG to portray state equations and differential equations. The significance of the state diagram is that it forms a close relationship among the state equations, computer simulation, and transfer functions. A state diagram is constructed following all the rules of the SFG using the Laplace-transformed state equations.

The basic elements of a state diagram are similar to the conventional SFG, except for the **integration** operation. Let the variables $x_1(t)$ and $x_2(t)$ be related by the first-order differentiation:

$$\frac{dx_1(t)}{dt} = x_2(t) \tag{10-20}$$

Integrating both sides of the last equation with respect to t from the initial time t_0, we get

$$x_1(t) = \int_{t_0}^{t} x_2(\tau)d\tau + x_1(t_0) \tag{10-21}$$

Because the SFG algebra does not handle integration in the time domain, we must take the Laplace transform on both sides of Eq. (10-20). We have

$$X_1(s) = \mathcal{L}\left[\int_{t_0}^{t} x_2(\tau)d\tau\right] + \frac{x_1(t_0)}{s} = \mathcal{L}\left[\int_{0}^{t} x_2(\tau)d\tau - \int_{0}^{t_0} x_2(\tau)d\tau\right] + \frac{x_1(t_0)}{s}$$

$$= \frac{X_2(s)}{s} - \mathcal{L}\left[\int_{0}^{t_0} x_2(\tau)d\tau\right] + \frac{x_1(t_0)}{s} \tag{10-22}$$

Because the past history of the integrator is represented by $x_1(t_0)$, and the state transition is assumed to start at $\tau = t_0$, $x_2(\tau) = 0$ for $0 < \tau < t_0$. Thus, Eq. (10-22) becomes

$$X_1(s) = \frac{X_2(s)}{s} + \frac{x_1(t_0)}{s} \quad \tau \geq t_0 \tag{10-23}$$

Eq. (10-23) is now algebraic and can be represented by an SFG, as shown in Fig. 10-3. Fig. 10-3 shows that *the output of the integrator is equal to s^{-1} times the input, plus the initial condition $x_1(t_0)/s$*. An alternative SFG with fewer elements for Eq. (10-23) is shown in Fig. 10-4.

Before embarking on several illustrative examples on the construction of state diagrams, let us point out the important uses of the state diagram.

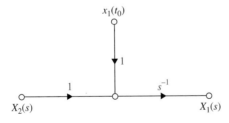

Figure 10-3 Signal-flow graph representation of $X_1(s) = [X_2(s)/s] + [x_1(t_0)/s]$.

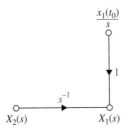

Figure 10-4 Signal-flow graph representation of $X_1(s) = [X_2(s)/s] + [x_1(t_0)/s]$.

1. A state diagram can be constructed directly from the system's differential equation. This allows the determination of the state variables and the state equations.
2. A state diagram can be constructed from the system's transfer function. This step is defined as the **decomposition** of transfer functions (Section 10-10).
3. The state diagram can be used to program the system on an analog computer or for simulation on a digital computer.
4. The state-transition equation in the Laplace transform domain may be obtained from the state diagram by using the SFG gain formula.
5. The transfer functions of a system can be determined from the state diagram.
6. The state equations and the output equations can be determined from the state diagram.

The details of these techniques will follow.

10-2-4 From Differential Equations to State Diagrams

When a linear system is described by a high-order differential equation, a state diagram can be constructed from these equations, although a direct approach is not always the most convenient. Consider the following differential equation:

$$\frac{d^n y(t)}{dt^n} + a_n \frac{d^{n-1} y(t)}{dt^{n-1}} + \cdots + a_2 \frac{dy(t)}{dt} + a_1 y(t) = r(t) \qquad (10\text{-}24)$$

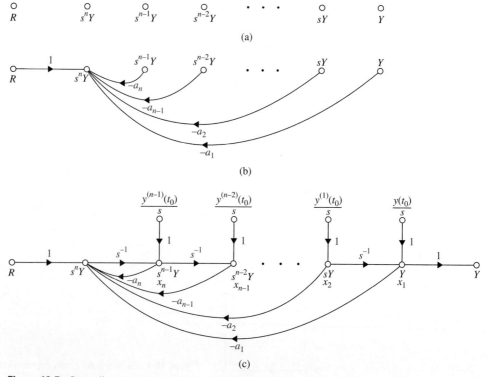

Figure 10-5 State-diagram representation of the differential equation of Eq. (10-24).

To construct a state diagram using this equation, we rearrange the equation as

$$\frac{d^n y(t)}{dt^n} = -a_n \frac{d^{n-1} y(t)}{dt^{n-1}} - \cdots - a_2 \frac{dy(t)}{dt} - a_1 y(t) + r(t) \tag{10-25}$$

As a first step, the nodes representing $R(s), s^n Y(s), s^{n-1} Y(s), \ldots, sY(s)$, and $Y(s)$ are arranged from left to right, as shown in Fig. 10-5(a). Because $s^i Y(s)$ corresponds to $d^i y(t)/dt^i$, $i = 0, 1, 2, \ldots, n$, in the Laplace domain, as the next step, the nodes in Fig. 10-5(a) are connected by branches to portray Eq. (10-25), resulting in Fig. 10-5(b). Finally, the integrator branches with gains of s^{-1} are inserted, and the initial conditions are added to the outputs of the integrators, according to the basic scheme in Fig. 10-3. The complete state diagram is drawn as shown in Fig. 10-5(c). *The outputs of the integrators are defined as the state variables*, $x_1, x_2, \ldots, x_n$. This is usually the natural choice of state variables once the state diagram is drawn.

When the differential equation has derivatives of the input on the right side, the problem of drawing the state diagram directly is not as straightforward as just illustrated. We will show that, in general, it is more convenient to obtain the transfer function from the differential equation first and then arrive at the state diagram through decomposition (Section 10-10).

▶ **EXAMPLE 10-2-2** Consider the differential equation

$$\frac{d^2 y(t)}{dt^2} + 3 \frac{dy(t)}{dt} + 2y(t) = r(t) \tag{10-26}$$

Equating the highest-ordered term of the last equation to the rest of the terms, we have

$$\frac{d^2 y(t)}{dt^2} = -3 \frac{dy(t)}{dt} - 2y(t) + r(t) \tag{10-27}$$

Following the procedure just outlined, the state diagram of the system is drawn as shown in Fig. 10-6. The state variables x_1 and x_2 are assigned as shown.

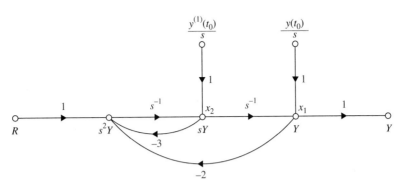

Figure 10-6 State diagram for Eq. (10-26).

10-2-5 From State Diagrams to Transfer Functions

The transfer function between an input and an output is obtained from the state diagram by using the gain formula and setting all other inputs and initial states to zero. The following example shows how the transfer function is obtained directly from a state diagram.

680 ▶ Chapter 10. State Variable Analysis

▶ **EXAMPLE 10-2-3** Consider the state diagram of Fig. 10-6. The transfer function between $R(s)$ and $Y(s)$ is obtained by applying the gain formula between these two nodes and setting the initial states to zero. We have

$$\frac{Y(s)}{R(s)} = \frac{1}{s^2 + 3s + 2} \tag{10-28}$$

◀

10-2-6 From State Diagrams to State and Output Equations

The state equations and the output equations can be obtained directly from the state diagram by using the SFG gain formula. The general form of a state equation and the output equation for a linear system is described in Chapter 2 and presented here.

State equation:

$$\frac{dx(t)}{dt} = ax(t) + br(t) \tag{10-29}$$

Output equation:

$$y(t) = cx(t) + dr(t) \tag{10-30}$$

where $x(t)$ is the state variable; $r(t)$ is the input; $y(t)$ is the output; and a, b, c, and d are constant coefficients. Based on the general form of the state and output equations, the following procedure of deriving the state and output equations from the state diagram are outlined:

1. Delete the initial states and the integrator branches with gains s^{-1} from the state diagram, since the state and output equations do not contain the Laplace operator s or the initial states.

2. For the state equations, regard the nodes that represent the derivatives of the state variables as output nodes, since these variables appear on the left-hand side of the state equations. The output $y(t)$ in the output equation is naturally an output node variable.

3. Regard the state variables and the inputs as input variables on the state diagram, since these variables are found on the right-hand side of the state and output equations.

4. Apply the SFG gain formula to the state diagram.

▶ **EXAMPLE 10-2-4** Fig. 10-7 shows the state diagram of Fig. 10-6 with the integrator branches and the initial states eliminated. Using $dx_1(t)/dt$ and $dx_2(t)/dt$ as the output nodes and $x_1(t)$, $x_2(t)$, and $r(t)$ as input nodes, and applying the gain formula between these nodes, the state equations are obtained as

$$\frac{dx_1(t)}{dt} = x_2(t) \tag{10-31}$$

$$\frac{dx_2(t)}{dt} = -2x_1(t) - 3x_2(t) + r(t) \tag{10-32}$$

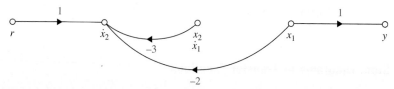

Figure 10-7 State diagram of Fig. 10-6 with the initial states and the integrator branches left out.

◀

Applying the gain formula with $x_1(t)$, $x_2(t)$, and $r(t)$ as input nodes and $y(t)$ as the output node, the output equation is written

$$y(t) = x_1(t) \tag{10-33}$$

▶ **EXAMPLE 10-2-5** As another example on the determination of the state equations from the state diagram, consider the state diagram shown in Fig. 10-8(a). This example will also emphasize the importance of applying the gain formula. Fig. 10-8(b) shows the state diagram with the initial states and the integrator branches deleted. Notice that, in this case, the state diagram in Fig. 10-8(b) still contains a loop. By applying the gain formula to the state diagram in Fig. 10-8(b) with $\dot{x}_1(t)$, $\dot{x}_2(t)$, and $\dot{x}_3(t)$ as output-node variables and $r(t)$, $x_1(t)$, $x_2(t)$, and $x_3(t)$ as input nodes, the state equations are obtained as follows in vector-matrix form:

$$\begin{bmatrix} \dfrac{dx_1(t)}{dt} \\ \dfrac{dx_2(t)}{dt} \\ \dfrac{dx_3(t)}{dt} \end{bmatrix} = \begin{bmatrix} 0 & 1 & 0 \\ \dfrac{-(a_2+a_3)}{1+a_0a_3} & -a_1 & \dfrac{1-a_0a_2}{1+a_0a_3} \\ 0 & 0 & 0 \end{bmatrix} \begin{bmatrix} x_1(t) \\ x_2(t) \\ x_3(t) \end{bmatrix} + \begin{bmatrix} 0 \\ 0 \\ 1 \end{bmatrix} r(t) \tag{10-34}$$

The output equation is

$$y(t) = \frac{1}{1+a_0a_3} x_1(t) + \frac{a_0}{1+a_0a_3} x_3(t) \tag{10-35}$$

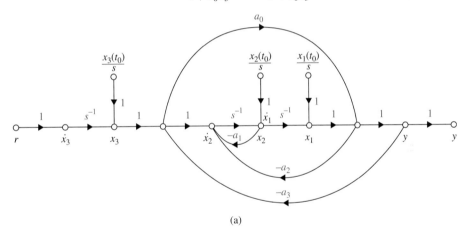

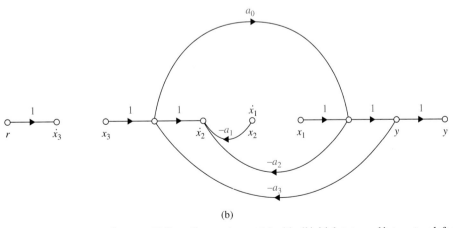

Figure 10-8 (a) State diagram. (b) State diagram in part (a) with all initial states and integrators left out.

10-3 VECTOR-MATRIX REPRESENTATION OF STATE EQUATIONS

Let the n state equations of an nth-order dynamic system be represented as

$$\frac{dx_1(t)}{dt} = f_i\big[x_1(t), x_2(t), \ldots, x_n(t), u_1(t), u_2(t), \ldots, u_p(t), w_1(t), w_2(t), \ldots, w_v(t)\big] \tag{10-36}$$

where $i = 1, 2, \ldots, n$. The ith state variable is represented by $x_i(t)$; $u_j(t)$ denotes the jth input for $j = 1, 2, \ldots, p$; and $w_k(t)$ denotes the kth disturbance input, with $k = 1, 2, \ldots, v$.

Let the variables $y_1(t), y_2(t), \ldots, y_q(t)$ be the q output variables of the system. In general, the output variables are functions of the state variables and the input variables. The **output equations** can be expressed as

$$y_j(t) = g_j\big[x_1(t), x_2(t), \ldots, x_n(t), u_1(t), u_2(t), \ldots, u_p(t), w_1(t), w_2(t), \ldots, w_v(t)\big] \tag{10-37}$$

where $j = 1, 2, \ldots, q$.

The set of n state equations in Eq. (10-36) and q output equations in Eq. (10-37) together form the **dynamic equations**. For ease of expression and manipulation, it is convenient to represent the dynamic equations in vector-matrix form. Let us define the following vectors:

State vector:

$$\mathbf{x}(t) = \begin{bmatrix} x_1(t) \\ x_2(t) \\ \vdots \\ x_n(t) \end{bmatrix} \quad (n \times 1) \tag{10-38}$$

Input vector:

$$\mathbf{u}(t) = \begin{bmatrix} u_1(t) \\ u_2(t) \\ \vdots \\ u_p(t) \end{bmatrix} \quad (p \times 1) \tag{10-39}$$

Output vector:

$$\mathbf{y}(t) = \begin{bmatrix} y_1(t) \\ y_2(t) \\ \vdots \\ y_q(t) \end{bmatrix} \quad (q \times 1) \tag{10-40}$$

Disturbance vector:

$$\mathbf{w}(t) = \begin{bmatrix} w_1(t) \\ w_2(t) \\ \vdots \\ w_v(t) \end{bmatrix} \quad (v \times 1) \tag{10-41}$$

By using these vectors, the n state equations of Eq. (10-36) can be written

$$\frac{d\mathbf{x}(t)}{dt} = \mathbf{f}[\mathbf{x}(t), \mathbf{u}(t), \mathbf{w}(t)] \tag{10-42}$$

where $\mathbf{f}$ denotes an $n \times 1$ column matrix that contains the functions $f_1, f_2, \ldots, f_n$ as elements. Similarly, the q output equations in Eq. (10-37) become

$$\mathbf{y}(t) = \mathbf{g}[\mathbf{x}(t), \mathbf{u}(t), \mathbf{w}(t)] \tag{10-43}$$

where $\mathbf{g}$ denotes a $q \times 1$ column matrix that contains the functions $g_1, g_2, \ldots, g_q$ as elements.

For a linear time-invariant system, the dynamic equations are written as

State equations:

$$\frac{d\mathbf{x}(t)}{dt} = \mathbf{A}\mathbf{x}(t) + \mathbf{B}\mathbf{u}(t) + \mathbf{E}\mathbf{w}(t) \tag{10-44}$$

Output equations:

$$\mathbf{y}(t) = \mathbf{C}\mathbf{x}(t) + \mathbf{D}\mathbf{u}(t) + \mathbf{H}\mathbf{w}(t) \tag{10-45}$$

where

$$\mathbf{A} = \begin{bmatrix} a_{11} & a_{12} & \cdots & a_{1n} \\ a_{21} & a_{22} & \cdots & a_{2n} \\ \vdots & \vdots & \ddots & \vdots \\ a_{n1} & a_{n2} & \cdots & a_{nn} \end{bmatrix} \quad (n \times n) \tag{10-46}$$

$$\mathbf{B} = \begin{bmatrix} b_{11} & b_{12} & \cdots & b_{1p} \\ b_{21} & b_{22} & \cdots & b_{2p} \\ \vdots & \vdots & \ddots & \vdots \\ b_{n1} & b_{n2} & \cdots & b_{np} \end{bmatrix} \quad (n \times p) \tag{10-47}$$

$$\mathbf{C} = \begin{bmatrix} c_{11} & c_{12} & \cdots & c_{1n} \\ c_{21} & c_{22} & \cdots & c_{2n} \\ \vdots & \vdots & \ddots & \vdots \\ c_{q1} & c_{q2} & \cdots & c_{qn} \end{bmatrix} \quad (q \times n) \tag{10-48}$$

$$\mathbf{D} = \begin{bmatrix} d_{11} & d_{12} & \cdots & d_{1p} \\ d_{21} & d_{22} & \cdots & d_{2p} \\ \vdots & \vdots & \ddots & \vdots \\ d_{q1} & d_{q2} & \cdots & d_{qp} \end{bmatrix} \quad (q \times p) \tag{10-49}$$

$$\mathbf{E} = \begin{bmatrix} e_{11} & e_{12} & \cdots & e_{1v} \\ e_{21} & e_{22} & \cdots & e_{2v} \\ \vdots & \vdots & \ddots & \vdots \\ e_{n1} & e_{n2} & \cdots & e_{nv} \end{bmatrix} \quad (n \times v) \tag{10-50}$$

$$\mathbf{H} = \begin{bmatrix} h_{11} & h_{12} & \cdots & h_{1v} \\ h_{12} & h_{22} & \cdots & h_{2v} \\ \vdots & \vdots & \ddots & \vdots \\ h_{q1} & h_{q2} & \cdots & h_{qv} \end{bmatrix} \quad (q \times v) \tag{10-51}$$

▶ 10-4 STATE-TRANSITION MATRIX

Once the state equations of a linear time-invariant system are expressed in the form of Eq. (10-44), the next step often involves the solutions of these equations given the initial state vector $\mathbf{x}(t_0)$, the input vector $\mathbf{u}(t)$, and the disturbance vector $\mathbf{w}(t)$, for $t \geq t_0$. The first term on the right-hand side of Eq. (10-44) is known as the homogeneous part of the state equation, and the last two terms represent the forcing functions $\mathbf{u}(t)$ and $\mathbf{w}(t)$.

The **state-transition matrix** is defined as a matrix that satisfies the linear homogeneous state equation:

$$\frac{d\mathbf{x}(t)}{dt} = \mathbf{A}\mathbf{x}(t) \tag{10-52}$$

Let $\boldsymbol{\phi}(t)$ be the $n \times n$ matrix that represents the state-transition matrix; then it must satisfy the equation

$$\frac{d\boldsymbol{\phi}(t)}{dt} = \mathbf{A}\boldsymbol{\phi}(t) \tag{10-53}$$

Furthermore, let $\mathbf{x}(0)$ denote the initial state at $t = 0$; then $\boldsymbol{\phi}(t)$ is also defined by the matrix equation

$$\mathbf{x}(t) = \boldsymbol{\phi}(t)\mathbf{x}(0) \tag{10-54}$$

which is the solution of the homogeneous state equation for $t \geq 0$.

One way of determining $\boldsymbol{\phi}(t)$ is by taking the Laplace transform on both sides of Eq. (10-52); we have

$$s\mathbf{X}(s) - \mathbf{x}(0) = \mathbf{A}\mathbf{X}(s) \tag{10-55}$$

Solving for $\mathbf{X}(s)$ from Eq. (10-55), we get

$$\mathbf{X}(s) = (s\mathbf{I} - \mathbf{A})^{-1}\mathbf{x}(0) \tag{10-56}$$

where it is assumed that the matrix $(s\mathbf{I} - \mathbf{A})$ is nonsingular. Taking the inverse Laplace transform on both sides of Eq. (10-56) yields

$$\mathbf{x}(t) = \mathcal{L}^{-1}\left[(s\mathbf{I} - \mathbf{A})^{-1}\right]\mathbf{x}(0) \quad t \geq 0 \tag{10-57}$$

By comparing Eq. (10-54) with Eq. (10-57), the state-transition matrix is identified to be

$$\boldsymbol{\phi}(t) = \mathcal{L}^{-1}\left[(s\mathbf{I} - \mathbf{A})^{-1}\right] \tag{10-58}$$

An alternative way of solving the homogeneous state equation is to assume a solution, as in the classical method of solving linear differential equations. We let the solution to Eq. (10-52) be

$$\mathbf{x}(t) = e^{\mathbf{A}t}\mathbf{x}(0) \tag{10-59}$$

for $t \geq 0$, where $e^{\mathbf{A}t}$ represents the following power series of the matrix $\mathbf{A}t$, and

$$e^{\mathbf{A}t} = \mathbf{I} + \mathbf{A}t + \frac{1}{2!}\mathbf{A}^2 t^2 + \frac{1}{3!}\mathbf{A}^3 t^3 + \cdots \tag{10-60}$$

It is easy to show that Eq. (10-59) is a solution of the homogeneous state equation, since, from Eq. (10-60),

$$\frac{de^{\mathbf{A}t}}{dt} = \mathbf{A}e^{\mathbf{A}t} \tag{10-61}$$

Therefore, in addition to Eq. (10-58), we have obtained another expression for the state-transition matrix:

$$\boldsymbol{\phi}(t) = e^{\mathbf{A}t} = \mathbf{I} + \mathbf{A}t + \frac{1}{2!}\mathbf{A}^2 t^2 + \frac{1}{3!}\mathbf{A}^3 t^3 + \cdots \tag{10-62}$$

Eq. (10-62) can also be obtained directly from Eq. (10-58). This is left as an exercise for the reader (Problem 10-5).

10-4-1 Significance of the State-Transition Matrix

Because the state-transition matrix satisfies the homogeneous state equation, it represents the **free response** of the system. In other words, it governs the response that is excited by the initial conditions only. In view of Eqs. (10-58) and (10-62), the state-transition matrix is dependent only upon the matrix $\mathbf{A}$ and, therefore, is sometimes referred to as the **state-transition matrix of A**. As the name implies, the state-transition matrix $\boldsymbol{\phi}(t)$ completely defines the transition of the states from the initial time $t = 0$ to any time t when the inputs are zero.

10-4-2 Properties of the State-Transition Matrix

The state-transition matrix $\boldsymbol{\phi}(t)$ possesses the following properties:

1. $\boldsymbol{\phi}(0) = \mathbf{I}$ (the identity matrix) \hfill (10-63)

   Proof: Eq. (10-63) follows directly from Eq. (10-62) by setting $t = 0$.

2. $\boldsymbol{\phi}^{-1}(t) = \boldsymbol{\phi}(-t)$ \hfill (10-64)

   Proof: Post-multiplying both sides of Eq. (10-62) by $e^{-\mathbf{A}t}$, we get

$$\boldsymbol{\phi}(t)e^{-\mathbf{A}t} = e^{\mathbf{A}t}e^{-\mathbf{A}t} = \mathbf{I} \tag{10-65}$$

Then, pre-multiplying both sides of Eq. (10-65) by $\boldsymbol{\phi}^{-1}(t)$, we get

$$e^{-\mathbf{A}t} = \boldsymbol{\phi}^{-1}(t) \tag{10-66}$$

Thus,

$$\boldsymbol{\phi}(-t) = \boldsymbol{\phi}^{-1}(t) = e^{-\mathbf{A}t} \tag{10-67}$$

An interesting result from this property of $\boldsymbol{\phi}(t)$ is that Eq. (10-59) can be rearranged to read

$$\mathbf{x}(0) = \boldsymbol{\phi}(-t)\mathbf{x}(t) \tag{10-68}$$

which means that the state-transition process can be considered as bilateral in time. That is, the transition in time can take place in either direction.

3. $\boldsymbol{\phi}(t_2 - t_1)\boldsymbol{\phi}(t_1 - t_0) = \boldsymbol{\phi}(t_2 - t_0) \quad$ for any t_0, t_1, t_2 (10-69)

Proof:

$$\begin{aligned} \boldsymbol{\phi}(t_2 - t_1)\boldsymbol{\phi}(t_1 - t_0) &= e^{\mathbf{A}(t_2-t_1)}e^{\mathbf{A}(t_1-t_0)} \\ &= e^{\mathbf{A}(t_2-t_0)} = \boldsymbol{\phi}(t_2 - t_0) \end{aligned} \tag{10-70}$$

This property of the state-transition matrix is important because it implies that a state-transition process can be divided into a number of sequential transitions. Fig. 10-9 illustrates that the transition from $t = t_0$ to $t = t_2$ is equal to the transition from t_0 to t_1 and then from t_1 to t_2. In general, of course, the state-transition process can be divided into any number of parts.

4. $[\boldsymbol{\phi}(t)]^k = \boldsymbol{\phi}(kt) \quad$ for $k =$ positive integer (10-71)

Proof:

$$\begin{aligned}{} [\boldsymbol{\phi}(t)]^k &= e^{\mathbf{A}t}e^{\mathbf{A}t}\ldots e^{\mathbf{A}t} \\ &= e^{k\mathbf{A}t} = \boldsymbol{\phi}(kt) \end{aligned} \tag{10-72}$$

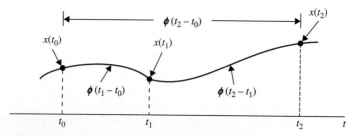

Figure 10-9 Property of the state-transition matrix.

10-5 STATE-TRANSITION EQUATION

The **state-transition equation** *is defined as the solution of the linear homogeneous state equation*. The linear time-invariant state equation

$$\frac{d\mathbf{x}(t)}{dt} = \mathbf{A}\mathbf{x}(t) + \mathbf{B}\mathbf{u}(t) + \mathbf{E}\mathbf{w}(t) \quad (10\text{-}73)$$

can be solved using either the classical method of solving linear differential equations or the Laplace transform method. The Laplace transform solution is presented in the following equations.

Taking the Laplace transform on both sides of Eq. (10-73), we have

$$s\mathbf{X}(s) - \mathbf{x}(0) = \mathbf{A}\mathbf{X}(s) + \mathbf{B}\mathbf{U}(s) + \mathbf{E}\mathbf{W}(s) \quad (10\text{-}74)$$

where $\mathbf{x}(0)$ denotes the initial-state vector evaluated at $t = 0$. Solving for $\mathbf{X}(s)$ in Eq. (10-74) yields

$$\mathbf{X}(s) = (s\mathbf{I} - \mathbf{A})^{-1}\mathbf{x}(0) + (s\mathbf{I} - \mathbf{A})^{-1}[\mathbf{B}\mathbf{U}(s) + \mathbf{E}\mathbf{W}(s)] \quad (10\text{-}75)$$

The state-transition equation of Eq. (10-73) is obtained by taking the inverse Laplace transform on both sides of Eq. (10-75):

$$\mathbf{x}(t) = \mathcal{L}^{-1}\left[(s\mathbf{I} - \mathbf{A})^{-1}\right]\mathbf{x}(0) + \mathcal{L}^{-1}\left\{(s\mathbf{I} - \mathbf{A})^{-1}[\mathbf{B}\mathbf{U}(s) + \mathbf{E}\mathbf{W}(s)]\right\}$$

$$= \boldsymbol{\phi}(t)\mathbf{x}(0) + \int_0^t \boldsymbol{\phi}(t-\tau)[\mathbf{B}\mathbf{u}(\tau) + \mathbf{E}\mathbf{w}(\tau)]d\tau \quad t \geq 0 \quad (10\text{-}76)$$

The state-transition equation in Eq. (10-76) is useful only when the initial time is defined to be at $t = 0$. In the study of control systems, especially discrete-data control systems, it is often desirable to break up a state-transition process into a sequence of transitions, so a more flexible initial time must be chosen. Let the initial time be represented by t_0 and the corresponding initial state by $\mathbf{x}(t_0)$, and assume that the input $\mathbf{u}(t)$ and the disturbance $\mathbf{w}(t)$ are applied at $t \geq 0$. We start with Eq. (10-76) by setting $t = t_0$, and solving for $\mathbf{x}(0)$, we get

$$\mathbf{x}(0) = \boldsymbol{\phi}(-t_0)\mathbf{x}(t_0) - \boldsymbol{\phi}(-t_0)\int_0^{t_0} \boldsymbol{\phi}(t_0-\tau)[\mathbf{B}\mathbf{u}(\tau) + \mathbf{E}\mathbf{w}(\tau)]d\tau \quad (10\text{-}77)$$

where the property on $\boldsymbol{\phi}(t)$ of Eq. (10-64) has been applied.

Substituting Eq. (10-77) into Eq. (10-76) yields

$$\mathbf{x}(t) = \boldsymbol{\phi}(t)\boldsymbol{\phi}(-t_0)\mathbf{x}(t_0) - \boldsymbol{\phi}(t)\boldsymbol{\phi}(-t_0)\int_0^{t_0} \boldsymbol{\phi}(t_0-\tau)[\mathbf{B}\mathbf{u}(\tau) + \mathbf{E}\mathbf{w}(\tau)]d\tau$$

$$+ \int_0^t \boldsymbol{\phi}(t-\tau)[\mathbf{B}\mathbf{u}(\tau) + \mathbf{E}\mathbf{w}(\tau)]d\tau \quad (10\text{-}78)$$

Now by using the property of Eq. (10-69) and combining the last two integrals, Eq. (10-78) becomes

$$\mathbf{x}(t) = \boldsymbol{\phi}(t-t_0)\mathbf{x}(t_0) + \int_0^t \boldsymbol{\phi}(t-\tau)[\mathbf{B}\mathbf{u}(\tau) + \mathbf{E}\mathbf{w}(\tau)]d\tau \quad t \geq t_0 \quad (10\text{-}79)$$

It is apparent that Eq. (10-79) reverts to Eq. (10-77) when $t_0 = 0$.

Once the state-transition equation is determined, the output vector can be expressed as a function of the initial state and the input vector simply by substituting $\mathbf{x}(t)$ from Eq. (10-79) into Eq. (10-45). Thus, the output vector is

$$\mathbf{y}(t) = \mathbf{C}\boldsymbol{\phi}(t - t_0)\mathbf{x}(t_0) + \int_{t_0}^{t} \mathbf{C}\boldsymbol{\phi}(t - \tau)[\mathbf{B}\mathbf{u}(\tau) + \mathbf{E}\mathbf{w}(\tau)]d\tau \qquad (10\text{-}80)$$
$$+ \mathbf{D}\mathbf{u}(t) + \mathbf{H}\mathbf{w}(t) \quad t \geq t_0$$

The following example illustrates the determination of the state-transition matrix and equation.

▶ **EXAMPLE 10-5-1** Consider the state equation

$$\begin{bmatrix} \dfrac{dx_1(t)}{dt} \\ \dfrac{dx_2(t)}{dt} \end{bmatrix} = \begin{bmatrix} 0 & 1 \\ -2 & -3 \end{bmatrix} \begin{bmatrix} x_1(t) \\ x_2(t) \end{bmatrix} + \begin{bmatrix} 0 \\ 1 \end{bmatrix} u(t) \qquad (10\text{-}81)$$

The problem is to determine the state-transition matrix $\boldsymbol{\phi}(t)$ and the state vector $\mathbf{x}(t)$ for $t \geq 0$ when the input is $u(t) = 1$ for $t \geq 0$. The coefficient matrices are identified to be

$$\mathbf{A} = \begin{bmatrix} 0 & 1 \\ -2 & -3 \end{bmatrix} \quad \mathbf{B} = \begin{bmatrix} 0 \\ 1 \end{bmatrix} \quad \mathbf{E} = 0 \qquad (10\text{-}82)$$

Therefore,

$$s\mathbf{I} - \mathbf{A} = \begin{bmatrix} s & 0 \\ 0 & s \end{bmatrix} - \begin{bmatrix} 0 & 1 \\ -2 & -3 \end{bmatrix} = \begin{bmatrix} s & -1 \\ 2 & s+3 \end{bmatrix} \qquad (10\text{-}83)$$

The inverse matrix of $(s\mathbf{I} - \mathbf{A})$ is

$$(s\mathbf{I} - \mathbf{A})^{-1} = \dfrac{1}{s^2 + 3s + 2} \begin{bmatrix} s+3 & 1 \\ -2 & s \end{bmatrix} \qquad (10\text{-}84)$$

The state-transition matrix of $\mathbf{A}$ is found by taking the inverse Laplace transform of Eq. (10-84). Thus,

$$\boldsymbol{\phi}(t) = \mathcal{L}^{-1}\left[(s\mathbf{I} - \mathbf{A})^{-1}\right] = \begin{bmatrix} 2e^{-t} - e^{-2t} & e^{-t} - e^{-2t} \\ -2e^{-t} + 2e^{-2t} & -e^{-t} + 2e^{-2t} \end{bmatrix} \qquad (10\text{-}85)$$

The state-transition equation for $t \geq 0$ is obtained by substituting Eq. (10-85), $\mathbf{B}$, and $u(t)$ into Eq. (10-76). We have

$$\mathbf{x}(t) = \begin{bmatrix} 2e^{-t} - e^{-2t} & e^{-t} - e^{-2t} \\ -2e^{-t} + 2e^{-2t} & -e^{-t} + 2e^{-2t} \end{bmatrix} \mathbf{x}(0) \qquad (10\text{-}86)$$
$$+ \int_0^t \begin{bmatrix} 2e^{-(t-\tau)} - e^{-2(t-\tau)} & e^{-(t-\tau)} - e^{-2(t-\tau)} \\ -2e^{-(t-\tau)} + e^{-2(t-\tau)} & -e^{-(t-\tau)} + 2e^{-2(t-\tau)} \end{bmatrix} \begin{bmatrix} 0 \\ 1 \end{bmatrix} d\tau$$

or

$$\mathbf{x}(t) = \begin{bmatrix} 2e^{-t} - e^{-2t} & e^{-t} - e^{-2t} \\ -2e^{-t} + 2e^{-2t} & -e^{-t} + 2e^{-2t} \end{bmatrix} \mathbf{x}(0) + \begin{bmatrix} 0.5 - e^{-t} + 0.5e^{-2t} \\ e^{-t} - e^{-2t} \end{bmatrix} \quad t \geq 0 \qquad (10\text{-}87)$$

As an alternative, the second term of the state-transition equation can be obtained by taking the inverse Laplace transform of $(s\mathbf{I} - \mathbf{A})^{-1}\mathbf{B}U(s)$. Thus, we have

$$\mathcal{L}^{-1}\left[(s\mathbf{I} - \mathbf{A})^{-1}\right]\mathbf{B}U(s) = \mathcal{L}^{-1}\left(\dfrac{1}{s^2 + 3s + 2}\begin{bmatrix} s+3 & 1 \\ -2 & s \end{bmatrix}\begin{bmatrix} 0 \\ 1 \end{bmatrix}\dfrac{1}{s}\right)$$
$$= \mathcal{L}^{-1}\left(\dfrac{1}{s^2 + 3s + 2}\begin{bmatrix} \frac{1}{s} \\ 1 \end{bmatrix}\right) = \begin{bmatrix} 0.5 - e^{-t} + 0.5e^{-2t} \\ e^{-t} - e^{-2t} \end{bmatrix} \quad t \geq 0 \qquad (10\text{-}88)$$

◀

10-5-1 State-Transition Equation Determined from the State Diagram

Eqs. (10-75) and (10-76) show that the Laplace transform method of solving the state equations requires obtaining the inverse of matrix $(s\mathbf{I} - \mathbf{A})$. We shall now show that the state diagram described earlier in Section 10-2-3 and the SFG gain formula (Chapter 3) can be used to solve for the state-transition equation in the Laplace domain of Eq. (10-75). Let the initial time be t_0; then Eq. (10-75) is rewritten as

$$\mathbf{X}(s) = (s\mathbf{I} - \mathbf{A})^{-1}\mathbf{x}(t_0) + (s\mathbf{I} - \mathbf{A})^{-1}[\mathbf{B}U(s) + \mathbf{E}\mathbf{W}(s)] \quad t \geq t_0 \quad (10\text{-}89)$$

The last equation can be written directly from the state diagram using the gain formula, with $X_i(s)$, $i = 1, 2, \ldots, n$ as the output nodes. The following example illustrates the state-diagram method of finding the state-transition equations for the system described in Example 10-2-1.

▶ **EXAMPLE 10-5-2** The state diagram for the system described by Eq. (10-81) is shown in Fig. 10-10 with t_0 as the initial time. The outputs of the integrators are assigned as state variables. Applying the gain formula to the state diagram in Fig. 10-10, with $X_1(s)$ and $X_2(s)$ as output nodes and $x_1(t_0)$, $x_2(t_0)$, and $U(s)$ as input nodes, we have

$$X_1(s) = \frac{s^{-1}(1 + 3s^{-1})}{\Delta}x_1(t_0) + \frac{s^{-2}}{\Delta}x_2(t_0) + \frac{s^{-2}}{\Delta}U(s) \quad (10\text{-}90)$$

$$X_2(s) = \frac{-2s^{-2}}{\Delta}x_1(t_0) + \frac{s^{-1}}{\Delta}x_2(t_0) + \frac{s^{-1}}{\Delta}U(s) \quad (10\text{-}91)$$

where

$$\Delta = 1 + 3s^{-1} + 2s^{-2} \quad (10\text{-}92)$$

After simplification, Eqs. (10-90) and (10-91) are presented in vector-matrix form:

$$\begin{bmatrix} X_1(s) \\ X_2(s) \end{bmatrix} = \frac{1}{(s+1)(s+2)}\begin{bmatrix} s+3 & 1 \\ -2 & s \end{bmatrix}\begin{bmatrix} x_1(t_0) \\ x_2(t_0) \end{bmatrix} + \frac{1}{(s+1)(s+2)}\begin{bmatrix} 1 \\ s \end{bmatrix}U(s) \quad (10\text{-}93)$$

The state-transition equation for $t \geq t_0$ is obtained by taking the inverse Laplace transform on both sides of Eq. (10-93).

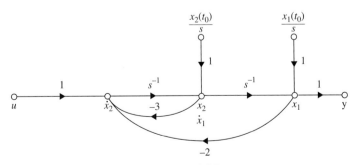

Figure 10-10 State diagram for Eq. (10-81).

Consider that the input $u(t)$ is a unit-step function applied at $t = t_0$. Then the following inverse Laplace transform relationships are identified:

$$\mathcal{L}^{-1}\left(\frac{1}{s}\right) = u_s(t - t_0) \quad t \geq t_0 \tag{10-94}$$

$$\mathcal{L}^{-1}\left(\frac{1}{s+a}\right) = e^{-a(t-t_0)} u_s(t - t_0) \quad t \geq t_0 \tag{10-95}$$

Because the initial time is defined to be t_0, the Laplace transform expressions here do not have the delay factor $e^{-t_0 s}$. The inverse Laplace transform of Eq. (10-93) is

$$\begin{bmatrix} x_1(t) \\ x_2(t) \end{bmatrix} = \begin{bmatrix} 2e^{-(t-t_0)} - e^{-2(t-t_0)} & e^{-(t-t_0)} - e^{-2(t-t_0)} \\ -2e^{-(t-t_0)} + 2e^{-2(t-t_0)} & -e^{(t-t_0)} + 2e^{-2(t-t_0)} \end{bmatrix} \begin{bmatrix} x_1(t_0) \\ x_2(t_0) \end{bmatrix}$$
$$+ \begin{bmatrix} 0.5 u_s(t-t_0) - e^{-(t-t_0)} + 0.5 e^{-2(t-t_0)} \\ e^{-(t-t_0)} - e^{-2(t-t_0)} \end{bmatrix} t \geq t_0 \tag{10-96}$$

The reader should compare this result with that in Eq. (10-87), which is obtained for $t \geq 0$.

▶ **EXAMPLE 10-5-3** In this example, we illustrate the utilization of the state-transition method to a system with input discontinuity. An *RL* network is shown in Fig. 10-11. The history of the network is completely specified by the initial current of the inductance, $i(0)$ at $t = 0$. At time $t = 0$, the voltage $e_{in}(t)$ with the profile shown in Fig. 10-12 is applied to the network. The state equation of the network for $t \geq 0$ is

$$\frac{di(t)}{dt} = -\frac{R}{L} i(t) + \frac{1}{L} e_{in}(t) \tag{10-97}$$

Comparing the last equation with Eq. (10-44), the scalar coefficients of the state equation are identified to be

$$A = -\frac{R}{L} \quad B = \frac{1}{L} \quad E = 0 \tag{10-98}$$

The state-transition matrix is

$$\phi(t) = e^{-At} = e^{-Rt/L} \tag{10-99}$$

The conventional approach of solving for $i(t)$ for $t \geq 0$ is to express the input voltage as

$$e(t) = E_{in} u_s(t) + E_{in} u_s(t - t_1) \tag{10-100}$$

where $u_s(t)$ is the unit-step function. The Laplace transform of $e(t)$ is

$$E_{in}(s) = \frac{E_{in}}{s} (1 + e^{-t_1 s}) \tag{10-101}$$

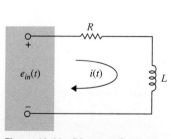

Figure 10-11 *RL* network.

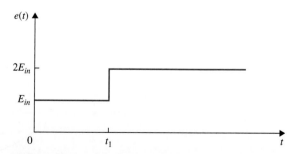

Figure 10-12 Input voltage waveform for the network in Fig. 10-3.

Then

$$(s\mathbf{I} - \mathbf{A})^{-1}\mathbf{B}U(s) = \frac{E_{in}}{Ls(s + R/L)}(1 + e^{-t_1 s}) \qquad (10\text{-}102)$$

By substituting Eq. (10-102) into Eq. (10-76), the state-transition equation, the current for $t \geq 0$ is obtained:

$$i(t) = e^{-Rt/L}i(0)u_s(t) + \frac{E_{in}}{R}\left(1 - e^{-Rt/L}\right)u_s(t) + \frac{E_{in}}{R}\left(1 - e^{-R(t-t_1)/L}\right)u_s(t - t_1) \qquad (10\text{-}103)$$

Using the state-transition approach, we can divide the transition period into two parts: $t = 0$ to $t = t_1$, and $t = t_1$ to $t = \infty$. First, for the time interval $0 \leq t \leq t_1$, the input is

$$e(t) = E_{in}u_s(t) \quad 0 \leq t < t_1 \qquad (10\text{-}104)$$

Then

$$(s\mathbf{I} - \mathbf{A})^{-1}\mathbf{B}U(s) = \frac{E_{in}}{Ls(s + R/L)} = \frac{E_{in}}{Rs[1 + (L/R)s]} \qquad (10\text{-}105)$$

Thus, the state-transition equation for the time interval $0 \leq t \leq t_1$ is

$$i(t) = \left[e^{-Rt/L}i(0) + \frac{E_{in}}{R}\left(1 - e^{-Rt/L}\right)\right]u_s(t) \qquad (10\text{-}106)$$

Substituting $t = t_1$ into Eq. (10-106), we get

$$i(t_1) = e^{-Rt_1/L}i(0) + \frac{E_{in}}{R}\left(1 - e^{-Rt_1/L}\right) \qquad (10\text{-}107)$$

The value of $i(t)$ at $t = t_1$ is now used as the initial state for the next transition period of $t_1 \leq t < \infty$. The amplitude of the input for the interval is $2E_{in}$. The state-transition equation for the second transition period is

$$i(t) = e^{-R(t-t_1)/L}i(t_1) + \frac{2E_{in}}{R}\left(1 - e^{-R(t-t_1)/L}\right) \quad t \geq t_1 \qquad (10\text{-}108)$$

where $i(t_1)$ is given by Eq. (10-107).

This example illustrates two possible ways of solving a state-transition problem. In the first approach, the transition is treated as one continuous process, whereas in the second, the transition period is divided into parts over which the input can be more easily presented. Although the first approach requires only one operation, the second method yields relatively simple results to the state-transition equation, and it often presents computational advantages. Notice that, in the second method, the state at $t = t_1$ is used as the initial state for the next transition period, which begins at t_1. ◀

▶ 10-6 RELATIONSHIP BETWEEN STATE EQUATIONS AND HIGH-ORDER DIFFERENTIAL EQUATIONS

In the preceding sections, we defined the state equations and their solutions for linear time-invariant systems. Although it is usually possible to write the state equations directly from the schematic diagram of a system, in practice the system may have been described by a high-order differential equation or transfer function. It becomes necessary to investigate how state equations can be written directly from the high-order differential equation or the transfer function. In Chapter 2 we illustrated how the state variables of an nth-order differential equation in Eq. (2-97) are intuitively defined, as shown in Eq. (2-105). The results are the n state equations in Eq. (2-106).

The state equations are written in vector-matrix form:

$$\frac{d\mathbf{x}(t)}{dt} = \mathbf{A}\mathbf{x}(t) + \mathbf{B}u(t) \qquad (10\text{-}109)$$

where

$$\mathbf{A} = \begin{bmatrix} 0 & 1 & 0 & \cdots & 0 \\ 0 & 0 & 1 & \cdots & 0 \\ \vdots & \vdots & \vdots & \ddots & \vdots \\ 0 & 0 & 0 & \cdots & 1 \\ -a_0 & -a_1 & -a_2 & \cdots & -a_{n-1} \end{bmatrix} \quad (n \times n) \qquad (10\text{-}110)$$

$$\mathbf{B} = \begin{bmatrix} 0 \\ 0 \\ \vdots \\ 0 \\ 1 \end{bmatrix} \quad (n \times 1) \qquad (10\text{-}111)$$

Notice that the last row of **A** contains the negative values of the coefficients of the homogeneous part of the differential equation in ascending order, except for the coefficient of the highest-order term, which is unity. **B** is a column matrix with the last row equal to one, and the rest of the elements are all zeros. The state equations in Eq. (10-109) with **A** and **B** given in Eqs. (10-110) and (10-111) are known as the **phase-variable canonical form** (**PVCF**), or the **controllability canonical form** (**CCF**).

The output equation of the system is written

$$y(t) = \mathbf{C}\mathbf{x}(t) = x_1(t) \qquad (10\text{-}112)$$

where

$$\mathbf{C} = \begin{bmatrix} 1 & 0 & 0 & \cdots & 0 \end{bmatrix} \qquad (10\text{-}113)$$

We have shown earlier that the state variables of a given system are not unique. In general, we seek the most convenient way of assigning the state variables as long as the definition of state variables is satisfied. In Section 10-9 we shall show that, by first writing the transfer function and then drawing the state diagram of the system by decomposition of the transfer function, the state variables and state equations of any system can be found very easily.

▶ **EXAMPLE 10-6-1** Consider the differential equation

$$\frac{d^3y(t)}{dt^3} + 5\frac{d^2y(t)}{dt^2} + \frac{dy(t)}{dt} + 2y(t) = u(t) \qquad (10\text{-}114)$$

Rearranging the last equation so that the highest-order derivative term is set equal to the rest of the terms, we have

$$\frac{d^3y(t)}{dt^3} = -5\frac{d^2y(t)}{dt^2} - \frac{dy(t)}{dt} - 2y(t) + u(t) \qquad (10\text{-}115)$$

The state variables are defined as

$$\begin{aligned} x_1(t) &= y(t) \\ x_2(t) &= \frac{dy(t)}{dt} \\ x_3(t) &= \frac{d^2y(t)}{dt^2} \end{aligned} \qquad (10\text{-}116)$$

Then the state equations are represented by the vector-matrix equation

$$\frac{d\mathbf{x}(t)}{dt} = \mathbf{A}\mathbf{x}(t) + \mathbf{B}u(t) \qquad (10\text{-}117)$$

where $\mathbf{x}(t)$ is the 2×1 state vector, $u(t)$ is the scalar input, and

$$\mathbf{A} = \begin{bmatrix} 0 & 1 & 0 \\ 0 & 0 & 1 \\ -2 & -1 & -5 \end{bmatrix} \quad \mathbf{B} = \begin{bmatrix} 0 \\ 0 \\ 1 \end{bmatrix} \qquad (10\text{-}118)$$

The output equation is

$$y(t) = x_1(t) = [\,1 \quad 0\,]\mathbf{x}(t) \qquad (10\text{-}119)$$

◀

▶ 10-7 RELATIONSHIP BETWEEN STATE EQUATIONS AND TRANSFER FUNCTIONS

We have presented the methods of modeling a linear time-invariant system by transfer functions and dynamic equations. We now investigate the relationship between these two representations.

Consider a linear time-invariant system described by the following dynamic equations:

$$\frac{d\mathbf{x}(t)}{dt} = \mathbf{A}\mathbf{x}(t) + \mathbf{B}\mathbf{u}(t) + \mathbf{E}\mathbf{w}(t) \qquad (10\text{-}120)$$

$$\mathbf{y}(t) = \mathbf{C}\mathbf{x}(t) + \mathbf{D}\mathbf{u}(t) + \mathbf{H}\mathbf{w}(t) \qquad (10\text{-}121)$$

where

$\mathbf{x}(t) = n \times 1$ state vector
$\mathbf{u}(t) = p \times 1$ input vector
$\mathbf{y}(t) = q \times 1$ output vector
$\mathbf{w}(t) = v \times 1$ disturbance vector

and **A, B, C, D, E,** and **H** are coefficient matrices of appropriate dimensions.

Taking the Laplace transform on both sides of Eq. (10-120) and solving for $\mathbf{X}(s)$, we have

$$\mathbf{X}(s) = (s\mathbf{I} - \mathbf{A})^{-1}\mathbf{x}(0) + (s\mathbf{I} - \mathbf{A})^{-1}[\mathbf{B}\mathbf{U}(s) + \mathbf{E}\mathbf{W}(s)] \qquad (10\text{-}122)$$

The Laplace transform of Eq. (10-121) is

$$\mathbf{Y}(s) = \mathbf{C}\mathbf{X}(s) + \mathbf{D}\mathbf{U}(s) + \mathbf{H}\mathbf{W}(s) \qquad (10\text{-}123)$$

Substituting Eq. (10-122) into Eq. (10-123), we have

$$\mathbf{Y}(s) = \mathbf{C}(s\mathbf{I} - \mathbf{A})^{-1}\mathbf{x}(0) + \mathbf{C}(s\mathbf{I} - \mathbf{A})[\mathbf{B}\mathbf{U}(s) + \mathbf{E}\mathbf{W}(s)] + \mathbf{D}\mathbf{U}(s) + \mathbf{H}\mathbf{W}(s) \qquad (10\text{-}124)$$

Because the definition of a transfer function requires that the initial conditions be set to zero, $\mathbf{x}(0) = \mathbf{0}$; thus, Eq. (10-124) becomes

$$\mathbf{Y}(s) = \left[\mathbf{C}(s\mathbf{I} - \mathbf{A})^{-1}\mathbf{B} + \mathbf{D}\right]\mathbf{U}(s) + [\mathbf{C}(s\mathbf{I} - \mathbf{A})\mathbf{E} + \mathbf{H}]\mathbf{W}(s) \qquad (10\text{-}125)$$

Let us define

$$\mathbf{G}_u(s) = \mathbf{C}(s\mathbf{I} - \mathbf{A})^{-1}\mathbf{B} + \mathbf{D} \qquad (10\text{-}126)$$

$$\mathbf{G}_w(s) = \mathbf{C}(s\mathbf{I} - \mathbf{A})^{-1}\mathbf{E} + \mathbf{H} \qquad (10\text{-}127)$$

where $\mathbf{G}_u(s)$ is a $q \times p$ transfer-function matrix between $\mathbf{u}(t)$ and $\mathbf{y}(t)$ when $\mathbf{w}(t) = \mathbf{0}$, and $\mathbf{G}_w(s)$ is a $q \times v$ transfer-function matrix between $\mathbf{w}(t)$ and $\mathbf{y}(t)$ when $\mathbf{u}(t) = \mathbf{0}$.

Then, Eq. (10-125) becomes

$$\mathbf{Y}(s) = \mathbf{G}_u(s)\mathbf{U}(s) + \mathbf{G}_w(s)\mathbf{W}(s) \qquad (10\text{-}128)$$

▶ **EXAMPLE 10-7-1** Consider that a multivariable system is described by the differential equations

$$\frac{d^2 y_1(t)}{dt^2} + 4\frac{dy_1(t)}{dt} - 3y_2(t) = u_1(t) + 2w(t) \qquad (10\text{-}129)$$

$$\frac{dy_1(t)}{dt} + \frac{dy_2(t)}{dt} + y_1(t) + 2y_2(t) = u_2(t) \qquad (10\text{-}130)$$

The state variables of the system are assigned as:

$$x_1(t) = y_1(t)$$
$$x_2(t) = \frac{dy_1(t)}{dt} \qquad (10\text{-}131)$$
$$x_3(t) = y_2(t)$$

These state variables are defined by mere inspection of the two differential equations, because no particular reasons for the definitions are given other than that these are the most convenient. Now equating the first term of each of the equations of Eqs. (10-129) and (10-130) to the rest of the terms and using the state-variable relations of Eq. (10-131), we arrive at the following state equations and output equations in vector-matrix form:

$$\begin{bmatrix} \frac{dx_1(t)}{dt} \\ \frac{dx_2(t)}{dt} \\ \frac{dx_3(t)}{dt} \end{bmatrix} = \begin{bmatrix} 0 & 1 & 0 \\ 0 & -4 & 3 \\ -1 & -1 & -2 \end{bmatrix} \begin{bmatrix} x_1(t) \\ x_2(t) \\ x_3(t) \end{bmatrix} + \begin{bmatrix} 0 & 0 \\ 1 & 0 \\ 0 & 1 \end{bmatrix} \begin{bmatrix} u_1(t) \\ u_2(t) \end{bmatrix} + \begin{bmatrix} 0 \\ 2 \\ 0 \end{bmatrix} w(t) \qquad (10\text{-}132)$$

$$\begin{bmatrix} y_1(t) \\ y_2(t) \end{bmatrix} = \begin{bmatrix} 1 & 0 & 0 \\ 0 & 0 & 1 \end{bmatrix} \begin{bmatrix} x_1(t) \\ x_2(t) \\ x_3(t) \end{bmatrix} = \mathbf{C}\mathbf{x}(t) \qquad (10\text{-}133)$$

To determine the transfer-function matrix of the system using the state-variable formulation, we substitute the $\mathbf{A}$, $\mathbf{B}$, $\mathbf{C}$, $\mathbf{D}$, and $\mathbf{E}$ matrices into Eq. (10-125). First, we form the matrix $(s\mathbf{I} - \mathbf{A})$:

$$(s\mathbf{I} - \mathbf{A}) = \begin{bmatrix} s & -1 & 0 \\ 0 & s+4 & -3 \\ 1 & 1 & s+2 \end{bmatrix} \qquad (10\text{-}134)$$

The determinant of $(s\mathbf{I} - \mathbf{A})$ is

$$|s\mathbf{I} - \mathbf{A}| = s^3 + 6s^2 + 11s + 3 \qquad (10\text{-}135)$$

Thus,

$$(s\mathbf{I} - \mathbf{A})^{-1} = \frac{1}{|s\mathbf{I} - \mathbf{A}|} \begin{bmatrix} s^2 + 6s + 11 & s+2 & 3 \\ -3 & s(s+2) & 3s \\ -(s+4) & -(s+1) & s(s+4) \end{bmatrix} \qquad (10\text{-}136)$$

The transfer-function matrix between $\mathbf{u}(t)$ and $\mathbf{y}(t)$ is

$$\mathbf{G}_u(s) = \mathbf{C}(s\mathbf{I} - \mathbf{A})^{-1}\mathbf{B} = \frac{1}{s^3 + 6s^2 + 11s + 3}\begin{bmatrix} s+2 & 3 \\ -(s+1) & s(s+4) \end{bmatrix} \quad (10\text{-}137)$$

and that between $\mathbf{w}(t)$ and $\mathbf{y}(t)$ is

$$\mathbf{G}_w(s) = \mathbf{C}(s\mathbf{I} - \mathbf{A})^{-1}\mathbf{E} = \frac{1}{s^3 + 6s + 11s + 3}\begin{bmatrix} 2(s+2) \\ -2(s+1) \end{bmatrix} \quad (10\text{-}138)$$

Using the conventional approach, we take the Laplace transform on both sides of Eqs. (10-129) and (10-130) and assume zero initial conditions. The resulting transformed equations are written in vector-matrix form as

$$\begin{bmatrix} s(s+4) & -3 \\ s+1 & s+2 \end{bmatrix}\begin{bmatrix} Y_1(s) \\ Y_2(s) \end{bmatrix} = \begin{bmatrix} U_1(s) \\ U_2(s) \end{bmatrix} + \begin{bmatrix} 2 \\ 0 \end{bmatrix}W(s) \quad (10\text{-}139)$$

Solving for $\mathbf{Y}(s)$ from Eq. (10-139), we obtain

$$\mathbf{Y}(s) = \mathbf{G}_u(s)\mathbf{U}(s) + \mathbf{G}_w(s)\mathbf{W}(s) \quad (10\text{-}140)$$

where

$$\mathbf{G}_u(s) = \begin{bmatrix} s(s+4) & -3 \\ s+1 & s+2 \end{bmatrix}^{-1} \quad (10\text{-}141)$$

$$\mathbf{G}_w(s) = \begin{bmatrix} s(s+4) & -3 \\ s+1 & s+2 \end{bmatrix}^{-1}\begin{bmatrix} 2 \\ 0 \end{bmatrix} \quad (10\text{-}142)$$

which will give the same results as in Eqs. (10-137) and (10-138), respectively, when the matrix inverses are carried out.

10-8 CHARACTERISTIC EQUATIONS, EIGENVALUES, AND EIGENVECTORS

Characteristic equations play an important role in the study of linear systems. They can be defined with respect to differential equations, transfer functions, or state equations.

10-8-1 Characteristic Equation from a Differential Equation

Consider that a linear time-invariant system is described by the differential equation

$$\frac{d^n y(t)}{dt^n} + a_{n-1}\frac{d^{n-1} y(t)}{dt^{n-1}} + \cdots + a_1\frac{dy(t)}{dt} + a_0 y(t)$$
$$= b_m\frac{d^m u(t)}{dt^m} + b_{m-1}\frac{d^{m-1} u(t)}{dt^{m-1}} + \cdots + b_1\frac{du(t)}{dt} + b_0 u(t) \quad (10\text{-}143)$$

where $n > m$. By defining the operator s as

$$s^k = \frac{d^k}{dt^k} \quad k = 1, 2, \ldots, n \quad (10\text{-}144)$$

Eq. (10-143) is written

$$(s^n + a_{n-1}s^{n-1} + \cdots + a_1 s + a_0)y(t) = (b_m s^m + b_{m-1}s^{m-1} + \cdots + b_1 s + b_0)u(t) \quad (10\text{-}145)$$

The **characteristic equation** of the system is defined as

$$s^n + a_{n-1}s^{n-1} + \cdots + a_1 s + a_0 = 0 \qquad (10\text{-}146)$$

which is obtained by setting the homogeneous part of Eq. (10-145) to zero.

▶ **EXAMPLE 10-8-1** Consider the differential equation in Eq. (10-114). The characteristic equation is obtained by inspection,

$$s^3 + 5s^2 + s + 2 = 0 \qquad (10\text{-}147)$$

◀

10-8-2 Characteristic Equation from a Transfer Function

The transfer function of the system described by Eq. (10-143) is

$$G(s) = \frac{b_m s^m + b_{m-1}s^{m-1} + \cdots + b_1 s + b_0}{s^n + a_{n-1}s^{n-1} + \cdots + a_1 s + a_0} \qquad (10\text{-}148)$$

The characteristic equation is obtained by equating the denominator polynomial of the transfer function to zero.

▶ **EXAMPLE 10-8-2** The transfer function of the system described by the differential equation in Eq. (10-114) is

$$\frac{Y(s)}{U(s)} = \frac{1}{s^3 + 5s^2 + s + 2} \qquad (10\text{-}149)$$

The same characteristic equation as in Eq. (10-147) is obtained by setting the denominator polynomial of Eq. (10-149) to zero.

◀

10-8-3 Characteristic Equation from State Equations

From the state-variable approach, we can write Eq. (10-126) as

$$\begin{aligned}\mathbf{G}_u(s) &= \mathbf{C}\frac{\text{adj}(s\mathbf{I} - \mathbf{A})}{(s\mathbf{I} - \mathbf{A})}\mathbf{B} + \mathbf{D} \\ &= \frac{\mathbf{C}[\text{adj}(s\mathbf{I} - \mathbf{A})]\mathbf{B} + |s\mathbf{I} - \mathbf{A}|\mathbf{D}}{|s\mathbf{I} - \mathbf{A}|}\end{aligned} \qquad (10\text{-}150)$$

Setting the denominator of the transfer-function matrix $\mathbf{G}_u(s)$ to zero, we get the characteristic equation

$$|s\mathbf{I} - \mathbf{A}| = 0 \qquad (10\text{-}151)$$

which is an alternative form of the characteristic equation but should lead to the same equation as in Eq. (10-146). *An important property of the characteristic equation is that, if the coefficients of $\mathbf{A}$ are real, then the coefficients of $|s\mathbf{I} - \mathbf{A}|$ are also real.*

▶ **EXAMPLE 10-8-3** The matrix $\mathbf{A}$ for the state equations of the differential equation in Eq. (10-114) is given in Eq. (10-128). The characteristic equation of $\mathbf{A}$ is

$$|s\mathbf{I} - \mathbf{A}| = \begin{vmatrix} s & -1 & 0 \\ 0 & s & -1 \\ 2 & 1 & s+5 \end{vmatrix} = s^3 + 5s^2 + s + 2 = 0 \qquad (10\text{-}152)$$

◀

10-8-4 Eigenvalues

The roots of the characteristic equation are often referred to as the eigenvalues of the matrix **A**.

Some of the important properties of eigenvalues are given as follows.

1. If the coefficients of **A** are all real, then its eigenvalues are either real or in complex-conjugate pairs.
2. If $\lambda_1, \lambda_2, \ldots, \lambda_n$ are the eigenvalues of **A**, then

$$\text{tr}(\mathbf{A}) = \sum_{i=1}^{n} \lambda_i \qquad (10\text{-}153)$$

That is, the trace of **A** is the sum of all the eigenvalues of **A**.

3. If λ_i, $i = 1, 2, \ldots, n$, is an eigenvalue of **A**, then it is an eigenvalue of $\mathbf{A}'$.
4. If **A** is nonsingular, with eigenvalues λ_i, $i = 1, 2, \ldots, n$, then $1/\lambda_i$, $i = 1, 2, \ldots, n$, are the eigenvalues of $\mathbf{A}^{-1}$.

▶ **EXAMPLE 10-8-4** The eigenvalues or the roots of the characteristic equation of the matrix **A** in Eq. (10-118) are obtained by solving for the roots of Eq. (10-152). The results are

$$s = -0.06047 + j0.63738 \quad s = -0.06047 - j0.63738 \quad s = -4.87906 \qquad (10\text{-}154)$$

◀

10-8-5 Eigenvectors

Eigenvectors are useful in modern control methods, one of which is the similarity transformation, which will be discussed in a later section.

Any nonzero vector $\mathbf{p}_i$ that satisfies the matrix equation

$$(\lambda_i \mathbf{I} - \mathbf{A})\mathbf{p}_i = \mathbf{0} \qquad (10\text{-}155)$$

where λ_i, $i = 1, 2, \ldots, n$, denotes the ith eigenvalue of **A**, called the eigenvector *of* **A** *associated with the eigenvalue* λ_i. If **A** has distinct eigenvalues, the eigenvectors can be solved directly from Eq. (10-155).

▶ **EXAMPLE 10-8-5** Consider that the state equation of Eq. (10-44) has the coefficient matrices

$$\mathbf{A} = \begin{bmatrix} 1 & -1 \\ 0 & -1 \end{bmatrix} \quad \mathbf{B} = \begin{bmatrix} 1 \\ 1 \end{bmatrix} \quad \mathbf{E} = \mathbf{0} \qquad (10\text{-}156)$$

The characteristic equation of **A** is

$$|s\mathbf{I} - \mathbf{A}| = s^2 - 1 \qquad (10\text{-}157)$$

The eigenvalues are $\lambda_1 = 1$ and $\lambda_2 = -1$. Let the eigenvectors be written as

$$\mathbf{p}_1 = \begin{bmatrix} p_{11} \\ p_{21} \end{bmatrix} \quad \mathbf{p}_2 = \begin{bmatrix} p_{12} \\ p_{22} \end{bmatrix} \qquad (10\text{-}158)$$

Substituting $\lambda_1 = 1$ and $\mathbf{p}_1$ into Eq. (10-155), we get

$$\begin{bmatrix} 0 & 1 \\ 0 & 2 \end{bmatrix} \begin{bmatrix} p_{11} \\ p_{21} \end{bmatrix} = \begin{bmatrix} 0 \\ 0 \end{bmatrix} \qquad (10\text{-}159)$$

Thus, $p_{21} = 0$, and p_{11} is arbitrary, which in this case can be set equal to 1.
Similarly, for $\lambda_2 = -1$, Eq. (10-155) becomes

$$\begin{bmatrix} -2 & 1 \\ 0 & 0 \end{bmatrix} \begin{bmatrix} p_{12} \\ p_{22} \end{bmatrix} = \begin{bmatrix} 0 \\ 0 \end{bmatrix} \quad (10\text{-}160)$$

which leads to

$$-2p_{12} + p_{22} = 0 \quad (10\text{-}161)$$

The last equation has two unknowns, which means that one can be set arbitrarily. Let $p_{12} = 1$, then $p_{22} = 2$. The eigenvectors are

$$\mathbf{p}_1 = \begin{bmatrix} 1 \\ 0 \end{bmatrix} \quad \mathbf{p}_2 = \begin{bmatrix} 1 \\ 2 \end{bmatrix} \quad (10\text{-}162)$$

◀

10-8-6 Generalized Eigenvectors

It should be pointed out that if **A** has multiple-order eigenvalues and is nonsymmetric, not all the **eigenvectors** can be found using Eq. (10-155). Let us assume that there are $q(<n)$ distinct eigenvalues among the n eigenvalues of **A**. The eigenvectors that correspond to the q distinct eigenvalues can be determined in the usual manner from

$$(\lambda_i \mathbf{I} - \mathbf{A})\mathbf{p}_i = 0 \quad (10\text{-}163)$$

where λ_i denotes the ith distinct eigenvalue, $\mathbf{i} = 1, 2, \ldots, q$. Among the remaining high-order eigenvalues, let λ_j be of the mth order ($m \leq n - q$). The corresponding eigenvectors are called the **generalized eigenvectors** and can be determined from the following m vector equations:

$$(\lambda_j \mathbf{I} - \mathbf{A})\mathbf{p}_{n-q+1} = \mathbf{0}$$
$$(\lambda_j \mathbf{I} - \mathbf{A})\mathbf{p}_{n-q+2} = -\mathbf{p}_{n-q+1}$$
$$(\lambda_j \mathbf{I} - \mathbf{A})\mathbf{p}_{n-q+3} = -\mathbf{p}_{n-q+2} \quad (10\text{-}164)$$
$$\vdots$$
$$(\lambda_j \mathbf{I} - \mathbf{A})\mathbf{p}_{n-q+m} = -\mathbf{p}_{n-q+m-1}$$

▶ **EXAMPLE 10-8-6** Given the matrix

$$\mathbf{A} = \begin{bmatrix} 0 & 6 & -5 \\ 1 & 0 & 2 \\ 3 & 2 & 4 \end{bmatrix} \quad (10\text{-}165)$$

The eigenvalues of **A** are $\lambda_1 = 2$, $\lambda_2 = \lambda_3 = 1$. Thus, **A** is a second-order eigenvalue at 1. The eigenvector that is associated with $\lambda_1 = 2$ is determined using Eq. (10-163). Thus,

$$(\lambda_1 \mathbf{I} - \mathbf{A})\mathbf{p}_1 = \begin{bmatrix} 2 & -6 & 5 \\ -1 & 2 & -2 \\ -3 & -2 & -2 \end{bmatrix} \begin{bmatrix} p_{11} \\ p_{21} \\ p_{31} \end{bmatrix} = 0 \quad (10\text{-}166)$$

Because there are only two independent equations in Eq. (10-166), we arbitrarily set $p_{11} = 2$, and we have $p_{21} = -1$ and $p_{31} = -2$. Thus,

$$\mathbf{p}_1 = \begin{bmatrix} 2 \\ -1 \\ -2 \end{bmatrix} \quad (10\text{-}167)$$

For the generalized eigenvectors that are associated with the second-order eigenvalues, we substitute $\lambda_2 = 1$ into the first equation of Eq. (10-164). We have

$$(\lambda_2 \mathbf{I} - \mathbf{A})\mathbf{p}_2 = \begin{bmatrix} 1 & -6 & 5 \\ -1 & 1 & -2 \\ -3 & -2 & -3 \end{bmatrix} \begin{bmatrix} p_{12} \\ p_{22} \\ p_{32} \end{bmatrix} = \mathbf{0} \quad (10\text{-}168)$$

Setting $p_{12} = 1$ arbitrarily, we have $p_{22} = -\frac{3}{7}$ and $p_{32} = -\frac{5}{7}$. Thus,

$$\mathbf{p}_2 = \begin{bmatrix} 1 \\ -\frac{3}{7} \\ -\frac{5}{7} \end{bmatrix} \quad (10\text{-}169)$$

Substituting $\lambda_3 = 1$ into the second equation of Eq. (10-164), we have

$$(\lambda_3 \mathbf{I} - \mathbf{A})\mathbf{p}_3 = \begin{bmatrix} 1 & -6 & -5 \\ -1 & 1 & -2 \\ -3 & -2 & -3 \end{bmatrix} \begin{bmatrix} p_{13} \\ p_{23} \\ p_{33} \end{bmatrix} = -\mathbf{p}_2 = \begin{bmatrix} -1 \\ \frac{3}{7} \\ \frac{5}{7} \end{bmatrix} \quad (10\text{-}170)$$

Setting p_{13} arbitrarily to 1, we have the generalized eigenvector

$$\mathbf{p}_3 = \begin{bmatrix} 1 \\ -\frac{22}{49} \\ -\frac{46}{49} \end{bmatrix} \quad (10\text{-}171)$$

◀

10-9 SIMILARITY TRANSFORMATION

The dynamic equations of a single-input, single-output (SISO) system are

$$\frac{d\mathbf{x}(t)}{dt} = \mathbf{A}\mathbf{x}(t) + \mathbf{B}u(t) \quad (10\text{-}172)$$

$$y(t) = \mathbf{C}\mathbf{x}(t) + \mathbf{D}u(t) \quad (10\text{-}173)$$

where $\mathbf{x}(t)$ is the $n \times 1$ state vector, and $u(t)$ and $y(t)$ are the scalar input and output, respectively. When carrying out analysis and design in the state domain, it is often advantageous to transform these equations into particular forms. For example, as we will show later, the controllability canonical form (CCF) has many interesting properties that make it convenient for controllability tests and state-feedback design.

Let us consider that the dynamic equations of Eqs. (10-172) and (10-173) are transformed into another set of equations of the same dimension by the following transformation:

$$\mathbf{x}(t) = \mathbf{P}\overline{\mathbf{x}}(t) \quad (10\text{-}174)$$

Chapter 10. State Variable Analysis

where $\mathbf{P}$ is an $n \times n$ nonsingular matrix, so

$$\overline{\mathbf{x}}(t) = \mathbf{P}^{-1}\mathbf{x}(t) \tag{10-175}$$

The transformed dynamic equations are written

$$\frac{d\overline{\mathbf{x}}(t)}{dt} = \overline{\mathbf{A}}\overline{\mathbf{x}}(t) + \overline{\mathbf{B}}u(t) \tag{10-176}$$

$$\overline{y}(t) = \overline{\mathbf{C}}\overline{\mathbf{x}}(t) + \overline{\mathbf{D}}u(t) \tag{10-177}$$

Taking the derivative on both sides of Eq. (10-175) with respect to t, we have

$$\frac{d\overline{\mathbf{x}}(t)}{dt} = \mathbf{P}^{-1}\frac{d\mathbf{x}(t)}{dt} = \mathbf{P}^{-1}\mathbf{A}\mathbf{x}(t) + \mathbf{P}^{-1}\mathbf{B}u(t)$$
$$= \mathbf{P}^{-1}\mathbf{A}\mathbf{P}\overline{\mathbf{x}}(t) + \mathbf{P}^{-1}\mathbf{B}u(t) \tag{10-178}$$

Comparing Eq. (10-178) with Eq. (10-176), we get

$$\overline{\mathbf{A}} = \mathbf{P}^{-1}\mathbf{A}\mathbf{P} \tag{10-179}$$

and

$$\overline{\mathbf{B}} = \mathbf{P}^{-1}\mathbf{B} \tag{10-180}$$

Using Eq. (10-174), Eq. (10-177) is written

$$\overline{y}(t) = \mathbf{C}\mathbf{P}\mathbf{x}(t) + \overline{\mathbf{D}}u(t) \tag{10-181}$$

Comparing Eq. (10-181) with Eq. (10-173), we see that

$$\overline{\mathbf{C}} = \mathbf{C}\mathbf{P} \quad \overline{\mathbf{D}} = \mathbf{D} \tag{10-182}$$

The transformation just described is called a **similarity transformation**, because in the transformed system such properties as the characteristic equation, eigenvectors, eigenvalues, and transfer function are all preserved by the transformation. We shall describe the controllability canonical form (CCF), the observability canonical form (OCF), and the diagonal canonical form (DCF) transformations in the following sections. The transformation equations are given without proofs.

10-9-1 Invariance Properties of the Similarity Transformations

One of the important properties of the similarity transformations is that the characteristic equation, eigenvalues, eigenvectors, and transfer functions are invariant under the transformations.

Characteristic Equations, Eigenvalues, and Eigenvectors
The characteristic equation of the system described by Eq. (10-176) is $|s\mathbf{I} - \overline{\mathbf{A}}| = 0$ and is written

$$|s\mathbf{I} - \overline{\mathbf{A}}| = |s\mathbf{I} - \mathbf{P}^{-1}\mathbf{A}\mathbf{P}| = |s\mathbf{P}^{-1}\mathbf{P} - \mathbf{P}^{-1}\mathbf{A}\mathbf{P}| \tag{10-183}$$

Because the determinant of a product matrix is equal to the product of the determinants of the matrices, the last equation becomes

$$|s\mathbf{I} - \overline{\mathbf{A}}| = |\mathbf{P}^{-1}||s\mathbf{I} - \mathbf{A}||\mathbf{P}| = |s\mathbf{I} - \mathbf{A}| \qquad (10\text{-}184)$$

Thus, the characteristic equation is preserved, which naturally leads to the same eigenvalues and eigenvectors.

Transfer-Function Matrix
From Eq. (10-126), the transfer-function matrix of the system of Eqs. (10-176) and (10-177) is

$$\begin{aligned}\overline{\mathbf{G}}(s) &= \overline{\mathbf{C}}(s\mathbf{I} - \overline{\mathbf{A}})\overline{\mathbf{B}} + \overline{\mathbf{D}} \\ &= \mathbf{CP}(s\mathbf{I} - \mathbf{P}^{-1}\mathbf{AP})\mathbf{P}^{-1}\mathbf{B} + \mathbf{D}\end{aligned} \qquad (10\text{-}185)$$

which is simplified to

$$\overline{\mathbf{G}}(s) = \mathbf{C}(s\mathbf{I} - \mathbf{A})\mathbf{B} + \mathbf{D} = \mathbf{G}(s) \qquad (10\text{-}186)$$

10-9-2 Controllability Canonical Form (CCF)

Consider the dynamic equations given in Eqs. (10-172) and (10-173). The characteristic equation of $\mathbf{A}$ is

$$|s\mathbf{I} - \mathbf{A}| = s^n + a_{n-1}s^{n-1} + \cdots + a_1 s + a_0 = 0 \qquad (10\text{-}187)$$

The dynamic equations in Eqs. (10-172) and (10-173) are transformed into CCF of the form of Eqs. (10-176) and (10-177) by the transformation of Eq. (10-174), with

$$\mathbf{P} = \mathbf{SM} \qquad (10\text{-}188)$$

where

$$\mathbf{S} = \begin{bmatrix} \mathbf{B} & \mathbf{AB} & \mathbf{A}^2\mathbf{B} \ldots \mathbf{A}^{n-1}\mathbf{B} \end{bmatrix} \qquad (10\text{-}189)$$

and

$$\mathbf{M} = \begin{bmatrix} a_1 & a_2 & \cdots & a_{n-1} & 1 \\ a_2 & a_3 & \cdots & 1 & 0 \\ \vdots & \vdots & \ddots & \vdots & \vdots \\ a_{n-1} & 1 & \cdots & 0 & 0 \\ 1 & 0 & \cdots & 0 & 0 \end{bmatrix} \qquad (10\text{-}190)$$

Then,

$$\bar{\mathbf{A}} = \mathbf{P}^{-1}\mathbf{A}\mathbf{P} = \begin{bmatrix} 0 & 1 & 0 & \cdots & 0 \\ 0 & 0 & 1 & \cdots & 0 \\ \vdots & \vdots & \vdots & \ddots & \vdots \\ 0 & 0 & 0 & \cdots & 1 \\ -a_0 & -a_1 & -a_2 & \cdots & -a_{n-1} \end{bmatrix} \quad (10\text{-}191)$$

$$\bar{\mathbf{B}} = \mathbf{P}^{-1}\mathbf{B} = \begin{bmatrix} 0 \\ 0 \\ \vdots \\ 0 \\ 1 \end{bmatrix} \quad (10\text{-}192)$$

The matrices $\bar{\mathbf{C}}$ and $\bar{\mathbf{D}}$ are given by Eq. (10-182) and do not follow any particular pattern. The CCF transformation requires that $\mathbf{P}^{-1}$ exists, which implies that the matrix $\mathbf{S}$ must have an inverse, because the inverse of $\mathbf{M}$ always exists because its determinant is $(-1)^{n-1}$, which is nonzero. The $n \times n$ matrix $\mathbf{S}$ in Eq. (10-189) is later defined as the **controllability matrix**.

▶ **EXAMPLE 10-9-1** Consider the coefficient matrices of the state equations in Eq. (10-172):

$$\mathbf{A} = \begin{bmatrix} 1 & 2 & 1 \\ 0 & 1 & 3 \\ 1 & 1 & 1 \end{bmatrix} \quad \mathbf{B} = \begin{bmatrix} 1 \\ 0 \\ 1 \end{bmatrix} \quad (10\text{-}193)$$

The state equations are to be transformed to CCF.

The characteristic equation of $\mathbf{A}$ is

$$|s\mathbf{I} - \mathbf{A}| = \begin{vmatrix} s-1 & -2 & -1 \\ 0 & s-1 & -3 \\ -1 & -1 & s-1 \end{vmatrix} = s^3 - 3s^2 - s - 3 = 0 \quad (10\text{-}194)$$

Thus, the coefficients of the characteristic equation are identified as $a_0 = -3$, $a_1 = -1$, and $a_2 = -3$. From Eq. (10-190),

$$\mathbf{M} = \begin{bmatrix} a_1 & a_2 & 1 \\ a_2 & 1 & 0 \\ 1 & 0 & 0 \end{bmatrix} = \begin{bmatrix} -1 & -3 & 1 \\ -3 & 1 & 0 \\ 1 & 0 & 0 \end{bmatrix} \quad (10\text{-}195)$$

The controllability matrix is

$$\mathbf{S} = \begin{bmatrix} \mathbf{B} & \mathbf{AB} & \mathbf{A}^2\mathbf{B} \end{bmatrix} = \begin{bmatrix} 1 & 2 & 10 \\ 0 & 3 & 9 \\ 1 & 2 & 7 \end{bmatrix} \quad (10\text{-}196)$$

We can show that $\mathbf{S}$ is nonsingular, so the system can be transformed into the CCF. Substituting $\mathbf{S}$ and $\mathbf{M}$ into Eq. (10-188), we get

$$\mathbf{P} = \mathbf{SM} = \begin{bmatrix} 3 & -1 & 1 \\ 0 & 3 & 0 \\ 0 & -1 & 1 \end{bmatrix} \quad (10\text{-}197)$$

Thus, from Eqs. (10-191) and (10-192), the CCF model is given by

$$\bar{\mathbf{A}} = \mathbf{P}^{-1}\mathbf{A}\mathbf{P} = \begin{bmatrix} 0 & 1 & 0 \\ 0 & 0 & 1 \\ 3 & 1 & 3 \end{bmatrix} \quad \bar{\mathbf{B}} = \mathbf{P}^{-1}\mathbf{B} = \begin{bmatrix} 0 \\ 0 \\ 1 \end{bmatrix} \quad (10\text{-}198)$$

which could have been determined once the coefficients of the characteristic equation are known; however, the exercise is to show how the CCF transformation matrix **P** is obtained. ◀

10-9-3 Observability Canonical Form (OCF)

A dual form of transformation of the CCF is the **observability canonical form** (OCF). The system described by Eqs. (10-172) and (10-173) is transformed to the OCF by the transformation

$$\mathbf{x}(t) = \mathbf{Q}\bar{\mathbf{x}}(t) \qquad (10\text{-}199)$$

The transformed equations are as given in Eqs. (10-176) and (10-177). Thus,

$$\bar{\mathbf{A}} = \mathbf{Q}^{-1}\mathbf{A}\mathbf{Q} \quad \bar{\mathbf{B}} = \mathbf{Q}^{-1}\mathbf{B} \quad \bar{\mathbf{C}} = \mathbf{C}\mathbf{Q} \quad \bar{\mathbf{D}} = \mathbf{D} \qquad (10\text{-}200)$$

where

$$\bar{\mathbf{A}} = \mathbf{Q}^{-1}\mathbf{A}\mathbf{Q} = \begin{bmatrix} 0 & 0 & \cdots & 0 & -a_0 \\ 1 & 0 & \cdots & 0 & -a_1 \\ 0 & 1 & \cdots & 0 & -a_2 \\ \vdots & \vdots & \ddots & \vdots & \vdots \\ 0 & 0 & \cdots & 1 & -a_{n-1} \end{bmatrix} \qquad (10\text{-}201)$$

$$\bar{\mathbf{C}} = \mathbf{C}\mathbf{Q} = \begin{bmatrix} 0 & 0 & \cdots & 0 & 1 \end{bmatrix} \qquad (10\text{-}202)$$

The elements of the matrices $\bar{\mathbf{B}}$ and $\bar{\mathbf{D}}$ are not restricted to any form. Notice that $\bar{\mathbf{A}}$ and $\bar{\mathbf{C}}$ are the transpose of the $\bar{\mathbf{A}}$ and $\bar{\mathbf{B}}$ in Eqs. (10-191) and (10-192), respectively.

The OCF transformation matrix **Q** is given by

$$\mathbf{Q} = (\mathbf{M}\mathbf{V})^{-1} \qquad (10\text{-}203)$$

where **M** is as given in Eq. (10-190), and

$$\mathbf{V} = \begin{bmatrix} \mathbf{C} \\ \mathbf{C}\mathbf{A} \\ \mathbf{C}\mathbf{A}^2 \\ \vdots \\ \mathbf{C}\mathbf{A}^{n-1} \end{bmatrix} \quad (n \times n) \qquad (10\text{-}204)$$

The matrix **V** is often defined as the **observability matrix**, and $\mathbf{V}^{-1}$ must exist in order for the OCF transformation to be possible.

▶ **EXAMPLE 10-9-2** Consider that the coefficient matrices of the system described by Eqs. (10-172) and (10-138) are

$$\mathbf{A} = \begin{bmatrix} 1 & 2 & 1 \\ 0 & 1 & 3 \\ 1 & 1 & 1 \end{bmatrix} \quad \mathbf{B} = \begin{bmatrix} 1 \\ 0 \\ 1 \end{bmatrix} \quad \mathbf{C} = \begin{bmatrix} 1 & 1 & 0 \end{bmatrix} \quad \mathbf{D} = \mathbf{0} \qquad (10\text{-}205)$$

Because the matrix $\mathbf{A}$ is identical to that of the system in Example 10-9-1, the matrix $\mathbf{M}$ is the same as that in Eq. (10-195). The observability matrix is

$$\mathbf{V} = \begin{bmatrix} \mathbf{C} \\ \mathbf{CA} \\ \mathbf{CA}^2 \end{bmatrix} = \begin{bmatrix} 1 & 1 & 0 \\ 1 & 3 & 4 \\ 5 & 9 & 14 \end{bmatrix} \quad (10\text{-}206)$$

We can show that $\mathbf{V}$ is nonsingular, so the system can be transformed into the OCF. Substituting $\mathbf{V}$ and $\mathbf{M}$ into Eq. (10-203), we have the OCF transformation matrix,

$$\mathbf{Q} = (\mathbf{MV})^{-1} = \begin{bmatrix} 0.3333 & -0.1667 & 0.3333 \\ -0.3333 & 0.1667 & 0.6667 \\ 0.1667 & 0.1667 & 0.1667 \end{bmatrix} \quad (10\text{-}207)$$

From Eq. (10-191), the OCF model of the system is described by

$$\overline{\mathbf{A}} = \mathbf{Q}^{-1}\mathbf{AQ} = \begin{bmatrix} 0 & 0 & 3 \\ 1 & 0 & 1 \\ 0 & 1 & 3 \end{bmatrix} \quad \overline{\mathbf{C}} = \mathbf{CQ} = \begin{bmatrix} 0 & 0 & 1 \end{bmatrix} \quad \overline{\mathbf{B}} = \mathbf{Q}^{-1}\mathbf{B} = \begin{bmatrix} 3 \\ 2 \\ 1 \end{bmatrix} \quad (10\text{-}208)$$

Thus, $\overline{\mathbf{A}}$ and $\overline{\mathbf{C}}$ are of the OCF form given in Eqs. (10-201) and (10-202), respectively, and $\overline{\mathbf{B}}$ does not conform to any particular form. ◀

10-9-4 Diagonal Canonical Form (DCF)

Given the dynamic equations in Eqs. (10-172) and (10-173), if $\mathbf{A}$ has distinct eigenvalues, there is a nonsingular transformation

$$\mathbf{x}(t) = \mathbf{T}\overline{\mathbf{x}}(t) \quad (10\text{-}209)$$

which transforms these equations to the dynamic equations of Eqs. (10-176) and (10-177), where

$$\overline{\mathbf{A}} = \mathbf{T}^{-1}\mathbf{AT} \quad \overline{\mathbf{B}} = \mathbf{T}^{-1}\mathbf{B} \quad \overline{\mathbf{C}} = \mathbf{CT} \quad \overline{\mathbf{D}} = \mathbf{D} \quad (10\text{-}210)$$

The matrix $\overline{\mathbf{A}}$ is a diagonal matrix,

$$\overline{\mathbf{A}} = \begin{bmatrix} \lambda_1 & 0 & 0 & \cdots & 0 \\ 0 & \lambda_2 & 0 & \cdots & 0 \\ 0 & 0 & \lambda_3 & \cdots & 0 \\ \vdots & \vdots & \vdots & \ddots & \vdots \\ 0 & 0 & 0 & \cdots & \lambda_n \end{bmatrix} \quad (n \times n) \quad (10\text{-}211)$$

where $\lambda_1, \lambda_2, \ldots, \lambda_n$ are the n distinct eigenvalues of $\mathbf{A}$. The coefficient matrices $\overline{\mathbf{B}}, \overline{\mathbf{C}},$ and $\overline{\mathbf{D}}$ are given in Eq. (10-210) and do not follow any particular form.

It is apparent that one of the advantages of the DCF is that the transformed state equations are *decoupled* from each other and, therefore, can be solved individually.

We show in the following that the DCF transformation matrix $\mathbf{T}$ can be formed by use of the eigenvectors of $\mathbf{A}$ as its columns; that is,

$$\mathbf{T} = \begin{bmatrix} \mathbf{p}_1 & \mathbf{p}_2 & \mathbf{p}_3 & \cdots & \mathbf{p}_n \end{bmatrix} \quad (10\text{-}212)$$

where $\mathbf{p}_i$, $i = 1, 2, \ldots, n$, denotes the eigenvector associated with the eigenvalue λ_i. This is proved by use of Eq. (10-155), which is written as

$$\lambda_i \mathbf{p}_i = \mathbf{A}\mathbf{p}_i \quad i = 1, 2, \ldots, n \tag{10-213}$$

Now, forming the $n \times n$ matrix,

$$[\lambda_1 \mathbf{p}_1 \quad \lambda_2 \mathbf{p}_2 \quad \cdots \quad \lambda_n \mathbf{p}_n] = [\mathbf{A}\mathbf{p}_1 \quad \mathbf{A}\mathbf{p}_2 \quad \cdots \quad \mathbf{A}\mathbf{p}_n]$$
$$= \mathbf{A}[\mathbf{p}_1 \quad \mathbf{p}_2 \quad \cdots \quad \mathbf{p}_n] \tag{10-214}$$

The last equation is written

$$[\mathbf{p}_1 \quad \mathbf{p}_2 \quad \cdots \quad \mathbf{p}_n]\overline{\mathbf{A}} = \mathbf{A}[\mathbf{p}_1 \quad \mathbf{p}_2 \quad \cdots \quad \mathbf{p}_n] \tag{10-215}$$

where $\overline{\mathbf{A}}$ is as given in Eq. (10-211). Thus, if we let

$$\mathbf{T} = [\mathbf{p}_1 \quad \mathbf{p}_2 \quad) \quad \mathbf{p}_n] \tag{10-216}$$

Eq. (10-215) is written

$$\overline{\mathbf{A}} = \mathbf{T}^{21}\mathbf{A}\mathbf{T} \tag{10-217}$$

If the matrix $\mathbf{A}$ is of the CCF and $\mathbf{A}$ has distinct eigenvalues, then the DCF transformation matrix is the Vandermonde matrix,

$$\mathbf{T} = \begin{bmatrix} 1 & 1 & 1 & \cdots & 1 \\ \lambda_1 & \lambda_2 & \lambda_3 & \cdots & \lambda_n \\ \lambda_1^2 & \lambda_2^2 & \lambda_3^2 & \cdots & \lambda_n^2 \\ \vdots & \vdots & \vdots & \ddots & \vdots \\ \lambda_1^{n-1} & \lambda_2^{n-1} & \lambda_3^{n-1} & \cdots & \lambda_n^{n-1} \end{bmatrix} \tag{10-218}$$

where $\lambda_1, \lambda_2, \ldots, \lambda_n$ are the eigenvalues of $\mathbf{A}$. This can be proven by substituting the CCF of $\mathbf{A}$ in Eq. (10-110) into Eq. (10-155). The result is that the ith eigenvector $\mathbf{p}_i$ is equal to the ith column of $\mathbf{T}$ in Eq. (10-218).

▶ **EXAMPLE 10.9.3** Consider the matrix

$$\mathbf{A} = \begin{bmatrix} 0 & 1 & 0 \\ 0 & 0 & 1 \\ -6 & -11 & -6 \end{bmatrix} \tag{10-219}$$

which has eigenvalues $\lambda_1 = -1$, $\lambda_2 = -2$, and $\lambda_3 = -3$. Because $\mathbf{A}$ is CCF, to transform it into DCF, the transformation matrix can be the Vandermonde matrix in Eq. (10-218). Thus,

$$\mathbf{T} = \begin{bmatrix} 1 & 1 & 1 \\ \lambda_1 & \lambda_2 & \lambda_3 \\ \lambda_1^2 & \lambda_2^2 & \lambda_3^2 \end{bmatrix} = \begin{bmatrix} 1 & 1 & 1 \\ -1 & -2 & -3 \\ 1 & 4 & 9 \end{bmatrix} \tag{10-220}$$

Thus, the DCF of $\mathbf{A}$ is written

$$\overline{\mathbf{A}} = \mathbf{T}^{-1}\mathbf{A}\mathbf{T} = \begin{bmatrix} -1 & 0 & 0 \\ 0 & -2 & 0 \\ 0 & 0 & -3 \end{bmatrix} \tag{10-221}$$

◀

10-9-5 Jordan Canonical Form (JCF)

In general, when the matrix $\mathbf{A}$ has multiple-order eigenvalues, unless the matrix is symmetric with real elements, it cannot be transformed into a diagonal matrix. However, there exists a similarity transformation in the form of Eq. (10-217) such that the matrix $\overline{\mathbf{A}}$ is almost diagonal. The matrix $\overline{\mathbf{A}}$ is called the **Jordan canonical form (JCF)**. A typical JCF is shown below.

$$\overline{\mathbf{A}} = \begin{bmatrix} \lambda_1 & 1 & 0 & 0 & 0 \\ 0 & \lambda_1 & 1 & 0 & 0 \\ 0 & 0 & \lambda_1 & 0 & 0 \\ 0 & 0 & 0 & \lambda_2 & 0 \\ 0 & 0 & 0 & 0 & \lambda_3 \end{bmatrix} \quad (10\text{-}222)$$

where it is assumed that $\mathbf{A}$ has a third-order eigenvalue λ_1 and distinct eigenvalues λ_2 and λ_3.

The JCF generally has the following properties:

1. The elements on the main diagonal are the eigenvalues.
2. All the elements below the main diagonal are zero.
3. Some of the elements immediately above the multiple-order eigenvalues on the main diagonal are 1s, as shown in Eq. (10-222).
4. The 1s together with the eigenvalues form typical blocks called the **Jordan blocks**. As shown in Eq. (10-222), the Jordan blocks are enclosed by dashed lines.
5. When the nonsymmetrical matrix $\mathbf{A}$ has multiple-order eigenvalues, its eigenvectors are not linearly independent. For an $\mathbf{A}$ that is $n \times n$, there are only r (where r is an integer that is less than n and is dependent on the number of multiple-order eigenvalues) linearly independent eigenvectors.
6. The number of Jordan blocks is equal to the number of independent eigenvectors r. There is one and only one linearly independent eigenvector associated with each Jordan block.
7. The number of 1s above the main diagonal is equal to $n - r$.

To perform the JCF transformation, the transformation matrix $\mathbf{T}$ is again formed by using the eigenvectors and generalized eigenvectors as its columns.

▶ **EXAMPLE 10-9-4** Consider the matrix given in Eq. (10-165). We have shown that the matrix has eigenvalues 2, 1, and 1. Thus, the DCF transformation matrix can be formed by using the eigenvector and generalized eigenvector given in Eqs. (10-167), (10-169), and (10-171), respectively. That is,

$$\mathbf{T} = [\mathbf{p}_1 \ \mathbf{p}_2 \ \mathbf{p}_3] = \begin{bmatrix} 2 & 1 & 1 \\ -1 & -\frac{3}{7} & \frac{22}{49} \\ -2 & \frac{5}{7} & -\frac{46}{49} \end{bmatrix} \quad (10\text{-}223)$$

Thus, the DCF is

$$\overline{\mathbf{A}} = \mathbf{T}^{-1}\mathbf{A}\mathbf{T} = \begin{bmatrix} 2 & 0 & 0 \\ 0 & 1 & 1 \\ 0 & 0 & 1 \end{bmatrix} \quad (10\text{-}224)$$

Note that in this case there are two Jordan blocks, and there is one element of 1 above the main diagonal. ◀

10-10 DECOMPOSITIONS OF TRANSFER FUNCTIONS

Up to this point, various methods of characterizing linear systems have been presented. To summarize, it has been shown that the starting point of modeling a linear system may be the system's differential equation, transfer function, or dynamic equations; all these methods are closely related. Furthermore, the state diagram is also a useful tool that can not only lead to the solutions of state equations but also serve as a vehicle of transformation from one form of description to the others. The block diagram of Fig. 10-13 shows the relationships among the various ways of describing a linear system. For example, the block diagram shows that, starting with the differential equation of a system, one can find the solution by the transfer-function or state-equation method. The block diagram also shows that the majority of the relationships are bilateral, so a great deal of flexibility exists between the methods.

One subject remains to be discussed, which involves the construction of the state diagram from the transfer function between the input and the output. The process of going from the transfer function to the state diagram is called **decomposition**. In general, there are three basic ways to decompose transfer functions. These are **direct decomposition**, **cascade decomposition**, and **parallel decomposition**. Each of these three schemes of decomposition has its own merits and is best suited for a particular purpose.

10-10-1 Direct Decomposition

Direct decomposition is applied to an input–output transfer function that is not in factored form. Consider the transfer function of an nth-order SISO system between the input $U(s)$

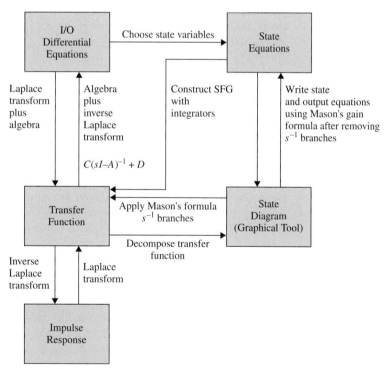

Figure 10-13 Block diagram showing the relationships among various methods of describing linear systems.

and output $Y(s)$:

$$\frac{Y(s)}{U(s)} = \frac{b_{n-1}s^{n-1} + b_{b-2}s^{n-2} + \cdots + b_1 s + b_0}{s^n + a_{n-1}s^{n-1} + \cdots + a_1 s + a_0} \quad (10\text{-}225)$$

where we have assumed that the order of the denominator is at least one degree higher than that of the numerator.

We next show that the direct decomposition can be conducted in at least two ways, one leading to a state diagram that corresponds to the CCF and the other to the OCF.

Direct Decomposition to CCF

The objective is to construct a state diagram from the transfer function of Eq. (10-225). The following steps are outlined:

1. Express the transfer function in negative powers of s. This is done by multiplying the numerator and the denominator of the transfer function by s^{-n}.

2. Multiply the numerator and the denominator of the transfer function by a dummy variable $X(s)$. By implementing the last two steps, Eq. (10-225) becomes

$$\frac{Y(s)}{U(s)} = \frac{b_{n-1}s^{-1} + b_{n-2}s^{-2} + \cdots + b_1 s^{-n+1} + b_0 s^{-n}}{1 + a_{n-1}s^{-1} + \cdots + a_1 s^{-n+1} + a_0 s^{-n}} \frac{X(s)}{X(s)} \quad (10\text{-}226)$$

3. The numerators and the denominators on both sides of Eq. (10-226) are equated to each other, respectively. The results are:

$$Y(s) = (b_{n-1}s^{-1} + b_{n-2}s^{-2} + \cdots + b_1 s^{-n+1} + b_0 s^{-n})X(s) \quad (10\text{-}227)$$

$$U(s) = (1 + a_{n-1}s^{-1} + \cdots + a_1 s^{-n+1} + a_0 s^{-n})X(s) \quad (10\text{-}228)$$

4. To construct a state diagram using the two equations in Eqs. (10-227) and (10–228), they must first be in the proper cause-and-effect relation. It is apparent that Eq. (10-227) already satisfies this prerequisite. However, Eq. (10-228) has the input on the left-hand side of the equation and must be rearranged. Eq. (10-228) is rearranged as

$$X(s) = U(s) - (a_{n-1}s^{-1} + a_{n-2}s^{-2} + \cdots + a_1 s^{-n+1} + a_0 s^{-n})X(s) \quad (10\text{-}229)$$

The state diagram is drawn as shown in Fig. 10-14 using Eqs. (10-227) and (10-228). For simplicity, the initial states are not drawn on the diagram. The state variables $x_1(t)$, $x_2(t)$, ..., $x_n(t)$ are defined as the outputs of the integrators and are arranged in order from the right to the left on the state diagram. The state equations are obtained by applying the SFG gain formula to Fig. 10-14 with the derivatives of the state variables as the outputs and the state variables and $u(t)$ as the inputs, and overlooking the integrator branches. The

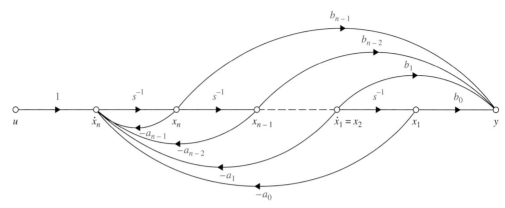

Figure 10-14 CCF state diagram of the transfer function in Eq. (10-225) by direct decomposition.

output equation is determined by applying the gain formula among the state variables, the input, and the output $y(t)$. The dynamic equations are written

$$\frac{d\mathbf{x}(t)}{dt} = \mathbf{A}\mathbf{x}(t) + \mathbf{B}u(t) \tag{10-230}$$

$$y(t) = \mathbf{C}\mathbf{x}(t) + Du(t) \tag{10-231}$$

where

$$\mathbf{A} = \begin{bmatrix} 0 & 1 & 0 & \cdots & 0 \\ 0 & 0 & 1 & \cdots & 0 \\ \vdots & \vdots & \vdots & \ddots & \vdots \\ 0 & 0 & 0 & 0 & 1 \\ -a_0 & -a_1 & -a_2 & \cdots & -a_{n-1} \end{bmatrix} \quad \mathbf{B} = \begin{bmatrix} 0 \\ 0 \\ \vdots \\ 0 \\ 1 \end{bmatrix} \tag{10-232}$$

$$\mathbf{C} = \begin{bmatrix} b_0 & b_1 & \cdots & b_{n-2} & b_{n-1} \end{bmatrix} \quad D = 0 \tag{10-233}$$

Apparently, $\mathbf{A}$ and $\mathbf{B}$ in Eq. (10-232) are of the CCF.

Direct Decomposition to OCF

Multiplying the numerator and the denominator of Eq. (10-225) by s^{-n}, the equation is expanded as

$$(1 + a_{n-1}s^{-1} + \cdots + a_1 s^{-n+1} + a_0 s^{-n})Y(s)$$
$$= (b_{n-1}s^{-1} + b_{n-2}s^{-2} + \cdots + b_1 s^{-n+1} + b_0 s^{-n})U(s) \tag{10-234}$$

or

$$Y(s) = -(a_{n-1}s^{-1} + \cdots + a_1 s^{-n+1} + a_0 s^{-n})Y(s)$$
$$+ (b_{n-1}s^{-1} + b_{n-2}s^{-2} + \cdots + b_1 s^{-n+1} + b_0 s^{-n})U(s) \tag{10-235}$$

Fig. 10-15 shows the state diagram that results from using Eq. (10-235). The outputs of the integrators are designated as the state variables. However, unlike the usual convention, the state variables are assigned in *descending order* from right to left. Applying the SFG gain

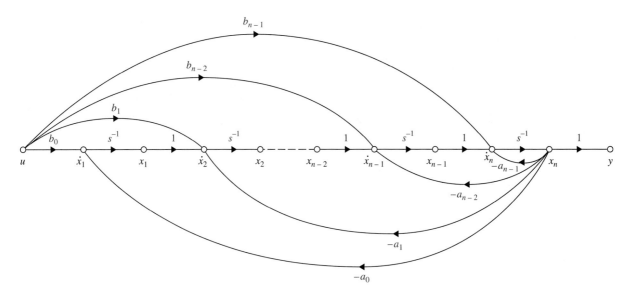

Figure 10-15 CCF state diagram of the transfer function in Eq. (10-225) by direct decomposition.

formula to the state diagram, the dynamic equations are written as in Eqs. (10-230) and (10-231), with

$$\mathbf{A} = \begin{bmatrix} 0 & 0 & \cdots & 0 & -a_0 \\ 1 & 0 & \cdots & 0 & -a_1 \\ 0 & 1 & \cdots & 0 & -a_2 \\ \vdots & \vdots & \ddots & \vdots & \vdots \\ 0 & 0 & \cdots & 1 & -a_{n-1} \end{bmatrix} \quad \mathbf{B} = \begin{bmatrix} b_0 \\ b_1 \\ b_2 \\ \vdots \\ b_{n-1} \end{bmatrix} \quad (10\text{-}236)$$

and

$$\mathbf{C} = \begin{bmatrix} 0 & 0 & \cdots & 0 & 1 \end{bmatrix} \quad \mathbf{D} = 0 \quad (10\text{-}237)$$

The matrices $\mathbf{A}$ and $\mathbf{C}$ are in OCF.

It should be pointed out that, given the dynamic equations of a system, the input–output transfer function is unique. However, given the transfer function, the state model is not unique, as shown by the CCF, OCF, and DCF, and many other possibilities. In fact, even for any one of these canonical forms (for example, CCF), while matrices $\mathbf{A}$ and $\mathbf{B}$ are defined, the elements of $\mathbf{C}$ and $\mathbf{D}$ could still be different depending on how the state diagram is drawn, that is, how the transfer function is decomposed. In other words, referring to Fig. 10-14, whereas the feedback branches are fixed, the feedforward branches that contain the coefficients of the numerator of the transfer function can still be manipulated to change the contents of $\mathbf{C}$.

▶ **EXAMPLE 10-10-1** Consider the following input–output transfer function:

$$\frac{Y(s)}{U(s)} = \frac{2s^2 + s + 5}{s^3 + 6s^2 + 11s + 4} \quad (10\text{-}238)$$

The CCF state diagram of the system is shown in Fig. 10-16, which is drawn from the following equations:

$$Y(s) = (2s^{-1} + s^{-2} + 5s^{-3})X(s) \quad (10\text{-}239)$$

$$X(s) = U(s) - (6s^{-1} + 11s^{-2} + 4s^{-3})X(s) \quad (10\text{-}240)$$

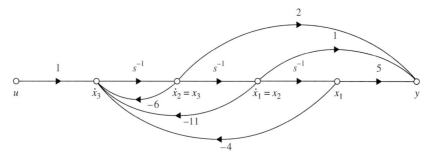

Figure 10-16 CCF state diagram of the transfer function in Eq. (10-238).

The dynamic equations of the system in CCF are

$$\begin{bmatrix} \dfrac{dx_1(t)}{dt} \\ \dfrac{dx_2(t)}{dt} \\ \dfrac{dx_3(t)}{dt} \end{bmatrix} = \begin{bmatrix} 0 & 1 & 0 \\ 0 & 0 & 1 \\ -4 & -11 & -6 \end{bmatrix} \begin{bmatrix} x_1(t) \\ x_2(t) \\ x_3(t) \end{bmatrix} + \begin{bmatrix} 0 \\ 0 \\ 1 \end{bmatrix} u(t) \qquad (10\text{-}241)$$

$$y(t) = \begin{bmatrix} 5 & 1 & 2 \end{bmatrix} \mathbf{x}(t) \qquad (10\text{-}242)$$

For the OCF, Eq. (10-238) is expanded to

$$Y(s) = (2s^{-1} + s^{-2} + 5s^{-3})U(s) - (6s^{-1} + 11s^{-2} + 4s^{-3})Y(s) \qquad (10\text{-}243)$$

which leads to the OCF state diagram shown in Fig. 10-17. The OCF dynamic equations are written

$$\begin{bmatrix} \dfrac{dx_1(t)}{dt} \\ \dfrac{dx_2(t)}{dt} \\ \dfrac{dx_3(t)}{dt} \end{bmatrix} = \begin{bmatrix} 0 & 0 & -4 \\ 1 & 0 & -11 \\ 0 & 1 & -6 \end{bmatrix} \begin{bmatrix} x_1(t) \\ x_2(t) \\ x_3(t) \end{bmatrix} + \begin{bmatrix} 5 \\ 1 \\ 2 \end{bmatrix} u(t) \qquad (10\text{-}244)$$

$$y(t) = \begin{bmatrix} 0 & 0 & 1 \end{bmatrix} \mathbf{x}(t) \qquad (10\text{-}245)$$

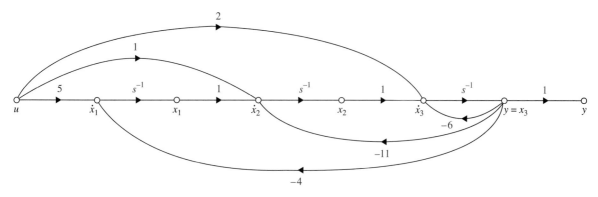

Figure 10-17 OCF state diagram of the transfer function in Eq. (10-238).

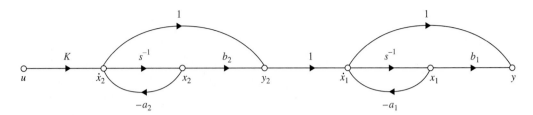

Figure 10-18 State diagram of the transfer function in Eq. (10-246) by cascade decomposition.

10-10-2 Cascade Decomposition

Cascade compensation refers to transfer functions that are written as products of simple first-order or second-order components. Consider the following transfer function, which is the product of two first-order transfer functions.

$$\frac{Y(s)}{U(s)} = K\left(\frac{s+b_1}{s+a_1}\right)\left(\frac{s+b_2}{s+a_2}\right) \quad (10\text{-}246)$$

where a_1, a_2, b_1, and b_2 are real constants. Each of the first-order transfer functions is decomposed by the direct decomposition, and the two state diagrams are connected in cascade, as shown in Fig. 10-18. The state equations are obtained by regarding the derivatives of the state variables as outputs and the state variables and $u(t)$ as inputs and then applying the SFG gain formula to the state diagram in Fig. 10-18. The integrator branches are neglected when applying the gain formula. The results are

$$\begin{bmatrix} \frac{dx_1(t)}{dt} \\ \frac{dx_2(t)}{dt} \end{bmatrix} = \begin{bmatrix} -a_1 & b_2 - a_2 \\ 0 & -a_2 \end{bmatrix} \begin{bmatrix} x_1(t) \\ x_2(t) \end{bmatrix} + \begin{bmatrix} K \\ K \end{bmatrix} u(t) \quad (10\text{-}247)$$

The output equation is obtained by regarding the state variables and $u(t)$ as inputs and $y(t)$ as the output and applying the gain formula to Fig. 10-18. Thus,

$$y(t) = [b_1 - a_1 \quad b_2 - a_2]\mathbf{x}(t) + Ku(t) \quad (10\text{-}248)$$

When the overall transfer function has complex poles or zeros, the individual factors related to these poles or zeros should be in second-order form. As an example, consider the following transfer function:

$$\frac{Y(s)}{U(s)} = \left(\frac{s+5}{s+2}\right)\left(\frac{s+1.5}{s^2+3s+4}\right) \quad (10\text{-}249)$$

where the poles of the second term are complex. The state diagram of the system with the two subsystems connected in cascade is shown in Fig. 10–19. The dynamic equations of the system are

$$\begin{bmatrix} \frac{dx_1(t)}{dt} \\ \frac{dx_2(t)}{dt} \\ \frac{dx_3(t)}{dt} \end{bmatrix} = \begin{bmatrix} 0 & 1 & 0 \\ -4 & -3 & 3 \\ 0 & 0 & -2 \end{bmatrix} \begin{bmatrix} x_1(t) \\ x_2(t) \\ x_3(t) \end{bmatrix} + \begin{bmatrix} 0 \\ 1 \\ 1 \end{bmatrix} u(t) \quad (10\text{-}250)$$

$$y(t) = [1.5 \quad 1 \quad 0]\mathbf{x}(t) \quad (10\text{-}251)$$

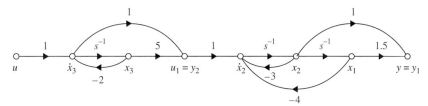

Figure 10-19 State diagram of the transfer function in Eq. (10-249) by cascade decomposition.

10-10-3 Parallel Decomposition

When the denominator of the transfer function is in factored form, the transfer function may be expanded by partial-fraction expansion. The resulting state diagram will consist of simple first- or second-order systems connected in parallel, which leads to the state equations in DCF or JCF, the latter in the case of multiple-order eigenvalues.

Consider that a second-order system is represented by the transfer function

$$\frac{Y(s)}{U(s)} = \frac{Q(s)}{(s+a_1)(s+a_2)} \tag{10-252}$$

where $Q(s)$ is a polynomial of order less than 2, and a_1 and a_2 are real and distinct. Although, analytically, a_1 and a_2 may be complex, in practice, complex numbers are difficult to implement on a computer. Eq. (10-253) is expansion by partial fractions:

$$\frac{Y(s)}{U(s)} = \frac{K_1}{s+a_1} + \frac{K_2}{s+a_2} \tag{10-253}$$

where K_1 and K_2 are real constants.

The state diagram of the system is drawn by the parallel combination of the state diagrams of each of the first-order terms in Eq. (10-253), as shown in Fig. 10-20. The dynamic equations of the system are

$$\begin{bmatrix} \dfrac{dx_1(t)}{dt} \\ \dfrac{dx_2(t)}{dt} \end{bmatrix} = \begin{bmatrix} -a_1 & 0 \\ 0 & -a_2 \end{bmatrix} \begin{bmatrix} x_1(t) \\ x_2(t) \end{bmatrix} + \begin{bmatrix} 1 \\ 1 \end{bmatrix} u(t) \tag{10-254}$$

$$y(t) = [K_1 \quad K_2]\mathbf{x}(t) \tag{10-255}$$

Thus, the state equations are of the DCF.

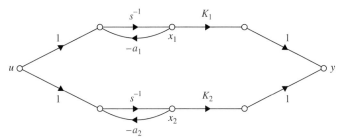

Figure 10-20 State diagram of the transfer function of Eq. (10-252) by parallel decomposition.

714 ▶ Chapter 10. State Variable Analysis

The conclusion is that, for transfer functions with distinct poles, parallel decomposition will lead to the DCF for the state equations. For transfer functions with multiple-order eigenvalues, parallel decomposition to a state diagram with a minimum number of integrators will lead to the JCF state equations. The following example will clarify this point.

▶ **EXAMPLE 10-10-2** Consider the following transfer function and its partial-fraction expansion:

$$\frac{Y(s)}{U(s)} = \frac{2s^2 + 6s + 5}{(s+1)^2(s+2)} = \frac{1}{(s+1)^2} + \frac{1}{s+1} + \frac{1}{s+2} \qquad (10\text{-}256)$$

Note that the transfer function is of the third order, and, although the total order of the terms on the right-hand side of Eq. (10-256) is four, only three integrators should be used in the state diagram, which is drawn as shown in Fig. 10-21. The minimum number of three integrators is used, with one integrator being shared by two channels. The state equations of the system are written directly from Fig. 10-21.

$$\begin{bmatrix} \dfrac{dx_1(t)}{dt} \\ \dfrac{dx_2(t)}{dt} \\ \dfrac{dx_3(t)}{dt} \end{bmatrix} = \begin{bmatrix} -1 & 1 & 0 \\ 0 & -1 & 0 \\ 0 & 0 & -2 \end{bmatrix} \begin{bmatrix} x_1(t) \\ x_2(t) \\ x_3(t) \end{bmatrix} + \begin{bmatrix} 0 \\ 1 \\ 1 \end{bmatrix} u(t) \qquad (10\text{-}257)$$

which is recognized to be the JCF.

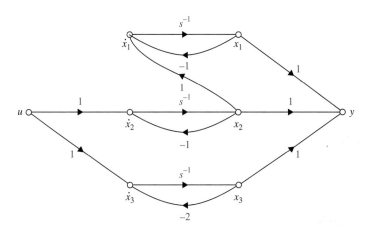

Figure 10-21 State diagram of the transfer function of Eq. (10-256) by parallel decomposition.

◀

▶ **10-11 CONTROLLABILITY OF CONTROL SYSTEMS**

The concepts of **controllability** and **observability**, introduced first by Kalman [3], play an important role in both theoretical and practical aspects of modern control. The conditions on controllability and observability essentially govern the existence of a solution to an optimal control problem. This seems to be the basic difference between optimal control theory and classical control theory. In the classical control theory, the design techniques are dominated by trial-and-error methods so that given a set of design specifications the designer at the outset does not know if any solution exists. Optimal control theory, on the

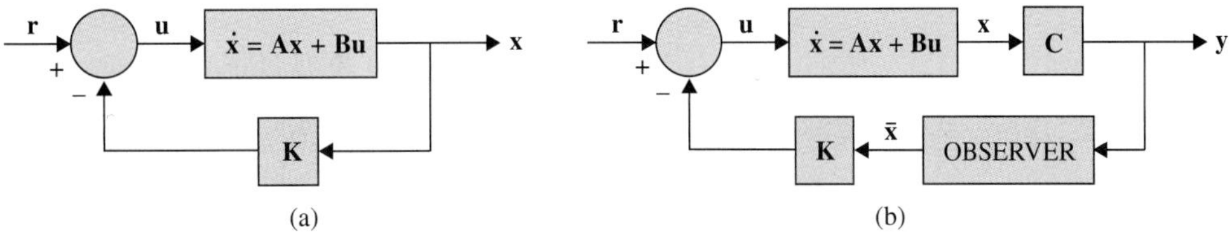

Figure 10-22 (a) Control system with state feedback. (b) Control system with observer and state feedback.

other hand, has criteria for determining at the outset if the design solution exists for the system parameters and design objectives.

We shall show that the condition of controllability of a system is closely related to the existence of solutions of state feedback for assigning the values of the eigenvalues of the system arbitrarily. The concept of *observability* relates to the condition of observing or estimating the state variables from the output variables, which are generally measurable.

The block diagram shown in Fig. 10-22 illustrates the motivation behind investigating controllability and observability. Figure 10-22(a) shows a system with the process dynamics described by

$$\frac{d\mathbf{x}(t)}{dt} = \mathbf{A}\mathbf{x}(t) + \mathbf{B}\mathbf{u}(t) \quad (10\text{-}258)$$

The closed-loop system is formed by feeding back the state variables through a constant feedback gain matrix **K**. Thus, from Fig. 10-22,

$$\mathbf{u}(t) = -\mathbf{K}\mathbf{x}(t) + \mathbf{r}(t) \quad (10\text{-}259)$$

where **K** is a $p \times n$ feedback matrix with constant elements. The closed-loop system is thus described by

$$\frac{d\mathbf{x}(t)}{dt} = (\mathbf{A} - \mathbf{B}\mathbf{K})\mathbf{x}(t) + \mathbf{B}\mathbf{r}(t) \quad (10\text{-}260)$$

This problem is also known as the **pole-placement design** through state feedback. The design objective in this case is to find the feedback matrix **K** such that the eigenvalues of $(\mathbf{A} - \mathbf{B}\mathbf{K})$, or of the closed-loop system, are of certain prescribed values. The word *pole* refers here to the poles of the closed-loop transfer function, which are the same as the eigenvalues of $(\mathbf{A} - \mathbf{B}\mathbf{K})$.

We shall show later that the existence of the solution to the pole-placement design with arbitrarily assigned pole values through state feedback is directly based on the controllability of the states of the system. The result is that *if the system of Eq. (10-225) is controllable, then there exists a constant feedback matrix* **K** *that allows the eigenvalues of* $(\mathbf{A} - \mathbf{B}\mathbf{K})$ *to be arbitrarily assigned.*

Once the closed-loop system is designed, the practical problems of implementing the feeding back of the state variables must be considered. There are two problems with implementing state feedback control: First, the number of state variables may be excessive, which will make the cost of sensing each of these state variables for feedback prohibitive. Second, not all the state variables are physically accessible, and so it may be necessary to design and construct an **observer** that will estimate the state vector from the output vector $\mathbf{y}(t)$. Fig. 10-22(b) shows the block diagram of a closed-loop system with an observer. The

716 ▶ Chapter 10. State Variable Analysis

Figure 10-23 Linear time-invariant system.

observed state vector $\bar{\mathbf{x}}(t)$ is used to generate the control $\mathbf{u}(t)$ through the feedback matrix **K**. *The condition that such an observer can be designed for the system is called the observability of the system.*

10-11-1 General Concept of Controllability

The concept of controllability can be stated with reference to the block diagram of Fig. 10-22(a). *The process is said to be **completely controllable** if every state variable of the process can be controlled to reach a certain objective in finite time by some unconstrained control* $\mathbf{u}(t)$, as shown in Fig. 10-23. Intuitively, it is simple to understand that, if any one of the state variables is independent of the control $\mathbf{u}(t)$, there would be no way of driving this particular state variable to a desired state in finite time by means of a control effort. Therefore, this particular state is said to be uncontrollable, and, as long as there is at least one uncontrollable state, the system is said to be not completely controllable or, simply, uncontrollable.

As a simple example of an uncontrollable system, Fig. 10-24 illustrates the state diagram of a linear system with two state variables. Because the control $\mathbf{U}(t)$ affects only the state $x_1(t)$, the state $x_2(t)$ is uncontrollable. In other words, it would be impossible to drive $x_2(t)$ from an initial state $x_2(t_0)$ to a desired state $x_2(t_f)$ in finite time interval $t_f - t_0$ by the control $\mathbf{U}(t)$. Therefore, the entire system is said to be uncontrollable.

The concept of controllability given here refers to the states and is sometimes referred to as **state controllability**. Controllability can also be defined for the outputs of the system, so there is a difference between state controllability and output controllability.

10-11-2 Definition of State Controllability

Consider that a linear time-invariant system is described by the following dynamic equations:

$$\frac{d\mathbf{x}(t)}{dt} = \mathbf{A}\mathbf{x}(t) + \mathbf{B}\mathbf{u}(t) \tag{10-261}$$

$$\mathbf{y}(t) = \mathbf{C}\mathbf{x}(t) + \mathbf{D}\mathbf{u}(t) \tag{10-262}$$

where $\mathbf{x}(t)$ is the $n \times 1$ state vector, $\mathbf{u}(t)$ is the $r \times 1$ input vector, and $\mathbf{y}(t)$ is the $p \times 1$ output vector. **A**, **B**, **C**, and **D** are coefficients of appropriate dimensions.

The state $\mathbf{x}(t)$ is said to be controllable at $t = t_0$ if there exists a piecewise continuous input $\mathbf{u}(t)$ that will drive the state to any final state $\mathbf{x}(t_f)$ for a finite time $(t_f - t_0) \geq 0$. If

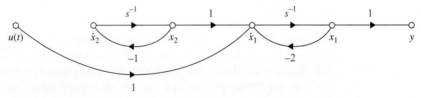

Figure 10-24 State diagram of the system that is not state controllable.

every state $\mathbf{x}(t_0)$ of the system is controllable in a finite time interval, the system is said to be completely state controllable or, simply, controllable.

The following theorem shows that the condition of controllability depends on the coefficient matrices $\mathbf{A}$ and $\mathbf{B}$ of the system. The theorem also gives one method of testing for state controllability.

■ **Theorem 10-1.** *For the system described by the state equation of Eq. (10-261) to be* **completely state controllable**, *it is necessary and sufficient that the following* $n \times nr$ **controllability matrix** *has a rank of n:*

$$\mathbf{S} = \begin{bmatrix} \mathbf{B} & \mathbf{AB} & \mathbf{A}^2\mathbf{B} & \cdots & \mathbf{A}^{n-1}\mathbf{B} \end{bmatrix} \qquad (10\text{-}263)$$

Because the matrices $\mathbf{A}$ *and* $\mathbf{B}$ *are involved, sometimes we say that the pair* $[\mathbf{A}, \mathbf{B}]$ *is controllable, which implies that* $\mathbf{S}$ *is of rank n.*

The proof of this theorem is given in any standard textbook on optimal control systems. The idea is to start with the state-transition equation of Eq. (10-79) and then proceed to show that Eq. (10-263) must be satisfied in order that all the states are accessible by the input.

Although the criterion of state controllability given in Theorem 10-1 is quite straightforward, manually, it is not very easy to test for high-order systems and/or systems with many inputs. If $\mathbf{S}$ is nonsquare, we can form the matrix $\mathbf{SS}'$, which is $n \times n$; then, if $\mathbf{SS}'$ is nonsingular, $\mathbf{S}$ has rank n.

10-11-3 Alternate Tests on Controllability

There are several alternate methods of testing controllability, and some of these may be more convenient to apply than the condition in Eq. (10-263).

■ **Theorem 10-2.** *For a single-input, single-output (SISO) system described by the state equation of Eq. (10-261) with* $r = 1$, *the pair* $[\mathbf{A}, \mathbf{B}]$ *is completely controllable if* $\mathbf{A}$ *and* $\mathbf{B}$ *are in CCF or transformable into CCF by a similarity transformation.*

The proof of this theorem is straightforward, since it was established in Section 10-9 that the CCF transformation requires that the controllability matrix $\mathbf{S}$ be nonsingular. Because the CCF transformation in Section 10-9 was defined only for SISO systems, the theorem applies only to this type of system.

■ **Theorem 10-3.** *For a system described by the state equation of Eq. (10-261), if* $\mathbf{A}$ *is in DCF or JCF, the pair* $[\mathbf{A}, \mathbf{B}]$ *is completely controllable if all the elements in the rows of* $\mathbf{B}$ *that correspond to the last row of each Jordan block are nonzero.*

The proof of this theorem comes directly from the definition of controllability. Let us assume that $\mathbf{A}$ is diagonal and that it has distinct eigenvalues. Then, the pair $[\mathbf{A}, \mathbf{B}]$ is controllable if $\mathbf{B}$ does not have any row with all zeros. The reason is that, if $\mathbf{A}$ is diagonal, all the states are decoupled from each other, and, if any row of $\mathbf{B}$ contains all zero elements, the corresponding state would not be accessed from any of the inputs, and that state would be uncontrollable.

For a system in JCF, such as the $\mathbf{A}$ and $\mathbf{B}$ matrices illustrated in Eq. (10-264), for controllability only the elements in the row of $\mathbf{B}$ that correspond to the last row of the Jordan block cannot all be zeros. The elements in the other rows of $\mathbf{B}$ need not all be nonzero, since the corresponding states are still coupled through the 1s in the Jordan blocks of $\mathbf{A}$.

$$\mathbf{A} = \begin{bmatrix} \lambda_1 & 1 & 0 & 0 \\ 0 & \lambda_1 & 1 & 0 \\ 0 & 0 & \lambda_1 & 0 \\ 0 & 0 & 0 & \lambda_2 \end{bmatrix} \quad \mathbf{B} = \begin{bmatrix} b_{11} & b_{12} \\ b_{21} & b_{22} \\ b_{31} & b_{32} \\ b_{41} & b_{42} \end{bmatrix} \quad (10\text{-}264)$$

Thus, the condition of controllability for the $\mathbf{A}$ and $\mathbf{B}$ in Eq. (10-264) is $b_{31} \neq 0$, $b_{32} \neq 0$, $b_{41} \neq 0$, and $b_{42} \neq 0$.

▶ **EXAMPLE 10-11-1** The following matrices are for a system with two identical eigenvalues, but the matrix $\mathbf{A}$ is diagonal.

$$\mathbf{A} = \begin{bmatrix} \lambda_1 & 0 \\ 0 & \lambda_1 \end{bmatrix} \quad \mathbf{B} = \begin{bmatrix} b_{11} \\ b_{21} \end{bmatrix} \quad (10\text{-}265)$$

The system is uncontrollable, since the two state equations are dependent; that is, it would not be possible to control the states independently by the input. We can easily show that in this case $\mathbf{S} = [\mathbf{B} \ \mathbf{AB}]$ is singular. ◀

▶ **EXAMPLE 10-11-2** Consider the system shown in Fig. 10-24, which was reasoned earlier to be uncontrollable. Let us investigate the same system using the condition of Eq. (10-263). The state equations of the system are written in the form of Eq. (10-263) with

$$\mathbf{A} = \begin{bmatrix} -2 & 1 \\ 0 & -1 \end{bmatrix} \quad \mathbf{B} = \begin{bmatrix} 1 \\ 0 \end{bmatrix} \quad (10\text{-}266)$$

Thus, from Eq. (10-263), the controllability matrix is

$$\mathbf{S} = [\mathbf{B} \ \mathbf{AB}] = \begin{bmatrix} 1 & -2 \\ 0 & 0 \end{bmatrix} \quad (10\text{-}267)$$

which is singular, and the system is uncontrollable. ◀

▶ **EXAMPLE 10-11-3** Consider that a third-order system has the coefficient matrices

$$\mathbf{A} = \begin{bmatrix} 1 & 2 & -1 \\ 0 & 1 & 0 \\ 1 & -4 & 3 \end{bmatrix} \quad \mathbf{B} = \begin{bmatrix} 0 \\ 0 \\ 1 \end{bmatrix} \quad (10\text{-}268)$$

The controllability matrix is

$$\mathbf{S} = [\mathbf{B} \ \mathbf{AB} \ \mathbf{A}^2\mathbf{B}] = \begin{bmatrix} 0 & -1 & -4 \\ 0 & 0 & 0 \\ 1 & 3 & 8 \end{bmatrix} \quad (10\text{-}269)$$

which is singular. Thus, the system is not controllable.

The eigenvalues of **A** are $\lambda_1 = 2$, $\lambda_2 = 2$, and $\lambda_3 = 1$. The JCF of **A** and **B** are obtained with the transformation $\mathbf{x}(t) = \mathbf{T}\bar{\mathbf{x}}(t)$, where

$$\mathbf{T} = \begin{bmatrix} 1 & 0 & 0 \\ 0 & 0 & 1 \\ -1 & 1 & 2 \end{bmatrix} \quad (10\text{-}270)$$

Then,

$$\bar{\mathbf{A}} = \mathbf{T}^{-1}\mathbf{A}\mathbf{T} = \begin{bmatrix} 2 & -1 & 0 \\ 0 & 2 & 0 \\ 0 & 0 & 1 \end{bmatrix} \quad \bar{\mathbf{B}} = \mathbf{T}^{-1}\mathbf{B} = \begin{bmatrix} 0 \\ -1 \\ 0 \end{bmatrix} \quad (10\text{-}271)$$

Because the last row of $\bar{\mathbf{B}}$, which corresponds to the Jordan block for the eigenvalue λ_3, is zero, the transformed state variable $\bar{x}_3(t)$ is uncontrollable. From the transformation matrix **T** in Eq. (10-235), $x_2 = \bar{x}_3$, which means that x_2 is uncontrollable in the original system. It should be noted that the minus sign in front of the 1 in the Jordan block does not alter the basic definition of the block. ◀

▶ 10-12 OBSERVABILITY OF LINEAR SYSTEMS

The concept of observability was covered earlier in Section 10-11 on controllability and observability. Essentially, *a system is completely observable if every state variable of the system affects some of the outputs.* In other words, it is often desirable to obtain information on the state variables from the measurements of the outputs and the inputs. If any one of the states cannot be observed from the measurements of the outputs, the state is said to be unobservable, and the system is not completely observable or, simply, unobservable. Fig. 10-25 shows the state diagram of a linear system in which the state x_2 is not connected to the output $y(t)$ in any way. Once we have measured $y(t)$, we can observe the state $x_1(t)$, since $x_1(t) = y(t)$. However, the state x_2 cannot be observed from the information on $y(t)$. Thus, the system is unobservable.

10-12-1 Definition of Observability

*Given a linear time-invariant system that is described by the dynamic equations of Eqs. (10-261) and (10-262), the state $\mathbf{x}(t_0)$ is said to be observable if given any input $\mathbf{u}(t)$, there exists a finite time $t_f \geq t_0$ such that the knowledge of $\mathbf{u}(t)$ for $t_0 \leq t < t_f$, matrices **A**, **B**, **C**, and **D**; and the output $\mathbf{y}(t)$ for $t_0 \leq t < t_f$ are sufficient to determine $\mathbf{x}(t_0)$. If every state of the system is observable for a finite t_f, we say that the system is completely observable, or, simply, observable.*

The following theorem shows that the condition of observability depends on the matrices **A** and **C** of the system. The theorem also gives one method of testing observability.

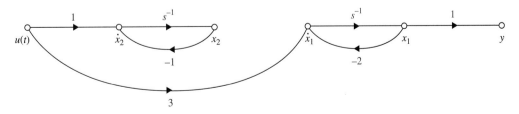

Figure 10-25 State diagram of a system that is not observable.

Theorem 10-4. *For the system described by Eqs. (10-261) and (10-262) to be completely observable, it is necessary and sufficient that the following $n \times np$ **observability matrix** has a rank of n:*

$$\mathbf{V} = \begin{bmatrix} \mathbf{C} \\ \mathbf{CA} \\ \mathbf{CA}^2 \\ \vdots \\ \mathbf{CA}^{n-1} \end{bmatrix} \qquad (10\text{-}272)$$

The condition is also referred to as the pair [**A**, **C**] being observable. In particular, if the system has only one output, **C** is a $1 \times n$ row matrix; **V** is an $n \times n$ square matrix. Then the system is completely observable if **V** is nonsingular.

The proof of this theorem is not given here. It is based on the principle that Eq. (10-272) must be satisfied so that $\mathbf{x}(t_0)$ can be uniquely determined from the output $\mathbf{y}(t)$.

10-12-2 Alternate Tests on Observability

Just as with controllability, there are several alternate methods of testing observability. These are described in the following theorems.

Theorem 10-5. *For an SISO system, described by the dynamic equations of Eqs. (10-261) and (10-262) with $r = 1$ and $p = 1$, the pair [**A**, **C**] is completely observable if **A** and **C** are in OCF or transformable into OCF by a similarity transformation.*

The proof of this theorem is straightforward, since it was established in Section 10-8 that the OCF transformation requires that the observability matrix **V** be nonsingular.

Theorem 10-6. *For a system described by the dynamic equations of Eqs. (10-261) and (10-262), if **A** is in DCF or JCF, the pair [**A**, **C**] is completely observable if all the elements in the columns of **C** that correspond to the first row of each Jordan block are nonzero.*

Note that this theorem is a dual of the test of controllability given in Theorem 10-3. If the system has distinct eigenvalues, **A** is diagonal, then the condition on observability is that none of the columns of **C** can contain all zeros.

▶ **EXAMPLE 10-12-1** Consider the system shown in Fig. 10-25, which was earlier defined to be unobservable. The dynamic equations of the system are expressed in the form of Eqs. (10-261) and (10-262) with

$$\mathbf{A} = \begin{bmatrix} -2 & 0 \\ 0 & -1 \end{bmatrix} \quad \mathbf{B} = \begin{bmatrix} 3 \\ 1 \end{bmatrix} \quad \mathbf{C} = \begin{bmatrix} 1 & 0 \end{bmatrix} \qquad (10\text{-}273)$$

Thus, the observability matrix is

$$\mathbf{V} = \begin{bmatrix} \mathbf{C} \\ \mathbf{CA} \end{bmatrix} = \begin{bmatrix} 1 & 0 \\ -2 & 0 \end{bmatrix} \qquad (10\text{-}274)$$

which is singular. Thus, the pair [**A**, **C**] is unobservable. In fact, because **A** is of DCF and the second column of **C** is zero, this means that the state $x_2(t)$ is unobservable, as conjectured from Fig. 10-24. ◀

10-13 RELATIONSHIP AMONG CONTROLLABILITY, OBSERVABILITY, AND TRANSFER FUNCTIONS

In the classical analysis of control systems, transfer functions are used for modeling of linear time-invariant systems. Although controllability and observability are concepts and tools of modern control theory, we shall show that they are closely related to the properties of transfer functions.

Theorem 10-7. *If the input–output transfer function of a linear system has pole–zero cancellation, the system will be uncontrollable or unobservable, or both, depending on how the state variables are defined. On the other hand, if the input–output transfer function does not have pole–zero cancellation, the system can always be represented by dynamic equations as a completely controllable and observable system.*

The proof of this theorem is not given here. The importance of this theorem is that, if a linear system is modeled by a transfer function with no pole–zero cancellation, then we are assured that it is a controllable and observable system, no matter how the state-variable model is derived. Let us amplify this point further by referring to the following SISO system.

$$\mathbf{A} = \begin{bmatrix} -1 & 0 & 0 & 0 \\ 0 & -2 & 0 & 0 \\ 0 & 0 & -3 & 0 \\ 0 & 0 & 0 & -4 \end{bmatrix} \quad \mathbf{B} = \begin{bmatrix} 1 \\ 1 \\ 0 \\ 0 \end{bmatrix} \quad \mathbf{C} = \begin{bmatrix} 1 & 0 & 1 & 0 \end{bmatrix} \quad \mathbf{D} = 0 \quad (10\text{-}275)$$

Because $\mathbf{A}$ is a diagonal matrix, the controllability and observability conditions of the four states are determined by inspection. They are as follows:

x_1: Controllable and observable (C and O)

x_2: Controllable but unobservable (C but UO)

x_3: Uncontrollable but observable (UC but O)

x_4: Uncontrollable and unobservable (UC and UO)

The block diagram of the system in Fig. 10-26 shows the DCF decomposition of the system. Clearly, the transfer function of the controllable and observable system should be

$$\frac{Y(s)}{U(s)} = \frac{1}{s+1} \quad (10\text{-}276)$$

whereas the transfer function that corresponds to the dynamics described in Eq. (10-275) is

$$\frac{Y(s)}{U(s)} = \mathbf{C}(s\mathbf{I} - \mathbf{A})^{-1}\mathbf{B} = \frac{(s+2)(s+3)(s+4)}{(s+1)(s+2)(s+3)(s+4)} \quad (10\text{-}277)$$

which has three pole–zero cancellations. This simple-minded example illustrates that a "minimum-order" transfer function without pole–zero cancellation is the only component that corresponds to a system that is controllable and observable.

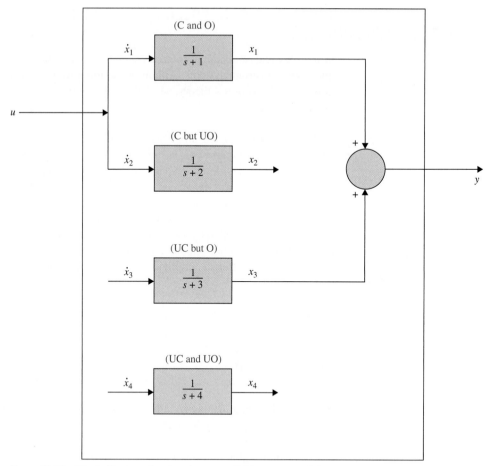

Figure 10-26 Block diagram showing the controllable, uncontrollable, observable, and unobservable components of the system described in Eq. (10-275).

▶ **EXAMPLE 10-13-1** Let us consider the transfer function

$$\frac{Y(s)}{U(s)} = \frac{s+2}{(s+1)(s+2)} \tag{10-278}$$

which is a reduced form of Eq. (10-277). Eq. (10-278) is decomposed into CCF and OCF as follows:

CCF:

$$\mathbf{A} = \begin{bmatrix} 0 & 1 \\ -2 & -3 \end{bmatrix} \quad \mathbf{B} = \begin{bmatrix} 0 \\ 1 \end{bmatrix} \quad \mathbf{C} = \begin{bmatrix} 1 & 1 \end{bmatrix} \tag{10-279}$$

Because the CCF transformation can be made, the pair [**A**, **B**] of the CCF is controllable. The observability matrix is

$$\mathbf{V} = \begin{bmatrix} \mathbf{C} \\ \mathbf{CA} \end{bmatrix} = \begin{bmatrix} 1 & 1 \\ -2 & -2 \end{bmatrix} \tag{10-280}$$

which is singular, and the pair [**A**, **C**] of the CCF is unobservable.

OCF:

$$\mathbf{A} = \begin{bmatrix} 0 & -2 \\ 1 & -3 \end{bmatrix} \quad \mathbf{B} = \begin{bmatrix} 1 \\ 1 \end{bmatrix} \quad \mathbf{C} = \begin{bmatrix} 0 & 1 \end{bmatrix} \qquad (10\text{-}281)$$

Because the OCF transformation can be made, the pair [**A, C**] of the OCF is observable. However, the controllability matrix is

$$\mathbf{S} = \begin{bmatrix} \mathbf{B} & \mathbf{AB} \end{bmatrix} = \begin{bmatrix} 1 & -2 \\ 1 & -2 \end{bmatrix} \qquad (10\text{-}282)$$

which is singular, and the pair [**A, B**] of the OCF is uncontrollable.

The conclusion that can be drawn from this example is that, given a system that is modeled by transfer function, the controllability and observability conditions of the system depend on how the state variables are defined. ◀

▶ 10-14 INVARIANT THEOREMS ON CONTROLLABILITY AND OBSERVABILITY

We now investigate the effects of the similarity transformations on controllability and observability. The effects of controllability and observability due to state feedback will be investigated.

■ **Theorem 10-8.** *Invariant theorem on similarity transformations:* Consider the system described by the dynamic equations of Eqs. (10-261) and (10-262). The similarity transformation $\mathbf{x}(t) = \mathbf{P}\bar{\mathbf{x}}(t)$, where **P** is nonsingular, transforms the dynamic equations to

$$\frac{d\bar{\mathbf{x}}(t)}{dt} = \bar{\mathbf{A}}\bar{\mathbf{x}}(t) + \bar{\mathbf{B}}\mathbf{u}(t) \qquad (10\text{-}283)$$

$$\bar{\mathbf{y}}(t) = \bar{\mathbf{C}}\mathbf{x}(t) + \bar{\mathbf{D}}\mathbf{u}(t) \qquad (10\text{-}284)$$

where

$$\bar{\mathbf{A}} = \mathbf{P}^{-1}\mathbf{AP} \quad \bar{\mathbf{B}} = \mathbf{P}^{-1}\mathbf{B} \qquad (10\text{-}285)$$

The controllability of $[\bar{\mathbf{A}}, \bar{\mathbf{B}}]$ and the observability of $[\bar{\mathbf{A}}, \bar{\mathbf{C}}]$ are not affected by the transformation.

In other words, controllability and observability are preserved through similar transformations. The theorem is easily proven by showing that the ranks of $\bar{\mathbf{S}}$ and $\mathbf{S}$ and the ranks of $\bar{\mathbf{V}}$ and $\mathbf{V}$ are identical, where $\bar{\mathbf{S}}$ and $\bar{\mathbf{V}}$ are the controllability and observability matrices, respectively, of the transformed system.

■ **Theorem 10-9.** *Theorem on controllability of closed-loop systems with state feedback:* If the open-loop system

$$\frac{d\mathbf{x}(t)}{dt} = \mathbf{A}\mathbf{x}(t) + \mathbf{B}\mathbf{u}(t) \qquad (10\text{-}286)$$

is completely controllable, then the closed-loop system obtained through state feedback,

$$\mathbf{u}(t) = \mathbf{r}(t) - \mathbf{K}\mathbf{x}(t) \qquad (10\text{-}287)$$

so that the state equation becomes

$$\frac{d\mathbf{x}(t)}{dt} = (\mathbf{A} - \mathbf{BK})\mathbf{x}(t) + \mathbf{B}\mathbf{r}(t) \tag{10-288}$$

is also completely controllable. On the other hand, if [**A, B**] *is uncontrollable, then there is no* **K** *that will make the pair* [**A** − **BK, B**] *controllable. In other words, if an open-loop system is uncontrollable, it cannot be made controllable through state feedback.*

Proof: The controllability of [**A, B**] implies that there exists a control $\mathbf{u}(t)$ over the time interval $[t_0, t_f]$ such that **the initial state** $\mathbf{x}(t_0)$ is driven to the final state $\mathbf{x}(t_f)$ over the finite time interval $t_f - t_0$. We can write Eq. (10-252) as

$$\mathbf{r}(t) = \mathbf{u}(t) + \mathbf{Kx}(t) \tag{10-289}$$

which is the control of the closed-loop system. Thus, if $\mathbf{u}(t)$ exists that can drive $\mathbf{x}(t_0)$ to any $\mathbf{x}(t_f)$ in finite time, then we cannot find an input $\mathbf{r}(t)$ that will do the same to $\mathbf{x}(t)$, because otherwise we can set $\mathbf{u}(t)$ as in Eq. (10-287) to control the open-loop system.

■ **Theorem 10-10.** *Theorem on observability of closed-loop systems with state feedback: If an open-loop system is controllable and observable, then state feedback of the form of Eq. (10-287) could destroy observability. In other words, the observability of open-loop and closed-loop systems due to state feedback is unrelated.*

The following example illustrates the relation between observability and state feedback.

▶ **EXAMPLE 10-14-1** Let the coefficient matrices of a linear system be

$$\mathbf{A} = \begin{bmatrix} 0 & 1 \\ -2 & -3 \end{bmatrix} \quad \mathbf{B} = \begin{bmatrix} 1 \\ 1 \end{bmatrix} \quad \mathbf{C} = \begin{bmatrix} 1 & 2 \end{bmatrix} \tag{10-290}$$

We can show that the pair [**A, B**] is controllable and [**A, C**] is observable.
Let the state feedback be defined as

$$u(t) = r(t) - \mathbf{Kx}(t) \tag{10-291}$$

where

$$\mathbf{K} = \begin{bmatrix} k_1 & k_2 \end{bmatrix} \tag{10-292}$$

Then the closed-loop system is described by the state equation

$$\frac{d\mathbf{x}(t)}{dt} = (\mathbf{A} - \mathbf{BK})\mathbf{x}(t) + \mathbf{B}r(t) \tag{10-293}$$

$$\mathbf{A} - \mathbf{BK} = \begin{bmatrix} -k_1 & 1 - k_2 \\ -2 - k_1 & -3 - g_2 \end{bmatrix} \tag{10-294}$$

The observability matrix of the closed-loop system is

$$\mathbf{V} = \begin{bmatrix} \mathbf{C} \\ \mathbf{C}(\mathbf{A} - \mathbf{BK}) \end{bmatrix} = \begin{bmatrix} 1 & 2 \\ -k_1 - 4 & -3k_2 - 5 \end{bmatrix} \tag{10-295}$$

The determinant of **V** is

$$|V| = 6k_1 - 3k_2 + 3 \qquad (10\text{-}296)$$

Thus, if k_1 and k_2 are chosen so that $|V| = 0$, the closed-loop system would be uncontrollable. ◀

▶ 10-15 CASE STUDY: MAGNETIC-BALL SUSPENSION SYSTEM

As a case study to illustrate some of the material presented in this chapter, let us consider the magnetic-ball suspension system shown in Fig. 10-27. The objective of the system is to regulate the current of the electromagnet so that the ball will be suspended at a fixed distance from the end of the magnet. The dynamic equations of the system are

$$M\frac{d^2x(t)}{dt^2} = Mg - \frac{ki^2(t)}{x(t)} \qquad (10\text{-}297)$$

$$v(t) = Ri(t) + L\frac{di(t)}{dt} \qquad (10\text{-}298)$$

where Eq. (10-262) is nonlinear. The system variables and parameters are as follows:

$v(t)$ = input voltage (V)
$i(t)$ = winding current (A)
R = winding resistance = 1 Ω
M = ball mass = 1.0 kg

$x(t)$ = ball position (m)
k = proportional constant = 1.0
L = winding inductance = 0.01 H
g = gravitational acceleration = 32.2 m/sec^2

The state variables are defined as

$$\begin{aligned} x_1(t) &= x(t) \\ x_2(t) &= \frac{dx(t)}{dt} \\ x_3(t) &= i(t) \end{aligned} \qquad (10\text{-}299)$$

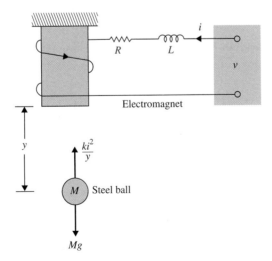

Figure 10-27 Ball-suspension system.

The state equations are

$$\frac{dx_1(t)}{dt} = x_2(t) \tag{10-300}$$

$$\frac{dx_2(t)}{dt} = g - \frac{k}{M}\frac{x_3^2(t)}{x_1(t)} \tag{10-301}$$

$$\frac{dx_3(t)}{dt} = -\frac{R}{L}x_3(t) + \frac{v(t)}{L} \tag{10-302}$$

These nonlinear state equations are linearized about the equilibrium point, $x_1(t) = x(t) = 0.5$ m, using the method described in Section 4-9. After substituting the parameter values, the linearized equations are written

$$\Delta \dot{\mathbf{x}}(t) = \mathbf{A}^* \Delta \mathbf{x}(t) + \mathbf{B}^* \Delta v(t) \tag{10-303}$$

where $\Delta \mathbf{x}(t)$ denotes the state vector, and $\Delta v(t)$ is the input voltage of the linearized system. The coefficient matrices are

$$\mathbf{A}^* = \begin{bmatrix} 0 & 1 & 0 \\ 64.4 & 0 & -16 \\ 0 & 0 & -100 \end{bmatrix} \quad \mathbf{B}^* = \begin{bmatrix} 0 \\ 0 \\ 100 \end{bmatrix} \tag{10-304}$$

All the computations done in the following section can be carried out with the MATLAB Toolbox appearing later in this chapter. To show the analytical method, we carry out the steps of the derivations as follows.

The Characteristic Equation

$$|s\mathbf{I} - \mathbf{A}^*| = \begin{bmatrix} s & -1 & 0 \\ -64.4 & s & 16 \\ 0 & 0 & s+100 \end{bmatrix} = s^3 + 100s^2 - 64.4s - 6440 = 0 \tag{10-305}$$

Eigenvalues: The eigenvalues of $\mathbf{A}^*$, or the roots of the characteristic equation, are

$$s = -100 \quad s = -8.025 \quad s = 8.025$$

The State-Transition Matrix: The state-transition matrix of $\mathbf{A}^*$ is

$$\phi(t) = \mathcal{L}^{-1}\left[(s\mathbf{I} - \mathbf{A}^*)^{-1}\right] = \mathcal{L}^{-1}\left(\begin{bmatrix} s & -1 & 0 \\ -64.4 & s & 16 \\ 0 & 0 & s+100 \end{bmatrix}^{-1}\right) \tag{10-306}$$

or

$$\phi(t) = \mathcal{L}^{-1}\left(\frac{1}{(s+100)(s+8.025)(s-8.025)}\begin{bmatrix} s(s+100) & s+100 & -16 \\ 64.4(s+100) & s(s+100) & -16s \\ 0 & 0 & s^2 - 64.4 \end{bmatrix}\right)$$

$$\tag{10-307}$$

By performing the partial-fraction expansion and carrying out the inverse Laplace transform, the state-transition matrix is

$$\phi(t) = \begin{bmatrix} 0 & 0 & -0.0016 \\ 0 & 0 & 0.16 \\ 0 & 0 & 1 \end{bmatrix} e^{-100t} + \begin{bmatrix} 0.5 & -0.062 & 0.0108 \\ -4.012 & 0.5 & -0.087 \\ 0 & 0 & 0 \end{bmatrix} e^{-8.025t}$$

$$+ \begin{bmatrix} 0.5 & 0.062 & -0.0092 \\ 4.012 & 0.5 & -0.074 \\ 0 & 0 & 0 \end{bmatrix} e^{8.025t} \quad (10\text{-}308)$$

Because the last term in Eq. (10-308) has a positive exponent, the response of $\phi(t)$ increases with time, and the system is unstable. This is expected, since without control, the steel ball would be attracted by the magnet until it hits the bottom of the magnet.

Transfer Function: Let us define the ball position $x(t)$ as the output $y(t)$; then, given the input, $v(t)$, the input–output transfer function of the system is

$$\frac{Y(s)}{V(s)} = \mathbf{C}^*(s\mathbf{I} - \mathbf{A}^*)^{-1}\mathbf{B}^* = \begin{bmatrix} 1 & 0 & 0 \end{bmatrix}(s\mathbf{I} - \mathbf{A}^*)^{-1}\mathbf{B}^*$$
$$= \frac{-1600}{(s+100)(s+8.025)(s-8.025)} \quad (10\text{-}309)$$

Controllability: The controllability matrix is

$$\mathbf{S} = \begin{bmatrix} \mathbf{B}^* & \mathbf{A}^*\mathbf{B}^* & \mathbf{A}^{*2}\mathbf{B}^* \end{bmatrix} = \begin{bmatrix} 0 & 0 & -1{,}600 \\ 0 & -1{,}600 & 160{,}000 \\ 100 & -10{,}000 & 1{,}000{,}000 \end{bmatrix} \quad (10\text{-}310)$$

Because the rank of $\mathbf{S}$ is 3, the system is completely controllable.

Observability: The observability of the system depends on which variable is defined at the output. For state-feedback control, which will be discussed later in Chapter 10, the full controller requires feeding back all three state variables, x_1, x_2, and x_3. However, for reasons of economy, we may want to feed back only one of the three state variables. To make the problem more general, we may want to investigate which state, if chosen as the output, would render the system unobservable.

1. $y(t) = $ ball position $= x(t)$: $\mathbf{C}^* = \begin{bmatrix} 1 & 0 & 0 \end{bmatrix}$
   The observability matrix is

$$\mathbf{V} = \begin{bmatrix} \mathbf{C}^* \\ \mathbf{C}^*\mathbf{A}^* \\ \mathbf{C}^*\mathbf{A}^{*2} \end{bmatrix} = \begin{bmatrix} 1 & 0 & 0 \\ 0 & 1 & 0 \\ 64.4 & 0 & -16 \end{bmatrix} \quad (10\text{-}311)$$

which has a rank of 3. Thus, the system is completely observable.

2. $y(t) = $ all velocity $= dx(t)/dt$: $\mathbf{C}^* = \begin{bmatrix} 0 & 1 & 0 \end{bmatrix}$
   The observability matrix is

$$\mathbf{V} = \begin{bmatrix} \mathbf{C}^* \\ \mathbf{C}^*\mathbf{A}^* \\ \mathbf{C}^*\mathbf{A}^{*2} \end{bmatrix} = \begin{bmatrix} 0 & 1 & 0 \\ 64.4 & 0 & -16 \\ 0 & 64.4 & 1600 \end{bmatrix} \quad (10\text{-}312)$$

which has a rank of 3. Thus, the system is completely observable.

3. $y(t) =$ winding current $= i(t)$: $\mathbf{C}^* = [0\ 0\ 1]$
   The observability matrix is

$$\mathbf{V} = \begin{bmatrix} \mathbf{C}^* \\ \mathbf{C}^*\mathbf{A}^* \\ \mathbf{C}^*\mathbf{A}^{*2} \end{bmatrix} = \begin{bmatrix} 0 & 0 & 1 \\ 0 & 0 & -100 \\ 0 & 0 & -10{,}000 \end{bmatrix} \quad (10\text{-}313)$$

which has a rank of 1. Thus, the system is unobservable. The physical interpretation of this result is that, if we choose the current $i(t)$ as the measurable output, we would not be able to reconstruct the state variables from the measured information.

The interested reader can enter the data of this system into any available computer program and verify the results obtained.

▶ 10-16 STATE-FEEDBACK CONTROL

A majority of the design techniques in modern control theory is based on the state-feedback configuration. That is, instead of using controllers with fixed configurations in the forward or feedback path, control is achieved by feeding back the state variables through real constant gains. The block diagram of a system with state-feedback control is shown in Fig. 9-2(c). A more detailed block diagram is shown in Fig. 10-28.

We can show that the PID control and the tachometer-feedback control discussed earlier are all special cases of the state-feedback control scheme. In the case of tachometer-feedback control, let us consider the second-order prototype system described in Eq. (5-87). The process is decomposed by direct decomposition and is represented by the state diagram of Fig. 10-29(a). If the states $x_1(t)$ and $x_2(t)$ are physically accessible, these variables may be fed back through constant real gains $-k_1$ and $-k_2$, respectively, to form the control $u(t)$, as shown in Fig. 10-29(b). The transfer function of the system with state feedback is

$$\frac{Y(s)}{R(s)} = \frac{\omega_n^2}{s^2 + (2\zeta\omega_n + K_2)s + K_1} \quad (10\text{-}314)$$

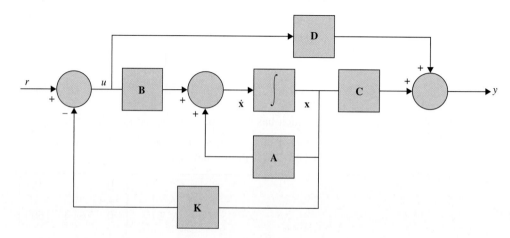

Figure 10-28 Block diagram of a control system with state feedback.

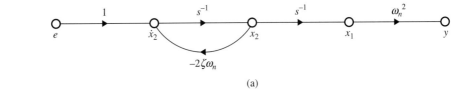

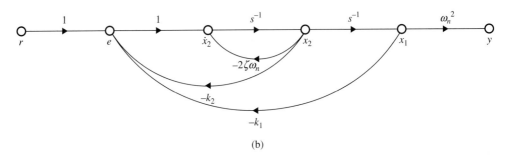

Figure 10-29 Control of a second-order system by state feedback.

For comparison purposes, we display the transfer functions of the systems with tachometer feedback and with PD control as follows:

Tachometer feedback:

$$\frac{Y(s)}{R(s)} = \frac{\omega_n^2}{s^2 + (2\zeta\omega_n + K_t\omega_n^2)s + \omega_n^2} \qquad (10\text{-}315)$$

PD control:

$$\frac{Y(s)}{R(s)} = \frac{\omega_n^2(K_P + K_D s)}{s^2 + (2\zeta\omega + K_D\omega_n^2)s + \omega_n^2 K_P} \qquad (10\text{-}316)$$

Thus, tachometer feedback is equivalent to state feedback if $k_1 = \omega_n^2$ and $k_2 = K_t\omega_n^2$. Comparing Eq. (10-314) with Eq. (10-316), we see that the characteristic equation of the system with state feedback would be identical to that of the system with PD control if $k_1 = \omega_n^2 K_P$ and $k_2 = \omega_n^2 K_D$. However, the numerators of the two transfer functions are different.

The systems with zero reference input, $r(t) = 0$, are commonly known as **regulators**. When $r(t) = 0$, the control objective is to drive any arbitrary initial conditions of the system to zero in some prescribed manner, for example, "as quickly as possible." Then a second-order system with PD control is the same as state-feedback control.

It should be emphasized that the comparisons just made are all for second-order systems. For higher-order systems, the PD control and tachometer-feedback control are equivalent to feeding back only the state variables x_1 and x_2, while state-feedback control feeds back all the state variables.

Because PI control increases the order of the system by one, it cannot be made equivalent to state feedback through constant gains. We show in Section 10-18 that if we combine state feedback with integral control we can again realize PI control in the sense of state-feedback control.

10-17 POLE-PLACEMENT DESIGN THROUGH STATE FEEDBACK

When root loci are utilized for the design of control systems, the general approach may be described as that of **pole placement**; the poles here refer to that of the closed-loop transfer function, which are also the roots of the characteristic equation. Knowing the relation between the closed-loop poles and the system performance, we can effectively carry out the design by specifying the location of these poles.

The design methods discussed in the preceding sections are all characterized by the property that the poles are selected based on what can be achieved with the fixed-controller configuration and the physical range of the controller parameters. A natural question would be: *Under what condition can the poles be placed arbitrarily?* This is an entirely new design philosophy and freedom that apparently can be achieved only under certain conditions.

When we have a controlled process of the third order or higher, the PD, PI, single-stage phase-lead, and phase-lag controllers would not be able to control independently all the poles of the system, because there are only two free parameters in each of these controllers.

To investigate the condition required for arbitrary pole placement in an nth-order system, let us consider that the process is described by the following state equation:

$$\frac{d\mathbf{x}(t)}{dt} = \mathbf{A}\mathbf{x}(t) + \mathbf{B}u(t) \quad (10\text{-}317)$$

where $\mathbf{x}(t)$ is an $n \times 1$ state vector, and $u(t)$ is the scalar control. The state-feedback control is

$$u(t) = -\mathbf{K}\mathbf{x}(t) + r(t) \quad (10\text{-}318)$$

where $\mathbf{K}$ is the $1 \times n$ feedback matrix with constant-gain elements. By substituting Eq. (10-318) into Eq. (10-317), the closed-loop system is represented by the state equation

$$\frac{d\mathbf{x}(t)}{dt} = (\mathbf{A} - \mathbf{B}\mathbf{K})\mathbf{x}(t) + \mathbf{B}r(t) \quad (10\text{-}319)$$

It will be shown in the following that if the pair [$\mathbf{A}, \mathbf{B}$] is completely controllable, then a matrix $\mathbf{K}$ exists that can give an arbitrary set of eigenvalues of $(\mathbf{A} - \mathbf{B}\mathbf{K})$; that is, the n roots of the characteristic equation

$$|s\mathbf{I} - \mathbf{A} + \mathbf{B}\mathbf{K}| = 0 \quad (10\text{-}320)$$

can be arbitrarily placed. To show that this is true, that if a system is completely controllable, it can always be represented in the controllable canonical form (CCF); that is, in Eq. (10-317),

$$\mathbf{A} = \begin{bmatrix} 0 & 1 & 0 & \cdots & 0 \\ 0 & 0 & 1 & \cdots & 0 \\ \vdots & \vdots & \vdots & \ddots & \vdots \\ 0 & 0 & 0 & \cdots & 1 \\ -a_0 & -a_1 & -a_2 & \cdots & -a_{n-1} \end{bmatrix} \quad \mathbf{B} = \begin{bmatrix} 0 \\ 0 \\ \vdots \\ 0 \\ 1 \end{bmatrix} \quad (10\text{-}321)$$

The feedback gain matrix **K** is expressed as

$$\mathbf{K} = [k_1 \quad k_2 \quad \cdots \quad k_n] \tag{10-322}$$

where $k_1, k_2, \ldots, k_n$ are real constants. Then,

$$\mathbf{A} - \mathbf{BK} = \begin{bmatrix} 0 & 1 & 0 & \cdots & 0 \\ 0 & 0 & 1 & \cdots & 0 \\ \vdots & \vdots & \vdots & \ddots & \vdots \\ 0 & 0 & 0 & \cdots & 1 \\ -a_0 - k_1 & -a_1 - k_2 & -a_2 - k_3 & \cdots & -a_{n-1} - k_n \end{bmatrix} \tag{10-323}$$

The eigenvalues of $\mathbf{A} - \mathbf{BK}$ are then found from the characteristic equation

$$|s\mathbf{I} - (\mathbf{A} - \mathbf{BK})| = s^n + (a_{n-1} + k_n)s^{n-1} + (a_{n-2} + k_{n-1})s^{n-2} + \cdots + (a_0 + k_1) = 0 \tag{10-324}$$

Clearly, the eigenvalues can be arbitrarily assigned, because the feedback gains $k_1, k_2, \ldots k_n$ are isolated in each coefficient of the characteristic equation. Intuitively, it makes sense that a system must be controllable for the poles to be placed arbitrarily. If one or more state variables are uncontrollable, then the poles associated with these state variables are also uncontrollable and cannot be moved as desired. The following example illustrates the design of a control system with state feedback.

▶ **EXAMPLE 10-17-1** Consider the magnetic-ball suspension system analyzed in Section 10-15. This is a typical regulator system for which the control problem is to maintain the ball at its equilibrium position. It is shown in Section 10-15 that the system without control is unstable.

The linearized state model of the magnetic-ball system is represented by the state equation

$$\frac{d\Delta \mathbf{x}(t)}{dt} = \mathbf{A}^* \Delta \mathbf{x}(t) + \mathbf{B}^* \Delta v(t) \tag{10-325}$$

where $\Delta \mathbf{x}(t)$ denotes the linearized state vector, and $\Delta v(t)$ is the linearized input voltage. The coefficient matrices are

$$\mathbf{A}^* = \begin{bmatrix} 0 & 1 & 0 \\ 64.4 & 0 & -16 \\ 0 & 0 & -100 \end{bmatrix} \quad \mathbf{B}^* = \begin{bmatrix} 0 \\ 0 \\ 100 \end{bmatrix} \tag{10-326}$$

The eigenvalues of $\mathbf{A}^*$ are $s = -100, -8.025,$ and 8.025. Thus, the system without feedback control is unstable.

Let us give the following design specifications:

1. The system must be stable.
2. For any initial disturbance on the position of the ball from its equilibrium position, the ball must return to the equilibrium position with zero steady-state error.
3. The time response should settle to within 5% of the initial disturbance in not more than 0.5 sec.
4. The control is to be realized by state feedback

$$\Delta v(t) = -\mathbf{K}\Delta \mathbf{x}(t) = -[k_1 \quad k_2 \quad k_3]\Delta \mathbf{x}(t) \tag{10-327}$$

where $k_1, k_2,$ and k_3 are real constants.

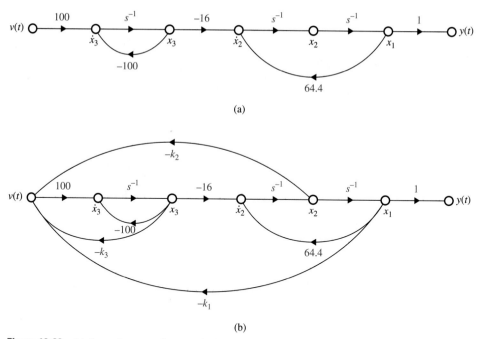

Figure 10-30 (a) State diagram of magnetic-ball-suspension system. (b) State diagram of magnetic-ball-suspension system with state feedback.

A state diagram of the "open-loop" ball-suspension system is shown in Fig. 10-30(a), and the same of the "closed-loop" system with state feedback is shown in Fig. 10-30(b).

We must select the desired location of the eigenvalues of $(s\mathbf{I} - \mathbf{A}^* + \mathbf{B}^*\mathbf{K})$ so that requirement 3 in the preceding list on the time response is satisfied. Without entirely resorting to trial and error, we can start with the following decisions:

1. The system dynamics should be controlled by two dominant roots.
2. To achieve a relatively fast response, the two dominant roots should be complex.
3. The damping that is controlled by the real parts of the complex roots should be adequate, and the imaginary parts should be high enough for the transient to die out sufficiently fast.

After a few trial-and-error runs, using the **ACSYS/MATLAB** tool, we found that the following characteristic equation roots should satisfy the design requirements:

$$s = -20 \quad s = -6 + j4.9 \quad s = -6 - j4.9$$

The corresponding characteristic equation is

$$s^3 + 32s^2 + 300s + 1200 = 0 \tag{10-328}$$

The characteristic equation of the closed-loop system with state feedback is written

$$|s\mathbf{I} - \mathbf{A}^* + \mathbf{B}^*\mathbf{K}| = \begin{vmatrix} s & -1 & 0 \\ -64.4 & s & 16 \\ 100k_1 & 100k_2 & s + 100 + 100k_3 \end{vmatrix}$$

$$= s^3 + 100(k_3 + 1)s^2 - (64.4 + 1600k_2)s - 1600k_1 - 6440(k_3 + 1) = 0 \tag{10-329}$$

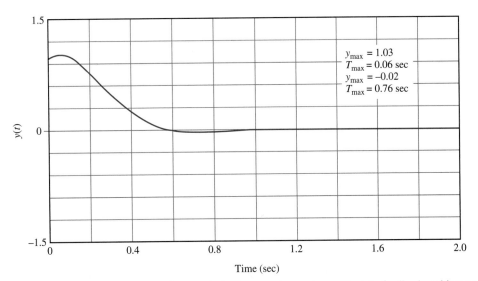

Figure 10-31 Output response of magnetic-ball-suspension system with state feedback, subject to initial condition $y(0) = x_1(0) = 1$.

which can also be obtained directly from Fig. 10-30(b) using the SFG gain formula. Equating like coefficients of Eqs. (10-328) and (10-329), we get the following simultaneous equations:

$$100(k_3 + 1) = 32$$
$$-64.4 - 1600k_2 = 300 \quad (10\text{-}330)$$
$$-1600k_1 - 6440(k_3 + 1) = 1200$$

Solving the last three equations, and being assured that the solutions exist and are unique, we get the feedback-gain matrix

$$\mathbf{K} = [k_1 \quad k_2 \quad k_3] = [-2.038 \quad -0.22775 \quad -0.68] \quad (10\text{-}331)$$

Fig. 10.31 shows the output response $y(t)$ when the system is subject to the initial condition

$$\mathbf{x}(0) = \begin{bmatrix} 1 \\ 0 \\ 0 \end{bmatrix} \quad (10\text{-}331)$$

◀

▶ **EXAMPLE 10-17-2** In this example, we shall design a state-feedback control for the second-order sun-seeker system treated in Example 4-11-2 and throughout Chapter 9. The CCF state diagram of the process with $K = 1$ is shown in Fig. 10-32(a). The problem involves the design of a state-feedback control with

$$\theta_e(t) = -\mathbf{K}\mathbf{x}(t) = -[k_1 \quad k_2]\mathbf{x}(t) \quad (10\text{-}332)$$

The state equations are represented in vector-matrix form as

$$\frac{d\mathbf{x}(t)}{dt} = \mathbf{A}\mathbf{x}(t) + \mathbf{B}\theta_e(t) \quad (10\text{-}333)$$

where

$$\mathbf{A} = \begin{bmatrix} 0 & 1 \\ 0 & -25 \end{bmatrix} \quad \mathbf{B} = \begin{bmatrix} 0 \\ 1 \end{bmatrix} \quad (10\text{-}334)$$

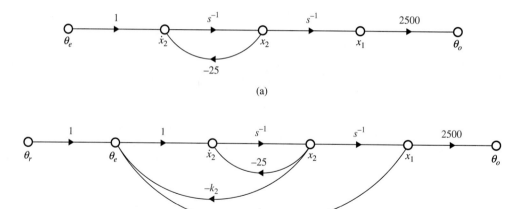

Figure 10-32 (a) State diagram of second-order sun-seeker system. (b) State diagram of second-order sun-seeker system with state feedback.

The output equation is

$$\theta_o(t) = \mathbf{C}\mathbf{x}(t) \tag{10-335}$$

where

$$\mathbf{C} = \begin{bmatrix} 1 & 0 \end{bmatrix} \tag{10-336}$$

The design objectives are as follows:

1. The steady-state error due to a step function input should equal 0.
2. With the state-feedback control, the unit-step response should have minimum overshoot, rise time, and settling time.

The transfer function of the system with state feedback is written

$$\frac{\Theta_o(s)}{\Theta_r(s)} = \frac{2500}{s^2 + (25 + k_2)s + k_1} \tag{10-337}$$

Thus, for a step input, if the output has zero steady-state error, the constant terms in the numerator and denominator must be equal to each other—that is, $k_1 = 2500$. This means that, while the system is completely controllable, we cannot arbitrarily assign the two roots of the characteristic equation, which is now

$$s^2 + (25 + k_2)s + 2500 = 0 \tag{10-338}$$

In other words, only one of the roots of Eq. (10-338) can be arbitrarily assigned. The problem is solved using **ACSYS**. After a few trial-and-error runs, we found out that the maximum overshoot, rise time, and settling time are all at a minimum when $k_2 = 75$. The two roots are $s = -50$ and -50. The attributes of the unit-step response are

$$\text{maximum overshoot} = 0\% \quad t_r = 0.06717 \text{ sec} \quad t_s = 0.09467 \text{ sec}$$

The state-feedback gain matrix is

$$\mathbf{K} = \begin{bmatrix} 2500 & 75 \end{bmatrix} \tag{10-339}$$

The lesson that we learned from this illustrative example is that state-feedback control generally produces a system that is type 0. For the system to track a step input without steady-state error, which requires a type 1 or higher-type system, the feedback gain k_1 of the system in the CCF state diagram cannot be assigned arbitrarily. This means that, for an nth-order system, only $n - 1$ roots of the characteristic equation can be placed arbitrarily. ◀

▶ 10-18 STATE FEEDBACK WITH INTEGRAL CONTROL

The state-feedback control structured in the preceding section has one deficiency in that it does not improve the type of the system. As a result, the state-feedback control with constant-gain feedback is generally useful only for regulator systems for which the system does not track inputs, if all the roots of the characteristic equation are to be placed at will.

In general, most control systems must track inputs. One solution to this problem is to introduce integral control, just as with PI controller, together with the constant-gain state feedback. The block diagram of a system with constant-gain state feedback and integral control feedback of the output is shown in Fig. 10-33. The system is also subject to a noise input $n(t)$. For a SISO system, the integral control adds one integrator to the system. As shown in Fig. 10-33, the output of the $(n + 1)$st integrator is designated as x_{n+1}. The dynamic equations of the system in Fig. 10-33 are written as

$$\frac{d\mathbf{x}(t)}{dt} = \mathbf{A}\mathbf{x}(t) + \mathbf{B}u(t) + \mathbf{E}n(t) \qquad (10\text{-}340)$$

$$\frac{dx_{n+1}(t)}{dt} = r(t) - y(t) \qquad (10\text{-}341)$$

$$y(t) = \mathbf{C}\mathbf{x}(t) + Du(t) \qquad (10\text{-}342)$$

where $\mathbf{x}(t)$ is the $n \times 1$ state vector; $u(t)$ and $y(t)$ are the scalar actuating signal and output, respectively; $r(t)$ is the scalar reference input; and $n(t)$ is the scalar disturbance input. The coefficient matrices are represented by $\mathbf{A}, \mathbf{B}, \mathbf{C}, D$ and $\mathbf{E}$, with appropriate dimensions. The actuating signal $u(t)$ is related to the state variables through constant-state and integral feedback,

$$u(t) = -\mathbf{K}\mathbf{x}(t) - k_{n+1}x_{n+1}(t) \qquad (10\text{-}343)$$

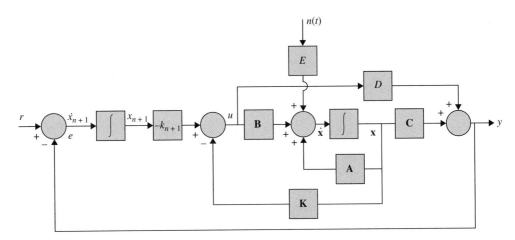

Figure 10-33 Block diagram of a control system with state feedback and integral output feedback.

where

$$\mathbf{K} = \begin{bmatrix} k_1 & k_2 & k_3 & \cdots & k_n \end{bmatrix} \qquad (10\text{-}344)$$

with constant real gain elements, and k_{n+1} is the scalar integral-feedback gain.

Substituting Eq. (10-343) into Eq. (10-340) and combining with Eq. (10-341), the $n + 1$ state equations of the overall system with constant-gain and integral feedback are written

$$\frac{d\overline{\mathbf{x}}(t)}{dt} = (\overline{\mathbf{A}} - \overline{\mathbf{B}}\,\overline{\mathbf{K}})\overline{\mathbf{x}}(t) + \begin{bmatrix} 0 \\ 1 \end{bmatrix} r(t) + \overline{\mathbf{E}}n(t) \qquad (10\text{-}345)$$

where

$$\overline{\mathbf{x}}(t) = \begin{bmatrix} \dfrac{d\mathbf{x}(t)}{dt} \\ \dfrac{dx_{n+1}(t)}{dt} \end{bmatrix} \qquad (n+1) \times 1 \qquad (10\text{-}346)$$

$$\overline{\mathbf{A}} = \begin{bmatrix} \mathbf{A} & \mathbf{0} \\ -\mathbf{C} & 0 \end{bmatrix} \quad (n+1) \times (n+1) \qquad \overline{\mathbf{B}} = \begin{bmatrix} \mathbf{B} \\ D \end{bmatrix} \quad (n+1) \times 1 \qquad (10\text{-}347)$$

$$\overline{\mathbf{K}} = \begin{bmatrix} \mathbf{K} & k_{n+1} \end{bmatrix} = \begin{bmatrix} k_1 & k_2 & \cdots & k_n & k_{n+1} \end{bmatrix} \quad 1 \times (n+1) \qquad (10\text{-}348)$$

$$\overline{\mathbf{E}} = \begin{bmatrix} \mathbf{E} \\ 0 \end{bmatrix} \quad [(n+1) \times 1] \qquad (10\text{-}349)$$

Substituting Eq. (10-343) into Eq. (10-342), the output equation of the overall system is written

$$y(t) = \overline{\mathbf{C}}\,\overline{\mathbf{x}}(t) \qquad (10\text{-}350)$$

where

$$\overline{\mathbf{C}} = \begin{bmatrix} \mathbf{C} - D\mathbf{K} & D\mathbf{K} \end{bmatrix} \quad [1 \times (n+1)] \qquad (10\text{-}351)$$

The design objectives are as follows:

1. The steady-state value of the output $y(t)$ follows a step-function input with zero error; that is,

$$e_{ss} = \lim_{t \to \infty} e(t) = 0 \qquad (10\text{-}352)$$

2. The $n + 1$ eigenvalues of $(\overline{\mathbf{A}} - \overline{\mathbf{B}}\,\overline{\mathbf{K}})$ are placed at desirable locations. For the last condition to be possible, the pair $[\overline{\mathbf{A}}, \overline{\mathbf{B}}]$ must be completely controllable.

The following example illustrates the applications of state-feedback with integral control.

▶ **EXAMPLE 10-18-1** We have shown in Example 10-17-2 that, with constant-gain state-feedback control, the second-order sun-seeker system can have only one of its two roots placed at will for the system to track a step input without steady-state error. Now let us consider the same second-order sun-seeker system in Example 10-17-2, except that an integral control is added to the forward path. The state diagram of the overall system is shown in Fig. 10-34. The coefficient matrices are

$$\mathbf{A} = \begin{bmatrix} 0 & 1 \\ 0 & -25 \end{bmatrix} \quad \mathbf{B} = \begin{bmatrix} 0 \\ 1 \end{bmatrix} \quad \mathbf{C} = [2500 \ 0] \quad D = 0 \quad (10\text{-}353)$$

From Eq. (10-347),

$$\overline{\mathbf{A}} = \begin{bmatrix} \mathbf{A} & \mathbf{0} \\ -\mathbf{C} & 0 \end{bmatrix} = \begin{bmatrix} 0 & 1 & 0 \\ 0 & -25 & 0 \\ -2500 & 0 & 0 \end{bmatrix} \quad \overline{\mathbf{B}} = \begin{bmatrix} \mathbf{B} \\ D \end{bmatrix} = \begin{bmatrix} 0 \\ 1 \\ 0 \end{bmatrix} \quad (10\text{-}354)$$

We can show that the pair $[\overline{\mathbf{A}}, \overline{\mathbf{B}}]$ is completely controllable. Thus, the eigenvalues of $(s\mathbf{I} - \overline{\mathbf{A}} + \overline{\mathbf{B}}\overline{\mathbf{K}})$ can be arbitrarily placed. Substituting $\overline{\mathbf{A}}$, $\overline{\mathbf{B}}$, and $\overline{\mathbf{K}}$ in the characteristic equation of the closed-loop system with state and integral feedback, we have

$$|s\mathbf{I} - \overline{\mathbf{A}} + \overline{\mathbf{B}}\overline{\mathbf{K}}| = \begin{vmatrix} s & -1 & 0 \\ k_1 & s + 25 + k_2 & k_3 \\ -2500 & 0 & s \end{vmatrix}$$
$$= s^3 + (25 + k_2)s^2 + k_1 s + 2500 k_3 = 0 \quad (10\text{-}355)$$

which can also be found from Fig. 10-34 using the SFG gain formula.

The design objectives are as follows:

1. The steady-state output must follow a step function input with zero error.
2. The rise time and settling time must be less than 0.05 sec.
3. The maximum overshoot of the unit-step response must be less than 5%.

Because all three roots of the characteristic equation can be placed arbitrarily, it is not realistic to require minimum rise and settling times, as in Example 10-17-2.

Again, to realize a fast rise time and settling time, the roots of the characteristic equation should be placed far to the left in the s-plane, and the natural frequency should be high. Keep in mind that *roots with large magnitudes will lead to high gains for the state-feedback matrix.*

The **ACSYS/MATLAB** software was used to carry out the design. After a few trial-and-error runs, the design specifications can be satisfied by placing the roots at

$$s = -200 \quad -50 + j50 \quad \text{and} \quad -50 - j50$$

The desired characteristic equation is

$$s^3 + 300s^2 + 25,000s + 1,000,000 = 0 \quad (10\text{-}356)$$

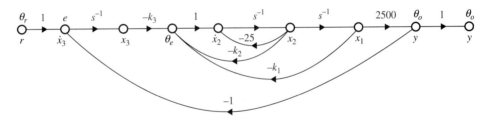

Figure 10-34 Sun-seeker system with state feedback and integral control in Example 10-18-1.

Equating like coefficients of Eqs. (10-355) and (10-356), we get

$$k_1 = 25,000 \quad k_2 = 275 \quad \text{and} \quad k_3 = 400$$

The attributes of the unit-step response are as follows:

Maximum overshoot = 4%
$t_r = 0.03247$ sec
$t_s = 0.04667$ sec

Notice that the high feedback gain of k_1, which is due to the large values of the roots selected, may pose physical problems; if so, the design specifications may have to be revised. ◀

▶ **EXAMPLE 10-18-2** In this example we illustrate the application of state-feedback with integral control to a system with a disturbance input.

Consider a dc-motor control system that is described by the following state equations:

$$\frac{d\omega(t)}{dt} = \frac{-B}{J}\omega(t) + \frac{K_i}{J}i_a(t) - \frac{1}{J}T_L \tag{10-357}$$

$$\frac{di_a(t)}{dt} = \frac{-K_b}{L}\omega(t) - \frac{R}{L}i_a(t) + \frac{1}{L}e_a(t) \tag{10-358}$$

where

$i_a(t)$ = armature current, A
$e_a(t)$ = armature applied voltage, V
$\omega(t)$ = motor velocity, rad/sec
B = viscous-friction coefficient of motor and load = 0
J = moment of inertia of motor and load = 0.02 N-m/rad/sec^2
K_i = motor torque constant = 1 N-m/A
K_b = motor back-emf constant = 1 V/rad/sec
T_L = constant load torque(magnitude not known), N-m
L = armature inductance = 0.005 H
R = armature resistance = 1 Ω

The output equation is

$$y(t) = \mathbf{Cx}(t) = [1 \quad 0]\mathbf{x}(t) \tag{10-359}$$

The design problem is to find the control $u(t) = e_a(t)$ through state feedback and integral control such that

1. $\lim_{t \to \infty} i_a(t) = 0$ and $\lim_{t \to \infty} \frac{d\omega(t)}{dt} = 0$ \hfill (10-360)

2. $\lim_{t \to \infty} \omega(t)$ = step input $r(t) = u_s(t)$ \hfill (10-361)

3. The eigenvalues of the closed-loop system with state feedback and integral control are at $s = -300, -10 + j10,$ and $-10 - j10$.

Let the state variables be defined as $x_1(t) = \omega(t)$ and $x_2(t) = i_a(t)$. The state equations in Eqs. (10-357) and (10-358) are written in vector-matrix form:

$$\frac{d\mathbf{x}(t)}{dt} = \mathbf{Ax}(t) + \mathbf{B}u(t) + \mathbf{E}n(t) \tag{10-362}$$

where $n(t) = T_L u_s(t)$.

$$\mathbf{A} = \begin{bmatrix} -\dfrac{B}{J} & \dfrac{K_i}{J} \\ -\dfrac{K_b}{L} & -\dfrac{R}{L} \end{bmatrix} = \begin{bmatrix} 0 & 50 \\ -200 & -200 \end{bmatrix} \quad (10\text{-}363)$$

$$\mathbf{B} = \begin{bmatrix} 0 \\ \dfrac{1}{L} \end{bmatrix} = \begin{bmatrix} 0 \\ 200 \end{bmatrix} \quad (10\text{-}364)$$

$$\mathbf{E} = \begin{bmatrix} -\dfrac{1}{J} \\ 0 \end{bmatrix} = \begin{bmatrix} -50 \\ 0 \end{bmatrix} \quad (10\text{-}365)$$

From Eq. (10-347),

$$\overline{\mathbf{A}} = \begin{bmatrix} \mathbf{A} & 0 \\ -\mathbf{C} & 0 \end{bmatrix} = \begin{bmatrix} 0 & 50 & 0 \\ -200 & -200 & 0 \\ -1 & 0 & 0 \end{bmatrix} \quad \overline{\mathbf{B}} = \begin{bmatrix} \mathbf{B} \\ 0 \end{bmatrix} = \begin{bmatrix} 0 \\ 200 \\ 0 \end{bmatrix} \quad (10\text{-}366)$$

$$\overline{\mathbf{C}} = \begin{bmatrix} \mathbf{C} & 0 \end{bmatrix} = \begin{bmatrix} 1 & 0 & 0 \end{bmatrix} \quad \overline{\mathbf{E}} = \begin{bmatrix} \mathbf{E} \\ 0 \end{bmatrix} = \begin{bmatrix} -50 \\ 0 \\ 0 \end{bmatrix} \quad (10\text{-}367)$$

The control is given by

$$u(t) = -\mathbf{K}\mathbf{x}(t) - k_{n+1}x_{n+1}(t) = \overline{\mathbf{K}}\overline{\mathbf{x}}(t) \quad (10\text{-}368)$$

where

$$\overline{\mathbf{K}} = \begin{bmatrix} k_1 & k_2 & k_3 \end{bmatrix} \quad (10\text{-}369)$$

Fig. 10-35 shows the state diagram of the overall designed system. The coefficient matrix of the closed-loop system is

$$\overline{\mathbf{A}} - \overline{\mathbf{B}}\,\overline{\mathbf{K}} = \begin{bmatrix} 0 & 50 & 0 \\ -200 - 200k_1 & -200 - 200k_2 & -200k_3 \\ -1 & 0 & 0 \end{bmatrix} \quad (10\text{-}370)$$

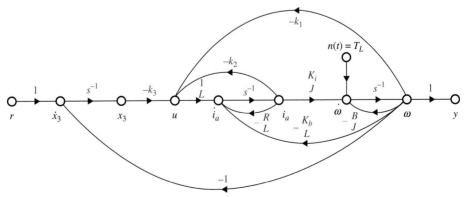

Figure 10-35 DC-motor control system with state feedback and integral control and disturbance torque in Example 10-18-2.

The characteristic equation is

$$|s\mathbf{I} - \overline{\mathbf{A}} + \overline{\mathbf{B}}\overline{\mathbf{K}}| = s^3 + 200(1 + k_2)s^2 + 10,000(1 + k_1)s - 10,000k_3 = 0 \quad (10\text{-}371)$$

which is more easily determined by applying the gain formula of SFG to Fig. 10-35.
For the three roots assigned, the last equation must equal

$$s^3 + 320s^2 + 6,200s + 60,000 = 0 \quad (10\text{-}372)$$

Equating the like coefficients of Eqs. (10-371) and (10-372), we get

$$k_1 = -0.38 \quad k_2 = 0.6 \quad k_3 = -6.0$$

Applying the SFG gain formula to Fig. 10-35 between the inputs $r(t)$ and $n(t)$ and the states $\omega(t)$ and $i_a(t)$, we have

$$\begin{bmatrix} \Omega(s) \\ I_a(s) \end{bmatrix} = \frac{1}{\Delta_c(s)} \begin{bmatrix} -\frac{1}{J}\left(s^2 + \frac{R}{L}s + \frac{k_2}{L}s\right) & -\frac{k_3 K_i}{JL} \\ -\frac{1}{J}\left(-\frac{K_b}{L}s - \frac{k_1}{L}s + \frac{k_3}{L}\right) & -\frac{k_3}{L}\left(s + \frac{B}{J}\right) \end{bmatrix} \begin{bmatrix} \frac{T_L}{s} \\ \frac{1}{s} \end{bmatrix} \quad (10\text{-}373)$$

where $\Delta_c(s)$ is the characteristic polynomial given in Eq. (10-372).

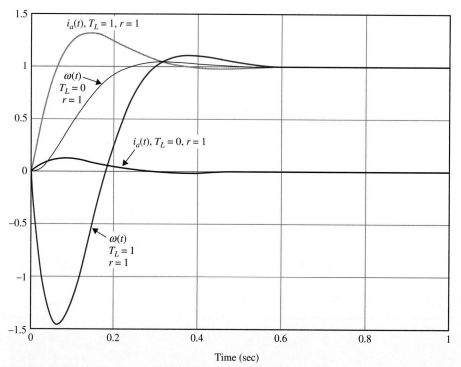

Figure 10-36 Time responses of dc-motor control system with state feedback and integral control and disturbance torque in Example 10-18-2.

Applying the final-value theorem to the last equation, the steady-state values of the state variables are found to be

$$\lim_{t \to \infty} \begin{bmatrix} \omega(t) \\ i_a(t) \end{bmatrix} = \lim_{s \to 0} s \begin{bmatrix} \Omega(s) \\ I_a(s) \end{bmatrix} = \begin{bmatrix} 0 & K_i \\ 1 & B \end{bmatrix} \begin{bmatrix} T_L \\ 1 \end{bmatrix} = \begin{bmatrix} 1 \\ T_L \end{bmatrix} \quad (10\text{-}374)$$

Thus, the motor velocity $\omega(t)$ will approach the constant reference input step function $r(t) = u_s(t)$ as t approaches infinity, independent of the disturbance torque T_L. Substituting the system parameters into Eq. (10-373), we get

$$\begin{bmatrix} \Omega(s) \\ I_a(s) \end{bmatrix} = \frac{1}{\Delta_c(s)} \begin{bmatrix} -50(s+320)s & 60,000 \\ 6200s + 60,000 & 1,200s \end{bmatrix} \begin{bmatrix} T_L \\ s \\ \dfrac{1}{s} \end{bmatrix} \quad (10\text{-}375)$$

Fig. 10-36 shows the time responses of $\omega(t)$ and $i_a(t)$ when $T_L = 1$ and $T_L = 0$. The reference input is a unit-step function. ◀

▶ 10-19 MATLAB TOOLS AND CASE STUDIES

In this section we present a MATLAB tool to solve most problems addressed in this chapter. The reader is encouraged to apply this tool to all the problems identified by a MATLAB Toolbox in the left margin of the text throughout this chapter. We use MATLAB's Symbolic Tool to solve some of the initial problems in this chapter involving inverse Laplace transformations. We will also use MATLAB to convert from transfer functions to state-space representation. These programs allow the user to conduct the following tasks:

- Enter the state matrices.
- Find the system's characteristic polynomial, eigenvalues, and eigenvectors.
- Find the similarity transformation matrices.
- Examine the system controllability and observability properties.
- Obtain the step, impulse, and natural (response to initial conditions) responses, as well as the time response to any function of time.
- Use MATLAB Symbolic Tool to find the state-transition matrix using the inverse Laplace command.
- Convert a transfer function to state-space form or vice versa.

To better illustrate how to use the software, let us go through some of the steps involved in solving earlier examples in this chapter.

10-19-1 Description and Use of the State-Space Analysis Tool

The State-Space Analysis Tool (statetool) consists of a number of m-files and GUIs for the analysis of state-space systems. The statetool can be invoked from the **A**utomatic **C**ontrol **S**ystems launch applet (**ACSYS**) by clicking on the appropriate button. You will then see the window pictured in Fig. 10-37. We use the example in Section 10-16 and Examples 10-5-1 and 10-5-2 to describe how to use the statetool.

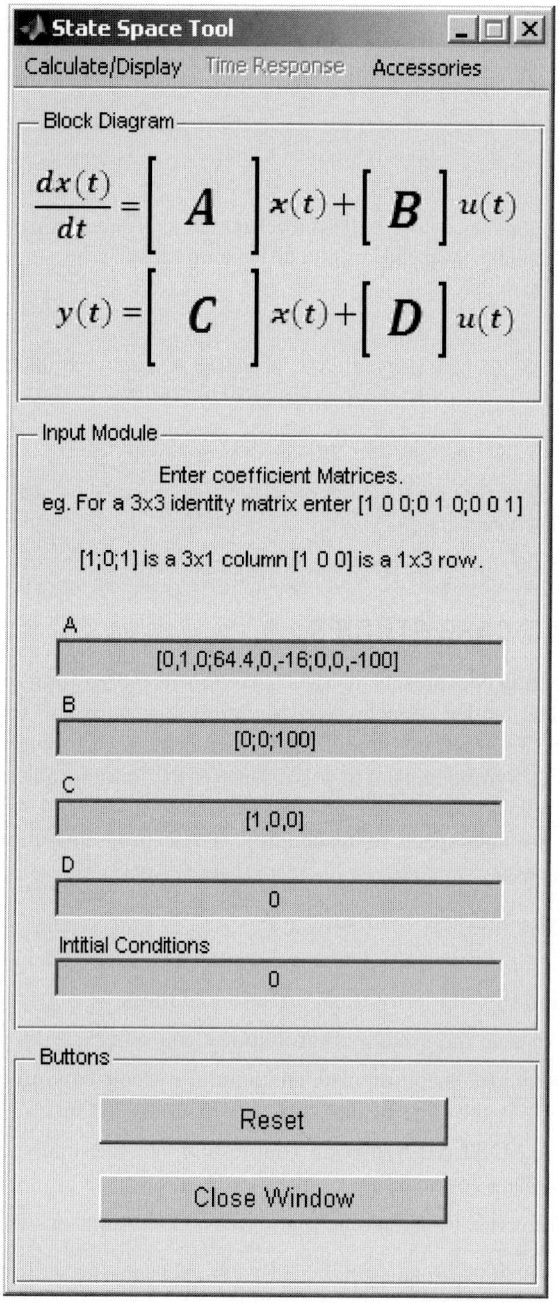

Chapter 10. State Variable Analysis

Figure 10-37 The State-Space Analysis window.

First consider the example in Section 10-14. To enter the following coefficient matrices,

$$\mathbf{A}^* = \begin{bmatrix} 0 & 1 & 0 \\ 64.4 & 0 & -16 \\ 0 & 0 & -100 \end{bmatrix} \quad \mathbf{B}^* = \begin{bmatrix} 0 \\ 0 \\ 100 \end{bmatrix} \quad \mathbf{C}^* = \begin{bmatrix} 1 & 0 & 0 \end{bmatrix} \quad (10\text{-}378)$$

enter the values in the appropriate edit boxes. Note that the default value of initial conditions is set to zero and you do not have to adjust it for this example. Follow the instructions on the screen very carefully. The elements in the row of a matrix may be separated by a space or a comma, while the rows themselves must be separated by a semicolon. For example, to enter matrix **A**, enter [0,1,0;64.4,0,-16;0,0,-100] in the **A** edit box, and to enter matrix **B**, enter [0;0;100] in the **B** edit box, as shown in Fig. 10-38. In this case, the **D** matrix is set to zero (default value). To find the characteristic Eq. (10-270),

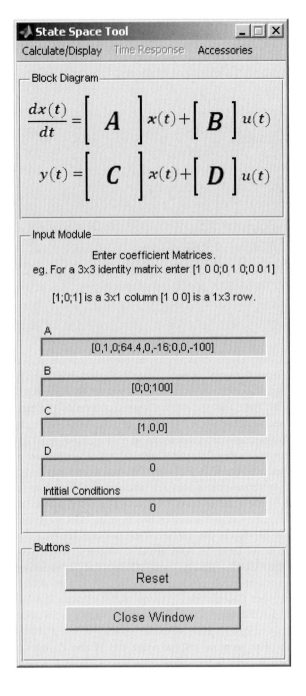

Figure 10-38 Inputing values in the State-Space window.

> The A matrix is:
>
> Amat =
>
> | | | |
> |-------:|-------:|---------:|
> | 0 | 1.0000 | 0 |
> | 64.4000| 0 | -16.0000 |
> | 0 | 0 |-100.0000 |
>
> Characteristic polynomial:
>
> ans =
>
> s^3+100*s^2-2265873562520787/35184372088832*s-6440
>
> Eigenvalues of A = diagonal canonical form of A is:
>
> Abar =
>
> | | | |
> |-------:|--------:|---------:|
> | 8.0250 | 0 | 0 |
> | 0 | -8.0250 | 0 |
> | 0 | 0 |-100.0000 |
>
> Eigenvectors are
>
> T =
>
> | | | |
> |-------:|--------:|--------:|
> | 0.1237 | -0.1237 | -0.0016 |
> | 0.9923 | 0.9923 | 0.1590 |
> | 0 | 0 | 0.9873 |

Figure 10-39 The MATLAB command window display after clicking the "Eigenvals & vects of A" button.

eigenvalues, and eigenvectors, choose the "Eigenvals & vects of A" option from the Calculate/Display menu. The detailed solution will be displayed on the MATLAB command window. The **A** matrix, eigenvalues of **A**, and eigenvectors of **A** are shown in Fig. 10-39. Note that the matrix representation of the eigenvalues corresponds to the diagonal canonical form (DCF) of **A**, while matrix **T**, representing the eigenvectors, is the DCF transformation matrix discussed in Section 10-9-4. To find the state-transition matrix $\phi(t)$, you must use the tfsym tool, which will be discussed in Section 10-19-2.

The choice of the **C** in Eq. (10-376) makes the ball position the output $y(t)$ for input $v(t)$. Then the input–output transfer function of the system can be obtained by choosing the "State-Space Calculations" option. The final output appearing in the MATLAB command window is the transfer function in both polynomial and factored forms, as shown in Fig. 10-40. As you can see, there is a small error due to numerical simulation. You may set the small terms to zero in the resulting transfer function to get Eq. (10-309).

Click the "Controllability" and "Observability" menu options to determine whether the system is controllable or observable. Note these options are only enabled after pressing the "State-Space Calculations" button. After clicking the "Controllability" button, you get the MATLAB command window display, shown in Fig. 10-41. The **S** matrix in this case is the same as Eq. (10-310) with the rank of 3. As a result, the system is completely controllable. The program also provides the **M** and **P** matrices and the system controllability canonical form (CCF) representation as defined in Section 10-9.

State-space model is:

a =

```
         x1     x2     x3
x1        0      1      0
x2      64.4    0     -16
x3        0     0    -100
```

b =

```
         u1
x1        0
x2        0
x3      100
```

c =

```
         x1    x2    x3
y1        1     0     0
```

d =

```
         u1
y1        0
```

Continuous-time model.
Characteristic polynomial:

ans =

s^3+100*s^2-2265873562520787/35184372088832*s − 6440

Equivalent transfer-function model is:

Transfer function:
$$\frac{4.263e - 014\ s^2 + 8.527e-014\ s - 1600}{s^3 + 100s^2 - 64.4s - 6440}$$

Pole, zero form:

Zero/pole/gain:
$$\frac{4.2633e - 014\ (s+1.937e008)(s - 1.937e008)}{(s+100)(s+8.025)(s - 8.025)}$$

Figure 10-40 The MATLAB command window after clicking the "State-Space Calculations" button.

The Controllibility matrix [B AB A^2B ...] is =

Smat =

```
     0          0      -1600
     0      -1600     160000
   100     -10000    1000000
```

The system is therefore controllable, rank of S matrix is =

rankS =

 3

Mmat =

```
  -64.4000   100.0000    1.0000
  100.0000     1.0000         0
    1.0000         0         0
```

The controllability canonical form (CCF) transformation matrix is:

Ptran =

```
  -1600         0      0
      0     -1600      0
  -6440         0    100
```

The transformed matrices using CCF are:

Abar =

1.0e+003 *

```
       0    0.0010         0
       0         0    0.0010
  6.4400    0.0644   -0.1000
```

Bbar =

0
0
1

Cbar =

-1600   0   0

Dbar =

0

Figure 10-41 The MATLAB command window after clicking the "Controllability" button.

The observability matrix (transpose:[C CA CA^2 ...]) is =

Vmat =

1.0000	0	0
0	1.0000	0
64.4000	0	-16.0000

The system is therefore observable, rank of V matrix is =

rankV =

   3

Mmat =

-64.4000	100.0000	1.0000
100.0000	1.0000	0
1.0000	0	0

The observability canonical form (OCF) transformation matrix is:

Qtran =

0	0	1.0000
0	1.0000	-100.0000
-0.0625	6.2500	-625.0000

The transformed matrices using OCF are:

Abar =

1.0e+003 *

0.0000	-0.0000	6.4400
0.0010	-0.0000	0.0644
0	0.0010	-0.1000

Bbar =

  -1600
     0
     0

Cbar =

  0 0 1

Dbar =

  0

Figure 10-42 The MATLAB command window after clicking the "Observability" button.

Once you choose the "Observability" option, the system observability is assessed in the MATLAB command window, as shown in Fig. 10-42. The system is completely observable, since the **V** matrix has a rank of 3. Note the **V** matrix in Fig. 10-42 is the same as in Eq. (10-311). The program also provides the **M** and **Q** matrices and the system observability canonical form (OCF) representation as defined in Section 10-9. As an exercise, the user is urged to reproduce Eqs. (10-312) and (10-313) using this software.

You may obtain the output $y(t)$ natural time response (response to initial conditions only), the step response, the impulse response, or the time response to any other input function by choosing the appropriate option from the Time Response menu.

The statetool program may be used on all the examples identified by a MATLAB Toolbox in the left margin of the text throughout this chapter, except problems involving inverse Laplace transformations and closed-form solutions. To address the analytical solutions, we need to use the tfsym tool, which requires the Symbolic Tool of MATLAB.

10-19-2 Description and Use of tfsym for State-Space Applications

You may run the Transfer Function Symbolic Tool by clicking the "Transfer Function Symbolic" button in the **ACSYS** window. You should get the window in Fig. 10-43. For

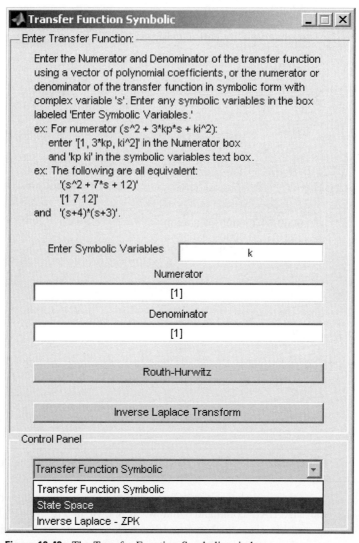

Figure 10-43 The Transfer Function Symbolic window.

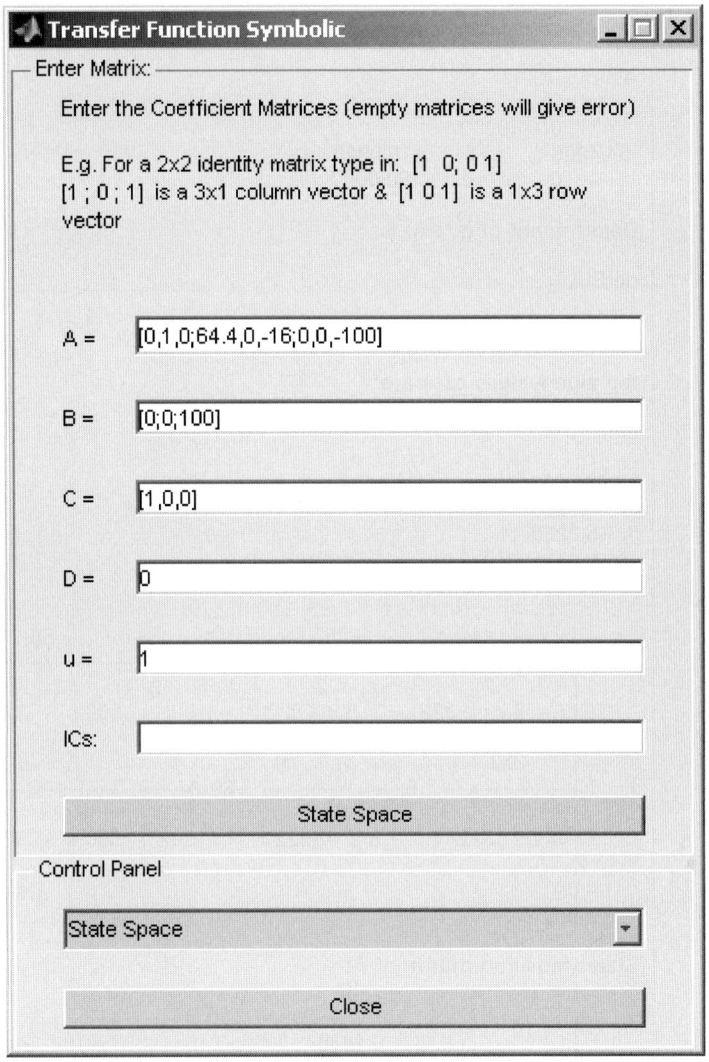

Figure 10-44 Inputting values into the Transfer Function Symbolic window.

this example we will use the State-Space mode. Choose the appropriate option from the drop-down menu as shown in Fig. 10-43.

Let us continue our solution to the example in Section 10-16. Fig. 10-44 shows the input of the matrices for this example into the state-space window. The input and output displays in the MATLAB command window are selectively shown in Fig. 10-45. Note that at first glance the $(s\mathbf{I}-\mathbf{A})^{-1}$ and $\phi(t)$ matrices may appear different from Eqs. (10-366) and (10-367). However, after minor manipulations, you may be able to verify that they are the same. This difference in representation is because of MATLAB symbolic approach. You may further simplify these matrices by using the "simple" command in the MATLAB command window. For example, to simplify $\phi(t)$, type "simple(phi)" in the MATLAB command window. If the desired format has not been achieved, you may have reached the software limit.

```
Enter A = [0 1 0;64.4 0 -16;0 0 -100]

Asym =

         0    1.0000         0
   64.4000         0   -16.0000
         0         0  -100.0000

Determinant of (s*I-A) is:

detSIA =

s^3+100*s^2-322/5*s-6440

the eigenvalues of A are:

eigA =

 -100.0000
    8.0250
   -8.0250

Inverse of (s*I-A) is:

[          s                 5                      80                          ]
[5 ----------------- , ----------------- , ----------------------------------- ]
[         2                 2                  3        2                      ]
[    5 s  - 322         5 s  - 322         5 s  + 500 s  - 322 s  - 32200]
[                                                                              ]
[         322                s                          s                      ]
[----------------- , 5 ----------------- , -80 ---------------------------------]
[       2                   2                      3        2                  ]
[  5 s  - 322           5 s  - 322             5 s  + 500 s  - 322 s  - 32200]
[                                                                              ]
[                                                                           1  ]
[0 ,                     0 ,                                         ----------]
[                                                                       s + 100]

State transition matrix of A:

[              40                        2000              40        ]
[%2 , 1/322 %1, - ----------- exp (-100 t) - ---------------- %1 + ---------- %2]
[                24839                      3999079              24839        ]
[                                                                              ]
[                      4000                       4000                         ]
[1/5 %1 ,    %2 ,   ----------- exp (-100 t) - ---------- %2 + 8/24839 %1]
[                    24839                       24839                         ]
[                                                                              ]
[0 ,                     0 ,                                    exp (-100 t)]

                              1/2              1/2
                  %1 : = 1610    sinh (1/5 1610    t)

                                     1/2
                    %2 : = cosh (1/5 1610    t)

Transfer function between u(t) and y(t) is:

                         8000
             - -----------------------------
                  3        2
               5 s  + 500 s  - 322 s  - 32200
```

Figure 10-45 Selective display of the MATLAB command window for the tfsym tool.

10-20 SUMMARY

This chapter was devoted to the state-variable analysis of linear systems. The fundamentals on state variables and state equations were introduced in Chapters 2 and 3, and formal discussions on these subjects were covered in this chapter. Specifically, the state-transition matrix and state-transition equations were introduced and the relationship between the state equations and transfer functions was established. Given the transfer function of a linear system, the state equations of the system can be obtained by decomposition of the transfer function. Given the state equations and the output equations, the transfer function can be determined either analytically or directly from the state diagram.

Characteristic equations and eigenvalues were defined in terms of the state equations and the transfer function. Eigenvectors of **A** were also defined for distinct and multiple-order eigenvalues. Similarity transformations to controllability canonical form (CCF), observability canonical form (OCF), diagonal canonical form (DCF), and Jordan canonical form (JCF) were discussed. State controllability and observability of linear time-invariant systems were defined and illustrated, and a final example, on the magnetic-ball-suspension system, summarized the important elements of the state-variable analysis of linear systems.

The MATLAB software tools statetool, tfsym, and tfcal were described in the last section. The program functionality was discussed with two examples. Together these tools can solve most of the homework problems and examples in this chapter.

REVIEW QUESTIONS

1. What are the components of the dynamic equations of a linear system?

2. Given the state equations of a linear system as

$$\frac{d\mathbf{x}(t)}{dt} = \mathbf{A}\mathbf{x}(t) + \mathbf{B}u(t)$$

give two expressions of the state-transition matrix $\phi(t)$ in terms of **A**.

3. List the properties of the state-transition matrix $\phi(t)$.

4. Given the state equations as in Review Question 2, write the state-transition equation.

5. List the advantages of expressing a linear system in the controllability canonical form (CCF). Give an example of **A** and **B** in CCF.

6. Given the state equations as in Review Question 2, give the conditions for **A** and **B** to be transformable into CCF.

7. Express the characteristic equation in terms of the matrix **A**.

8. List the three methods of decomposition of a transfer function.

9. What special forms will the state equations be in if the transfer function is decomposed by direct decomposition?

10. What special form will the state equations be in if the transfer function is decomposed by parallel decomposition?

11. What is the advantage of using cascade decomposition?

12. State the relationship between the CCF and controllability.

13. For controllability, does the magnitude of the inputs have to be finite?

14. Give the condition of controllability in terms of the matrices **A** and **B**.

15. What is the motivation behind the concept of observability?

16. Give the condition of observability in terms of the matrices **A** and **C**.

17. What can be said about the controllability and observability conditions if the transfer function has pole–zero cancellation?
18. State the relationship between OCF and observability.

Answers to these review questions can be found on this book's companion Web site: www.wiley.com/college/golnaraghi.

▶ REFERENCES

State Variables and State Equations

1. B. C. Kuo, *Linear Networks and Systems*, McGraw-Hill, New York, 1967.
2. R. A. Gabel and R. A. Roberts, *Signals and Linear Systems*, 3rd ed., John Wiley & Sons, New York, 1987.

Controllability and Observability

3. R. E. Kalman, "On the General Theory of Control Systems," *Proc. IFAC*, Vol. 1, pp. 481–492, Butterworths, London, 1961.
4. W. L. Brogan, *Modern Control Theory*, 2nd Ed., Prentice Hall, Englewood Cliffs, NJ, 1985.

▶ PROBLEMS

10-1. The following differential equations represent linear time-invariant systems. Write the dynamic equations (state equations and output equations) in vector-matrix form.

(a) $\dfrac{d^2y(t)}{dt^2} + 4\dfrac{dy(t)}{dt} + y(t) = 5r(t)$

(b) $2\dfrac{d^3y(t)}{dt^3} + 3\dfrac{d^2y(t)}{dt^2} + 5\dfrac{dy(t)}{dt} + 2y(t) = r(t)$

(c) $\dfrac{d^3y(t)}{dt^3} + 5\dfrac{d^2y(t)}{dt^2} + 3\dfrac{dy(t)}{dt} + y(t) + \displaystyle\int_0^t y(\tau)d\tau = r(\tau)$

(d) $\dfrac{d^4y(t)}{dt^4} + 1.5\dfrac{d^3y(t)}{dt^3} + 2.5\dfrac{dy(t)}{dt} + y(t) = 2r(t)$

10-2. The following transfer functions show linear time-invariant systems. Write the dynamic equations (state equations and output equations) in vector-matrix form.

(a) $G(s) = \dfrac{s+3}{s^2 + 3s + 2}$

(b) $G(s) = \dfrac{6}{s^3 + +6s^2 + 11s + 6}$

(c) $G(s) = \dfrac{s+2}{s^2 + 7s + 12}$

(d) $G(s) = \dfrac{s^3 + 11s^2 + 35s + 250}{s^2(s^3 + 4s^2 + 39s + 108)}$

10-3. Repeat Problem 10-2 by using MATLAB.

10-4. Write the state equations for the block diagrams of the systems shown in Fig. 10P-4.

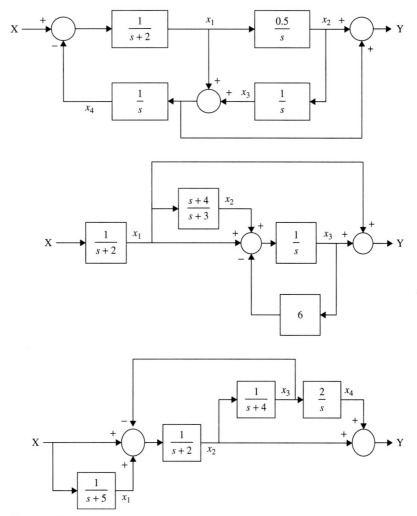

Figure 10P-4

10-5. By use of Eq. (10-58), show that

$$\phi(t) = \mathbf{I} + \mathbf{A}t + \frac{1}{2!}\mathbf{A}^2 t^2 + \frac{1}{3!}\mathbf{A}^3 t^3 + \cdots$$

10-6. The state equations of a linear time-invariant system are represented by

$$\frac{d\mathbf{x}(t)}{dt} = \mathbf{A}\mathbf{x}(t) + \mathbf{B}\mathbf{u}(t)$$

Find the state-transition matrix $\phi(t)$, the characteristic equation, and the eigenvalues of $\mathbf{A}$ for the following cases:

(a) $\mathbf{A} = \begin{bmatrix} 0 & 1 \\ -2 & -1 \end{bmatrix}$ $\mathbf{B} = \begin{bmatrix} 0 & 1 \\ 1 & 0 \end{bmatrix}$

(b) $\mathbf{A} = \begin{bmatrix} 0 & 1 \\ -4 & -5 \end{bmatrix}$ $\mathbf{B} = \begin{bmatrix} 1 \\ 1 \end{bmatrix}$

(c) $\mathbf{A} = \begin{bmatrix} -3 & 0 \\ 0 & -3 \end{bmatrix}$ $\mathbf{B} = \begin{bmatrix} 0 \\ 1 \end{bmatrix}$

(d) $\mathbf{A} = \begin{bmatrix} 3 & 0 \\ 0 & -3 \end{bmatrix}$ $\mathbf{B} = \begin{bmatrix} 0 \\ 1 \end{bmatrix}$

(e) $\mathbf{A} = \begin{bmatrix} 0 & 2 \\ -2 & 0 \end{bmatrix}$ $\mathbf{B} = \begin{bmatrix} 0 \\ 1 \end{bmatrix}$

(f) $\mathbf{A} = \begin{bmatrix} -1 & 0 & 0 \\ 0 & -2 & 1 \\ 0 & 0 & -2 \end{bmatrix}$ $\mathbf{B} = \begin{bmatrix} 0 \\ 1 \\ 0 \end{bmatrix}$

(g) $\mathbf{A} = \begin{bmatrix} -5 & 1 & 0 \\ 0 & -5 & 1 \\ 0 & 0 & -5 \end{bmatrix}$ $\mathbf{B} = \begin{bmatrix} 0 \\ 0 \\ 1 \end{bmatrix}$

10-7. Find $\phi(t)$ and the characteristic equation of the state variables in Problem 10-6 using a computer program.

10-8. Find the state-transition equation of each of the systems described in Problem 10-6 for $t \geq 0$. Assume that $\mathbf{x}(0)$ is the initial state vector, and the components of the input vector $\mathbf{u}(t)$ are all unit-step functions.

10-9. Find out if the matrices given in the following can be state-transition matrices. [Hint: check the properties of $\phi(t)$.]

(a) $\begin{bmatrix} -e^{-t} & 0 \\ 0 & 1-e^{-t} \end{bmatrix}$

(b) $\begin{bmatrix} 1-e^{-t} & 0 \\ 1 & e^{-t} \end{bmatrix}$

(c) $\begin{bmatrix} 1 & 0 \\ 1-e^{-t} & e^{-t} \end{bmatrix}$

(d) $\begin{bmatrix} e^{-2t} & te^{-2t} & t^2 e^{-2t}/2 \\ 0 & e^{-2t} & te^{-2t} \\ 0 & 0 & e^{-2t} \end{bmatrix}$

10-10. Find the time response of the following systems:

(a) $\begin{bmatrix} \dot{x}_1 \\ \dot{x}_2 \end{bmatrix} = \begin{bmatrix} 0 & 1 \\ -2 & -3 \end{bmatrix} \begin{bmatrix} x_1 \\ x_2 \end{bmatrix} + \begin{bmatrix} 0 \\ 1 \end{bmatrix} u$

(b) $\begin{bmatrix} \dot{x}_1 \\ \dot{x}_2 \end{bmatrix} = \begin{bmatrix} -1 & -0.5 \\ 1 & 0 \end{bmatrix} \begin{bmatrix} x_1 \\ x_2 \end{bmatrix} + \begin{bmatrix} 0.5 \\ 0 \end{bmatrix} u$ $y = \begin{bmatrix} 1 & 0 \end{bmatrix} \begin{bmatrix} x_1 \\ x_2 \end{bmatrix}$

10-11. Given a system described by the dynamic equations:

$$\frac{d\mathbf{x}(t)}{dt} = \mathbf{A}\mathbf{x}(t) + \mathbf{B}u(t) \quad y(t) = \mathbf{C}\mathbf{x}(t)$$

(a) $\mathbf{A} = \begin{bmatrix} 0 & 1 & 0 \\ 0 & 0 & 1 \\ -1 & -2 & -3 \end{bmatrix}$ $\mathbf{B} = \begin{bmatrix} 0 \\ 0 \\ 1 \end{bmatrix}$ $\mathbf{C} = \begin{bmatrix} 1 & 0 & 0 \end{bmatrix}$

(b) $\mathbf{A} = \begin{bmatrix} -1 & 1 \\ 0 & -1 \end{bmatrix}$ $\mathbf{B} = \begin{bmatrix} 0 \\ 1 \end{bmatrix}$ $\mathbf{C} = \begin{bmatrix} 1 & 1 \end{bmatrix}$

(c) $\mathbf{A} = \begin{bmatrix} 0 & 1 & 0 \\ 0 & 0 & 1 \\ 0 & -1 & -2 \end{bmatrix}$ $\mathbf{B} = \begin{bmatrix} 0 \\ 0 \\ 1 \end{bmatrix}$ $\mathbf{C} = \begin{bmatrix} 1 & 1 & 0 \end{bmatrix}$

(1) Find the eigenvalues of $\mathbf{A}$. Use the **ACSYS** computer program to check the answers. You may get the characteristic equation and solve for the roots using tfsym or tcal components of **ACSYS**.

(2) Find the transfer-function relation between $\mathbf{X}(s)$ and $U(s)$.

(3) Find the transfer function $Y(s)/U(s)$.

10-12. Given the dynamic equations of a time-invariant system:

$$\frac{d\mathbf{x}(t)}{dt} = \mathbf{A}\mathbf{x}(t) + \mathbf{B}u(t) \quad y(t) = \mathbf{C}\mathbf{x}(t)$$

where

$$\mathbf{A} = \begin{bmatrix} 0 & 1 & 0 \\ 0 & 0 & 1 \\ -1 & -2 & -3 \end{bmatrix} \quad \mathbf{B} = \begin{bmatrix} 0 \\ 0 \\ 1 \end{bmatrix} \quad \mathbf{C} = \begin{bmatrix} 1 & 1 & 0 \end{bmatrix}$$

Find the matrices $\mathbf{A}_1$ and $\mathbf{B}_1$ so that the state equations are written as

$$\frac{d\bar{\mathbf{x}}(t)}{dt} = \mathbf{A}_1 \bar{\mathbf{x}}(t) + \mathbf{B}_1 u(t)$$

where

$$\bar{\mathbf{x}}(t) = \begin{bmatrix} x_1(t) \\ y(t) \\ \frac{dy(t)}{dt} \end{bmatrix}$$

10-13. Given the dynamic equations

$$\frac{d\mathbf{x}(t)}{dt} = \mathbf{A}\mathbf{x}(t) + \mathbf{B}u(t) \quad y(t) = \mathbf{C}\mathbf{x}(t)$$

(a) $\mathbf{A} = \begin{bmatrix} 0 & 2 & 0 \\ 1 & 2 & 0 \\ -1 & 0 & 1 \end{bmatrix} \quad \mathbf{B} = \begin{bmatrix} 0 \\ 1 \\ 1 \end{bmatrix} \quad \mathbf{C} = \begin{bmatrix} 1 & 0 & 1 \end{bmatrix}$

(b) $\mathbf{A} = \begin{bmatrix} 0 & 2 & 0 \\ 1 & 2 & 0 \\ -1 & 1 & 1 \end{bmatrix} \quad \mathbf{B} = \begin{bmatrix} 1 \\ 1 \\ 0 \end{bmatrix} \quad \mathbf{C} = \begin{bmatrix} 1 & 0 & 1 \end{bmatrix}$

(c) $\mathbf{A} = \begin{bmatrix} -2 & 1 & 0 \\ 0 & -2 & 0 \\ -1 & -2 & -3 \end{bmatrix} \quad \mathbf{B} = \begin{bmatrix} 1 \\ 1 \\ 1 \end{bmatrix} \quad \mathbf{C} = \begin{bmatrix} 1 & 0 & 0 \end{bmatrix}$

(d) $\mathbf{A} = \begin{bmatrix} -1 & 1 & 0 \\ 0 & -1 & 1 \\ 0 & 0 & -1 \end{bmatrix} \quad \mathbf{B} = \begin{bmatrix} 0 \\ 1 \\ 1 \end{bmatrix} \quad \mathbf{C} = \begin{bmatrix} 1 & 0 & 1 \end{bmatrix}$

(e) $\mathbf{A} = \begin{bmatrix} 1 & 1 \\ -2 & -3 \end{bmatrix} \quad \mathbf{B} = \begin{bmatrix} 0 \\ 1 \end{bmatrix} \quad \mathbf{C} = \begin{bmatrix} 1 & 0 \end{bmatrix}$

Find the transformation $\mathbf{x}(t) = \mathbf{P}\bar{\mathbf{x}}(t)$ that transforms the state equations into the controllability canonical form (CCF).

10-14. For the systems described in Problem 10-13, find the transformation $\mathbf{x}(t) = \mathbf{Q}\bar{\mathbf{x}}(t)$ so that the state equations are transformed into the observability canonical form (OCF).

10-15. For the systems described in Problem 10-13, find the transformation $\mathbf{x}(t) = \mathbf{T}\bar{\mathbf{x}}(t)$ so that the state equations are transformed into the diagonal canonical form (DCF) if $\mathbf{A}$ has distinct eigenvalues and Jordan canonical form (JCF) if $\mathbf{A}$ has at least one multiple-order eigenvalue.

10-16. Consider the following transfer functions. Transform the state equations into the controllability canonical form (CCF) and observibility canonical form (OCF).

(a) $\dfrac{s^2 - 1}{s^2(s^2 - 2)}$ (b) $\dfrac{2s + 1}{s^2 + 4s + 4}$

10-17. The state equation of a linear system is described by

$$\frac{d\mathbf{x}(t)}{dt} = \mathbf{A}\mathbf{x}(t) + \mathbf{B}u(t)$$

The coefficient matrices are given as follows. Explain why the state equations cannot be transformed into the controllability canonical form (CCF).

(a) $\mathbf{A} = \begin{bmatrix} -2 & 0 \\ 0 & -1 \end{bmatrix}$ $\mathbf{B} = \begin{bmatrix} 0 \\ 1 \end{bmatrix}$
(b) $\mathbf{A} = \begin{bmatrix} -1 & 0 & 0 \\ 0 & -1 & 0 \\ 0 & 0 & -1 \end{bmatrix}$ $\mathbf{B} = \begin{bmatrix} 1 \\ 2 \\ 3 \end{bmatrix}$

(c) $\mathbf{A} = \begin{bmatrix} 1 & 2 \\ 1 & 1 \end{bmatrix}$ $\mathbf{B} = \begin{bmatrix} 2 \\ \sqrt{2} \end{bmatrix}$
(d) $\mathbf{A} = \begin{bmatrix} -2 & 1 & 0 \\ 0 & -2 & 0 \\ -1 & -2 & -3 \end{bmatrix}$ $\mathbf{B} = \begin{bmatrix} 1 \\ 0 \\ 1 \end{bmatrix}$

10-18. Check the controlibility of the following systems:

(a) $\begin{bmatrix} \dot{x}_1 \\ \dot{x}_2 \end{bmatrix} = \begin{bmatrix} -1 & 0 \\ 0 & -2 \end{bmatrix} \begin{bmatrix} x_1 \\ x_2 \end{bmatrix} + \begin{bmatrix} 2 \\ 5 \end{bmatrix} u$

(b) $\begin{bmatrix} \dot{x}_1 \\ \dot{x}_2 \end{bmatrix} = \begin{bmatrix} -1 & 0 \\ 0 & -2 \end{bmatrix} \begin{bmatrix} x_1 \\ x_2 \end{bmatrix} + \begin{bmatrix} 2 \\ 0 \end{bmatrix} u$

(c) $\begin{bmatrix} \dot{x}_1 \\ \dot{x}_2 \end{bmatrix} = \begin{bmatrix} -1 & 1 & 0 \\ 0 & -1 & 0 \\ 0 & 0 & -2 \end{bmatrix} \begin{bmatrix} x_1 \\ x_2 \\ x_3 \end{bmatrix} + \begin{bmatrix} 4 & 2 \\ 0 & 0 \\ 3 & 0 \end{bmatrix} \begin{bmatrix} u_1 \\ u_2 \end{bmatrix}$

(d) $\begin{bmatrix} \dot{x}_1 \\ \dot{x}_2 \end{bmatrix} = \begin{bmatrix} -1 & 1 & 0 \\ 0 & -1 & 0 \\ 0 & 0 & -2 \end{bmatrix} \begin{bmatrix} x_1 \\ x_2 \\ x_3 \end{bmatrix} + \begin{bmatrix} 0 \\ 4 \\ 3 \end{bmatrix} u$

(e) $\begin{bmatrix} \dot{x}_1 \\ \dot{x}_2 \\ \dot{x}_3 \\ \dot{x}_4 \\ \dot{x}_5 \end{bmatrix} = \begin{bmatrix} -2 & 1 & 0 & 0 & 0 \\ 0 & -2 & 1 & 0 & 0 \\ 0 & 0 & -2 & 0 & 0 \\ 0 & 0 & 0 & -5 & 1 \\ 0 & 0 & 0 & 0 & -5 \end{bmatrix} \begin{bmatrix} x_1 \\ x_2 \\ x_3 \\ x_4 \\ x_5 \end{bmatrix} + \begin{bmatrix} 0 & 1 \\ 0 & 0 \\ 3 & 0 \\ 0 & 0 \\ 2 & \end{bmatrix} \begin{bmatrix} u_1 \\ u_2 \end{bmatrix}$

(f) $\begin{bmatrix} \dot{x}_1 \\ \dot{x}_2 \\ \dot{x}_3 \\ \dot{x}_4 \\ \dot{x}_5 \end{bmatrix} = \begin{bmatrix} -2 & 1 & 0 & 0 & 0 \\ 0 & -2 & 1 & 0 & 0 \\ 0 & 0 & -2 & 0 & 0 \\ 0 & 0 & 0 & -5 & 1 \\ 0 & 0 & 0 & 0 & -5 \end{bmatrix} \begin{bmatrix} x_1 \\ x_2 \\ x_3 \\ x_4 \\ x_5 \end{bmatrix} + \begin{bmatrix} 4 \\ 2 \\ 1 \\ 3 \\ 0 \end{bmatrix} u$

10-19. Check the observability of the following systems:

(a) $\begin{bmatrix} \dot{x}_1 \\ \dot{x}_2 \end{bmatrix} = \begin{bmatrix} -1 & 0 \\ 0 & -2 \end{bmatrix} \begin{bmatrix} x_1 \\ x_2 \end{bmatrix}$ $y = \begin{bmatrix} 1 & 3 \end{bmatrix} \begin{bmatrix} x_1 \\ x_2 \end{bmatrix}$

(b) $\begin{bmatrix} \dot{x}_1 \\ \dot{x}_2 \end{bmatrix} = \begin{bmatrix} -1 & 0 \\ 0 & -2 \end{bmatrix} \begin{bmatrix} x_1 \\ x_2 \end{bmatrix}$ $y = \begin{bmatrix} 0 & 1 \end{bmatrix} \begin{bmatrix} x_1 \\ x_2 \end{bmatrix}$

(c) $\begin{bmatrix} \dot{x}_1 \\ \dot{x}_2 \\ \dot{x}_3 \end{bmatrix} = \begin{bmatrix} 2 & 1 & 0 \\ 0 & 2 & 1 \\ 0 & 0 & 2 \end{bmatrix} \begin{bmatrix} x_1 \\ x_2 \\ x_3 \end{bmatrix}$ $\begin{bmatrix} y_1 \\ y_2 \end{bmatrix} = \begin{bmatrix} 0 & 1 & 3 \\ 0 & 2 & 4 \end{bmatrix} \begin{bmatrix} x_1 \\ x_2 \\ x_3 \end{bmatrix}$

(d) $\begin{bmatrix} \dot{x}_1 \\ \dot{x}_2 \\ \dot{x}_3 \end{bmatrix} = \begin{bmatrix} 2 & 1 & 0 \\ 0 & 2 & 1 \\ 0 & 0 & 2 \end{bmatrix} \begin{bmatrix} x_1 \\ x_2 \\ x_3 \end{bmatrix}$ $\begin{bmatrix} y_1 \\ y_2 \end{bmatrix} = \begin{bmatrix} 3 & 0 & 0 \\ 4 & 0 & 0 \end{bmatrix} \begin{bmatrix} x_1 \\ x_2 \\ x_3 \end{bmatrix}$

(e) $\begin{bmatrix} \dot{x}_1 \\ \dot{x}_2 \\ \dot{x}_3 \\ \dot{x}_4 \\ \dot{x}_5 \end{bmatrix} = \begin{bmatrix} 2 & 1 & 0 & 0 & 0 \\ 0 & 2 & 1 & 0 & 0 \\ 0 & 0 & 2 & 0 & 0 \\ 0 & 0 & 0 & -3 & 1 \\ 0 & 0 & 0 & 0 & -3 \end{bmatrix} \begin{bmatrix} x_1 \\ x_2 \\ x_3 \\ x_4 \\ x_5 \end{bmatrix} \qquad \begin{bmatrix} y_1 \\ y_2 \end{bmatrix} = \begin{bmatrix} 1 & 1 & 1 & 0 & 0 \\ 0 & 1 & 1 & 1 & 0 \end{bmatrix} \begin{bmatrix} x_1 \\ x_2 \\ x_3 \\ x_4 \\ x_5 \end{bmatrix}$

(f) $\begin{bmatrix} \dot{x}_1 \\ \dot{x}_2 \\ \dot{x}_3 \\ \dot{x}_4 \\ \dot{x}_5 \end{bmatrix} = \begin{bmatrix} 2 & 1 & 0 & 0 & 0 \\ 0 & 2 & 1 & 0 & 0 \\ 0 & 0 & 2 & 0 & 0 \\ 0 & 0 & 0 & -3 & 1 \\ 0 & 0 & 0 & 0 & -3 \end{bmatrix} \begin{bmatrix} x_1 \\ x_2 \\ x_3 \\ x_4 \\ x_5 \end{bmatrix} \qquad \begin{bmatrix} y_1 \\ y_2 \end{bmatrix} = \begin{bmatrix} 1 & 1 & 1 & 0 & 0 \\ 0 & 1 & 1 & 0 & 0 \end{bmatrix} \begin{bmatrix} x_1 \\ x_2 \\ x_3 \\ x_4 \\ x_5 \end{bmatrix}$

10-20. The equations that describe the dynamics of a motor control system are

$$e_a(t) = R_a i_a(t) + L_a \frac{di_a(t)}{dt} + K_b \frac{d\theta_m(t)}{dt}$$

$$T_m(t) = K_i i_a(t)$$

$$T_m(t) = J \frac{d^2\theta_m(t)}{dt^2} + B \frac{d\theta_m(t)}{dt} + K\theta_m(t)$$

$$e_a(t) = K_a e(t)$$

$$e(t) = K_s[\theta_r(t) - \theta_m(t)]$$

(a) Assign the state variables as $x_1(t) = \theta_m(t)$, $x_2(t) = d\theta_m(t)/dt$, and $x_3(t) = i_a(t)$. Write the state equations in the form of

$$\frac{d\mathbf{x}(t)}{dt} = \mathbf{A}\mathbf{x}(t) + \mathbf{B}\theta_r(t)$$

Write the output equation in the form $y(t) = \mathbf{C}\mathbf{x}(t)$, where $y(t) = \theta_m(t)$.

(b) Find the transfer function $G(s) = \Theta_m(s)/E(s)$ when the feedback path from $\Theta_m(s)$ to $E(s)$ is broken. Find the closed-loop transfer function $M(s) = \Theta_m(s)/\Theta_r(s)$.

10-21. Given the matrix $\mathbf{A}$ of a linear system described by the state equation

$$\frac{d\mathbf{x}(t)}{dt} = \mathbf{A}\mathbf{x}(t) + \mathbf{B}u(t)$$

(a) $\mathbf{A} = \begin{bmatrix} 0 & 1 \\ -1 & 0 \end{bmatrix}$

(b) $\mathbf{A} = \begin{bmatrix} -1 & 0 \\ 0 & -2 \end{bmatrix}$

(c) $\mathbf{A} = \begin{bmatrix} 0 & 1 \\ 1 & 0 \end{bmatrix}$

Find the state-transition matrix $\phi(t)$ using the following methods:

(1) Infinite-series expansion of $e^{\mathbf{A}t}$, expressed in closed form
(2) The inverse Laplace transform of $(s\mathbf{I} - \mathbf{A})^{-1}$

10-22. The schematic diagram of a feedback control system using a dc motor is shown in Fig. 10P-22. The torque developed by the motor is $T_m(t) = K_i i_a(t)$, where K_i is the torque constant.

The constants of the system are

$K_s = 2 \qquad R = 2\,\Omega \qquad R_s = 0.1\,\Omega$
$K_b = 5$ V/rad/sec $\qquad K_i = 5$ N-m/A $\qquad L_a \cong 0$ H
$J_m + J_L = 0.1$ N-m-sec$^2 \qquad B_m \cong 0$ N-m-sec

Assume that all the units are consistent so that no conversion is necessary.

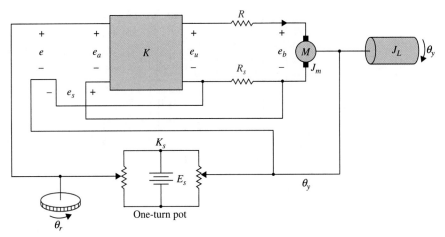

Figure 10P-22

(a) Let the state variables be assigned as $x_1 = \theta_y$ and $x_2 = d\theta_y/dt$. Let the output be $y = \theta_y$. Write the state equations in vector-matrix form. Show that the matrices **A** and **B** are in CCF.

(b) Let $\theta_r(t)$ be a unit-step function. Find $\mathbf{x}(t)$ in terms of $\mathbf{x}(0)$, the initial state. Use the Laplace transform table.

(c) Find the characteristic equation and the eigenvalues of **A**.

(d) Comment on the purpose of the feedback resistor R_s.

10-23. Repeat Problem 10-22 with the following system parameters:

$K_s = 1$ $K = 9$ $R_a = 0.1\,\Omega$
$R_s = 0.1\,\Omega$ $K_b = 1$ V/rad/s $K_i = 1$ N-m/A
$L_a \cong 0$ H $J_m + J_L = 0.01$ N-m-sec^2 $B_m \cong 0$ N-m-sec

10-24. Consider that matrix **A** can be diagonalized. Show that $e^{\mathbf{A}t} = \mathbf{P}e^{\mathbf{D}t}\mathbf{P}^{-1}$, where **P** transforms **A** into a diagonal matrix, and $\mathbf{P}^{-1}\mathbf{A}\mathbf{P} = \mathbf{D}$, where **D** is a diagonal matrix.

10-25. Consider that matrix **A** can be transformed to the Jordan canonical form, then $e^{\mathbf{A}t} = \mathbf{S}e^{\mathbf{J}t}\mathbf{S}^{-1}$, where **S** transforms **A** into a Jordan canonical form and **J** is in a Jordan canonical form.

10-26. The block diagram of a feedback control system is shown in Fig. 10P-26.

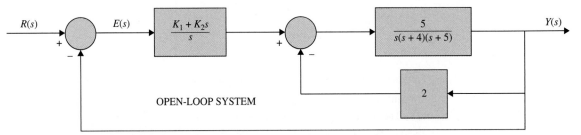

Figure 10P-26

(a) Find the forward-path transfer function $Y(s)/E(s)$ and the closed-loop transfer function $Y(s)/R(s)$.

(b) Write the dynamic equations in the form of

$$\frac{d\mathbf{x}(t)}{dt} = \mathbf{A}\mathbf{x}(t) + \mathbf{B}r(t) \quad y(t) = \mathbf{C}\mathbf{x}(t) + \mathbf{D}r(t)$$

Find **A**, **B**, **C**, and **D** in terms of the system parameters.

(c) Apply the final-value theorem to find the steady-state value of the output $y(t)$ when the input $r(t)$ is a unit-step function. Assume that the closed-loop system is stable.

10-27. For the linear time-invariant system whose state equations have the coefficient matrices given by Eqs. (10-191) and (10-192) (CCF), show that

$$\text{adj}(s\mathbf{I} - \mathbf{A})\mathbf{B} = \begin{bmatrix} 1 \\ s \\ s^2 \\ \vdots \\ s^{n-1} \end{bmatrix}$$

and the characteristic equation of $\mathbf{A}$ is

$$s^n + a_{n-1}s^{n-1} + \cdots + a_1 s + a_0 = 0$$

10-28. A linear time-invariant system is described by the differential equation

$$\frac{d^3 y(t)}{dt^3} + 3\frac{d^2 y(t)}{dt^2} + 3\frac{dy(t)}{dt} + y(t) = r(t)$$

(a) Let the state variables be defined as $x_1 = y$, $x_2 = dy/dt$, and $x_3 = d^2 y/dt^2$. Write the state equations of the system in vector-matrix form.

(b) Find the state-transition matrix $\phi(t)$ of $\mathbf{A}$.

(c) Let $y(0) = 1$, $dy(0)/dt = 0$, $d^2 y(0)/dt^2 = 0$, and $r(t) = u_s(t)$. Find the state-transition equation of the system.

(d) Find the characteristic equation and the eigenvalues of $\mathbf{A}$.

10-29. A spring-mass-friction system is described by the following differential equation:

$$\frac{d^2 y(t)}{dt^2} + 2\frac{dy(t)}{dt} + y(t) = r(t)$$

(a) Define the state variables as $x_1(t) = y(t)$ and $x_2(t) = dy(t)/dt$. Write the state equations in vector-matrix form. Find the state-transition matrix $\phi(t)$ of $\mathbf{A}$.

(b) Define the state variables as $x_1(t) = y(t)$ and $x_2(t) = y(t) + dy(t)/dt$. Write the state equations in vector-matrix form. Find the state-transition matrix $\phi(t)$ of $\mathbf{A}$.

(c) Show that the characteristic equations, $|s\mathbf{I} - \mathbf{A}| = 0$, for parts (a) and (b) are identical.

10-30. Given the state equations $d\mathbf{x}(t)/dt = \mathbf{A}\mathbf{x}(t)$, where σ and ω are real numbers:

(a) Find the state transition matrix of $\mathbf{A}$.

(b) Find the eigenvalues of $\mathbf{A}$.

10-31. (a) Show that the input–output transfer functions of the two systems shown in Fig. 10P-31 are the same.

(b) Write the dynamic equations of the system in Fig. 10P-31(a) as

$$\frac{d\mathbf{x}(t)}{dt} = \mathbf{A}_1 \mathbf{x}(t) + \mathbf{B}_1 u_1(t) \qquad y_1(t) = \mathbf{C}_1 \mathbf{x}(t)$$

and those of the system in Fig. 10-31(b) as

$$\frac{d\mathbf{x}(t)}{dt} = \mathbf{A}_2 \mathbf{x}(t) + \mathbf{B}_2 u_2(t) \qquad y_2(t) = \mathbf{C}_2 \mathbf{x}(t)$$

10-32. Draw the state diagrams for the following systems.

$$\frac{d\mathbf{x}(t)}{dt} = \mathbf{A}\mathbf{x}(t) + \mathbf{B}u(t)$$

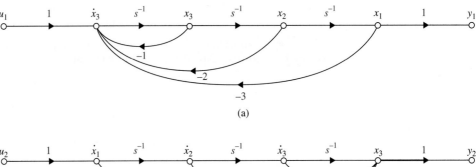

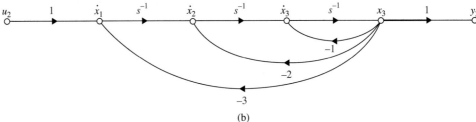

Figure 10P-31

(a) $\mathbf{A} = \begin{bmatrix} -3 & 2 & 0 \\ -1 & 0 & 1 \\ -2 & -3 & -4 \end{bmatrix} \quad \mathbf{B} = \begin{bmatrix} 0 \\ 0 \\ 1 \end{bmatrix}$

(b) Same **A** as in part (a), but with

$$\mathbf{B} = \begin{bmatrix} 0 & 1 \\ 1 & 0 \\ 1 & 0 \end{bmatrix}$$

10-33. Draw state diagrams for the following transfer functions by direct decomposition. Assign the state variables from right to left for $x_1, x_2, \ldots$. Write the state equations from the state diagram and show that the equations are in CCF.

(a) $G(s) = \dfrac{10}{s^3 + 8.5s^2 + 20.5s + 15}$

(b) $G(s) = \dfrac{10(s+2)}{s^2(s+1)(s+3.5)}$

(c) $G(s) = \dfrac{5(s+1)}{s(s+2)(s+10)}$

(d) $G(s) = \dfrac{1}{s(s+5)(s^2+2s+2)}$

10-34. Draw state diagrams for the systems described in Problem 10-33 by parallel decomposition. Make certain that the state diagrams contain a minimum number of integrators. The constant branch gains must be real. Write the state equations from the state diagram.

10-35. Draw the state diagrams for the systems described in Problem 10-33 by using cascade decomposition. Assign the state variables in ascending order from right to left. Write the state equations from the state diagram.

10-36. The block diagram of a feedback control system is shown in Fig. 10P-36.

(a) Draw a state diagram for the system by first decomposing $G(s)$ by direct decomposition. Assign the state variables in ascending order, $x_1, x_2, \ldots$, from right to left. In addition to the state-variable-related nodes, the state diagram should contain nodes for $R(s)$, $E(s)$, and $C(s)$.

(b) Write the dynamic equations of the system in vector-matrix form.

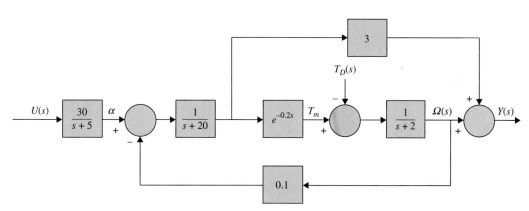

Figure 10P-36

(c) Find the state-transition equations of the system using the state equations found in part (b). The initial state vector is $\mathbf{x}(0)$, and $r(t) = u_s(t)$.

(d) Find the output $y(t)$ for $t \geq 0$ with the initial state $\mathbf{x}(0)$, and $r(t) = u_s(t)$.

10-37. **(a)** Find the closed-loop transfer function $Y(s)/R(s)$, and draw the state diagram.
(b) Perform a direct decomposition to $Y(s)/R(s)$, and draw the state diagram.
(c) Assign the state variables from right to left in ascending order, and write the state equations in vector-matrix form.
(d) Find the state-transition equations of the system using the state equations found in part (c). The initial state vector is $\mathbf{x}(0)$, and $r(t) = u_s(t)$.
(e) Find the output $y(t)$ for $t \geq 0$ with the initial state $\mathbf{x}(0)$, and $r(t) = u_s(t)$.

10-38. The block diagram of a linearized idle-speed engine-control system of an automobile is shown in Fig. 10P-38. (For a discussion on linearization of nonlinear systems, refer to Section 4-9.) The system is linearized about a nominal operating point, so all the variables represent linear-perturbed quantities. The following variables are defined: $T_m(t)$ is the engine torque; T_D, the constant load-disturbance torque; $\omega(t)$, the engine speed; $u(t)$, the input-voltage to the throttle actuator; and α, the throttle angle. The time delay in the engine model can be approximated by

Figure 10P-38

$$e^{-0.2s} \cong \frac{1 - 0.1s}{1 + 0.1s}$$

(a) Draw a state diagram for the system by decomposing each block individually. Assign the state variables from right to left in ascending order.

(b) Write the state equations from the state diagram obtained in part (a), in the form of

$$\frac{d\mathbf{x}(t)}{dt} = \mathbf{A}\mathbf{x}(t) + \mathbf{B}\begin{bmatrix} u(t) \\ T_D(t) \end{bmatrix}$$

(c) Write $Y(s)$ as a function of $U(s)$ and $T_D(s)$. Write $\Omega(s)$ as a function of $U(s)$ and $T_D(s)$.

10-39. The state diagram of a linear system is shown in Fig. 10P-39.

(a) Assign state variables on the state diagram from right to left in ascending order. Create additional artificial nodes if necessary so that the state-variable nodes satisfy as "input nodes" after the integrator branches are deleted.

(b) Write the dynamic equations of the system from the state diagram in part (a).

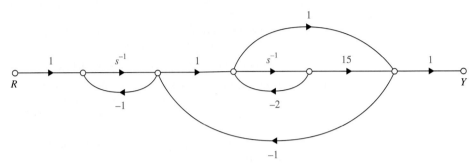

Figure 10P-39

10-40. The block diagram of a linear spacecraft-control system is shown in Fig. 10P-40.

(a) Determine the transfer function $Y(s)/R(s)$.

(b) Find the characteristic equation and its roots of the system. Show that the roots of the characteristic equation are not dependent on K.

(c) When $K = 1$, draw a state diagram for the system by decomposing $Y(s)/R(s)$, using a minimum number of integrators.

(d) Repeat part (c) when $K = 4$.

(e) Determine the values of K that must be avoided if the system is to be both state controllable and observable.

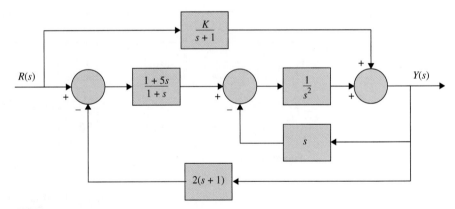

Figure 10P-40

10-41. A considerable amount of effort is being spent by automobile manufacturers to meet the exhaust-emission-performance standards set by the government. Modern automobile-power-plant systems consist of an internal combustion engine that has an internal cleanup device called a catalytic converter. Such a system requires control of such variables as the engine air–fuel (A/F) ratio, ignition-spark timing, exhaust-gas recirculation, and injection air. The control-system problem considered in this problem deals with the control of the A/F ratio. In general, depending on fuel composition and other factors, a typical stoichiometric A/F is 14.7:1, that is, 14.7 grams of air to each gram of fuel. An A/F greater or less than stoichiometry will cause high hydrocarbons, carbon monoxide, and nitrous oxides in the tailpipe emission. The control system shown in Fig. 10P-41 is devised to control the air–fuel ratio so that a desired output is achieved for a given input command.

The sensor senses the composition of the exhaust-gas mixture entering the catalytic converter. The electronic controller detects the difference or the error between the command and the error and computes the control signal necessary to achieve the desired exhaust-gas composition. The output $y(t)$ denotes the effective air–fuel ratio. The transfer function of the engine is given by

$$G_p(s) = \frac{Y(s)}{U(s)} = \frac{e^{-T_d s}}{1 + 0.5s}$$

where $T_d = 0.2$ sec is the time delay and is approximated by

$$e^{-T_d s} = \frac{1}{e^{T_d s}} = \frac{1}{1 + T_d s + T_d^2 s^2/2! + \cdots} \cong \frac{1}{1 + T_d s + T_d^2 s^2/2!}$$

The gain of the sensor is 1.0.

(a) Using the approximation for $e^{-T_d s}$ given, find the expression for $G_p(s)$. Decompose $G_p(s)$ by direct decomposition, and draw the state diagram with $u(t)$ as the input and $y(t)$ as the output. Assign state variables from right to left in ascending order, and write the state equations in vector-matrix form.

(b) Assuming that the controller is a simple amplifier with a gain of 1, i.e., $u(t) = e(t)$, find the characteristic equation and its roots of the closed-loop system.

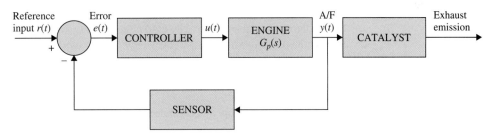

Figure 10P-41

10-42. Repeat Problem 10-41 when the time delay of the automobile engine is approximated as

$$e^{-T_d s} \cong \frac{1 - T_d s/3}{1 + \tfrac{2}{3}T_d s + \tfrac{1}{6}T_d^2 s^2} \quad T_d = 0.2 \text{ sec}$$

10-43. The schematic diagram in Fig. 10P-43 shows a permanent-magnet dc-motor-control system with a viscous-inertia damper. The system can be used for the control of the printwheel of an electronic word processor. A mechanical damper such as the viscous-inertia type is sometimes used in practice as a simple and economical way of stabilizing a control system. The damping effect is achieved by a rotor suspended in a viscous fluid. The differential and algebraic equations that describe the dynamics of the system are as follows:

$e(t) = K_s[\omega_r(t) - \omega_m(t)]$ $K_s = 1$ V/rad/sec

$e_a(t) = Ke(t) = R_a i_a(t) + e_b(t)$ $K = 10$

$e_b(t) = K_b \omega_m(t)$ $K_b = 0.0706$ V/rad/sec

$T_m(t) = J\dfrac{d\omega_m(t)}{dt} + K_D[\omega_m(t) - \omega_D(t)]$ $J = J_h + J_m = 0.1$ oz-in.-sec^2

$T_m(t) = K_i i_a(t)$ $K_i = 10$ oz-in./A

$K_D[\omega_m(t) - \omega_D(t)] = J_R \dfrac{d\omega_D(t)}{dt}$ $J_R = 0.05$ oz-in.-sec^2

$R_a = 1\,\Omega$ $K_D = 1$ oz-in.-sec

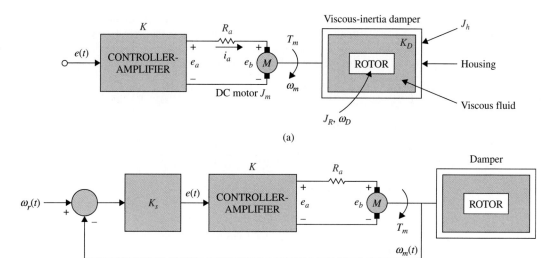

Figure 10P-43

(a) Let the state variables be defined as $x_1(t) = \omega_m(t)$ and $x_2(t) = \omega_D(t)$. Write the state equations for the open-loop system with $e(t)$ as the input. (*Open-loop* refers to the feedback path from ω_m to e being open.)

(b) Draw the state diagram for the overall system using the state equations found in part (a) and $e(t) = K_s[\omega_r(t) - \omega_m(t)]$.

(c) Derive the open-loop transfer function $\Omega_m(s)/E(s)$ and the closed-loop transfer function $\Omega_m(s)/\Omega_r(s)$.

10-44. Determine the state controllability of the system shown in Fig. 10P-44.

(a) $a = 1$, $b = 2$, $c = 2$, and $d = 1$.

(b) Are there any nonzero values for a, b, c, and d such that the system is uncontrollable?

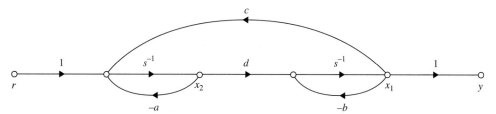

Figure 10P-44

10-45. Determine the controllability of the following systems:

(a) $\mathbf{A} = \begin{bmatrix} -1 & 0 & 0 \\ 0 & -1 & 0 \\ 0 & 0 & -1 \end{bmatrix}$ $\mathbf{B} = \begin{bmatrix} 1 \\ 1 \\ 1 \end{bmatrix}$

(b) $\mathbf{A} = \begin{bmatrix} -1 & 0 & 0 \\ 0 & -2 & 0 \\ 0 & 0 & -3 \end{bmatrix}$ $\mathbf{B} = \begin{bmatrix} 1 \\ 1 \\ 1 \end{bmatrix}$

10-46. Determine the controllability and observability of the system shown in Fig. 10P-46 by the following methods:

(a) Conditions on the **A**, **B**, **C**, and **D** matrices

(b) Conditions on the pole–zero cancellation of the transfer functions

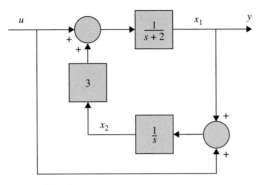

Figure 10P-46

10-47. The transfer function of a linear control system is

$$\frac{Y(s)}{R(s)} = \frac{s+\alpha}{s^3 + 7s^2 + 14s + 8}$$

(a) Determine the value(s) of α so that the system is either uncontrollable or unobservable.

(b) With the value(s) of α found in part (a), define the state variables so that one of them is uncontrollable.

(c) With the value(s) of α found in part (a), define the state variables so that one of them is unobservable.

10-48. Consider the system described by the state equation

$$\frac{d\mathbf{x}(t)}{dt} = \mathbf{A}\mathbf{x}(t) + \mathbf{B}u(t)$$

where

$$\mathbf{A} = \begin{bmatrix} 0 & 1 \\ -1 & a \end{bmatrix} \quad \mathbf{B} = \begin{bmatrix} 1 \\ b \end{bmatrix}$$

Find the region in the a–b plane such that the system is completely controllable.

10-49. Determine the condition on b_1, b_2, c_1, and c_2 so that the following system is completely controllable and observable.

$$\frac{d\mathbf{x}(t)}{dt} = \mathbf{A}\mathbf{x}(t) + \mathbf{B}u(t) \quad y(t) = \mathbf{C}\mathbf{x}(t)$$

$$\mathbf{A} = \begin{bmatrix} 1 & 1 \\ 0 & 1 \end{bmatrix} \quad \mathbf{B} = \begin{bmatrix} b_1 \\ b_2 \end{bmatrix} \quad \mathbf{C} = [d_1 \ d_2]$$

10-50. The schematic diagram of Fig. 10P-50 represents a control system whose purpose is to hold the level of the liquid in the tank at a desired level. The liquid level is controlled by a float whose position $h(t)$ is monitored. The input signal of the open-loop system is $e(t)$. The system parameters and equations are as follows:

Motor resistance R_a	= 10 Ω	Motor inductance L_a	= 0 H
Torque constant K_i	= 10 oz-in./A	Rotor inertia J_m	= 0.005 oz-in.-sec^2
Back-emf constant K_b	= 0.0706 V/rad/sec	Gear ratio n	= N_1/N_2 = 1/100
Load inertia J_L	= 10 oz-in.-sec^2	Load and motor friction	= negligible
Amplifier gain K_a	= 50	Area of tank A	= 50 ft^2

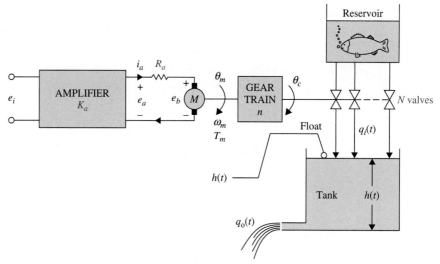

Figure 10P-50

$$e_a(t) = R_a i_a(t) + K_b \omega_m(t) \qquad \omega_m(t) = \frac{d\theta_m(t)}{dt}$$

$$T_m(t) = K_i i_a(t) = (J_m + n^2 J_L)\frac{d\omega_m(t)}{dt} \qquad \theta_y(t) = n\theta_m(t)$$

The number of valves connected to the tank from the reservoir is $N = 10$. All the valves have the same characteristics and are controlled simultaneously by θ_y. The equations that govern the volume of flow are as follows:

$q_i(t) = K_I N \theta_y(t)$ $K_I = 10\,\text{ft}^3/\text{sec-rad}$

$q_o(t) = K_o h(t)$ $K_o = 50\,\text{ft}^2/\text{sec}$

$$h(t) = \frac{\text{volume of tank}}{\text{area of tank}} = \frac{1}{A}\int [q_i(t) - q_o(t)]dt$$

(a) Define the state variables as $x_1(t) = h(t)$, $x_2(t) = \theta_m(t)$, and $x_3(t) = d\theta_m(t)/dt$. Write the state equations of the system in the form of $d\mathbf{x}(t)/dt = \mathbf{A}\mathbf{x}(t) + \mathbf{B}e_i(t)$. Draw a state diagram for the system.

(b) Find the characteristic equation and the eigenvalues of the $\mathbf{A}$ matrix found in part (a).

(c) Show that the open-loop system is completely controllable; that is, the pair $[\mathbf{A}, \mathbf{B}]$ is controllable.

(d) For reasons of economy, only one of the three state variables is measured and fed back for control purposes. The output equation is $y = \mathbf{C}\mathbf{x}$, where $\mathbf{C}$ can be one of the following forms:
(1) $\mathbf{C} = [1\ 0\ 0]$ (2) $\mathbf{C} = [0\ 1\ 0]$ (3) $\mathbf{C} = [0\ 0\ 1]$
Determine which case (or cases) corresponds to a completely observable system.

10-51. The "broom-balancing" control system described in Problem 4-21 has the following parameters:

$$M_b = 1\,\text{kg} \quad M_c = 10\,\text{kg} \quad L = 1\,\text{m} \quad g = 32.2\,\text{ft/sec}^2$$

The small-signal linearized state equation model of the system is

$$\Delta \dot{\mathbf{x}}(t) = \mathbf{A}^* \Delta \mathbf{x}(t) + \mathbf{B}^* \Delta r(t)$$

where

$$\mathbf{A}^* = \begin{bmatrix} 0 & 1 & 0 & 0 \\ 25.92 & 0 & 0 & 0 \\ 0 & 0 & 0 & 1 \\ -2.36 & 0 & 0 & 0 \end{bmatrix} \qquad \mathbf{B}^* = \begin{bmatrix} 0 \\ -0.0732 \\ 0 \\ 0.0976 \end{bmatrix}$$

(a) Find the characteristic equation of $\mathbf{A}^*$ and its roots.

(b) Determine the controllability of $[\mathbf{A}^*, \mathbf{B}^*]$.

(c) For reason of economy, only one of the state variables is to be measured for feedback.

The output equation is written

$$\Delta y(t) = \mathbf{C}^* \Delta \mathbf{x}(t)$$

where

(1) $\mathbf{C}^* = \begin{bmatrix} 1 & 0 & 0 & 0 \end{bmatrix}$ (2) $\mathbf{C}^* = \begin{bmatrix} 0 & 1 & 0 & 0 \end{bmatrix}$
(3) $\mathbf{C}^* = \begin{bmatrix} 0 & 0 & 1 & 0 \end{bmatrix}$ (4) $\mathbf{C}^* = \begin{bmatrix} 0 & 0 & 0 & 1 \end{bmatrix}$

Determine which $\mathbf{C}^*$ corresponds to an observable system.

10-52. The double-inverted pendulum shown in Fig. 10P-52 is approximately modeled by the following linear state equation:

$$\frac{d\mathbf{x}(t)}{dt} = \mathbf{A}\mathbf{x}(t) + \mathbf{B}u(t)$$

where

$$\mathbf{x}(t) = \begin{bmatrix} \theta_1(t) \\ \dot{\theta}_1(t) \\ \theta_2(t) \\ \dot{\theta}_2(t) \\ x(t) \\ \dot{x}(t) \end{bmatrix}$$

$$\mathbf{A} = \begin{bmatrix} 0 & 1 & 0 & 0 & 0 & 0 \\ 16 & 0 & -8 & 0 & 0 & 0 \\ 0 & 0 & 0 & 1 & 0 & 0 \\ -16 & 0 & 16 & 0 & 0 & 0 \\ 0 & 0 & 0 & 0 & 0 & 1 \\ 0 & 0 & 0 & 0 & 0 & 0 \end{bmatrix} \quad \mathbf{B} = \begin{bmatrix} 0 \\ -1 \\ 0 \\ 0 \\ 0 \\ 1 \end{bmatrix}$$

Determine the controllability of the states.

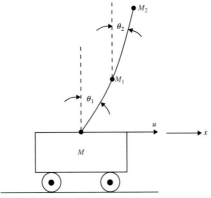

Figure 10P-52

10-53. The block diagram of a simplified control system for the large space telescope (LST) is shown in Fig. 10P-53. For simulation and control purposes, model the system by state equations and by a state diagram.

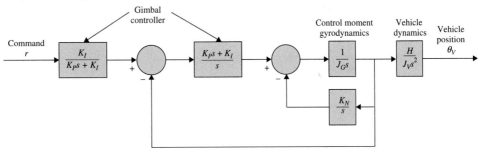

Figure 10P-53

(a) Draw a state diagram for the system and write the state equations in vector-matrix form. The state diagram should contain a minimum number of state variables, so it would be helpful if the transfer function of the system is written first.

(b) Find the characteristic equation of the system.

10-54. The state diagram shown in Fig. 10P-54 represents two subsystems connected in cascade.

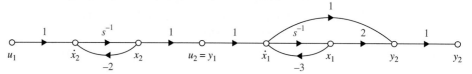

Figure 10P-54

(a) Determine the controllability and observability of the system.

(b) Consider that output feedback is applied by feeding back y_2 to u_2; that is, $u_2 = -ky_2$, where k is a real constant. Determine how the value of k affects the controllability and observability of the system.

10-55. Given the system

$$\frac{d\mathbf{x}(t)}{dt} = \mathbf{A}\mathbf{x}(t) + \mathbf{B}u(t) \quad y(t) = \mathbf{C}\mathbf{x}(t)$$

where

$$\mathbf{A} = \begin{bmatrix} 0 & 1 \\ -1 & -3 \end{bmatrix} \quad \mathbf{B} = \begin{bmatrix} 1 \\ 2 \end{bmatrix} \quad \mathbf{C} = \begin{bmatrix} 1 & 1 \end{bmatrix}$$

(a) Determine the state controllability and observability of the system.

(b) Let $u(t) = -\mathbf{K}\mathbf{x}(t)$, where $\mathbf{K} = [k_1 k_2]$, and k_1 and k_2 are real constants. Determine if and how controllability and observability of the closed-loop system are affected by the elements of $\mathbf{K}$.

10-56. The torque equation for part (a) of Problem 10-21 is

$$J\frac{d^2\theta(t)}{dt^2} = K_F d_1 \theta(t) + T_s d_2 \delta(t)$$

where $K_F d_1 = 1$ and $J = 1$. Define the state variables as $x_1 = \theta$ and $x_2 = d\theta/dt$. Find the state-transition matrix $\phi(t)$ using any available computer program.

10-57. Starting with the state equation $d\mathbf{x}(t)/dt = \mathbf{A}\mathbf{x}(t) + \mathbf{B}\theta_r$ obtained in Problem 10-22, use ACSYS/MATLAB or any other available computer program to do the following:

(a) Find the state-transition matrix of $\mathbf{A}$, $\phi(t)$.

(b) Find the characteristic equation of $\mathbf{A}$.

(c) Find the eigenvalues of $\mathbf{A}$.

(d) Compute and plot the unit-step response of $y(t) = \theta_y(t)$ for 3 seconds. Set all the initial conditions to zero.

10-58. The block diagram of a control system with state feedback is shown in Fig. 10P-58. Find the real feedback gains k_1, k_2, and k_3 so that:

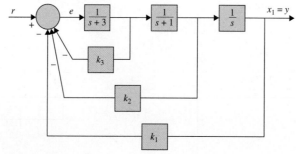

Figure 10P-58

The steady-state error e_{ss} [$e(t)$ is the error signal] due to a step input is zero.
The complex roots of the characteristic equation are at $-1 + j$ and $-1 - j$.

Find the third root. Can all three roots be arbitrarily assigned while still meeting the steady-state requirement?

10-59. The block diagram of a control system with state feedback is shown in Fig. 10P-59(a). The feedback gains k_1, k_2, and k_3 are real constants.

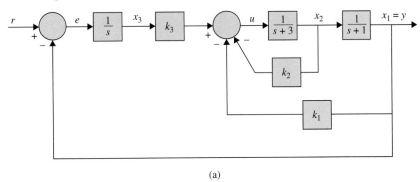

(a)

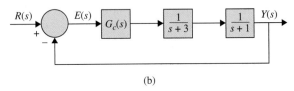

(b)

Figure 10P-59

(a) Find the values of the feedback gains so that:

The steady-state error e_{ss} [$e(t)$ is the error signal] due to a step input is zero.
The characteristic equation roots are at $-1 + j$, $-1 - j$, and -10.

(b) Instead of using state feedback, a series controller is implemented, as shown in Fig. 10P-59(b). Find the transfer function of the controller $G_c(s)$ in terms of k_1, k_2, and k_3 found in part (a) and the other system parameters.

10-60. Problem 9-39 has revealed that it is impossible to stabilize the broom-balancing control system described in Problems 4-21 and 10-51 with a series PD controller. Consider that the system is now controlled by state feedback with $\Delta r(t) = -\mathbf{K}\mathbf{x}(t)$, where

$$\mathbf{K} = [k_1 \quad k_2 \quad k_3 \quad k_4]$$

(a) Find the feedback gains k_1, k_2, k_3, and k_4 so that the eigenvalues of $\mathbf{A}^* - \mathbf{B}^*\mathbf{K}$ are at $-1+j, -1-j, -10$, and -10. Compute and plot the responses of $\Delta x_1(t), \Delta x_2(t), \Delta x_3(t)$, and $\Delta x_4(t)$ for the initial condition, $\Delta x_1(0) = 0.1$, $\Delta \theta(0) = 0.1$, and all other initial conditions are zero.

(b) Repeat part (a) for the eigenvalues at $-2 + j2, -2 - j2, -20$, and -20. Comment on the difference between the two systems.

10-61. The linearized state equations of the ball-suspension control system described in Problem 4-57 are expressed as

$$\Delta \dot{\mathbf{x}}(t) = \mathbf{A}^* \Delta \mathbf{x}(t) + \mathbf{B}^* \Delta i(t)$$

where

$$\mathbf{A}^* = \begin{bmatrix} 0 & 1 & 0 & 0 \\ 115.2 & -0.05 & -18.6 & 0 \\ 0 & 0 & 0 & 1 \\ -37.2 & 0 & 37.2 & -0.1 \end{bmatrix} \quad \mathbf{B}^* = \begin{bmatrix} 0 \\ -6.55 \\ 0 \\ -6.55 \end{bmatrix}$$

Let the control current $\Delta i(t)$ be derived from the state feedback $\Delta i(t) = -\mathbf{K}\Delta\mathbf{x}(t)$, where
$$\mathbf{K} = [k_1 \quad k_2 \quad k_3 \quad k_4]$$

(a) Find the elements of $\mathbf{K}$ so that the eigenvalues of $\mathbf{A}^* - \mathbf{B}^*\mathbf{K}$ are at $-1+j, -1-j, -10$, and -10.

(b) Plot the responses of $\Delta x_1(t) = \Delta y_1(t)$ (magnet displacement) and $\Delta x_3(t) = \Delta y_2(t)$ (ball displacement) with the initial condition

$$\Delta\mathbf{x}(0) = \begin{bmatrix} 0.1 \\ 0 \\ 0 \\ 0 \end{bmatrix}$$

(c) Repeat part (b) with the initial condition

$$\Delta\mathbf{x}(0) = \begin{bmatrix} 0 \\ 0 \\ 0.1 \\ 0 \end{bmatrix}$$

Comment on the responses of the closed-loop system with the two sets of initial conditions used in (b) and (c).

10-62. The temperature $x(t)$ in the electric furnace shown in Fig. 10P-62 is described by the differential equation

$$\frac{dx(t)}{dt} = -2x(t) + u(t) + n(t)$$

where $u(t)$ is the control signal, and $n(t)$ the constant disturbance of unknown magnitude due to heat loss. It is desired that the temperature $x(t)$ follows a reference input r that is a constant.

(a) Design a control system with state and integral control so that the following specifications are satisfied:

$$\lim_{t \to \infty} x(t) = r = \text{constant}$$

The eigenvalues of the closed-loop system are at -10 and -10.
Plot the responses of $x(t)$ for $t \geq 0$ with $r = 1$ and $n(t) = -1$, and then with $r = 1$ and $n(t) = 0$, all with $x(0) = 0$.

(b) Design a PI controller so that

$$G_c(s) = \frac{U(s)}{E(s)} = K_P + \frac{K_I}{s}$$

$$E(s) = R(s) - X(s)$$

where $R(s) = R/s$.
Find K_P and K_I so that the characteristic equation roots are at -10 and -10. Plot the responses of $x(t)$ for $t \geq 0$ with $r = 1$ and $n(t) = -1$, and then with $r = 1$ and $n(t) = 0$, all $x(0) = 0$.

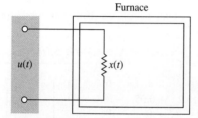

Figure 10P-62

10-63. The transfer function of a system is given by

$$G(s) = \frac{10}{(s+1)(s+2)(s+3)}$$

Find the state-space model of the system if:

$x_1 = y$
$x_2 = \dot{x}_1$
$x_3 = \dot{x}_2$

Design a state control feedback $u = -Kx$ so that the closed-loop poles are located at $s = -2 + j2\sqrt{3}$, $s = -2 - j2\sqrt{3}$, and $s = -10$.

10-64. Fig. 10P-64 shows an inverted pendulum on a moving platform.
Assuming $M = 2$ kg, $m = 0.5$ kg, and $l = 1$ m.

(a) Find the state-space model of the system if $x_1 = \theta$, $x_2 = \dot{\theta}$, $x_3 = x$, $x_4 = \dot{x}$, $y_1 = x_1 = \theta$, and $y_2 = x_3 = x$

(b) Design a state feedback control with gain $-K$ so that the closed-loop poles are located at $s = -4 + 4j$, $s = -4 - 4j$, $s = 210$, and $s = 210$

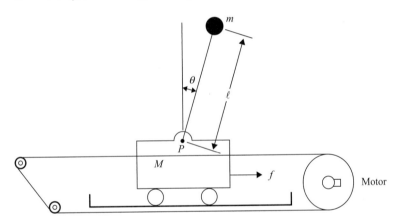

Figure 10P-64

10-65. Consider the following state-space equation of a system:

$$\begin{bmatrix} \dot{x}_1 \\ \dot{x}_2 \end{bmatrix} = \begin{bmatrix} 0 & 1 \\ -6 & -5 \end{bmatrix} \begin{bmatrix} x_1 \\ x_2 \end{bmatrix} + \begin{bmatrix} 0 \\ 1 \end{bmatrix} u$$

(a) Design a state feedback controller so that:
  (i) The damping ratio is $\zeta = 0.707$.
  (ii) Peak time of the unit-step response is 3 sec.

(b) Use MATLAB to plot the step response of the system and show how your design meets the specification in part (a).

10-66. Consider the following state-space equation of a system:

$$\begin{bmatrix} \dot{x}_1 \\ \dot{x}_2 \\ \dot{x}_3 \end{bmatrix} = \begin{bmatrix} -1 & -2 & -2 \\ 0 & -1 & 1 \\ 1 & 0 & -1 \end{bmatrix} \begin{bmatrix} x_1 \\ x_2 \\ x_3 \end{bmatrix} + \begin{bmatrix} 2 \\ 0 \\ 1 \end{bmatrix} u$$

(a) Design a state feedback controller so that:
  (i) Settling time is less than 5 sec (1% settling time).
  (ii) Overshoot is less than 10%.

(b) Use MATLAB to verify your design.

10-67. Fig. 10P-67 shows an RLC circuit.

(a) Find the state equation for the circuit when $v(t)$ is an input, $i(t)$ is an output, and capacitor voltage and the inductor current are the state variables.

(b) Find the condition that the system is controllable.

(c) Find the condition that the system is observable.

(d) Repeat parts (a), (b), and (c) when $v(t)$ is an input, the voltage of the R_2 is output, and capacitor voltage and the inductor current are the state variables.

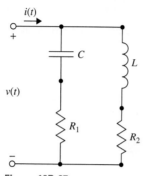

Figure 10P-67

INDEX

A

Absolute acceleration control system (Quarter Car Modeling Tool), 361
Absolute stability, 72
ac control systems, 12–13
  with potentiometers, 192, 193
ac motors, 12, 193, 198. *See also* dc motors
ac signals, 191, 192
Acceleration. *See also* Translational motion
  angular, 157, 159
  symbol/units, 151
  translational motion and, 148
Accelerometer, 242–243
Accumulators, 182. *See also* Fluid capacitance
ACSYS (Automatic Control Systems software), 488. *See also* MATLAB; SIMLab; Virtual Lab
  aircraft attitude-control system and, 632–644
  applet, 341
  frequency-domain plots and, 26
  purpose of, 223, 319, 320
  sun-seeker control system and, 645–647
  tfcal, 86
  tfsym tool, 54, 77, 84–85, 86, 744, 748–750, 754
Active filter
  minor-loop feedback controller with, 603–604
  second-order, 579–580
Actuating signals, 2
Actuators, 105, 107, 108
  double-acting single rod, 605–606
  ideal, 606
  linear, modeling, 605–606
  models of, 289
Addition of poles to $G(s)H(s)$, root-locus diagrams and, 385–387
Addition of zeros to $G(s)H(s)$, root-locus diagrams and, 387–393
Adiabatic process, reversible, 184
Air, as ideal gas, 215
Air flow, through pipe with orifice, 187
Aircraft. *See also* Attitude-control system; Position-control system
  fly-by-wire control system of, 216
  motion equations of, 229–230
Aircraft turboprop engine, 135
  signals of, coupling between, 143, 144
Air-flow system, temperature control of, 243–244
Algebra/manipulation rules, for SFGs, 123–124
Amplifiers. *See also* Operational amplifiers
  dc motor model and, 340
  dead zone and, 272
  saturation and, 11, 216, 272, 337, 340, 346, 347, 348, 349, 351, 352, 368, 369
Analogies, 213–216
  mechanical/electrical components, 213–214
  mechanical/thermal/fluid systems and electrical equivalents, 215–216
  pneumatic system, 214–215
  RLC network and, 213–216
  single-tank liquid-level system and, 215–216
Analytic function, 20
Angles of asymptotes, of root loci, 378–379, 381
Angles of departure/arrival, of root loci, 380, 382
Angular acceleration, 157, 159
Angular displacement, 157, 159
Angular velocity, 157, 159
Antenna control system, of solar collector field, 137
Anticipatory control, 496. *See also* PD (proportional-derivative) controllers
Antiwindup protection, 177
Armature-controlled dc motor, 338–339, 358–359
  block diagram of, 222, 290
  position control of, 292–293
  speed control of, 291–292
Asymptotes, 378
  angles of, of root loci, 378–379, 381
  intersect of, 379, 381
Asymptotic plots. *See* Bode plots
Asymptotic stability, 75–76
Attitude-control system (aircraft), 216–217, 295–296, 498–505. *See also* Position-control system
  ACSYS/MATLAB tools and, 632–644
  block diagram of, 217
  forward-path transfer function of, 498
  frequency-domain design, 503–505, 509–511, 520–523, 526–528, 531–532
  PD controller and, 498–505, 506–511
  PI controller and, 516–528
  PID controller and, 529–532
  root loci of, 302
  steady-state response of, 304
  system parameters, 216
  third-order, 300–304
  time response of, 300–304
  time-domain design, 498–503, 506–508, 516–520, 529–530
  transfer-function block diagram of, 217
  transient response of, 301
  unit-ramp responses of, 299
  unit-step responses of, 295, 303, 501
Attitude-control system (guided missile), 232–233
Automatic Control Systems software. *See* ACSYS
Automobiles. *See also* Quarter-car model
  high performance real-time control of, 4
  idle-speed control system, 4, 239–240
  intelligent systems in, 3–4
  steering control of, 4
Auxiliary equation, 81

B

Back emf (back electromotive force), 199, 202, 203, 204
  electric friction and, 204
Back emf constant, 202, 204, 216, 223, 245, 246, 247, 344, 599, 659, 666, 738, 765
  torque constant $v.$, 204–205
Backlash, 11–12, 164
  gear, 164, 216, 340, 368
  input-output characteristic of, 164
  physical model of, 164
  Quarter Car Modeling Tool and, 361
Ball and beam system, 228–229
  lever arm, 228, 483, 665
Ball bearings, in spacecraft systems, 151
Ball-suspension control system, 250, 251
Bandwidth (BW)
  feedback and, 11, 15
  prototype second-order system and, 416–418
  specification, 412–413
Belt and pulley, 161, 162
BIBO (bounded-input, bounded-output) stability, 73–74
Block diagrams, 104–118
  antenna control system (of solar collector field), 137
  armature-controlled dc motor, 222, 290
  attitude-control system of aircraft, 217
  closed-loop idle-speed control system, 7
  control system with conditional feedback, 144
  dc-motor control system, 105, 203
  digital autopilot for guided missile control, 14
  electric train control, 138

773

Block diagrams (*Continued*)
  elements of, 106–107
  feedback control systems, 109, 136–137
  gain formula and, 128–129
  general control system, 106
  heating system, 104
  idle-speed control system, 5
  mass-spring-friction system, 152, 153, 155
  mathematical equations and, 109–113
  MATLAB tools and, 129–132
  motor-control system with tachometer feedback, 102
  of multi-input systems (with disturbance), 115–117
  of multivariable systems, 117–118, 674–676
  open-loop control system, 6
  potentiometer, 191
  problems/exercises for, 134–138
  reduction, 113–115
  references for, 133, 134
  RLC network, 166
  sampled-data control system, 13–14
  sensing devices (of control systems), 107
  SFG, of control system, 129
  SFGs *v.*, 119, 125
  speed-control system, 581
  sums/differences of signals in, 173–174
  sun-seeker control system, 220, 537
  transfer functions from, MATLAB and, 129–132
  of transfer functions in parallel, 108
  of transfer functions in series, 108
Blocks, 106
Bode diagrams, 77–78
  of aircraft position-control system, 467
  of phase-lag controller, 565
  of PI controller, 515
"bode" function, 33, 41, 43
Bode plots (corner plots/asymptotic plots), 26, 32–43
  advantages of, 455
  of constant K, 34, 35
  disadvantages of, 456
  gain margin on, 456
  gain-crossover point and, 47
  of $L(s)$, 457
  of $(jw)^p$, 36
  PD controllers and, 497
  phase curve of, 48
  phase margin on, 456
  phase-crossover point and, 47
  of phase-lag system, 571
  of phase-lead controller, 535–536, 552
  poles/zeros at origin and, 34–36
  pure time delays and, 42–43, 458–459
  quadratic poles/zeros and, 39–41
  simple pole, $1/(1+jwT)$, 39
  simple zero, $1+jwT$, 37–38
  slope of the magnitude curve of, 459–462
  stability analysis with, 455–459
  of sun-seeker system, 552, 557
  time delays and, 42–43, 458–459
Bounded-input, bounded-output stability. *See* BIBO stability

Branch point, 108
  relocation, 113, 114
Branches, on root loci, 378, 381
Breakaway points (saddle points), on root loci, 380–381, 382
British units, 148, 204
Broom-balancing system, 233–234
Brushless PM dc motors, 201
Bulk modulus, 182

C

Capacitance, 182. *See also* RLC network
  electrical, 172
  fluid, 181
  heat transfer and, 177
  pneumatic systems, 182–183
  RC circuit systems and, 63, 171–172
  units for, 172, 179, 189
Capacitors, 165
Cart, inverted pendulum on, 227–228
Cartesian coordinate frame, 16, 17
Cascade compensation, 489, 490
Cascade decomposition, 712–713
Causal system, 53
Cause-and-effect relationships, 8. *See also* Feedback
Centroid, 379. *See also* Asymptotes
Characteristic equations, 71, 695
  from differential equations, 695–696
  roots of, stability and, 74
  from state equations, 696
  from transfer functions, 696
Charge, units for, 172
Circles, constant-M, 464, 465
Circuit systems. *See* RC circuit systems
Closed-loop acceleration control (quarter-car model), 359–360, 366–367
Closed-loop control systems. *See* Feedback control systems
Closed-loop frequency response, of aircraft position-control system, 469
Closed-loop idle-speed control system, 7
Closed-loop position control (quarter-car model), 221–222, 359
Closed-loop relative position control (quarter-car model), 365–366
Closed-loop transfer function
  poles added to, 307–308
  steady-state error and, 266–270
  zero added to, 308–309
Comparators, 105, 106, 107, 109
  relocation, 113, 114
Compensated phase-lag system
  Bode plot of, 571
  root loci of, 564
Compensated sun-seeker system
  Bode plots of, 557
  unit-step responses of, 568
Compensation, 489
  cascade, 489, 490
  feedback, 489, 490
  feedforward, 490–491
  series, 489, 490
  series-feedback, 489, 490
  state-feedback, 489, 490
Completely controllable process, 716

Completely state controllable, 717
Complex convolution (real multiplication), 56, 57
Complex multiplication. *See* Real convolution
Complex numbers, 16–18
  polar form of, 17
  properties of, 18
  rectangular form of, 16, 17
Complex s-plane, 18, 19
Complex shifting (theorem), 56
Complex variables, 18–26
  functions of, 19–26
  imaginary component of, 18
  MATLAB commands and, 25
  problems/exercises for, 93
  real component, 18
  references for, 92
Complex-conjugate poles, 61
Conditional feedback (block diagram), control system with, 144
Conditionally stable system, 459–462, 555
Conduction, 178
Configuration Parameters, 343
Conjugate, of complex number, 17
Conservation of mass, 181, 182, 185, 188, 606
  pneumatic systems and, 182
Conservation of volume, 181, 182
Constant K, 34, 35
Constant pressure, 183
Constant temperature, 183
Constant volume, 183
Constant-M circles, 464, 465
Constant-M loci, 463–470
Constants. *See also* Back emf constant; Time constant
  electrical time, amplifier-motor system and, 217, 294
  error, 265
  low-time, 198
  mechanical time, motor-load system and, 217, 294
  motor electric-time, 290, 359
  motor-mechanical time, 291
  multiplication by a constant (theorem), 54, 56
  spring, 149, 151, 159
  tachometer, 195
  torque, 202, 204, 216, 223, 245, 247, 249, 339, 599, 659, 666, 668, 738, 757, 765
  torsional spring, 158
Continuous-data control systems, 12–13
  time response of, 253–254
Control lab, 337–371
  SIMLab, 223, 340–344
  virtual experimental system, 338–340
  Virtual Lab, 223, 340–344
Control Parameters window, 363
Control system design, 487–672. *See also* Frequency-domain design; Time-domain design
  controller configurations, 489–491
  design specifications, 487–489
  Experiment 5, 355–356
  feedforward controllers and, 588–590
  forward controllers and, 588–590
  fundamental principles of, 491–492

Index ◀ 775

hydraulic control system, 605–617
lead-lag controller and, 574–576
minor-loop feedback control, 601–604
PD controller and, 492–511
phase-lag controller and, 561–574
phase-lead controller and, 532–561
PI controller and, 511–528
PID controller and, 528–532
pole-zero-cancellation design in, 576–588
robotic-arm-joint (controller design), 617–630
robust, 590–601
three steps for, 487
Control System Toolboxes. *See* Toolboxes
Control systems. *See also* Attitude-control system; Feedback control systems; Frequency-domain analysis; Linear control systems; Nonlinear control systems; Sun-seeker control system; Time-domain analysis
ac, 12–13, 192, 193
applications of, 2–6
ball-suspension, 250, 251
block diagram, 106
block diagram/SFG of, 129
components of, 2
with conditional feedback (block diagram), 144
continuous-data, 12–13
controllability of, 714–719, 721–725
dc, 12–13, 105, 137, 192
definition of, 1
digital, 14
discrete-data, 13–14
frequency-domain analysis of, 409–486
importance of, 1
objectives of, 1, 2
open-loop, 5–7, 109
with PI controller, 317
printwheel, 231, 246, 329
robust, 590–601
rotary-to-linear motion, 161
sampled-data, 14
sensitivity and, 9–10
SFG/block diagram of, 129
with state feedback, 715
sun-tracking, 4–6
time-domain analysis of, 253–336
Control volume, net mass flow rate and, 181
Controllability, 714–719, 721–725
control system with state feedback, 715
general concept of, 716
invariant theorems on, 723–725
observability and, 714, 715
state, 716–717
testing methods for, 717–719
transfer functions/observability, relationship among, 721–723
Controllability canonical form (CCF), 692, 701–703
direct decomposition to, 708–709
Controllability matrix, 702
Controllable process, completely, 716
Controlled process, 6
Controlled variables, 2
Controller design projects

quarter-car model, 357–367
robotic arm, 354–357
Controller Design Tool, 86, 631
Controllers, 7, 107. *See also* PD controllers; PI controllers; PID controllers
feedforward, 588–590
forward, 588–590
lead-lag, 574–576
minor-loop feedback, 601–604
notch, 581–588
P, 617–621
phase-lag, 561–574
phase-lead, 532–561
robust, 590–601
Convection, 178–179
fluid-boundary, 178
Conversion between translational/rotational motions, 161
Conversion factors
mechanical system properties and, 151
thermal system properties and, 179
Corner frequency, 37
Corner plots. *See* Bode plots
Coulomb friction, 150, 158
coefficient, 150
steady-state errors and, 273–274
Critical point, 432, 434
Cross-section view
brushless PM dc motor, 201
hot oil forging in quenching vat, 238
iron core PM dc motor, 200
moving-coil PM dc motor, 201
surface-wound PM dc motor, 200
Current, units for, 172
Current Law, 165
Cutoff rate, 413
Cylindrical container, fluid flow into, 182

D

Damping ratio/damping factor (prototype second-order system), 277–278
Damping term, 315, 499
Dashpot, for viscous friction, 149, 150
dB (decibels), 26, 32, 33, 452
dc motor model (virtual experiments). *See also* Robotic arm
amplifier in, 340
interfacing, 340
open-loop sine input and, 347–350
open-loop speed, 345–347
position control and, 352–354
position sensor/speed sensor in, 339–340
simulation and, 345–354
speed control and, 350–352
system parameters, 339
dc (direct-current) motors, 198–205. *See also* Permanent-magnet dc motors
armature-controlled, 222, 289–293, 338–339, 358–359
operational principles of, 199
sun-seeker control system and, 218, 220
torque production in, 199
dc signals, 191
dc-motor control systems, 12–13, 245, 246, 247, 248
block diagram, 105, 203

with integral controller, 738–740
with nonzero initial conditions, SFG of, 203
with potentiometers, 192
problems/exercises for, 137
with transfer functions and amplifier characteristics, 105
Dead zone
amplifier with, 272
gear trains and, 164
Simulink Library Browser and, 349
Decades, 34
Decibels (dB), 26, 32, 33, 452
Decompositions (of transfer functions), 678, 707–714
cascade, 712–713
direct, 707–712
parallel, 713–714
Delay time, 257, 283–285, 488
Derivative control, 176, 496. *See also* PID controllers
Design aspects of root loci, 385–393
Design projects. *See also* Control system design
quarter-car model, 357–367
robotic arm, 354–357
Diagonal canonical form (DCF), 704–705
Diagrams. *See also* Block diagrams; Free-body diagrams; Modeling; Signal-flow graphs
Bode, 77–78
gear trains, 162
motor-load system, 232
Nyquist, 29
op-amps, 173
rotary incremental encoder, 196
rotational mechanical system, 159
state, 133, 673, 676–681
state-flow, 138, 145, 231, 234
sun-seeker control system, 218
Differential equations, 49–52
characteristic equations from, 695–696
first-order, 50
high-order, state equations and, 691–693
integro-, 49
linear ordinary, 49, 62–67
nonlinear, 49
for pendulum, 49
problems/exercises for, 94
RC circuit and, 171–172
RLC network and, 49
second-order, 49
state diagrams from, 678–679
Differentiation (theorem), 54, 56
Digital autopilot, for guided missile control, 14
Digital control systems, 14
Dirac delta function, 68
Direct decomposition, 707–712
to CCF, 708–709
to OCF, 709–712
Direct-current motors. *See* dc motors
Discrete-data control systems, 14
Displacement, 148. *See also* Translational motion
angular, 157
Distance (symbol/units), 151
Distributed mass systems, 148

Disturbance, 105
  heat loss, 104
  multi-input systems (block diagram) with, 115–117
  noise and, 10–11
Disturbance input, system with, 260
Disturbance rejection, 487
Disturbance vector, 682
Disturbance-open loop response, 289
Divider, voltage, 171–172
Dominant poles/zeros of transfer functions, 311–314
Dominant roots, 301
Double-tank liquid-level system, 186–187, 242
Drive-by-wire technology, 3
Driver assist systems, 3
Dry friction, rolling, 150–151
Dual-channel incremental encoder, 197–198
  one cycle of output signals of, 198
  signals, in quadrature, 196, 197
Dynamic equations, 682
Dynamic systems, 147. *See also* Electrical systems; Fluid systems; Mechanical systems; Modeling; Pneumatic systems; Thermal systems
  with transportation lag, 205–207

E
Eigenvalues, 697
Eigenvectors, 697–698
  generalized, 698–699
Electric circuit representation, of potentiometer, 190
Electric friction, 204. *See also* Back emf
Electric train control (block diagram), 138
Electrical circuits, 235
Electrical elements
  active, 172–177
  passive, 165
Electrical networks
  examples, 165–172
  modeling of, 165–172
  problems, SFGs and, 141–142
  state equations of, 167, 168–169
  state variable analysis of, 134
Electrical schematics, RLC network, 166
Electrical systems. *See also* Operational amplifiers
  equivalents, fluid systems and, 215–216
  equivalents, mechanical systems and, 215
  equivalents, thermal systems and, 216
  modeling of, 165–172
  op-amps, 172–177
  properties, 172
  simple, 165–172
Electrical time constant, amplifier-motor system and, 217, 294
Electro-hydraulic valve, four-way, 606–611
Electromechanical systems, 235–236, 289
Electromechanical transducer, 189. *See also* Potentiometers
Electromotive force. *See* Back emf
Elementary heat transfer properties, 177–180
  capacitance and, 177
  rectangular object and, 180
  references for, 224

Encirclements
  defined, 428
  number of, 429
Enclosures
  defined, 428
  number of, 429
Encoders, 189. *See also* Sensors
  incremental, 195–198
  sensors and, 189
Energy (units/symbols)
  electrical system, 172
  heat stored, 179, 189
  rotational mechanical system, 159
Engine, turboprop, 135
  signals of, coupling between, 143, 144
Equation of state, 182. *See also* State equations
Error, quantization, 273
Error constants, 265. *See also* Steady-state errors
Error signal, 108
Estimator, 489
Euhler formula, 17
Evans, W. R., 372, 377. *See also* Root-locus technique
Experiment control window, 342
Experiment menu, 342
Experiments. *See also* dc motor model
  1: Speed Control, 350–352
  2: Position Control, 352–354
  3: Open-Loop Speed, 345–347
  4: Open-Loop Sine Input, 347–350
  5: Control System Design, 355–356
External disturbance, 10–11

F
Feedback, 8–11, 105. *See also* State feedback
  bandwidth and, 11, 15
  frequency responses and, 11
  impedance and, 11
  negative, 8
  noise and, 10–11
  overall gain and, 8–9
  positive, 109
  stability and, 9–10, 15
  transient responses and, 11
Feedback compensation, 489, 490
Feedback control systems (closed-loop control systems), 7, 108–109. *See also* Linear control systems
  block diagram of, 109, 136–137
  classifications of, 11–14
  conditions for, 108
  configuration, 8
  frequency response of, 410–412
  gain-phase characteristics of, 412
  linear *v.* nonlinear, 11–12
  with noise signal, 10
  nonunity, 258, 265–272
  open-loop control systems *v.*, 7
  with PD controller, 314–316, 493
  SFG of, 124
  state, 715, 728–729
  time-invariant *v.* time-varying, 12–14
  torque-angle curve of, 273
  with two feedback loops, 9
  unity, 260–261

Feedback controller, minor-loop, 601–604
Feedback loops
  negative, 109
  positive, 109
  unity, 109
Feedback-path transfer function matrix, of multivariable feedback control system, 676
Feedforward compensation, 490–491
Feedforward compensators, sun-seeker system and, 589–590
Feedforward controllers, 588–590
Fin positions, of aircraft. *See* Position-control system
Final-value theorem, 55–56
First-order differential equations, 50
First-order linear system, 62
First-order op-amp configurations, 174–177
First-order prototype systems, 63–64
  time constant and, 274
  time response of, 274–275
  unit-step response of, 274
Fixed-configuration design, 489. *See also* Compensation
Flow rate
  fluid volume, 185
  heat, 177, 178, 179
  mass, 181, 182, 184, 185, 187, 189
  volume, 181, 182, 189
Flows, laminar, 184, 185
Fluid capacitance, 181, 182
Fluid continuity equation, 181
Fluid forced through frictionless pipe, 184
Fluid inductance, 184
Fluid inertance, 184
Fluid systems, 180–189. *See also* Pneumatic systems
  electrical equivalents for, 215–216
  incompressible, 181
  parameters in, 180–181
  properties, 180–189
  references, 224
Fluid viscosity, 185
Fluid volume flow rate, 185
Fluid-boundary heat convection, 178
Fly-by-wire control system, 216
Force (symbol/units), 151
Force-mass system, 149
Force-spring system, 149
Forward compensators, sun-seeker system and, 589–590
Forward controllers, 588–590
Forward path, 122
Forward-path gain, 122
Forward-path transfer function
  Bode plots of, in third-order sun-seeker control system, 552
  pole added to, 305–307, 424–426
  of second-order aircraft attitude control system, 498
  third-order system with, 425
  unit-step responses of second-order system with, 423
  zero added to, 309–311, 418–424
Forward-path transfer function matrix, of multivariable feedback control system, 676

Index ◀ 777

Four-way electro-hydraulic valve, 606–611
  input voltage/main spool displacement relationship in, 610–611
  liberalized flow equations for, 608–610
  main valve schematic in, 608
  orifice equation and, 607–608
  rectangular valve-port geometry in, 609
  two-stage valve block diagram, 611
Free-body diagrams (FBDs)
  broom-balancing system, 233
  electrical circuits (exercise), 235, 236
  grain scale, 251
  inverted pendulum on cart, 227, 228
  mass-spring-damper system, 154
  mass-spring-friction system, 151
  motor-load system, 160
  Newton's law of motion and, 148
  rotational system, 158, 159
  spring-supported pendulum and, 209
  train exercise, 226
  2-DOF spring-mass system, 156
Frequency
  gain-crossover, 453
  phase-crossover, 451
Frequency response function, 26
Frequency responses, 487
  of closed-loop systems, 410–412
  feedback and, 11
  step responses/Nyquist plots, correlation among, 450–451
Frequency-domain analysis, 409–486
  crux of, 410
  frequency response of closed-loop systems, 410–412
  sensitivity studies and, 470–472
  time-domain analysis v., 409
  transfer function and, 409–410
Frequency-domain design
  frequency-domain characteristics in, 492
  notch controller and, 586–588
  PD controllers and, 496–497, 503–511
  performance specifications, 487, 488
  phase-lag controllers and, 563–566, 569–572
  phase-lead controllers and, 535–537, 543–548, 551–554
  PI controllers and, 514–516, 526–528
  pole-zero-cancellation and, 580–581
  second-order aircraft attitude control system, 503–505, 520–523
  third-order aircraft attitude control system, 509–511, 526–528, 531–532
  third-order sun-seeker control system and, 551–554
  time-domain design v., 488, 492
Frequency-domain plots, 26–48
  computer-aided construction of, 26–27
  problems/exercises for, 93–94
Frequency-domain specifications, 412–413
  bandwidth, 412–413
  cutoff rate, 413
  resonant frequency w_r, 412, 413–416
  resonant peak M_r, 412, 413–416
Friction. *See also* Mass-spring-friction system
  Coulomb, 150, 158, 273–274
  electric, 204

gear train with, 163
rolling dry, 150–151
for rotational motion, 158–159
static, 150, 158
for translational motion, 149–150
viscous, 149–150, 158
Functions (of complex variable), 19–26. *See also* Poles; Transfer functions; Zeros
  analytic, 20
  frequency response, 26
  single-valued, 19
  singularities of, 20

G

$G(s)$
  frequency response function of, 26
  polar plots of, 27–32
  polar representation of, 22–24
$G(s)H(s)$
  addition of poles to, 385–387
  addition of zeros to, 387–393
  Nyquist criterion and, 434–435
$G(s)$-plane, 19
$G_2(s)H_2(s)$, 394, 396, 399, 400
Gain crossover, 453
Gain formula, 124–127
  block diagrams and, 128–129
  output nodes/noninput nodes and, 127–128
  SFGs and, 124–128
  simplified, 129
Gain margin (GM), 46, 86, 451–453, 487, 488
  on Bode plot, 456
  definition of, 451–452
  of nonminimum-phase systems, 452–453
  physical significance of, 452
Gain and Phase Calculator, 350
Gain-crossover frequency, 453
Gain-crossover point, 46, 47
Gain-phase characteristics
  of feedback control system, 412
  of ideal low-pass filter, 411
Gain-phase plots
  of aircraft position-control system, 468
  of $L(s)$, 462
Gains, 109
Gas flow, into rigid container system, 187–188
Gas law, perfect, 183, 187
Gas systems. *See* Pneumatic systems
Gear backlash, 164, 216, 340, 368
Gear trains, 162–164
  backlash in, 164, 216, 340, 368
  dead zone in, 164
  diagram, 162
  with friction/inertia, 163
  motor-load system and, 230
Generalized eigenvectors, 698–699
Grain scale, 251, 252
Guided missile
  attitude control of, 232–233
  control, digital autopilot for, 14

H

Hardware in the loop simulation, 2
Heat conduction flow, one-directional, 178
Heat convection. *See* Convection

Heat exchanger system, 239, 250
Heat flow rate, 177, 178, 179
Heat loss, 104
Heat radiation system, with directly opposite ideal radiators, 179
Heat transfer problem, between fluid/insulated solid object, 180
Heat transfer properties (elementary), 177–180
  capacitance and, 177
  rectangular object and, 180
  references for, 224
Heating system (block diagram), 104
High-order differential equations, state equations v., 691–693
High-pass filter characteristics, of PD controllers, 496, 497
Horsepower (hp), 204
Hurwitz criterion. *See* Routh-Hurwitz criterion
Hybrid powertrains, 3–4
Hydraulic capacitance (symbol/units), 189
Hydraulic control system, 605–617. *See also* Robot-arm-joint system
  applications, 613–617
  double-acting single rod actuator in, 605–606
  four-way electro-hydraulic valve in, 606–611
  modeling, 612–613
  P controllers for, 617–621
  PD controllers for, 621–626
  PI controllers for, 626–628
  PID controllers for, 628–630
  rotational system and, 615–616
  translational motion and, 613–615
  variable load and, 616–617
Hydraulic diameter, 185
Hydraulic generator system, 240–241
Hydraulic resistance (symbol/units), 189
Hydraulic servomotor, 242

I

Ideal gas, air as, 215
Ideal linear actuators, 606, 612, 613, 618
Ideal low-pass filter, 411
Ideal op-amp, 173
Ideal radiators, heat radiation system and, 179
Idealized models, linear feedback control systems as, 11
Idle-speed control system, 5
  automobile, 4, 239–240
  block diagram, 5
  closed-loop, 7
  open-loop, 7
ilaplace command, 85
Imaginary axis, intersection of root loci with, 380, 382
Imaginary component, of complex variable, 18
Impedance, feedback and, 11
Impulse response, 67–69
Incompressible fluid systems, 181–187
  inductance and, 184
  open-top cylindrical container and, 182
  resistance and, 184–185

Incremental encoders, 195–198. *See also* Encoders; Sensors
  dual-channel, 196–198
  linear, 196
  rotary, 196
  single-channel, 196, 197
Inductance. *See also* RLC network
  fluid, 184
  incompressible fluids and, 184
  units for, 172
Inductors, 165
Inertance, fluid, 184
Inertia, 157. *See also* Load inertia
  gear train with, 163
  symbol/units, 159
Initial states, 51
Initial-value theorem, 55, 56
Input node, 121
Input vector, 682
Input voltage/main spool displacement relationship (four-way valve), 610–611
Inputs, 2, 104
Insignificant poles, steady-state response, 313–314
Insulated solid object/fluid, heat transfer problem between, 180
Integral controllers, 176. *See also* PI controllers; PID controllers
  dc-motor control system with, 738–740
  state feedback with, 735–741
  sun-seeker system and, 737–738
Integration (theorem), 55, 56
Integration operation, state diagrams and, 677
Integrator output magnitude, 177
Integrodifferential equation, 49
Intelligent systems, 2
  in automobiles, 3–4
Interfacing, dc motor model and, 340
Intersect of asymptotes, 379, 381
Intersection, of root loci with imaginary axis, 380, 382
Invariance properties, of similarity transformations, 700–701
Invariant theorems, on controllability/observability, 723–725
Inventory-control system, 103
Inverse Laplace transform, 54
  MATLAB and, 64
  by partial-fraction expansion, 57–62
  problems/exercises for, 97–99
Inverted pendulum, on cart, 227–228
Inverting op-amp configuration, 175
Inverting op-amp transfer functions, 175–176
Iron-core PM dc motors, 199–200
Isentropic process, 184
Isobaric process, 183
Isothermal process, 183
Isovolumetric process, 183

J
$(jw)^p$, 34–36
Jerk function, 256
Jordan blocks, 706
Jordan canonical form (JCF), 706
Junction points. *See* Nodes

K
$K = \pm \infty$ points, on root loci, 377–378, 381
$K = 0$ points, on root loci, 377–378, 381
K values on root loci, calculation of, 382
Kalman, E., 714
Kirchoff's laws, 147, 165

L
$L(s)$ plot
  Bode plot and, 457
  gain-phase plot and, 462
  Nyquist criterion and, 434–435
  Nyquist plot and, 444–449
  poles added to, 445–448
  zeros added to, 448–449
Laminar flows, equations of resistance for, 184, 185
Laplace operator, 52
Laplace transform, 52–57. *See also* Inverse Laplace transform
  definition of, 52–53
  features of, 52
  linear ordinary differential equations and, 52, 62–67
  MATLAB and, 53
  one-sided, 52–53
  problems/exercises for, 94–97
  references for, 92
  theorems of, 54–57
Laplace transform table, 53, 54, 57, 62, 65, 97, 276, 295, 298, 758
Laser printers, 289
Lead screw, 161
Lead-lag controller, 574–576
Lever
  gear train and, 162
  thermal, 237, 238
  throttle, 242
Lever arm (ball and beam system), 228, 483, 665
Library Browser, Simulink, 349
Light source (rotary incremental encoder), 196
Linear actuators
  force balance equation for, 612
  modeling, 605–606, 612–613
Linear control systems, 11–12. *See also* Nonlinear control systems
  block diagram of, 109
  characteristic equation of, 71, 74
  as idealized models, 11
  mathematical foundations for, 16–90
  nonlinear control systems v., 11–12, 15
  observability of, 714, 715, 719–725
  rotary-to-linear motion control systems, 161
  stability of, 72–73
Linear incremental encoder, 196
Linear motion potentiometer, 189, 190
Linear ordinary differential equations, 49
  first-order prototype systems and, 63–64
  Laplace transform and, 52, 62–67
  procedure for solving, 62
  second-order prototype systems and, 64–67
Linear spring, 149
Linear variable differential transformer (LVDT), 359, 360
Linearization (of nonlinear systems), 206–213

state space approach and, 207–213
Taylor series and, 207, 208
Liquid-level system
  double-tank, 186–187
  single-tank, 185–186
Load inertia, 160, 244, 246, 247, 290, 344, 765
  armature-controlled dc motor and, 289, 338
  printwheels and, 599
  variable, position-control system and, 599
Load torque, 4, 5
Loop gain, 122
Loop Method, 165
Loops, 122, 123. *See also* Feedback loops
  nontouching, 122
  phase-locked, 245
Low-pass filter, ideal, 411
Low-time-constant properties, 198
Lumped mass models, 148
LVDT (linear variable differential transformer), 359, 360

M
Magnetic-ball-suspension system, 211–213, 725–728
  state feedback and, 731–733
Magnification v. normalized frequency, of prototype second-order system, 415
Magnitude phase, 9
Magnitude-phase plane, constant-M loci in, 463–470
Magnitude-phase plot, 26, 44–46
  gain-crossover point and, 47
  phase-crossover point and, 47
  stability analysis with, 462–463
Main spool displacement/input voltage relationship (four-way valve), 610–611
Manipulation rules/algebra, for SFGs, 123–124
Marginally stable/unstable, 76
Mason, S. J., 119. *See also* Gain formula; Signal-flow graphs
Mass
  conservation of, 181, 182, 185, 188, 626
  defined, 148
  distributed mass systems, 148
  lumped mass models, 148
  in polytropic process, 183
  symbol/units, 151
Mass flow rate, 181, 182, 184, 185, 187
  symbol/units, 189
  volume flow rate v., 182
Mass-spring-friction system, 151–153, 224
  block diagrams, 152, 153, 155, 156
  FBD of, 151
  SFG, 155
Mathematical equations, block diagrams and, 109–113
Mathematical foundations (for linear control systems), 16–90
Mathematical modeling. *See* Modeling
MATLAB. *See also* Toolboxes
  aircraft attitude-control system and, 632–644
  block diagrams/SFGs and, 129–132
  "bode" function, 33, 41, 43
  complex variables and, 25

development/availability of, 488
frequency-domain plots and, 26
inverse Laplace transform and, 64
Laplace transforms and, 53
"nichols" function, 45
Nyquist diagram and, 29
partial-fraction expansion and, 58–61
phase/gain margins and, 46
rise time and, 288
role of, 2
settling time and, 288
SISO Design Tool, 363, 366, 367, 371, 634, 635, 636, 638, 642
stability tools, 85–90
sun-seeker control system and, 645–647
Symbolic Tool, 53, 64, 83, 84, 741, 748
tfsym tool, 54, 77, 84–85, 86, 744, 748–750, 754
time response and, 67
unit impulse response and, 69
velocity-control system and, 259
zero-pole-gain models and, 21
Matrices
  controllability, 702
  feedback-path transfer function matrix, 676
  forward-path transfer function matrix, 676
  state-transition matrix, 684–686
  vector-matrix representation of state equations, 682–684
Matrix algebra, 16
Maximum overshoot, 256–257, 280–283, 487, 488
Mechanical systems
  conversion between translational/rotational motions in, 161
  electrical equivalents, 215
  gear trains, 162–164
  modeling of, 148–164
  Newton's second law of motion and, 147, 148, 156, 157, 184
  rotational motion in, 157–160
  symbols/units/conversion factors, 151
  translational motion in, 148–157
Mechanical time constant, motor-load system and, 217, 294
Microradians, 258
Minimal set, of variables, 51
Minimum-phase transfer functions, 47–48
  Nyquist criterion for, 435–437, 440–444
Minor-loop feedback controller, 601–604
  with active filter, 603–604
  rate-feedback and, 601–602
  sun-seeker system and, 603–604
  tachometer-feedback control and, 601–602
Model Parameters window, 362
Modeling. See also Block diagrams; dc motor model; Signal-flow graphs
  of actuators, 289
  electrical elements (active), 172–177
  electrical elements (passive), 165
  electrical networks, 165–172
  electrical systems, 165–177
  fluid systems, 180–189
  hydraulic control system, 612–613
  linear actuators, 605–606, 612–613
  mechanical systems, 148–164

of PM dc motors, 201–205
references, 224
tachometers, 195
thermal systems, 177–180
Modulus, bulk, 182
Moment equation, 158, 210
Morning sickness, 9–10
Motion
  Newton's second law of, 147, 148, 156, 157, 184
  rotational, 157–161
  translational, 148–157, 161
Motion equations, of aircraft, 229–230
Motor blocks, SIMLab, 344
Motor electric-time constant, 290, 359
Motor-control system
  open-loop, 244
  with tachometer feedback, 102
  torque-angle curve of, 273
Motor-load system, 159–160
  gear train and, 230
  schematic diagram of, 232
Motor-mechanical time constant, 291
Motors. See also dc motors
  ac, 12, 193, 198
  servomotors, 13, 198, 242, 328, 331, 477
  voice-coil, 247
Moveable-plate capacity, 235, 236
Moving-coil PM dc motors, 200–201
M_r. See Resonant peak M_r
Multi-input systems with disturbance (block diagram), 115–117
Multiple-order poles, 59–60
Multiple-parameter variation. See Root contours
Multiplication by a constant (theorem), 54, 56
Multistage phase-lead controller, 555. See also Two-stage phase-lead controller
Multivariable feedback control system
  block diagram of, 675
  feedback-path transfer function matrix of, 676
  forward-path transfer function matrix of, 676
Multivariable systems, 4, 71–72
  block diagrams of, 117–118, 674–676
  transfer functions of, 4, 71–72, 117–118, 673–674

N
Natural undamped frequency, 278–280
Negative feedback, 8
Negative feedback loop, 109
Net mass flow rate, control volume and, 181
Newton's second law of motion, 147, 148, 156, 157, 184
Nichols chart, 44, 463–470
  of aircraft position-control system, 468
  nonunity feedback control systems and, 469–470
"nichols" function, 45
Node Method, 165–166, 170
Nodes, 119
  input, 121
  output, 121, 127–128
Noise, feedback and, 10–11

Nonfeedback systems. See Open-loop control systems
Noninput nodes/output nodes, gain formula and, 127–128
Nonlinear control systems, 49, 223. See also Linear control systems
  linear control systems v., 11–12, 15
  linearization of, 206–213
Nonlinear differential equations, 49
Nonlinear system elements, steady-state error and, 272–274
Nonminimum-phase systems, GM of, 452–453
Nonminimum-phase transfer functions, 47–48
Nontouching loops, 122
Nonunity feedback control systems, 258, 265–266
  Nichols chart applied to, 469–470
  steady-state errors and, 266–272
Normalized frequency v. magnification, of prototype second-order system, 415
Notation. See Units/symbols
Notch controllers, 581–588. See also Speed-control system
  frequency-domain design and, 586–588
  pole-zero-cancellation design with, 582–584
Notch filters, 576–588
Number of branches, on root loci, 378, 381
Numerical control machines, 289
Nyquist diagram, 29
Nyquist path, 433–434
Nyquist plots
  advantages of, 455
  disadvantage of, 455
  gain crossover on, 453
  $L(s)$ plot and, 444–449
  phase crossover on, 451
  root loci and, 437–439
  step responses/frequency responses, correlation among, 450–451
Nyquist stability criterion, 29, 48, 77, 426–444. See also Root-locus technique; Routh-Hurwitz criterion
  critical point and, 432, 434
  fundamentals, 426–435
  generalized, 437
  $G(s)H(s)$ plot and, 434–435
  $L(s)$ plot and, 434–435
  minimum-phase transfer functions and, 435–437, 440–444
  origination of, 429
  principles of the argument, 429–433
  root-locus technique v., 426, 437–439
  stability problem and, 427–428

O
Objectives, 1, 2
Observability, 714, 715, 719–725
  controllability and, 714, 715
  definition of, 719–720
  invariant theorems on, 723–725
  testing methods for, 720
  transfer functions/controllability, relationship among, 721–723

Observability canonical form (OCF), 703–704
  direct decomposition to, 709–712
Observer, 489
Octaves, 35
Ohms, 172, 289, 339
Ohm's law, 165
Oil well system, 241
One degree of freedom (1-DOF) quarter-car model, 220, 221, 252, 357, 358
One-directional heat conduction flow, 178
One-sided Laplace transform, 52–53
One-tank system. See Single-tank liquid-level system
One-to-one mapping, 19
Open-loop base excitation (quarter-car model), 221
Open-loop control systems (nonfeedback systems), 5–7, 109. See also Feedback control systems
Open-loop motor-control system, 244
Open-loop response, disturbance-, 289
Open-loop sine input (virtual experiment), 347–350
Open-loop speed (virtual experiment), 345–347
Open-top cylindrical container, fluid flow into, 182
Operational amplifiers (op-amps), 172–177
  configuration, inverting, 175
  exercises, 236–237
  first-order, 174–177
  ideal, 173
  input-output relationship for, 173
  issues with, 173
  PD controller and, 493
  phase-lead controller and, 543
  PI controller and, 512–513
  realization, of transfer function, 176–177
  schematic diagram of, 173
  and sums/differences of signals, 173–174
  transfer functions, inverting, 175–176
  uses for, 172–173, 177
Orifice equation, 607–608
Output equations, 51–52
  state diagrams from, 680–681
Output nodes, 121
  noninput nodes, gain formula and, 127–128
Output sensor, 108
Output vector, 682
Outputs, 2, 104
  state variables v., 51
Overall gain, feedback and, 8–9
Overshoot. See Maximum overshoot

P

P (proportional) controllers, 617–621
Pade approximation, 206
Parabolic-function input, 256
  steady-state error and, 263–264, 268
Parallel decomposition, 713–714
Parameter variations, sensitivity to, 487
Partial-fraction expansion
  inverse Laplace transform by, 57–62
  references for, 92
Passive electrical elements, modeling of, 165

Passive suspension, quarter car model and, 364–365
Path gain, 122
Paths, 122
Payload, of space-shuttle-pointing control system, 102, 103
PD (proportional-derivative) controllers, 314–316
  as anticipatory control, 496
  Bode plot and, 497
  design principle of, 496
  design with, 492–511
  disadvantage of, 496
  feedback control system with, 314–316, 493
  frequency-domain interpretation of, 496–497, 503–511
  high-pass filter characteristics of, 496, 497
  op-amp circuit realization of, 493
  for robot-arm-joint system, 621–626
  second-order aircraft attitude control system and, 498–505
  summary effects of, 497
  third-order aircraft attitude control system and, 506–511
  time-domain interpretation of, 494–496, 498–505
Pendulum
  differential equation for, 49
  inverted, on cart, 227–228
  spring-supported, 209–211
Perfect gas law, 183, 187
Permanent-magnet (PM) dc motors
  brushless, 201
  classifications of, 199–201
  control system, 763–764
  iron-core, 199–200
  modeling of, 201–205
  moving-coil, 200–201
  surface-wound, 200
Permanent-magnet technology, 198
Phase crossover, 451
Phase margin (PM), 46, 86, 453–455, 487, 488. See also Gain margin
  on Bode plot, 456
  definition of, 453
Phase-crossover frequency, 451
Phase-crossover point, 46, 47
Phase-lag controller, 533, 561–574
  Bode diagram of, 565
  compensated system, 564, 571
  design strategies for, 562
  frequency-domain design of, 563–566, 569–572
  pole-zero configuration of, 561
  speed-control system and, 584
  sun-seeker system and, 566–572
  third-order sun-seeker system and, 572–574
  time-domain design of, 561–563, 566–569
  uncompensated system, 564, 571
Phase-lead controller, 532–561
  Bode plot of, 535–536
  effects of, 554
  frequency-domain design of, 535–537, 543–548, 551–554
  limitations of, 555
  multistage, 555

op-amp-circuit realization of, 543
pole-zero configuration of, 534
single-stage, 555
sun-seeker control system and, 537–548
third-order sun-seeker control system and, 548–554
time-domain design of, 534–535, 537–543, 548–551
two-stage, 555–559
Phase-locked loops, 245
Phase-variable canonical form (PVCF), 692
Physically realizable system, 53
PI (proportional-integral) controllers, 316–319
  advantages/disadvantages of, 516
  Bode diagram of, 515
  control system with, 317
  design with, 511–528
  frequency-domain design of, 514–516, 520–523
  op-amp circuit realization of, 512–513
  pole-zero configuration of, 317–318, 514
  prototype second-order system with, 511–512
  for robot-arm-joint system, 626–628
  second-order attitude control system and, 516–523
  speed-control system and, 584–585
  third-order attitude-control system and, 523–528
  time-domain design of, 513–514, 516–520, 523–528
PID (proportional, integral, derivative) controllers, 176, 492
  design with, 528–532
  implementation of, 176–177
  for robot-arm-joint system, 628–630
  role of, 304–305
  third-order attitude-control system and, 529–532
Pinion, rack and, 161
Pipe
  with fluid resistor, incompressible fluid flow through, 184
  frictionless, fluid through, 184
  with orifice, air flow through, 187
Piston system, spring-loaded, 182–183
Plant, 107
Plotting tutorial, 647–648
Pneumatic systems. See also Fluid systems
  capacitance in, 182–183
  conservation of mass and, 182
  conservation of volume and, 182
  gas flow into rigid container system, 187–188
  perfect gas law and, 183, 187
  properties, 180–189, 224
  resistance and, 187
  spring-loaded piston system, 182–183
  time constant of, 214–215
  with valve and spherical rigid tank (analogy), 214–215
Polar form, of complex numbers, 17
Polar plots, 26, 27–32
  gain-crossover point and, 47
  phase curve of, 48
  phase-crossover point and, 47

Polar representation, of $G(s)$, 22–24
Pole placement, 730
Pole-placement design, 715
　through state feedback, 730–735
Pole-zero configuration
　of $G2(s)H2(s)$, 394, 396, 399, 400
　of phase-lag controller, 561
　of phase-lead controller, 534
　of PI controller, 317–318, 514
Pole-zero-cancellation design, 576–588
　exact cancellation, 576
　frequency-domain design and, 580–581
　inexact cancellations, 577, 578
　with notch controller, 582–584
　second-order active filter, 579–580
　speed-control system and, 581–588
Poles, 20
　added to $L(s)$ plot, 445–448
　closed-loop transfer function with, 307–308
　definition of, 20
　dominant, of transfer functions, 311–313
　forward-path transfer function with, 305–307, 424–426
　graphical representation of, 21
　at origin, 34–36
　quadratic, 39–41
　simple, 20
　simple, $1/(1 + jwT)$, 39
Polytropic exponent, 183
Polytropic process, 183, 189
Position control
　of armature-controlled dc motor, 292–293
　dc motor model and, 352–354
Position Control (virtual experiment 2), 352–354
Position indicator, potentiometer as, 191
Position sensor, dc motor model and, 339–340
Position-control system (aircraft). *See also*
　Attitude-control system
　Bode diagrams of, 467
　case study, 216–217
　closed-loop frequency response of, 469
　gain-phase plots of, 468
　Nichols chart of, 468
　steady-state response and, 298
　time domain analysis of, 293–304
　unit-step response and, 294–297
Position-control systems
　of electronic word processor, 136
　robust controllers and, 599–601
　with tachometer feedback, 195
　variable load inertia and, 599
Positive feedback, 109
Potentiometers, 189–194
　ac control system with, 192, 193
　block diagram representation of, 191
　dc-motor position-control system with, 192
　electric circuit representation of, 190
　linear motion, 189, 190
　as position indicator, 191
　rotary, 189, 190
Power, units for, 172
Power supply, within enclosure, 238
Powertrains, hybrid, 3–4
Pressure drop, 185
Principles of the argument, 429–433

equation form of, 431
Nyquist criterion and, 429–433
statement of, 430
summary of outcomes of, 433
Printers, laser, 289
Printwheels, 197, 198
　control system, 231, 246, 329
　load inertia and, 599
　permanent-magnet dc-motor-control system and, 763, 764
　velocity of, 198
Proper transfer functions, 71
Proportional control, 492
Proportional controllers. *See* P controllers
Proportional gain, 176
Proportional-derivative controllers. *See* PD controllers
Proportional-integral controllers. *See* PI controllers
Prototype first-order systems. *See* First-order prototype systems
Prototype second-order systems. *See* Second-order prototype systems
Pulley, belt and, 161, 162
Pure time delays, Bode plots and, 42–43, 458–459

Q

Quadratic poles/zeros, 39–41
Quadrature, dual-channel encoder signals in, 196, 197
Quantization error, 273
Quantizer, 272, 273
Quarter Car Modeling Tool, 360–364
　absolute acceleration control system, 361
　backlash and, 361
　control window, 362
　Model Parameters window, 362
　saturation and, 361
Quarter-car model, 220–222, 335
　closed-loop acceleration control, 359–360, 366–367
　closed-loop position control, 221–222, 359
　closed-loop relative position control, 365–366
　design project 2, 357–367
　introduction to, 357–359
　1-DOF, 220, 221, 252, 357, 358
　open-loop base excitation, 221
　parameter values, 357
　passive suspension and, 364–365
　2-DOF, 220, 221, 358

R

Rack and pinion, 161
Radiation, 179
Ramp function, 255
Ramp-function input, 255–256
　steady-state error with, 262–263, 268
Rate-feedback, 601–602
Rational functions, time-delay function and, 206
RC circuit systems
　capacitance and, 63, 171–172
　differential equation of, 170–171
　unit response of, 63

Reactor tank, 97
Real axis, root loci on, 380, 381
Real component, of complex variable, 18
Real convolution (complex multiplication), 56, 57
Real multiplication. *See* Complex convolution
Rectangular form, of complex numbers, 16, 17
Rectangular object, heat transfer problem and, 180
Rectangular output waveform, of single-channel encoder device, 196, 197
Rectangular valve-port geometry, 609
References
　block diagrams, 133, 134
　complex variables, 92
　elementary heat transfer properties, 224
　fluid/gas system properties, 224
　Laplace transform, 92
　modeling, 224
　partial-fraction expansion, 92
　SFGs, 133, 134
　stability, 93
　state variable analysis, of electric networks, 134
Regulator system, 7
Relative damping ratio, 313
Relative Position Time Response plot, 366
Relative stability, 72, 449–455, 487
　gain margin and, 46, 86, 451–453
　phase margin and, 46, 86, 453–455
　slope of the magnitude curve of Bode plot and, 459–462
Relocation
　branch point, 113, 114
　comparator, 113, 114
Resistance. *See also* RLC network
　equations, for laminar flows, 184, 185
　incompressible fluids and, 184–185
　pneumatic systems and, 187
　RC circuit systems and, 63, 170–171
　thermal, 178
　turbulent, 185
　units for, 172, 179
Resistance-inductance-capacitance network.
　See RLC network
Resistors, 165
Resonant frequency w_r
　prototype second-order system, 413–416
　specification, 412
Resonant peak M_r, 487, 488
　prototype second-order system, 413–416
　specification, 412
Results. *See* Outputs
Reversible adiabatic process, 184
Rigid container system, gas flow into, 187–188
Rise time, 257, 283–285, 288, 487, 488
RLC (resistance-inductance-capacitance) network, 49, 165–166, 213, 772
　analogies and, 213–216
　block diagram representation, 166
　differential equation for, 49
　electrical schematics, 166
　modeling of, 165–166
　SFG representation, 166

Robot-arm-joint system, 249
  P controllers for, 617–621
  PD controllers for, 621–626
  PI controllers for, 626–628
  PID controllers for, 628–630
  schematic diagram, 605, 612
Robotic arm (design project 1), 354–357
  control of, 354
  side view of, 355
Robotics, 1, 2
Robots, 289
Robust controllers, 590–601
  position-control system and, 591–601
  sun-seeker system and, 591–599
Robustness, 487, 590
Rolling dry friction, 150–151
Root contours (RC), 373, 393–400
  of sun-seeker control system, 540, 541, 570
Root loci (RL), 86, 372
  angles of asymptotes of, 378–379, 381
  angles of departure/arrival of, 380, 382
  basic properties of, 373–377
  breakaway points (saddle points) on, 380–381, 382
  calculation of values of K on, 382
  of compensated phase-lag system, 564
  design aspects of, 385–393
  graphical construction of, 375–377
  and intersect of asymptotes, 379, 381
  intersection of, with imaginary axis, 380, 382
  $K = \pm \infty$ points on, 377–378, 381
  $K = 0$ points on, 377–378, 381
  number of branches on, 378, 381
  Nyquist plot and, 437–439
  properties of, 377–385
  on real axis, 380, 381
  summarization of properties, 381–382
  of sun-seeker system, 567
  symmetry of, 379
  of third-order attitude-control system, 302
  of uncompensated phase-lag system, 564
Root sensitivity, 382–385
Root-locus diagrams
  addition of poles to $G(s)H(s)$, 385–387
  addition of zeros to $G(s)H(s)$, 387–393
Root-locus technique, 206, 372–408. *See also*
    Nyquist stability criterion; Routh-
    Hurwitz criterion
  Nyquist stability criterion v., 426, 437–439
  Routh-Hurwitz criterion and, 426
ROOTS command, 99
Rotary disk, 196
Rotary incremental encoders
  diagram, 196
  parts in, 196
Rotary potentiometer, 189, 190
Rotary-to-linear motion control systems, 161
  belt and pulley, 161
  lead screw, 161
  rack and pinion, 161
Rotational mechanical system
  diagram, 159
  hydraulic control system and, 615–616
  motor-load system, 159–160
  properties, 159
  SFG representation, 160

Rotational motion, 157–160
  friction for, 158–159
  translational motion, conversion between, 161
Routh-Hurwitz criterion, 77, 78–84. *See also*
    Nyquist stability criterion; Root-locus
    technique
  problems/exercises for, 99, 100, 101, 102, 103
  root-locus technique and, 426
Routh-Hurwitz stability routine. *See* tfrouth
Routh's tabulation, 79, 80, 81

S

Saddle points (breakaway points), on root loci, 380–381, 382
Sampled-data control systems, 14
Sampler, 14
Saturation, 11–12, 216, 272, 337, 340, 346, 347, 348, 349, 351, 352, 368, 369
  dc motor model and, 346
  Quarter Car Modeling Tool and, 361
Schematic diagrams. *See* Diagrams
Screw, lead, 161
Second-order active filter, 579–580
Second-order attitude control system
    (aircraft), 295–296, 498–505. *See also*
    Attitude-control system
  forward-path transfer function of, 498
  frequency-domain design, 503–505, 520–523
  PD controller and, 498–505
  PI controller and, 516–523
  time-domain design, 498–503, 516–520
Second-order differential equations, 49
Second-order linear system, 62
Second-order prototype function, 61
Second-order prototype systems, 64–67
  BW and, 416–418
  damping ratio/damping factor and, 277–278
  delay time/rise time and, 283–285
  magnification v. normalized frequency of, 415
  maximum overshoot and, 280–283
  natural undamped frequency and, 278–280
  with PI controller, 511–512
  resonant peak/resonant frequency, 413–416
  settling time and, 285–288
  transient response of, 275–288
  unit-step responses of, 276
Second-order system, 49
  with forward-path transfer function, unit-step responses of, 423
Sensing devices of control systems (block diagram), 107
Sensitivity
  control systems and, 9–10
  to parameter variations, 487
  speed-control system and, 585–586
  studies, in frequency domain, 470–472
  third-order sun-seeker system and, 559–561
Sensitivity function, 9–10, 470, 471, 472, 474, 559
Sensors, 104, 107, 189–195. *See also* Encoders
  encoders and, 189
  output, 108

potentiometers, 189–194
  in rotary incremental encoder, 196
  tachometers, 194–195
Series compensation, 489, 490
Series-feedback compensation, 489, 490
Servomechanisms, 289
Servomotors, 13, 198, 242, 328, 331, 477
Settling time, 257, 285–288, 487, 488
Shift in time (theorem), 55, 56
SI units. *See* Units/symbols
Signal-flow graphs (SFGs), 119–129. *See also*
    State diagrams
  algebra/manipulation rules for, 123–124
  block diagram, of control system, 129
  block diagrams v., 119, 125
  of dc-motor system with nonzero initial conditions, 203
  electric network problems and, 141–142
  elements of, 119–120
  of feedback control system, 124
  gain formula and, 124–128
  mass-spring-friction system, 155
  MATLAB tools and, 129–132
  problems/exercises for, 138–146
  properties of, 120
  references for, 133, 134
  RLC network, 166
  rotational system, 160
  state diagrams and, 673, 676
  step-by-step construction, 121
  sums/differences of signals in, 173–174
  terminology for, 121–122
Signals
  dual-channel encoder, in quadrature, 196, 197
  sums/differences, op-amps and, 173–174
  suppressed-carrier-modulated, 193–194
  test, for time-domain analysis, 254–256
Similarity transformations, 699–706
  CCF and, 692, 701–703
  DCF and, 704–705
  invariance properties of, 700–701
  JCF and, 706
  OCF and, 703–704
SIMLab, 223
  experiment control window, 342
  Experiment menu, 342
  motor blocks, adjustable parameters for, 344
  Speed Control Simulink model, 343
Simple pole, 20
  $1/(1+jwT)$, 39
Simple zero, $1+jwT$, 37–38
Simplified gain formula, 129. *See also* Gain formula
Simulation, virtual experiments and, 345–354. *See also* dc motor
Simulink
  Library Browser, 349
  role of, 2
  Speed Control Simulink model, 343
Single channel incremental encoder
  rectangular output waveform of, 196, 197
  sinusoidal output waveform of, 196, 197
Single-input, single-output systems. *See* SISO systems
Single-stage phase-lead controller, 555

Single-tank liquid-level system, 184–185
  analogies for, 215–216
Single-valued function, 19
Single-valued mapping, from s-plane to $G(s)$-plane, 19
Singularities, of function, 20
Sinusoidal output waveform, of single-channel encoder device, 196, 197
SISO Design Tool, 363, 366, 367, 371, 634, 635, 636, 638, 642
SISO (single-input, single-output) systems, 70–71
Slope of the magnitude curve, of Bode plots, 459–462
Smart transportation systems, 3–4, 5
Solar collector field, 4–5
  antenna control system, block diagram of, 137
Solar power, water extraction and, 4–5, 6
Space state form. *See* State space form
Space-shuttle pointing control system, 102, 103
Spacecraft systems, ball bearings in, 151
Speed control
  of armature-controlled dc motor, 291–292
  dc motor model and, 350–352
Speed Control (virtual experiment), 350–352
Speed Control Simulink model, 343
Speed sensor, dc motor model and, 339–340
Speed-control system, 581–588
  block diagram of, 581
  notch controller and, 581–588
  phase-lag controller for, 584
  PI controller for, 584–585
  pole-zero-cancellation design with notch controller, 582–584
  sensitivity and, 585–586
  time-domain performance attributes, 584
  unit-step responses of, 585
s-plane
  complex, 19
  $G(s)$-plane and, 19
Spring, 149. *See also* Mass-spring-friction system
  force-spring system and, 149
  linear, 149
  torsional, 158
Spring constant
  defined, 149
  symbol/units, 151, 159
Spring system, torque torsional, 158
Spring-loaded piston system, 182–183
Spring-mass system, 2-DOF, 156
Spring-mass-damper, 22, 110
Spring-supported pendulum, 209–211
Stability. *See also* Bode plots; Nyquist stability criterion; Relative stability
  absolute, 72
  asymptotic, 75–76
  BIBO, 73–74
  feedback and, 9–10, 15
  of linear control systems, 72–73
  magnitude-phase plot and, 462–464
  marginal, 76
  MATLAB tools for, 85–90
  methods for determining, 77–78

Nyquist criterion and, 427–428
  references for, 93
  roots of characteristic equation v., 74
  zero-input, 74–75
Stable system, conditionally, 459–462, 555
State controllability, 716–717
State controllable, completely, 717
State diagrams, 133, 676–681
  differential equations to, 678–679
  integration operation and, 677
  output equations from, 680–681
  SFGs and, 673, 676
  state equations from, 680–681
  state-transition equation from, 689–691
  transfer functions to, 679–680
  uses of, 678
State equations, 50, 51
  characteristic equations from, 696
  of electrical networks, 167, 168–169
  equation of state, 182
  high-order differential equations v., 691–693
  state diagrams from, 680–681
  transfer functions v., 693–695
  vector-matrix representation of, 682–684
State feedback, 489
  control system, 715, 728–729
  with integral controller, 735–741
  magnetic-ball suspension system and, 731–733
  pole-placement design through, 730–735
  sun-seeker system and, 733–735
State feedback compensation, 489, 490
State space approach
  linearization and, 207–213
  magnetic-ball-suspension system and, 211–213
  spring-supported pendulum and, 209–211
State space form, 51, 152, 167
State space systems, 16, 50, 741
State variable analysis, 673–772
  of electric networks, 134
  transfer functions and, 673–674
State variables, 50–51
  conditions for, 51
  outputs v., 51
State vector, 682
State-flow diagrams, 138, 145, 231, 234
State-Space Analysis Tool (statetool), 741–748
  inputting values in window, 744
  window, 742
State-transition equation, 687–691
  definition, 687
  from state diagram, 689–691
State-transition matrix, 684–686
  properties of, 685–686
  significance of, 685
Statetool. *See* State-Space Analysis Tool
Static friction, 150, 158
Stationary mask (rotary incremental encoder), 196
Steady-state accuracy, 487
Steady-state errors, 257–274, 487
  closed-loop transfer function and, 266–270
  Coulomb friction and, 273–274

defined, 257, 258
  error constants and, 265
  linear continuous-data control systems and, 258–272
  nonlinear system elements and, 272–274
  nonunity feedback and, 266–272
  parabolic-function input and, 263–264, 268
  ramp-function input and, 262–263, 268
  step-function input and, 261–262, 267
  system configuration and, 258–259
  unity feedback systems and, 260–266
Steady-state responses, 253, 254
  insignificant poles and, 313–314
  position-control system and, 298
  of third-order attitude-control system, 304
Steel-rolling process, 231–232
Steering control, of automobile, 4
Step responses. *See* Unit-step responses
Step-function input, 255
  steady-state error with, 261–262, 267
Stephan-Boltzmann law, 179
Stiffness. *See* Spring constant
Sum and difference (theorem), 54, 56
Sums/differences of signals, 173–174
Sun-seeker control system, 217–220, 548–554
  ACSYS/MATLAB tools and, 645–647
  block diagram of, 220, 537
  Bode plots of forward-path transfer function of, 552
  Bode plots of phase-lead controller in, 552
  compensated, 557, 568
  coordinate system of, 218, 219
  dc motor in, 218, 220
  error discriminator of, 218–219
  feedforward compensators and, 589–590
  forward compensators and, 589–590
  frequency-domain design, 551–554
  integral controller and, 737–738
  lead-lag controller and, 574–576
  minor-loop feedback controller and, 603–604
  phase-lag controller and, 566–574
  phase-lead controller and, 537–554
  robust controllers and, 591–599
  root contours of, 540, 541, 570
  root loci of, 567
  schematic diagram of, 218
  sensitivity considerations, 559–561
  servoamplifier of, 218, 220
  state feedback and, 733–735
  tachometer of, 218, 220
  third-order, 548–554
  time-domain design, 548–551
  two-stage phase-lead controller and, 556–559
  uncompensated, 557, 568
  unit-step responses of, 538, 551, 560, 568
Sun-tracking control systems, 4–5
Superposition principle, 11, 71, 73, 115, 173, 176, 222, 291, 358, 674
Suppressed-carrier-modulated signal, 193–194
Surface-wound PM dc motors, 200

Symbolic Tool
  MATLAB, 53, 64, 83, 84, 741, 748
  Transfer Function, 54, 77, 84–85, 86, 744, 748, 750, 754
Symbols. *See* Units/symbols
Symmetry of root loci, 378
System with disturbance input, 260
System error, 258. *See also* Steady-state errors

T

Tachometer constant, 195
Tachometer-feedback control, 601–602
Tachometers, 194–195
  modeling of, 195
  position-control system with, 195
  sun-seeker control system and, 218, 220
  transfer function of, 195
  velocity-control system with, 194
Taylor series, 17, 182
  liberalized flow equations and, 608
  linearization and, 207, 208
Temperature, symbol/units for, 179, 189
Temperature control, of air-flow system, 243–244
Ten-turn rotary potentiometer, 189, 190
Testing methods
  for controllability, 717–719
  for observability, 720
tfcal (Transfer Function Calculator), 86, 741, 751
tfrouth (Routh-Hurwitz Stability Routine), 77, 86–90
tfsym (Transfer Function Symbolic Tool), 54, 77, 84–85, 86, 744, 748–750, 754
Theorems
  complex convolution, 56, 57
  complex shifting, 56
  on controllability/observability, 723–725
  differentiation, 54, 56
  final-value, 55–56
  initial-value, 55, 56
  integration, 55, 56
  of Laplace transform, 54–57
  multiplication by a constant, 54, 56
  real convolution, 56, 57
  shift in time, 55, 56
  sum and difference, 54, 56
Thermal expansion coefficient, 182
Thermal lever, 237, 238
Thermal resistance, 178
Thermal systems, 177–180
  conduction in, 178
  convection in, 178–179
  electrical equivalents for, 216
  properties, 179
  radiation in, 179
Third-order attitude-control system (aircraft), 300–304. *See also* Attitude-control system
  frequency-domain design, 509–511, 526–528, 531–532
  PD controller and, 506–511
  PI controller and, 523–528
  PID controller and, 529–532
  root loci of, 302
  steady-state response of, 304
  time response of, 300–304
  time-domain design, 506–508, 529–530
  transient response of, 301
  unit-step responses of, 303
Third-order system, with forward-path transfer function
  magnification of, 425
  unit-step responses of, 425
Throttle angle, 5
Throttle lever, 242
Time constant, 63, 171, 188
  electrical, amplifier-motor system and, 217, 294
  low-time-constant properties and, 198
  mechanical, motor-load system and, 217, 294
  motor electric, 290, 359
  motor-mechanical, 291
  of pneumatic system, 214–215
  prototype first-order system and, 274
  symbol/unit for, 180, 189
  transient response and, 320
Time delays. *See also* Delay time
  Bode plots and, 42–43, 458–459
  systems with, 205–207
Time responses, 64, 67, 253
  continuous data systems, 253–254
  of prototype first-order system, 274–275
  test signals for, 254–256
  of third-order attitude-control system, 300–304
  to unit-ramp input, 298–300
Time-delay function, approximation of, 206
Time-domain analysis, 253–336
  of aircraft position-control system, 293–304
  frequency-domain analysis *v.*, 409
  parabolic-function input and, 256
  ramp-function input and, 255–256
  step-function input and, 255
  test signals and, 254–256
  unit-step response and, 256–257
Time-domain design
  frequency-domain design *v.*, 488, 492
  PD controllers and, 494–496, 498–505
  performance specifications, 487, 488
  phase-lag controller and, 561–563, 566–569
  phase-lead controller and, 534–535, 537–543, 548–554
  PI controllers and, 513–514, 516–520, 523–528
  second-order aircraft attitude control system, 498–503, 516–520
  speed-control system and, 584
  third-order aircraft attitude control system, 506–508, 529–530
  third-order sun-seeker control system, 548–551
  time-domain characteristics in, 492
Time-invariant feedback control systems, 12–13
Time-varying feedback control systems, 12–13
Timing belt, over pulley, 161, 162
Toolboxes. *See also* MATLAB
  2-1-1, 21
  2-1-2, 25
  2-2-1, 29
  2-2-2, 33
  2-2-3, 42
  2-2-4, 43
  2-2-5, 45
  2-2-6, 46
  2-4-1, 54
  2-5-1, 58
  2-5-2, 60
  2-6-1, 64
  2-6-2, 67
  2-7-1, 69
  2-13-1, 80
  2-13-2, 83
  3-3-1, 130–131
  3-3-2, 132
  4-1-1, 153–154
  4-2-1, 167–168
  4-2-2, 169–170
  5-4-1, 259
  5-4-2, 265–266
  5-4-3, 271–272
  5-6-1, 277
  5-6-2, 288
  5-8-1, 296
  5-8-2, 300
  5-8-3, 303
  5-9-1, 306
  5-9-2, 308
  5-9-3, 311
  5-11-1, 316
  5-11-2, 319
  7-3-1, 380
  7-3-2, 384
  7-4-1, 387
  7-4-2, 389
  7-4-3, 391
  7-5-1, 395
  7-5-2, 397
  7-5-3, 400
  8-2-1, 414–415
  8-2-2, 417
  8-3-1, 422
  8-3-2, 424
  8-3-3, 424
  8-4-1, 426
  8-8-1, 441
  8-8-2, 442
  8-8-3, 443
  8-9-1, 448
  8-10-1, 449
  8-11-1, 457
  9-2-1, 498
  9-2-2, 502
  9-2-3, 502–503
  9-2-4, 503
  9-2-5, 506
  9-2-6, 510
  9-2-7, 511
  9-3-1, 519
  9-3-2, 520
  9-3-3, 522
  9-3-4, 527
  9-4-1, 529
  9-5-1, 539
  9-5-2, 540

Index 785

9-5-3, 542–543
9-5-4, 545
9-5-5, 546
9-5-6, 550
9-5-7, 553
9-5-8, 556
9-5-9, 559
9-10-1, 592
9-10-2, 594
9-10-3, 594–595
9-13-1, 620
9-13-2, 624
9-13-3, 625
9-13-4, 628
9-13-5, 630
Torque, 157, 159
Torque constant, 202, 204, 216, 223, 245, 247, 249, 339, 599, 659, 666, 668, 738, 757, 765
  back emf constant v., 204–205
Torque production, in dc motor, 199
Torque torsional spring system, 158
Torque-angle curve of motor/closed-loop system, 273
Torque-inertia system, 157
Torsional spring, 158
Torsional spring constant, 158
Traction system, train in, 248–249
Trailer, vehicle and, 227
Train, in traction system, 248–249
Train controller
  block diagram, 138
  exercise, 226
Trains, gear. *See* Gear trains
Transfer Function Calculator (tfcal), 86, 741, 751
Transfer Function Symbolic Tool (tfsym), 54, 77, 84–85, 86, 744, 748–750, 754
Transfer functions, 26, 67, 68, 70–72, 106. *See also* PID controllers
  from block diagrams, MATLAB and, 129–132
  characteristic equations from, 696
  controllability/observability, relationship among, 721–723
  decompositions of, 678, 707–714
  definition of, 70
  dominant poles/zeros of, 311–314
  frequency-domain analysis and, 409–410
  impulse response and, 67–69
  inverting op-amp, 175–177
  minimum-phase, 47–48, 435–437
  multivariable systems, 4, 71–72, 117–118, 673–674
  nonminimum-phase, 47–48
  op-amp realization of, 176–177
  in parallel, block diagram of, 108
  proper, 71
  properties of, 71
  in series, block diagram of, 107, 108
  single-input, single-output systems, 70–71
  state diagrams from, 679–680
  state equations v., 693–695
  state variable analysis and, 673–674
  of tachometers, 195
Transient responses, 253, 254, 487

feedback and, 11
  of prototype second-order system, 275–288
  of third-order attitude-control system, 301
  time constant and, 320
Translational mechanical system properties, 151
Translational motion, 148–157
  acceleration and, 148
  displacement and, 148
  force-mass system, 149
  friction for, 149–150
  hydraulic control system and, 613–615
  rotational motion, conversion between, 161
  velocity and, 148
Transportation lags, dynamic systems with, 205–207
Turboprop engine, 135
  signals of, coupling between, 143, 144
Turbulent resistance, 185
Tutorial, plotting, 647–648
Two degrees of freedom (2-DOF)
  quarter-car model, 220, 221, 358
  spring-mass system, 156
Two-stage phase-lead controller, 555–559
Two-tank system. *See* Double-tank liquid-level system

U

U tube manometer, 240
Uncompensated phase-lag system
  Bode plot of, 571
  root loci of, 564
Uncompensated sun-seeker system
  Bode plots of, 557
  unit-step responses of, 568
Unit impulse response, 68, 69. *See also* Dirac delta function
Unit-ramp input, 298–300
  attitude-control system and, 299
  time responses to, 298–300
Units/symbols
  acceleration, 151
  angular acceleration, 159
  angular displacement, 159
  angular velocity, 159
  British, 148, 204
  capacitance (electrical system), 172
  capacitance (fluid/pneumatic system), 189
  capacitance (hydraulic), 189
  capacitance (thermal system), 179
  charge, 172
  distance, 151
  electrical system properties, 172
  energy (electrical system), 172
  energy (heat stored), 179, 189
  energy (rotational mechanical system property), 159
  force, 151
  heat flow rate, 179
  inductance, 172
  inertia, 159
  mass, 151
  mass flow rate, 189
  power, 172
  resistance (electrical system), 172
  resistance (hydraulic), 189

  resistance (thermal system), 179
  spring constant, 151, 159
  temperature, 179, 189
  thermal system properties, 179
  time constant, 180, 189
  torque, 159
  translational mechanical system properties, 151
  velocity, 151
  viscous friction coefficient, 159
  voltage, 172
  volume flow rate, 189
Unit-step responses
  of attitude-control system, 295, 501
  delay time and, 257
  frequency responses/Nyquist plots, correlation among, 450–451
  maximum overshoot and, 256–257, 280–283
  position-control system and, 294–297
  of prototype first-order system, 274
  of prototype second-order system, 276
  rise time and, 257
  of second-order system with forward-path transfer function, 423
  settling time and, 257
  of speed-control system, 585
  of sun-seeker system, 538, 568
  of third-order attitude-control system, 303
  of third-order sun-seeker control system, 551, 560
  time-domain specifications and, 256–257
Unity feedback loop, 109
Unity feedback systems
  poles added to forward-path transfer function, 305–307
  steady-state error and, 260–266
Unstable, 76

V

Variable load, hydraulic control system and, 616–617
Vector-matrix representation, of state equations, 682–684
Vehicle, with trailer, 227
Vehicle suspension system, 225, 336
Velocity
  angular, 157, 159
  symbol/units, 151
  translational motion and, 148
Velocity-control system
  MATLAB and, 259
  with tachometer feedback, 194
Vibration absorber, 234
Virtual experiments. *See* dc motor model; Experiments
Virtual ground, 173
Virtual Lab, 12, 223, 340–344
  Experiment menu for, 342
Virtual prototyping, 2
Virtual short, 173
Viscous friction, 149–150, 158
  dashpot for, 149, 150
  graphical representation, 150
Viscous friction coefficient, 151, 159
Voice-coil motor, 247

Voltage, units for, 172
Voltage divider, 171–172
Voltage law, 165–166, 170
Volume, conservation of, 181
Volume flow rate, 181
    mass flow rate v., 182
    symbol/units, 189

W

Washing machine, 6
Water extraction, solar power and, 4–5, 6
Word processor

position-control system of, 136
printwheel of, 197, 198

Z

Zero initial conditions, 70
Zero-input response, 73
Zero-input stability, 74–75
Zero-pole-gain models, 21. *See also* Pole-zero configuration
Zero-state response, 72
Zeros
    added to $L(s)$ plot, 448–449
    closed-loop transfer function and, 308–309
    definition of, 20
    dominant, of transfer functions, 311–314
    forward-path transfer function and, 309–311, 418–424
    graphical representation of, 21
    at origin, 34–36
    quadratic, 39–41
    simple, $1+jwT$, 37–38
zpk, 21, 45, 131, 259, 265, 457